Vertebrate paleozoology

VERTEBRATE PALEOZOOLOGY

EVERETT C. OLSON

Department of Zoology,
University of California at Los Angeles

WILEY-INTERSCIENCE, *a Division of John Wiley & Sons, Inc.*

New York · London · Sydney · Toronto

Library of Congress Catalog Card Number: 71-127667

ISBN 0-471-65364-0

Printed in the United States of America

10 9 8 7 6 5 4 3 2 1

Preface

This book brings together some of the results of my studies of extinct life—studies that began in the early 1930s at the University of Chicago and that have continued with only minor interruptions since that time. I have chosen to call the book *Vertebrate Paleozoology* because it is more than anything else a discussion of the problems of the fossil vertebrates and the roles they play in our understanding of life and its evolution. Questions rather than answers carry the line of development throughout the four sections.

Many excellent books are currently available to give authoritative and extensive information about the morphology, systematics, geological distributions, and phylogeny of the fossil vertebrates and there is little need for additional books in these areas. On the other hand, there is need for a book that presents the problems and viewpoints of vertebrate paleozoology in the light of the kinds of materials that are available for study and the philosophical bases of the approaches that are currently prevalent. Conclusions of paleontology and paleozoology sometimes receive much more credence than they merit and other times tend to be rejected out of hand without adequate reason. One of the aims of this book is to consider the reliability of various kinds of data and to examine the different ways of approaching paleontological materials and the consequences of these approaches.

The format of the text is basically topical, using selected materials from paleozoology and related fields to develop fundamental concepts. The questions that stimulate research on fossils and the adequacy of the materials and methods to provide answers are constantly examined. To some of these questions definitive answers are given in the text. To many, however, specific answers are not attempted, although preferences among alternatives may be included. Perhaps this absence of firm conclusions will prove unsatisfying to readers. My justification is that it expresses the actual state of affairs and presents the field as one of challenge, which it is, rather than one in which many of the crucial questions have been answered, which it is not.

To meet the various aims the book is divided into four sections. The first is short and deals with the nature of the vertebrate record, the aims of research in vertebrate paleozoology, and the extent to which the record and aims are compatible. The nature of the samples of ancient life available to the paleozoologist and the adequacy of these samples for interpretation are central questions.

The second section presents a relatively brief consideration of comparative and functional anatomy in studies of extinct vertebrates. How much can be known of body systems in the various groups of vertebrates from their extinct representatives and what methods of study have yielded pertinent information are among the questions raised. The different levels of inference used in such studies are examined.

The third section, which is much longer, treats matters of vertebrate classification. The questions in this section relate to the actual ways in which different groups of vertebrates have been classified and how the history of discovery and the nature of the fossil record have influenced the treatments. The aim has not been to arrive at a single, preferred classification, but rather to explore the nature of specific classifications in the light of the ways in which they came to be. In order to have a stable and widely available base, it was decided to accept A. S. Romer's (1966a) classification as presented in the third edition of his *Vertebrate Paleontology* as the terminal version. Occasionally, where they appear to be particularly instructive, more recent formulations have been included.

The fourth and final section of the book treats some of the major patterns of vertebrate evolution as revealed in the fossil record. Different aspects of this record are viewed from several perspectives. I have selected areas with which my work has been directly concerned as well as some others that I have found to be especially interesting and instructive as vehicles for study of these patterns. Little attention has been given to the understanding of evolution that has come from studies of the origins and radiations of placental mammals. Instead, the less well known patterns of evolution revealed by members of other classes of vertebrates have been emphasized.

Adaptation to aquatic life is the subject of the first two chapters of Section IV. This is an area of study that is being attacked on many fronts at the present time. It provides an opportunity to analyze jointly the adaptive modification of organisms whose entire evolutionary history was in the water and of those that, having moved onto land, returned once again to life in an aquatic medium. The two principal phases of evolution, the attainment of new levels of organization and the exploitation of these levels in adaptive radiation, are viewed in juxtaposition in the third chapter.

Successively subsequent chapters explore in some detail the nature of a major change in environment, from water to land; the exploitation of land, in a chronofaunal-ecological context; and the evolution of primitive mammals by way of their reptilian predecessors, treated in the dual perspective of adaptive radiation and retrospective analysis from the mammalian vantage point.

Finally, in a different vein, some of the problems of paleobiogeography are analyzed in the last chapter. They are viewed in the light of general biogeographic principles and models. The origin and deployment of the primitive therian mammals, continuing their introduction in the preceding chapter, is used as the vehicle for exposition. Into focus once again come the ever present problems of how adequately the available information answers questions about both the biological and physical events of the past.

The source of the materials and the ideas that form the heart of this book have, of course, been many. The most significant lie in my studies; my contacts with graduate students in classes, seminars, field parties, and "coffee sessions"; and the influences of various friends and associates who have made special impacts upon my way of looking at things.

Several persons from the many on whom I have depended have contributed to the background of this work in such a way that I should like to mention them specifically. First, of course, is my close friend and former teacher Professor Alfred S. Romer. Without his stimulation I would not have found my way into vertebrate paleonotology or found a secure and stimulating place in which to make my start. His continuing advice, criticism, and encouragement through the years have always been of the greatest value to me.

Professor Robert L. Miller, of the Department of the Geophysical Sciences at the University of Chicago—with whom I have worked on many projects, some of which found the culmination in the book *Morphological Integration* which we authored jointly—instilled in me a sense and understanding, if not mastery, of quantitative thinking in the problems of paleozoology and evolutionary biology. He too has contributed much to the concepts that form undercurrents in this book. Professor James R. Beerbower, as a graduate student and thereafter, was instrumental in stirring my interest in ecological approaches to the study of vertebrate fossils. A constant friend and critic in all aspects of the work that culminated in this volume has been Professor Ralph G. Johnson, a member of the Department of the Geophysical Sciences and the Committee on Evolutionary Biology of the University of Chicago.

There are many others to whom I owe a debt, too many to be mentioned individually here. As a critical reader of the manuscript whose suggestions have been invaluable, Dr. Matthew H. Nitecki of the Field Museum of Natural History cannot be left unnamed. Finally, major influences of my thinking as portrayed in this book have stemmed from two men. One is Professor David Kitts of the University of Oklahoma. Many of the underlying notions in the text have come from his stimulation. During three trips to Moscow over the last decade it was my pleasure and good fortune to come into close contact with Professor Ivan A. Efremov of that city and to benefit immensely from the friendship, counsel, and insights of this Soviet scientist, scholar, and writer. To these persons, and the many others, I express my appreciation for their aid while denying them any part in the shortcomings of this book.

During the decades of 1950 and 1960 my studies have been in part financed by the National Science Foundation. Both directly and indirectly this support has contributed to this book. Most recently NSF Grant B7070X has given direct support by providing funds for a research technician, Miss Eleanor Daly who has worked constantly with me in preparation of illustrations and in all phases of editing, Dr. Tibor Perenyi of the Field Museum of Natural History has drawn about half the figures, in particular many of those to be found in the second half of the book. To these two, to Mrs. Sharon Petraitis, who has carried much of the burden of manuscript preparation, and to the National Science Foundation, I express my sincere appreciation.

Los Angeles   Everett C. Olson
December 1970

Contents

Tables

LAND-FOWL are either such as have

Crooked Beak and Talons which are either

Carnivorous and **rapacious**, called **BIRDS OF PREY**, and these either

Diurnal, that prey in the day-time

The **Greater**, and these either

┌ The *more generous*, called EAGLES, The *Golden Eagle*, the *Sea-Eagle*, the *black Eagle*, &c. Part 1. Sect. 1. Chap. 3.

└ The *more cowardly and sluggish*, called VULTURES, Part 1. Sect. 1. Chap. 4.

The **Lesser**, called in Latine *Accipitres*,

The *more generous*, that are wont to be reclaimed and manned for fowling, called **Hawks**, which our Falconers distinguish into

┌ *Long-wing'd*, whose wings reach almost as far as the end of their Train, as the *Falcon, Lanner*, &c. Part 1. Sect. 1. Chap. 9.

└ *Short-wing'd*, whose Wings when closed fall much short of the end of their Trains, as the *Goshawk* and *Sparrow-hawk*, Part 1. Sect. 1. Chap. 10.

The *more cowardly and sluggish*, or else *indocile*, and therefore by our Falconers neglected and permitted to live at large,

┌ The *Greater*; The *common Buzzard, bald Buzzard*, &c. Part 1. Sect. 1. Chap. 8.

└ **The Lesser,**
  ┌ *European*; BUTCHER-BIRDS or *Shrikes*. Part 1. Sect. 1. Cha. 11.
  └ *Exotic*; BIRDS of PARADISE, Part 1. Sect. 1. Chap. 12.

Nocturnal, that fly and prey by night

┌ *Horned* or eared, as the *Eagle-Owl, Horn-Owl*, &c. Part 1. Sect. 2. Chap. 1.

└ *Without Horns*, as the **brown** *Owl*, grey *Owl*, &c. Part 1. Sect. 2. Chap. 2.

Frugivorous, called by a general name **Parrots**, distinguished by their bigness into the

┌ *Greatest* kind; called MACCAWS, Part 1. Sect. 3. Chap. 2.

├ *Middle-sized* and most common; called PARROTS and POPINJAYES, Part 1. Sect. 3. Chap. 3.

└ *Least* kind; called PARRAKEETS, Part 1. Sect. 3. Chap. 4.

More streight Bill and Claws, distinguishable into the

Greatest kind; which by reason of the bulk of their bodies and smalness of their Wings cannot fly at all, Exotic Birds of a singular nature; The *Ostrich*, the *Cassowary* and the *Dodo*. Part 2. Sect. 1. Chap. 8.

Middle-sized; which may be divided by their Bills into such as have

Large, thick, strong, and long ones; feeding either

┌ *Promiscuously* upon *Flesh, Insects*, and *Fruits*, distinguishable by their colour into
  ┌ *Wholly black*; The CROW-kind, Part 2. Sect. 1. Cha. 3.
  └ *Particoloured*; The PIE-kind, Part 2. Sect. 1. Cha. 2.

├ Upon *Fish*; as the KINGFISHER, &c. Part 2. Sect. 1. Chap. 7.

└ Upon *Insects* only; The WOODPECKER-*kind*, Part 2. Sect. 1. Chap. 5.

Smaller, and *shorter*; whose flesh is either

┌ *White*; The POULTRY-*kind*, Part 2. Sect. 1. Chap. 10, 11, 12.

└ the *Black*;
  ┌ *Greater*; the PIGEON-*kind*, Part 2. Sect. 1. Chap. 15.
  └ *Lesser*; the THRUSH-*kind*, Part 2. Sect. 1. Chap. 17, 18.

Least kind; called *small* **Birds** which are either

┌ *Soft-beak'd*, which have slender weight, and the most pretty-long Bills; which kind feeds chiefly upon Insects, Part 2. Sect. 2. Memb. 1.

└ *Hard-beak'd*, which have thick and short Bills, and feed most upon Seeds. Part 2. Sect. 2. Memb. 2.

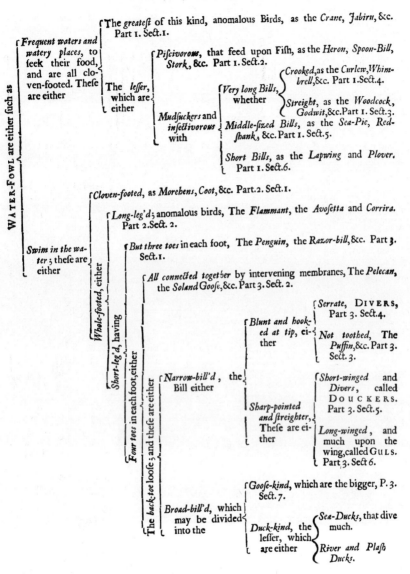

WATER-FOWL are either such as

Frequent waters and watery places, to seek their food, and are all cloven-footed. These are either

- The greatest of this kind, anomalous Birds, as the *Crane*, *Jabiru*, &c. Part 1. Sect.1.
- The *lesser*, which are either
  - *Piscivorous*, that feed upon Fish, as the *Heron*, *Spoon-Bill*, *Stork*, &c. Part 1. Sect.2.
    - *Very long Bills*, whether
      - *Crooked*, as the *Curlew*, *Whimbrell*, &c. Part 1. Sect.4.
      - *Streight*, as the *Woodcock*, *Godwit*, &c. Part 1. Sect.3.
  - *Mudsuckers* and *insectivorous* with
    - *Middle-sized Bills*, as the *Sea-Pie*, *Redshank*, &c. Part 1. Sect.5.
    - *Short Bills*, as the *Lapwing* and *Plover*. Part 1. Sect.6.

Swim in the water; these are either

- Cloven-footed, as *Morehens*, *Coot*, &c. Part.2. Sect.1.
- Whole-footed, either
  - Long-leg'd; anomalous birds, The *Flammant*, the *Avosetta* and *Corrira*. Part 2. Sect. 2.
  - Short-leg'd, having
    - But three toes in each foot, The *Penguin*, the *Razor-bill*, &c. Part 3. Sect.1.
    - Four toes in each foot, either
      - All connected together by intervening membranes, The *Pelecan*, the *Soland Goose*, &c. Part 3. Sect. 2.
      - The back-toe loose; and these are either
        - *Narrow-bill'd*, the Bill either
          - Blunt and hooked at tip, either
            - Serrate, DIVERS, Part 3. Sect.4.
            - Not toothed, The *Puffin*, &c. Part 3. Sect. 3.
          - Sharp-pointed and streighter, These are either
            - Short-winged and *Divers*, called DOUCKERS. Part 3. Sect.5.
            - Long-winged, and much upon the wing, called GULS. Part 3. Sect 6.
        - *Broad-bill'd*, which may be divided into the
          - Goose-kind, which are the bigger, P. 3. Sect. 7.
          - Duck-kind, the lesser, which are either
            - Sea-Ducks, that dive much.
            - River and Plash Ducks.

[†] The Wilughby classification of Aves, reproduced directly from Ray (1678).

section I
The vertebrate record and its study

1 Introductory considerations

THE SUBPHYLUM VERTEBRATA

Definitions and Relationships

The vertebrates form a subphylum of the phylum Chordata, which includes all animals that possess a notochord. This dorsal, axial, supporting structure is universally present among the vertebrates but is developed in relatively few adults, being confined in most groups to the early developmental stages. On the basis of this definitive structure, there is no question that the vertebrates are appropriately placed among the chordates, with demonstrable relationships to tunicates, acorn worms, and the cephalochordate *Amphioxus*. However, no known vertebrate is at all close to any of these, nor can any vertebrate be considered to have arisen directly from any of the groups they represent.

A substantial internal skeleton, in addition to the notochord, is a feature found throughout the vertebrates. It consists of two distinctive tissues, bone and cartilage, which by their several functions as supporting structures, sites for muscle origins and insertions, and protectors of vital organs have made possible the development of highly active animals and very extensive adaptive radiations. The skeleton and the radiations, which carried the vertebrates into many different environments, have resulted in an excellent fossil record. This is in sharp contrast with other chordates, which, except for the controversial graptolites, have essentially no record. The most commonly preserved tissues are bone and bonelike tissues, including teeth. Bone, as far as is known,

3

is unique to the vertebrates, and its occurrence in a specimen is considered sufficient grounds for assignment to the subphylum Vertebrata. Cartilage or very cartilagelike tissues, on the other hand, occur in members of some of the nonchordate phyla, apparently developed convergently. It is thus not a completely diagnostic feature, but as far as the fossil record is concerned, the presence of a substantial amount of cartilage can be considered as essentially conclusive evidence that the remains are those of a vertebrate.

The name of the subphylum Vertebrata implies that all of its members possessed some sort of vertebral structure, and this is almost certainly the case. The presence of vertebrae in some of the earliest and most primitive of the fishlike vertebrates has not, however, been demonstrated, and it is the presence of bone or bonelike tissues, plus general morphological resemblances to less primitive vertebrates, that has allowed assignments to the subphylum.

In addition to the notochord, vertebrae, and bone, vertebrates are characterized by a hollow dorsal longitudinal nerve chord that is expanded anteriorly to form a brain. Characteristic sense organs occur in the cephalic region, and these and the brain are partially encased in a distinctive series of cranial skeletal elements. The combination of these features is sufficient to separate members of the subphylum from all other major groups of animals. In

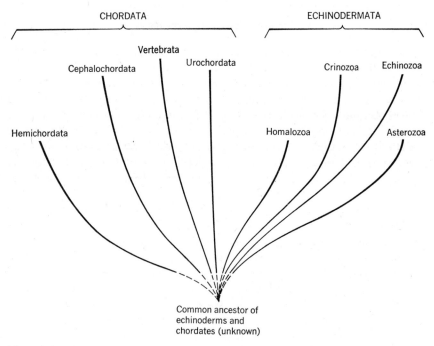

Figure 1. Suggested relationships of chordates and echinoderms shown in phylogenetic form.

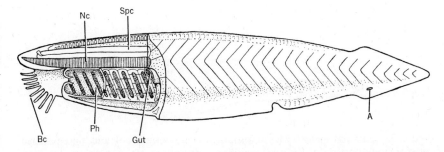

Figure 2. Lateral view of *Amphioxus* with the anterior portion cut away to show the internal features: gills, notochord, segmented musculature, pharynx, and so forth. A, anus; Bc, buccal cirri; Gut, general gut; Nc, notochord; Spc, spinal cord; Ph, pharynx. Redrawn from *A textbook of Zoology*, 7th edit., by T. J. Parker and W. A. Haswell, rev. by A. J. Marshall. Copyright © 1965 by Macmillan and Company, Publishers.

addition each of the systems has its own distinctive features that, for the most part, set vertebrates apart from all other animals, even from members of the other subphyla of the Chordata.

Neither the origin of the phylum Chordata nor the origin of the vertebrates within this phylum is known from direct evidence. Vertebrate ancestry has been sought in several groups of invertebrates, among them the Mollusca, Annelida, Arthropoda, Echinodermata, and of course Coelenterata. Part of the problem of origin has arisen from the fact that none of the well-known representatives of any of these lines, either living or fossil, can be shown to lie very near the chordate base. The resemblances between the larvae of some of the echinoderms and those of the chordates have for many years been taken as evidence of possible relationships between these two phyla, and more recently this has been strengthened by biochemical evidence. Currently most students agree that the chordates and echinoderms should be grouped together to form one of the major lines of descent within the animal kingdom (Figure 1). Presumably they had a common ancestor that diverged from stocks leading to other major lines before the beginning of the Ordovician period. The common ancestral line probably passed through a coelenterate level of organization, but the nature of the organisms at this level is almost entirely conjectural.

Amphioxus (Figure 2) is much the most vertebratelike of any of the known chordates that are not vertebrates. This small animal, although itself far removed from any vertebrate, possesses such features as a segmented body, a notochord as an axial support, and gill slits, all of which suggest a fairly intimate relationship. Both *Amphioxus* and tunicates are filter feeders. The mouth and pharyngeal structures of some of the very early fishlike vertebrates, although not known in great detail, suggest that this way of feeding

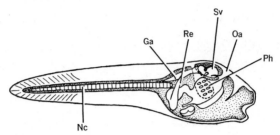

Figure 3.   Larval tunicate with a propulsive, segmented tail and unsegmented anterior body portion or "trunk" region. Ga, ganglion; Nc, notochord; Oa, oral aperture; Ph, pharynx; Re, rectum; Sv, sensory vesicle. Redrawn from *A Textbook of Zoology*, 7th edit., by T. J. Parker and W. A. Haswell, rev. by A. J. Marshall. Copyright © 1965 by Macmillan and Company, Publishers.

may have been the ancestral mode among the vertebrates as well. *Amphioxus* is mobile, like the vertebrates, whereas the tunicates, in their adult stage, are sessile. Tunicate larvae (Figure 3), however, are mobile, and their segmented tail and more anterior gill-basket structure have been the basis for suggestions that they may bear structural resemblances to vertebrate ancestors. Romer (1956c) has developed the theory of origin of the vertebrate body from the two sources suggested by the tunicate larva. The segmented propulsive portion is considered the source of the somatic portion of the vertebrate, whereas the gill basket and associated structures provided the visceral parts of the body. Persistence of the larval condition into an adult stage, perhaps paedogenetically, could have resulted in a mobile adult animal with at least some of the properties to be expected in a vertebrate ancestor. Such a creature may well have existed at some time in the remote past; however, there are no animals either among the known fossils or adults of living animals that approximate this hypothetical ancestor very closely.

Efforts to identify vertebrate ancestors among extinct groups have not been very successful. *Jaymoytius* (Figure 4), a soft-bodied, fishlike creature

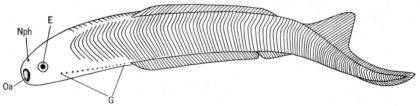

Figure 4.   *Jaymoytius.* One of several published reconstructions, which differ in many particulars. This one suggests a structure quite close to that of anaspid agnathans. Note round mouth, numerous gill openings, segmented body and nearly continuous fins. E, eye; G, external gill openings; Nhp, nasohypophyseal opening; Oa, oral aperture. (After Ritchie, 1968.)

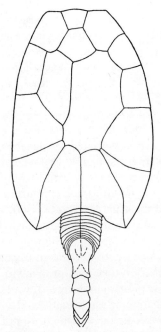

Figure 5. *Mitrocystella,* a carpoid echinoderm, showing the general resemblances to some of the early agnathan fishes. Note especially the anterior shield and segmented posterior portion. Length about 3 cm. (After Jeffries, 1968.)

from the late Silurian, has been cast in this role. Although its most usual association has been with anaspids, the creature has also been called a coelolepid and interpreted as a remote ancestor of *Amphioxus.* E. I. White (1959), noting a suggestion of Hardy (1954) that *Amphioxus* might be a paedomorphic cephalaspid, a modified ammocoete, returned tentatively to his original interpretation (E. I. White, 1946) that *Jaymoytius* might be a primitive ancestor of all of the Agnatha.

Jeffries (1968), following an earlier suggestion of Gislén (1930), made a case for chordate affinities of the stylophoran Carpoidea, usually considered to be echinoderms. These creatures range from Cambrian through Devonian and have typical echinoderm skeletons, some of which are rather reminiscent in outline of those of some agnathan fishes (Figure 5). Finding what he considered to be extensive homologies between chordate and echinoderm structures (notochord = chambered organ, dorsal nerve chord = peduncular nerve, brain = aboral nerve center, stem = fish tail, etc.), Jeffries concluded that the Cornutes and Mitrates (Stylophora), which he placed in a subphylum Calcichordata, were ancestral to the Urochordata, Cephalochordata, and Craniata (Vertebrata). The stem, shown in the vaguely ostracodermlike *Mitrocystella* in Figure 5, is considered to be equivalent on the one hand to the stem of crinoids and on the other to the tail portion of the ostracoderm. It is interpreted as containing muscle blocks posteriorly. That the carpoids or some part of their total occupy such an ancestral position has not been strongly favored by most students of vertebrate origins. The detailed analysis and unusual interpretations of Jeffries present the evidence for this proposition clearly but necessarily involve a great deal of speculation intermixed with some points of strong resemblance between the stocks in question.

The earliest known organisms universally accepted as vertebrates are the remains of fishlike creatures from the early and middle Ordovician. They are known mostly from small scraps of bone with some associated cartilagelike material (Denison, 1967c). A few larger pieces from the middle Ordovician and some assemblages of scales indicate that some of these creatures were shaped much like some of the heterostracans from beds of the late Silurian. They were jawless, in the sense that they lacked the usual vertebrate visceral jaws, were quite flat, and had a cephalic shield and presumably a propulsive tail. During the time from the middle Ordovician to the upper Silurian an immense amount of evolution of the vertebrates appears to have taken place, so that the first well-known creatures had departed in time, and probably in form, far from the very primitive vertebrates. The early fossil record, while giving some possible insights into the general nature of the primitive vertebrates and vertebrate ancestors, is deficient in specific information on both counts.

The unique features of the vertebrates, attained at least 400 million years ago, on the one hand imposed restrictions on the evolutionary radiation that followed and, on the other, opened up wide vistas of potential adaptive successes. It has been the latter aspect rather than the former that has been strikingly displayed over millions of years of vertebrate evolution.

This success was based primarily in the complexity and plasticity of the interrelated systems of the vertebrate body. For simplicity the body is commonly subdivided into a number of systems, such as the respiratory, digestive, reproductive, excretory, skeletal, integumentary, and muscular systems. None of these is unique to the vertebrates, for each finds analogous counterparts in various other groups in the animal kingdom, but each has many features that are peculiarly vertebrate. It is less the uniqueness and more the organization that has permitted the extensive modifications of the system interrelationships critical to the success of vertebrate radiations.

The actual history of the vertebrates, of course, must be read from the fossil record, even though most of the major groups do have representatives in the current world faunas. The history is recorded directly by the skeletal system, and much that is known pertains to the skeleton and the modest insights that it provides into the nature of the other systems. Functions directly related to the skeleton—for example, locomotion—can be fairly assessed. Some aspects of the muscular system and nervous system are evident because of the close relationships of their parts to the hard structures. To a lesser extent aspects of the vascular system and digestive system can be understood, the latter in particular as it is revealed in the mouth parts and teeth. The fossil record, however, provides very little information on the integumentary, reproductive, and endocrine systems or on organs of special sense other than their general location and size.

By means of interplay of data from existing animals at different stages of evolutionary development and the fossil record it is possible to arrive at a good understanding of the general evolutionary course of the various systems and to study their roles in vertebrate evolution. When this is done, the most distinctive feature of the subphylum Vertebrata becomes evident—namely, its capacity to adapt through evolutionary processes to an extremely broad spectrum of environmental circumstances and to a wide variety of modes of existence in these many circumstances.

THE AIMS AND PROBLEMS OF PALEOZOOLOGY

Paleozoology and Zoology

The studies of special interest to vertebrate paleontologists and paleozoologists are extremely varied, and seemingly the work is guided largely by the available materials. Although this is to a large degree true, almost all students are motivated by the common aim of revealing the history of vertebrate life within the general framework of evolutionary theory. In one way or another most of the diverse activities of collection, description, taxonomy, environmental analysis, and search for temporal relationships proceed under this unifying theme. Fundamental data are of two distinct kinds: morphological and temporal, or stratigraphic. On the basis of the former, animals are grouped into arrays that include similar and presumably related organisms; on the basis of the latter, these arrays are ordered in time. Subsidiary to the morphological data, which are directly obtained, are all of the aspects of biology that can be read into the record. Complementary to the stratigraphic data are the physical aspects of the rocks that aid in determination of the ecological setting and in making time determinations.

Considered under evolutionary theory, the temporal arrays of animals are interpreted as sequences of descent; that is, as phylogenetic or evolutionary sequences. The assessment of affinities on the basis of morphological features and the determination of temporal relationships are extremely intricate and require careful application of principles of biology, known from modern organisms, and of the geological sciences.

Because of the changing nature of the record of life with time, the levels of both taxonomic and morphological investigation differ notably as they are

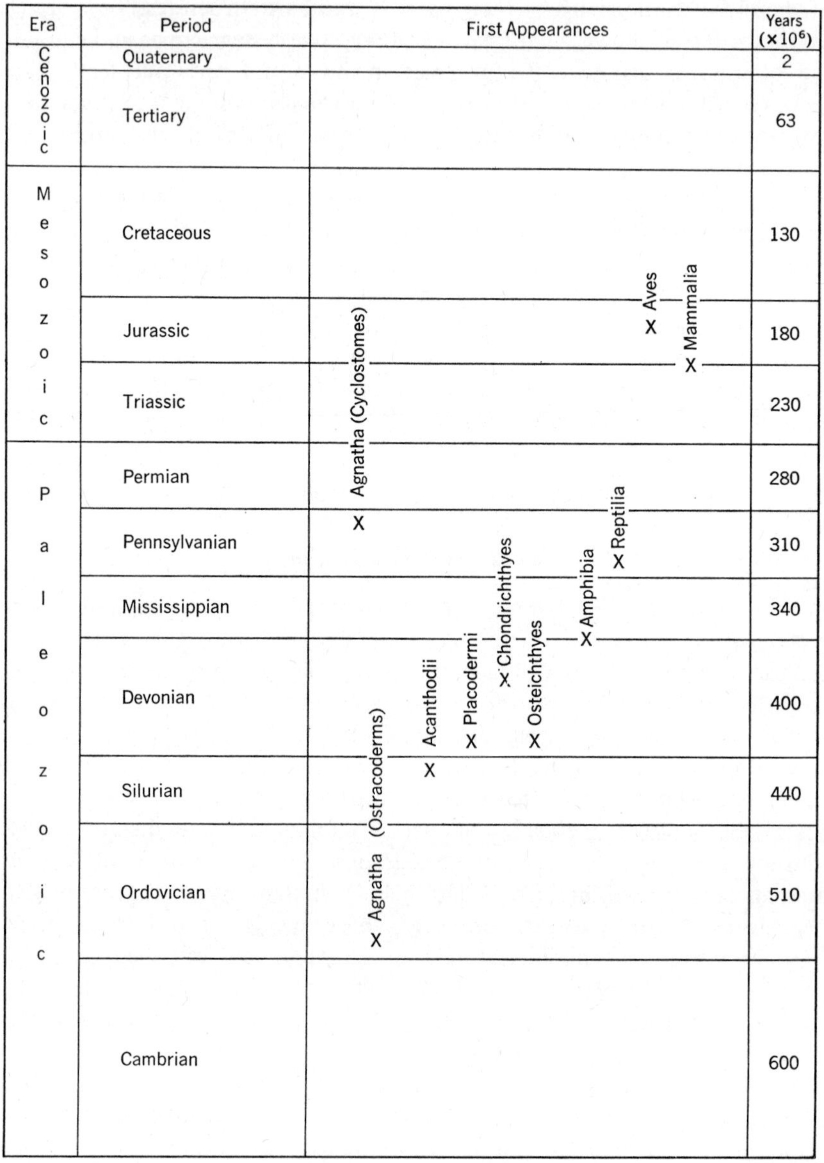

Era	Period	First Appearances	Years ($\times 10^6$)
C e n o z o i c	Quaternary		2
	Tertiary		63
M e s o z o i c	Cretaceous		130
	Jurassic	X Aves · X Mammalia	180
	Triassic		230
P a l e o z o i c	Permian	Agnatha (Cyclostomes) X	280
	Pennsylvanian	X Reptilia	310
	Mississippian	X Amphibia	340
	Devonian	Acanthodii X · X Placodermi · X Chondrichthyes · X Osteichthyes	400
	Silurian	Agnatha (Ostracoderms) X	440
	Ordovician	X	510
	Cambrian		600

Figure 6.   Diagram of the times of appearance of the major groups of vertebrates in the fossil record. Note that the cyclostomes and ostracoderms are entered separately, although it is quite possible that the roots of the former are to be found within the latter. Years is the time in millions of years since the beginning of the period.

applied to different periods of past time. The temporal framework likewise varies, with rather large units of the ancient past being replaced by finer and finer divisions as the present is approached. Time determination, which may be quite precise in the very recent past, becomes increasingly inaccurate the older the record is. Similarly, of course, knowledge of the totality of life both for the earth as a whole and for local sites decreases rapidly as the record becomes more ancient, and the taxonomic arrays that form the framework of evolutionary interpretations tend to become broader and more crudely conceived.

The classes of vertebrates appeared successively in geological time (Figure 6). First came the heterostracans, agnathous fishlike creatures. Then, in the later part of the Silurian, osteostracan and coelolepid agnathans appeared, accompanied by the first of the acanthodians. Shortly after, in the early Devonian, osteichthyan fishes appeared, then in succession came the chondrichthyan fishes (mid-Devonian), amphibians (late Devonian), reptiles (early Pennsylvanian), mammals (latest Triassic), and birds (mid-Jurassic). With the increasingly better record from ancient to modern times, it is inevitable that the levels at which different groups have been studied differ. Mammals, for example, are very recent in their major radiations and can be treated much more appropriately under concepts of evolution based on existing animals than can, say, amphibians, which reached their climax in the late Paleozoic and early Mesozoic, or, even more so, agnathous fishes, which, except for some recently discovered Pennsylvanian lampreys, did not extend beyond the Devonian as fossils.

All of these groups, however, are treated as if the same life processes applied and as if it were entirely appropriate to analyze their evolutionary histories under the broad tenets of evolutionary theory that have emerged from studies of modern organisms. This position is warranted by the morphological and temporal data. The decreasing capacity to make even approximately direct application of the population concepts of evolution, which derive from moderns, in times increasingly remote from the present, argues for some caution in acceptance of conclusions dependent on these concepts. At least the question must be raised as to whether phylogenetic sequences based on morphotypes well separated in time are in fact expressing something that is at all closely allied to evolutionary sequences displayed in successions of subspecies or species and exhibiting small genetic modifications whose perpetuation is expressed as natural selection.

However they may be considered in detail, evolution theory and the course of evolutionary history form central themes in paleozoological work, and the phylogenies that are constructed serve as one important end result of a large segment of paleozoological work. The phylogenies form scaffoldings on which the relationships of organisms are entered, and when depicted in the usual diagrams, they are graphic representations of the systematic interpretations of the investigators.

Phylogenies, however, are not the only goal toward which paleozoological research is aimed, for they themselves serve as a primary basis for deductions in studies of the modes of evolution as viewed within the temporal framework of geological history. The dynamic processes of evolution must be learned from living organisms, and it is under general theory so based that temporal arrays of fossils can be construed in a phylogenetic sense. Once so integrated, these constructions are generally considered legitimate data for determining "second level" laws of evolution. Understanding of the processes of evolution and recognition of its laws have been the aim of many of the outstanding students of paleozoology since the recognition in the nineteenth century that fossils did offer a picture of the coherent flow of life through time.

The need for intimate linkage of studies of paleontological and modern materials has been recognized by many students of biology over the years. It is through this recognition that the major contributions in the fields of paleozoology have been made. Successful syntheses characterize the work of such students as Cope, Osborn, Matthew, Abel, Watson, W. K. Gregory, Romer, and Simpson, to name a few who are both outstanding and diverse in their approaches. Cope early grasped and applied biological concepts to the fossil record and treated ancients and moderns under a single disciplinary concept. Osborn and Simpson, each in his own way, have synthesized evolutionary doctrines as appropriate to their times and under joint considerations of both the principles derived from neozoology and the information on history supplied by fossil remains. Abel and Watson, although very different in their basic approaches to problems, both arrived at biological interpretations of the fossil record by constant recourse to the data and principles of zoology. Gregory and Romer represent the fruitful school of comparative anatomy applied to the fossil record. Matthew, in one of his major fields of interest, bridged moderns and ancients in his approach to a synthesis of principles of zoogeography, a field also of interest to Simpson. Darwin, of course, sought ties between the past and present in his search for the concepts brought together in his theory of natural selection.

The primary link of biology with the past in all such studies is morphology, for therein lies the tangible common ground of materials between the actual creatures that live today and those that lived in times past. Morphology takes on the cloak of dynamics of the life processes from zoology, both with respect to the physiology and mechanics of living organisms and to the interactions between organisms and their environments. This is then cast in the perspective of time through interpretations of the morphology of the remains of creatures that have been found preserved in the rocks.

SPECIFIC PALEOZOOLOGICAL ACTIVITIES

Perusal of any representative collection of paleozoological research reports shows that the great bulk of work being done currently and carried out in the past is descriptive and systematic. The results of this unglamorous task form the mainstay of paleozoology and zoology, and the primary link between modern and ancient materials. A first step in the study of materials, once a collection has been made, is description and "alpha" taxonomy. This requires a full knowledge of the characteristics of the materials and ends with description and arrangement of these materials within a systematic framework.

The hundred and more years of active paleontological study of vertebrates have provided a wealth of materials that, when adequately described, form the basis for the many other kinds of study that can be made. Such studies tend to be grouped into three general categories:

1. Those that are concerned with particular sites or localities; that is, collections that have come from a limited area and are more or less contemporary in time.

2. Those that are concerned with events that took place more or less simultaneously over broad areas, continental or intercontinental in extent.

3. Those that are concerned with events spaced over appreciable amounts of time in the geological scale.

None of these, of course, is strictly definable, and the scope and limits tend to increase for each as older parts of the record are studied. It is studies of the last type, from which emerge the documented genetic reconstructions and their interpretations, which are peculiar to paleozoological studies and from which the major contributions have come.

Analyses of temporally and spatially localized events may be as restricted as descriptions and systematics, or they may be as extensive and complex as analyses of the ecology of the community represented by some collection of fossils. Within these contexts all aspects of the once living organisms become of interest. All the knowledge and techniques that can be brought to bear on the particular circumstances may come into play. Biologically such things as morphology, functional anatomy, physiology, food webs, ways of life of species, and species interactions are among those that are important. Physically the interpretation of the environment of deposition through geophysical analyses is of extreme importance, superimposed on more classical geological field observations. Bridging the biological and physical are the coordinated studies involved in taphonomic interpretations.

This type of study in general aims toward an understanding comparable to that sought by a student of modern organisms who investigates the many aspects of some particular locality. The accomplishments of the paleo-zoologist are of course much more limited. The knowledge of the organisms themselves and their systematics represents a positive increase in data on life not available otherwise. Interpretation of the nature of a community may have some intrinsic interest if it suggests a circumstance not found in modern times. Such studies also may serve to stimulate observations on living animals that have not been undertaken previously, to provide better models for paleoecological analyses. There is little chance that studies of particular sites will contribute much directly to an understanding of the biological or physical characteristics of species, interspecies populations, or ecology. The most important biological role of such studies is that they provide building blocks for syntheses of evolutionary history.

By inference, the information gained from a local site is often extended to a much broader area, serving as the basis for interpretation of the life and living conditions over a considerable range. This is one way of making generalizations concerning contemporary conditions over broader regions, the second type of study. Often several sites within some common province may be considered as contemporary and lead to more concrete understanding of the extent and variation of conditions in a more or less homogeneous environment of deposition. Coordination of studies of more extensively dispersed sites are directed toward the understanding of contemporary conditions over major parts of the earth. These, however, run into serious difficulties of temporal precision because of the vexing problems of correlation of the deposits.

There is potentially a wide and fascinating area of research on contem-poraneous faunas of the past, but for these studies the determination of the time relationships must rest on information that is not dependent on the organism. That a prime criterion of contemporaneity in the study of cor-relations of widely separated sites is the degree of similarity of the organisms poses a major problem. Physical criteria, independent of the life record, are used, but most of them are unsatisfactory for establishing the precise time relationships needed for studies of animal distributions. Roughly, time equivalences should be such that the effects of evolutionary change and modifications by faunal interchanges are negligible. The greatest chance for precision lies in the various methods of dating by radioactive materials. Only for the very recent past, however, are these accurate enough to bring studies of extinct organisms close to the sort of time references that are used in studies of modern zoogeography.

A result of this difficulty is that paleozoological studies of "contemporary faunas" from different areas actually involve considerable spans of time. Contributions to this error come partly from the difficulties of precise corre-lation and also, to a lesser degree, from the time spread involved in the accumulation of the fossils of a particular locality. Thus in older beds, say of late Paleozoic age, samples considered contemporary for practical purposes may have been taken from several feet of deposits or may, not uncommonly, have been lumped from sequences of as much as several hundred feet. Such practices are perfectly valid if it is recognized that the results and interpretations cannot be cast in the same framework as those from the more precise studies of modern organisms. At best even the most precise of the paleozoogeographic investigations that strive for contemporaneity involve an appreciable amount of time relative to the histories of the organisms they treat.

The contributions of these studies, even with their difficulties, are in themselves of interest in depicting distributions of life in the past and providing bases for comparing distributions at different times. They supply important criteria in studies of paleoclimatology and have been used extensively, if with some difficulties, in analyses of polar shifting and continental drift. Except for the time factor, the greatest stumbling block in the application of faunal distributions to such analyses is the very erratic nature of the geographical and ecological sampling at any given time in the past. This is of course an ever present problem in studies of modern distributions, but one that can in theory at least be overcome. Much of the potential record of the past, however, either was not preserved or, if preserved, was destroyed. In studies of distributions negative evidence—that is, the absence of some species or group of species—is critical. The absence of particular organisms from the fossil record is, however, a treacherous kind of evidence. Reliability is in part a function of time, with rapid decrease as remoteness from the present increases, and in part a function of the extent of sampling.

The third type of study is that which considers the time factor. It is in this sort of study that paleozoology makes its greatest contributions to an understanding of biology. Most paleozoogeographic studies are cast within a framework that uses broadly conceived "contemporary" faunas as a basis for analyses of changing distributions and interpretations of the meanings of these changes. The phylogenies that form the bulwark of evolutionary interpretation of past history are to some extent constructed with geographic distributions of their stages in mind, but the degree of dependence on this aspect of life history varies widely.

In Cenozoic deposits, with fair stratigraphic control and knowledge of the general disposition of continents and continental connections, the relationships between phylogenies and animal distributions may be drawn with some confidence, although this decreases even in the older portions of the Tertiary. In older deposits this is much less the case. Thus, for example, the phylogenetic sequence passing from primitive reptiles to very advanced mammal-like reptiles is based largely on an early segment from midcontinental United States, an intermediate one from European Russia, and a later segment largely from South Africa. It is true, of course, that parts of this record are known from other places. The most recent part, for example, is also known from other parts of Africa, South America, Southeast Asia, and, poorly, from Australia and North America. A strong tendency to think of the area in which fossils have actually been found as the primary center of evolution of the group involved crops up frequently in paleontological literature. Except for very recent deposits, where the record is sufficient to confirm or deny this idea, such a conclusion is rarely based on reliable evidence and can be very misleading.

The third type of paleozoological study provides not only the temporal sequences of evolutionary lines but also the potential for looking at these as parts of more complex structures as they change with time. Important among these are studies of complexes of animals, faunal analyses, of animals and plants together, community analyses, and finally of the total interacting complex of organisms and their physical environment—ecosystem analyses. Changes of these complex units with time are of interest in themselves and provide a meaningful framework within which the temporal changes of phyla may be interpreted.

Few paleozoologists, of course, work on their materials and their problems with a set, self-conscious goal within limits of one or another particular type of approach. What is done is guided by the opportunities that the materials at hand offer. The common theme, however—the theme that in general motivates paleozoological research—is organic evolution and its expression in a temporal framework by phylogenetic structures and their interpretations. We wish to know the course of history and what, in evolutionary context, caused it to be so.

THE NATURE OF THE RECORD

The General Nature of the Samples

Any collection of fossils can be viewed as a sample of some larger assemblage from which it was drawn, and this assemblage may be defined to suit the needs of particular studies. In this section it is designated as the universe of interest for ease of reference. In its broadest sense the total fossil record is a sample of the universe that comprises all the life that has ever existed on earth. Only some extremely small portion of this totality has persisted to form the sample, and there is no valid way to arrive either at the percentage of individuals present or the completeness with which constituent taxonomic units are recorded.

The adequacy of representation is primarily a function of the objectives of the use of the samples. As discussed in the preceding section, even if these are limited to primarily biological problems, they may be highly varied. Very different sample characteristics, for example, are necessary to one who wishes to determine the species composition of a fauna and to one who is interested in the properties of a particular species. Ideal to the former would be a sample with at least one representative of each species; but one individual, although better than none, falls far short of meeting the needs of the latter. The paleozoologist who wishes to construct a morphotypic sequence representing what he feels may be a major line of descent needs only a sampling that provides well-preserved individuals of the line at rather widely spaced time intervals, with little or no geographic control. If interest lies in lesser changes of a developing line, say in a chronofaunal context, closely spaced samples taken over a relatively restricted geographic range may best serve the needs.

If we take as major aims of the paleozoologist those outlined in the last section, then his success depends to a large extent on his ability to relate the characteristics of his materials to the properties exhibited by the living creatures from which they came. Inference from the dead to the living is a constant process in paleozoology, and the extent to which a sample can serve as a basis for such inferences, as appropriate to the special questions being asked, is a measure of its potential effectiveness.

The Formation of Samples

The remains that finally reach the hands of the investigator have travelled a long course from the living population, through taphonomic processes, diagenesis, and accidents of preservation of deposits—and then exposure, discovery, and collection. In the course of this journey the number of individuals is immensely reduced and, more importantly, preservation is highly selective. Many students have given this matter careful study, most recently Clark, Beerbower, and Kietzke (1967). Some of the principal steps are illustrated in Figure 7 to show important landmarks in the continuous process. Between these steps repeated samplings by various natural processes have taken place. Of the losses that occur, some are trivial, some important, and the significance can best be judged in context of the uses to which a sample is to be put. The sequence illustrated in Figure 7 applies to single species and to larger and more complex units, such as interspecies populations or even ecosystems.

The processes of taphonomy are those involved in the passage from life to burial and are very general in their effects. They are particularly pertinent to the first and second steps in the diagram. On burial the diagenetic processes take over and continue until the time that collections are made, through steps 3 to 6 inclusive. After burial, of course, erosion of deposits becomes an

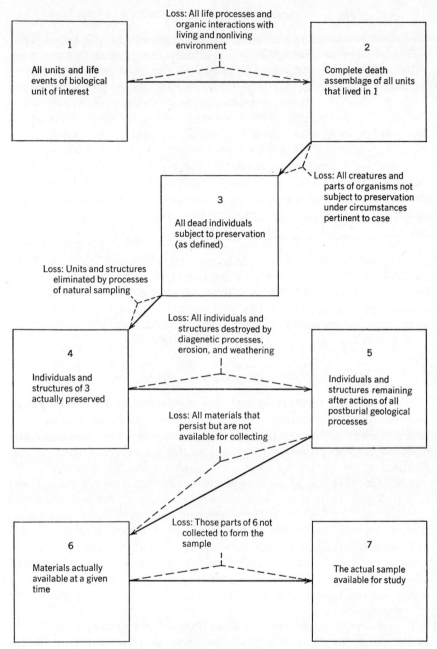

Figure 7. A diagram of the course from the living populations into the fossil record and to the final sample that provides the data for study of the nature of the population. In the diagram interest is focused on the losses incurred between the successive stages. See text (pages 17–24) for discussion of other aspects.

important factor both in determining what of the original materials persist and what part is exposed at any given time at the surface. Weathering, as fossils approach the surface, is a special aspect of diagenesis that may act at the very last step or at any intermediate time along the way as deposits intersect the surface of the earth.

Some of the important aspects of the steps in Figure 7 and some of the invervening processes are as follows:

Step 1. This includes all the life events of the biological unit of interest. During its life span some finite number of individuals lived and died, contributing to the total death assemblage of step 2. Ultimately, it would seem, it would be ideal to know all that can be known about the total population, for this is the final basis for all other interpretations. Yet much of this information, even if it were attainable, would be of little significance to paleozoological studies. It cannot be encompassed in a framework sufficiently broad to be of importance at the scale at which paleozoology can make contributions that are not merely confirmations of things better known from studies of moderns.

Step 2. This represents the complete death assemblage, or the universe of all individuals that lived in step 1. The life processes of individuals, interactions between individuals, and those between individuals and their biological and physical environments have been lost. If the living population of step 1 is to be reconstructed from the assemblage of step 2, understanding must come from samples of the universe of 2. Studies of such samples are an important activity of zoologists as analyses of anatomy, genetics, or some aspects of population structure are made to obtain information without direct observation of biological processes and interactions.

Step 3. This includes all of the dead individuals that are subject to preservation in the rocks. Whereas steps 1 and 2 are precisely definable on the basis of a fixed number of individuals, this is not the case for step 3, which is subject to interpretation. Technically all organisms are preservable, making step 3 a possible equivalent of step 2. Usually, however, only hard parts are preserved. In any event it is necessary to define for any given case in which inferences about the death assemblage are being made just what is meant by this assemblage. This is of prime importance in many studies, for this is perhaps the most common sample-to-universe inference that is made in paleozoological studies. In most studies of vertebrates, where samples are specifically concerned, step 3 is considered, either consciously or otherwise, to include those animals that had some sort of skeletal elements.

Step 4. This represents the universe of all individuals of some specified unit 3 that were actually preserved. It is thus the portion of step 3 which formed the initial burial assemblage and from which the fossil deposits have eventually come. The totality of this assemblage may not, of course, have been buried at the same time, and erosion may have removed parts of it before later contributions were made.

It is between stages 3 and 4, which bridge death and burial, that the critical roles of natural processes of sampling first come into play. The multitude of factors that affect the outcome—that is, the nature of the actual burial assemblage—include aspects of the living population, where, when, and how its members lived, the duration of the period or periods of contribution relative to the longevity and population structure, sorting and mixing in the processes of transportation and burial, and actions of other organisms on the dead. Each of these has been treated in detail in various treatises and papers on paleontology and paleoecology (e.g., Ager, 1963; Johnson, 1960; Beerbower, 1963; and Clark, Beerbower, and Kietzke, 1967).

The essential point that emerges from these considerations is that many "biases" are introduced and that these complicate immensely the study of stage 3 on the basis of stage 4 and all subsequent stages.

Step 5.   After burial, diagenetic processes act on the deposits and erosion may remove some or all of what was originally laid down. Although erosion and diagenesis are very different in character and have potentially different biasing effects, they comprise the sum of events that intervene between burial and the time of collection of the sample. What remains after they have acted constitutes the sample that exists at the present time, here illustrated as step 5.

Broadly, diagenetic change can be assessed in the context of the shifts in the chemical and physical relationships between the buried remains of organisms and their physical and biological environments. These may be quite stable or very unstable. As for taphonomic processes, the relationships of diagenetic processes to the preservation of fossils is a field of study in itself, one to which much attention has been given both by geologists and chemists. The results of the various processes are differential preservation and destruction of buried remains. In addition to outright destruction, various degrees of distortion and breakage may occur. These processes are highly selective with respect to materials, structures, and sizes of fossil remains. In general diagenetic effects are greater for older deposits, but the range of variables is so great that the age factor may be overshadowed.

Erosion may act during the time of deposition and at any time thereafter. The biases that it introduces tend to be less severe for samples of species or local faunas than for those representing a more complex universe with considerable geographic and temporal extent. Deposits formed under different conditions do not have equal chances of escaping destruction by erosion, so that biases occur that favor persistence of some types of life records over others. Deposits formed on stable continental platforms tend to suffer different fates from those formed in subsiding geosynclines, and those formed on highlands have little chance of survival as compared with contemporary deposits accumulated in lowlands. As in the case of diagenesis, the factor of the time of deposition is important, but other variables modify its effects. In general the spectrum of environments that persist is greatest for very recent deposits and narrows rapidly in the more remote geological past.

Step 6.   The materials that are actually available to the paleontologists at a given time constitute an important landmark. Essentially they are controlled or limited by what fossil-bearing beds, those of step 5, intersect the surface of the earth at the time collections are made. This is of course somewhat modified by man's activities, as his excavations may enlarge his site or as operations in connection with mining, quarrying, or highway construction may reveal otherwise unexposed fossiliferous beds. Access to what is preserved, step 5, is generally very limited. Just how this biases various samples with respect to step 5 and earlier steps is not always immediately evident. For a single-species population exposure of only a small part of a deposit may introduce only minor difficulties, since the outcrop may present a fairly random sample. If, however, there were segregating factors—for example, different life zones of young and old or differential transportation of size groups—the story might be quite different. For more complex populations, say an interspecies population, the distributions of the elements in the rocks may be such that considerable nonrandomness may be introduced if but some small part of the total is collected.

Present topography, geography, and climatology of continents have an immense effect on what parts of all preserved fossil materials are exposed. Vegetative cover is of course very important. The factors that determine what parts of the preserved record are available are extremely complex. Even within the recent past marked changes in distributions of climate have taken place, and hence it is possible to make only very broad generalizations. Earlier, when differences were much greater and possibly complicated by polar shifts and movements of continents, relationships to present distributions provide essentially no guide.

Step 7.   The final process, which supplies the sample for study, is collecting. Sampling by collectors can have two strongly biasing effects. One, which is rather easily overcome, is the selectivity of the collector. Any experienced field paleontologist is aware of the practical difficulties of sampling—how one generally takes what he can get under the circumstances and time available, how different objectives modify what is taken as well as what is seen, how the ease or difficulty of access may be important, and how any of almost numberless considerations of like kind alter the nature of the sample.

When the totality of exposures that could yield fossils today is considered, another important factor emerges—namely, the accessibility of the various areas of exposure to the collectors. This has had a major effect on the constitution of present-day collections. Many aspects of the record of life of the past have been strongly influenced by the accessibility of areas and the intensity of collecting. Added to this is the intangible factor of what has been popular, which has influenced the intensity of searches for one or another type of animal and deposits of one or another age. Concurrently with the increase of ease of travel and transportation and availability of more ample funds, explorations have spread rapidly to many places little studied or unstudied heretofore. A major change in knowledge of the past is being made by this recent increase in world coverage.

The final product of these series of events, the sample, is very different from the parent universe from which it came. The severity of this difference can be measured roughly by an assessment of the difficulties that arise in making the inferences to the living populations necessary for a particular study.

For some types of study, the processes of inference require only an assumption that a specimen, which may be considered the sample, does actually represent a biological entity from the past. It may then be considered taxonomically and many sorts of biological studies can be made on it. From this most simple case we may pass through a series of more complex situations, and in each the number of assumptions necessary for successful study increases.

The effects of sampling both by natural processes and by man are significant in all types of paleozoological work. The stringent restrictions of probability theory bring these sharply into focus in studies that involve inferential statistics. The requirement of randomness of a sample with respect to its universe is of course fundamental and satisfied, if at all, by very careful considerations of the definition of the universe in terms of the probable events of sampling (see Olson, 1957b). Determination of many biological properties of populations known by samples from the fossil record requires this kind of study, and results may be seriously hampered by the commonly unknown sample-universe relationship.

Some types of qualitative studies—for example, constructions of morphotypic series—may suffer little, since the major requirement is that some minimal part of the record be preserved, with sufficient spread to bring out the major steps in a sequence. The greatest difficulties arise because of near or complete loss of some significant part of the record, often, it seems, during the critical initial stage of a radiation. Even in such cases, however, although quantitative and probabilistic statements are not possible, the general sample universe context provides a profitable way of looking at the nature of samples and aids in arriving at logically derived conclusions.

Regardless of the nature of the problems treated and the methods used, information on the sample-universe relationship is fundamental. Whether conclusions are sought within a quantitative or qualitative framework, or by a combination of the two, the broad relationship of the part to the whole must be constantly considered. A danger of qualitative studies lies in the tendency to relax the stringency with which this relationship is considered.

2 The vertebrate record

GENERAL FEATURES

From its beginnings in the scant and fragmentary remains in the Ordovician period the vertebrate record follows an irregular course of expansion in kinds of animals, kinds of environments sampled, and in the spread of localities over the world. As representation from local sites improves and as spatial and temporal continuities increase, the kinds of problems that can be treated successfully multiply.

The fossil record of the very recent past merges with the present, with overlapping temporal and geographic ranges of living and extinct species of animals marking the transition. The results of extensive studies of this part of the record, exemplified by such detailed analyses as those of Beringia and the Bering Straits in recent years (see, for example, Hopkins, 1967), show the great potential of joint uses of zoological and paleozoological data. Distributions and physical characteristics of extant populations of both man and associated animals and plants can be neatly related to those of recently extinct populations. Minor evolutionary changes, routes of migrations, and, in the case of man, cultural changes can be viewed in the perspectives of but a few thousand years. The efforts of zoologists, archeologists, anthropologists, and paleozoologists merge in the treatments of common problems and materials.

Even in such studies problems of incompleteness and bias of the record occur, but since they stand out clearly, they may be recognized and evaluated. Thus such investigations provide models for interpretations of more ancient records where the discrepancies are less clear. As more ancient deposits are treated, however, the utility of these analogs becomes less valuable, for the kinds of animals are increasingly different, the record less complete and more biased, and the environments, both physical and biological, less like those recorded in the very recent past.

The differences between the recent records and those of earlier vertebrates result partly from the increasing inadequacies of the samples. Important as well, but more difficult to assess, are the changes that have occurred in vertebrate assemblages as new major groups and new adaptive types have arisen, creating new environments by exploitation of an ever wider spectrum of the physical attributes of the earth. These attributes themselves have altered through time, providing constantly changing circumstances for the vertebrate radiations.

One can hardly doubt that vertebrate evolution followed a course from shallow aquatic environments to the open waters of the seas and to the lakes and streams of the continents, over the strandline onto the land, to the uplands of the continents, into the air, and, many times, back to the water once more. Many of these steps were taken over and over again, by different groups with different levels of organization and at different times. This is the clear picture that has emerged from paleontological studies.

Yet many facets of this story are not supported by positive, first-order evidence from the fossil record. Strong but secondary support, which brings many details into a common fold, rests on the fact that the major steps in the occupations of the environments by successive radiations *can* be accounted for by the evolutionary sequences outlined in the last paragraph. This is not, of course, a proof that these were the actual courses of events. Essentially they represent the simplest satisfactory framework. For example, all of the tetrapods of the late Paleozoic could have come from a transition from water to land in the later part of the Devonian period. There is no need to postulate a substantially earlier emergence on land, multiple penetrations, or even more complex events. The record shows that amphibians had come into being in the very late Devonian. None are found in earlier deposits, some of which are of suitable kinds. As far as the evidence goes, it supports origin during the later part of the Devonian, but in view of the scantiness of the samples, this evidence shows positively only that tetrapod emergence was *no later* than latest Devonian.

Similarly the evolution of Permian reptiles and their Triassic descendants does not require postulation of unknown upland faunas. The only known deposits, however, were formed in the lowlands in rivers, swamps, and deltas, so there is no way of knowing that there were no upland faunas. However, in context of later events and the general framework of evolutionary thought, none is needed.

Earlier in the record, during the Ordovician and Silurian, the deposits formed under usual marine conditions contain no remains of fish or fishlike creatures. It seems reasonable to suppose that no such creatures inhabited these depositional environments, but this could be the result of sampling deficiencies. All inhabitants might, for example, have been soft bodied and unlikely to have been preserved. The fact that Devonian marine fishes do not require the existence of such Ordovician and Silurian ancestors lends support to the supposition that they did not exist. An equally confident conclusion that there were no early abyssal marine fishes has been reached, but of course there are no records whatsoever to test it.

In later parts of the record many major groups of vertebrates appear without any evidence of close, direct ancestors. One well-known case is that of the turtles, which appeared in the middle Triassic. Once again the general evolutionary framework, aided by a few cases in which origins of new groups are known from the record, provides a guide under which it is confidently assumed that the turtles and other similar "ancestorless" groups did have ancestors that bridged the gaps between fully matured known representatives and much more primitive stocks from which they are presumed to have come. The gaps are frequently filled, often graphically, by reconstructions of the presumed intermediate organisms.

The vertebrate record thus supplies a sample of past vertebrate life that shows very well part of what was actually present. Vertebrate history has been built on this fragmentary base, with heavy dependence on some of the patterns that the best parts of the record reveal, taken in conjunction with the flow of life in context of the general precepts of evolutionary theory. Explanations and interpretations are generally cast within the simplest explanatory framework commensurate with the factual information.

The nature of sampling by biological and geological processes seriously reduces the chances of finding critical evidence that might refute some part of the accepted story. That most of the available deposits formed during the Silurian or early Devonian would, for example, carry tetrapods, even had they been present on land, is unlikely in view of their predominantly marine nature. Most freshwater deposits appear to have been formed near the seas. The rather special conditions necessary for burial and preservation of vertebrates are generally at best realized by only a very small percentage of all deposits formed in environments in which such creatures lived. Thus, even with much greater expanses of terrestrial beds than actually exist, it is quite possible that land vertebrates could have been present and remain unrecorded.

Upland Permian reptiles, were they present, would have had to find their way into lowland deposits to have been preserved. Even had they made this unlikely transition, the chance of their being recognized as upland forms is slight. In general, organisms living in environments in which deposits are formed in abundance and preserved in quantity are those that contribute to the fossil record.

Important exceptions do occur, however, and these can be significant in understanding the ecological distributions and geological ranges of the animals of the past. The soft-bodied invertebrates preserved in the Burgess shales of Cambrian age provide a good example. Vertebrates from tree stumps in the middle Pennsylvanian of Nova Scotia represent another such case. Localities in the upper Pennsylvanian of Illinois, Kansas, and Colorado, atypical for this time in not being of the "coal measures" type, have produced evidences of the beginnings of Permian red-bed amphibians and reptiles little known otherwise. Fissure-fill deposits from the Triassic–Jurassic of the British Isles have yielded a wealth of materials of very primitive mammals and of early lizards little known elsewhere. Among more recent deposits, the Baltic ambers of Oligocene age have produced perfectly preserved insects in which even the color remains intact.

The role of such atypical samples, produced by some unusual circumstances, diminishes as the record becomes more complete. Even in times as recent as the later Tertiary, however, chances of burial, preservation, and discovery are vastly different for organisms living under different environmental circumstances. The kinds of problems that can be studied and the reliability of answers do improve as the present is approached, but attrition in the early phases of the sampling process persists to the level of subrecent deposits, right to the threshold of the present time.

THE TANGIBLE VERTEBRATE RECORD

Records of Fishes

The Earliest Vertebrates

Not all students of vertebrates agree as to precisely when in geological time the vertebrate record begins. There is no question that vertebrate remains are present in deposits of middle Ordovician age, for they occur in abundance in the Harding sandstone of Colorado and in similar deposits in Wyoming and South Dakota in the United States. Toothlike remains have come from the early Ordovician, from the glauconitic sands near Leningrad in the Soviet Union. The shape and histology of these fragments suggest that they came from vertebrates. The principal difficulties concern the problematical remains known as conodonts. These small fossils assumed a variety of shapes, some rather toothlike. Their morphology and histology have suggested that some of them at least may have come from vertebrates. If so, they extend the record a little farther back in time and also into types of marine rocks in which vertebrates are otherwise unknown. The evidence for vertebrate assignment is at best tentative, and conodonts are not included in most discussions of vertebrate origins or environments because of their equivocal nature.

The earliest verified vertebrates thus may be considered to be of early Ordovician age. The nature and relationships of these first representatives are unknown. The middle Ordovician vertebrates, however, are members of the heterostracan agnathans. Remains are mostly fragments of bone, but a few larger plates and aggregations of plates have been found. On the basis of these and of later heterostracans some idea of the general form and habits can be had. These were fishlike creatures, the anterior parts of whose rather flattened bodies were covered by an ossified shield. Plates in some were associated with a cartilagelike tissue.

What can be known from these earliest remains is limited, but it is of greatest importance. They give some information on the morphology at a very early stage of vertebrate development, in particular on the histology of the bony shield. From this has arisen speculation on the probable nature of the initial vertebrates and the origin of the subphylum as a whole. A problem concerning the primitiveness of bone in the vertebrates has originated from various sources, and, of course, the first vertebrates have been pertinent to it. A more prominent role, however, has been in studies of the environment of the origin of the vertebrates. Proponents of both a freshwater and a saltwater origin have used the evidence of the Ordovician heterostracans to support their views. The beds in which these creatures occur originated in a near-shore, marine environment, but the question is whether this was the life environment of the animals or whether, as some argue on the basis of their fragmentary nature, they were washed in.

Whether there were other kinds of vertebrates living during the Ordovician is an open question. Marine beds are widespread, and marginal marine deposits occur in several places, but there are no known, clearly defined non-marine deposits. If there was any life on the continents, either plant or animal, its record would have to come from marine deposits. As far as animals are concerned no trace of such evidence has come to light. Unless conodonts be considered vertebrates, no records of vertebrates living in open seas have been found.

The next representatives come from many millions of years later, in beds of middle and late Silurian age. Vertebrates at that time were much more diversified than those of the middle Ordovician, but the degree of change is not such that it *requires* a greater diversification than is known from the Ordovician, if it be granted that the known Ordovician forms might represent a common ancestral stock. If this is not the case, and there appear to be good reasons for considering it unlikely, then some other source must be sought. What it may have been is wide open to speculation. It is unlikely that the few Ordovician vertebrate sites did tap the totality of vertebrate types of that time. It has been argued, for example, that the main line or lines of early vertebrates were unossified creatures and that the origin of the major stocks is obscured by this condition. If so, the fossils from the Harding sandstone might be considered as an early offshoot that acquired ossification. In the absence of "exotic" finds in rocks of Ordovician or Silurian ages the questions cannot be answered and their discussion, although interesting and useful, must remain largely speculative.

The Silurian Expansion

After a long hiatus, from the middle Ordovician to the late middle Silurian, the geological record once again includes remains of vertebrate organisms. The fauna of the later parts of the Silurian is rich in comparison with that of the middle Ordovician. All remains pertain to agnathans and gnatho-stomatous acanthodians. Some of the agnathans, the heterostracans, are fairly close to the known Ordovician vertebrates, but others—osteostracans, anaspids, and coelolepids—are members of different major groups, sometimes given class distinctions. Nowhere are unmistakably recognizable ancestors of Placodermi, Osteichthyes, or Chondrichthyes to be found, although sources of each group have been at one time or another sought within the acanthodians.

The widespread marine limestones and shales of the upper Ordovician and the lower and middle Silurian have yielded no remains of vertebrates. Deposits that appear to have been formed under near-shore conditions and possibly even on land are known from the margins of the ancient land of Appalachia, to the west of the highlands. Only in beds formed during the very late part of the middle Silurian have any traces of bone been found, although vertebrates evidently were in existence. There must have been continuity between those of the middle Ordovician and late Silurian unless multiple origin of the vertebrates be granted. Arguing strongly against this as an explanation is not only the constancy of structure in all vertebrates but also the fact that some of the late Silurian forms are fairly closely related to those known from the Ordovician.

During this long, unrecorded interval a major expansion seems to have taken place. Not only are several different kinds of vertebrates present in the Silurian, but there is considerable evidence of assemblages that are adapted to different environmental conditions. Gross (1951), Romer (1955a), Denison (1956), and E. I. White (1959) have all considered the composition and life environments of the late Silurian vertebrates in detail in their assessments of the probable habitats of the initial vertebrates. Dension recognized an osteostracan–anaspidan–eurypterid assemblage, living in fresh to brackish waters, and a heterostracan–acanthodian assemblage in marine waters, with acanthodians at least largely marine, and the soft-bodied coelolepids euryhaline.

Even for this early stage in the vertebrate record, paleoecology is important to the study of vertebrate evolution. In Europe a temporal sequence passing from essentially marine to nonmarine deposits has been traced in a single section along with accompanying changes in the contained vertebrate faunas. Remains are well enough preserved for it to be possible to learn much about the morphology of some of the fishes and to employ this both in studies of functional adaptation to the environments and in interpretation of phylogenies based on morphological resemblances and differences of the animals from these and later beds (Figure 8).

One other major step forward, made possible by the greater spread of vertebrate sites in the Silurian, is the analysis of similarities and differences between assemblages from different geographic regions. Classic sites occur in the British Isles, around the Baltic, extending east to the Urals, and in North America, from Canada south along the Appalachian chain. Assemblages from these areas differ in details but bear such strong resemblances to each other that close connections between them seem quite certain.

The positive aspects of the upper Silurian are so important that they tend to overshadow the relative incompleteness of the environmental and geographic sampling. Actually much of the world remains a blank.

At this time, as at others, the negative aspect of the record is of two types. The first involves known deposits in which the organisms of concern have not been found, and the second, those areas in which there are no deposits of the age in question. The second type can have only very indirect significance relative to the distributions themselves but, of course, presents possibilities of extensive manipulation in explanation of known distributions from contemporary fossiliferous areas.

The first type of negative evidence has very wide use, but the usage has some curious aspects. Absence of one or another group of organisms is often taken as evidence that it did not exist in the area of deposition. The absence of vertebrates from typically marine Ordovician and Silurian deposits is a case in point. The credibility of such conclusions is difficult to assess and has been the focal point of many of the disputes among paleontologists about the fossil record. No reliable yardstick exists to test the level of credibility or to indicate when a condition of virtual certainty is reached, as in the case of typically marine Silurian fishes.

Clearly an important aspect is the number of instances. In the Silurian, for example, but one case of marine beds lacking fish would mean essentially nothing, 100 cases much more, and 1000 cases might be convincing. But an important factor in this case is that marine fishes are not expected, since they are not necessary to explain what occurred before or after. Many barren sites

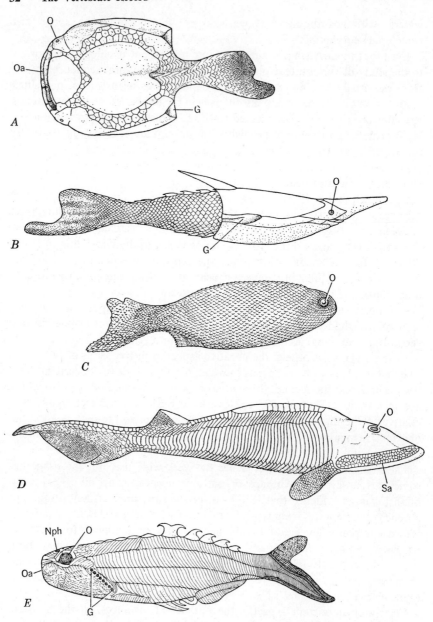

Figure 8.   Various agnathans. *A, Drepanaspis*, a heterostracan; *B, Pteraspis*, a heterostracan; *C, Phlebolepis*, a coelolepid; *D, Hemicyclaspis*, an osteostracan; *E, Birkenia*, an anaspid. G, gill opening(s); Nhp, nasohypophyseal opening; O, orbit; Oa, oral aperature; Sa, sensory area; (*A* and *C* after Obruchev, 1943, 1964; *B* after E. I. White, 1935; *D* after Stensiö, 1964; *E* after Heintz, 1958.)

in red beds formed under deltaic conditions in the Triassic could not similarly be taken as evidence that tetrapods did not inhabit the area of deposition, for it is to be expected that they were in fact there. Yet the tangible evidence from the deposits in the two cases is identical, and it is the general reading of the evolutionary record that makes for differences in assessment of the evidence.

Even though they are much broader and varied than their Ordovician predecessors, the vertebrate faunas of the Silurian do not provide a sufficient evolutionary base for the vertebrates of the early Devonian. The absence of osteichthyans and placoderms has been noted. The relative diversification of each in beds of the lowest Devonian epoch implies that they were present, unless it is assumed that rates of evolutionary diversification were inordinately rapid. Both appear to have been initially freshwater groups if their environment at their first appearance is a trustworthy criterion. Although late Silurian freshwater deposits are not abundant, they do exist. Most seem to have been formed near the sea, and some may have been deposited in moderately saline waters. It is very possible that the unrepresented groups of fish were present in kinds of environments for which there is no record, but at present there is no possibility of testing such a conclusion.

The kinds of postulates necessary to reconcile the Silurian and early Devonian records run throughout the course of vertebrate history. It is only by the use of some model, such as that of vertebrate evolution, gradually built up over the years under evolutionary theory, that such postulates have meaning.

The Devonian: Age of Fishes

The expansion of vertebrates first encountered in beds of late Silurian age continues into and through the Devonian. By the middle of the period all classes of fishes have come into the record, and only the class Chondrichthyes is absent from rocks of early Devonian age. The actinopterygian Osteichthyes are only questionably represented in the lower Devonian, but both Dipnoi and Crossopterygii are present and morphologically quite distinct.

The end of the Devonian is marked by two significant events: the disappearance of ostracoderms and placoderms, except for a few remnants, and the appearance of the first tetrapods.

The classic lower Devonian rocks occur in the British Isles, and there the transition from Silurian to Devonian is such that no marked break occurs in the sediments. This sequence includes the Ludlow beds, which include a widespread bone bed. On the basis of the appearance of large numbers of vertebrates, E. I. White (1950) has taken this bed to mark the base of the lower Devonian. Denison (1956), however, has stressed the continuity with the Silurian on the basis of the resemblances of the vertebrate assemblages. The Ludlow series is the oldest in which vertebrates have come to play a role, albeit a controversial one, in the establishment of the major stratigraphic break.

By the middle and late parts of the Devonian period rapid expansions were taking place within the classes of fishes and fishlike animals. This involved three more or less separate but interdependent kinds of events. First there was a sharp increase in the kinds of organisms, both with respect to major groups and to the diversity within groups. The most important additions were the Chondrichthyes and Actinopterygii, both in the middle Devonian. Sharks appeared in marine deposits and very soon after their initial appearance are represented by a diversity of kinds. The source of the sharks is not known in the record, and this has occasioned speculation both as to their phylogenetic and environmental source. It has been suggested on the basis of the soft anatomy and physiology of modern sharks that the origin was in fresh water. If this were the case the presumed soft-bodied ancestors might well have been missed in the relatively few freshwater beds known from earlier times. Or it is possible that the ancestors lived in types of freshwater environments for which there is no record. The record itself gives no basis for supporting one sort of speculation or another.

Actinopterygians, rare at first appearance, expanded rapidly to begin the complex radiation of ray-finned fishes that was to dominate the scene and culminate in the teleosts. Their early appearances are in both freshwater and marine deposits, and at the outset they are quite distinct from Dipnoi and Crossopterygii, which had appeared somewhat earlier. Their source is uncertain, although derivation from the acanthodians seems a possibility.

The second important aspect of Devonian fishes was their appearance in an increasingly varied array of deposits. Most striking was the successful occupancy of typical marine environments, away from the strandline. Similarly they appear to have expanded into a variety of freshwater environments, although it is more difficult to assess how much of this was the result of actual radiations and how much the result of a better sampling of these environments.

The third important aspect of the record is its increasingly wider geographic spread during the Devonian. Fish-producing localities of middle and upper Devonian age are not only dispersed over North America and Europe but occur as well on the far side of the earth, in Australia.

The result of the expansion of the record is that by the Devonian a major vertebrate radiation is well enough represented to permit most of the kinds of study possible on the fossil record as a whole. Some of the most detailed morphology of extinct vertebrates has come from agnathans and placoderms of this time. Phylogenies of various lines, based it is true on rather widely spaced morphotypes, have provided insights into the course of evolution in these groups. Ecological studies have played an important role in efforts to explain the courses of various lines of evolution. For the first time it has been possible to determine something concerning worldwide distributions of vertebrates and make at least some advances in the understanding of ancient zoogeography.

Despite all these successes, studies of this period have been less effective than those of the present and the recent past. Localities, while covering a broad spectrum of time and space, are sporadic in their distribution, cover relatively small parts of the total earth area, and are difficult to correlate in time. Preservation of the material is often poor and preparation is difficult and time consuming. Many types of organisms, in particular the agnathans and placoderms, are very different from any modern fish, so that interpretations of their physiology and habits pose problems.

Some of these difficulties have been overcome by the many paleontologists who have specialized in fishes of the Paleozoic. Many, of course, cannot be mastered. Much of the understanding of life and evolution of this time of great radiation of the fishes must be considered in very general terms. Nevertheless, the first major success of vertebrate evolution, beginning in the record of the early Ordovician and continuing to the end of the Devonian, has been well enough preserved for the record to provide a penetrating insight into the evolution of groups of organisms not known in other parts of the vertebrate record.

Fish After the Devonian

It is convenient to make a somewhat artificial subdivision of the vertebrate record beginning with the end of the Devonian. Near the end of this period the first tetrapods appeared and began to pass through their early history, which, as far as the record is concerned, somewhat resembles that of the early fishes. The fishes, on the other hand, after the Devonian radiation, have a record dominated by the actinopterygian Osteichthyes. Most of the prominent Devonian groups are either absent or very much reduced (Figure 9). It is thus useful to treat the post-Devonian record of fishes as a unit, rather in relationship to the accompanying record of the tetrapods. This does, however, tend to divert emphasis from the important fact that the interdependence of the tetrapods and fishes at various times after the Devonian had important impacts on the evolution of both groups.

Acanthodians, which had first appeared in the Silurian, persisted past the Devonian into the Permian and then disappeared. Only in a limited number of assemblages are they an important element. Chondrichthyes are widely scattered and not uncommon in both marine and nonmarine deposits. In large part they are represented by teeth and spines, which are notably unsatisfactory for either systematic or evolutionary studies. Throughout the post-Devonian, as in the Devonian, a few exceptional sites have preserved

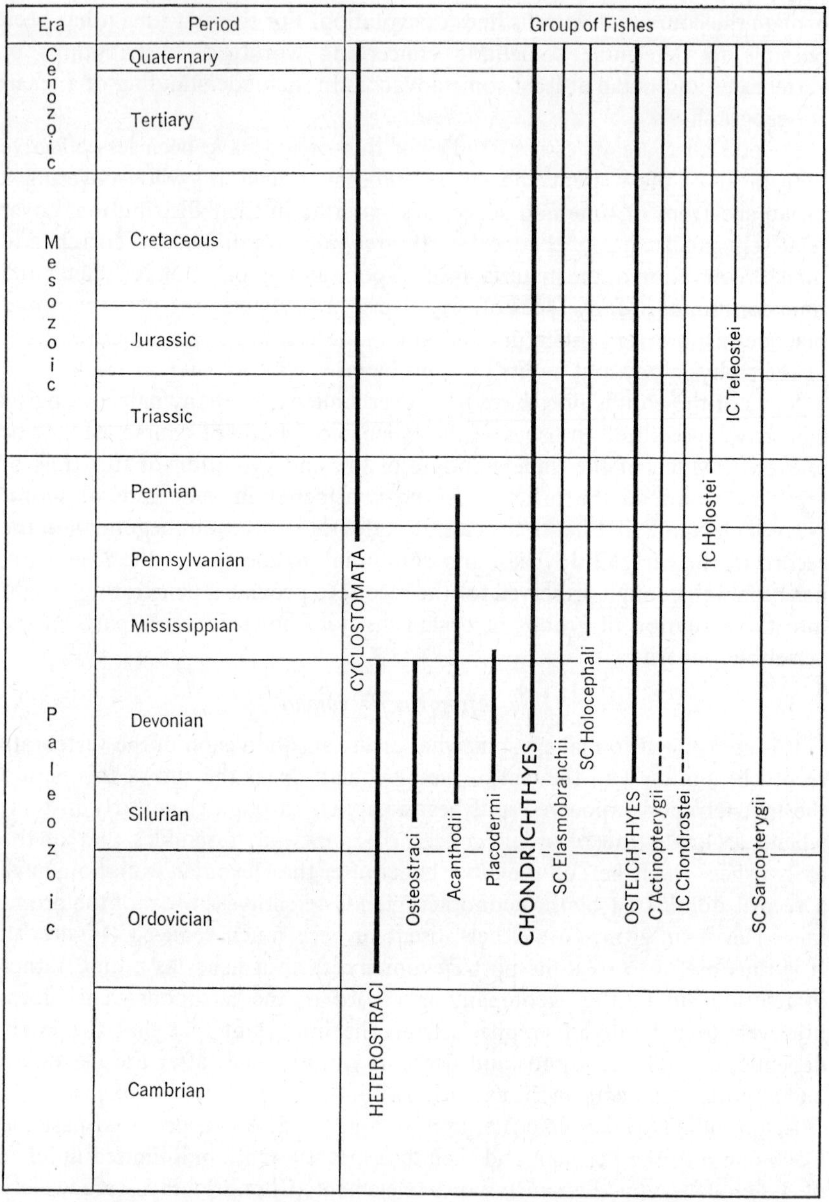

Era	Period	Group of Fishes

Figure 9.   The ranges of the major groups of fishes. Heavy lines represent classes, medium lines subclasses (SC), light lines infraclasses (IC) (for Osteichthyes only). Cyclostomes entered as a class, but may not form a coherent unit. The ranges are illustrated as continuous, but many of them have major gaps, and the connection is based on the assumption that there was actual continuity.

skeletons of sharks. These have given a basis for a rather general reconstruction of shark phylogeny, but one that is at present in a far from satisfactory state. The problems are considered in the classification of the Chondrichthyes in Section III, Chapter 4.

Among the Osteichthyes only the actinopterygians are consistently abundant in post-Devonian deposits, and their record, of course, fluctuates markedly during the different periods. Crossopterygians reached their peak in the Devonian. One stock, the rhipidistians, continued on in fresh water into the Permian; the other, the coelacanths, invaded marine waters and persisted to the present time, leaving a rather scant record. Lungfishes, the Dipnoi, also underwent a major radiation in the Devonian and thereafter recorded a slow evolutionary course in sporadic occurrences from the Mississippian to the present. Very early and several times during their history lungfishes occur in marine deposits, suggesting the possibility that some may have lived in this medium, although they are predominantly inhabitants of fresh waters. Two groups of lungfishes exist today. The course of evolution suggested by the record, that of relatively little change since the Paleozoic and no important adaptive radiation after the Devonian, is sufficient to explain their structure, dispersal, and modes of existence.

Actinopterygians reveal a very different pattern, one of successive radiations after attainment of new levels of organization. The first evidence of the earliest radiation is in deposits of middle Devonian age. This radiation continued into the late Devonian, with a fair record, and on into the next period, the Mississippian, which has yielded a wide range of adaptive types. It involved the highly diverse group of fishes classed as paleoniscoids. The variety of Mississippian paleoniscoids indicates that acceleration of the deployment began in the Devonian, but the actual record is limited and comes from a relatively few localities in the British Isles and the Appalachian region of North America. How representative it is of the full range of paleoniscoids of the time cannot be determined from firsthand evidence.

The paleoniscoid and subsequent actinopterygian radiations are treated in detail in Section IV, Chapter 3, and the problems of classification that have resulted from the complexities of the radiations are taken up in Section III, Chapter 6. A brief résumé of the history of occurrences will be sufficient to round out the general account of the nature of the vertebrate record being treated in this section.

Paleoniscoids occur in abundance in widespread Pennsylvanian deposits, which were formed primarily in coal swamps. Permian sites are more restricted, although geographically fairly extensive. End members of the various radiations are prominent in the Triassic and continue on into the lower Cretaceous in reduced numbers. Replacement by a new radiation, at the holostean grade, began with the latest Permian. Transitional genera are abundant in the early and middle Triassic, and the late Triassic and Jurassic deposits are characterized by the major holostean expansion.

The records of the early phase of actinopterygian evolution are well enough preserved to show the great complexities involved. Somewhat similar, although less complex, radiations probably occurred among earlier groups of fishes. The scantier records, however, tend to obscure the patterns and to give a false sense of simplicity. Even the record of the paleoniscoids has so many temporal, geographical, and environmental discontinuities that it has been impossible, to date at least, to put together more than a very broad outline of its evolutionary history.

From the late Jurassic on, the third, or teleost, phase of actinopterygian radiation is dominant. Worldwide distributions of well-preserved assemblages of fishes from many kinds of environments have opened the way for extensive and intensive studies on many facets of teleost evolution. What has emerged, however, is not a well-understood evolutionary picture, but rather immense confusion about the multitudes of fishes arising from the extreme intricacy and complexity of the adaptive radiations.

Discussions of the problems of classification of the teleosts in Section III, Chapter 6, illustrate the difficulties that arise when efforts are made to reconcile classifications based on living and extinct animals. This has been one of the major sources of confusion in understanding the teleosts. Two other factors are also involved. One is merely the fact that the scope of the radiation has been very great—much greater, it would seem, than in other actinopterygian groups. A second is the relative completeness of the fossil record. The latter, however, always influences estimations of the actual complexity of a radiation, and the presumed simplicity of earlier radiations may be in some part the result of the decreasing completeness of the fossil record.

Tetrapod Records

General Patterns

The geological time scale of the Phanerozoic was developed largely from studies of marine rocks and their invertebrate faunas. Boundaries between time units indicate breaks in the biological and physical aspects of these deposits. Although this scale is the temporal framework under which the record of terrestrial tetrapods is usually treated, no close correspondence between marine and nonmarine records actually exists. To some extent they are reciprocal. At times of widespread epicontinental seas, when continents were low records of nonmarine deposits tend to be few but the radiations of the marine invertebrates appear to reach full expression. As the seas were expelled and lands became higher, producing breaks between one period of marine deposition and the next, terrestrial sedimentation became more wide- spread, and the record of tetrapods tends to increase.

The composition of faunas and the evolutionary courses that they follow have been strongly affected by the physical events of the earth's history, although it is difficult to assess the effects except in very broad terms. Much of what may appear to be the result of an impact of physical change on evolutionary processes can also be explained as the result of the sampling patterns induced by the physical changes. Notable exceptions do occur, as at the end of the Cretaceous period, but unless detailed documentation is possible, decisions on the relative weight to be given to the factors are difficult.

The best and most widespread records of tetrapods occur in the Permian– Triassic and in the late Cretaceous and Cenozoic. Both of these time spans were marked by extensive orogenic activity of long duration. The records are only fair in the Carboniferous, very poor in the lower Carboniferous, the Mississippian, and somewhat improved but very biased in the upper part, the Pennsylvanian. With the beginning of the late Paleozoic orogeny, more wide- spread and varied environments began to be represented, reaching a high point in the late Triassic.

Jurassic and early Cretaceous records have come from relatively few areas and from rather special environmental circumstances. With the beginning of the late Mesozoic Laramide revolution, records improve, to become excellent for the late Cretaceous. With a short hiatus during the very early Tertiary, the terrestrial record continues to be good and to improve steadily through the Cenozoic.

The selective preservation of the vertebrates during this time is dependent partly on the general geological history. Also important and in some degree related is the evolutionary status of the vertebrates at successive stages. The problems of interpretation are like those of earlier parts of the record, but in less remote times the various influences stand out somewhat more clearly.

The record shows successively a major radiation of the amphibians, be- ginning in the Devonian and climaxing in the Pennsylvanian, a dual reptilian radiation including the birds, and finally a mammalian radiation. The reptilian radiation began in the Pennsylvanian with the early anapsid–synapsid lineage that persisted into middle Triassic and was continued to the end of the Cretaceous by the archosaur–lepidosaur–sauropterygian complex.

Each of the major radiations found its source within faunal complexes of an earlier one and traces of these beginnings are commonly found among the constituents of the previous radiation. To some undeterminable degree, however, it appears that important elements that led to succeeding radiations arose in environments other than those from which the dominant faunal aspects of any one time have been determined. Although some elements of each of the major radiations have some known antecedents from times prior to their flowering, many do not—and this fact has resulted in some of the most controversial and interesting interpretations of the fossil record.

Another feature of the geological and paleontological record of tetrapods that biases its interpretation is the tendency for a greater percentage of the sediments formed to be preserved from the more recent deposits. This applies not only to the total amounts of sediments formed but also to the percentages deposited in various environments. Only in most recent times, for example, are sediments formed in restricted depositional areas within a generally eroding province preserved in the record. To complicate matters still more, the percentages of highly preservable and readily destroyed beds actually formed vary widely and irregularly through geological time.

These aspects of the terrestrial record play important roles in the interpretations of tetrapod history. What seems to be a very satisfactory overall concept of evolutionary history has resulted. Yet this understanding has come only from bits and pieces that remain from an almost infinitely greater universe of past life. The record has mostly been gleaned from relatively small samples widely spaced in locale and time. Detailed understanding of many of the isolated localities is excellent, but connections between them, as a rule, are tenuous and must be made under some general integrating concepts of the history of life.

The Devonian–Carboniferous Record—Amphibians

A skull roof, presumably amphibian, from Nova Scotia and skull and skeletal remains of at least two distinct types of amphibians from Greenland form the upper Devonian record of tetrapods. Most of the deposits of the next period, the Mississippian, are marine, but in the British Isles and along the Appalachian region of North America are freshwater and estuarine deposits of this age. From them, as well as associated marine deposits in the British Isles, have come some remains of amphibians. The bone-bearing horizons range from middle to late Mississippian in age and consist of sediments ranging from coarse clastics through fine, carbonaceous shale to limestones.

Only a few genera of amphibians have been obtained from the Mississippian, but significantly they fall into five orders and include members from the two major subclasses of ancient amphibians. A major adaptive dispersal was well under way, but its beginning and early stages, presumably middle and late Devonian in age, are as yet unknown (Figure 10).

The Pennsylvanian situation is different. Many parts of North America and Europe lay near the sea level and were subjected to repeated flooding by shallow seas. The periods of inundation alternated with periods during which vast amounts of vegetation were deposited in coal swamps. These cyclical

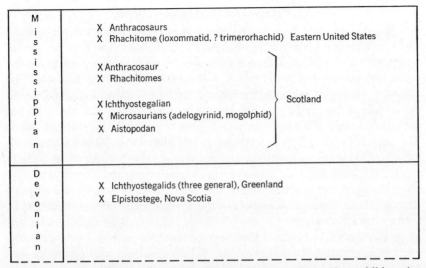

| M
i
s
s
i
s
s
i
p
p
i
a
n | X Anthracosaurs
X Rhachitome (loxommatid, ? trimerorhachid) Eastern United States

X Anthracosaur
X Rhachitomes
X Ichthyostegalian
X Microsaurians (adelogyrinid, mogolphid) } Scotland
X Aistopodan |
| D
e
v
o
n
i
a
n | X Ichthyostegalids (three general), Greenland
X Elpistostege, Nova Scotia |

Figure 10.   The Devonian–Mississippian (Lower Carboniferous) amphibians (see also Figure 135). The groups are arranged in approximate order of appearance. Most are represented by very few specimens. Correlations in the Mississippian are somewhat uncertain so that placement of various genera as lower, middle, or upper is open to some question. Correlation between the European and American fresh-water deposits in particular is uncertain.

series of deposits are generally called coal measures, and from them have come many vertebrates, often referred to as coal-measures faunas. Amphibians and freshwater fish predominate, but remains of a few reptiles have been found.

The variety of amphibians is immense in comparison with the few known from the Mississippian. The Pennsylvanian is usually considered to mark the peak of radiation of this group. The record, however, has a strong bias toward aquatic and semiaquatic animals. Strictly terrestrial vertebrates would occur in coal-measures formations only if transported into the en-vironment of deposition. The records indicate that this happened rarely.

Reptiles and amphibians from the early part of the Permian suggest that faunal elements not preserved in the coal measures and not part of the common coal-measures complex existed during the Pennsylvanian, somewhat dispelling the impression from the swamp deposit that the Pennsylvanian amphibians were strictly aquatic. Some bits of evidence from the Pennsylvanian itself support this conclusion. From beds deposited during the middle part of the period in Nova Scotia have come remains of reptiles and am-phibians that are of a more terrestrial character than those from typical coal measures. From a few sites in the later Pennsylvanian, in Illinois, Kansas, and Colorado, have come reptiles and amphibians that correspond more closely to those found in the early Permian.

The Permo–Triassic Record—Reptiles

Red beds dominate the terrestrial deposits of Permian and Triassic age and carry with them a variety of red-bed faunal complexes that become successively advanced and complicated with time. Coal measures persist into the lower Permian, especially in Europe, but soon disappear from the scene. The record of tetrapods is almost exclusively semiterrestrial and terrestrial until the middle of the Triassic, when marine reptiles appear, and the localities are rather limited and widely separated geographically during this time span. The early Permian is best represented in midcontinental United States, with lesser records in the more eastern and western parts of this country and minor records in South America, central and eastern Europe, and Siberia. The next stage, the basal part of the upper Permian, is displayed in deposits in European Russia. These overlap the section in the United States, below, and that in South Africa, above. By the early Triassic, additional productive sites are present in central and eastern Europe, mostly rather poor in fossils, in India, southeast Asia, and in South America. Vertebrates from South America and Africa are very similar, and, although collections are less good and not thoroughly worked, the same appears to apply to faunas of the other areas from which tetrapods have been obtained, except perhaps central Europe, where facies are quite different. The middle Triassic sees a continuation of much the same picture, but with the important addition of marine beds in which important tetrapod invasions are recorded. By the late Triassic terrestrial deposits are very widespread and even include a site recently discovered in Antarctica.

Deposits of the Permian and early Triassic were formed largely on the floodplains and in the channels of large streams. Much of the deposition took place as the rivers approached the seas and formed systems of major deltas. Lowland deposition was extensive. Presumably deposits also were formed far from the sea in higher lands, but few, if any, have been preserved. Faunal complexes existing during this time consisted of integrated units of reptiles, amphibians and freshwater fishes. Successively three more or less recognizable complexes appeared, and added to these in mid-Triassic was the marine complex of fish and reptiles. The first may be termed the *anapsid–pelycosaur* complex. It consisted of reptiles of these two groups closely associated with a large array of terrestrial, semiterrestrial and aquatic amphibians, a wide variety of osteichthyan fishes, and freshwater sharks. This is the typical early Permian red-bed assemblage, best known from the lower Permian beds of the United States. Its roots lay deep in the Pennsylvanian, the time when the adaptive deployment of most of its reptilian elements must have occurred.

By the beginning of the upper Permian (Kazanian of the Russian section), this complex began to give way to what may be called an *anapsid–therapsid* complex. Pelycosaurs were replaced by their more advanced descendants, therapsids, and new anapsids, procolophons and pareiasaurs, replaced the captorhinomorphs, dominant earlier. Labyrinthodont amphibians, mostly aquatic, remained important, as did paleoniscoid fishes. In its early phases, in North America and Russia, this anapsid–therapsid complex, like its forerunner, was closely associated with aquatic environments, but in the later Permian and early Triassic, as seen especially in South Africa, this was lessened and exploitation of more upland conditions commenced.

The anapsid–therapsid complex reached its peak in the early Triassic and continued vigorously into the middle Triassic. Very likely, as the rather sketchy record suggests, it had a worldwide distribution. Scant representation in the Northern Hemisphere, however, has given rise to the notion of a southern distribution, over a continental mass known as Gondwanaland. The distribution has been used in support of the idea that Africa and South America, India, Southeast Asia, and perhaps Australia formed the major portion of this continent. That such a continent may have existed seems highly likely, but the absence of information on faunas of the Northern Hemisphere and the worldwide distribution of faunas of the late Triassic argue for caution in applying the faunal data as evidence concerning the distribution of land masses in the early and middle Triassic.

The third phase of Permo-Triassic tetrapod radiation consists of the complex *archosaur–lepidosaur–sauropterygian* phase. It includes those reptiles that give rise to the major reptilian groups of the later Mesozoic, the last remnants of labyrinthodont amphibians, and highly specialized, persistent, mammal-like members of the therapsids. In the very late Triassic the first mammals appeared. This faunal complex replaced the anapsid–therapsid complex and in so doing marked a major change in the direction of tetrapod evolution.

The roots of this new radiation, in part, are found earlier among some minor elements of the anapsid–therapsid complex. Most, although not all, of the genera occur in close association with the members of the anapsid–therapsid complex. Much the earliest are the lizardlike eosuchians of the Permian. Close to them, probably derived from them, are the rhynchocephalians. One group of this order, rhynchosaurids, underwent a strong expansion in early Triassic to become an important element of the anapsid–therapsid complex in some places.

Archosaurs have a problematical Permian representation but are known in the early Triassic from a fair variety of extremely primitive members, grouped as proterosuchians. Many are very poorly known and assignment is difficult. They are widely scattered geographically and in most cases occur as minor constituents of the anapsid–therapsid complex.

In absence of anything but the most sketchy data, two points of view may be taken with respect to the relationships of the anapsid–therapsid complex and the base of the great expansion of the archosaur–lepidosaur–sauropterygian complex. One recognizes in the scant and scattered remains that are found associated with therapsids and anapsids the ancestors of a number of the lines in the later radiations. The other considers these minor constituents of the dominant complexes in the record as offshoots of other contemporary complexes. Much of the ancestry of the later radiations is conceived to have been in the unknown "central" structure of these complexes. Both explanations have merit and difficulties and are much more fully explored in the fourth section of this book. Relevance of the record to this general survey is found in the fact that the record allows such multiple interpretations, without pointing to a clear decision between them, and that the consequences of following one or another line lead to very different readings of the meanings of later parts of the record.

In central Europe and southwestern United States marine beds of middle Triassic age contain a variety of marine and semimarine reptiles. Best known are the ichthyosaurs, nothosaurs, and placodonts. Earlier records show almost nothing about the ancestry of these groups, although it is possible that some of the puzzling areosceloids may be related to the nothosaurs. Generally, however, it appears that the lines that led to these marine animals developed in regions or environments for which no record has been found. Similarly lacking a record of ancestry are the chelonians, which appeared as "good turtles" in the very latest middle or earliest late Triassic.

The record up to and into the middle Triassic appears to be confined largely to one major aspect of tetrapod evolution, that of the anapsid–synapsid complexes, with evidences of other radiations, which must be postulated in view of the later record, appearing sporadically within these complexes, very occasionally in other circumstances, and, for the most part, not at all.

Later Mesozoic Records—Reptiles

The great Mesozoic radiations of the reptiles, with their multiple phases and intricate faunal complexes on land and in the water, as well as in the air, have become so much a part of the common lore that to dispute the fullness of our knowledge verges on heresy. The land radiations produced the age of dinosaurs and the marine radiations the time of the sea monsters. Birds came into being, and the mammals followed a tortuous course of existence awaiting the demise of their betters.

But the record in fact is less than one to inspire confidence in much of what has been written in a popular and semipopular vein. The distribution of terrestrial vertebrate-bearing beds (Figure 11) shows great gaps in the geographic distributions at most times and little continuity between the times during which remains of vertebrates were preserved. Only vertebrate-bearing

Period	E.Eu.	C.,W. Eu.	B.I.	E.NA.	W. NA.	S.A.	Mongolia, China	India	S.Af.	E.Af.	N.Af.	Aust.
Upper Cretaceous		x	x	M	(X) (M)	(X)	(X)	(X)		x	x	x
Lower Cretaceous			x	(o)	(X)		x		x			(o)
Upper Jurassic		(M)	(M)		(X)					(X)		
Middle Jurassic			x							x		
Lower Jurassic		(M)	(M)		?		(X)					x
Upper Triassic	?	(X)	(X)	(X)	(X) (M)	(X)	(X)	(X)	(X)	(X)	x	x
Middle Triassic	x	(M)			x	(X)			(X)	(X)		x
Lower Triassic	X	x (M)			x	(X)	(X)	(X)	(X)	x		x

Figure 11. Geographic distributions of occurrences of vertebrate remains during the Mesozoic era. (X), major terrestrial site; x, minor terrestrial site; (o), small terrestrial site, few specimens; (M), major marine site: (M), partly marine site; ?, age in question; – – – –, site being developed.

deposits of the late Triassic and the late Cretaceous are well distributed over the earth and include a variety of depositional types. Otherwise producing localities are restricted both geographically and environmentally at any one time.

As a result of this pattern of preservation, records from local sites are in many instances excellent, but relationships between sites are poorly established. Detailed patterns of descent, zoogeography, and the interrelations of different radiations are difficult to assess.

The wide dispersal of upper Triassic beds, however, comes at a critical time, one which marks the initiation of the archosaur and lepidosaur radiations, the final phases of the therapsid radiation, and the initiation of lines that eventually led to the much later radiations of the mammals. This appears to have been one of the crossroads in the history of tetrapod life.

After this time, during the early and middle Jurassic, the vertebrate record, except for a few bright spots, is very poor. Marine reptiles are better known than nonmarine forms, for a considerable part of the evidence of the latter comes from creatures that were accidentally carried into areas of marine deposition. One site in India, being currently developed, has yielded a truly terrestrial early Jurassic fauna.

From the Morrison beds of the western United States and the Tendaguru beds of east Africa have come the principal finds of upper Jurassic terrestrial reptiles. Both appear to have tapped lowland, swampy environments and have yielded similar faunas. The marine record continues to be somewhat better. The few sites in the basal part of the lower Cretaceous, in England, in the adjacent continent, and in Asia, have produced faunas not greatly different. Some good collections, partly unstudied, have come from the upper part of the lower Cretaceous of North America. For the most part, however, there is a marked hiatus between the well-known tetrapods of the upper Jurassic and those known from the upper Cretaceous.

The late Cretaceous record comes from widespread beds of many types in which a wealth of well-preserved vertebrates has been found. A truly spectacular radiation of reptiles on land, in fresh and marine waters, and in the air is well recorded from localities spread over several continents. Terrestrial reptiles, in large part, although broadly related to those that have been found in earlier deposits, lack close antecedents in the earlier radiations. To a lesser degree this applies to the marine reptiles and the pterosaurs, for continuity is somewhat better in marine deposits of the Jurassic and Cretaceous.

Many of the remains in both the terrestrial and marine deposits are those of very large creatures, much as they are in the deposits of the upper Jurassic. This aspect of the record, particularly well shown at this time, has a very strong bearing on its utility. Large skeletons are difficult and expensive to collect, prepare, study, and store. They make spectacular exhibits in museums but are far from ideal for scientific study. Sampling of the preserved remains tends to be severely restricted, and the small samples, plus the major difficulties of study, tend to limit investigations to the simplest kinds of taxonomy and morphotypic phylogeny and to rather crude studies of morphology and function. Not all studies have been at this level, but even the most sophisticated are hampered by the very limited numbers that can be assembled. The large mammals of the Cenozoic have posed similar, although usually less severe, problems.

Various kinds of reptiles that had been prominent earlier appear to have decreased in numbers during the Cretaceous and, in some cases, to have become extinct well before the end of the period. Prior to the end of the period the radiations had passed their peaks. Many lines, both on land and in the sea, however, are well represented late in the record and, it seems, disappeared nearly simultaneously at what has been designated as the end of the Cretaceous period. This constitutes one of the great extinctions of history; unlike some others, it involved many kinds of terrestrial animals and both vertebrates and invertebrates of the oceans.

Data on extinctions must, of course, always depend on negative evidence, on the absence of remains of the animals in question from the deposits that follow the presumed time of final death. It becomes virtually impossible, in view of the incompleteness of the record to pinpoint the time when some groups or arrays of groups did in fact terminate. Apparent suddenness may be a function of the accidents of sampling.

The apparent extinction at the end of the Cretaceous, however, although surely subject to this difficulty, involves so many groups in many environments, with cross-references between the events, that its reality and relative rapidity can hardly be doubted. As far as the admittedly difficult world corelation of beds formed under different circumstances can be carried, it does indicate that the disappearance of several major groups of vertebrates and invertebrates from the record was essentially contemporaneous. In no case have the members of any of these groups, dinosaurs, marine reptiles, or ammonite cephalopods, been found in beds that on other grounds can be considered to have been formed during the Cenozoic.

That a widespread and rather rapid extinction occurred at the end of the Cretaceous now seems amply demonstrated by the record. The reasons for it are much more difficult, perhaps impossible, to determine from what the record shows. Of the multitudes of suggestions that have been made to explain part or all of the extinction, some may contain elements of a solution. But this is hard to demonstrate. It seems certain that any extinction, whether it be that of a species or of a series of ecological complexes, must ultimately find its causes in disruptions of the dynamics of ecological interactions. Unless such interactions can be fully understood for any case under consideration, and for extinct groups this must be speculative, discussions of factors involved in the disruption can at best be tentative.

It is possible to understand the basic aspects of processes of extinction, but to explain a particular extinction calls for data that cannot be supplied from the record. The recorded physical and biological events may suggest the nature of ecological interactions between organisms and their physical environments and indicate possible changes that might have altered these interactions. Most reasonable proposals about extinctions have been made through this type of interpretation.

By analogies with moderns the probable biological properties of the participants may be determined, and sometimes their interactions may be assessed from their structure and the relationships of their remains. Physical or biological changes for which there is some evidence can then be evaluated with respect to the effects that they might have had on this biological system. In the Cretaceous, for example, modifications of the Laramide revolution and subsequent changes of such things as temperature, rainfall, soil contents, and floral distributions are worthy of consideration. Many specific instances of abrupt modifications run into difficulties in applications to coordinated biological events in both marine and terrestrial conditions.

Ubiquitous changes, such as modifications of the oxygen content of the atmosphere, have recently received study as explanations of major times of extinctions. Both very general and specific interpretations rest on the proposition that answers lie in the joint use of analogies and data from the record. It is quite possible, of course, that the critical events left no trace whatsoever. Disease might be one such factor. Whatever approach may be taken, it is in the very nature of the record that firm conclusions cannot be reached, but in the course of speculation and study much of interest comes to light.

The Mesozoic–Cenozoic Record—Mammals

Mammals dominate the record of terrestrial and aquatic tetrapods from the Cenozoic. Present-day faunal complexes involve, in addition to mammals, remnants of the reptilian radiations of the Mesozoic, members of the lacertilian–ophidian reptilian radiation, which has had great success in the Cenozoic; members of the lissamphibians; and great numbers and varieties of birds. Aquatic complexes and those that have both terrestrial and aquatic members also incorporate fish in their structure. The terrestrial records, however, both as a result of the nature of formation of the samples and the predilections of collectors over the years are strongly biased toward the mammalian aspects of this part of tetrapod history. Generally, in contrast, marine tetrapods make up only a small part of the record from marine waters.

As for each of the earlier major radiations for which we have records, that of the mammals reaches deep into the earlier radiations for its beginnings. The first record of mammals, or very mammal-like creatures, is from the latest Triassic. The roots are clearly within the therapsid reptiles, and the transition is such that definitions of class relationships are blurred. Representatives of these first mammals are found at approximately the same time in the British Isles, East Africa, central Europe, and China. Somewhat similar remains have been found in the southeastern United States. This remarkable distribution for small animals whose skulls are little more than a centimeter in length is an expression of the excellence of the upper Triassic.

Through the rest of the Mesozoic, until the later Cretaceous, the record affords only a few glimpses of the course of mammalian development, from the middle Jurassic of England, the upper Jurassic of England, Portugal and the United States, and the early Cretaceous of midcontinental United States and Mongolia. The very late Cretaceous, however, has produced many sites in North America and in Asia. With rare exceptions the materials of mammals from the Mesozoic are teeth and occasional jaw fragments. Skulls and jaws have come primarily from the very late Triassic and from the very late Cretaceous of Asia.

The remains of these early mammals have been studied intensively, and a broad picture of the history and phylogeny of Mesozoic mammals has emerged. The seriousness of the inadequacies of the samples, both in numbers and in composition, is fully realized, yet these materials are the source of all of the information on the origin, early radiations, and initiation of the major stocks of mammals. Theories of the origin of mammalian cusp patterns, so important in studies of mammalian phylogenies, depend on these few samples. Before the late Cretaceous a total of ten primary collections, four of them from the latest Triassic, span a period of 80 to 100 million years.

All of the mammals from the late Cretaceous are quite small and with a few exceptions are known only from teeth and jaws. Generally similar remains, as far as preservation is concerned, also come from the early Paleocene. Inevitably much of the interpretation of the initial phases of radiation of the marsupial and placental mammals has become centered around the properties and changes of dentition and the information that can be gained from teeth and related jaw elements. Throughout the record the teeth of mammals tend to be excellently preserved, and studies of mammalian systematics and phylogeny rely heavily on them. Diet is a center of critical interest in studies of the adaptive radiation of mammals.

The relatively great stability of the components of the mammalian skeleton and the intimate association of changes in dentitions with those in other parts of the skeleton, particularly the skulls and jaws and locomotor structures, reduce considerably the loss of information that might otherwise result from this predilection for dental studies. By the late Paleocene and increasingly thereafter, more complete remains form an important part of the mammalian record. Although dentitions remain supreme in systematic studies, broader studies of evolution involving skull and jaws and postcranial elements have been made for a number of groups, and special studies of such critical features as brains and locomotor adaptations have added much to an understanding of the origins and histories of modern groups of mammals.

Late Paleocene deposits that carry mammals are known in Europe, North America, Asia, and South America. Earlier Paleocene remains have come only from North America and there from rather limited facies. It was during this time, it would seem, that the differentiation of many of the later dominant lines of placental mammals was taking place. The restricted deposits have left many gaps at this critical stage. Even the more widespread and varied upper

Paleocene beds have failed to produce initial stages of some stocks that the Eocene faunas indicate were differentiated at this time. Notable among these are the perissodactyls. Early samples of the Cenozoic thus seem to have failed to tap areas and ecologies in which many events of mammalian evolution were taking place.

The remainder of the Cenozoic record tends to become increasingly representative and widespread. The progress, of course, is irregular as the multiplicity of factors that influence the end result fluctuate in their relative importance. Records of terrestrial vertebrates for each of the epochs are known on all continents except Antarctica, with those from Australia being poor and subject to considerable uncertainty in dating relative to the history of the rest of the world. Gaps, both temporal and geographical, do occur, but in comparison with those of earlier eras they are insignificant in their effects on reconstruction of the general history of the vertebrates.

The Cenozoic vertebrates are of unsurpassed importance as the vehicle for study of patterns of evolution, which are seen less clearly in earlier parts of the record. Directional change over significantly long periods of geological time can be documented over and over again. Overlaps of the old and the new, with the roots of the new stemming from earlier phases of the old, are well displayed in the Eocene and Oligocene epochs. The multiple sources and different times of origin of differing elements of the new radiations are clear, and at least some of the possible motivating factors of change can be deduced. Relationships of changing climate and evolution were well documented. Interrelationships of zoogeographic distributions and evolutionary change are evident in the patterns of development of different faunal complexes in the various continents and the impacts of contacts of elements of these complexes as migrations alter distribution.

The Cenozoic record, though excellent, was not drawn randomly from the total vertebrate life of the time. The earlier parts in particular suffer many biases. These show up in the difficulties that attend attempts to unravel the histories of particular lines of evolution, among both those producing the earlier primitive complexes and those leading to the more advanced creatures of the later complexes. Studies of individual groups, which must strive for completeness over limited ranges of time, may suffer severely from complications of even minor biases in the records. At even a lesser level, taxonomic analyses that depend on randomness of the samples face the problem that randomness is rarely, if ever, attained. As the present is approached such difficulties in one sense are reduced, but the ever greater detail results in complementary refinement of the levels of investigation and similar problems at a lesser order of magnitude.

In its most recent phases the fossil record merges with the Recent, and the studies of the paleontologist become one with those of the anthropologist, archaeologist, and historian. The bridge from existing to extinct is made at this level and has opened the way for profitable studies of evolution expressed in minor morphological modifications under interpretable modifications of the life environments. The final bridge is supplied by the entirely new way of sampling the record of the recent past provided by the written documents of man, following his earlier, less complete, portrayal of life in paintings and ideographs.

section II
Vertebrate morphology and function

I General and skeletal system

INTRODUCTION

The primary information on the morphology and function of vertebrates derived from paleozoology comes in very large part from the relatively small portion of the total animal that is preserved. Except in relatively rare instances this consists of the hard skeletal materials. In addition impressions such as tracks and skin, coprolites, and similar evidences of life are of value. Interpretations of the skeletal materials of fossils depend on knowledge of the relationships of morphology and function that can be gained from living animals; by studies of their comparative anatomy, physiology, embryology, and genetics; as well as by an understanding of their activities and interrelationships gained through studies of ecology.

The excellent books dealing with anatomy, physiology, and embryology cited in this and later sections provide an adequate background for general paleozoological analyses. In addition there exists a vast literature concerned with particular systems and special groups of vertebrates. Neither of these sources is entirely adequate to fill the total needs of the paleozoologist, because the general references cannot supply the detail necessary for special studies and the anatomical and physiological literature is rather limited in its sampling of the totality of living vertebrates.

The second section of this book is concerned primarily with the systems of the vertebrate body as they are preserved in the fossil record. It includes information on the systems and parts of the systems that are not preserved to

55

the extent that is necessary to an understanding of their roles in evolution. It is not intended as a survey of comparative anatomy but rather assumes that the reader has a fair knowledge of this field.

The vertebrate body is of course an integrated system, and no subdivision can separate out units that can be considered as strictly independent in their form or function. It can, however, be divided by various schemes into systems that have some useful characteristics of autonomy. For our purpose a basically functional division serves best. The individual systems, of course, interact extensively. To the paleozoologist the most important interactions are between the soft systems and the skeleton, for these interactions, as they leave records on the skeleton, make possible interpretation of the soft anatomy from the more or less direct evidence of hard parts.

The skeletal system supplies direct information on its own form and function, indirect information about closely associated systems, and serves in a third way as well. The skeletal organization in modern animals is generally associated with distinctive levels of organization of other body systems. If the assumption is made that such relationships hold generally, it is possible to draw broad inferences about the levels and general conditions of non-preserved systems, even though they have left no trace in the record. This is a common procedure, so common as to be almost automatic.

The most evident example is the reproductive system, which in tetrapods is distinctive at the class level among living representatives. Recognition among fossils of, say, a reptilian skeletal organization, even from a small part of the skeleton, is generally taken to be sufficient evidence that the reproduction, and hence the reproductive system, was of the reptilian type. This general approach is extended with little hesitation to both the morphology and function, including physiological function.

Dangers exist both in this method and in the more direct approach in which muscles scars, fossae, foramina, and similar features of the hard skeleton are taken as the basis for recognition of the existence and, perhaps, condition of function of some soft structure. Both depend on analogies with existing animals, and as extinct forms depart increasingly from living relatives the chances of correct analysis decrease. Problems exist in particular in instances where a feature in question has no adaptive counterpart in living animals and at transitions between major organizational levels, where the reproductive criteria used for separation were themselves undergoing change. The claws of large extinct, presumably herbivorous, mammals, such as the chalicotheres, have proven very difficult to interpret, for no living herbivorous mammals have such features. At the transition from reptiles to mammals the intermediate structure of the skeleton offers no sound basis for interpreting reproduction as "reptilian," "mammalian," or transitional.

Yet, for all such problems, an immense amount of understanding of vertebrate evolution has come from studies of the record of fossil vertebrates, and by these studies the knowledge of vertebrate morphology and its evolutionary history has been greatly enlarged.

Each of the systems of the vertebrate body is treated in this section, with special attention to the major features of its evolutionary change and the ways in which it contributes to the fossil record. This focus leads to an emphasis of skeletal, muscular, and nervous systems. No implication that these are more important to the evolutionary development of vertebrates than are other systems is intended, because this certainly is not the case. The following breakdown of the systems of the vertebrate body, which is fairly standard, is used:

1. Skeletal	6. Nervous
2. Integumentary	7. Digestive
3. Urogenital	8. Endocrine
4. Respiratory	9. Muscular
5. Circulatory	

ADAPTATION AND EVOLUTION OF THE SKELETON

Functions

The vertebrate skeleton forms a support for all other systems of the body and provides a stable base for the integration of all physical movements. The intimacy of relationships to different systems varies greatly, but in its roles of support and coordination the skeleton is sensitive to adaptive changes throughout the body, whether the involvement is direct or through the media of other systems.

The functional and morphological relationships of the different systems can be studied concurrently among living animals. Fossils, of course, allow direct study of the skeleton alone, and dependence on analogies with living animals for interpretation is inevitable. The main activities of vertebrates, however, can be grouped conveniently into a few broad categories, each of which involves more than a single body system, providing insight into relationships needed for interpretation from less than full information. Some of the activities are strongly reflected in gross morphology, in particular such dynamic functions as locomotion, feeding, respiration, and reproduction. Others, which are more passive, such as sensing of the environment or coordination of the several systems, are generally less clearly portrayed in morphology but involve organs that are morphologically distinct. The extent to which the soft systems and the skeleton are related in the performance of the various activities is the major determinant of how much can be inferred about nonskeletal systems from the fossil record.

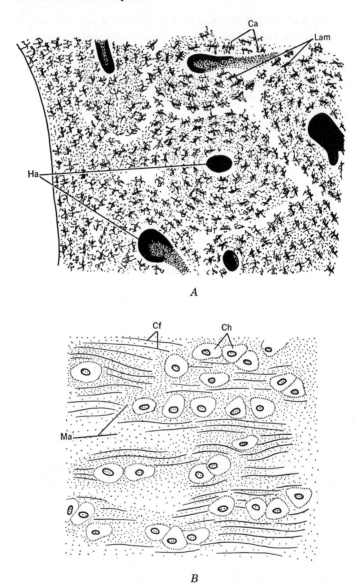

Figure 12.   Drawings of microscopic sections showing the principal features of *A*, bone and *B*, cartilage. Ca, canaliculi; Cf, collagenous fibers: Ch, chondrocytes; Ha, Haversian canals; Lam, lamellae; Ma, matrix.

Locomotion involves the whole postcranial skeleton and the related musculature in such a way that much can be inferred from the skeleton about the nature and functions of the musculature. Feeding and digestive activities, on the other hand, are reflected mostly by the skulls, jaws, and teeth, so that limits of interpretation are greater. Reproductive processes involve the skeleton but little and in secondary ways. Respiration, on the other hand, as it has undergone marked evolutionary change has become related to the skeleton in very different ways at different stages, and these modifications can partly be seen in the fossil record. Sensory and coordinating activities of the nervous system have retained a fairly constant and mostly passive relationship to the skull, vertebral column, and appendages.

Evolution, as seen in the record, has two rather distinct phases: one of adaptive radiation and the other of the development of new levels of organization. The fossil record tends to document the morphological changes during the adaptive phases well, but the modifications that took place during transitions between major groups are more obscure. The latter, however, are critically important for an understanding of the broad patterns of vertebrate evolution and are but rather dimly portrayed by living animals. It is in providing information on such changes and the factors that control them, even though the record is often sketchy, that the skeletons of extinct vertebrates have made their most important contributions to our knowledge of the evolution of the subphylum.

Composition of the Skeleton

The vertebrate skeleton is made up of connective tissue of which two types, cartilage and bone, predominate. Each of these has its own distinctive, although highly varied, histology (Figure 12) and can readily be recognized in the fossil record. In addition to them, the notochord, with its own distinctive histology, is an important skeletal element in a small number of adult vertebrates and occurs universally during ontogenetic stages. Many vertebrates also have a fourth skeletal component, horny or corneal tissues that cover the body or form special structures, such as hoofs, beaks, and horns. This tissue too has a distinctive composition and histology, expressed in a variety of forms, but is rarely preserved except in very recent fossils. Unlike the other three skeletal components, corneal tissues rarely play an important role in the structural, supporting parts of the skeleton.

Among very primitive vertebrates several bonelike tissues—aspidine (an acellular bone), several types of dentine, and enamel—are parts of the coherent and often solid external skeleton. It is at this level that histological studies have been most interesting and instructive. Beyond this primitive level most histological research has concerned teeth and, in the cases of some fishes, the bony scales. Teeth, in both structure and form, have been one of the major sources of information on taxonomy of the vertebrates of the past. An excellent summary of studies on teeth is to be found in *Comparative Odontology* by Peyer (1968). Considerable study has also been devoted to the histology of bone among fossil vertebrates. Preservation tends to be good so that details are available. The effectiveness of such studies, however, is restricted by the fact that, although there are many variations in the histology of bone, variations between different parts of animals are about as extensive as those between major categories. It is primarily in investigations of functional aspects of the skeleton that bone structure has proven useful, because its value in taxonomy is minimal.

Of the two principal major skeletal components only cartilage occurs in all vertebrates, being a universal embryological tissue that persists into adults in varying amounts in different groups. The adult skeleton of cyclostomes includes some cartilage, although much of its skeletal tissue is in a "precartilaginous" condition. The skeletons of adult sharks are formed of cartilage only except for teeth and dermal structures. Cartilage generally is poorly preserved in the fossil record. In special circumstances, especially in very-fine-grained black shales, cartilage may be well enough preserved to show many features of the skeleton in good detail, and calcification of the cartilage during the life of some sharks produces a tissue that can leave an excellent fossil record.

In spite of such instances most of the concrete record of the fossil vertebrates comes from bones and teeth. Ontogenetically bone is of two basic types: endochondral, or cartilage-replacement, bone and dermal bone, formed in the dermal tissue without a cartilage precursor. Histologically the two are indistinguishable, and hence their differentiation in fossils depends on an understanding of their relationships to each other and to other structures as determined in recent animals. Relatively few problems of identification actually occur, because the cartilaginous skeleton and its ossification have been highly stable throughout vertebrate history. It is mainly in the head region, where dermal and cartilage-replacement bones are intimately associated, that problems occasionally arise.

On a gross basis the skeleton may be conveniently divided into five subdivisions: axial, appendicular, visceral, dermal, and skull and jaws. The principal usefulness of this subdivision is that the different parts have a fair degree of functional integrity. They do, however, overlap, with parts of one system entering into one or more of the others to a greater or lesser extent at different levels of organization.

The skull and jaws form the most complex unit, including unique elements— the sensory capsules, lateral and dorsal cartilages, connecting structures— and visceral, axial, dermal, and appendicular components in some instances.

In gill-breathing animals the visceral skeleton is primarily involved in respiration, but in air-breathing forms this unique function is lost. The axial and appendicular portions of the skeleton have a fair integrity, and their various functions can be moderately well defined. The dermal skeleton is more complex. If the dermal bones that enter into the structural skeleton are included, the dermal bones to be considered take part in formation of the skull and jaws and appendicular skeleton in many vertebrates. As a rule the phrase is applied only to dermal elements that are not directly incorporated into the structural skeleton. Both bony and corneous tissues fall within this definition, the former including scales, scutes, and armorplate and the latter horny scales, hoofs, horns, and feathers. Teeth may be considered as a specialized part of this dermal skeleton, but as a rule they are treated separately because of their peculiar value in studies of morphology, function, and taxonomy of fossil vertebrates.

The Axial Skeleton

The principal components of the axial skeleton are present throughout much of the subphylum. The notochord, presumably the primitive element, persists in developmental stages throughout and occurs somewhat sporadically in adult stages. Its presence in the adult cyclostomes may be a primary feature, but in other lines the unrestricted notochord is probably secondary. Although the notochord itself is not preserved in the fossil record, its existence can often be detected by the development of canals through the vertebral centra or in the total absence of centra when some other vertebral structures, such as the arches, have been preserved.

The situation in primitive agnathans, acanthodians, and most placoderms is unclear. Somewhat restricted notochords persist into adult stages in various fishes, Chondrichthyes and Crossopterygia among others, and in amphibians rather generally. Among fishes, full formation of the central elements, accompanied by a marked reduction of the notochord, occurs in some Chondrichthyes and in teleosts.

Vertebrae are known in all groups of gnathostomes and presumably were present, if only in rudimentary form, in agnathans, since they are present in specialized and degenerate form even in living cyclostomes. In full development vertebrae consist of central elements, neural and haemal arches and spines, intervertebral articular processes, the zygapophyses and supplementary facets, and transverse processes for attachment of the ribs. In many fishes only the arches and spines are developed. Centra are fully formed only in a few groups, and articular processes are absent or rudimentary. Haemal arches and spines, if present at all, are confined to the region posterior to the termination of the alimentary canal at the anus. In some fishes, however,

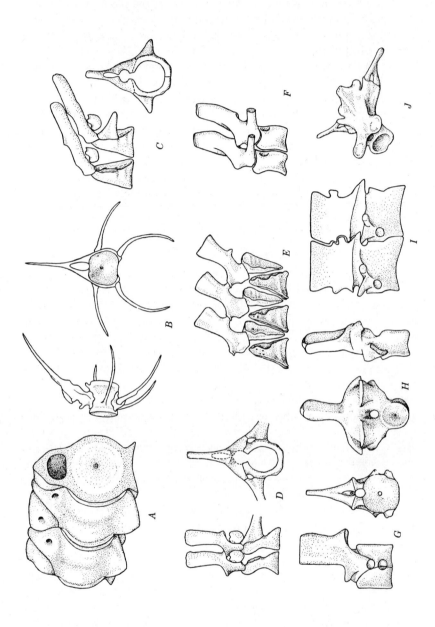

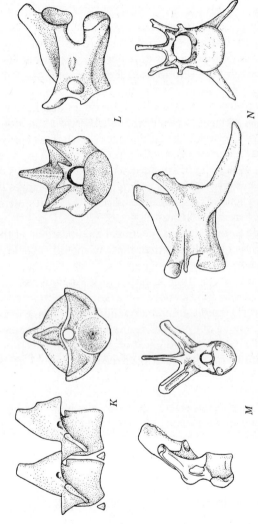

Figure 13.   Representative vertebrae. *A, Squalus,* a shark; *B, Salmo,* a teleost; *C, Eusthenopteron,* a rhipidistian crossopterygian; *D, Ichthyostega,* a primitive labyrinthodont amphibian; *E, Neldasaurus,* a rhachitomous labyrinthodont amphibian; *F, Mastodonsaurus,* a stereospondylous labyrinthodont amphibian; *G, Archeria,* an embolomerous anthracosaur amphibian; *H, Diadectes,* a problematic amphibian or reptile; *I, Crossotelos,* a nectridean, lepospondylous amphibian; *J, Necturus,* a caudate amphibian; *K, Labidosaurus,* a captorhinomorph reptile; *L, Iguana,* a lizard; *M, Anteosaurus,* a dinosaur; *N, Felis,* a mammal. (*A, J,* and *N* redrawn from *Atlas and Dissection Guide for Comparative Anatomy* by Saul Wischnitzer. W. H. Freeman and Company, copyright © 1967; *B* after Montagna, 1959; *C* after W. K. Gregory and Raven, 1941; *D* after Jarvik, 1955; *E* after Chase, 1965; *G* after Williston, 1918; *H* after Olson, 1936a; *K* after Williston, 1918; *M* after Lull and Wright, 1942.)

ventral ribs in the abdominal region represent a more anterior expression of these structures.

Representative vertebrae are illustrated in Figure 13. In addition to differences between major groups, as shown in the figure, regional differentiations along the column occur in many tetrapods as segments of the column assume specialized functions.

Ribs are almost universally present in gnathostomes, but in a few they have been lost in the course of specialization. As a rule they are not particularly instructive either taxonomically or functionally among fossil vertebrates. The major change that takes place in the evolution of the ribs is the loss of the ventral (pleural) ribs during the transition from aquatic to terrestrial life. The dorsal, or intramuscular, ribs, both in gnathostome fishes and in tetrapods, have undergone various specializations, including differentiation along the column in conjunction with vertebral modifications. A sternum occurs in the thoracic region of many amniote tetrapods, accompanied by ventral extensions of the dorsal ribs, usually poorly ossified. The heads, the shapes of the necks and shafts, and histology of the ribs undergo moderate changes in the course of evolution, but in general ribs, like other parts of the axial system, are conservative throughout vertebrate history.

This conservatism is related to the fact that functional modifications of the axial system have been fewer and less drastic than those of the other skeletal systems. Axial structures became involved in functions not related to support and locomotion, mainly as they have attained respiratory functions among tetrapods. The extension of neural spines above the axial musculature mass in such creatures as *Platyhystrix* among amphibians and *Dimetrodon, Ctenospondylus*, and *Spinosaurus* among the reptiles is another example of some interest. The functional significance of this modification, however, is uncertain.

The most critical modifications of function in the axial skeleton took place during the period of transition from life in water to life on land. In fishes, with modest exceptions, the primary dynamic role of the axial skeleton is its participation in locomotion. The development of thrust in swimming comes largely from body flexures that involve the trunk musculature and the cartilage, bone, and soft connective tissue of the axial skeleton. Axial support is a minor function. As terrestrial locomotion developed, the importance of the axial skeleton in generating thrust became reduced. This function was transferred in large part to the appendicular skeleton. Propulsive forces were generated by the hind limbs and transmitted to the body through the pelvis, sacrum, and vertebral column.

Axial conservatism in fishes reflects the general constancy of axial function. It is, however, unrealistic to ignore the many fundamental changes in locomotor patterns that have occurred in the various groups of fishes, for these have been of great importance both in the origin of major groups and in adaptive radiations within the groups. Largely they were attained by slowly evolving accommodations to somewhat modified ways of life in the water and usually occurred along with new feeding adaptations that increased the adaptive advantage of some particular form of locomotion.

Fundamental changes of axial structure may well have been critical in the early deployment of the gnathostomes into major groups, and the differences between mature representatives of these groups suggest that this did occur. The record of the transitions, however, is too incomplete to lend any positive support to this conclusion. The best documented case in which locomotion has played a key role is found among the teleost fishes. The full ossification of the vertebral centra is the most evident innovation, but this is only one aspect of the generally high integration of the axial structures. Development of extreme mobility of various teleosts depends in part on axial differentiation and integration, but also, of course, on the full array of related changes of other systems, in particular the appendicular structures, feeding mechanisms, hydrostatic mechanisms, and the scale covering (see Section IV, Chapter 1 for a full consideration of this matter).

Various degrees of ossification of the axial skeleton occur in the rhipidistian crossopterygians, which lie close to the ancestry of the tetrapods. In the relatively few specimens in which the nature of the axial skeleton is well known the vertebral arches and spines are well ossified, but the centra are quite varied and consist of one or two small osseous elements. The notochord was an important and persistent supporting structure, even in the adults (see Thomson, 1969, for discussion).

Support provided by this type of axial structure depended on the notochord and the axial muscles and tendons that extended between the vertebral spines, arches, the ribs, and the transverse body septa. The role of centra probably was insignificant. This structural situation evidently was satisfactory to many types of fishes, for in the course of their evolution members of several groups tended to reduce or lose the central portions of the vertebrae. Sharks and teleosts, however, proceeded in the opposite direction, increasing either calcification or ossification of the central cartilages.

The history of the long, slow emancipation of the vertebrates from the water is treated in some detail in Section IV. As far as the axial skeleton is concerned, the important point is the modification of its function from a primary role in locomotion to a primary role in body suspension. That this was a slow process is indicated by the fact that it lagged considerably as compared with changes in the appendages. The vertebral structure of the oldest known tetrapods was very close to that of the rhipidistian ancestors, whereas the tetrapod limbs were fully developed. Articulation between the vertebrae was slightly greater, and attachment of the pelvis to the column, though weak, was improved. Little change in central structure or differentiation along the column took place.

Once tetrapod limbs had been developed, the potential for changes in the axial skeleton was present, and within the amphibians many modifications took place as the effects of this change in habitat were brought to maturity. The modifications of axial structure did not, however, follow a single course. In fact the differences in the presumably adaptive changes of the vertebral centra have provided the primary basis for the development of the major taxonomy of the Amphibia. Here, as elsewhere, the functional and adaptive values of changes in the vertebral centra are difficult to interpret. In contrast to the central elements, the arches and spines do not show any consistent adaptive divergences.

Several schemes of amphibian classification based on centra have been proposed, as discussed in Section III. The general tendencies in evolution, regardless of what the actual sequence of events may have been, were toward the development of a single central ossification, whether formed by a fusion of two elements or by reduction or loss of one or the other of the primitive pair of elements, intercentrum and pleurocentrum, and emphasis of the other. The result was a unified axial column in which the articulation of the ossified centra formed a strong but flexible base for the "suspension bridge" activities of the axis. The capacity for dorsal flexure was reduced, but considerable lateral flexure was retained. The thrust provided by the hind limbs could be transmitted through the pelvis and the sacrum to a strong longitudinally and vertically resistant structure.

Truly terrestrial habits have been attained by a few amphibians—some anurans and probably some of the dissorophid amphibians of the late Paleozoic. In large part, however, it was the advent of the reptilian stage that witnessed the final loss of the direct impact of aquatic life on the axial skeleton. The vertebral modifications leading to reptiles were marked by development of a single solid central element, the pleurocentrum, with intercentra present as small wedges (see Figures 13 and 73). As in many amphibians, but generally better expressed in reptiles, the sacrum was modified, first by the possession of specialized sacral ribs and then by increasing modification of the sacral vertebrae. The tail, no longer primarily involved in locomotion, became a distinct structure, in contrast to the condition in many fishes, and the caudal vertebrae and ribs assumed patterns rather different from those anterior to the sacrum. The presacral vertebrae, however, tended to remain little differentiated, as did the ribs. The primary modifications took place in the two anteriormost vertebrae, the atlas and axis, as mobility of the head relative to the rest of the body increased.

The axial column, composed of vertebrae with a unified centrum, well defined arches, strong zygapophyses, and fully ossified transverse processes, was a structure eminently suitable for life on land. Although the various lines of reptiles, birds, and mammals, in keeping with their special adaptations, developed a wide range of modifications of the basic pattern, rarely did radical innovations appear. Studies of the evolution of the axial skeleton among tetrapods deal in large part with rather subtle changes in which few general trends can be detected.

The Appendicular Skeleton

The parts of the appendicular skeleton, like those of the axis, underwent a major transition in form and function as the vertebrates passed from aquatic to terrestrial existence. They reveal the functional modifications much more vividly. The function of propulsion in locomotion was transferred from primitive association with the axial system to the paired appendages, which in aquatic habitats were initially concerned primarily with balance, planing, and positioning. As this change of function took place, the median appendages were lost. Except for the caudal fin, these too had been primarily involved in control rather than propulsion in primitive fishes. The caudal fin, on the other hand, was a primary organ in propulsion, and its loss was directly related to the diminished participation of the trunk in this activity.

Median fins appear to have been present in the most primitive vertebrates and persisted in nearly all fishes. Some embryological evidence, in particular from studies of sharks, suggests that these fins may have originated as a continuous fold (finfold) that passed along the dorsal midline of the body around the tail to a position just behind the anal opening. The various patterns of dorsal, caudal, and anal fins are presumed to have arisen by differentiation from this primitive structure. Paleontological evidence has given little support to this theory, although it cannot argue strongly against it. Continuous median fins are not known among the most primitive fin-bearing fossil fishes.

Dorsal and anal fins have undergone many changes in the course of the evolution of different groups of fishes. In general their functions did not alter drastically, although in eel-like fishes, flatfishes, and some specialized teleosts they did take on a propulsive function. The skeletal structure of median and paired appendages in most individuals conforms to a common ground plan, which suggests that the various parts of the appendicular skeleton developed under a common genetic control that remained dominant even when the functions of the various fins diverged markedly. The caudal fins, being a vital element of the propulsive system, were somewhat more independent. In primitive fishes of all classes a heterocercal tail fin was present (Figure 14). In the radiations of the classes evolution toward a symmetrical fin was the rule, usually producing either a homocercal or diphycercal condition.

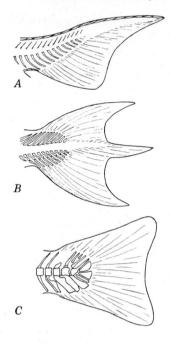

Figure 14. Diagrammatic sketches of the three principal types of caudal fins. *A*, heterocercal; *B*, diphycercal; *C*, homocercal.

As the transition to terrestrial life took place, all median fins were reduced and eventually lost. At just what stage complete loss occurred is uncertain. A caudal fin was present in the earliest amphibians, but it was absent in most. The median fins of aquatic reptiles and mammals were, of course, new developments.

The paired fins of each of the major groups of fishes are characterized by a distinctive structural pattern and are of considerable taxonomic use. The conditions in agnathans do not compare well with those found in any group of gnathostomes. In some agnathans, such as *Cephalaspis*, scale-covered flaps form a pectoral fin of sorts, but there is no evident genetic relationship with the pectorals of the gnathostome fishes. Some agnathans had no paired fins whatsoever.

The paired fins of gnathostomes were separate and localized, and in primitive members of each of the major groups of gnathostome fishes they have common features (Figure 15). Notable are the broad bases, more or less parallel distal rays, and separation of pectorals and pelvics, with one pair back of the gill chambers and the other at about the level of the anus. Except for the acanthodians, two pairs of fins is the usual number. In acanthodians as many as seven pairs of lateroventral spines were present in positions that suggest they may represent fins. The fins in all of these primitive stocks appear to have been relatively immobile.

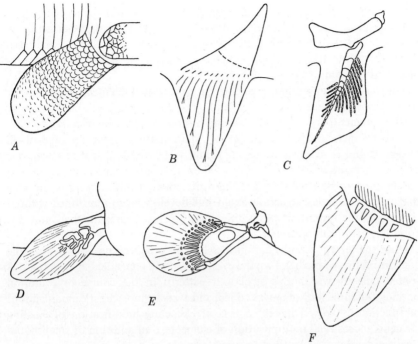

Figure 15.   Diagrammatic sketches of some types of skeletal structures of paired fins. *A*, cephalaspid, an osteostracan, internal structure not known; *B*, broad-based fin, cladoselachian type; *C*, fin with central axis, in some sharks and some osteichthyans, here based on *Xenacanthus; D*, osteolepid type, with complex internal ossifications; *E*, *Polypterus*, showing a well-developed, osseous internal skeleton in the actinopterygians; *F*, actinopterygian type, with fin rays.

Somewhat similar adaptive trends occurred within each of the groups as fins tended to change in parallel ways. As this took place, however, the initial structural differences tended to be emphasized. Narrowing of the fin bases occurred in each of the major lines, and development and emphasis of a central axis are found in several groups.

These changes were closely associated with locomotor modifications, evident in the shift of the tail away from the heterocercal condition and, in some lines at least, in the modification of body shape from a somewhat ventrally flattened form to a fusiform type. In each group many lesser modifications accompanied adaptive responses to special circumstances, and in some lines changes directly opposite to the general trends occurred.

One such specialization that arose a number of times in different groups was the development of fins and body form that were suitable for "bottom-walking." Among the Osteichthyes such fins are found in lungfishes, crossopterygians, *Polypterus*, probably some of the paleoniscoids, and in teleosts.

Increase in internal appendicular support, development of strong but mobile external elements, and differentiation of musculature are characteristic changes related to this habit. Fishes of this sort were to some extent pre-adapted to terrestrial life, and several crossings of the strand took place, as discussed in detail on pages 632 and 633.

The basic tetrapod limb consists of a proximal element (humerus or femur), two distal elements (radius and ulna or tibia and fibula), a number of small elements (carpals or tarsals), and a series of radiating elements (meta-carpals or metatarsals and phalanges). This pattern is the heritage of all tetrapods and is so constant among primitive tetrapods that it has been one of the principal bases for the proposition that all tetrapods arose from a common or single source among the fishes. Behind this concept is the idea that such a complex structure could not develop more than once without showing some important basic differences in the nature, number, and distribution of its components.

Once the tetrapod limbs and girdles had become well developed, with the concurrent loss of median fins, the evolution of the appendicular skeleton was rooted entirely within this structural pattern. In the course of evolution nothing fundamentally new was added, and even the most drastic modifications of functions involved merely a loss of elements, modification of existing elements, occasional multiplication of elements, and addition of small bones (sesamoids) that frequently developed in tendons.

The locomotor function of the paired appendages was in large part re-tained, even with such innovations as flight among the reptiles, birds, and mammals. In some terrestrial amphibians and reptiles, however, limbs were lost and the axial skeleton once again became the system most innately involved on locomotion. Among swimming "tetrapods" the tendency for the axial structures to take over locomotor functions also was strong, and this happened in amphibians, reptiles, and mammals. But various swimming reptiles, such as the plesiosaurs and turtles, and aquatic birds retained the loco-motor function of the paired appendages, carrying aquatic specializations of this function to a high level of perfection.

The adaptive radiations of the tetrapods have been strongly reflected in both internal and external modifications of the appendages. The close association of the appendages with the external environment as well as with the functions of other body systems has made the osseous elements of the system important in interpretations of the habits of animals and important markers of taxonomy, especially at ordinal and familial ranks. The absence of major reorganizations in the composition of the skeletons, with loss of elements representing the dominant compositional change, has meant that homologies of elements can be traced through the great majority of tetrapods with little difficulty. Thus the appendicular system has become one of the prime targets in studies of functional morphological changes of homologous structures.

Skull and Visceral Skeleton

Roles in Morphology, Evolution, and Taxonomy

The skeletal structures of the head are by far the most complex and the most studied of any of the vertebrate body. The skull and jaws in adult amniotes form a coherent complex of elements derived from axial, visceral, dermal, and cranial parts of the skeleton. In primitive vertebrates the intimacy of the components is less developed than in advanced groups, but in all an amalgamation of at least some elements of the visceral, axial, and cranial skeletons has taken place. One of the most intricate and fascinating histories revealed by the joint efforts of comparative anatomy, embryology, and paleozoology portrays the development of this complex structure from a synthesis of once discrete entities. Intimate relationships of the functional modifications and morphological changes are seen with a clarity not evident from other parts of the body.

Parts of the substantive discussion of Sections III and IV deal with the skulls and jaws of particular groups of vertebrates, and in later parts of this section attention is given to the relationships of the skull and jaws to soft anatomy. Here we are concerned primarily with the general nature of this complex and its interpretability as a key to the general evolutionary history of the vertebrates.

The adult skulls and jaws of all gnathostomatous vertebrates have arisen through ontogenetic processes that involve a cartilaginous stage. In some adults the cartilaginous structure, chondrocranium, and visceral cartilages persist, but in most they are at least partially replaced by bone. These osseous structures are generally augmented by dermal bones to form the complete skeleton of the head. Although cartilage and bone are histologically distinct (see Figure 12), they play generally similar roles in the support and protection of the soft structures of the head and in the feeding mechanisms. Bones of the skull formed in cartilage generally approximate the form of the tissue that they replace, whereas dermal bones take their initial form without preformation in other skeletal tissue. The ultimate origin of these two types of bone is not clear, but the embryological separation is distinct, and the two may be treated somewhat separately in a description of the structure of the skull and jaws. In adults, however, increasingly in higher categories, the two types are intimately associated in the functions of the head.

Although many parts of the skull are largely passive in their relationships to function, they reflect with considerable accuracy the modifications of the associated soft systems. The jaws, in their relationships to the more passive portions of the skull, provide insight into the dynamics of feeding and have figured prominently in analyses of the structural changes associated with major evolutionary stages. Both skulls and jaws are important as taxonomic tools and are critical in analyses of the functional significance of morphological change both within adaptive radiations of major groups and in the transitions between major categories. The chondrocranium and the bones that replace it tend to be stable during evolution and form a solid foundation on which homologies may be traced through both ontogeny and phylogeny.

The dermal structures are much less constant. Major problems arise from difficulties in determining homologies between dermal elements in major vertebrate groups, since identities must be established if changes are to be traced. Adding to the confusion, of course, is the fact that the concept of homology itself is open to a wide variety of interpretations. Position in the skull and jaws and relationships of the bones are the most commonly used criteria of homology. These are augmented by relationships to soft parts that have left traces on the bones, such structures as lateral-line canals, foramina, and canals that carried nerves and blood vessels, muscle scars, and included such organs as the brain and organs of special sense. The chondrocranium and its ossifications tend to retain the primitive relationships of organs of the head and the skeleton. Dermal bones, being more superficial and less a fundamental part of the initial skeleton, show much less constant relationships. Their patterns have for the most part arisen independently within the major categories of vertebrates, with the result that even close morphological resemblances may be suspect as indicators of homologies. The special evolutionary sequences treated in Section IV and some of the taxonomic problems discussed in Section III give detailed consideration to the sorts of problems that have arisen.

The Chondrocranium and Endochondral Bone in Gnathostomes

The chondrocranium and its ossifications in gnathostomes are sufficiently similar to make generalizations possible. Although the differences between the jawed and jawless animals are considerable, they are not so great that continuity of structure cannot be established. The greater independence of the cranial and visceral elements, however, introduces complexities that make formulation of a single, simple covering model impossible. Inasmuch as only the ossifications of the cartilaginous skull and jaws are preserved in extinct animals, except in some unusual cases, interpretations of the nature of the chondrocranium must come from reconstructions based on the osteocranium.

The chondrocranium, here considered to include the palatoquadrate and Meckel's cartilages, derives its parts from the axial and visceral skeleton and from independent cartilages of the head region. Its fundamental features can most easily be understood by reference to Figures 16, 17, and 18. Basically the developing chondrocranium can be divided into a posterior part that forms around the notochord, the parachordal portion, and a prechordal part.

BASAL ASPECT.   Viewed ventrally (Figure 16*A*), the parachordal and prechordal portions of the developing chondrocranium are clear, the former including a pair of parachordal cartilages and the other a pair of prechordal cartilages. The former flank the notochord, being of axial origin, whereas the latter, termed trabeculae cranii, appear to have been ultimately derived from the visceral skeleton. In the subsequent ontogeny the stage diagrammed in Figure 16*B* is reached. The order of development of different structures varies considerably and details are different among the many gnathostomes, but the diagrammatic stages portray the usual elements in typical relationships.

In Figure 16*B* the capsules of the three principal organs of special sense— nasal, optic, and otic—are included, the parachordal cartilages have formed a basal plate, with a basicranial fenestra, the trabeculae cranii have joined anteriorly to form a trabecula communis and have joined posteriorly to the basal plate, leaving a hypophyseal fenestra. Nasal capsules are shown here, and throughout, in very general form, joined to the rest of the chondrocranium by a median septum developed in the internasal region above the trabecula communis. Conditions in this region are quite variable, but they do not enter importantly into consideration of the crania of most fossil groups, because ossification tends to be poor. The optic capsules do not become connected to the rest of the chondrocranium.

The otic capsules are shown as joined to the basal plate by a single basicapsular commissure and a more anterior prefacial commissure. Once again there is a great deal of variation in the details of these connections, but the simple arrangement shown is about the limit to which interpretation of fossils generally can be carried. Behind the basicapsular commissure, between the capsule and the basal plate, lies a metotic fissure. This is a complex conduit because it marks the point of issuance of cranial nerves IX, X, and XI and a jugular vein from the cranial cavity and houses the recessus scala tympani, which carries the perilymphatic duct, giving access both to the cranial cavity medially and the tissue of the throat region laterally or posteriorly. Intricate changes alter this structure of the inner ear as hearing in the atmosphere is perfected. Some of these modifications can be traced in the delicate bony structure of the otic region, but, except in mammals, the record is relatively poor because cartilage tends to persist and obscure the details.

A fenestra ovalis, shown in the otic capsule in Figure 16*B*, is developed in most tetrapods, where hearing in the atmosphere is critical, but is generally absent in nontetrapods. More anteriorly, behind the prefacial commissure, a facial foramen for nerve VII is present. This is quite constant in tetrapods and

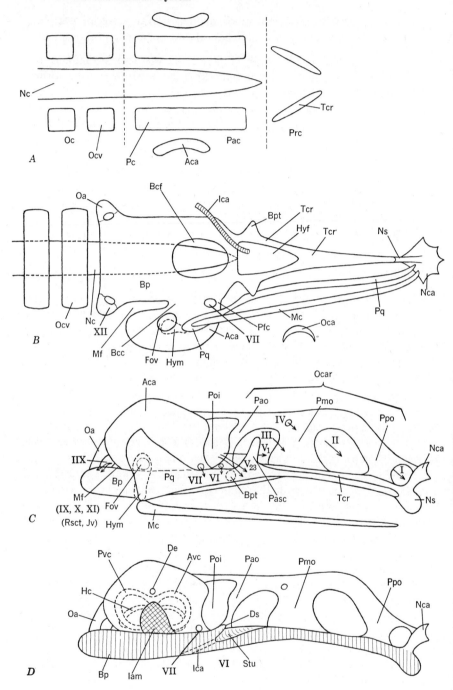

present in many fishes, but additional exits of the facial nerve occur in many of the latter. At the posterior end of the basal plate occipital arches of the vertebrae that have been incorporated into the plate are present, and they isolate one or more foramina through which nerve XII exits.

The hypophyseal fenestra, between the trabeculae cranii, is walled posteriorly by a crista sellaris. The fenestra itself remains open for the period in ontogeny during which the formation of the infundibulum and hypophysis, from brain and epithelium of the oral cavity, respectively, takes place. During this time the internal carotid arteries make entrance through the fenestra. With closing of the fenestra, a recess, usually termed the pituitary fossa, is defined and the internal carotids enter through paired carotid foramina. Laterally to the pituitary fossa are the basitrabecular processes, one on each side. These, as shown, form an articulation with the palatoquadrate cartilage. This important joint between the brain case and the palate persists in many adults as a freely movable articulation and figures prominently in cranial kinetics, especially as related to feeding.

In Figure 16B the palatoquadrate and Meckel's cartilages also are included. These are cartilages of the mandibular visceral arch, a free gill support in agnathans; they form the primary jaws in gnathostomes. Elements of the hyoidean arch, representing the next most posterior gill support in agnathans, occur variously in gnathostomes, as a supporting structure of the palatoquadrate in many fishes and as the stapes in most tetrapods. The hyomandibula

---

Figure 16.   Diagrams of chondrocrania of a generalized vertebrate at about the level of a primitive tetrapod. Posterior to the left. A, an early stage in ontogeny, showing the primary cartilages in ventral view. Note in particular the parachordal and prechordal parts of the cranium; B, a more advanced ontogenetic stage, with cartilages largely joined and in fairly mature condition, seen in ventral view; C, approximately the same stage as B, in right-lateral view, showing exits of cranial nerves; D, medial view of chondrocranium cut along a midline section, same developmental stage as C. Aca, auditory capsule; Avc, anterior vertical semicircular canal; Bcc, basicapsular commissure; Bcf, basicranial fenestra; Bp, basal plate; Bpt, basipterygoid or basitrabecular process; De, opening for endolymphatic duct; Ds, dorsum sellae; Fov, fenestra ovalis; Hc, horizontal semicircular canal; Hym, hyomandibula; Hyf, hypophyseal fenestra; Iam, internal auditory meatus; Ica, internal carotid canal, artery, or foramen for artery; Jv, jugular vein; Mc, Meckel's cartilage; Mf, metotic fissure; Nc, notochord; Nca, nasal capsule; Ns, nasal septum; Oa, occipital arch; Oc, occipital segment of chondrocranium; Oca, optic capsule; Ocar, orbital cartilage; Ocv, occipital vertebra; Pac, parachordal part of chondrocranium; Pao, pila antotica; Pasc, ascending process of palatoquadrate cartilage; Pc, parachordal cartilage; Pfc, prefacial commisure; Pmo, pila metoptica; Poi, proötic incisure; Ppo, pila preoptica; Pq, palatoquadrate cartilage; Prc, prechordal part of chondrocranium; Pvc, posterior vertical semicircular canal; Rsct, recessus scala tympani; Stu, sella turcica; Tcr, trabecula cranii; I–XII, cranial nerves or the foramina through which they pass.

is shown in dashed lines in Figure 16*B*, in its usual position passing from the otic capsule (here from the fenestra ovalis) to the posterior end of the palatoquadrate cartilage. Other hyoidean cartilages are not shown.

LATERAL AND MEDIAL ASPECTS. Figure 16*C* shows about the same ontogenetic stage of the chondrocranium as Figure 16*B*, but in lateral view, and Figure 16*D* represents a median section, viewed from the midline laterally. Various additional features are included, and those described earlier are seen in different aspects as far as they are visible. The palatoquadrate has three basic connections to the chondrocranium: one on the otic capsule, one to the basitrabecular process, and one in the ethmoidal region. Rising above the main bar of the palatoquadrate in the basitrabecular region is an ascending process. It isolates medially an extracranial space, the cavum epipterycum, which has figured prominently in studies of the transition from reptiles to mammals (see pages 711–712).

In medial view, Figure 16*D*, the position of the semicircular canals in the otic capsule, the opening for the endolymphatic duct into the cranial cavity, and a wide-open internal auditory meatus are shown. Ossification in this region tends to be good, so that details are often available in fossils.

The most important additional feature in Figure 16*C* and *D* is the orbital cartilage and its roots. Like the capsules, this is an independent cartilage of the head skeleton. It takes on many forms, from nearly solid to highly fenestrated, and what is shown is somewhat intermediate between the extremes. The cartilage consists of a main dorsal element and three roots, which together form the primary sidewall of the preotic part of the cranial cavity. The most posterior root is the pila antotica (pro-otica) which joins the basicranial part of the chondrocranium adjacent to the crista sellaris (or its derivative, the dorsum sellae). In this region the polar cartilage, a small paired element not shown in the diagrams, may be present. It has been thought to have important implications with respect to interpretations of sources of parts of the chondrocranium, but its presence and significance are undetectable among fossils.

Posterior to the pila antotica is a space through which the important trigeminal cranial nerve (nerve V) issues from the cranium. When broadly open, this is often called the proötic incisure. Two parts of the compound fifth nerve, the second and third branches (or maxillary and mandibular) pass posterior to the ascending process of the palatoquadrate, and the first branch passes medially to it. The constancy of this relationship among extinct forms is not certain, but the usual practice has been to extrapolate from moderns on the assumption that it holds. The abducens cranial nerve (nerve VI) may issue through the basicranium, near the base of the foot of the pila antotica as shown, or it may not penetrate hard tissue. It is only occasionally identifiable in fossils.

Anterior to the pila antotica and posterior to a second root, the pila metoptica, is a space through which the third cranial nerve, the oculomotor, issues. The fourth nerve, the trochlear, generally penetrates the orbital cartilage proper. Unless the wall of the cranial cavity is very well ossified, the exits of these nerves cannot be identified in the osseous skull. Anterior to the pila metoptica and posterior to the third root of the orbital cartilage, the pila preoptica, lies the foramen for the optic nerve (nerve II). Anterior to the third root the olfactory tracts pass toward the nasal capsule, issuing through a foramen and passing extracranially for a short distance. In mammals, and in some reptiles, amphibians, and fishes, the courses of these two nerves can be identified, but on the whole ossification is such that this is difficult. The stability of optic and nasal organs is such that their positions and innervation rarely contribute materially to studies of morphological changes in the evolution of the skull.

DORSAL AND POSTERIOR ASPECTS.   The cranial roof is well developed in cases in which the chondrocranium forms the adult skull, but in most vertebrates it is not well formed and its structure is highly varied. Ossification does not tend to preserve much of the roof, even if it is formed in cartilage, and hence this part of the chondrocranium does not figure importantly in fossils. Posteriorly two important tecti occur, the tectum posterius between the occipital arches, and the tectum synoticum between the otic capsules. The identity of these between members of major groups is not fully established, and the discreteness of the two in some groups is questionable. Either or both may be absent.

Figure 17 shows the posterior view of the chondrocranium, including the tecti and the occipital structures shown in other views in the preceding figures.

THE HYOID COMPONENT.   The hyoid component of the chondrocranium was noted briefly and shown diagrammatically in Figure 16. More detail is shown in Figure 18A, B, and C. This part of the skull has played an important role in many studies of the evolution of vertebrates, since it is involved in the transition from the agnathous to the gnathostomatous stage, in systems of jaw suspension among fishes, in the development of hearing in the atmosphere, and in the transition from reptiles to mammals. Each of these figures prominently in later parts of this book as applied to particular problems. The basic structures and their relationships to each other and to associated soft structures are shown in the figure. In Figure 18A the condition in *Amia* illustrates the hyomandibula in its role as a support of the palatoquadrate. Figure 18C (*Sphenodon*) illustrates the full complement of attachments of the stapes in a tetrapod. In more primitive animals, even among the fishes, where the functions are very different, these same general connections of the hyomandibula exist. As discussed on pages 607 and 612, this constancy of

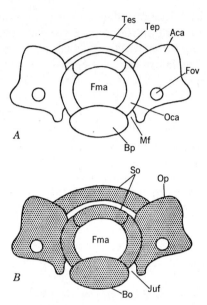

Figure 17. Diagram of posterior view of the occiput. *A*, chondrocranium; *B*, ossifications in chondrocranium. Bo, basioccipital; Fma, foramen magnum; Juf, jugular foramen; Op, opisthotic bone; So, supraoccipital bone; Tep, posterior tectum; Tes, synotic tectum. Other abbreviations as in Figure 16.

relationships has made possible very detailed analyses of the evolutionary course of vertebrate auditory structures from water to land and from reptiles to mammals.

OSSIFICATIONS. Ossification of the cartilaginous parts of the skull and jaws has tended to be stable throughout vertebrate history. The principal components of the osteocranium as they form in their cartilaginous precursors are shown in Figure 19. Only a few points need special mention. The posterior parts of the basicranium are consistently ossified by the basioccipital and basisphenoid, unpaired bones. More anteriorly a presphenoid forms, but this is often associated, sometimes indistinguishably in the adult, with other ossifications in the orbital cartilages, in its anterior roots and the internasal septum, whether this is of independent origin or derived from the orbital cartilage. The complex that forms is commonly termed sphenethmoid, and it may be a single bone or have several parts. It appears to carry the equivalents of the presphenoid, orbitosphenoid, and various ethmoidal bones. Ossifications in the nasal capsule, not shown, also tend to be highly varied.

If the pila antotica ossifies, the element formed in it is termed either the laterosphenoid or the pleurosphenoid. The extent of ossification is highly varied. Otic capsules generally ossify from two main centers to produce a posterior opisthotic and anterior proötic. These may fuse to form a single element sometimes called periotic. The terminology of the bones, however, tends to be different for the major groups of vertebrates.

The palatoquadrate rarely ossifies fully, and in some tetrapods it does not form a single, coherent cartilaginous unit. The posterior portion ossifies to form the quadrate bone, the primary articular element of the upper jaw, lost only in mammals in which the quadrate becomes the incus of the middle ear. More anteriorly, ossification occurs in the vicinity of the palatocranial joint to form the epipterygoid bone, with its ascending process ossified in the ascending ramus of the palatoquadrate bar. In mammals this element is termed

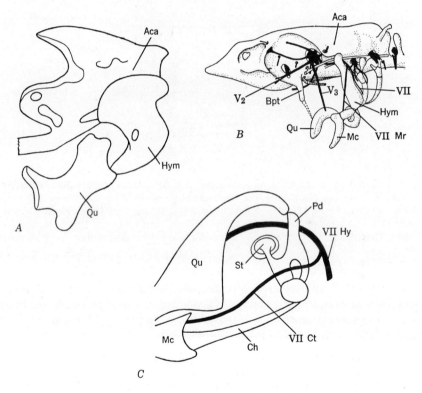

Figure 18.   Lateral views of the suspensory apparatus and associated structures in *A, Amia; B, Acipenser;* and *C, Sphenodon.* The relationships of the hyomandibula and stapes are shown along with the position of the facial nerve (VII) in *B* and *C.* Pd, dorsal process of stapes; Qu, quadrate; St, stapes; V_2, V_3, maxillary and mandibular branches of the trigeminal nerve, respectively; VII$_{ct}$, VII$_{hy}$, and VII$_{mr}$—chorda tympani, hyoidean, and mandibular branches of the facial nerve, respectively. Other abbreviations as in Figure 16. (Modified after De Beer, 1937.)

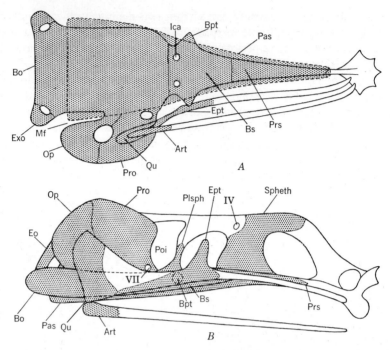

Figure 19. Diagrams of ossifications of the chondrocranium with the dermal parasphenoid bone added. *A*, ventral aspect; *B*, right lateral aspect. Art, articular bone; Bo, basioccipital bone; Bs, basisphenoid bone; Ept, epipterygoid bone; Exo, exoccipital bone; Op, opisthotic bone; Pas, parasphenoid bone; Plsph, pleurosphenoid ossification; Poi, proöticincisure; Pro, proötic bone; Prs, presphenoid bone; Qu, quadrate bone; Spheth, sphenethmoid ossification. Other abbreviations as in Figure 16.

the alisphenoid. The principal ossification of Meckel's cartilage is the articular, which forms the articulation of the lower jaw with the skull. Various more anterior centres of ossification may form, but they are, by and large, of little phylogenetic or functional significance.

The Dermal Bones

Dermal bones, in contrast to those formed by replacement of cartilage, exist in a wide range of patterns. Being external, they are the parts of the skull and jaw most generally available for study in fossils, but in occupying this position they tend to be less intimately involved in the primary functions of the skull, especially in more primitive vertebrates. Dermal bones occur in all vertebrates except cyclostomes and sharks and thus have been one of the major sources of information on the evolution of the skulls and jaws both within and between major categories of vertebrates.

Within the major groups dermal patterns of the skulls and jaws are moderately consistent, and hence changes may be followed with considerable confidence. Between major groups, however, the contrary is true. Differences in patterns may be so great that no interrelationships can be seen, and even where some resemblances of pattern do appear it is unsafe to infer homologies of the elements unless the bones in question can be, at least to some extent, traced phylogenetically. The dermal patterns of the head shields and jaws in agnathans, placoderms, and acanthodians differ so much from one another and from the patterns among the osteichthyans that no valid homologies between the elements can be drawn and probably none exist. Within the bony fishes the patterns in the dipnoans are so different from those of crossopterygians and actinopterygians that a completely different nomenclature is necessary. Somewhat greater resemblances are to be found between crossopterygians and actinopterygians, but although common names are used for some bones, the homologies are at best very uncertain.

In general tetrapods, which stem from a crossopterygian ancestry, show a greater constancy of dermal bones of the skulls and jaws than do the various classes of fishes. In spite of strong controversives on some critical points, the dermal patterns have proven very useful in interpretations of major evolutionary trends. In many tetrapods, as in some lines of advanced fishes, the dermal elements have become intimately involved in the masticatory mechanisms, and their functional roles have made studies of adaptive changes possible and illuminating.

The problems of the resemblances and interpretations of interrelationships of the dermal elements among major groups arise in part from the fact that these various patterns were developed independently within the groups. Among primitive vertebrates there is some evidence that dermal bones have arisen from coalescence of very small elements into the larger plates. In its full development this process has been formulated into the lepidomorial theory by Stensiö. Small placoidlike denticles are considered to be the source of dermal elements, coalescing to form larger denticles, which in turn provided the elements that fused to form dermal bones.

If this were in fact the basic process, much of it has been lost in more advanced vertebrates since dermal bones have only one or a limited number of centers of origin. On the other hand, some of the agnathans, sharks, acanthodians, and primitive members of the Osteichthyes give at least some evidence of this sort of origin in the many small elements in the dermal covering of the head. If carried to an extreme, with the dermal bones in each major group having origin in the coalescence of small plates, it must be concluded that there are no homologies between such elements. The consequences in particular cases are taken up in some detail on pages 617–618.

The most important force opposing this potential heterogeneity is found in the presumed organizing capacities of the organs of the lateral-line sensory system. Although this capacity is the subject of continuing argument, it has been the basis for establishment of homologies of important dermal bones through which parts of the lateral-line system pass, based on the thought that the elements induced by homologous organizers are themselves to be considered homologous.

Studies made within major categories are much less prone to these difficulties. Patterns of dermal bones tend to be pervasive and to change in regular and determinable ways in the course of evolution within the groups. The presence or absence of various elements thus has come to have considerable taxonomic and phylogenetic signficance. In general, trends have been toward loss of elements, with resulting simplification of patterns and the assumption of additional functions by the remaining bones. The contacts of elements and their relationships to other major skull features, such as the apertures for the organs of special sense, figure in evolutionary analyses, predicated on the proposition that these relationships are so constant that they may be used as guides in tracing the course of development.

In most major groups of vertebrates the composition of the dermal covering of the skull and jaws changes materially in the course of evolution, with some elements becoming smaller and eventually dropping out. In the mammals and the birds, however, the composition has essentially stabilized and changes are those of proportions and form of the constituent bones in adaptation to changes in functions.

Evolutionary Stages

The most fundamental step in the evolution of the skull and visceral skeleton saw the emergence of gnathostomes from their presumed agnathous ancestors. Unfortunately there is no recognized record of this transition in the fossil history of vertebrates, and it is not known from precisely what sort of agnathous ancestors the gnathostomes may have come. None of the known agnathans is necessarily at all close to this ancestry.

Embryological origins of the hard parts, patterns of innervation of muscles associated with skeletal structures, and patterns of circulation provide conclusive evidence that a transition from agnathous to gnathostomatous forms did occur and that the agnathous forms were the primitive vertebrates. They offer very little information on the precise nature of the ancestral organisms.

The head structures of the cephalaspid agnathans formed a unified cephalic shield, consisting of dermal and internal ossification. It appears to have included an extensive visceral skeleton, extending nearly to the anterior end of the head, as well as ossifications enclosing the sense organs, the brain, and other cranial structures. The dermal skeleton included a complex and highly ramified system of canals and pores that housed the lateral-line sensory

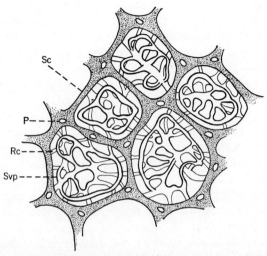

Figure 20.   Superficial view of sensory and vascular systems of *Tremataspis*. The lateral lines are only slightly differentiated, parts of the sensory network covering the whole body. P, pore of sensory canal system; Rc, "radiating" vascular canals; Sc, sensory canal; Svp, subepidermal vascular plexus. (After Denison, 1947.)

system (Figure 20). It can reasonably be assumed that most of the vital organs of circulation, digestion, and reproduction lay in the region encompassed by the cephalic shield and that the postshield body was mostly concerned with the function of propulsion.

There were no jaws that incorporated visceral elements. There appears to have been a premandibular arch, and a mandibular arch, with gill pouches, which have been lost in gnathostomes. The cephalic shield in the best known forms is a little differentiated unit, and hence it is difficult to assess the contributions of the various skeletal systems to it. *Petromyzon* has been studied from this point of view, and general similarities are clearly present. The high degree of specialization has produced some difficulties in detailed interpretations and in comparisons with the extinct creatures, which did not persist beyond the Devonian.

Members of other agnathous groups, anaspid osteostracans, heterostracans, and coelolepids do not offer much aid in comparisons with gnathostomes except very generally, because their internal structures were very little impressed on the hard parts of the skeleton.

The agnathans represent an extensive radiation of fishlike creatures that probably stemmed from the very base of the vertebrate stock. Differentiation of the somatic and visceral portions of the body appears to have been quite distinct. The latter, along with the mouth, sensory organs, and enlarged anterior portion of the central nervous system included structures related to the skeletal elements that eventually formed the skull and jaws in tetrapods.

With the advent of the gnathostomes, regardless of their precise origin, the way for a vast radiation was opened by the presence of jaws. Feeding mechanisms lay at the heart of much of what was to follow, with the means of locomotion under varied circumstances of different media changing concurrently in intimate relationship to dietary modifications. The radiations took place in the two primary media, water and air, and in each followed somewhat parallel courses. The potentials of each of these major habitats were exploited with increasing intensity as new life forms, both animal and plant, became established, augmenting the more stable physical environments. Radiation on land, of course, awaited the development of aquatic vertebrates with potentials for life out of water, but this stage was reached long before aquatic forms had fully exploited the opportunities available to gill-breathing vertebrates.

With the initiation of the complex radiations, fundamental shifts in the organization of skulls and visceral skeletons occurred. Within the fishes major changes related primarily to modifications of the feeding and respiratory functions and were tied closely to locomotor activities. Structurally these modifications have been difficult to isolate and analyze. This is due partly to difficulties in the study of living aquatic creatures among living animals, as compared with the ease of studying air breathers, and partly to the relatively poor preservation of the fossil fishes that occupied the critical positions in the major transitions.

The most striking radiation of the gnathostome fishes is that of the teleosts, which is explored in greater detail later in this book. An initial modification in locomotion appears to have set the stage for introduction of highly significant changes in feeding mechanisms, followed by an almost incredible burst of adaptive radiation.

Less spectacular and with clouded beginnings is the radiation of paleoniscoids in the latter part of the Paleozoic and early part of the Mesozoic eras. The basic skull, jaw, and gill structures that set the stage for this radiation were fully developed when the paleoniscoids first appeared in the fossil record in the Devonian (see Section IV, Chapter 2 for a full discussion).

Other radiations—those of the acanthodians, sharks, and placoderms— also probably were based on some threshold combination of feeding, locomotor, and respiratory functions, but the initial stages are not known for any of them.

Dipnoans and crossopterygians are generally thought to have stemmed from a common source with the actinopterygians (paleoniscoids), but once again extensive divergence had taken place prior to the time they appeared in the record. By this time the characteristic skull-and-jaw patterns had been established for each. Once the characteristic dental modifications of the lung-fishes had been established, the skull and jaws were restricted, or so it would appear, to very limited evolutionary changes.

The crossopterygian skull pattern, like that of the other Osteichthyes, was distinctive in the earliest creatures in which it has been found. The very early patterns suggest an origin of the dermal pattern from an ancestor in which there were many small plates, as seems to have been true for dipnoans and primitive actinopterygians as well. Modest changes in the surficial dermal skeleton occurred in the course of radiation of the crossopterygians, and within their patterns can be seen the structures that led to the tetrapods.

In contrast to both the axial and appendicular skeletons, the skulls and jaws did not undergo a marked reorganization during transition from water to land. The most fundamental modification at this stage appears to have been in respiration and involved the loss of the visceral skeleton as a supporting mechanism for gills. Gill bars, well developed in rhipidistians, are absent in primitive amphibians, although their homologs presumably were partially represented in the hyoidean skeleton, essentially unknown in transitional creatures.

The structure of skull-and-jaw patterns of primitive amphibians was largely set within the rhipidistians, and changes to terrestrial life appear to have been in large part modifications related to feeding. Such changes were gradual and did not result in fundamental alterations that can be considered critical for the initiation of the ensuing terrestrial radiations.

Modifications of the skulls and jaws during the terrestrial phases of vertebrate radiation—after suitable locomotion, respiration, sensing, and capacities to withstand desiccation had been attained—were related most importantly to feeding. The initial tetrapods arose, it would appear, from predacious ancestors and most, if not all, of the major groups within the tetrapods originated in predacious stocks. Herbivores have, of course, been extremely successful, but largely as evolutionary offshoots that did not lead to new major levels of organization. In the reconstruction of this history much reliance has been put on the information from the skulls and jaws, and much less on the postcranial skeleton.

Only the basic changes leading to mammals are well enough known for the course to be traced. Most of the steps in other major transitions are hypothetical. In the history of mammals attention has centered on a variety of coordinated modifications in the course from primitive (pelycosaurian) ancestors through a fairly-well-graded series of morphological stages to the mammalian level. These are considered in detail in Section IV, Chapter 6.

A rather curious sequence of events took place in the course of the evolution of amphibians' skulls and jaws. The very early and primitive labyrinthodonts were characterized by skulls that were fairly narrow and deep and had distinct otic notches, small interpterygoidal vacuities, an open joint between the palate and braincase, and a well-ossified endocranium. Essentially all of these features were lost in most of the evolving lines of labyrinthodonts. The resulting skulls were very flat, akinetic, with little internal ossification, and with widely open palates. In a few groups, such as dissorophids, fairly deep skulls persisted. The changes in genera, however, represent common persistent directional changes that took place throughout most, although not all, of the subclass and occurred repeatedly in distinct lines. Adaptively many of the animals were aquatic, and in some lines at least they became increasingly so as time went by. Thus these features could in large part be associated with adaptation to a common life pattern. But this poses problems, because many of the same sorts of changes took place in a wide array of different adaptive types, creatures as distinct in habitat as the dissorophids and trimerorhachids.

One group, the anthracosaurs, although probably common in origin with the labyrinthodonts, did not undergo this series of changes, even though some members were distinctly aquatic. On the other hand, many of the trends are found among the lepospondyls, in large part fully matured in the first well-known skulls. Loss of kinetic structure and internal ossification, however, are not fully realized in some lepospondyls, such as the microsaurs.

Reptiles appear to have commenced their radiation with the captorhinomorphs, from an anthracosaur base. The initial skull was evidently anapsid, lacking any temporal fenestration, rather deep, and with a mobile joint between the palate and basicranium. A fundamental modification that occurred during a reptilian radiation was the development of temporal fenestration. With the exception of one persistent group, the Chelonia, all modern reptiles have fenestrated skulls.

Temporal fenestration has been very important in formulating classifications of the reptiles. In modern forms it is clearly related to the adductor musculature of the jaws. In Chelonia, in which fenestration has not developed, much the same relationships have been attained by emarginations of the temporal bones. In some rather poorly understood way the major radiations of reptiles seem to have been related to the differentiations of musculature as reflected in the temporal regions. The temporal structure evidently is related to jaw musculature, but how the musculature relates to the different types of fenestration is not clear, nor are the reasons for the apparent fact that each of several important lines is characterized by a particular type. It is quite possible that a variety of mechanical factors is involved and that the adductor-muscle system represents but one of these.

In spite of all of the changes among the reptiles, some underlying organizational features did not alter materially. Important among these were the general features of the brain, which tended to remain relatively small, to occupy only part of the cranial cavity, and to have relatively little specialization of its various components. Undoubtedly this correlated in large measure with the maintenance of "reptilian" features of much of the soft anatomy and physiology.

Some breaks with the conservative pattern are found in each of the divergent lines. In the archosaurian line that led to birds and the synapsid line that led to mammals marked changes took place. In both a dominant feature is the relative increase in the size and complexity of organization of the brain. Many other anatomical and physiological changes were associated with this, some of which are at least potentially detectable in the fossil record whereas others can be known only from modern animals.

The modifications leading to birds appear to be related in very large part to a locomotor shift, as flight replaced terrestrial locomotion. This was a different step from that which led to flying reptiles, which on the whole were much more conservative in the retention of reptilian features. The history of development of neither group, however, has any appreciable evolutionary documentation.

In the line leading to mammals, however, the relationships of structural change and function have proven to be very complex and to involve the interactions of all body systems. Among the mammal-like reptiles the modifications of the structure of the braincase have given evidence of increased size and specialization of the brain. The covering of the cranial cavity became more complete, especially with the expansion of the epipterygoid into a mammalian alisphenoid. Rather than being early innovations, changes in the brain and braincase were subsequent to the establishment of many of the other "mammal-like" features. Like other changes, modifications of the brain took place independently in different phyletic lines. The mammalian brain, as reflected in the braincase, had departed far from the reptilian condition in size and complexity by the time of origin of marsupials and placentals. Braincases are little known for animals that occupy intermediate stages leading to therians. It is thus difficult to assess the significance of the brain and associated parts of the nervous system in the early development of mammals, but it seems that changes were slow relative to many other structures.

Studies based on fossils of the evolution of mammals have centered on dentition, although, of course, other parts of the organisms have not been neglected. This focus is essentially unavoidable in the early stages of mammalian radiation, because, with some notable exceptions, only teeth and associated jaws are known. Mammalian dental patterns were anticipated in a number of reptilian lines, especially by some of the cynodonts. The fully matured pattern, with no more than two sets of teeth, fully differentiated into incisors, canines, premolars, and molars, was attained only after considerable evolution at an essentially mammalian level. Feeding mechanisms, reflected in the dentitions, jaws, and skulls, have been intimately involved in the sorting out of the various mammalian lines, and their role in evolution is emphasized by the fact that taxonomy is based on them.

Studies of morphology and adaptation of skulls and jaws in mammalian radiations have been based largely on teeth and related structures. These have been directed to the establishment of taxonomies and phylogenies as well as to the understanding of evolution in the light of dietary modifications. Relationships between dietary and locomotor changes have not been ignored, but the correlations have been generally rather loosely drawn. Other aspects of skulls and jaws have been studied as they have been particularly relevant to special groups. Thus among the horned and antlered ruminants much attention has been paid to these parts of the skeleton. Special studies of endocranial casts have been made where interest in the brain has been foremost, as in the well-known analysis of the evolution of the horse brain by Edinger. Auditory structures have come in for special attention where they show special features, as among the South American notoungulates. Probably the greatest departures from the fundamentally basic mammalian patterns are reflected in the enormous increase in cranium in some primates and great reduction of the facial and masticatory structures in proboscideans, whales, and some of the edentates. In the latter cases, however, the modifications are keyed largely to high specialization of feeding, whereas in *Homo* the modifications, irrespective of their initial motivation, have opened vast new thresholds of ways of life, comparable in scope to others that were fundamental to the initiation of new major radiations.

2 Integumentary, urogenital, respiratory, and circulatory systems

INTEGUMENTARY SYSTEM

The integumentary system, in forming the surface of the body has intimate contact with the external environment and is a sensitive indicator of the kind of environment in which an animal lives and the ways in which it interacts with the surrounding medium. Its functions, of course, are many and highly varied. In the course of evolution the integument has been modified notably in accommodation to major habitat shifts. The skin thus is an important structure both in interpretation of functions of organisms and in differentiation of major adaptive types and taxonomic divisions.

The skin and associated dermal structures form the integumentary system, which technically includes various skeletal elements, such as fish scales, dermal bone, teeth, as well as horny scales, hoofs, antlers, and similar structures. Except for the horny or corneous scales, these hard structures, although dermal in origin, are most conveniently treated as parts of the skeleton.

The integument proper is composed of an epidermis, derived from the ectoderm, and an underlying dermis of mesodermal origin. The skin is variously equipped with glands. Variations in glandular structure, in the thickness and complexity of the epidermis and dermis, and the presence or absence of a corneous layer constitute the major integumentary differences among the principal categories of vertebrates. Various specializations related to particular adaptations occur throughout the classes. The skin in mammals, for example, is intimately related to thermal regulation. In fast-swimming animals, such as porpoises, the skin is specialized to reduce drag, and in fast-swimming fishes the scales serve a similar function. Skin respiration by utilization of a highly developed vascular network is found in various fishes and amphibians.

Skin glands are well developed in amphibians, at least some reptiles, and in mammals, but except for a few specialized structures, they are not prominent or highly varied in other classes of vertebrates. Mucous glands are well developed in some fishes, but they are simple in structure, and the skin of the fishes is relatively uncomplicated. Amphibians possess both mucous and poison glands, and some members of the group have highly vascular skins that play an important role in respiration. A corneous layer is developed in some.

The skin of many reptiles is marked by a well-developed horny layer and a fairly simple glandular system. The feathers of birds, stemming from reptilian scales, are distinctive, and the skin has few glands, the most important being the uropygial oil gland, present in the majority of birds. The skin of mammals is much more complex than that of any other major group of vertebrates (Figure 21). Its complex system of glands has many functions. The sweat glands are primarily related to temperature control, a part of the complex system of regulatory devices contributing to the stable homoiothermy of mammals. Modified sweat glands have produced another characteristic mammalian feature, the mammary glands.

Although the integumentary system has broad and important evolutionary significance and a complex evolutionary history, fossil materials have contributed relatively little to the understanding of its functions and history. As a rule the nature of the skin is inferred from the evolutionary level indicated by the skeleton.

Cases of actual preservation of the soft tissue of the integument are very rare and found only under most unusual conditions in materials of little age. Frozen carcasses and mummified tissue are examples. Occasionally in older materials carbonized remains of skin occur in fine-grained sediments. Ichthyosaurs are among the best known examples. Such remains have been important in determinations of the limits of the skin and in recognition of the presence of soft structures, such as unsupported fins and webs between bones, but they have not given much information about the nature of the skin itself, indicating at most the presence or absence of horny scales. Recently Chudinov (1968) has decribed preserved skin in the dinocephalian *Estemmenosuchus*. Although the somewhat complex structure is difficult to interpret, it appears

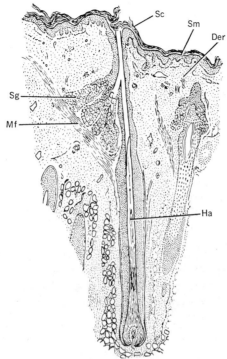

Figure 21.   A section of mammalian skin with hair follicle, showing the general complexity of structure characteristic of mammals. Der, dermis; Ha, a hair; Mf, muscle fibers; Sc, scaly outer surface of epidermis; Sg, sebaceous gland; Sm, Malpighian layer of epidermis. Redrawn from *A Textbook of Histology*, 8th ed., by W. Bloom and D. Fawcett. W. B. Saunders and Company, copyright © 1962.

that in this primitive therapsid the skin was not typicaly reptilian but rather closer to the structure that can be considered premammalian. This is one of the few cases in which any extensive detail has been found preserved in fossilized skin.

Skin impressions often give some evidence about the nature of the integument in fossils. The most common of these are impressions of skins that carried denticles, scales, osteoderms, extensive dermal armour, and various horny scales or their derivatives. There is a fine and indistinct line between impressions, which are commonly spoken of as skin impressions, and the actual scales and bones themselves, generally considered as skeletal elements.

Skin impressions have proven of some value in revealing the nature of the integument in one or another extinct animal and, in some instances, in showing features in which extinct representatives of classes or lesser groups differ from their modern counterparts. For example, many fossil amphibians are

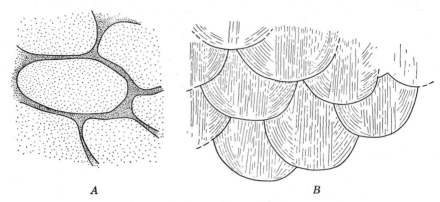

A   *B*

Figure 22.   The skin pattern in two Permian amphibians. *A, Eryops*, in which bony plates were embedded in the skin; *B, Trimerorhachis*, with overlapping, complex, bony scales. (*A* after Romer, 1941*a*; *B* after Colbert, 1955.)

known to have had bony scales (Figure 22), whereas among moderns remnants are present only in the skin folds of gymnophionans. The wing membranes of some of the flying reptiles appear to show fine hairlike structures that, regardless of their somewhat uncertain nature, are not like anything found among modern reptiles. The impressions of feathers in *Archaeopteryx* are of course classic as an indicator of the existence of these birdlike structures in an animal that in most respects is distinctly reptilian.

Informative as such structures are, they give at best only indirect evidence on the histology of the skin itself and the degree and nature of development of skin glands. In various fishes and many amphibians and reptiles rugose, pitted patterns occur on the dermal bones in areas where the skin was closely applied to the bone. The patterns appear to develop under strict genetic control and are, in many instances, taxonomically useful. Occasionally, in addition, some evidence of the vascularity of the overlying skin is to be found in the vascular character of skin-covered bone. A fin between the long neural spines of *Dimetrodon* has been postulated, in part on the evidence of an extensive vascularization of the surface of the bone of the long spines. This finding complements data on healing of bone fractures in place, which also suggest a webbing.

All these kinds of evidence contribute in one way or another to an understanding of the history of the integumentary system. The evidence is spotty in its taxonomic distribution among vertebrates and of relatively minor significance in contributing to understanding of the details of the evolution of the skin.

For the most part it appears that the integument of fishes, even very early in their history, was little different from that found among the various modern groups, and a group-by-group comparison of fishes, as far as it can be made, suggests similar correspondence. This conclusion, of course, is based very largely on resemblances of scales and not of direct observation of the soft tissue.

How close the resemblances may have been between the extinct ostracoderms and living cyclostomes is an open question. Most of the ostracoderms that are well known had very heavy dermal ossification, some very different from that known among any other groups of vertebrates. Others, such as the coelolepids, however, had a soft skin set with small denticles. This skin could have been quite similar to that of the lampreys or hagfish.

Many extinct amphibians carried bony scales in their skin, in contrast to anurans and urodeles. Some, such as the dissorophids and some of the seymouriamorphs, developed bony armor, which has been interpreted as a means of reducing dehydration through skin. It appears to be generally supposed that the skin of extinct amphibians was much like that of many moderns, lacking a corneous layer and set with numerous glands. The actual conditions are not very clear. As far as the rather scant evidence of early amphibians shows, development of extensive horny scales did not take place. The glandular structure, of course, is not known from any direct evidence.

Skin impressions of reptiles in general suggest that the range of structure shown by moderns, by lizards, snakes, turtles, *Sphenodon*, and crocodiles covers most of the types that have existed. Marine forms, in at least some instances (e.g., the ichthyosaurs), appear to have had soft, scaleless skins. Terrestrial forms appear to have had either a keratinized or osseous covering, or both. Very little is known, however, about the actual history of the development of the reptilian types of integument, because information from very primitive forms of reptiles and initial members of most major groups is essentially nonexistent.

Interesting questions arise about the integumentary history of the mammals and their reptilian ancestors. Whether mammalian ancestors went through a "reptilian" phase of skin type, with a heavy horny layer, or whether, perhaps, they remained in a more "amphibianlike" condition, with no extensive covering to prevent dehydration, is uncertain. Skin impressions of some small pelycosaurs suggest that horny scales were present, but *Estemmenosuchus*, with its "nonreptilian" skin, seems to show no evidence of scales. If present in the ancestry, they were lost very early. When hair developed is another related question of interest with no answer.

A similar question pertains to the origin of the skin of modern amphibians. Is it perhaps largely secondary with extensive development of mucous glands and respiratory functions coming late in amphibian history? It is quite possible that many amphibians did develop a corneous system similar to that of reptiles. If, for example, the dissorophids led to anurans, as has been suggested, was there possibly not only a bony but also a corneous integument in anuran ancestry? If so, the skin of frogs may be somewhat less like that of the ancestors than the more corneous skin of some of the toads.

Answers to questions of these sorts may eventually come from the record, because occasionally preservation is suitable. That more detailed information on the histology of the soft structures of the integument can come from it appears to be essentially out of the question.

UROGENITAL SYSTEM

Organs of the excretory and reproductive system, although basically different in origin and separate in very primitive vertebrates, are closely related structurally and often are treated together for convenience. Evolutionary trends are in general toward more complex interrelationships, so that treatment as separate systems in advanced vertebrates poses problems. Both the functions and structures undergo very important changes, with major reorganizational patterns characteristic of the different classes of vertebrates.

The most important evolutionary modifications of the excretory organs take place in the kidneys. These, as is known from successively more advanced modern groups, show successive changes related to attainment of increasingly higher levels of organization. Essentially, in addition to increase in complexity, changes involve successive development of functional kidneys from the anterior to posterior (Figure 23). The evolutionary history of the kidneys and related structures and its relationships to the embryological development in higher vertebrates make an interesting and fascinating story. Undoubtedly its events have played an important role in the adaptive evolution of the major categories of vertebrates as they have become adjusted to new environments and attained increasingly complex ontogenies. The parallelism of phylogenetic and ontogenetic stages in particular emphasizes the significance of embryological adaptation and the ways in which successive stages in both types of development may resemble and differ from each other. A lucid account of this is to be found in Romer's The Vertebrate Body (1956c).

The history, however, is one that has been gleaned from comparative anatomy and embryology of living animals, because there is essentially no interpretable record among fossils of any age. The end members of the lines that have persisted to the present time present a good morphological series, but they do not show most of the transitional stages between the major groups.

Some speculations on the roles of the kidneys in osmoregulation have been made in connection with investigations into the origin of vertebrates and of some of the major groups of fishes. These have come into prominence particularly in discussions of the origins of sharks in fresh or marine waters

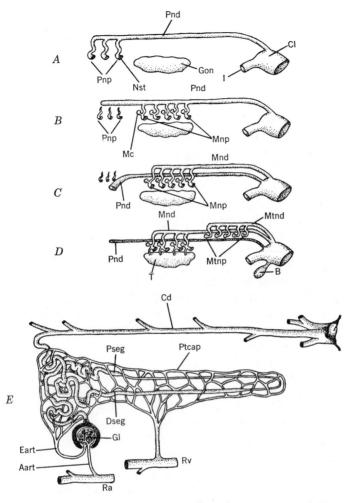

Figure 23. The developmental stages (*A* through *D*) of the urogenital organs in vertebrates, showing the successive stages of pronephros, mesonephros, and metanephros along with associated structures. The fine structure of an adult kidney is shown in *E*. Aart, afferent arteriole; Cd, collecting duct; Cl, cloaca; Dseg, distal segment; Eart, efferent arteriole; Gl, glomerulus; Gon, gonad; I, intestine; Mc, Malpighian capsule; Mnd, Mesonephric duct; Mnp, mesonephros; Mtnd, metanephric duct; Mtnp, metanephros; Nst, nephrostome; Pdis, distal segment; Pnd, pronephric duct; Pnp, pronephros; Pseg, proximal segment; Ptcap, capillary; Ra, renal artery; Rv, renal vein. (*A* through *D* redrawn from *A Textbook of Zoology*, 7th edit., by T. J. Parker and W. A. Haswell, rev. by A. J. Marshall. Copyright © 1963 by Macmillan and Company, Publishers. *E* redrawn from *From Fish to Philosopher* by H. J. Smith. Little, Brown and Company, Copyright © 1953.)

(see, for example, Carter, 1967). This matter is taken up in context later (pages 511–512) and, as noted there, requires extrapolation over a period in excess of 300 million years without any intervening data. Although interesting, such speculations are ineffective as bases for interpretation of events so remote.

Reproductive parts of the urogenital system are extremely important in the evolution of the vertebrates, since the major steps in reorganization of the functions of these organs relate directly to major changes in environments and the divergence of the major lines of vertebrate adaptive radiation. Separation of three of the four classes of tetrapods, amphibians, reptiles, and mammals is based primarily on the structure and function of this system, as is the division of all vertebrates into anamniotes and amniotes.

Comparative studies of reproductive systems have yielded an immense amount of information about their structures and functions in the different groups of vertebrates. Not only are there clear distinctions between major categories, but widely diverse adaptive changes have occurred within groups (see, for example, Figure 24).

Throughout the subphylum the gonads are the principal sex organs. Both ovaries and testes show rather broad evolutionary changes, evident particularly in the increasing complexity and subdivision of the component parts. Such changes are related mainly to patterns of egg formation in the case of the ovaries. Accessory structures of the reproductive systems follow much the same patterns of general evolutionary advancement and display rather marked differences among the major groups. Males and females show interesting patterns of complementary evolution related to the development of different patterns of insemination. Intensive studies have made the courses of evolution very well known, as far as this is possible from an understanding of existing animals.

From studies of skeletons it appears that some of the stages of evolutionary development of the reproductive system are not represented by living vertebrates. These are the stages intermediate between major categories, and they can be determined only by interpolation, because the record preserves very little that is directly interpretable.

The fossil record, however, does give a few glimpses into one or another aspect of the reproductive system. Some secondary sexual characters are exhibited directly by the skeletons and thus, occasionally, give some insight into the nature of sexual dimorphism and behavioral patterns between the sexes. Most evidence, however, comes neither from the skeleton as such nor from preserved reproductive organs, but from reproductive products, such as eggs and larvae. Reptile eggs, though not abundant in the record, are known for several groups, including turtles, lizards, and dinosaurs. The earliest fossil egg known is from the early Permian, showing the existence of a

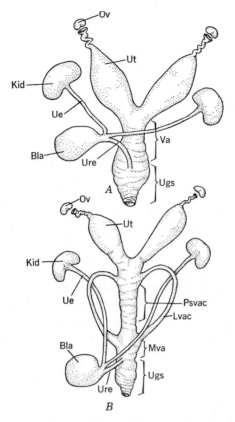

Figure 24. Diagram of the female urogenital systems in ventral view. *A*, placental (Eutheria); *B*, marsupial (Metatheria). Bla, bladder; Kid, kidney; Lvac, lateral vaginal canal, Mva, median vaginal canal; Ov, ovary; Psvac, pseudovaginal canal; Ue, ureter; Ugs, urogenital sinus; Ure, urethra; Ut, uterus; Va, vagina. (After Lillegraven, 1959.)

reptilian type of reproduction at that time. It is not known, however, to what particular animal this egg pertains.

Larval stages are known for some extinct amphibians. Some of these from the Paleozoic show the existence of external gills, much like those found in some of the larvae of extant amphibians. Discovery of gilled larval stages in one family of seymouriamorphs indicates an amphibian type of reproduction for this group, or at least for one of its families, and seems to have settled the long standing controversy about the reptilian or amphibian nature of this suborder.

A few bits of osteological evidence on the reproductive system do come from the record. Males of various extinct sharks have been found to possess clasping organs, indicating fertilization similar to that found among moderns. The presence of vestigial grooves for lateral-line organs in *Seymouria* (T. E. White, 1939) suggests an aquatic larval stage in its ontogeny. On the other hand, T. E. White also has indicated the possible existence of a sexual dimorphism in the haemal arches of the anterior caudal region, with the first arch farther back from the pelvis in some (?females) than in others (?males). This resembles a condition known in some lizards, which, of course, have relatively large eggs. These bits of evidence on *Seymouria* appear to be somewhat contradictory.

Because the function of reproduction is so important in the classification of vertebrates and is a key to the major adaptive shifts in tetrapods, it has been the subject of much speculation. The nature of reproduction during the transition between amphibians and reptiles, and reptiles and mammals, has been argued without end and without agreement.

The skeleton serves mainly as a secondary source of data in such speculations. Particular skeletal features occur in association with definable levels of reproductive structure and function in living animals. Quite evidently these relationships do not hold in animals that mark transitions between the major groups, and like the skeletal features that are known, the reproductive structures were undergoing modifications from one pattern to another. The skeletons do not supply the evidence necessary to judge the stage of transition of the reproductive systems reached by representatives of the adaptive shifts preserved in the record.

RESPIRATORY SYSTEM

The function of respiration, the gaseous exchange between the organism and its environment, is carried out by various parts of the body in different groups of vertebrates; hence the term "system," though convenient, does not refer to the same structures in various large categories of vertebrates. Gills, lungs, the integument, the pharynx, the gut, and the cloacal or anal region all may take part in this exchange. Gills, furthermore, are of two general types—external and internal. The latter, it would appear, are the most primitive vertebrate respiratory device. They alone are present in the cyclostomes and, apparently, in their ancient ancestors, the agnathans.

Lungs go far back in vertebrate history, having been identified in one group of placoderms, the Antiarcha, by Denison (1941) (see Figure 25). Evidence of their existence has been found in all classes of vertebrates except the agnathans and Chondrichthyes. In the latter a rudimentary structure detected in the embryos has been considered to represent a vestige of a gas bladder, but this has been seriously questioned. Most existing fishes possess

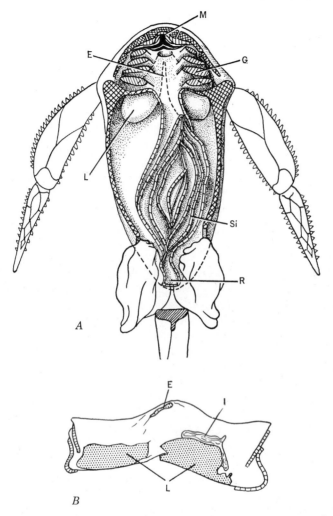

Figure 25. *Bothriolepis*, an antiarch placoderm. *A*, reconstruction of the body systems; *B*, cross section through the posterior lung region. E, esophagus; G, gills; I, intestine; L, lung; M, mouth; R, rectum; Si, possibly spiral valve of intestine. After Denison (1941).

gas bladders rather than lungs, but there is considerable evidence to indicate their phylogenetic source in the lungs proper. Lungs are characteristic of tetrapods, although they may be absent in some amphibians, replaced by gills or by skin breathing.

Exchange of gases through the skin is developed in a number of types of amphibians and occurs as well in some teleost fishes. Anal respiration is a localized expression of this type of gaseous exchange, found in a few fishes and in turtles, mainly as a supplementary device. Any highly vascular tissue that brings capillaries close to the surface of the animal has the potential for the development of gaseous exchange with the environment, and such tissues have been so used in a variety of ways in the course of evolution of the vertebrates.

Respiratory functions thus involve tissues with appropriate vascular structure and are so positioned that fluids containing accessible gases and capable of accepting gases from tissues can be passed over their surfaces. Air and water are such media, and vertebrates have developed a multitude of special structures and functions to ensure their passage over respiratory tissues. Development of the system can be assessed in terms of these two aspects: the development and positioning of appropriate tissue and mechanisms related to producing currents that pass fluids over the surfaces of the tissue.

Fossil vertebrates provide some evidence concerning these functions, which, however, is largely indirect. The intimate association of internal gills with visceral arches account for a considerable body of information on the placement and general dynamics of the gill mechanism. Tendencies for development of various kinds of gill coverings add data. It is known, for example, that at least many agnathans, like the cyclostomes, possessed a larger number of functional gills than did other fishes and that major differences between agnathan groups were present in the ways that circulating waters made their exit from the gill chambers. Individual openings were present in osteostracans, and a single common opening in heterostracans (see Figure 5).

In the presumed transition from agnathous to gnathostomatous forms the anterior visceral arches became incorporated into the skull in the form of the trabeculae cranii (according to usual interpretations), the mandibular arch, which formed the palatoquadrate and Meckel's cartilages, and the hyoidean arch, which became a part of the jaw-support apparatus. In the course of this change some modifications of the structure of gills can be reconstructed and modifications of the gill supports provide some rather direct insight into the ways in which functions were modified.

Gill structures in both lampreys and hagfish appear to be highly specialized and may provide no more than very general insight into the nature of the soft tissue in the extinct agnathans. Some interpretations of anaspids, however, have implied the existence of structures very similar to those of lampreys, and recent discoveries of fossil lampreys in the later Pennsylvanian of Illinois show that such "specializations" did at least have a very long history.

The skeletal remains of other fossil fishes show close correspondence to those known from comparable groups of living fishes and add little structural information. They do, on the other hand, indicate that there has been notable stability in this supporting apparatus of the gills since it attained the level found in gnathostomatous, water-breathing vertebrates.

Gills in living chondrichthyans range from five to seven in number, with a spiracle generally present. All known fossil sharks fall within this range. Very little is known of the gills of most placoderms. One group of ancient fishes, acanthodians, variously classed with placoderms or Osteichthyes, or as a separate class, was the basis for a condition termed aphetohyoidean, on the presumption of the existence of a free hyoidean gill support and fully developed hyoidean gill (Watson, 1937). This condition was furthermore considered to be characteristic of the placoderms, with which the acanthoidians were generally associated. It has since been proven that this interpretation was incorrect, and what appeared to have been a major addition to knowledge of the gill system from the fossil record is known to have been misleading.

In higher gnathostome fishes the gills are covered by an operculum and lack the interbranchial supports found in sharks. Gills range from five to three in number, and, except for some obvious specializations related to specific conditions found in a few groups, they conform closely to a central pattern. Evidence in the fossil record amply confirms the stability insofar as hard structures of supports and covers of gills are concerned but adds nothing of importance to the knowledge of soft anatomy.

Embryological gill pouches are present in tetrapods, but for the most part they are modified to perform nonrespiratory functions in adult individuals. Gill supports likewise are modified in ontogeny in many ways to perform various functions related to feeding, sound production, sound reception, and, secondarily in some instances, to aid in movements connected with respiration.

Modern amphibians generally have external gills in the larval stages, and in some instances these persist into stages at which reproduction occurs, especially where there is partial or full neotony. These gills are filamentous and arise from the bases of the first three gill arches (Figure 26). A similar structure has been found preserved in various amphibians in the late part of the Paleozoic era. It is particularly distinctive of a group termed branchiosaurs, interpreted by some (e.g., Romer) as larval stages of labyrinthodonts,

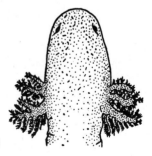

Figure 26. External gills in a modern amphibian, showing the three arches involved in support. (Redrawn from *Anatomy of the Chordates*, 3rd edit., McGraw-Hill Book Company. Copyright © 1962.)

but by others as members of a separate group, the Phyllospondyli. Similar structures occur in the larval stages of one family of the Seymouriamorpha, the Discosauriscidae. Such gills are of great interest, but they serve more to indicate the larval nature of the creatures bearing them than to give information on the nature of the gills themselves. As in so many instances related to soft systems in the fossil record, the contribution to understanding of the particular system lies in the information that the kinds of systems found today have been in existence for a long time.

The more useful aspect of information from the gills is that it has aided in determining something about the life habits and modes of development of some extinct creatures and thus has aided in an understanding of the reproductive systems and in placement of the organisms taxonomically.

The fossil record has provided only rather minimal direct information about lungs. The structure identified as a lung in an antiarch by Denison is one of the few instances in which such a determination has been made. Calcified gas bladders, such as that in Cretaceous coelacanths, also show the presence of lung-derived structures, and there are other similar isolated bits of information.

If the evidence on the antiarch is correct, as it certainly seems to be, then it is clear that lungs are very ancient in the vertebrates. There is no evidence of them in agnathans or sharks, and the situation in placoderms (except the antiarchs) and acanthodians is unknown.

All large groups of higher fishes, Osteichthyes, and all major groups of tetrapods are known from modern materials to have had lungs, structures derived from lungs, or to have lost lungs once present in their ancestry. The structures in most fishes are classified as gas bladders, but some of these have respiratory functions, and the line between lung and gas bladder cannot be clearly drawn.

Lungs, as seen in modern animals, show considerable evolutionary development. The most primitive known stage appears to be the ventral, bilobed structure found in *Polypterus*. Various other fishes have functional lungs that are also simple in structure but lie dorsal to the digestive tract, with connections variously ventral, lateral, and dorsal (Figure 27). In more advanced fishes the connection of the gas bladder and digestive tract tends to be lost, and in some the gas bladder does not develop. In such forms, of course, all lung-related respiratory function has disappeared.

Increasing complexity of lungs obtains in the successive stages of tetrapod evolution, although various reductions and special modifications occur along the way. A high level of complexity is reached in birds, in which air passages pass into various recesses, including the hollow portions of the long bones. This feature, of course, is readily recognized in the fossil record. In connection with this highly developed condition a special complex of mechanisms for circulating air is established. Lungs are generally complex in mammals,

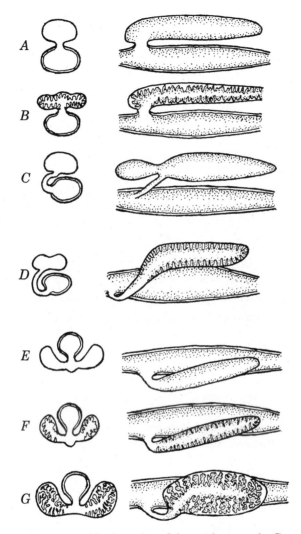

Figure 27.   Lungs or gas bladder in various fishes and tetrapods. Cross sections to the left, lateral views to the right. *A*, Sturgeon and some teleosts; *B*, *Lepisosteus* and *Amia; C, Erythrinus: D, Ceratodus; E, Polypterus* and *Calamoichthys; F, Lepidosiren* and *Protopterus; G*, reptiles, birds, and mammals. (Mostly from Dean, 1895, after Wilder.)

reflecting the high rate of metabolism. The emphasis is on a relative increase of the exposed surface of vascular tissue, as compared with the condition in reptiles.

Fossil vertebrates add nothing to what can be known of the structure of lungs from moderns. Some rather vague ideas of the course of lung evolution can be gained from analysis of associated hard parts. The development of internal nares, giving passage of air into the mouth from external nares, the development of secondary palatal elements, and related modifications of the mouth give something of a history of development of air breathing from the choanate fishes through amphibians to reptiles and alternately to birds and mammals.

Correlated with such changes undoubtedly were modifications in the mechanisms of air transfer. It is known from moderns that many modifications have been related to these processes. The most striking is the development of the muscular diaphragm in mammals. In the development of the therapsids, along with increase in the secondary palate, a differentiation between dorsal and lumbar ribs takes place, suggesting the beginning of muscular modification, perhaps related to breathing.

Some dinosaurs, like birds, have hollow long bones, suggesting the possibility of birdlike modifications. This also was possibly the case in pterosaurs. This sort of information, like that for gills, gives some additional insight into the times of origin and extent of distribution of some aspects of the respiratory system.

Other types of respiration involving modifications of the larynx and gut, outpouching of the pharyngeal chambers, and the skin do not leave an obvious record on the hard anatomy. As far as the amphibians are concerned, it may be supposed that a scale covering such as found in primitive forms precluded effective cutaneous respiration. Highly vascular dermal bone to which skin was closely applied, as found in some fossil fishes, amphibians, and reptiles (e.g., the aquatic Permian reptile *Dictybolos*), may indicate the existence of skin respiration, but this is at best speculative. It is essentially impossible to tell what sorts of innovations may have been made among extinct animals. A fairly straightforward account of the transition to land that results in the class Amphibia is possible, as outlined in detail in Section IV, Chapter 4, but even this must be based largely on secondary sources.

CIRCULATORY SYSTEM

The circulatory system is made up of two distinct parts: the blood vascular system and the lymphatic system. The former is well defined and often thought of as *the* circulatory system. The lymphatic system, of course, is of great importance but virtually impossible to detect in fossil vertebrates. For this reason this section is devoted exclusively to the blood vascular system.

The basic pattern of this system of vertebrates is rather simple, but superimposed on the primary pattern are rather extensive evolutionary changes and, in each group, complex elaborations of the ground plan. The changes

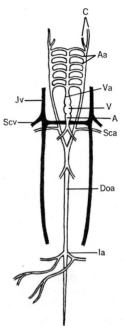

Figure 28.   Diagram of vascular system in an air-breathing vertebrate embryo, showing principal structures. Veins black, arteries white. A, auricle; Aa, aortic arches; C, carotid arteries; Doa, dorsal aorta; Ia, iliac artery; Ju, jugular vein; Sca, subclavian artery; Scv, subclavian vein; V, ventricle; Va, ventral artery. Redrawn from *A Textbook of Zoology*, 7th edit., by T. J. Parker and W. A. Haswell, rev. by A. J. Marshall. Copyright © 1963 by Macmillan and Company, Publishers.

between the classes have been the subject of intensive morphological study and have produced some of the most elegant results attained in the field of developmental and comparative morphology. Except for minor contributions from the fossil record, the information on the details of the system has come from living animals.

Essentially the circulatory system consists of the pumping organ, the heart, which receives blood from a few major veins and pumps it through the system in arteries in which blood is oxygenated and sent to the various parts of the body (Figure 28). In branchiate forms the heart pumps blood through a ventral artery into the gill region, with a branch to each of the gill pouches. Here the blood is oxygenated and continues in the arterial system to the dorsal aorta, in which it is carried back to the body and distributed to the various organs and muscles. It is carried to the head by the carotid system.

In lunged vertebrates there is in addition a pulmonary artery and vein leading to the lungs and to the heart from the lungs, respectively. The artery develops from the sixth aortic arch (Figure 28). Typical examples of circulatory systems are shown in Figures 29 and 30. Major evolutionary advances have taken place in the heart, leading in general to a greater separation of oxygenated and unoxygenated blood, in the system of aortic arches, as gills have been modified and lost, and in the circulation of the head as its functions

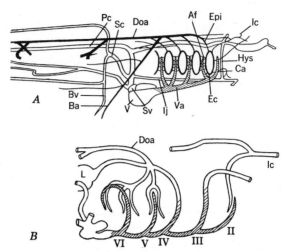

Figure 29.   Two diagrams of circulation in fishes. *A*, shark in which only gill breathing occurs; *B*, lungfish, with both gill and lung breathing. Af, afferent aorta; Ba, brachial artery; Bv, brachial vein; Ca, carotid artery; Doa, dorsal aorta; Ec, efferent collector loop; Ep_1, epibranchial artery 1; Hys, hyoidean sinus; Ic, internal carotid artery; Ij, inferior jugular vein; L, lung; Pc, posterior cardinal sinus; Sc, subclavian vein; Sv, sinus venosus; V, ventricle; Va, ventral aorta. Redrawn from *A Textbook of Zoology*, 7th edit, by T. J. Parker and W. A. Haswell, rev. by A. J. Marshall. Copyright © 1963 by Macmillan and Company, Publishers.

and structures have changed. The two most profound steps have come with the development of gnathostomes from agnathous forms and of lung-breathing animals from gill breathers.

As for the other systems, the fossil vertebrates supply information on blood circulation only to the extent that parts of the system are recorded on or in the hard tissue. The parts of the circulatory system that undergo the most

Figure 30.   Examples of major elements of the circulatory systems in tetrapods, seen ventrally. *A* and *B*, *Rana*, an amphibian, arterial and venous systems, respectively, in separate diagrams; *C*, *Columba*, representing the arterial and venous conditions in birds and reptiles; *D*, *Lepus* (*Oryctolagus*), a mammal. Abd, abdominal vein; Bra, brachial artery; Brv, brachial vein; Cart, conus arteriosus; Cc, common carotid artery; Dao, dorsal aorta; Ec, external carotid artery; Eju, external jugular vein; Fma, femoral artery; Fmv, femoral vein; Hp, hepatic portal vein; Ic, internal carotid artery; Iju, internal jugular vein; Ila, iliac artery; Iv, iliac vein; Ju, jugular vein; Lau, left auricle; Lpa, left pulmonary artery; Lvn, left ventricle; Pa, pulmonary artery; Pcv, postcaval vein; Prev, precaval vein; Pv, pelvic vein; Rau, right auricle; Rav, renal artery and vein; Rnv, renal vein, Rna, renal artery; Rnpt, renal portal vein; Rpa, right pulmonary artery; Sclv, subclavian vein; Sv, sinus venosus; Syt, systemic trunk artery; V, ventricle. Redrawn from *A Textbook of Zoology*, 7th edit., by T. J. Parker and W. A. Haswell, rev. by A. J. Marshall. Copyright © 1963 by Macmillan and Company, Publishers.

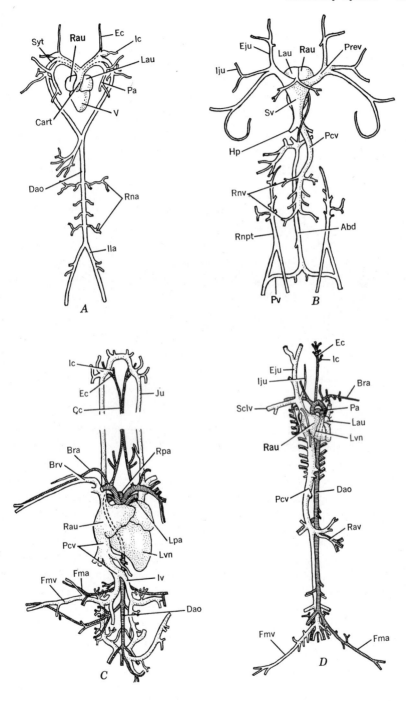

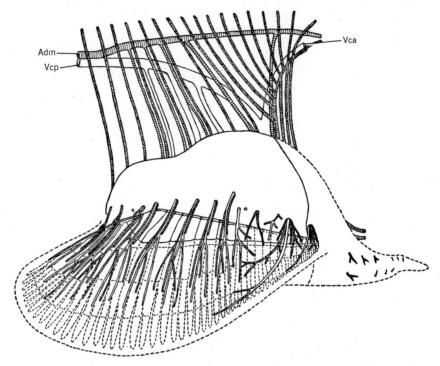

Figure 31.   Reconstructions of the circulatory and nervous systems of the fin of an arthrodire (*Enseostius jaekali* Gross). Note that the only actual reference points are the canals and foramina in the hard tissues. All the rest depends very strongly on analogies with modern fishes. Veins white, arteries with annular pattern, nerves stippled. Adm, unpaired postcephalic division of dorsal aorta; Vca, Vcp, anterior and posterior veins. (After Stensiö, 1959.)

extensive and significant evolutionary changes are not recorded on the bone. This is completely true as far as the heart is concerned. Although the presence or absence of gill bars indirectly can give some information on the aortic arches, the lack of intimate association between the arches and the hard gill parts has precluded anything but the most general interpretations. Vertebral structures, likewise, give only minimal information on arteries. In some particular cases girdles and limb bones record some details about the arteries and veins that accompanied them. Among tetrapods standard associations of certain vessels and foramina are such that the soft structures can be confidently inferred for the fossils. Occasionally rather intricate patterns of foramina have been found, and interpretations of the circulation have been made. In keeping with their efforts at very detailed analyses, the Swedish paleontologists have been foremost in taking advantage of such circumstances. Figure 31 shows interpretations of arthrodires as made by Stensiö. All such reconstructions must be based in some large part on extension from knowledge of modern vertebrates, but starting from this base they may become significant in showing the nature of development in one or another particular group of animals. Their role in a broad understanding of the evolutionary course of circulatory systems is necessarily very limited.

It is mainly in the head region that valuable detail of the circulatory system is available in fossils. As for the nervous system, the most detailed information comes from primitive vertebrates, Osteostraci and arthrodires, and from mammals. The former have the greatest intrinsic interest, for there is much less that can be determined from modern materials concerning agnathans than mammals. The degree to which reconstructions have been possible are shown in Figure 32. Much of this work has stemmed from Stensiö and his associates and has been based on study of prepared specimens and serial sections from which reconstructions have been made.

As shown in the figures, an immense amount of data has been gathered. This has added importantly to the rather meager information on circulation of primitive vertebrates available from the living cyclostomes. It also shows the rather remarkable persistence of this level of development from the Devonian to the present. The fossil record does not, however, give any evidence on the transition from agnathous to gnathostomatous fishes, and this is lacking from moderns as well. The stages can be interpolated, but these are hypothetical, and even the assumption that there was a transition from any known type of agnathous condition to agnathostome condition is, of course, speculative.

For the higher levels of fishes, in general, relatively little data on head circulation is available from fossils. In cases where detailed studies have been made most of the information has come from ossified or chondrified remains of the endocranium. Moderate success has attended efforts to analyze various conduits and foramina interpretable as carrying particular vessels.

Among the fishes some arthrodires, xenacanth sharks, crossopterygians, and a few ray-finned fishes have yielded information on head circulation. Minor amounts have come from amphibians and reptiles, but within these groups only relatively rare exceptions show patterns of circulation much different from those that can be studied much more effectively among moderns. In some instances, however, such as the caseids, conditions without any modern counterparts are encountered. These are potentially most interesting, but the problem arises that without a guide among living animals it is extremely difficult or impossible to make an accurate assessment of what the condition of the soft anatomy actually was.

Mammals, by virtue of the high degree of ossification of the skull, preserve a better record of the circulatory system than do any other tetrapods. Very primitive mammals, along with advanced mammal-like reptiles, could be of

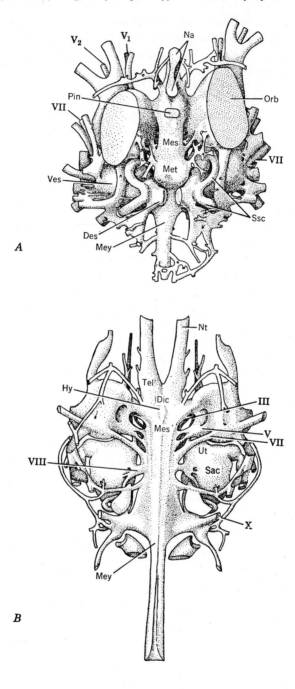

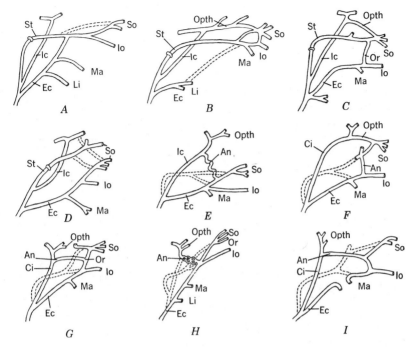

Figure 33.   Diagram of representative carotid circulation patterns in mammals, showing the extent of variation and some of the major groups represented by characteristic examples. This variable part of the circulatory system is but little reflected in the foramina and canals of the skull. *A, Ornithorhynchus; B, Erinaceus; C, Lemur; D, Vespertilio; E, Viverra; F, Homo; G, Phoca; H,* ruminant; *I, Pedetes.* All lettering refers to arteries. An, anastomosing ramus; Ec, external carotid; Ic, internal carotid; Io, infraorbital; Li, lingual; Ma, mandibular; Opth, ophthalmic; Or, orbital ramus; So, supraorbital; St, stapedial. (After an unpublished compilation by James R. Beerbower.)

considerable interest in tracing the origin of the mammalian circulatory system in the head region. The fact is, of course, that except for the multituberculates, which appear to be a side branch, none of the available fossil materials is satisfactory for such detailed analyses as are necessary.

Figure 32.   Reconstruction of two cranial casts showing positions of nerves and blood vessels on the basis of the spaces available for their occupancy. *A, Mimetaspis,* an osteostracan, dorsal; *B, Macropetalichthys,* an arthrodire, ventral. Des, heavy nerves to dorsal, dermal, sensory areas; Dic, diencephalon; Hy, hypophysis; Mes, mesencephalon; Met, metencephalon; Mey, myelencephalon; Na, nasal structures; Nt, nasal tract; Orb, orbit; Pin, pineal; Sac, sacculus; Ssc, semicircular canals; Tel, telencephalon; Ut, utriculus; Ves, vestibule of ear; III, oculomotor nerve; V, V_1, V_2, trigeminal nerve and its branches; VII, branch of facial nerve; VIII, acoustic nerve; X, vagus nerve. (After Stensiö, 1963a.)

Within the mammals an interesting and significant series of evolutionary steps occurs in the carotid system, with both the internal and external carotids undergoing extensive modifications (Figure 33). Even in an extreme case, such as the Felidae, in which a rete has replaced the internal carotid entrance to the cranium, the cranial anatomy gives little indication of what has occurred. It would appear that, with the possible exception of extremely primitive mammals, most of the patterns that have existed are still in existence in one form or another and that the head circulation of mammals can be rather fully understood from modern examples. The patterns leading to the different types, of course, cannot be bridged in true phylogenetic series, but only by comparisons of their end members.

3 The nervous system

Although the nervous system consists entirely of soft tissue, which is almost never itself preserved directly in the fossil record, some of the critical parts are closely related to the skeletal elements and provide considerable information about the condition of the system in some groups of extinct vertebrates. For this reason treatment is more extensive than for other systems. Extensive work has been done on the nervous system of fossil vertebrates, but this is necessarily irregular with respect to coverage of both the major taxonomic categories and the parts of the system.

The most primitive and the most advanced of the major categories have supplied the greatest amount of data. In some of the agnathans and the mammals, more than in most other groups, the cranial parts of the nervous system are encased by bone, in which have been preserved impressions that to a greater or lesser degree tell something about the structure of the related brain and nerves.

The nervous system can conveniently be divided into two parts, central and peripheral. The two are, of course, intimately related at points of contact but are generally topographically and morphologically distinct. Each division is complex and can readily be subdivided into meaningful subunits. The peripheral system, however, is but poorly recorded in fossils, and its subdivisions, which even in moderns are more functional than morphologic, are of little practical consequence.

113

The central nervous system consists of the brain and the spinal cord, both of which do, under some circumstances, have rather close relationships to the bones of encasing structures. Of the two, the brain has yielded much the more extensive and informative data.

THE CENTRAL NERVOUS SYSTEM

The Brain

Representative brains at several evolutionary levels are illustrated in Figure 34. From this figure it is clear that the brain has undergone extensive evolution from the primitive to advanced states. There has been a general increase in complexity, better expressed in changes of some portions than of others, and in specializations closely related to the ways of life of the major groups and to the special expressions of these ways by particular organisms.

Throughout the vertebrates three major portions of the brain can be recognized: forebrain, midbrain, and hindbrain. The first two tend toward additional subdivision so that a total of five parts is generally recognized:

Prosencephalon	1. Telencephalon
(forebrain)	2. Diencephalon
Mesencephalon	3. Mesencephalon
(midbrain)	
Rhombencephalon	4. Metencephalon
(hindbrain)	5. Myelencephalon

Even in the most primitive known vertebrates the five subdivisions are partially developed. It appears that each of the three major divisions was initially related to a major sensory receptor—olfaction, sight, and hearing, respectively—with the last involving a very broad interpretation of hearing as reception of a wide range of relatively long wave impulses. An immense amount of specialization followed on this early state. The capsular parts of the head developed in association with these major functions and thus are very directly related to the process of cephalization.

The most anterior part of the brain, the telencephalon, has undergone more extensive evolutionary changes than any of the other subdivisions. In very primitive forms, such as the lampreys and hagfishes, the telencephalon consists of an anterior and posterior swelling, both related to olfactory functions. An olfactory lobe persisted in the course of evolution, but the cerebral lobe has undergone vast change. The dorsal portion, the roof, became enlarged and highly differentiated. Important changes were initiated at the reptilian level, with the development in some reptiles of a neopallium, added to the archipallium and paleopallium that are present in amphibians

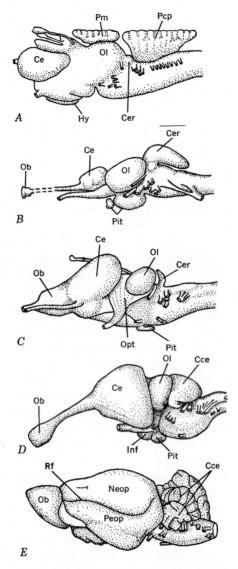

Figure 34.   Brains of representative vertebrates in left lateral view, to show the general levels of organization. *A*, *Petromyzon*, lamprey; *B*, *Gadus*, fish; *C*, *Rana*, amphibian; *D*, *Alligator*, reptile; *E*, *Gymnura*, mammal. Cce, corpus cerebelli; Ce, cerebrum; Cer, cerebellum; Hy, hypophysis; Inf, infundibulum; Neop, neopallium; Ob, olfactory bulb; Ol, olfactory lobe; Opt, optic tract; Pcp, posterior choroid plexus; Peop, paleopallium; Pm, mesencephalic plexus; Pit, pituitary; Rf, rhinal fissure. (After Romer, 1956*c*, from various sources.)

and less advanced vertebrates. The neopallium became well developed in mammals, reaching its greatest expression in man. Various convolutions that increase the surface area have resulted in the formation of a series of lobes, or gyri, and these are impressed, in spite of intervening meningeal tissue, on the inner surface of the cranial bones in many mammals. The extent of the record varies widely among the mammals, in general being best in the advanced forms. Where the patterns are clear, they provide considerable insight into the nature of the surface of the brain.

In the course of evolution the parts of the telencephalon have assumed functions remote from olfaction. In mammals the cerebral cortex is associated with "higher" mental activity, which reaches a peak in primates and man. There is an areal localization of many functions, such as acoustic, tactile, and visual, and the areas of parts of the surfaces of the lobes show correspondence with the degree of development of the functions with which they are intimately associated. As recent work by Radinsky has shown, information on degree of development of various functions among fossil mammals can be attained by use of this characteristic.

The diencephalon includes the optic vesicles in the embryonic state, and from these arise the eyes proper. Three main subdivisions, from dorsal to ventral, respectively, are present: epithalamus, thalamus, and hypothalamus. From the epithalamus arise the parietal body, or parapineal, and epiphysis, or pineal body. These appear to have been primitively paired light receptors, as suggested by impressions on the internal surface of the crania of some fishes (Figure 35). The parapineal possesses rudimentary optic structures in some vertebrates, whereas the pineal, which appears to lack them or possibly has them in extremely rudimentary form, is possibly of endocrine nature. Many vertebrates below the level of mammals and birds possess a "parietal" foramen, variously accompanied by related modification of the skull, and through this have provided some provocative if puzzling information about the distribution and size of the associated pineal system in extinct groups.

The thalamus is lateral and serves in the integration of sensory and motor impulses. The ventral hypothalamus acts as an integrator of the autonomic, or sympathetic, nervous system with other parts of the system and relates in this way to the control of a variety of physiological actions. At its anterior end is an important landmark of the brain, the optic chiasma. Ventrally the infundibulum, which provides the posterior lobe of the pituitary, forms in sufficiently close relationships to the basicranial part of the skull that it sometimes leaves an impression on the bone and thus a record of its general characteristics.

The metencephalon has undergone only moderate evolution during vertebrate history. Its dorsal part forms the cerebellum, a neuromuscular-activity coordinator. Its sensory functions are related to those of the inner

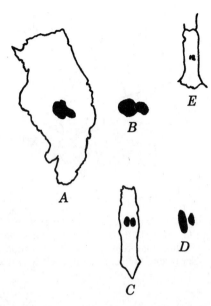

Figure 35.   Internal surface of the crania of various fishes, showing impressions of pineal body and epiphysis arranged laterally, suggesting the possibility that these were once paired organs. *A* through *D*, arthrodires; *E*, a "stegoselachian." (After Edinger, 1956.)

ear, as this structure plays its role in equilibrium. Auricular lobes provide centers of equilibrium and in some reptiles, birds, and mammals are specialized to form floccular lobes that are highly developed in active forms, with a three-dimensional aspect to their way of life. The floccular lobes are variously impressed on the otic portion of the inner surface of the braincase so that it is possible in some instances, for example, in mammal-like reptiles (Figure 36), to assess their dimensions roughly and make at least very crude estimates of the extent and nature of activities of extinct animals.

In birds and mammals a central feature of the metencephalon, the vermis, is highly developed. Under favorable circumstances its external features are recorded on the inner surface of the osseous cranium. Between the cerebellar lobes in some birds and in mammals a ventral bridge, the pons, is developed, and this too may leave impressions on the adjacent bone.

The mesencephalon includes dorsal processes, the optic lobes, which are primarily centers for conduction of sensory impulses from the organs of sight but also, in advanced vertebrates, may serve to some extent for auditory reception and coordination. Nuclei related to cranial nerves III and IV also lie in this part of the brain.

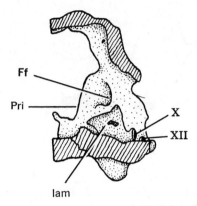

Figure 36.   Medial view of the inner surface of the otic part of the brain case of a dicynodont, therapsid reptile, with skull halved along the vertical median plane. Based on serial sections. Ff, floccular fossa; Iam, internal auditory meatus; Pri, proötic incisure; X, XII, foramina for tenth and twelfth cranial nerves. (After Olson, 1944.)

The myelencephalon, or medulla oblongata, is associated with various physiological regulators. Cranial nerves V through XII lead to this area of the brain, with their somatic visceral motor and sensory parts associated with different portions. An important portion of the medulla oblongata, from the standpoint of evolutionary change, is the acousticolateralis area, which is related to the lateral-line sensory system, where present, and to the equilibrial and auditory functions of the inner ear. As the lateral-line system decreases and disappears with the advent of terrestrial life, marked changes occur in this section of the medulla.

The brain proper is covered by membranes, the meninges, which support and protect it. They intervene between the brain and hard tissues and even in the most favorable cases reduce materially the actual correspondence in topography between the brain surface and the adjacent bone. All so-called fossil brains are, of course, casts, and these in no case correspond in any great detail to the actual surface of the brain.

The Spinal Cord

The spinal cord has undergone much less extensive morphological change than the brain in the course of vertebrate evolution. What has occurred is not for the most part evident in the neural canal first because the cord is not closely encased by the canal and, second, because most changes appear to have been in internal organization, not reflected in shape. The only records of the canal, and the nerves leading to and from it, are found in the vertebrae.

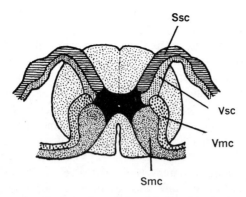

Figure 37.   Cross section of a spinal cord showing the principal nerve tracts, to be used for comparison with cranial nerves in Figure 38. Smc, somatic motor column; Ssc, somatic sensory column; Vmc, visceral motor column; Vsc, visceral sensory column. Redrawn from *Anatomy of the Chordates*, 3rd ed., by C. K. Weichert. McGraw-Hill Book Company, copyright © 1965.

Among tetrapods some enlargement of the cord in the regions that supply the fore and hind limbs does occur. In extreme cases, especially in very large animals, expansion of the cord may be sufficient to be reflected in the neural canal. Thus in some dinosaurs, for example, there is evidence of a very much enlarged complex in the sacral region, which supplies the hind limbs.

Figure 37 shows a diagrammatic cross section of the cord, the location and relationships of the spinal nerves relative to the column and the vertebrae. Little or none of this is commonly available for fossils, but the basic distribution is important for an understanding of the cranial nerves, most of which primitively appear to have been of this nature, being segmental. Various of these nerves (see chart on page 121) can be interpreted as part of one or more of the four fundamental divisions of a typical spinal nerve.

THE PERIPHERAL NERVOUS SYSTEM

Much of the extremely complex pattern of the peripheral parts of the nervous system must inevitably be lost in all fossil vertebrates. The cranial nerves and spinal nerves, and occasionally their ganglia may be to some extent impressed on the bone. Various foramina and canals in the appendicular girdles and in the elements of the paired appendages show the passage of structures which, on comparison with modern animals, may be interpreted as parts of the peripheral system. The remainder of the system, being entirely soft in composition and not having any association with skeletal elements, cannot be traced and even under the very best circumstances can be only very generally inferred as passing between positions where indications of some identifiable nerve or nerve complex are present.

In spite of these obvious restrictions the literature on fossils does have many reconstructions of parts of this system (see, for example, Figures 32 and 43). Although these are interesting and the result of deductions based on often very extensive knowledge of anatomy, it must be recognized that they depend on the anatomy of existing animals and that at best only relatively minor differences from the closest counterparts among existing animals can be inferred and interpreted with any likelihood of correctness among extinct types.

Of the total peripheral system, then, only a relatively small part is closely associated with hard tissue sufficiently to leave an interpretable record. Of this portion the cranial nerves are much the best represented and fortunately do play an important part in the functions of the system and portray some instructive changes in the course of evolution. Nerves associated with vertebrae and limb elements are generally much less important in this regard.

Considerable attention has been paid to the cranial nerves in the studies of fossil vertebrates, both in efforts to make reconstructions of the nerves themselves and in analyses and descriptions of the foramina and canals that mark their courses through or along bones. Throughout the subphylum Vertebrata

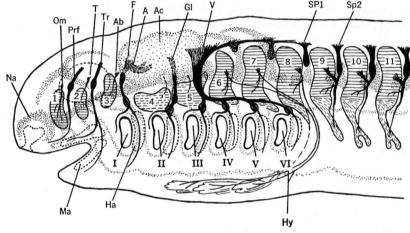

Figure 38. A schematic representation of head segmentation showing the cranial nerves in relationship to segments and gill pouches. A, acoustic nerve (VIII); Ab, abducens nerve (VI); Ac, auditory capsule; F, facial nerve (VII); Gl, glossopharyngeal nerve (IX); Ha, hyoid arch; Hy, hypoglossal nerve; Ma, mandibular arch; Na, nasal capsule; Om, oculomotor nerve (III); Prf, profundus nerve; Sp1, first spinal nerve; Sp2, second spinal nerve; T, trochlear nerve (IV); Tr, trigeminal nerve (V); V, vagus nerve (X); 1–11, myomeres; I–IV, gill pouches. (Modified after De Beer, 1937, from Goodrich, 1930.)

there is a consistent pattern of cranial nerves. Some of the nerves change immensely in function and in their extra cranial distribution, but a common basic structure is found throughout modern representatives and, as far as can be judged, among extinct groups as well. It would appear that no radical modifications have occurred among extinct forms, and all of the changes that have been observed have persisted to the present in one group of vertebrates or another.

As the anatomy of the head was worked out, the structures that were to form the suite of cranial nerves, as usually defined, were numbered I to X, with two additional, XI and XII, present in many groups. Discovery of a nerve anterior to I gave rise to the designation of nerve 0. Each of the nerves has a name that is very generally applied. The list as now accepted, at least for convenience, is as follows:

Number	Name
0	terminal
I	olfactory
II	optic
III	oculomotor
IV	trochlear
V	trigeminal
VI	abducens
VII	facial
VIII	acoustic (auditory)
IX	glossopharyngeal
X	vagus
XI	accessory
XII	hypoglossal

In some fishes and some of the living amphibian groups only nerves 0 through X are present, with XI not separable from X and XII represented by spinal nerves. The simple arrangement is based on a generally serial pattern of the nerves as they issue from the cranium (Figure 38). It has been long known that this arrangement, though descriptively convenient, does not properly express the relationships of the nerves to each other or to the brain.

The cranial nerves appear to have been originally segmental, as are the spinal nerves today. Similarly, primitively there was a dorsal and ventral root with structures and functions similar to those shown in Figure 37. Unlike the spinal nerves, except those of the lamprey, the dorsal and ventral roots did not unite. Primitively, it seems likely, both somatic and visceral sensory and motor elements were present. Technically the cranial nerves represent only one root, either dorsal or ventral. The arrangement now very generally accepted is as follows:

Segment	Arch	Dorsal Root	Ventral Root
Premandibular	Trabecula	Ramus ophthalmicus profundus V	Oculomotorius III
Mandibular	Palato-pterygo-quadrate bar and Meckel's cartilage	Ramus ophthalmicus superficialis, maxillaris, and mandibularis V	Trochlearis IV
Hyoid	Hyoid	Facialis VII Acousticus VIII	Abducens VI
First branchial	First branchial	Glossopharyngeus IX	(Absent)
Second branchial	Second branchial		
Third branchial	Third branchial	Vagus X plus Accessorius XI	Hypoglossus XII
Fourth branchial	Fourth branchial		
Fifth branchial	Fifth branchial		

The dorsal roots primitively (see Figure 38) innervated the visceral arches, as shown in the chart. The ophthalmic profundus rami of nerve V were closely related to the trabeculae cranii, which appear to represent a premandibular arch. The maxillary and mandibular branches of nerve V innervated the mandibular arch, the jaws in gnathostomes, and nerve VII innervated the hyoid arch. The branchials proper in gnathostomes are served by nerves IX, X, and XI. Of the ventral roots, with that of the first branchial absent, the first three, III, IV, and VI, served muscles of the eye, and nerve XII, which arises from several segments (6, 7, 8 in *Squalus*, Figure 38), passed to the ventral (hypoglossal) muscles (Figure 39).

Not entered in the chart are nerves 0, I, and II. The last, the optic "nerve," is not a true nerve, but is composed of fibers of the brain that pass to the optic cup through the optic chiasma. Nerve I is the olfactory nerve. Nerve 0 is a somatic sensory nerve that is well developed in some fishes and has been distinguished in primitive agnathans as a rather prominent structure. It appears to be a remnant of an anterior branchial nerve, significant in ancient agnathans but now long without an ascertainable function.

A major modification of nerve patterns and functions appears to have taken place at the shift from agnathous to gnathostomatous fishes, although this transition is unknown either among moderns or fossils. Nerves associated with the anterior branchials came to serve the structures of the jaws, both the mandibular and hyoidean structures. With the development of tetrapods other major changes took place with the shift from a strictly aquatic to an initially partially terrestrial way of life. The basic patterns,

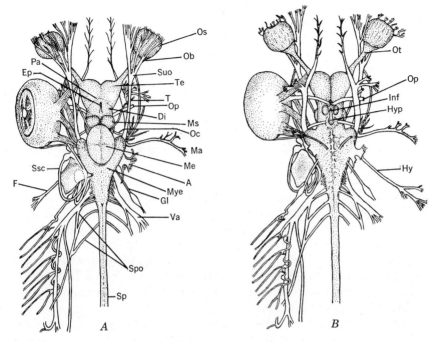

Figure 39.   Brain and sensory organs of *Squalus*, showing the generally primitive proportions and arrangements of their principal structures. *A*, dorsal; *B*, ventral. A, acoustic nerve (VIII); Di, diencephalon; Ep, epiphysis; F, facial nerve (VII); Gl, glossopharyngeal nerve (IX); Hy, hyomandibular nerve (branch of nerve VII); Hyp, hypophysis; Inf, infundibulum; Ma, mandibular nerve (branch of nerve V); Me, metencephalon (cerebellum); Mye, myelencephalon (medulla oblongata); Ms, mesencephalon (optic lobes); Ob, olfactory bulb; Oc, oculomotor nerve (III); Op, optic nerve (II); Os, olfactory sac; Ot, olfactory tract; Pa, paraphysis; Sp, spinal cord; Spo, spino-occipital nerves; Scc, semicircular canals; Suo, superficial ophthalmic nerve (V and VII); T, trigeminus nerve; Te, telencephalon; Va, vagus nerve (X). Redrawn from *Anatomy of the Chordates*, 3rd edit., by C. K. Weichert. McGraw-Hill Book Company, copyright ©, 1965.

however, appear to have been extremely persistent. Where changes have occurred, the association of nerves with muscles is assumed to have remained constant and is used as a principal criterion in the establishment of muscle homologies.

Nerves I, II, and VIII are basically sensory nerves, with "nerve" II being a sensory tract of the brain. They pass from receptors of external stimuli to the brain. These nerves did not undergo any important reorganizations in the course of the major steps noted above. Nerve VIII is closely associated with the facial nerve, VII, and probably had a common origin related to the acousticoequilibrial system. Its functions relate in part to hearing and in part to the maintenance of equilibrium. In its association with nerve VII it is intimately related to the extensive changes that this nerve has undergone.

Nerves III, IV, and VI are extremely conservative in evolution, servicing the muscles of the eyes at all stages. The principal evolutionary changes within the gnathostome level are recorded in nerves V, VII, and X, primitively associated with branchial arches. Each in its primitive stage (see, for example, shark, Figure 38) included a root in the cranial cavity, a ganglion, and a pretrematic and post-trematic branch.

The trigeminal nerve (V), as indicated in the chart, arises embryologically in two segments, with the ophthalmic profundus branch from the premandibular and the other branches from the mandibular segment. The profundus is a separate nerve in the cyclostomes, but it is closely associated with nerve V in higher types and is generally called a branch of V. It is a variable nerve, usually with two branches, which come into very close association with the ophthalmic branches of nerves V and VII, making separation very difficult.

Nerve V has two main branches, which arise from the gasserian ganglion. One, the maxillary, passes to the upper jaw and is in general cutaneous. The other, the mandibular, supplies the lower jaw, being the principal supplier of the jaw adductor musculature.

The facial nerve, VII, carries a large geniculate ganglion. In fishes and aquatic amphibians it has a large somatic sensory component that serves part of the lateral-line system. As this system is reduced or lost, this part of the nerve VII becomes insignificant. Visceral sensory portions persist, forming the chorda tympani, which passes through the middle ear to the tongue in common with parts of nerve V. The hyoidean branch persists but is reduced, and the palatine branch likewise persists. Most widely distributed however, are the motor visceral components that supply the constrictor muscles and derivatives, such as the depressor mandibuli and sphincter colli in reptiles and part of the digastric, platysma, and facial muscles in mammals. With the advent of mammals with mobile facial regions this portion of the nerve enlarges.

The glossopharyngeal nerve, IX, supplies the first branchial in fishes and various derived muscles in tetrapods. Following it is the very important vagus nerve, X, which supplies, through its branchiovisceral branch, branchial arches 3 through 6 in the fishes. It is a highly complex nerve with a somewhat obscure history. Through elements of the autonomic system it controls such things as heartbeat, peristalsis, and respiration (in lung-breathing vertebrates). Great changes have occurred in the course of its evolution, with the disappearance of the gills and the alteration of functions of various structures served by this nerve. Few of these changes, however, are of such a nature that direct evidence on the modifications of nerve X can come from fossils.

Nerve XI, accessory, is closely allied to nerve X, derived in large part from it but having somatic fibers that appear to represent a spinal component. The hypoglossal, nerve XII, is strictly a somatic nerve. It has several roots and is involved in supply of the muscles of the tongue and lower jaw.

Nerves XI and XII are absent in some amphibians, and this has sometimes been considered to be a primitive condition. Foramina in various of the ancient amphibians, however, show the presence of nerve XII. In the course of amphibian evolution reduction of the number of segments contributing to the occipital part of the chondrocranium occurred, and with this a reduction in the number of cranial nerves took place.

RECORDS OF THE NERVOUS SYSTEM IN FOSSILS

Except for the neural canal, vertebral foramina, and foramina and canals in limbs and girdles, the record of the nervous system in fossil vertebrates comes almost entirely from records left by the brain and cranial nerves. In some exceptional cases (see Figure 31) some detailed interpretations have been made of nerves supplying the appendages, but generally data are meager and the presence or absence of particular nerve and associated foramina are used more for taxonomic purposes than for anatomical interpretations.

The development and accuracy of the record of the brain and cranial nerves are in general proportional to the degree of ossification of the skull and jaws, and the intimacy of association of nervous tissue to the hard parts. As a result great variation exists both within and between major categories with regard to the amount of information potentially available. In addition, some groups of vertebrates have attracted more attention than others. These various factors contribute to a very irregular knowledge of the condition of brains and cranial nerves in fossil vertebrates.

External evidence concerning the brain and cranial nerves is available when the braincase itself forms an exposed surface of the skull, as in birds and mammals, and when one or more of the nerves have left a record on externally exposed bones. In the first instance the external shape of the brain-case may give some indications of the general size of the brain if the brain approximately fills the cranial cavity. In such creatures as sharks, however, even though the cartilaginous braincase forms the surface of the skull, the shape of this surface indicates nothing about the shape or size of the brain, which lies deeply embedded in tissue within the braincase. In birds and mammals the opposite is the case, and major differences in brain proportions can be judged externally.

Aquatic animals, except those derived from tetrapods, possess a lateral-line sensory system on the surface of the head and body, involving nerves

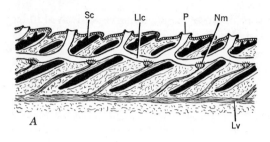

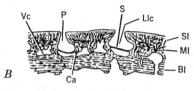

Figure 40.   Lateral-line system. *A*, shown in relationship to scales on the body of *Perca: B*, dispersed in bone in a heterostracan. Bl, basal layer; Ca, canal connecting mesh canal with vascular canal; Llc, lateral-line canal; Lv, lateral-line ramus of vagus nerve; Ml, median layer of exoskeleton; Nm, neuromast; P, pore of canal system; Sc, scale; S, septum dividing lateral line canal; Sl, superficial layer of exoskeleton (see also Figure 20.) (*A* redrawn from *A Textbook of Zoology*, 7th edit., by T. J. Parker and W. A. Haswell, rev. by A. J. Marshall. Copyright © 1963 by Macmillan and Company, Publishers; *B* after Denison, 1966.)

VII, IX, and X. In such animals, if there is well-developed bone near the surface of the head or a well-developed scale covering, the bone surface may show the presence and course of the lateral-line canals. This may be in the form of pit canals (Figure 40) if the lateral lines are housed within the bone or merely in the form of grooves on the surface of the bone (or the scales). In such cases, which are common in the record, the patterns of the lateral-line system can be worked out.

Lateral-line impressions are important in evaluation of the evolution of the system itself. It appears, for example, that in some very primitive agnathans, the heterostracans, the system was not organized entirely into well-formed lines but was dispersed throughout the bone (Figure 40 and also Figure 20). A second important role of the information from this system stems from the general organizational relationship between the lateral-line organs and ossifications. The apparent constancy of relationships between these two is used as a basis for the determination of bone homologies.

The functional aspects of the lateral-line system, which is sensory, has led to its use in interpretation of ecological aspects of the life of different sorts of aquatic animals. Systematic uses have been made as well. Since these organs are developed only in aquatic animals, their presence first gives evidence of the way of life. Vestiges in adults have suggested that there were larval stages that lived in the water. Thus T. White (1939) found traces in adult *Seymouria* and suggested the existence of an amphibianlike larva. This leads to a placement of seymouriamorphs with amphibians, a position now strongly supported on other grounds.

The somatic sensory components of nerves VII, IX, and X thus can to some extent be studied in external manifestations of various groups of aquatic animals. In large part it is the facial nerve VII that is best seen, since it is the predominant supplier of the head region.

The various cranial foramina that may be seen in the skulls and jaws of extinct vertebrates are a source of additional evidence on the cranial nerves. As for other parts of the nervous system, the amount of information and the detail available differ greatly between classes. Generally an exit for nerves IX, X, and XI is found between ossifications formed in the basal plate and in the otic capsule. When nerve XII is present, foramina for its exit are likewise readily identified in the posterior part of the skull. Foramina for other cranial nerves are much more variable in their preservation.

Many indications of the positions of exits of cranial nerves from the cranium are available in mammals. To a lesser extent this is true for birds, in which the fragile nature of the skull reduces the possibility of preservation. Within mammals some attention has been paid to the condition of cranial nerves as determined from their foramina and from their courses on the bones of the skull surface. Work on the felids and canids by Hough (1948, 1953), following Flower (1869), is an example of this sort of study.

By far the greatest amount of information on the nervous system of the head, except for that of the lateral-line system, has come from studies of the internal structures of the cranium. Such studies have been made for all the major groups of vertebrates. Outstanding results have come from ostracoderms and placoderms in which there is a very high level of ossification of the bone around the brain and nerve passages. Mammal skulls also provide excellent opportunities for studies based on internal features of the cranium. Between the agnathans and the mammals, in the other classes of vertebrates, opportunities for obtaining detailed information are less good, although there are great differences both between and within groups. For the most part, except in birds (for which fossil skulls are rare), the brain and passages of the cranial nerves within the cranial cavity are not impressed on the surrounding bone. The floor of the braincase is generally more closely associated with nervous tissues than are the sidewalls and top of the cranium. Thus in many instances it is possible to get some information on the base of the brain, whereas data on the lateral and dorsal parts are completely absent.

There are many ways of getting at the information available from the

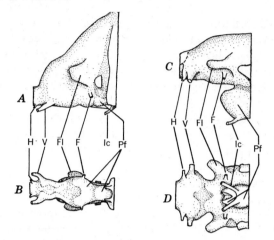

Figure 41.   Cranial "casts" of two therapsids, based on wax reconstructions from thin sections. *A*, *B*, a dicynodont, herbivorous therapsid; *C*, *D*, a gorgonopsian, carnivorous therapsid; lateral and ventral, respectively. F, facial nerve (VII); Fl, floccular portion of the brain; H, hypoglossal nerve (XII); Ic, internal carotid artery; Pf, cast of pituitary fossa; V, vagus nerve (X). (After Olson, 1944.)

internal structure of the cranium. Serial sections of skulls provide one means of access. Reconstructions from closely spaced sections may give good replicas of the internal cavities. They are more or less equivalent to another useful source, the natural brain cast, but do not give as much detail or accuracy. Figure 41 shows some results from this process. Serial sections also are used as a basis for interpretation of the structures of the brains and cranial nerves, without direct references to the nature of the cranial cavity in toto. Prepared skull materials can be used in much the same way, and frequently the two kinds of evidence are used in a complementary fashion.

It is from this type of work that many of the "brains" of fossil vertebrates have become known. Reconstructions tend, however, to engender much more confidence than is warranted in the accuracy with which they depict the brain and cranial nerves. Actually, for most vertebrates, the cranium provides a limited number of landmarks—for example, the foramen magnum; the parietal opening; positions of the exits of cranial nerves VII, VIII, and X; and the position of olfactory tracts. From these few solid bits of information reconstructions proceed with very heavy dependence on information from modern counterparts.

In many respects the most reliable data on fossil brains and associated cranial nerves come from cranial casts. The amount of information, as for all kinds of data, depends entirely on the intimacy of association of nervous

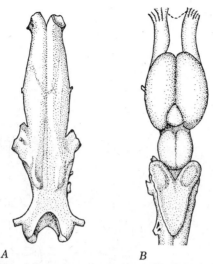

A   B

Figure 42.   Comparison of the actual brain and a cranial cast of *Cryptobranchus*, showing the extent to which the outer structure of the brain of an amphibian is reflected in the cranial cast. *A*, cranial cast; *B*, the brain. Both in dorsal aspect. (After Romer and Edinger, 1942.)

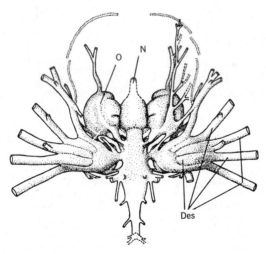

Figure 43.   Cast of cavities and canals in the endocranial and visceral portions of the cephalic shield of *Procephalaspis*, an osteostracan. Especially designed to show the large structures associated with the dorsal, dermal sensory organs, not known in other groups of vertebrates. Ventral aspect. Des, cast of nervous structures associated with sensory organs; N, naris; O, orbital cast. (After Stensiö, 1963.)

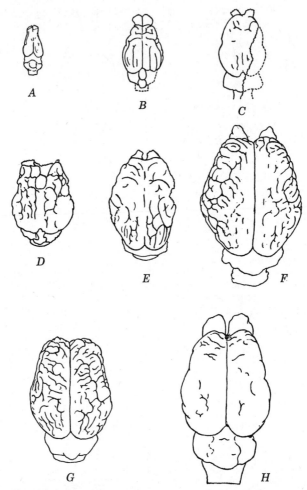

Figure 44. Sketches of the cranial casts of members of the Equidae, showing changes in size and proportions of the various parts of the brain. *A, Hyracotherium,* lower Eocene; *B, Mesohippus,* Oligocene; *C,* primitive *Merychippus,* middle Miocene; *D,* advanced *Merychippus,* upper Miocene; *E, Pliohippus,* Pliocene; *F, Equus occidentalis,* Pleistocene; *G, Equus caballus,* a mature pony, Recent; *H, Equus caballus,* a normal-sized horse, Recent. All to same (greatly reduced) scale. The last two, *G* and *H,* show some of the effects of size on structure in animals at the same level of evolutionary development. (After Edinger, 1948.)

tissue and bone (or cartilage) and the degree of perfection of representation in the material of which the cast is formed. Under this general heading can be included both natural casts and casts of crania made in some such medium as latex after the matrix has been removed. The sorts of information available are indicated in some typical examples of cranial casts shown in Figures 42 and 43A. Tilly Edinger, more than any other student, has been responsible for the development of use of this kind of information. She has emphasized repeatedly the point that the cranial cast is just that and is not to be considered an accurate representation of the surface of a brain, even under the best of circumstances.

Cranial casts are the most effective for studies of mammals, for in these alone among the commonly preserved vertebrates is the brain reasonably reflected in the cast. Figure 42 shows the endocranial cast of an amphibian compared with the actual brain and indicates the great discrepancy between the two at this level.

In comparison with the evidence on many of the soft systems from fossils, that for some parts of the nervous system is excellent. Much has been possible in the analysis of evolution of some of the gross features of the brain and cranial nerves. At the very primitive levels of vertebrate evolution, understanding of the initiation of features found in more advanced lines has been witnessed among agnathans. Knowledge of the presence of oddly specialized structures, such as the massive nerve tracts to areas on the surfaces of the skulls of osteostracans (Figure 43), is one type of contribution. Evidence on the beginning of some features of mammalian brains among therapsid reptiles, suggesting parallel origin of some of them, has come to light. Specializations of the brain and enlargements of the sacral complex of the neural cord have been determined from studies of dinosaurs. Finally, a great deal has emerged about the evolution of mammalian brains from the work of many students, led by Edinger. A classic example is her study of the evolution of the brain of the horse (Figure 44), and well known, but less detailed, are studies of the development of the brain in man, based in large part on analyses of cranial capacities and very crude estimates of the parts emphasized.

This is a field that offers much unrealized potential. As the nature and functions of brains among wider arrays of modern mammals become better known, studies of relationships of areas of the brain surfaces and functions of areas to which they are related become increasingly possible. This avenue of approach, as being explored by Radinsky, can be of particular interest in special cases.

4 The digestive and endocrine systems

THE DIGESTIVE SYSTEM

Diet is an extremely important factor in the evolution of vertebrates. Very crudely a subdivision of all vertebrates can be made with respect to their predilection for animal or plant food. In all major groups both types of feeders have developed, along with a large number of mixed feeders intermediate between the two extremes. Diet, and hence the digestive system, thus plays a cardinal role in the adaptation within major groups. It is also, however, involved in the more general progressive aspects of evolution, being one of the most important factors in adaptive acquisition of features that have brought some lines of developing vertebrates to the threshold of new levels of body organization. In general it has been through carnivorous rather than herbivorous lines that major advances have taken place, related it would seem to the rigorous requirements of locomotion, agility, and perception involved in finding, apprehending, and subduing active prey.

In all vertebrates the digestive system includes special structures concerned with the intake of food, incorporation of it into the body proper, and the elimination of waste. In general, in the course from the most primitive vertebrates to the advanced mammals, each of these parts of the system has undergone increasing differentiation and developed greater complexities. Under this very general trend, however, exists immense variation in the degree of complexity, related primarily to the particular adaptation of

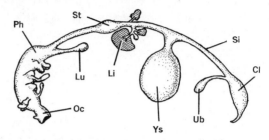

Figure 45. The generalized vertebrate digestive system based on the digestive tract in an amniote embryo. Cl, cloaca; Li, liver; Lu, lung; Oc, oral cavity; Ph, pharynx; Si, small intestine; St, stomach; Ub, urinary bladder; Ys, yolk sac. (Modified after Romer, 1956c, from Turner.)

the system in particular cases. The challenge of interpretation of diet and the nature of the digestive system in paleozoological systems is both a rewarding and frustrating one. Although much interpretation can and has been done, it generally is so highly inferential as to be more tantalizing than satisfying.

A mouth, pharynx, esophagus, stomach, intestine, and anal opening are the primary elements in the system (Figure 45). The mouth and the anal or cloacal termini of the system are usually lined with epithelium, but the rest is strictly of endodermal origin. Associated with the major tract are glands that secrete fluids, as well as such structures as lungs and gills, parts of the respiratory system, jaws, from the visceral skeleton, dermal bones, and teeth.

The primary functions of the system are the procurement of food, its preparation for digestion, provision for absorption into the body, and elimination of wastes.

Much of the information on anatomy, function, and evolution of the digestive system must necessarily come from studies of animals now in existence. It is such studies that have resulted in the conclusion that there has been a general increase in the complexity and differentiation of function of the parts of the system in the course of vertebrate evolution. It is possible to recognize a series of morphological and functional levels, each more or less characteristic of a major group of vertebrates, and to see superimposed on this level the various specializations related to particular adaptations. In general, however, the modern record does not provide much evidence on actual stages intermediate between the levels of development.

Most parts of the digestive system are related only very indirectly to the hard structures customarily preserved in the fossil record. Relatively unusual instances of preservation of stomach contents and very rare cases of impressions of some portions of the digestive tract or associated organs themselves give some information about the system in particular cases. In large part, however, these amount to curiosities rather than being generally informative. Coprolites, fossil feces, are a common item in the fossil record and can provide some evidence on diet if the animal and the coprolite can be associated. Most coprolites appear to have come from carnivorous vertebrates, and if recognizable elements of the food are present, they are in the form of bones and teeth. Association with the animal is in many cases impossible. In special instances some feature of the animal may be stamped on the coprolite—for example, the spiral valve in the intestine of sharks. In others it may be possible to make association by a process of elimination. Thus in the case of the very large reptile *Dimetrodon* of the early Permian, association of very large, bone-carrying coprolites seems possible on the basis of size, since no other carnivores of comparable dimensions are known to have been in existence in the environments in which the remains are found. Gastroliths, or stomach stones, occasionally tell something about the digestive system.

All such cases as these are, however, unusual. The actual nature of much of the digestive system for most extinct animals cannot be known from direct evidence and must be inferred from comparison with the closest living relatives, determined on the basis of skeletal evidence.

Two regions of the digestive system are rather intimately associated with the osseous or cartilaginous structures, the mouth and the pharynx. In the case of the former, jaws, palatal bones, and teeth are closely related to functions of procurement and initial preparation of food. The cartilages and bones of the laryngeal region have a primarily respiratory function in fishes and some amphibians but become incorporated into the functional complex of the digestive system as they come to form skeletal elements of the jaws and palatoquadrate, with the development of a gnathostomatous condition, and of the tongue and throat as gill breathing is reduced and abandoned with the development of terrestrial animals.

Mouth structure provides considerable insight into the evolutionary history of the processes of food procurement and preparation. This is written partly by the major changes of the structure of hard parts as new levels of organization are reached in the development of gnathostomes from agnathous forms, incorporation of dermal elements into the jaws, and shifts in respiration from gills, involving the gill arches, to lungs. Partly, and functionally more important, are the records of adaptive changes within major groups as radiation carried evolving stocks into different dietary provinces.

Evidence for the latter sort of change comes largely from the structure of the jaw; inferences concerning jaw musculature, based on the shapes and sizes of bones of the skulls and jaws, and evidence of muscle origins and insertions; and dentitions. Indications of nerve and blood-vessel supplies are of much less importance, although occasionally useful. Of potential value, but as yet little developed, is the possibility of using the data of the external molds of brains in mammals as a key to some aspects of functions of the

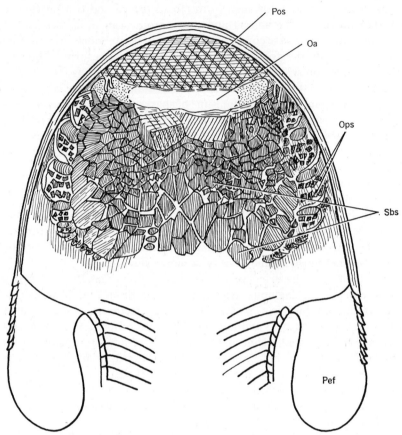

Figure 46.   Ventral view of the anterior part of *Hirella*, a hemicyclaspid osteostracan, showing the structure of the oral region and associated features. Oa, oral opening; Ops, scales on opercular fold; Pef, pectoral fin; Pos, pre-oral scales; Sbs, sub-branchial scales and bony plates. (After Stensiö, 1968.)

mouth parts. It is thus, through understanding of the skeletal systems that are related to the mouth, but in varying ways, that the possibility of gaining information on the digestive system is realized.

The mouth area in agnathous vertebrates includes a true mouth, but this lies behind a buccal funnel in cyclostomes. The location in extinct forms is difficult to determine, although in most at least there was no cyclostomelike funnel, and, from what is known, the mouth appears to have been either anterior, slightly dorsal, or moderately to fully ventral (Figure 46; see also Figure 8). In these animals the mouth lacks any intimate association with the visceral cartilages or teeth. In some of the osteostracans and heterostracans the mouth appears to have been associated with more or less labile dermal plates. The function of these mouth parts, however, is not very clear, because there are no counterparts among living vertebrates. Very probably there was a reasonably wide diversity of feeding habits among the agnathans, but there is little evidence as to the nature of food habits other than that foodstuffs must have been mainly small and relatively soft. Algal feeding has been suggested, and it has also been thought that these may have been largely filter-feeding organisms.

Teeth, jaws, areas of muscle attachment, palatal structures, and cartilages of the hyoid system supply considerable evidence on the function of the mouth region in gnathostome vertebrates. They give, however, little information on soft tissues other than that of gross distribution and function. It is very largely by reasoning from extant animals, on the basis of the degree of similarity of hard parts, that conclusions on the presence of particular glands, lips, types of tongue, and similar soft features can be reached. Functions of taste, glandular activities, and other features of this sort are not subject to any sort of analysis.

Much can be learned of the dynamics of mouth action. In more primitive swimming vertebrates the mouth served in part as a passage for water, and although it functioned in food intake, it was equally important as a part of the respiratory system. With the advent of internal nares only part of the mouth retained this function, and eventually, as hard and soft palatal separation of the mouth and narial passages was attained, the breathing function was very largely lost. This sequence, with its many ramifications, is representative of the kind that can be traced in the fossil record.

Teeth, of course, have been a primary source of evidence of diet and, through this evidence, a basis for inferences concerning various functions of the digestive system. Teeth alone, however, are only moderately reliable indicators of the principal food sources, and even of this in only a broad way. Even among mammals it is possible to specify only rather generally that animals tend to be carnivorous, omnivorous, or herbivorous on the basis of dentitions alone. Within these categories, in some instances, it may be possible to go farther. Among herbivores, for example, separation of grazing and browsing types is generally possible.

As long as there are living relatives or reasonable living analogs, the general diets of extinct forms can be inferred from the existence of common dental patterns. The farther the extinct forms depart from patterns that can be found in moderns, the less confidence can be placed in the teeth as indicators of food preference. Some of the difficulties can be seen among captorhinomorph and pelycosaurian reptiles. A common pattern in captorhinomorphs (Figure 47) involves the presence of multiple tooth rows, and the increase of this feature in evolution suggests that it has adaptive significance. Yet,

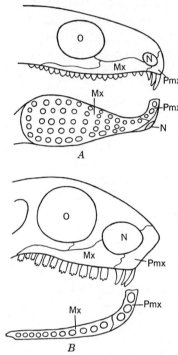

Figure 47.   Diagrams of two types of dentition in herbivorous reptiles: the crushing type, *A*, and the cutting type, *B*. *A* based on *Labidosaurikos*, a large captorhinid reptile; *B* based on *Casea*, a moderate-sized pelycosaur. Both showing lateral and crown (ventral) views, respectively. Mx, maxilla; N, naris; O, orbit; Pmx, premaxilla.

common as these animals are, there is no good basis for determining their diets, and they have been variously considered to be herbivorous, insectivorous, and molluscivorous. Caseid pelycosaurs are indicative of the problems that plague efforts to assess diets of presumed herbivorous reptiles. The teeth have a rather common pattern, including somewhat elongated anterior marginal teeth, posterior teeth with slight to moderate longitudinally arranged cuspules, and a well-ordered array of sharp palatal teeth. Presumably these were herbivores, for certainly the structure of the teeth and body features are not those of carnivores. But what the food was, even with a fair knowledge of associated plants, is at best highly speculative.

Related structures, other than teeth, do aid to some extent in the assessment of probable diet. The physics of jaw action, involving the lever systems and muscle attachments, are important. The kinetics of skulls and lower jaws provide another kind of key to probable dynamics. The kinds and disposition of the palatal bones and the teeth they bear can give some indications of the kinds of materials, especially their size, that were swallowed.

Beyond the mouth and pharynx only very little direct evidence of the digestive morphology and function comes from fossils. The general body structure is of some aid. Overall proportions are generally different in terrestrial herbivores and carnivores. The very large ribcase and the short,

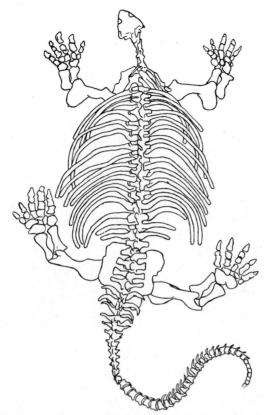

Figure 48. Outline drawing of the skeleton of *Cotylorhynchus romeri*, a large, herbivorous, edaphosaurian pelycosaur. The drawing shows the animal from above in the position in which it is mounted and close to the position in which it lay in the rock on discovery. The very extensive rib cage, strong limbs, claws, and small head are significant features in such reptilian herbivores, with dentitions of the type shown in Figure 47B. Based on a specimen in the Stovall Museum, University of Oklahoma, Norman. (After photograph in Olson, 1968.)

heavy limbs in caseids, whose teeth were noted above, are clearly those of a herbivore (Figure 48). The same criteria of body proportions serve well to differentiate herbivorous and carnivorous dinosaurs. Many kinds of mammals (e.g., uintatheres or ancylopods) could hardly be mistaken for carnivores, even were the dentitions unknown. Among mammals, teeth readily confirm such determinations, but among reptiles this is less true.

Body form is much less useful in aquatic animals, although in general it is not difficult to separate predacious from nonpredacious forms. Conditions of occurrence, or in general ecological associations, may also be brought into the picture for interpretation of food habits and the nature of the digestive system. Sources are, of course, limited, and it is only under the most favorable circumstances that evidence sufficiently good to supplement morphological data is available.

Contributions of the fossil record to an understanding of the evolution of the digestive system must, from a general point of view, be considered to be relatively slight. The contributions relate largely to interpretations of the nature of the diets of one or another type of extinct animal. In this way some increase in the understanding of adaptation and evolution in terms of dietary change comes about, and this perhaps is the most important aspect of this sort of analysis. Through such studies it is possible to gain a better understanding of the history of development of the sorts of digestive systems found among existing stocks, to see their divergence from common bases, and to evaluate the significance of dietary change in adaptive development of the various systems to the body. What is known of the anatomy and functioning of the system, however, must come almost entirely from modern animals.

THE ENDOCRINE SYSTEM

The endocrine system is one of the two that function primarily to interrelate body activities, the other being the nervous system. The endocrine system operates through secretions of hormones from glandular cells into the bloodstream and lymph, and communication is by means of the circulation of these media.

The great importance of the endocrine system has been increasingly realized over the years with proliferation of knowledge of the chemical structure and activities of the hormones. New information on their sources, kinds, and effects is continuously emerging. The greatest body of information pertains to mammals, in particular to man and a limited number of kinds of experimental animals used in research. Information on hormone activities in the lower vertebrate classes, both with regard to sampling of taxa and coverage of the activities of the various glands, is relatively poor.

Endocrine glands are present in all classes of vertebrates known from living representatives, and presumably this was true for extinct groups as well. The principal glands are the following.

> Ovaries
> Testes
> Thyroid
> Parathyroid
> Adrenal
> Pancreas (Islets of Langerhans)
> Pituitary

In addition the pineal and thymus are sometimes included, but whether they have functions that class them as endocrines is still under investigation. The placenta also has an endocrine function, and some glands, such as the testes and ovaries, are exocrine as well as endocrine.

As far as the conditions are known the endocrines are more or less similar in all classes of vertebrates. Most of them occur, in somewhat rudimentary form, in the cyclostomes. Parathyroids, however, have not been identified as such in fish, although structures that may be their forerunners are known. The adrenal glands in mammals include two very closely associated parts, cortex and medulla, with the latter central and the former peripheral. Functionally these two parts are distinct. In less advanced classes the physical distinctness increases as one goes down the evolutionary scale. In fishes the two tend to be entirely separate, and in amphibians the first association occurs.

The evolutionary changes of the adrenal glands are more or less representative of the levels of morphological changes found in other glands. There are few striking evolutionary changes either in form or function, and the system is in general a constant one. On the other hand, as advances in complexities of the morphology and activities of the body have taken place, the functions of the glands have kept pace, and seemingly the complexity of the hormone arrays that they yield has become greater.

The pituitary gland is of extreme importance in that its secretions, more than those of other glands, affect and control the operations of other glands as well as act on nonglandular tissue. The gland itself is complex, deriving in part from epithelial tissue of the primitive mouth and in part from the diencephalon of the brain. It has three principal lobes in tetrapods, each with particular hormones and hormonal effects. In cyclostomes development is somewhat rudimentary, and in those fishes in which it has been studied the structure varies rather widely. In Selachii, for example, the posterior lobe is diffuse.

The pituitary is the only one of the endocrines that bears any direct relationship to the bony structure and is the only one for which there is any evidence in fossils (if the pineal be excluded). In tetrapods and in some fishes it is housed in the sella turcica (Figure 49), and the outlines of this structure often give a general concept of the size of the gland. It is known to have been relatively enormous in some dinosaurs and ratite birds. The sella turcica, and presumably the pituitary gland, is extremely large, relatively, in sphenacodont pelycosaurs and gorgonopsians. It has been suggested (Edinger, 1942) that gigantism may be related to the size of the gland in dinosaurs. There are, however, rather extensive differences in the structures of pituitaries of modern animals that cannot be documented by the crude reflections of the gland in extinct forms. These differences are functionally significant. Thus, although it is tempting to relate the large pituitary to particlar functions—for example,

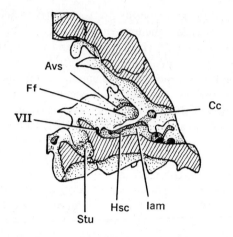

Figure 49.   Medial view of the posterior part of a skull cut on median vertical plane and showing the otic portion of the brain case and basicranial region in a primitive therocephalian. Note the very large sella turcica (Stu) and the deep floccular fossa (Ff). Avs, anterior vertical semicircular canal; Cc, opening of crus communis of utriculus; Ff, floccular fossa; Hsc, horizontal semicircular canal; Iam, open internal auditory meatus; Stu, sella turcica; VII, foramen for facial nerve. (After Olson, 1944.)

to modifying reproductive functions in mammal-like reptiles—the evidence is in fact not a sufficient basis for such speculation.

Except in most speculative or very general ways, such as inferences concerning the effects of hormones on hard anatomy, analyses of the conditions of the endocrines in fossil vertebrates is impossible. T. E. White (1959 and earlier) has made the most ambitious efforts in this direction. He has attempted to relate features of dentition, hypsodonty, and cement formation to modifications in endocrine activities. He treated only mammals, for which close analogs are available for study among extant forms.

Edinger (1942, 1964) made studies to relate the large pituitary fossa to gigantism, as noted, and also has made studies of the pineal and parietal organs, which some have considered to have endocrine functions. As she notes, such studies have intrinsic interest, but there are many pitfalls, such as the conclusion that the size of recesses in fact does reflect the size of the associated organ or its general proportions.

Among fossil fishes occasional information on the development of the hypophyseal cleft has some bearing on the pituitary. Once again, insofar as overall evolution of the pituitary is concerned, this information is relatively trivial.

In general paleozoological studies have added, and probably can add, but little to an understanding of the nature and evolution of the endocrine system. In view of the relative stability of the system and the apparently rather gradual advance in its complexities and functions, studies of living vertebrates can give a reasonably satisfactory understanding of the evolution of the system.

5 The musculature

The movements of all parts of the vertebrate body are dependent on the capacities of muscular tissue to expand and contract. The critical functions of feeding and locomotion are accomplished by joint actions of the muscular and skeletal systems. In this intimate relationship areas of attachment of muscles on the hard tissue often leave some sort of evidence, in the form of rugosities, processes, fossae, and so forth, and these tend to be preserved in fossils. Gross muscle patterns of extinct vertebrates thus often can be worked out and in some instances are of great interest in interpretation of the evolution of both the muscular and skeletal systems and in analyses of function.

As for the other soft systems, interpretation of muscles from the skeleton depends heavily on modern analogs. In contrast to most others, however, an appreciable part of the muscular system is available for study on fossils. Paleontologists have taken advantage of these circumstances and have attempted many analyses and reconstructions of muscles of the gill arches, skulls, and jaws; of the axis; and of the appendages for some members of all classes of vertebrates represented in the fossil record. Within the limitations imposed by the nature of the record, a great deal of information on myology and extensive analysis of its functional roles has been produced. All of this, as for all systems, tends to be conservative, since in any cases of doubt there is a strong tendency to interpret in terms of what is known from moderns rather than to make some opposite interpretation for which an analog does not exist.

Studies are very irregular with respect to classes of vertebrates, partly because of the differential suitability of materials and partly because of the areas of interest of those who have studied myology. The most comprehensive studies have come from the work of W. K. Gregory and a group of students who worked closely with him and under his inspiration, in particular C. L. Camp, A. S. Romer, R. W. Miner, and L. A. Adams. Many others have made important contributions both under the influence of the work of these men and independently, although exploitation of comparative anatomy, functional anatomy, and the synthesis of extinct and modern vertebrates in such studies have found their primary inspiration in Gregory and his school.

ORIGIN AND SUBDIVISION OF MUSCULATURE

Embryologically the muscular system of the vertebrate body arises from two principal sources. Much of the system that is amenable to paleontological study is of the somatic musculature derived from myotomes. This forms the segmented, striated, voluntary system that has more or less intimate association with the skeleton (Figure 50). In particular it is the axial and appendicular components of this system that are important in vertebrate evolution, since they relate especially to the important functions of locomotion. Other parts of the system, eye muscles, hypobranchial muscles, and dermal muscles play important, although less evident and less pervasive, functional roles.

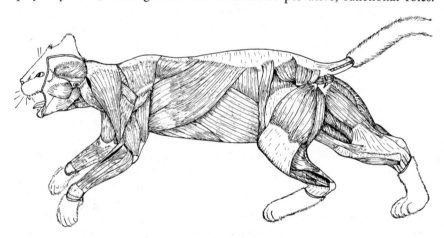

Figure 50. The striated muscle system as exemplified by a superficial dissection of the cat. The differentiation of the skull-and-jaw, trunk, and locomotor muscle groups, along with the interrelationships are evident. Composite based on *Vertebrate Dissection* by W. F. Walker, Jr., W. B. Saunders and Company, Copyright ©, 1965.

Generally their record is not well cast on the skeleton. Most important of these lesser groups insofar as evolutionary change is concerned is the hypobranchial system that functions in the gill system of fishes and comes to form various muscles of the hyoid apparatus, tongue, and lower jaws in tetrapods. Nerve supply is by cranial nerve XII or its equivalents in the spinal nerves.

The other major subdivision of the muscular system arises in mesenchyme from the peritoneum, not from myotomes. It forms the visceral system and is in large part composed of the nonstriated, or smooth, involuntary muscles associated with the digestive tract and other internal organs. Although cardiac and branchial muscles are derivatives of this system, they are striated and the latter are voluntary.

Most of the visceral muscles do not make close contact with the skeleton and cannot be directly inferred from fossil remains. The branchial muscles, however, are a marked exception. Initially in vertebrates they participated in movement of the gill arches and branchial chambers, and this functional relationship has been maintained both with respect to arches that remain part of a gill system and those parts of the branchial arches that have changed form and function in the course of vertebrate evolution. Most important, of course, are the muscles of the jaws. Innervation of these muscles is by cranial nerves V, VII, IX, and X. It is primarily on homologies based on innervation that it has been possible to trace the course of evolution of these muscles among modern vertebrates.

The adductor and depressor muscles of the jaws have been derived from this system. It is this part that has received the greatest amount of study among fossil vertebrates, in particular the divisions of the adductor system, often called the trigeminal musculature by virtue of its innervation by parts of the fifth cranial, or trigeminal, nerve. Significant understanding of the evolution of this group has come from analysis dependent primarily on extinct vertebrates.

RECORDS OF MUSCLES IN FOSSIL VERTEBRATES

The areas of origin and insertion of muscles on bones are recorded in a number of ways (see Figures 51 and 52). These give not only insight into the positions and sizes of attachments but some indication as well concerning the nature of the attachment, whether tendinous or fleshy. Major processes, ridges, and trochanters mark the positions of muscles that played major roles in skeletal dynamics. On these prominent structures, as well as on otherwise unmarked bone surfaces, irregular pits, rugosities, and spurs may attest to the tendinous nature of the muscle attachment. Smooth surfaces, in

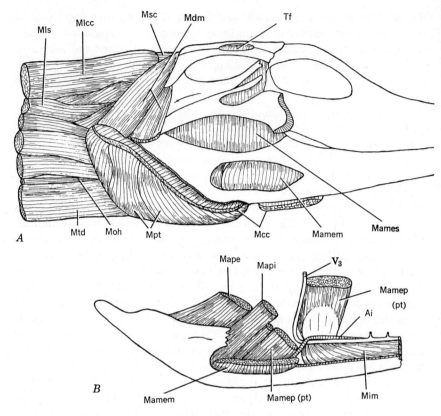

Figure 51. The alligator. *A*, superficial jaw and cervical musculature in lateral aspect; *B*, musculature of the lower jaw with part of the jaw removed. The patterns are somewhat specialized and particular to this animal, but representative of the major components in the reptiles as a whole. Ai, intermandibular artery; Mamem, Mamep (pt), Mames, adductor mandibulae externus medialis, profundus (part), and superficialis, respectively; Mape, Mapi, adductor posterius externus and internus, respectively; Mcc, constrictor colli; Mdm, depressor mandibulae; Mim, mandibularis externus; Mlcc, longissimus cervicus capitis; Mls, levator scapulae; Moh, omohyoideus; Mpt, pterygoideus; Msc, spinalis capitis; Mtd, tendinomandibularis; Tf, upper temporal fenestra; V_3, mandibular branch of trigeminal nerve.

areas where attachment is presumed to have occurred, indicate fleshy origins or insertions, with considerable intervening fascia presumably present. For such areas precise definition of the positions of attachment is frequently difficult.

Fossae and fenestrae, such as those in the temporal regions of many reptiles and mammals, may mark areas of origin and insertion of muscles, with osseous attachments around the margins. Pockets and tubular channels in bones in some instances house muscles, as in the case of the lower jaw of the alligator (Figure 51) and channels, separated by flanges or ridges, may show the passage of different muscle slips, as in the dorsal part of the ilium of *Diadectes* (Figure 52). The internal structure of bone is subject to stress by muscle action, and this may show up in the patterns of bone structure. Although it is of possible application to fossils, this feature of bone is difficult to use except in fresh materials, and very little use of it has been made in paleozoology.

Analyses of muscle patterns in fossils take advantage of all such indications. Muscle scars, processes, trochanters, fossae, and similar features are not in themselves, however, sufficient for reconstruction of the full musculature of a system. They may show the position of origin and insertion of muscles, which can be confidently identified if there are reasonably close analogs in modern animals. These areas, of course, give a minimal extent of the origin or insertion. They may show the degree of tendon development but can generally tell little about the fleshy belly of the muscle. If, however, some key muscles can be placed properly, others may be added to complete the system. This addition must, of course, rest on a concept of similarity of the system in the fossil in question and modern counterparts. In general, if the functional aspects of the system interpreted from the bone are comparable to those of the living analog, such interpretation is reasonable. Such a philosophy, however, makes for conservative interpretations, for differences that might

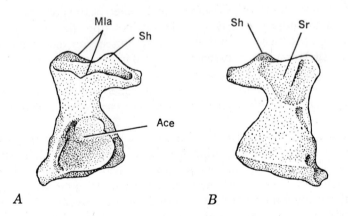

A   *B*

Figure 52.   Ilium and part of pubis and ischium of *Diadectes*, an amphibian or reptile?, showing the dorsal channels that carried part of the lateral axis musculature. *A*, lateral; *B*, medial. Ace, acetabulum; Mla, grooves for two parts of lateral axial musculature passing to the caudal region. Sh, shelf partially isolating the two grooves; Sr, area for attachment of sacral rib. (After Olson, 1936*a*.)

in fact persist in the unrecorded or poorly recorded muscles will not be included in reconstructions.

Muscle systems are often inferred, although less often reconstructed, on the basis of the existence of particular kinds of osseous systems that themselves give little direct information on muscles from scars and processes. It is reasonable to assume, for example, that a fin of some particular type in an extinct animal meets much the same conditions and performs in much the same way as one in a similar modern animal. It follows that the musculature probably was quite similar. The closer the taxonomic affinity, of course, the more reliable such a conclusion can be considered to be.

Difficulties of reconstruction arise, as in all systems, when there are no close counterparts among living vertebrates. In general these problems are less severe than for most other systems since the skeletal parts do allow some interpretation of the general dynamic situation and some direct indications of parts of the muscle system may be present on the bones. Although interpretations of muscle systems in fossils must always take much for granted, they can nevertheless start from tangible evidence and produce valuable results.

MUSCLES OF THE HEAD AND GILLS

The axial and appendicular musculature is rather clearly definable, but the musculature related to the anterior part of the body, the head region in general, comes from several sources and undergoes major changes in the course of evolution. Nevertheless, the muscles from this area do make a convenient if somewhat arbitrary unit for discussion and bring out the important relationships between the visceral arches and the skull and jaws.

Included in this suite are the eye muscles, derived from the proötic somites, branchial muscles, from the splanchnic mesoderm, hypobranchial muscles, dermal muscles, mostly in mammals, and cervical muscles from the axial system.

Of these, the branchial muscles are of much the greatest significance in studies of extinct animals, being associated with the critical functions of respiration and feeding and, in performing these functions, being closely associated with the skeleton. Axial muscles of the neck attach to the posterior part of the skull and often leave evidence of their presence in the form of processes, rugosities, and ridges. Hypobranchial muscles have close association with the bases of the gill arches, but their record on the whole is poor, since they do not tend to leave definitive impressions on the rather delicate gill elements in fishes, or in the derived hyoid apparatus in tetrapods. In addition, the hyoidean structures are but rarely preserved as fossils.

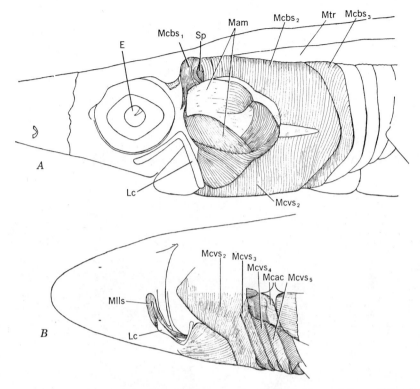

Figure 53.   Dorsal and ventral constrictors and mandibular musculature of *Squalus*.
A, lateral; *B*, ventral, one-half the scale of *A*. E, eye; Lc, labial cartilage; Mam,
adductor mandibulae; Mcac, coracoarculais communis; $Mcbs_1$–$Mcbs_3$, constrictor
basalis superficialis one through three; $Mcvs_2$–$Mcvs_5$, constrictor ventralis superficialis two through five; Mlls, levator labialis superior; Mtr, trapezius; Sp, spiracle.
(After D. Davis, unpublished.)

Records of the eye musculature and the dermal muscles are essentially nonexistent.

The branchial muscles of *Squalus*, as shown in Figure 53, represent a
common point of departure for tracing of the history of these muscles in the
gnathostomes. Patterns from this "generalized" type can more or less be
traced into more specialized gnathostomes, although of course the shark is
far from truly primitive. Much of the fundamental work on muscle modifications and homologies has been done on modern materials, and the
homologies have been based largely on innervation of the components as
they assume different roles in different systems (Figure 54).

The first dorsal constrictor has given rise to the adductor musculature.
From a primitive condition in which it is a single muscle serving the visceral

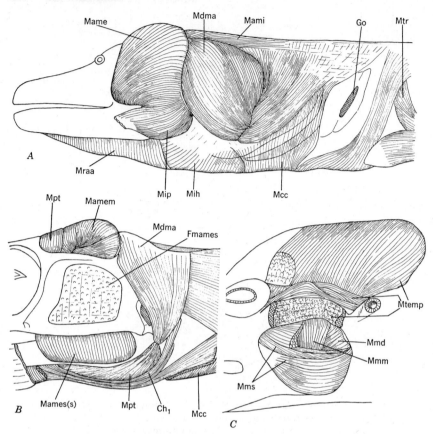

Figure 54. Head (and, in some, cervical) muscles of *A*, *Cryptobranchus*: *B*, *Sphenodon*: *C*, *Canis*. In lateral superficial aspect. Ch_1, first ceratohyal; Fmames, fossa with insertion of adductor mandibulae superficialis externus on covering fascia; Go, gill opening; Mamem, Mames(s), adductor mandibularis externus medialis and superficialis (superficial part), respectively; Mami, adductor mandibularis internus; Mcc, constrictor colli; Mdma, depressor mandibulae anterior; Mih, interhyoideus; Mip, intermandibularis posterius; Mmd, Mmm, Mms, masseter—deep, medial, and superficial, respectively; Mpt, pterygoideus; Mraa, retractor angulae oris; Mtemp, temporalis; Mtr, trapezius. (*A* after R. Zangerl, unpublished; *B* and *C* after Olson, unpublished.)

jaws, the adductor becomes differentiated in a variety of ways, related to a broad series of adaptive patterns. Generally it differentiates to form a complex consisting of a large adductor mass, which consists of an internal and external part. This passes from the outer surface of the cranium to the posterior part of the lower jaw. This mass becomes further differentiated in the classes of vertebrates, and homologies between the derived parts have proven difficult to trace. A second major division passes from the palatal region to the posterior part of the lower jaw, generally being termed the pterygoid series. This too differentiates into a series of different patterns in the various classes.

Depressor muscles of the jaw come from the musculature of the hyoidean arch, either as a whole or in part. To the extent that this is the source of the depressors, innervation is by cranial nerve VII. Various different patterns of muscles for depressing the jaws have, however, developed so that this functional system is much less subject to generalization than is the adductor system.

Paleontological studies of the head musculature have dealt primarily with muscles of the jaws and, of these, particularly with the adductor musculature. A wide taxonomic range has been covered, including placoderms, chondrichthyans, crossopterygians, actinopterygians, and all branches of the tetrapods except the birds.

Studies of jaw musculature, as well as that of other systems studied from fossils, can be divided into three broad, somewhat overlapping types. One involves reconstructions of musculature at a rather gross level, based largely on general information from living groups of animals. From these data muscles are superimposed on the fossil skulls and jaws, in the areas and the spaces available. The reconstructions of Adams (1919) are among the most effective of this general type. Variously in descriptions and more or less popular accounts of vertebrates (e.g., among dinosaurs) this kind of restoration may be encountered.

More detailed and more analytical studies have involved critical analysis of processes, fossae, rugosities, and other indications of the positions of muscle origins and insertions. These are correlated with the probable dynamics of the skulls, jaws, and dentition. Adams' studies, noted above, represent a beginning in this direction. Some reconstructions by R. C. Fox (1964) (Figure 55B) have gone somewhat further in analysis. One end purpose of this kind of study has been an understanding of the myology and function of some particular animal. This information is very often used comparatively in studies of the evolution of form and function.

Most studies of the second type have been made on tetrapods, although a good deal of attention has been paid to the structural meaning of the bones among various groups of fishes and, by implication, muscle systems are involved. In some instances reconstructions of muscles have been made, as that by Schaeffer and Rosen (1961), shown in Figure 55A. Many such studies have been made on amphibians and reptiles, and some work, such as that of Gregory (in Osborn, 1929) on titanotheres, has been done on mammals (Figure 56). A natural follow-up of such study has been the treatment

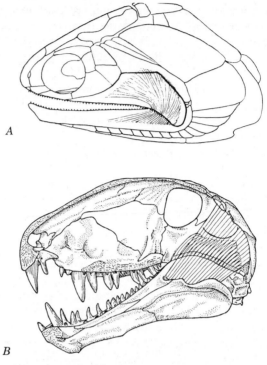

Figure 55.   Reconstructions of adductor musculature of the jaws for two extinct Paleozoic vertebrates. *A*, a paleoniscoid; *B*, *Dimetrodon*, a sphenacodont, pelycosaurian reptile. (*A* after Schaeffer and Rosen, 1961; *B* after R. C. Fox, 1964.)

of the masticatory apparatus as a mechanism subject to physical analysis. This third type of study is the one most commonly followed at the present time.

Some effort within this third framework has been directed toward understanding of general adaptive types. Figure 57 shows an example of a "phylogeny" of adaptive types in tetrapods, derived from a rhipidistian ancestry. This is based on the assumption of an undifferentiated adductor mass in the ancestral type and assessment of the functional patterns found among both extinct and living tetrapods. Data for the static-kinetic system have come largely from studies of type two, such as that of Wilson (1941) on *Buettneria* (Figure 58).

A great deal of effort has been devoted to analytical studies of the transformation of muscles and skulls and jaws from a primitive reptilian condition to that of mammals. The starting point has generally been among the captorhinomorph reptiles, and a series including sphenacodont pelycosaurs and a graded series of carnivorous therapsids has provided a series of type

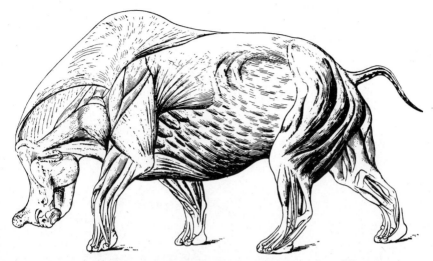

Figure 56.   Restoration of the full superficial musculature of an Oligocene titanothere, an extinct perissodactyl. (From Osborn, 1929.)

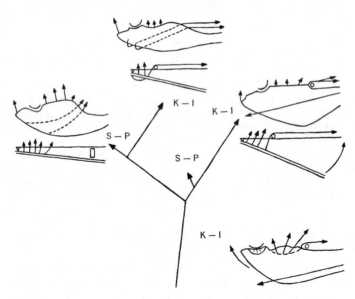

Figure 57.   Diagram of a "phylogeny" of patterns of the mechanical system of the adductor musculature of the lower jaw. K-I, kinetic-inertial system; S-P, static-pressure system. Arrows show the direction of major resolution of forces. (After Olson, 1961b.)

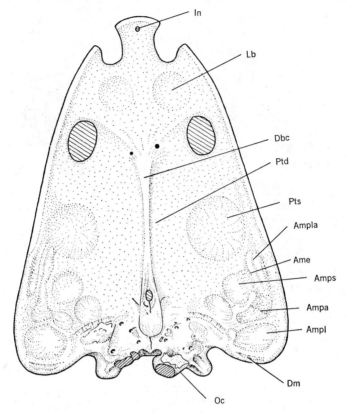

Figure 58. *Buettneria*, a large, flat-skulled, upper Triassic stereospondylous amphibian, in palatal view. Muscle scars and positions of some of the contacts of bone and cartilage are shown. Ame, Ampa, Ampl, Ampla, Amps, musculus adductor mandibulae externus, posterius articularis, posterius longus, posterius lateralis, and posterius subexternus, respectively; Dbc, dorsal side of cartilaginous braincase; Dm, musculus depressor mandibulae; In, internasal fenestra; Lb, musculus levator bulbae; Oc, occipital; Ptd, musculus pterygoideus, deep portion; Pts, musculus pterygoideus, superficial portion. (After Wilson, 1941.)

stages. In some instances offshoots of the main line, such as the dicynodont pattern, have been analyzed as well.

The fundamental model has been based largely on the musculature of such living reptiles as *Sphenodon* and some lizards (Figure 54*B*). This has followed from the close resemblance of the general morphology of the skull of captorhinomorphs to these living forms, with the one major difference being in the closed temporal region of the former. The "end" mammalian stage is based mostly on the condition of the marsupial *Didelphis* (Figure 59). Problems

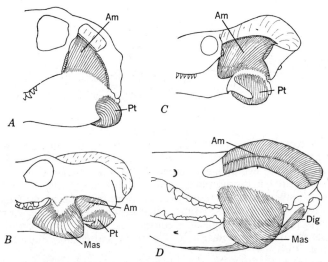

Figure 59. Comparisons of jaw-musculature reconstructions on the basis of extinct animals and the living opposum. The series is representative of a morphological sequence whose relationship to changing function is a key to the evolution of the morphotypic series shown. *A*, a sphenacodont pelycosaur, *Dimetrodon: B*, a primitive therapsid; *C*, an advanced therapsid; *D*, a primitive therian mammal, *Didelphis*. Am, adductor mandibulae; Dig, digastric; Mas, masseter; Pt, pterygoideus. (*A* through *C* after Barghusen, 1968; *D* based on Romer, 1956c.)

of muscle homologies between the jaw musculature of modern reptiles and living mammals have been difficult to solve, and one of the focal points in the evolutionary studies has been this problem.

For all the attention that has been devoted to this area, no firm answers to some of the problems have yet been produced. Problems remain in particular for structures in the evolutionary series of fossils that are not represented among moderns. Outstanding in this respect is the angular flange in sphenacodonts and therapsids (see Figure 59). This has been interpreted as a place of insertion of the masseter, and from this basis many conclusions on the evolution of this mammalian complex have been drawn. Barghusen (1968), on the other hand, has rejected this interpretation and, from a different base, reached distinctly different conclusions on the course of muscle evolution.

Studies such as those just described and figured represent the peak of myological studies among the fossil vertebrates. They illustrate the potentials and limitations of this kind of work. Mammals offer opportunities for reconstruction of muscles and studies of evolution, but as far as jaw muscles are concerned relatively little has been done. One reason is simply that the musculature for fossils in large part cannot be seen to be much different from that found in living adaptive equivalents. Turnbull (1970), in assessment

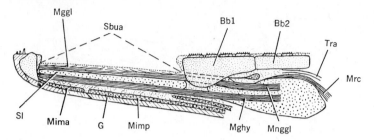

Figure 60.   Reconstruction of the musculature of the tongue in a rhipidistian fish. Much of the musculature is necessarily based on comparison with living forms and interpolated into the evolutionary scheme of tongue muscle on the basis of its approximate position in the morphotypic series. Little comes from the hard anatomy itself. $Bb_{1,2}$, basibranchials one and two; G, principal gular plate; Mggl, genio-glossus; Mghy, geniohyoideus; Mima, intermandibularis anterius; Mimp, inter-mandibularis posterius; Mnggl, hyogenioglossus; Mrc, rectus cervicis; Sbua, anterior portion of sub-branchial unit; Tra, truncus arteriosus. Sl, sublingual rod. (After Jarvik, 1963.)

of the principal masticatory types in mammals, has used fossil evidence as a part of the basis of his analysis of functional types. This has proven of considerable interest both in a general understanding of evolution of the principal types and in a knowledge of their distribution among mammals when not only recent but also ancient types are taken into consideration.

Other muscles of the head and gill region have posed different and difficult problems, and relatively few specific efforts have been made to study them. Although there are a number of reconstructions of gill musculature for various fish, these are very much of the first type noted above, in which muscles known from moderns are superimposed on the skeletal structure of the ancients. Jarvik (1954) has attempted a reconstruction of the tongue muscles in a rhipidistian (Figure 60), but this sort of attempt is rare and must be based on extremely scant firsthand evidence.

Cervical musculature, either as a reconstruction or in the form of analysis of areas of insertion on the skull (Figure 61A, C), has been studied sporadically in different groups. Special attention has been paid to it in some cases where structure and function have some special significance—for example, in the odd relationship of the cephalic and thoracic shields in some arthrodires (Figure 61A) and in such animals as saber-toothed cats, in which the head action is considered an important aspect of the feeding mechanism.

AXIAL MUSCULATURE

As compared with the work on musculature of the skull and jaws, very little has been done on the musculature of the axis in vertebrates. Fishes, for

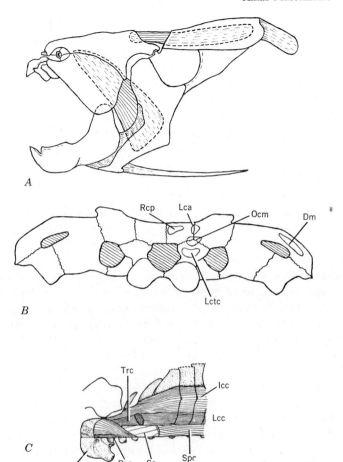

Figure 61.   Reconstruction of cervical musculature in several extinct vertebrates. *A,* *Dinichthys*, a large arthrodiran placoderm; *B, Buettneria,* showing areas of insertion of cervical musculature on the occiput in a stereospondylous amphibian; *C, Diadectes,* an amphibian or reptile? Dm, depressor mandibulae; Icc, iliocostalis capitis; Lca, longissimus capitis pars articuloparietalis; Lcc, longissimus cervico-capitis; Lctc, longissimus capitis pars transversalis capitis; Ocm, obliquus capitis magnus; Rcp, rectus capitis posterius; Sc, semispinalis capitis; Spc, spinalis capitis; Trc, transversalis capitis. (*A* after Adams, 1919; *B* after Wilson, 1941; *C* after Olson, 1936*a*.)

the most part, offer little opportunity for such studies, first because the preservation is seldom suitable and, second, because the system in fishes is generally stable and not especially subject to major changes in evolution which might be recorded on the bones.

In both fishes and tetrapods the dorsal, or epaxial, musculature and ventral, or hypaxial, musculature is developed. In fishes the epaxial musculature is essentially an unspecialized, segmented muscle mass, on which modest specializations are superimposed. External and internal portions may develop, but these are little if at all reflected on the bone. In all but the cyclostomes a horizontal septum separates this dorsal mass from the ventral, unified hypaxial mass.

The axial muscles have a relatively stable history in the tetrapods as well. In most the epaxial musculature tends to be divisible into a medial and lateral series. Anteriorly each of these is modified in the neck region to a cervico-occipital series. The more lateral muscles are interrupted by the shoulder girdle and to a greater or lesser extent specialized to aid in its support. The hypaxial muscles are simple in fishes but in tetrapods are divided into several subdivisions, some of which become very complex. Ventrally is a rectus series, which remains fairly simple in many animals but may become complex in mammals. A more lateral flank musculature is a complex basically formed of three sheets that, in their relationships to each other and hard structures, such as the ribs, may be highly complicated. In addition there is a small, subvertebral series that belongs to this complex. The hypaxial series is interrupted by the shoulder girdle, with muscles of the hypobranchial series occupying its general position anterior to the girdle in tetrapods.

The dorsal axial part of the axial musculature offers a better opportunity for study in extinct vertebrates than does the hypaxial musculature. Even for it, however, studies have been relatively restricted. The osseous contacts—except for those of the skull, shoulder girdle, and pelvis—are with the vertebrae and ribs, and these reflect only the innermost parts of the muscles. Various general reconstructions of musculature in which the axial muscles are included have been made, such as that of a dinosaur by Lull and Wright (1942) or the titanotheres by Gregory (in Osborn, 1929) (Figure 56). These are in large part extremely general and have not been based on detailed analyses of the origins and insertions on the vertebrae, ribs, and girdles. Reconstructions for three Permian tetrapods (Figure 62) were based on such an analysis and show about the extent to which such studies can be carried. In general, however, the evidence is not extensive, and a principal conclusion from the studies that have been done is that the musculature that is subject to some degree of analysis appears to be quite constant within very large segments of the vertebrates, showing the greatest differences between the fishes and tetrapods but relatively small differences within these groups.

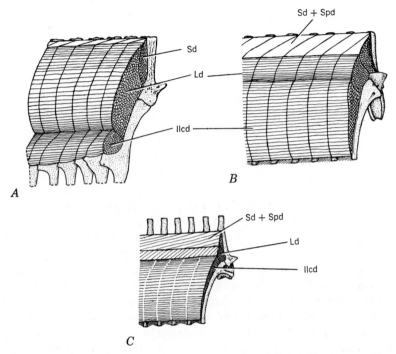

Figure 62.   Dorsal axial musculature as reconstructed for A, *Eryops*, a moderately advanced labyrinthodont amphibian; B, *Diadectes*, a possible amphibian with many reptilian features of the postcranium; C, *Dimetrodon*, a sphenacodont pelycosaur. The three very broadly represent a morphotypic series. Ilcd, iliocostalis dorsi; Ld, longissimus dorsi; Sd, semispinalis dorsi; Spd, spinalis dorsi. (After Olson, 1936a.)

The epaxial muscles continue into the tail without interruption. Initially part of the epaxial mass in tetrapods lay lateral to the iliac blade, but with enlargement of the limbs and increase in their locomotor functions, the muscles came to lie entirely internal to the ilium. *Diadectes* represents an intermediate stage (see Figure 52) and illustrates the transition in a way not found among living amphibians or reptiles.

The ventral musculature of the tail resembles the epaxial rather than hypaxial musculature in structure, for with specialization of the tail both dorsal and ventral muscles have tended to become differentiated and highly tendinous. The tendons in various forms, particularly among some of the dinosaurian groups, have tended to ossify and thus to preserve some actual vestige of the musculature. Such ossifications also occur in the precaudal regions of some of the erect, bipedal dinosaurs, such as the hadrosaurs. These give some added insight into musculature, but except for this the vertebrae of the caudal region provide only very general data on the massiveness and the degree of tendon development in the caudal axial muscles.

APPENDICULAR MUSCULATURE

The appendicular musculature of the vertebrates is ultimately of myotomic origin, although this is not clearly demonstrated in the embryology of the higher vertebrates. In general the musculature of the appendages consists of a dorsal and a ventral mass, related to elevation and depression of the external appendages. In many fishes these two masses are not differentiated to any great extent, although in some there has been a considerable amount of specialization related to special functions. In tetrapods, in contrast, the musculature of both the pectoral and pelvic limbs and girdles is highly complex, consisting of many intimately related separate muscles that produce the great variety of movement of which limbs are capable.

The fins of fish have played an important role in classification at both high and low categorical levels. The skeletal structure is generally well preserved in the fossil record, in particular that which is external to the body proper. The nature of muscle attachments and the usual delicate nature of the skeletal elements of the fins combine to make interpretation of the muscle systems extremely difficult. As a result relatively little has been made of the myology of the appendages of extinct fishes. Some general patterns from modern fishes are shown in Figure 63. It is evident that fossil types with similar skeletal structures had similar musculature, but this is as far as analyses can usually be carried.

Were it possible to study myology of fish appendages at some of the important stages in evolution, important results would be forthcoming. In particular some of the events intervening between the rhipidistian fishes and primitive amphibians could aid in solution of problems of bone homologies and relative limb orientations. The internal skeleton is, however, known in very few rhipidistians (Figure 64), and even in these there is no chance for more than highly speculative reconstructions of musculature.

The case of tetrapods is very different. Muscles of the pectoral and pelvic girdles and free appendages have offered opportunities for study over a wide range of taxonomic types. Extensive and detailed studies have been carried out, especially by W. K. Gregory and students working under his guidance' and inspiration. The approach has been one of detailed comparative anatomy, with continued interplay between extant and extinct animals. These studies have covered some amphibians, many groups of reptiles, and some selected mammals. In all of the studies inferences concerning the muscles of extinct animals have been based on penetrating knowledge of the morphology and

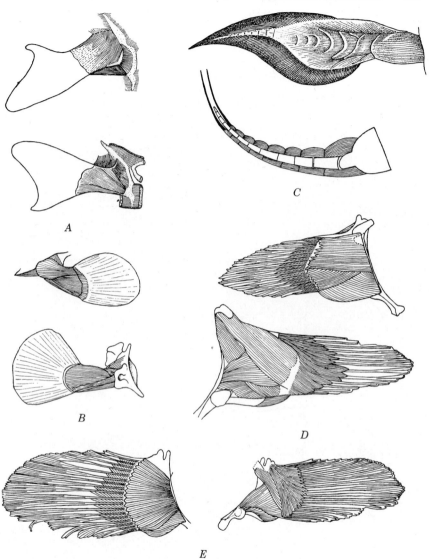

Figure 63.   Fin musculature of the pectoral appendages of some modern fishes emphasizing those in which the limb and girdle musculature is well developed. The nature of this musculature and the skeleton illustrates the immense difficulty of making reconstructions of these structures on the basis of fossil materials. *A*, *Squalus: B, Polypterus: C, Neoceratodus: D, Periophthalmus: E, Gobius.* (*B* after Shann, 1924; *C* after Ihle et al., 1927, modified after Braus; *D* and *E* after Bruno, 1929.)

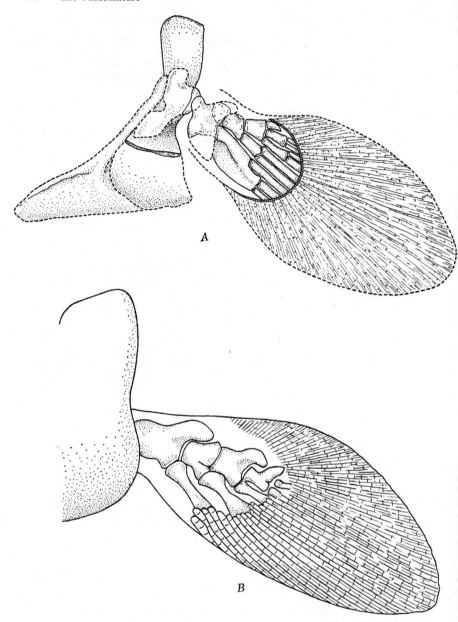

Figure 64. Reconstructed pectoral fins of two osteolepid crossopterygians, showing the extent to which reconstruction is possible with the best preserved fin materials of members of this group. *A, Sauripterus: B, Eusthenopteron. (A* after W. K. Gregory, 1935; *B* after W. K. Gregory and Raven, 1941.)

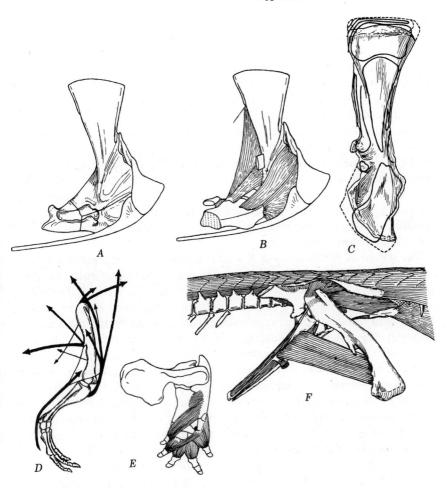

Figure 65.   Various methods of depicting aspects of the limbs and girdles and their reconstructed musculature in fossil vertebrates. *A*, the shoulder girdle in *Dimetrodon*, a sphenacodont pelycosaur, illustrating the basic, tangible evidence on which reconstructions must be based; *B*, the same girdle with some of the musculature to the humerus restored; *C*, the scapulocoracoid of a therapsid cynodont, *Cynognathus*, with areas of muscle origin and insertion outlined; *D*, diagram showing muscular forces acting in main propulsive movement in the hind limbs of a typical ornithischian; *E*, *Erypos*, a labyrinthodont amphibian, showing muscle reconstructions of the lower limbs based on the presence of unusually heavy processes; *F*, *Thescelosaurus*, a bipedal ornithischian, showing reconstruction of the deep musculature of the pelvis and femur, on the basis of sound osteological evidence and the inferred relationships to dorsal axial and posterior abdominal muscles. For the latter in particular data are scant. (*A* and *B* after Romer, 1922; *C* after W. K. Gregory and Camp, 1918; *D* and *F* after Romer, 1927; *E* after Miner, 1925.)

development of musculature of living animals, application of functional concepts, and detailed studies of the evidence of processes, fossae, and muscle scars on fossil bones.

The data in these studies are presented in the form of descriptions and illustrations of the bones, areas of muscle origin and insertion, in restorations of muscles at various levels, as would be seen in dissections, and in variously derived vectorial diagrams that show the principal directions of muscle actions and the general relative strength of the forces. Examples of the results are shown in Figure 65.

The studies have not been carried as far as have those of jaw musculature in the application of the physics of lever systems to the limb actions. Although this is possible, the extremely complicated actions make this much more difficult, as is evident from analyses of modern animals in which it has been possible to make experimental functional analyses (see, for example, Davis, 1964; Hildebrand, 1966).

The school of W. K. Gregory has carried studies of comparative myology using both living and extinct animals to the peak of its development, as far as general analysis and reconstruction are concerned. The aims of the studies are basically evolutionary. Particular attention has been paid to muscle homologies with the general conclusions that there has been a marked constancy in the general differentiation of muscles in the tetrapods and that homologies can, with some exceptions, be established between classes with considerable confidence. Attention was also directed specifically toward functional analyses, as exemplified by W. K. Gregory in his study of the myology of titanotheres and the functional analysis of locomotion in this and other ambulatory and cursorial types of mammals. It is this aspect of the studies that has carried over most extensively into modern studies in which functional morphology has become a principal area of research.

Transitional stages of morphology in evolution have played a dominant role in studies of limb musculature, just as they have in studies of jaw musculature. Finally, as seen in Romer's (1927) study of the myology of ornithischian dinosaurs, taxonomic conclusions have been drawn from studies of musculature of extinct animals. Saurischians and crocodilians and ornithischians and birds are found to have close relationships, and Romer has argued for four rather than three groups of ornithischians on the basis of the inferred patterns of musculature.

section III
Classification of the vertebrates

1 Bases of classification

GENERAL CONSIDERATIONS

The bases of classification are fundamentally similar throughout the whole field of biology. The particular problems that face the vertebrate taxonomist, or any specialist in his own field, arise from the special properties of the materials with which he is concerned. The groups that have a fossil record, as do the vertebrates, pose special problems in the relationships of the living and extinct organisms. Differences in the treatment of the fossils and extant animals arise in part because of the nature of preservation of fossils, the kinds of structures that are preserved, and the imperfections of the fossil record. Most important as well is the absence of information on life processes among the extinct organisms and the temporal aspect of the fossil record.

Vertebrates are all relatively large, active, complex organisms, and their morphology provides information of great use in interpretations of function. The recorded history shows that vertebrates possessed evolutionary mobility and the capacity to adapt to a wide range of circumstances and to change rapidly. It is to be expected that such a group would be a focal point for evolutionary studies and that classification of its constituents would be strongly oriented toward an evolutionary or phyletic approach. This, of course, has been the case since the beginning of the evolutionary phase of biological investigation.

In the everyday work with taxonomy of vertebrates, as in most special groups, the broader problems of classification tend to be ignored. Each area of study develops classifications with their own special features and peculiarities. These may be meaningful and suitable under particular circumstances but equally may not be well understood or properly interpreted by students in other areas. Later in this section the special aspects of vertebrate classification are considered, but in an analysis of the classification of this group as it is practiced, it is important to have a basis in the general tenets of biological classification, if for no other reason than to provide a rationale for interpretation of the special practices. The nature of taxonomy has been thoroughly discussed from many points of view in an extensive literature. Excellent treatments of the basic principles, problems and practices are available[1] so that a short summary of the most pertinent points will serve the needs of the later parts of this chapter. It is assumed that the reader has a fair understanding of taxonomic procedures which will serve as a basis for development of some of the rather difficult points of practice and procedure as they relate to the interactions of phyletic and phenetic principles.

Classification serves to reduce a complex array of objects or ideas to manageable proportions by introducing some sort of order. A classification must produce a convenient catalog, appropriate to the materials at hand and the kinds of study anticipated. This is generally accomplished by division of the universe of concern, some specified set, into subsets. Each such subset may in turn be considered a set and itself subdivided into subsets, each of which may be treated similarly. Thus there is established a hierarchical system such as that recognizable in biological classification in its descending levels, or categories, such as phylum, subphylum, class, order, family, genus, and species. Particular taxa—for example, Chordata, Vertebrata, Mammalia, Carnivora, Felidae, *Felis*, and *catus*—are concrete expressions of the categories. As usually conceived, in what will here be termed the classical sense, the subdivisions have no overlaps. Each element belongs to one and only one set at the categorical level of immediate concern.

Each set in a particular classification is defined by the characters required for membership in it. This provides a conceptual basis of which the actual category is an extension. Within this framework the suite of characters forms a necesary and sufficient criterion for membership in the set in question. This mode of classification is basic to various current practices in biological taxonomy. It is, in particular, essential to the structure of identification keys. It provides an approach to phylogenetic, or phyletic, classification that is widely used and functional. Biological classifications, in addition to acting as catalogs, bring together in succinct form biological information pertinent to the organisms involved and utilize this information in expressions of relationships. The classical form of classification fulfills this second role as each member of the small suite of defining characters represents but one of a complex of associated characters. In a sense one such defining character may be

[1] See Beckner, 1959; Hemple, 1965; Hennig, 1966; Mayr, 1963; Simpson, 1945; 1961*a*; Sokal and Sneath, 1963.

considered a key character of the correlated complex or as a representative of the complex. Associated arrays of characters, of course, may have intricate overlaps. An extensive amount of biological information thus is carried by a few characters that are convenient in classification.

Both theoretical and practical problems arise in the application of the classical approach. These have led to reevaluations of practices and suggestions of alternative practices and principles. The theoretical problems revolve around a central point, which will recur repeatedly in the discussions in this chapter. The strictest concept of classical taxonomy is not appropriate to the evolutionary structure of the organic world. Categories, both species and higher categories, are not strictly definable, as brought out in the following discussion. In addition, even if this difficulty were ignored to arrive at a working approximation, the establishment of a defining concept for a particular set raises technical difficulties. Selection of a few characters from many is necessary. This selection must be meaningful and objective, with reference to the goals of the classification that is being made. The process of selection and of category formation must proceed simultaneously. It is thus necessary to bring some theoretical consideration into these processes for guidance. Selection of characters with some particular end in mind and under some guiding concept may impair application of the criterion of objectivity, may not be repeatable by different students, and may, in extreme cases, lend itself to circularity. The fact that the procedures in classical taxonomy involve successive stages of approximation and that of biological principles are applied within a phyletic framework at the various stages provide reasonably strong safeguards against such errors.

In practice an overlapping, dual approach is generally used, involving both the classical approach and a less restrictive one that calls into play a principle called polytypic by Beckner.[2] The duality also involves a process of subdivision, or "classification from the top," and synthesis, or "classification from the bottom." The classical approach is applied in most cases in which subdivision is the principal operation. Synthesis, in which individuals are grouped into species, species into genera, genera into families, and so on, is more often carried out within a polythetic framework, although this need not be the case. There is, within this framework, no simple defining set of necessary and sufficient characters for a given aggregate. Rather there is a large

---

[2] Beckner (1959) has described and discussed this principle in detail. He termed the type of classification that we have here called classical, monotypic, and the type in which there is a large number of characters, no one of which is necessary and sufficient and none of which is possessed by all members of an aggregate, polytypic. Sokal and Sneath (1963) in discussion of this matter have suggested the substitution of "monothetic" and "polythetic" for Beckner's terms, citing the fact that his terms have other well-established meanings. For clarity the modification suggested by Sokal and Sneath is followed here.

suite of characters. Each individual of the aggregate in question possesses some large (undefined) part of this suite, each character is possessed by a large number of individuals, but none is possessed by all. No character is considered either necessary for membership or sufficient for inclusion.

Beckner has made a strong case for the polythetic nature of all categories with respect to the set of taxonomic characters. The basis of definition under these circumstances must depend in part on some available theory, for test procedures are formulated with the aid of such a theory. This, of course, is the pattern followed in the construction of phyletic classifications, in which the evolutionary theory is involved in testing results. Assignments made under the polythetic principle are not infallible, and group membership is established only with some, usually high, degree of probability.

The concept as developed by Beckner is related primarily to the principles expressed under the "new systematics," in particular to the fact that categories are not precisely definable. The problems posed by the incompatibility of the gradual nature of evolutionary processes and the discrete nature of categories are brought into sharp focus by this approach, and the probabilistic nature of assignments offers an avenue to solution of some of the difficulties. The result of polythetic formation of categories, however, is not one of sets that have no overlaps in the distributions of taxonomic characters, and if the approach is applied to its fullest extent, it requires complete knowledge of the characters of organisms—knowledge that is of course never available. In phyletic classifications a compromise between the multiple aims and alternative methods seems to be inevitable.

Problems of the phyletic approach prompted exploration of this principle by Beckner. However, elements of the polythetic principle lie close to the roots of some of the practices of numerical taxonomy. Here, in what is called a phenetic approach, some measure of taxonomic affinity is developed on the basis of the degree of resemblance of elements as revealed by comparisons based on large samples of the total characteristics. Biological theory, however, is not directly involved in the formulation of the defining concept. In this way the technical difficulties considered by many pheneticists to be inherent in character selection and emphasis under phyletic theory, as outlined earlier, are bypassed. Phenetic approaches produce definitive classifications and, on the whole, using generally similar raw materials, phenetic and phyletic methods produce classifications that do not differ markedly. Characters of organisms enter into both, and the information that they provide, even treated differently, produces meaningful results. Biology, in particular evolutionary biology, is considered *in the course* of phyletic classification and *after* the formation of classification under phenetic procedures.

The existence of these two systems raises a point that is of considerable interest—namely, the matter of the "naturalness" of classification developed under the different principles. Loosely it may be said that any classification that is meaningful with reference to the material classified is natural. Hemple (1965) stipulates that to be natural a classification must have scientific import—that is, it must be scientifically fruitful, lending itself to the formulation of general laws or uniformities that provide a basis for scientific understanding, for explanation, and for prediction. There can be little disagreement with this general formulation. Most taxonomists concerned with phylogeny consider that the natural classification is that which serves best for understanding, explanation, and prediction within the framework of evolutionary theory. Simply, as Mayr (1965) has put it, "the classification that has the highest predictive value is the best, that is the most natural."

Gilmour (1940, 1951, 1961), however, has suggested a broader concept to the effect that the measure of the naturalness of a classification is the number of propositions that can be made about its constituent classes. This is not restricted to biology or to any particular aspect of biology, such as evolutionary theory. Information content is important in any of the definitions of naturalness, but whether this is conceived to bear directly on one concept or construed as applying to a broader range of explanations, not in any particular field, is a matter open to argument and debate.

CRITERIA OF CLASSIFICATION

The term "character" has been used in the preceding section without definition, but with the general understanding that it stands for taxonomic character and is some attribute of an organism useful in classification. Characters, either singly or in combination, are the criteria of classification. All, or certainly nearly all, conceivable properties of organisms have been used in one way or another in taxonomic procedures, and most of them have played roles in the formulation of classifications. The concept of the character thus is extremely important, and a definition of the term would seem to be critical. On analysis, however, it becomes evident that, except in a loose, broad sense or in an overly restrictive sense, the term "character" is not subject to definition. As used currently, it must be understood in the context of its use and conception in particular situations. The more or less central idea of a taxonomic character as an observable morphological characteristic, still in common practice, fails to encompass many other uses of the term. Somewhat broader is the definition of "character" as some attribute that is the distinguishing feature of a taxon. Within this broader framework, which is itself more limited than current usage, can be found a wide range of application. For example, Meckel's cartilage, the lower jaw element of some vertebrates derived from the mandibular visceral arch, is a character of the gnathostomatous vertebrates; a lower temporal fenestra is a character of synapsid reptiles; a mesaxonic foot is a character of the perissodactyl ungulates; a particular call is a characteristic of a species of frog. The type of character use at one categorical level may be completely without meaning at another. The calls of several species of frogs, for example, though differing at the species level and serving as species characters, might all be of such similar pattern that they form together a character of a genus, encompassing several species, in contradistinction to other anuran genera. Within the ungulate mammals the mesaxonic foot has value at an ordinal level, delimiting the perissodactyls, but of course a similar foot pattern occurs widely among other groups of mammals.

Characters can thus be definitive within specified limits, such as a taxon at a defined categorical level, but meaningless if carried beyond the limits of this boundary. It would appear offhand that a character that is valid for a particular taxon at some categorical level must be ubiquitous in the taxons lower in the particular hierarchical structure. This does not, however, always follow. Four-footedness, in contrast to finnedness, among vertebrates is a character, and animals possessing it may be classified as tetrapods, sometimes formalized as Tetrapoda. But some animals—for example, snakes—generally grouped as tetrapods when a major division of the vertebrates is made on the basis of the nature of the paired apendages, lack limbs. They are placed with tetrapods because they have descended from animals that had four limbs. Historical and phyletic matters play an important role in categorization and may, as in the case must noted, give an odd twist to the use of characters.

At low categorical levels functional characters, such as capacities to interbreed, are often used. They, like various other kinds of characters that provide bases for very fine divisions, are not applicable at higher levels. Examples of this kind of limited applicability and use of the particular characters that "work" at different levels can be multiplied almost without limit. What has happened, of course, is that as time passed the meaning of the term "character," which has a fairly clear-cut dictionary definition, has been gradually extended into many realms to which this meaning is not appropriate. Procedures in classification are in some large part pragmatic, and the term "character" has become a convenient and generally understood expression for all sorts of criteria suitable for taxonomic differentiation.

Dissatisfaction with the results of this sort of practice and the need for precise definition in numerical work has led to the development of other concepts of the nature of the taxonomic character. One definition that has come into use in numerical work recognizes a character as a property or "feature which varies from one kind of organism to another" (Michener and Sokal, 1957). This leads into the concept of a character as some attribute of organisms that exists in two or more states, which may be merely the presence or absence of some attribute. For example, a character might be the nature of brachiopod shells with respect to ribs, and the states might be ribbed or not ribbed. Multiple states also exist, implicitly in the statement "two or more." For example, in mammalian molars in a particular family the nature of the molar with respect to a stylar cusp might be the character, with the states being weak, moderate, strong, and very strong—or perhaps some quantification of these four conditions. In another context a character might be the resistance of different strains or species of organisms to a specific infection, stated in appropriate terms that express degrees.

For multivariate studies in which characters are treated together, it is critical that there be some basis for assessing the amount of information supplied by the characters. Deliberate weighting can be used if it is desired to give emphasis to some and not to others. When, as in numerical taxonomy, weighting often is avoided in initial stages, a basis for unbiased coding is sought. This has led to the concept of the unit character. Sneath (1957) defined the unit character as one that yielded one bit of information; that is, about which a single statement could be made. This can, of course, be carried to limits that are presently totally inapplicable—for example, to the level of the genetic code. After an elaboration of this point Sokal and Sneath (1963) proposed a working definition that a unit character is "a taxonomic character of two or more states, which within the study at hand cannot be subdivided logically, except for subdivision brought about by changes in the method of coding." Statements such as this and others found in the language of numerical taxonomy do give precise definition to the term "character." They are, however, much too restrictive to cover the full scope of application outlined in this section.

Difficulties in definition thus stem in part from the lack of any unifying concept in applied taxonomy that is applicable at all levels, from the undefinable nature of categories, from the overlapping use of polythetic and monothetic principles in formulations of classifications, and in some major part from the wide variety of data that are used in classification and subsumed under the single, somewhat inappropriate, but useful appellation of character.

Criteria, or characters, that are used in some or all parts of biology fall under four general headings: physical, chemical, functional, and distributional.

Physical characters are largely, although not wholly, morphological, relating in one way or another to the form of the organism or the form of some one of its parts. Total form, per se, is not often expressible in complex organisms. Simpler ones, Protozoa or Algae, for example, may conform to some simple shape, such as a sphere, and occasionally more complex organisms have a total form that corresponds to some readily designated shape, as ball, thin disk, or cone. Total volume or weight are physical features expressing something about the whole, but they fall in the nonmorphological part of the physical array. Form, whether macroscopic or microscopic, is usually expressed by some abstraction consisting either of qualitative or quantitative statements. Readily observable macroscopic features of organisms represent one kind of character under this general heading, but histological features, chromosome shape and number, and even the structure of molecules are of the general physical type. They are linked by common kinds of expression and treatment. Qualitative descriptive statements (e.g., rough, short, green, transparent) can be used, although they may be subject to quantification or to coding. Measurements of various sorts (linear, angular, volumetric, weight) and counts represent common means of expressions of physical features.

Chemical characters relate to composition of organisms from the standpoint of the reactivity of their components rather than structure; for example, proteins have particular structures, which are physical attributes, and reactivity, which is a chemical property. Studies of reactivities, as an expression of the chemical composition of organisms, occupy an important and rapidly growing field of taxonomy, called chemotaxonomy. This is playing an increasingly important role in classification. The studies range over a large part of the categorical spread of organisms from demes to phyla. Many chemotaxonomic studies are grounded in phyletic theory, with major efforts devoted to establishment of phylogenetic relationships on the basis of chemical constituents and the use of the results in the formulation of classifications. However, one of the most thriving parts of numerical taxonomy is that dealing with reactivity patterns of microorganisms, for which few characters of other types tend to be available. Initially at least, in many such studies the aim is merely one of finding a basis for differentiation. Aims, once this is accomplished, vary widely from practical to theoretical and may include the establishment of phylogenesis and phyletic classifications.

Characters grouped under function pertain to all kinds of actions of organisms, both intrinsic and extrinsic. Physiological processes, patterns of cell differentiation, growth phenomena, mechanical body operations, and behavioral patterns all can be conveniently grouped. These are the attributes of organisms in terms of their kinetics. They enter into classifications in many different ways, from the level of species and lesser groups all the way up to the differentiation of the two kingdoms, plants and animals. They bear intimate relationship to both the static physical properties and the reactivity of chemical properties, because the functions depend on the existence of these properties for their performance.

Characters in these three categories all are in some large part a function of the genetic structure of the organisms in which they occur. From the phyletic or evolutionary points of view, genetic similarities and differences provide the ultimate measures of relationships. The characters are the observable criteria of this base. Direct genetic analysis is rarely possible at the present time, and hence classification must deal primarily with manifestations that are the consequence of genetic structure. Some kinds of characters lie close to the base—those related to aspects of proteins, nucleic acids, and so on. Studies of such characters, in theory at least, should provide the most reliable approach to the determination of relationships. It is this area that chemotaxonomy is probing, as are analytical studies of molecular structure by X-ray diffraction and related methods. The various other types of characters lie farther from the genetic base to varying degrees. Gross morphological features are relatively remote, and hence the chances of nongenetic influences playing a part in their final state are rather high. They are remote from the particular genetic base as well in that they are largely, if not entirely, polygenic in origin.

Distributional features, both spatial and temporal, play an important part in the formulation of classifications. Although such characters cannot be divorced from genetics and evolution, important aspects of the history of organisms which have led to the configurations of interest are not functions of the genetic composition and phyletic patterns, although they may have contributed to them. Geographic distributions of living populations often play important roles in decisions on classification. Temporal distributions of related members of phyla are of major concern in classifications of extinct organisms. The somewhat general, if arbitrary, practice of assigning samples to different species if they come from temporally different formations, even when there are no evident morphological differences, is an obvious case, similar to that of assigning samples of populations from well-separated areas to different species. Some key problems in the classification of vertebrates fall into this general area, and these will be the subject of discussion in several later parts of this book.

Phyletic classifications, in their attempts to give the greatest weight to those characters that carry the highest content of phyletic or genetic information, are making steady progress in their utilization of information that lies closest to the genetic system, especially in the fields of cytogenetics, chemotaxonomy, and molecular analysis. For the most part, among the vertebrates and most macroscopic organisms, such studies are currently in their infancy and are used in large part as an adjunct to the more usual and older analyses based on gross morphology and function. At low taxonomic levels and occasionally at higher ones, where morphology does not produce satisfying answers, the "newer" types of characters are being used more and more and, in some areas, are beginning to supplant the "older" kinds. At present, however, characters in the sense of "readily observable distinguishing characteristics" form the backbone of classifications. As a whole and at all categorical levels in both neozoological and paleozoological studies of vertebrates use of gross morphological and functional criteria predominates. Other criteria are variously used, as possible and profitable, but the overall structure of current vertebrate classification depends on them only to a very modest extent.

In summary, as the recognition and use of characters as criteria for classification are viewed in the current scene, especially with reference to vertebrates, it is clear that much of the present classification has been based on a strictly pragmatic and largely nonsystematic utilization of information. Some order and direction were introduced by the development of evolutionary theory and emphasis on phyletic history, but the major structure of formal classification was established on a nonevolutionary basis and has not been modified fundamentally by this shift in emphasis. Generally the policy has been and still is to use for classification "whatever will work," that is, whatever will provide a basis for subdivision or association of the materials of interest. Such a haphazard system could have led to chaos and certainly involves dangers of circularity. Both dangers have been largely avoided, partly because there has been general conformity to a set of taxonomic principles in day-to-day work and partly because the building of the classification has been a continuing process that has moved through a series of steps, with checks and balances at the various stages. There are still major problems, and the necessarily compromise character of classifications of biological materials suggests that there always will be unless there are drastic modifications of the fundamental bases. These do not stem, however, from the inadequacy of criteria or the way that they are used, but from the conflict of the evolutionary nature of the materials and the system of classification.

PRACTICES AND PROBLEMS IN VERTEBRATE CLASSIFICATION

Vertebrates form a coherent subphylum of the phylum Chordata, which, presumably as a result of a faulty record, can be readily characterized so that there is no overlap with any other group of equal rank. Classification practices within the subphylum have almost exclusively followed the phyletic mode since the introduction of evolutionary practices into taxonomy. The major outlines of the most accepted classifications of today, however, were fairly well established before the acceptance of evolutionary theory by the biological community. With relatively minor changes this framework has proven suitable for expression of the major phyletic relationships.

Historically the classification of vertebrates was based in large part on contemporary animals. As fossils were discovered, they were brought into the established framework and, at first, fitted into it as well as was possible.

Thus, for example, in the early days of discovery of fossil reptiles a great many extinct genera were placed in the Rhynchocephalia, conceived to be a very primitive group on the basis of *Sphenodon*. As time went on modifications that took into consideration increased knowledge of fossils and a better awareness of the scope of the reptiles resulted in many readjustments. A second result of the nature of the early history of vertebrate classification was the definition of major categories on the basis of the characters most evident in living animals. Methods and structures of reproduction, skin characters, metabolic properties, and clearly evident features of major organs and body systems provided the main criteria. These, of course, were for the most part not appropriate for the assignment of fossils.

The problem of the differences of criteria for extant and extinct animals is by no means completely historical, for neozoologists and paleozoologists still tend to differ in the characters they use, and this results in many cases in different values for taxons of comparable ranks among modern and fossil animals. Neozoologists, quite rightly, select from all of the information that is available and use the portion that best fits their needs. Very often this will not include the relatively restricted kinds of information available to the paleozoologist. Studies of relationships between the kinds of characters that are used and those of hard anatomy can to some extent alleviate the difficulty, at least to the extent of testing the homogeneity of the taxons that results from the different approaches. This is done in some instances, but it is a time-consuming task and one that often does not fit into the main line of research of the neozoologists. One who wishes to cross between the two sources of data usually must make such comparisons for himself.

Classification based on modern animals finds few problems in the establishment of discrete groups, except at the very lowest levels. Over any short period of time in the history of vertebrate life most phyletic lines will appear as discrete and readily separable, with confusion arising only in those that in retrospect appear to have undergone rapid differential branching. Thus among moderns, and over most short time intervals in the past, essentially monothetic classifications, with categories possessing clear-cut suites of defining characters, can be drawn. For the modern moment of time, with access to the full range of characters, construction of such a classification poses few insurmountable problems.

At very low taxonomic levels, however, some slight warning signals about the inadequacy of this kind of classification appear in such problems as those encountered in making species assignments for the populations that stretch continuously to form geographic clines. Here species intergrades may be found. Such problems are not, however, sufficient to upset the central plan, and it is essentially an operationally monothetic classification, with everything neatly in its own place, that the neozoologists present as a framework.

The paleozoologist, who follows his materials through time, is presented with the problem of conforming to such a scheme. It is true, of course, that breaks in the record usually remove the real problem of coordination, but this does nothing to alleviate the conceptual problem that paleontological materials continually bring to the forefront. The ever present time factor gives practical emphasis to the phyletic aspects that must be accommodated in classification and the inability of standard classifications, which are quite suitable to moderns and established on the basis of them, to meet this need.

Such differences in criteria and in temporal distribution underlie some of the problems of coordination of studies of ancients and moderns. They are basic and in part insurmountable. Beyond these there are features of the fossil record itself that add to the difficulties. Serious problems are raised by the facts that for increasingly remote times the record is less and less complete, sampling is increasingly biased with respect to ecological circumstances, and preservation is less and less good. Furthermore, some extinct major groups have no close living relatives. Although such creatures are among the most interesting of fossil animals, they are also most difficult to place in the established classification. The deterioration of the record in older times results in increasingly poorer articulation between the ancient groups and moderns. This must be recognized in the evaluation of any classifications that include materials from a spread of geological time.

Analysis of any standard classification in use today will reveal the effects of this difficulty. Attention to the special problems that relate primarily to paleozoological materials is important in understanding these classifications.

Species Problems

Throughout biology the species has been a focal point of study and controversy. The nature of the species in contemporary organisms is fairly clear and generally understood. Although there is no single criterion that will serve without fail to separate species and no general set of criteria that is necessary and sufficient for the assignment of individuals to species, the concept of the species as the largest interbreeding categorical unit is reasonably satisfactory and useful. As is well recognized, the time factor introduces problems into the species concept, but in a way little different from the problems that arise among organisms arrayed along interbreeding geographic clines. The technical difference is that members of a time series cannot be brought together, whereas those of a cline, in theory at least, can. Accommodation to the fact of continuity must be made if the important role of the species as a category in the taxonomic hierarchy is to be maintained. Arbitrary decisions are sometimes necessary in splitting both the populations of a geographic cline and of a temporal phyletic line. If the processes involved in evolution are recognized and the arbitrary nature of the decisions is understood, the ends of classification are met without harm. This does not appear as a serious problem in practical taxonomy, for a compromise solution is readily made.

Actually, of course, the temporal continuity and the problem it raises would be recognized whether there were a fossil record or not, for it is inherent in the evolutionary concept as generally held. Occurrences of actual and recognizable cases of continuity in the fossil record are rare enough for only very occasional difficulties in classification to arise. There are, however, other matters that have more serious consequences. For many reasons it is highly desirable to maintain a binomial system of genus and species for fossil vertebrates. Coordination with practices of neozoology is sufficient reason, but much more important is the fact that the species is the building block of evolution under almost all forms of evolutionary theory.

Species identification of fossils must be based primarily on the morphology of hard parts, with some aid from spatial and temporal distributions. Although the limitations that this restriction poses are serious, they are not prohibitive to proper species assignment, at least in theory. Several things, however, complicate the picture. First of all, fossils are subject to distortions so that features may not be displayed in "natural" or life form. Second, it is rare that the full array of hard parts is preserved. Third, the preparation of the specimen, that is, its removal from the rock, is often difficult; hence important characters may remain unrevealed and a less than adequate sample may be made available. All such problems tend to become increasingly important the more ancient the materials. They vary, moreover, in intensity in the different major animal groups and for different parts of animals in related groups. Fairly reliable species determinations can be made for many quite recent fossils, using morphological information, but this becomes increasingly difficult, other things being equal, for specimens from more remote times.

The occurrence of similar animals over appreciable time spans poses some interesting problems of species among fossils, in two ways in particular. If an assemblage of specimens that is thought to be composed of samples from a single species population has been gathered over a fairly wide area, there is generally no way of knowing that the temporal span was sufficiently limited for this population to have been a breeding population in the sense that a contemporary species is. As a result, except for special cases in which some independent criterion ensures contemporaneity, possibility of the existence of changes of magnitudes that would be recognized as of specific rank among moderns cannot be ignored. These changes may have affected the hard parts in question and be potentially detectable, but they may well have been in soft anatomy and completely lost. Species defined on the basis of such materials must always be somewhat suspect as being broader in scope than modern species, although this may not in fact have been the case. The general tendency for the coefficient of variability (V) to be higher for comparable parts in fossil samples than in modern ones—even in structures, such as mammalian teeth, where distortion tends to be low and growth of no consequence—is suggestive that greater breadth is a common thing.

Under other circumstances samples of similar animals may have come from sites that cover a time known to be quite long, say represented by the deposition of several hundred feet of fine sediments. The available morphological structures frequently fail to show any detectable differences over such spans. This varies, of course, depending on the kinds of features available for study. If relatively good indicators of minor modifications exist, such as mammalian cheek teeth, the situation is very different from one in which, say, only skull or limb proportions supply the information.

If a strict morphological basis of species assignment is used, the "species" in such a continuum may appear to have existed over a very long period of time, up to several million years or more. A somewhat similar problem arises in the case of widely separated samples of vertebrates thought to be more or less contemporary. Inability to separate on the basis of morphological grounds may result, if strictly morphological determination is employed, in samples being assigned to the same species. Zoogeographic inferences may follow from such assignments, without adequate justification. Such a case arises in the apparent identity of samples of fishes and small amphibians from the freshwater Carboniferous of Europe and North America. If there is, in fact, species identity, conclusions concerning continental connections may follow. This has been used as a basis for suggestions that North America and Europe were part of a single landmass at this time.

The various ways of handling these problems give practical solutions. None fulfills the goal of equating modern and extinct species. One method, cited in examples above, is the adoption of a strictly morphological basis for species assignment. If specimens thought to be closely related have no detectable morphological differences in the common structures, they are placed in the same species. With good materials this has some merit. Where the parts preserved tend to be less definitive or preservation is not good, say a series of somewhat crushed humeri of a suite of reptiles, such a method may result in the "species" actually including individuals that were members of two or more species populations. Reasonable discretion is the only safeguard in the use of this solution.

Another practice is quite different. This is to ignore apparent morphological identity and consider either temporal or geographic discontinuity as a basis for assigning two or more samples to different species. Such a practice has equally serious drawbacks and may lead to unwarranted conclusions.

A third approach might be to abandon species assignment altogether for fossils in which there is any question, which would be a large percentage of all that are known. The common occurrence in systematic lists of generic names followed by the abbreviation for species, "sp." (species indeterminate) or by "cf." and then a species name (meaning "compare with"), may suggest that this is a practice rather widely followed. Actually this is not the case. Whereever it is at all feasible, most paleontologists do make species assignments. The references "sp." or "cf." are used when for one reason or another no reasonable assignment is possible. In some instances the materials are too poorly preserved for assignment to an existing species or for naming a new one. In other cases the lack of assignment comes from the not uncommon situation that defined species, to which the material in question might belong, are based on characters not preserved in the specimens being studied. This has, however, by no means always acted as a deterrent to the naming of new species or assignment of materials to one already named. The literature is replete with such cases and, of course, with attempts to rectify the mistakes that have resulted.

If classification were the end in itself and it were simply a matter of pigeon-holing specimens for easy reference, no great damage would come from these practices. But the results should be useful for various kinds of study for which species are important—evolutionary studies, stratigraphy, paleozoogeography, and so on. They can be most misleading. Individual cases, where such information is to be used, must be carefully evaluated. The role of paleozoology is reduced by its incapacity to handle species assignments in a consistently meaningful way. This is less the case for Tertiary vertebrates than for others, and generally less so for mammals than for members of other classes. For example, for paleozoic fish, amphibians, and reptiles it is foolhardy to consider most species assignments as anything more than approximations, often very rough ones, of their counterparts in modern faunas.

Evolutionary studies relating to rates of change, modes of change, duration of species, and the like must be suspect unless they pertain to carefully studied materials from the very late part of the record. The same of course applies to paleozoogeographic studies that rely on species identifications. Students who have had close contact with materials of the fossil record, being aware of these difficulties, tend to use the genus as the basic unit, rather than the species. This involves loss of the particular properties of the species that make it uniquely valuable but avoids in some large part the very real danger of false conclusions that result from inability to apply the species concept reliably to most fossil materials.

Higher Categories

Whereas the great preponderance of taxonomic problems that concern the neozoologist relate to species and species groups, the vertebrate paleozoologist is more prone to find his problems of classifications at higher taxonomic levels. This follows from the fact that in large part his materials are sampled over long periods of time and represent significant parts of the adaptive radiation phase of evolution. Vertebrates reveal with unusual clarity the alternating phases of origin of higher levels of organization and subsequent adaptive radiations. Transitions to new levels are for the most part not well preserved in the record and, it would seem, difficult to recognize when they are preserved. The flowering of radiations, on the contrary, provides a wealth of materials. One obvious reason for this difference in the records of the two phases relates directly to the greater number of individuals in the radiating phase. Others, more subtle and requiring more detailed discussion, are considered explicitly in later parts of this book.

Many problems that the continuity of the record might pose are only infrequently encountered. The lack of information about most transitions "solves" many problems of classification that would exist were it available. Where such transitions are known, as between reptiles and mammals, difficulties with the formal system appear. From the standpoint of understanding evolution, multiplication of such problems would be highly welcome.

The other major difficulties come at and near the base of radiations. Following initial penetration of a new ecological zone, made possible by a newly attained organizational level, rapid adaptive deployment into the new zone tends to take place. This deployment is usually not accompanied by a very extensive morphological differentiation in the various lines. There results a complex of similar lines, some of which may persist to form the dominant phyla during the full flowering of the radiation.

Differentiation of the lines within this "basal" complex is generally difficult. Determination of relationships of the various phyletic lines to those that endured in the radiation is equally hard. The tendency in classifying such materials is to lump all of the lines together into a single category, say an order or a family. This process is sometimes called "horizontal" classification with reference to the pictorial process of cutting across the bases of vertical lines of descent in phylogenetic diagrams in which time is represented on the ordinate. The Condylarthra and Creodonta have represented such groups among early placentals. Hopefully these are interim groups that will be modified, molded, and perhaps totally eliminated as the phyletic lines are clarified and the relationships between them and with their descendants are worked out. In both the cited cases this process has gone on steadily over many years as new materials have been found and new concepts of relationships have been developed.

Another type of interim classification comes from the erection of a taxon to include possibly related animals that for one reason or another are not readily classifiable. This constitutes what is often called a "wastebasket," into which the "leftovers" are swept for neatness. The family Eothyrididae among the ophiacodont pelycosaur reptiles of the late Paleozoic is such a group. A number of small, rather poorly known, primitive ophiacodonts with short snouts and rather large orbits, among other things, have been so grouped. There is no assurance that this group does in fact represent a coherent radiation appropriate to the rank of family. Most comprehensive vertebrate classifications contain such assemblages, some less coherent than the Eothyrididae and others more so.

The attachment of genera of uncertain ordinal or familial affinities to an existing category is another practice that stems from the desire to see all animals fitted into some place in a classification. Placement of *Barytheria* of the early Tertiary of Egypt with the order Proboscidea is an illustration of such a procedure. The alternatives to such a practice are either to leave a great many loose ends throughout classifications or to erect a classification in which there are categories of appropriate level to accommodate all of the problematical forms. Each of these practices has been followed in the making of vertebrate classifications. Romer's (1945a) classification in the second edition of *Vertebrate Paleontology*, which has been widely accepted as a standard, follows the practice of finding a place, mostly in recognized categories, for all vertebrates, with very little use of the concept of "incertae sedis." Watson's classification of the reptiles in the *Encyclopaedia Britannica*, on the other hand, left several orders under the class Reptilia unassigned. In his more recent classification Romer (1966a) adopted a somewhat less rigorous practice in the assignment of problematical forms.

Some of the most puzzling problems in classification have come from instances in which some evidence of the transition from one major radiation to another has been preserved. These appear among the vertebrates, particularly within the osteichthyan fishes, and the transitions from amphibians to reptiles and reptiles to mammals, but they are also found to a lesser degree among other groups of fishes and in the transition from fishes to amphibians. Details of these cases are the subject matter of the fourth section of this book. Here attention is directed only to certain aspects that are pertinent directly to classification.

In several major vertebrate radiations it is possible, in retrospect, to identify a particular subradiation from which a new organizational level originated. Expressing this may be such terms as "mammal-like reptiles" or "reptilelike amphibians." Some of the steps in transition often can be followed in the history of such groups. Among the bony fishes, the rhipidistian crossopterygians reveal some of the steps leading toward amphibians; among amphibians, anthracosaurs have features that are suggestive of reptiles; and among reptiles synapsids represent mammal-like reptiles. If the course from any one of these groups to the next higher level of organization were fully known, assignment of some members to one category or the other would cause trouble. If we are committed to a set-type arrangement, as we are, this is inevitable.

Beyond the theoretical problems, which can be handled only by arbitrary solutions, the matter is complicated by the way in which the major groups of vertebrates were formed.

Amphibia, Reptilia, Aves, and Mammalia, as well as Osteichthyes and Chondrichthyes among the fishes, have been defined on the basis of modern animals; the class characters are those most readily recognized in moderns—to some extent recognizable only in moderns. As far as the tetrapod groups are concerned, the defining characters involve soft anatomy, reproductive processes and structures, and physiology. Integumentary features are important. Associated skeletal characteristics have been added, and many of these show unvarying association with the distinguishing soft features among present-day representatives. These skeletal features are used to assign extinct animals to their "proper" classes along with information on soft systems, reproduction, and ontogeny that can be inferred from the hard parts by analysis of their structures. Difficulties of placement arise only in a relatively few instances. For the most part the secondary criteria do produce a classification that conforms closely to the dictates of taxonomic procedure, allowing unequivocal assignments of the organisms to one or another class. But in some cases this breaks down. It does so for several reasons, some of which are trivial and some of which are fundamental.

Trivial cases result largely from lack of information. Additional data, which might someday be forthcoming, would allow solutions. A case in point will illustrate this problem and also bring out another relevant difficulty—that of assessing the validity of primary versus secondary or inferred characters in assignments. *Diadectes*, an extinct animal from the early part of the Permian, was long classified as a reptile on the basis of the generally reptilian cast of its skeleton and particularly because of skeletal resemblances to contemporary animals that no one doubted to be reptilian. However, evidence accumulated to show that another group of contemporary animals, the seymouriamorphs, also once considered reptilian, were amphibians. Also it became evident that they were related to *Diadectes*. The evidence for the assignment of seymouriamorphs—whose skeletons were not definitive, having both reptilian and amphibian characters—was in their ontogeny, which was found to have progressed through a larval stage, implying a mode of reproduction like that of modern amphibians. This, being a primary amphibian feature on the basis of moderns, has been considered widely as a proper basis for assigning seymouriamorphs to the Amphibia.

The ontogeny of *Diadectes* is unknown. Skeletal features most resemble those considered to be reptilian. However, since it is related to the seymouriamorphs, which on the basis of reproduction are considered amphibians, it too is an amphibian, according to an authority in this field, A. S. Romer. This is, of course, a very indirect approach to assignment. It is based on a number of assumptions, the most important being that this animal did actually have an amphibian type of ontogeny. This is not directly known but derives from the resemblance of some of the osteological features of *Diadectes* to members of another family, some of which at least did have an amphibian-like ontogeny. Assignment of *Diadectes* to the Amphibia makes a nice disposition of it, one that does not disrupt a useful system. In view of the lack of appropriate information, however, it must be considered mostly an assignment of convenience. The criterion of naturalness of the resulting classification has little applicability in such cases.

More fundamental in nature is the case of the reptilian-mammalian boundary. In this case, even with extensive information, it is possible to make only arbitrary assignments of some known animals to the Reptilia or Mammalia as defined on the basis of modern materials. One criterion commonly used, the presence or absence of a squamosal-dentary articulation of the upper and lower jaws, an absolute criterion of mammalian affinities among moderns, breaks down in the case of a number of creatures that, though mammalian in many respects, retained the quadrate-articular joint although the squamosal-dentary joint had been acquired and become functional. Some other systems of the animals with this dual joint also appear to be intermediate, whereas still others are distinctly either "reptilian" or "mammalian." This is precisely what evolutionary theory predicts, but not what taxonomic procedures demand.

What has happened, of course, is that some parts of the skeletal system have attained what, on the basis of associations in moderns of hard and soft parts, is considered a mammalian level, whereas other parts retained some features considered reptilian. It is futile to attempt to shift the reptile-mammal boundary in classification up or down to find a suitable—that is, definable—level at which assignment is possible so that results are un-equivocal. As De Beer (1954) has emphasized under the concept called mosaic evolution, systems of the body do not all progress to a new level at the same rate. Some retain ancestral conditions longer than others.

When viewed in fully evolved animals, well along in the radiative phase, a definitive array is seen. In the transition some systems are at the grade represented by the ancestral array, here reptilian, whereas others have attained the more advanced, here mammalian, level. There exists a broad zone of transition to which neither of the major categories as defined applies. Just as it is futile to try to find a level where a "real" break occurs, so is it futile to try to redefine the category in a way that permits precise nonarbitrary assignment of all animals to one or another major category.

Under the very strictest interpretation, that every species must belong to one and only one category at each categorical level (genus, family, order, class, and so on), the situation must be such that there existed a reptilian species that produced as its direct and immediate descendant a mammalian species. Under any modern concept of speciation this is ridiculous. If then the system of classification is not applicable to organisms under the patterns of evolution as we know them, it might seem that the system should be abandoned. To do this, in the face of the tremendous utility of the system, would be complete folly. Mere recognition of the problems, both practical and theoretical, in the incompatability of the system and the relationships of the elements that it classified should be sufficient to alert students to the one real danger, the utilization of the strict framework of the classification in support of theoretical studies where results depend primarily on the taxonomic position accorded the creatures in question.

As this has become clear, a concept of ordering based on grades of evolution has come to augment strict adherence to formal taxonomy in consideration of evolution at transition stages. This has generally involved a somewhat odd but readily comprehensible mixture of formal definition in the taxonomic sense with a recognition of grades of development with reference to the levels inherent in these definitions. Thus a mammal grade is recognized as that expressed in the definition of Mammalia, based on modern materials. Reptilian grade is the condition expressed by the several systems in the formally defined Reptilia. In a particular animal, as noted, one system may be at the mammalian level, another at the reptilian. Thus in *Tachyglossus*, the spiny echidna, parts of the skeletal system are at a reptilian level or grade, but the integument, including hair, is at the mammalian grade. Rudimentary mammary glands give a mammalian or submammalian cast. Reproduction is by means of eggs, which is certainly not a mammalian feature if the predominant mammals, marsupials and placentals, are used as the basis for definition. Classification of *Tachyglossus* as a mammal in the face of this evidence could certainly raise some theoretical questions, but no more so than its classification as a reptile. It is conveniently considered a mammal to the satisfaction of most students. This, however, need not cloak the fact that in an ordering based on its several systems this animal is at a lower grade in attainment of complete "mammalness" than is, say, another anteater, *Tamandua*, which has a full complement of completely "mammalian" features. Were only parts of the skeleton of *Tachyglossus* preserved in the fossil record and the modern representatives unknown, it would doubtless be considered reptilian without hesitation.

The case of the evolution of actinopterygian fishes, which is taken up in some detail in Section IV, Chapter 3, similarly illustrates the interaction of formal assignment, the recognition of what are called clades, and informal ordering which recognized the mosaic pattern of evolution in the existence of different grades in the same organisms. Here the once formal terms "Chondrostei," "Holostei," and Teleostei" have become primarily expressions of grade of development.

To provide a norm on which grade levels can be based, recourse is generally to formally defined taxonomic groups; hence the use of such grades as reptilian, mammalian, or holostean. The result is a useful and meaningful overlap of the two systems. It becomes necessary, as was seen in the case of the mammary glands of *Tachyglossus*, to involve the concept of partial attainment of a grade, but this can be done without altering the system materially. Many problems of evolutionary position can be clarified by the grade-ordering system. The highly functional structure of formal taxonomy can be preserved, but its artificiality must be kept in mind, with particular recognition that the artificiality comes into focus most clearly at transitions between major categories. Arbitrary decisions on assignment are necessary. The criteria for such assignments should be convenience and, as far as possible, consensus among the persons most directly concerned.

One other major point needs to be recognized in the utilization of classifications of the vertebrates. The development of the various classifications has been by the efforts of many workers in various fields at different times and under different theoretical concepts. This has resulted in a structure in which categorical meanings within various subgroups, even down to the lowest levels, are not necessarily commensurate. It is, for example, common knowledge that avian classifications include very finely drawn distinctions, so that an avian family tends to represent either a subfamilial or generic level in comparison with, say, reptilian designations. The class Aves, itself, of course is much more limited in the overall variation of its major structures than are other tetrapod classes. This sort of irregularity extends throughout, even when only modern animals are concerned. It is accentuated when fossil animals, requiring the use of different criteria, are added. The discrepancies at the species level were noted earlier. This extends to higher categories. In many segments of the vertebrate subphylum, at about the familial level there is an interaction of the modern and fossil representatives. At this stage both sets of evidence become the determinants of classification, with fossils in particular forming the concrete basis for actual phylogenies. This practice tends to produce consistencies within the classifications in which both types of evidence are used, but it also tends to widen the gulf between these and others for which only one or the other type of data is available.

As matters stand at present, the classification of vertebrates, like most classifications of major groups, is a highly complex structure which has been put together from many sources and which reflects this fact in its internal heterogeneity. In general the construction of classifications among vertebrates has followed the dictates of the formal system of taxonomy, superimposing on an originally nonevolutionary hierarchical system one that is phyletic and attempts to express as well as possible the phylogeny of the animals it includes. Although the formal system is not fully suitable to this end, it provides a useful and practical classification, both as a catalog and as an integration of the biology of the vertebrates. Few serious difficulties arise, and these mostly at transitions between major groups. Where such difficulties are acute, or where evolution is the principal focus of study, a grade system of ordering is useful, but it is not a fully satisfactory substitute for formal classification. The vertebrate classifications used today are then useful and valid, but they must be used with intelligence as conveniences and not as sole guides to the development of evolutionary concepts.

FORMATION OF HIGHER CATEGORICAL LEVELS

Introduction

The discussions to this point took it more or less for granted that there exists a stable and generally accepted classification of the vertebrates. The problems and suggested solutions were cast within this frame of reference. Broadly conceived, this is the case, and the major categories that have been discussed and used as examples do meet with general, although not universal, acceptance.

Important differences in classifications do exist with regard to some high categorical levels, and there are many rather sharp differences of opinion relative to many lesser categories. This section and the following chapters are devoted to analyses of some of the areas of agreement and disagreement, and through this means to exploration of the ways that vertebrates have been classified and the criteria that have been found effective.

Treatment in the various major groups is quite irregular. Those that illustrate one or another kind of approach or problems related to special kinds of materials are emphasized. A more or less systematic organization will be used, but the categorical levels will not be carried below the point necessary to understand the approaches applied in the higher category under consideration. It is assumed that the reader has a fair general grasp of general vertebrate taxonomy and of the characteristics of the main groups.

Major Subdivisions

Table 1 shows the framework of seven different classifications, which cover fairly well the types that have been in use over the last several decades. Conformity rather than diversity in general outline predominates. Only one of these, number VI, is markedly different in concept from the others.

Each of the classifications, with the exception of the first and last, which are conservative in this respect, makes some sort of division of the vertebrates into superclasses, either two or three. These are indicative of the existence of

TABLE I   Recent Classifications of the Vertebrates

I. *After Romer (1945a)*

  Class Agnatha
  Class Placodermi
    (Aphetohyoidea)
  Class Chondrichthyes
  Class Osteichthyes
  Class Amphibia
  Class Reptilia
  Class Aves
  Class Mammalia

II. *After Jarvik (1960)*

  Superclass Agnatha
  Superclass Gnathostomata
    Class Pisces
    Tetrapoda
    Class Amphibia
    Class Reptilia
    Class Aves
    Class Mammalia

III. *After Camp, Allison, and Nichols (1964)*

  Superclass Agnatha
    Class Ostracodermi
    Class Cyclostomi
  Superclass Pisces
    Class Placodermi
      (Aphetohyoidea)
    Class Elasmobranchii
      (Chondrichthyes)
    Class Teleostomi
      (Osteichthyes)
  Superclass Tetrapoda
    Class Amphibia
      (Batrachosauria)
    Class Reptilia
    Class Aves
    Class Mammalia

IV. *After Weichert (1965)*

  Superclass Pisces
    Class Agnatha
    Class Placodermi
    Class Chondrichthyes
    Class Osteichthyes
  Superclass Tetrapoda
    Class Amphibia
    Class Reptilia
    Class Aves
    Class Mammalia

V. *After Goodrich (1930)*

  Branch Monorhina
    Class Cyclostomata
  Branch and Class Ostracodermi
  Branch Gnathostomata
    Grade Ichthyopterygia
    Class Pisces
      Subgrade Chondrichthyes
      Subgrade Osteichthyes
    Grade Tetrapoda
    Subgrade Anamnia
    Class Amphibia
      (Batrachia)
    Subgrade Amniota
    Class Reptilia
    Class Aves
    Class Mammalia

VI. *After Von Huene and Kuhn (1940 to 1966)[a]*

  Tetrapoda
  Urodelomorpha
  Eutetrapoda
  Batrachomorpha
  Reptiliomorpha
  Theromorpha
  Synapsida
  Mammalia

(continued overleaf)

TABLE I   (continued)

Eutetrapoda (continued)	Class Chondrichthyes
Sauromorpha	Class Osteichthyes
Reptilia	Superclass Acanthodii
Aves	Superclass Actinopterygii
	Superclass Sarcopterygii
VII. After Romer (1966a)	Class Amphibia
Class Agnatha	Class Reptilia
Superclass Monorhina	Class Aves
Superclass Diplorhina	Class Mammalia
Class Placodermi	

[a] Ranks are not given in the classifications, and no specific treatment of fishes is present. Mainly taken from phylogeny diagrams.

strongly separated radiations. These major divisions are based on jaw structure or nature of the paired appendages, or both. A suggestion of a third kind of major division is found in classification V, presence or absence of an amnion, but there it is not used for highest level division. This use has been made in some classifications in recognition of the fact that the state of this structure is an important determinant in radiation.

Criteria

JAWS.   The vast majority of known vertebrates possesses jaw structures that have been derived from the mandibular arch of the visceral skeleton. In sharks the primary jaws are composed of elements from this source. In other fishes and many tetrapods some portion of the visceral jaw structures persists in the masticatory apparatus throughout life. Below the level of mammals, viscerally derived elements enter into the craniomandibular articulation. The participation in the jaws of these elements has disappeared in mammals, but since their ancestry has come through stages in which visceral jaws did exist, they can conveniently be classified with the rest as gnathostomes.

Among the living vertebrates, lampreys and hagfishes lack such jaws, the mandibular visceral arch persisting as part of the gill support system. These are referred to as agnathous forms, along with a large array of extinct Paleozoic fishlike animals that show the same conditions.

The division into gnathostomatous and agnathous vertebrates is fundamental. Two very distinct radiations have been followed by members of the two groups. One, the agnathous radiation, took place largely early in vertebrate history, from Ordovician to late Devonian. The only certain record thereafter is in the lampreys and hagfishes. Presumably this was the primitive group of vertebrates; that is, the one most closely related to prevertebrate ancestors. All other vertebrates belong to the gnathostomes. Their very complex radiation is known from the late Silurian through all periods to the present.

Recognition of these two radiations as equivalents in a classification produces a highly asymmetrical subdivision: one major group is small, with only a moderate radiation in limited aquatic conditions; the other is large and has exploited most facets of the waters of the earth as well as the land and air, and these in various ways from various sources at different times. This asymmetry can be avoided, as in the first classification in the list, by considering the agnathous groups as merely a class, comparable to the others. The arrangement, however, tends to underemphasize the very real differences between agnathous forms and all others, and the fact that their radiations are so strongly separated.

Whether the Agnatha are to be considered fish or fishlike creatures is a question posed by the different classifications, because both designations are made. For the most part classifiers have not put them in the fishes proper— that is, Pisces—although Weichert, in classification IV, does so place them.

Goodrich's classification is somewhat older than the others and reflects an intensely evolutionary point of view. Both the underlying concept and the time of formulation of a classification are important in understanding its structure. Goodrich has two branches: Monorhina for the cyclostomes and Ostracodermi for the extinct Agnatha. This is somewhat like a terminology used by Romer (1966a) but applied in a different way. Goodrich noted that these two will perhaps be combined on the basis of evidence then coming to the fore. Later classifications follow his anticipation. His use of "Branch" and other terms not found in other classifications among other things reflects the very fine divisions that he used, for which there are not sufficient categories in the usual hierarchical system.

PAIRED APPENDAGES.   Whether a vertebrate has fins or limbs is extremely important to its capacity for adaptive radiation. Limbs, as defined in this sense, are those of tetrapods, with a single proximal element, paired long distal elements, a series of small bones, and phalanges. Animals with this pattern and all derived from them are called tetrapods. Although some finned creatures have paired appendages that function somewhat as limbs in walking and some tetrapods have limbs that function as fins, no closely intermediate structural stage is known. These likenesses are convergent. Some rhipidistians show a very general and vague approach to the tetrapod pattern, but none known comes close to possessing a tetrapodlike limb. This is then within the current frame of knowledge an absolute character for differentiation of two major groups.

Basically the fin–limb differentiation separates the major aquatic radiations from the major land radiations, with, of course, some crossovers. Weichert's classification (IV) makes this the basis for two superclasses, Pisces and

Tetrapoda. Goodrich's classification (V) uses it as a basic division, but only after separating the jawed and not-jawed into the initial dual division. The other classifications treat this similarly except for Romer's, which uses no superclass division. The classification of Von Huene and Kuhn (VI) does not extend to this level technically but does tacitly make the distinction of Pisces and Tetrapoda. This is a useful and much used subdivision, but often it is employed only in a nonformal sense.

AMNION.   The amnion is a protective embryonic membrane, present in all tetrapods except the amphibians. It provides a sound, twofold basis for division of the vertebrates since its presence allows a type of land radiation not otherwise possible. Although it is used often in an informal way, it does not enter into many modern classifications. Of those cited here, it is present only in Goodrich's, and there for the division of the Tetrapoda, not the vertebrates as a whole.

Of the three pairs of characters, or better perhaps the three characters with two states each, only those that relate to hard anatomy are generally suitable for the fossil record. In some instances, however, it is possible to infer the presence or absence of an amnion even in rather early groups, from the ontogeny when evidence of this is preserved or from adult features that suggest the nature of the pattern of ontogeny. For example, in adult *Seymouria* from the Permian there is some trace of lateral-line canals, suggesting an aquatic development with larval stages. This has been confirmed in another seymouriamorph, *Discosauriscus*, in which the larval stages are known from well-preserved specimens. It is reasonable to assume no amnion for this array.

In general, however, osseous characters are most readily used. The combination of limb and jaw structures produces a threefold division, such as that used by Camp, Allison, and Nichols. From the standpoint of expressing radiation this is excellent, since as is now known, there was an agnathid radiation, a fish radiation, and a tetrapod radiation. On the other hand, such a classification tends to mask the dominant role of jaws in vertebrate success.

If the cyclostomes and ostracoderms are brought into a single group, Goodrich's classification handles the major criteria in a particularly useful fashion. His arrangement comes about as close as possible to recognizing the role of each in proper perspective. A problem arises when an effort is made to fit this into a more usual framework of categorical designations, in order to maintain the conventional use of the class level for major groups of fishes (placoderms, chondrichthyans, and osteichthyans) and tetrapods (amphibians, reptiles, birds, and mammals).

None of the classifications serves all purposes equally well. In each there is inequality of major groups and things that might be, but are not, expressed. None is necessarily right or wrong, but of the seven that have been illustrated and discussed, Goodrich's presents and relates the most information and in this sense is the most natural.

Superclasses, Classes, and Lower Categorical Levels

Division of all vertebrates into classes without higher groupings, such as that of Romer (Table 1), provides a useful framework, but it does ignore the fundamentally important differences between agnathans and gnathostomes. A Pisces–Tetrapoda division similarly has uses but ignores the importance of jaw structure. In many respects separation into the superclasses Agnatha and Gnathostomata is the most useful. At these highest levels under the subphylum Vertebrata, thus, an absolute assignment by one or two characters, with no overlap, is possible. Although there are problems about the most informative and most "natural" groupings, there is none in the assignment of any well-known vertebrate to one or another of the several possible superclasses.

The classes, on the other hand, if viewed over the full range of animals that each class is generally considered to include, must be deemed polythetic in basic structure. The definitions as used, however, are of the classical type, monothetic in concept. Conflicts in assignments of particular vertebrates develop between these two in a number of instances.

The monothetic character has emerged largely because the definitions of most of the classes were initially based on living materials and are still conceived in this way. This is not the case for class Placodermi, all of whose members are extinct, and does not hold for the extinct classes of Agnatha, if this rank is afforded the major subdivisions. Among the others it does hold in general. Mammary glands, for example, provide a single necessary and sufficient character for inclusion in Mammalia. Hair is another such character. Feathers provide an absolute basis for membership in the Aves. Cold-bloodedness and the presence of an amnion define membership in the Reptilia, and an anamniote tetrapod is a member of the Amphibia.

Monothetic definition is more difficult for the fish groups, inasmuch as a somewhat more complex array of defining characters is needed. Osteichthyes are fishes with lungs or derived structures, a specialized spiracle, distinctive hyoid support for the jaws, and bone. None of these is in a broad sense unique, but the combination is. Chondrichthyes are cartilaginous jawed fishes, totally lacking ossification of the internal skeleton.

If each of these rather simple definitions held throughout each of the class, the straightforward, monothetic characterization would suffice. No one can say with absolute certainty that this is not the case, at least for the classes of tetrapods. Here, however, we immediately encounter problems of assignment. If any animal that definitely cannot be shown to have the key features is excluded, then few fossils of tetrapods can be assigned to any of the classes.

Intergradations between the defining states in ancestor-descendant related classes are to be expected. It is known, of course, that gradations do occur in characters which, in moderns alone, show very different states, such as the middle-ear ossicles of reptiles and mammals. Such "secondary" defining characters strongly suggest that the same sort of processes applied to the "primary" characters. If this is the case, the definitions of the major groups, classes, are not ultimately monothetic but polythetic.

In the studies of the classes that follow much interplay of the two systems is evident. These studies are directed mostly to what has entered into the formulations that are in use and less concerned with the "best" classification. Each of the major subdivisions poses its own substantive problems and its own theoretical problems. Different students using different materials and with different concepts of classification procedures have arrived at divers interpretations.

A vast literature on classifications and phylogenies exists for each of the major groups. Although the history of classification is an interesting field in itself and somewhat pertinent to the conception of current classifications, it will be necessary in most cases to keep considerations of it to a minimum. introducing them only when needed to understand an interpretation.

2 Superclass Agnatha

GENERAL CONSTITUTION

The Agnatha are fishlike, aquatic, jawless vertebrates. They have long been known from rocks of the lower and middle Ordovician, from the upper half of the Silurian, and throughout the Devonian. Lampreys and hagfishes are modern representatives. Recently lampreylike cyclostomes have been obtained from upper Pennsylvanian beds of Illinois. The earliest agnathans, from the basal Ordovician, are extremely fragmentary, but a few specimens from the middle part of the period give some idea of the general shape of the animals. The Silurian remains are better preserved, and excellent records have come from widely scattered deposits of Devonian age.

The Agnatha have not been found in sediments deposited in the open marine waters of epicontinental seas and geosynclines during the Paleozoic. Rather, they occur in deposits formed in fresh water and in marine waters along coastal margins and in estuaries. Most of them appear to have been benthic animals, probably slow swimmers and not very active.

By the time that well-preserved representatives first occurred, the radiation of this superclass was well under way. This has reduced the utility of geological age as an aid in classification, and in some classifications age relationships have been essentially ignored. In others the information, even though limited, has been used to good advantage. A few typical agnathans are illustrated in Figure 8, and in these the defining features of the superclass are shown.

TABLE 2   Classification of the Agnatha

I. *Romer (1945a)*	III. *Camp, Allison, and Nichols (1964)*
Class Agnatha	Superclass Agnatha
Subclass Cephalaspidomorphi	Class Ostracodermi
Order Osteostraci	Subclass Cephalaspidomorpha
Order Anaspida	Superorder Osteostraci
Order Cyclostomata	Superorder Anaspida
Suborder Myxinoidea	Subclass Pteraspidomorpha
Suborder Petromyzontia	Subclass Thelodonti
Subclass Pteraspidomorphi	Class Cyclostomata
Order Heterostraci	Order Petromyzontiformes
Order Coelolepida	Order Myxiniformes
II. *Stensiö (1958)*	IV. *Romer (1966a)*
Superclass Cyclostomati	Class Agnatha
(Agnatha)	Subclass Monorhina
Class I. Cephalaspidomorphi	Order Osteostraci
Superorder Osteostraci	Order Anaspida
Superorder Anaspida	Order Cyclostomata
Superorder Petromyzontida	Suborder Petromyzontoidea
Class II. Pteraspidomorphi	Suborder Myxinoidea
Superorder Heterostraci	Subclass Diplorhina
Superorder Myxinoidea	Order Heterostraci
Class III. Thelodonti	Order Coelolepida
Superorder Phlebolepida	
Superorder Thelodontida	

There have been various approaches to the classification of the Agnatha, with fundamental differences centering on the relationships of the moderns and ancients. One, exemplified by classification III in Table 2, considers that the modern group, the cyclostomes, forms a separate class or superclass and places the ancients, the ostracoderms, in another equivalent category. This gives major emphasis to temporal separation and does not recognize morphological similarities and differences between the major extinct groups as a basis for assignment below superclass (or class) level. Another approach ties the moderns and ancients together. This may be done as in classifications I and IV in Table 2, in which all are placed in a single class. Below this level, at a subclass stage, the ancients may be divided into two groups, with lampreys assigned to one and hagfishes to the other. This is the system used in classification I of Table 2. Stensiö's classification (II in Table 2) is similar, but the Agnatha are considered to represent a superclass.[1]

In such classifications the lampreys, Petromyzontia (Petromyzontida), are associated with the extinct Cephalaspidomorpha and the hagfishes, Myxinoidea, with the Pteraspidomorpha. The implications concerning the ancient derivations of the two living forms are evident. The case for relationships of the lampreys to the cephalaspidomorphs is based on sound evidence, because the endocranial tissues in the fossil osteostracans are well ossified and show many characters in common with the living lampreys.

An alternative to this type of classification is shown in system IV of Table 2; here the major subdivisions are Monorhina and Diplorhina. The separation is based on interpretations of the condition of the nasal capsule. Only the Heterostraci (= Pteraspidomorpha) are placed in the Diplorhina, resulting in a very wide separation from all other Agnatha.

Much of the difficulty stems from the fact that evidence for the relationships of the heterostracans to other groups is weak, because there was no internal ossification under the cephalothoracic shield. In some instances, however, the cartilages of the endocranium and branchial region are lightly impressed on the underside of the dermal surface.

Until recently the rather vague evidence that the dermal bone gave concerning the position and nature of the narial opening was the primary center of attention. Thus, as in Romer's classification (IV in Table 2), all Agnatha that showed evidence of a single (coalesced) nasopharyngeal capsule, whether this was dorsal or terminal, were placed in one group, Monorhina. The heterostracans, for which there is some vague evidence of more widely spaced capsules, formed the Diplorhina. Using the criterion of the nasal structure, Stensiö had attempted to demonstrate the existence of a terminal capsule in the heterostracans, but had met with opposition based on evidence that the positions of the articulating plates in the snout region failed to provide the necessary space.

Stensiö (1968) made a very detailed analysis of the situation, listing a series of 11 points of agreement between the heterostracans and the myxinoids. He utilized the detail available for the heterostracans as the basis for very extensive hypothetical reconstructions of probable conditions. The reconstructions are extensively used in support of his position. The result is a fascinating but highly speculative fabric, consisting of an intricate series of interrelated inferences. For example, the apparent rigidity of the exoskeleton in the anterior regions is taken as evidence that the principal superficial constrictors were absent, suggesting in turn the presence of a powerful velum,

[1] As is evident in the text and tables, a great deal of variation in the use of terms occurs in the classifications. Much of this has little real taxonomic meaning, and reflects the individual preference of the particular author.

requiring a nasopharyngeal duct leading back from the prenasal sinus (reconstructed from other evidence). The existence of this duct is "proven" by its necessity in feeders on plankton and debris, inferred for this group from still other structures, in particular those of the mouth. Several other lines of evidence, such as the small eyes and presumed use of smell in search of food (as in hagfishes), led him in the same direction.

Throughout much of the history of the efforts to make major subdivisions among the Agnatha studies have been devoted to finding key characters, usually a single key character, and then to broaden the evidence for the division suggested by such a character with as many supplemental bits of data as possible. Attention has centered mostly on the morphology of the naso-hypophyseal organ and its capsule. This is the basis for the classifications shown in Table 2 and others of similar nature.

Stensiö, starting with a hypothesis of relationships whose roots go far back in his broad experience and certainly involved the nasopharyngeal structure, has attempted to assess relationships, in support of his hypothesis, on the basis of a wider range of features. He has used not only the available morphology, and many tenuous inferences based on it, but also interpretations of feeding and other functions that suggest similarities between some of the ancient and living groups. Without knowledge of structures of the modern agnathans, much of what exists in his reconstructions could not have been deduced. As in all such cases, it must be recognized the previous concepts on relationships cannot fail to influence what is "seen" among partially preserved fossil remains.

THE CLASSES AND THEIR SUBDIVISIONS

If the concept of the Agnatha as a superclass is followed, there are three classes of Agnatha, suggesting a status comparable to that of the Pisces or the Tetrapoda. Structurally, for at least two of the classes, this is reasonable. Clearly the separation of the fossil Cephalaspidomorpha and Pteraspidomorpha is as great, for example, as that between the Amphibia and Reptilia. In terms of numbers and kinds of animals known these two classes of Agnatha are very small, and their radiations, as far as exploitation of a variety of habitats and production of numbers of distinct kinds of body form are concerned, cannot compare with those of most other classes. On the other hand, diversity with regard to basic morphology is relatively great.

The fossil Cephalaspidomorpha and Pteraspidomorpha can be separated on a single key character, histology. The hard tissues of the latter consist of a basal layer of laminated acellular bone or bonelike material (aspidin), a more porous middle layer, and a surficial layer of dentine. The cephalaspidomorphs differ markedly in their possession of osteoblasts. There are many other characters that may be used—for example, the nasal capsules, the nature of opening of the gills to the exterior, and the position of the orbits.

The class Thelodonti (Coelolepida) is composed of soft-bodied creatures and may very well be an "artificial" class. It is variously treated as primitive (e.g., Obruchev places the members of this group as primitive pteraspidomorphs) or as advanced, on the idea that bone degeneration is typical of the evolution of the ostracoderms and that these are end members in this process. On the latter basis it is possible that they arose in part from the cephalaspidomorphs and in part from the pteraspidomorphs.

The phylogenetic relationships that the classifications are designed to depict clearly leave much to be learned. Presumably the ordering in classifications places the primitive members first and the more advanced forms after. The wide differences between Stensiö and Obruchev and Denison come out clearly in a simple side-by-side tabulation of their classification:

Stensiö (1958)	Obruchev (1964)	Denison (1951)
Superclass Cyclostomata (Agnatha)	Class Diplorhina (Pteraspidomorpha)	Subclass Diplorhina (Pteraspidomorpha)
Class Cephalaspidomorpha	Subclass Thelodonti	Order Heterostraci
Superorder Osteostraci	Subclass Heterostraci	?Order Coelolepida
Superorder Anaspida	Class Monorhina	Subclass Monorhina
Superorder Petromyzontida	(Cephalaspidomorpha)	Superorder Hyperotreti
Class Pteraspidomorpha	Subclass Osteostraci	Order Myxinoidea
Superorder Heterostraci	Subclass Anaspida	Superorder Hyperoarti
Superorder Myxinoidea	Subclass Cyclostomata	Order Petromyzontida
Class Thelodonti		Order Osteostraci
Superorder Phlebolepida		Order Anaspida
Superorder Thelodontida		

To the extent that these classifications may be considered ordered in terms of line of descent, almost complete opposites are represented. Geological age cannot be considered as an important contributor to the evidence, although it does provide some information. Stensiö's classification does not consider it, whereas Obruchev's and Denison's do, as far as possible.

Class Pteraspidomorpha (Diplorhina)

The principal component of this class, or subclass of some classifications, is the Heterostraci, a term often considered synonymous with Pteraspidomorpha. Although Thelodonti and Myxinoidea have been included, the dubiousness of their membership has been noted above, and there is no general agreement on how they should be handled. As far as Heterostraci are concerned, there is moderate consensus that this is a valid and homogeneous

TABLE 3   Groups, Orders, and Lineages of Heterostraci

Stensiö (1958)	Obruchev (1964) (Orders Only)	Tarlo (1962) (Lineages)
Group I	Astraspida	Base
	Eriptychia	
Group III	Cyathaspida	Lineage I
Group IV	Corvaspiformes: a family of Psammosteida, Tarlo; a family of Cyathaspida, Obruchev)	
Group II	Psammosteida	Lineage II
	Drepanaspida, Stensiö	
	Traquairaspida	Lineage III
	Pteraspida	
Group VI	Cardipeltida	
Group V	Amphiaspida	Lineage IV
	Hiberiaspida	
	Obliaspida (new order, Obruchev)	
Group VII:		
Turiniiformes (Thelodonti Obruchev, Tarlo)	Lyktaspida (new order, Heinz)	

group and that in large part the breakdown into the series of subgroups (orders) that is given in Tables 3 and 4 is reasonably valid. Some differences will be brought out, but it is less the groups than their phylogenetic arrangement that is controversial.

The Heterostraci illustrate particularly well some of the problems and procedures that attend attempts to study and classify fossil groups that lack close living relatives and whose preservation is at best only moderately good. The absence of internal ossification in the heterostracans means that superficial characters and characters of the histology of the exoskeleton must supply much of the information on morphology. Their relatively long geological history, from middle Ordovician to late Devonian, gives some chance for the use of stratigraphic distribution as a key to phylogeny and classification.

The number of ordinal groups, expressing the different kinds of hetero-stracans that are to be found in the record, ranges from 8 to 11 as determined by various classifiers. Table 3 shows the classifications of Tarlo, Stensiö, and Obruchev (mostly after Berg). In contrast to agreement on the general ordinal constitution is the diversity of the arrangements into lineages and groupings above the ordinal level.

Table 4 shows the kinds of characters that are used in subdivision and the general distributions in the orders, ten being used following the scheme of Obruchev. The futility of attempting to assess the nature of such groups from the ways they are arrayed in a formal classification is evident from the kind of information available. Historically the process has involved early recognition of a few types, attempts to assess their relationships, and the establishment of groups to include them. As more representatives have been found, they have been either placed in these groups or, if this was not thought possible, used as the basis for new groups. From time to time a major "overhaul" has been made, with some reshuffling and a new system of classification. In this process efforts have consistently been made to find "key" characters—that is, those that would allow simple and unquestionable definition of the groups. One or a combination of several characters has been used, as can be seen from the chart. Effort has thus been toward the monotypic kind of approach, with elimination of overlaps. This has been reasonably successful, as indicated by the fact that in large part there is no fundamental disagreement on the major ordinal groupings and little problem of intermediates. That this is the case results in part from the scanty record, which has left little in the way of well-known intermediates to cause trouble. Only the very prominent types have been preserved in any appreciable numbers.

Very probably the orders themselves are polythetic in structure, but again, the available materials are such that the necessary conditions cannot be assessed on what we have. There is no great number of characters and it cannot be known whether or not those that can be identified were possessed by most, or that none was possessed by all.

Although there is some agreement on the ordinal subdivisions, with exceptions that appear in Table 3 there is little on the phylogeny which is expressed in the ordering in the classifications. Obruchev's orders, in the sequence he gives them, are used in Table 3 to compare the systems of Stensiö and Tarlo. The differences between the two patterns are largely self-explanatory if cast in the framework of phylogeny, with the most primitive first and the derived groups following. Branching to produce two equivalent groups, of course, cannot be shown.

The differences illustrated result mainly from differences in the inter-pretation of the order of changes, both within the "divisions" and between them. This also results in conflicts in placement of some of the subgroups

TABLE 4   Characters Used in Ordir

	Order			
Character	Astraspida	Eriptychia	Cyathaspida	Psammosteida
Armor, how formed by plates	Discrete, polygonal, cyclomorial tesserae	Discrete, polygonal, cyclomorial tesserae	1. Two or four main plates: dorsal, ventral, and paired branchials 2. Some with paired suborbitals 3. Outlines of tesserae in some	Independent polygonal plates in some, formi into large, solid plate in families; include dorsal, ventral media plates, one paired or tal and branchials, plus rostral, pineal, cornual in some
Outline shape of shield	Long, subelliptical, posteriors truncated	Unknown	Elliptical with posterior truncated	Elliptical to ovoid, posterior to ovoid, posterior truncated
Cross-sectional shape of shield	Unknown	Unknown	Depressed fusiform to flat	Flat
Ornamentation	Tubercles, forming longitudinal ridges and scattered	Tubercles, scattered and in ?lateral ridges	Thin, parallel, long ridges, cyclomorial tubercle patterns; intricate, fine patterns	Varied, some irregular, interdigitating dermal "teeth," pustulae, polygonal tesserae
Branchial opening	?Posterior to corner of cornua	Unknown	Between branchial plate and dorsal plate	At posterior end of branchial plate, above cornual plate when plate is present
Mouth position	Unknown	Unknown	Ventroanterior	Terminal or dorsal
Orbital position	Unknown	Unknown	Lateral, separated	Dorsolateral in orbital plate
Pineal opening	Absent	Unknown	Present	Absent

fferentiation of the Heterostraci

		Order			
raquairaspida	Pteraspida	Cardipeltida	Amphiaspida	Hiberiaspida	Obliaspida
ht main lates: rostral, ineal, dorsal, entral, paired rbitals and ranchio-ornuals	Ten main plates: rostral, pineal, dorsal, ventral, paired orbital, branchial, and cornual	Large dorso-branchial plate, ?anterior tesserae.	Undivided dorsoventral armor; No rostrum	Fused dorsal, ventral and branchial plates	Undivided dorso-ven-tral armor.
adly ovoid runcated osterior	Elongated, with rostrum; some narrow, others broad	?Subelliptical, posterior broad	Subcircular	Subtriangular	Oval
oid	Fusiform to flattened	?Flattened	Ventrally flat, dorsal plate convex	Flattened	Dorsal plate ventral convex
rmal lenticles anged in ows, fol-owing contour f shield lements	Stellate, ribbed, or forming thin striae, never marked	Very little or none	Dentine tubercles	Flat, dentine ridges, marginal spines	Fine dentine scrolls in intricate patterns
losed in ranchio-ornual plate	Between cor-nual and branchial plate	Lateral, ?in branchial plate	Unknown	Unknown	Dorsal where known
tral, back f rostrum	Ventral, back of rostrum	Unknown	Ventral, anterior	Ventral, anterior, or at end of tube	Terminal
eral on rbital plates	Lateral on orbital plates	Unknown	Anterior margin of shield	Close to-gether, antero-dorsal, or none	Anterolateral or dorsal
ne	Present	None	Uncertain	None	Present

under different major units in the different plans. Geological distribution is also a factor, and the degree of credence given the sampling as an expression of actual age relationships enters in. In general Stensiö has paid less heed to it than has Tarlo. Both appear to agree that the Ordovician occurrences are primitive, being very early. Both phylogenies stem from this base. Beyond this there is little agreement.

As matters stand only rather vague guidelines exist for assessment of these relationships. The usual guidelines of general evolutionary theory are in large part invalidated by the fact that their application depends on kinds of information, especially that of the nature of the samples, that are not available. Classifications of this sort, then, are expressions of opinions on phylogeny and, in the absence of more than vaguest guides on the actual courses of evolution, in large part arbitrary. Their structure must be looked at in this light.

Class Cephalaspidomorpha (Monorhina)

The class level for this group, given in the classifications outlined in Table 5, is only one of several categorical levels at which it has been cast. In its most

TABLE 5   Classifications of the Cephalaspidomorpha

I. *Berg (1940)*
Class Cephalaspida
  Order Cephalaspidomorpha
    Family Cephalaspidae
    Family Boreaspidae
    Family Thyestidae
    Family Didymaspidae
    Family Sclerodidae
    Family Dartmuthiidae
  Order Tremataspidiformes
    Family Tremataspidae
    Family Oeselaspidae

II. *Stensiö (1958)*
Class Cephalaspidomorphi
  Subclass Osteostraci
    Order Orthobranchiata
    Order Oligobranchiata
    Order Nectaspiriformes

III. *Denison (1951)*
Subclass Cephalaspidomorpha
  Order Osteostraci
    Family Tremataspidae
      Subfamily Tremataspinae
      Subfamily Didymaspinae
    Family Cephalaspidae
    Family Kiaeraspidae
    Family Ateleaspidae

comprehensive sense Cephalaspidomorpha includes both the extinct Osteo-straci and Anaspida and the living cyclostomes, both Petromyzontia and Myxinoidea. There is little disagreement on inclusion of the former, but a wide diversity of opinions on the position of the myxinoids. Among these is the idea that they are related to the pteraspidomorphs and the contrary belief that they have no known close relatives in the record. Inclusion of the myxinoids in the Cephalaspidomorpha reduces the definitiveness of the characterization of the group but permits a definition as agnathans with a single nasal capsule, either terminal or dorsal. If the myxinoids are removed, the highly characteristic single dorsal nasohypophyseal capsule is a readily recognized character. Within the Agnatha, the presence of a discrete, external opening for each gill is a valid character, separating the cephalaspidomorphs from the myxinoids and pteraspidomorphs.

Some of the fossil representatives of the osteostracans have very well ossified internal cranial skeletons; determination of many details of the structure has been possible. This has had a distinct influence on classification, in particular playing a role in the classification of Stensiö, detailed later in this section. Both fossil groups, Osteostraci and Anaspida, have well-ossified exoskeletons, which provide a wealth of characters that serve to separate them into discrete groups (see Figure 8). Both have lacunae for bone cells, or osteocytes, which provide a basis for relating them to each other and for separation from the heterostracan pteraspidomorphs, which lack such cells.

Osteostraci

Although separation from the Anaspida poses no problems, classification within the Osteostraci has had a complex history and raises some interesting problems. Berg, Stensiö, Westoll, and Denison have given fairly recent consideration to this classification, building on the earlier studies of Heintz, Kiaer and Bryant, and others. Classifications by Berg (1940), Denison (1951), and Stensiö (1958) are shown in Table 5.

Berg's classification recognized two major groups, which are clearly distinct, and draws very fine lines on the basis of small features below these levels. Additional studies since the time of this work have shown that many of his characters are not fundamental; that is, they do not correlate with suites that have consistency over rather large segments of the Osteostraci and suggest radiative units. Denison's classification is based primarily on (a) the length of the head shield relative to the overall dimensions, (b) presence or absence of the pectoral fin, and (c) presence or absence of the pectoral spine. Temporal distribution also enters in. Resultant groupings show consistencies in a large variety of characters, suggesting that these three are fundamental.

Stensiö's classification is based on features that depend on the preservation of details, present in some but not in all. This lessens its practical use. His main groups, orders, are based on mouth form (transverse, elongated, and circular in Orthobranchiata, Oligobranchiata, and Nectaspiformes, respectively) and orientation of the gill chambers (transverse or oblique). How valid these structures may be for fundamental divisions is difficult to check, since they are not to be found in many specimens because of inadequate preservation. They do produce a grouping quite different from that of Denison, and this suggests, since his groups are related to presence of large suites of correlated characters, that Stensiö's features may be adaptive and perhaps the result of convergence.

Within these major groups Stensiö has drawn very fine divisions, recognizing 18 families, most of which have but a single genus. Monotypic genera, with a single species, and families with a single genus, are sometimes a necessity in the classification of extinct organisms, mainly, it would seem, because of lack of knowledge of the radiation at these levels in the majority of cases. That this applies to the large number of families in this case seems questionable, and this suggests that some type of systematic concept other than one that relies primarily on the idea of family as a unit of radiation has entered in. Only if each genus, or two in a small number of cases, is considered to be the only known representative of the familial radiation can such a scheme hold under the usual interpretation. To some extent this pattern suggests that proposed by Osborn in his classification of the Proboscidea (see pages 407– 411 in this volume).

Regardless of its validity as an expression of phyletic relationships, this classification is impractical, because confident major assignments can be made in only a relatively small number of cases. Denison's classification, on the other hand, makes use of features that appear to be fundamental in the radiation at the levels specified, and can be seen and evaluated for most specimens. It appears to be a biologically "natural" classification.

Classification of the Osteostraci, in some contrast to that of the Heterostraci, includes sufficient information for testing the validity of a proposed classification. There is a wealth of morphological detail and a host of specimens with fair geological distribution. It is possible, and to some degree this has been done, to assess positions in terms of suites of related characters, to classify by a polythetic approach, or to select from these suites "key" characters and employ the more usual, monotypic principle.

Anaspida

The anaspids are small, rather uncommon fishes, usually not very well preserved and hard to study. Their classification is thus not clearly defined, and the tendency has been to place each type, designated as a genus, in a separate family. In this instance, with the dearth of material, this practice may have some phyletic merit. It is, if nothing more, a matter of convenience and avoids the expression of relationships for which there is no basis.

Stensiö has recognized eight or nine families—nine if the controversial animal *Jaymoytius* (Figure 4) is included. Families are based on the states of several characters, the number of gill openings, the form of the scales of the dorsal "comb," the angular disposition of scale rows, and the nature of the anal fin.

Berg (1955) proposed a division into three orders. This was based on the concept that evolution in this group, as in various others, passed from a stage of heavy exoskeleton to a reduced exoskeleton. The state of the exoskeleton and the fin (spines) give the basis for the subdivision. His arrangement (as modified and used by Obruchev, 1964) is as follows:

> Subclass Anaspida
> Order Birkeniida (Barycnemata)
> (Five families plus one possibly assigned)
> Order Lasaniida (Oligocnemata)
> (One family)
> Order Endoiolepidida
> (One family)
> *Incertae sedis*
> (One family (Jaymoytiidae))

Little probably is to be gained from the use of orders at present, because information on relationships is very slight and little that has clear-cut phyletic meaning is known.

Class Thelodonti (Coelolepida)

Just where this group should be placed, what its status is, and even whether there is such a group, are open questions. It is conceived to include a number of very small to medium-sized soft-bodied creatures whose only hard parts are dermal denticles. Structures, of course, are poorly known and there is little to go on in classification.

Placement has depended primarily on two things. One relates to the interpretation of the soft condition as primitive or advanced. The other is the degree of credence given to the vague indications of structures, whether or not they are sufficient for assignment to one group or another.

Stensiö, among others, has formally placed the group as a class, thus avoiding any formal commitment on relationships. He has suggested, however, that the two superorders which he recognizes, Phlebolepida and Thelodonta, may have affinities with the Osteostraci and Heterostraci, respectively. The class is thus merely a convenience and a temporary repository until positions are clarified.

Obruchev, on the other hand, considers this to be a primitive group, a subclass under the Pteraspidomorpha, for reasons already noted. In addition to the two groups included by Stensiö, he places the Turiniida here, as does Tarlo. In spite of the very marked differences in positions in classifications, the comments of Obruchev and Stensiö suggest that their positions are not far apart. Both recognize affinities to both groups of ostracoderms. Obruchev is more definite in placing the group under a major heading.

Here, as in many cases encountered later in other groups, placement of known forms in a formal classification seems to convey much more than is actually justified by the evidence and much more than the classifier may have intended by his assignment, which can be extremely tentative.

3 Class Placodermi

INTRODUCTION

Since its introduction shortly before 1850, the term "Placodermi" has been widely used in a number of senses to designate assemblages of extinct fishes with heavy armor over the anterior part of the body. From the beginning the fishes generally known as arthrodires have been associated with the name. These are characterized by the presence of cephalic and thoracic shields that are joined by movable articulations, usually on the dorsolateral or lateral margins of the shields. Various other fishes have been included, and some that are now thought to be placoderms were earlier assigned to other groups.

Until the 1920s most of the work had been rather superficial, but since then many paleontologists have contributed to an understanding of the morphological detail and phylogenetic relationships of the class. Intensive studies are still continuing and new and better materials are being obtained, so that the classification remains somewhat in a state of flux. A fair consensus of what constitutes the placoderms exists today, representing stabilization of a situation that was very fluid not many years ago.

The state of affairs in the early 1930s is shown in Romer's (1933) classification in the first edition of his *Vertebrate Paleontology*. This classification (see Table 6) is quite similar to that of Heintz (1932) and reflects the general thinking of this period.

An outline of a current classification by Obruchev (1964) (Table 6) shows that a firm base for some of the recent practices had been established by 1933.

TABLE 6   Classifications of the Placodermi

I. *Romer (1933)*
    Class Placodermi
        Order Arthrodira
            Suborder Acanthaspidoidea
            Suborder Coccosteoidea
            Suborder Homosteoidea
            Suborder Ptyctodontoidea
        Order Antiarcha

II. *Obruchev (1964)*
    Class Placodermi
        Subclass Arthrodira (Coccostei)
            Superorder Rhenanida
                Order Radotinida
                Order Kolymaspida
                Order Palaeacanthaspida (Acanthothoraci)
                Order Stensioellida ("Stegoselachii")
                Order Gemuendinida
            Superorder Euarthrodira
                Order Arctolepidida (Acanthaspida, Dolichothoraci)
                    Suborder Arctolepidoidea
                    Suborder Williamsaspidoidea
                    Suborder Holomenatoidea
                Order Coccosteida (Brachythoraci, partim;
                    Coccosteoformes, partim)
                Order Pachyosteida
            Superorder Petalichthyida
                Order Macropetalichthyida (Anarthrodira)
                Order Ptyctodontoidea
                Order Phyllolepidida
        Subclass Antiarcha (Pterichthyes)
            Order Asterolepidida
            Order Remigolepidida

Some of the names, such as Homosteoidea, have been modified, and the content of some of the groups has been changed. Some animals now considered as placoderms were earlier placed among the Chondrichthyes; e.g., Stegoselachia (Petalichthyida) and Rhenanida.

External features have provided the principal criteria for establishing relationships. The two shields, resemblances of their plate patterns, and the nature of the mobile articulation—along with the presence of paired, osseous jaws, unlike those in other major groups—comprise the main characters. Only in a few instances has the endocranium been well preserved, and only occasionally is the internal postcranial skeleton available for study. The structure of the paired fins, especially the pectoral fins, and the shoulder girdle have recently become important in some classifications, and work on the cranium, still in progress, is having effects as well.

Expansion of the kind of classification used by Heintz and by Romer—with some modifications introduced by considerations of geological distribution, new finds, and inferences on the general course of evolutionary development—has resulted in such modern classifications as that of Obruchev and the various classifications shown in Table 7.

Obruchev's classification is based on morphology, especially as expressed in trends, but it also lays stress on environment, with the Arthrodira and the Antiarcha including members of marine and nonmarine radiations, respectively.

INTERIM CHANGES

Between the time of publication of Romer's 1933 classification and the modern stabilization of the general outlines of relationships, changes took place that are not evident in a comparison of the two ends of the history. Most influential for a period of 10 to 15 years was an organization proposed by Watson (1937). On the basis of work on the extinct fishes called Acanthodii, he reached the conclusion that a free hyoidean arch existed within this group; that is, there was a "normal" hyoidean gill and the gill arch was not incorporated into the jaw suspension. This condition was termed aphetohyoidean, or free hyoidean, and the acanthodians and placoderms were grouped together as supposed common possessors of this feature. Relationships between the acanthodians and arthrodires were based on supposedly common features and for the most part the free-hyoidean structure in placoderms was assumed on the basis of the relationship rather than sound morphological evidence of its presence. The classification that resulted is given in Table 7 (I) in outline form. One of its major effects was to consolidate the associations of various fishes that later became the mainstays of the Placodermi.

The influence of Watson's classification was widespread, being evident in such well-known classifications as that of Moy-Thomas (1939a) and Romer (1945a) (see Table 7, II). The internal structure of the classifications, below the main grouping on the basis of the aphetohyoidean nature, was strongly influenced by other studies, in particular by those of Gross (1937). Westoll (1945) recognized the aphetohypoidean characteristic but grouped on this basis only at a superclass level, expressed as a grade of evolution (see Table

TABLE 7   Classifications of the Aphetohyoidea and Placodermi

I. *Watson (1937)*
  Grade and Class Aphetohyoidea
  Order Acanthodii
  Order Arthrodira
  Order Antiarcha
  Order Petalichthyida
  Order Rhenanida

II. *Romer (1945a)*
  Class Placodermi (Aphetohyoidea
    in text)
  Order Acanthodii
  Order Arthrodira
    Suborder Euarthrodira
      Infraorder Arctolepida
        (Acanthaspida)
      Infraorder Brachythoraci
    Suborder Ptyctodontida
    Suborder Phyllolepida
  Order Macropetalichthyida
    (Petalichthyida)
  Order Antiarcha
  Order Stegoselachii
    Suborder Stensioellida
    Suborder Rhenanida
  Order Palaeospondyloidea

III. *Westoll (1945)*
  Grade Aphetohyoidea
  Class Acanthodii
    Order Acanthodii
  Class Placodermi
    Order Arthrodira
      Suborder Euarthrodira
        a. Arctolepida
        b. Brachythoraci
      Suborder Phyllolepida
      Suborder Ptyctodontoidea

Order Petalichthyida
  Suborder Macropetal-
    ichthyida
  Suborder Rhenanida
Order Antiarcha
Order *incertae sedis*
  Palaeospondyloidea

IV. *Stensiö (1959)*
  Arthrodira
    Division A. Euarthrodira
      Superorder a. Aspino-
        thoracidi
        Order I. Pachyosteo-
          morphi
      Superorder b. Spinothor-
        acidi
        Order II. Coccosteo-
          morphi
        Order III.
          Dolichothoraci
          (Acanthaspida, Arcto-
          lepida)
        Order IV. Acanthothoraci
        Order V. Radotinida
        Order VI. Rhenanida
        Order VII. Petalichthyida
        Order VIII. Stensiöellida
        Order IX. Phyllolepida
        Order X. Ptyctodontoidea
    Division B. Antiarcha
      Order I. Asterole-
        pidiformes
      Order II. Remigole-
        piformes

7, III). This did not imply a necessary close genetic relationship between the acanthodians and placoderms, but merely that both were at a level of development to be expected in the early evolution of gnathostomes.

Additional studies of the acanthodians by a number of students have shown quite conclusively that they were not in fact aphetohyoidean. This is a stage that theoretically should have existed but is not well documented in the record as it now stands. Thus the formal use of the term "Aphetohyoidea" and the grade level as used by Westoll have been abandoned, and with it the association of the acanthodians and placoderms. A resurgence of many of the concepts that were held prior to Watson's work has taken place, although a recent association of acanthodians and placoderms is found in Camp, Allison, and Nichols (1964), where four subclasses of the class Placodermi are recognized: Pterichthyes, Coccostei, Gemuendina, and Acanthodii.

Most classifications of the last decade and a half have considered the arthrodires and antiarchs to pertain to a single class, as two separate branches of a major radiation. Berg (1940, 1955), however, placed them as distinct classes, not recognizing the Placodermi as a valid unit. He used the terms "Pterichthyes" and "Coccostei" for his two classes.

Within the last decade a largely, but not altogether, new relationship has been recognized—that between the placoderms and Holocephali, the chimeras. The latter are usually classed with the Chondrichthyes. Gross

TABLE 8   Classifications of Placodermi, Elasmobranchiomorphi, and Holocephali

I. *Romer (1966a)*	Subclass II. Placodermata
Class Placodermi	Superorder Arthrodira
Order Petalichthyida	Superorder Holocephali
Order Rhenanida	Class Acanthodii
Order Arthrodira	
Suborder Arctolepida	III. *C. Patterson (1968)*
Suborder Brachythoraci	Class Holocephali
Order Phyllolepida	Order Chimaeriformes
Order Ptyctodontida	Suborder Chimaeroidei
Order Antiarcha	Suborder Squalorajoidei
Systematic position uncertain:	Suborder Myriacanthoidei
Palaeospondylus, Palaeomyzon	Suborder Menaspoidei
	Suborder Cochliodontoidei
II. *Stensiö (1963a)*	Suborder Helodontoidei
Class Elasmobranchiomorphi	Order Chondrenchelyiformes
Subclass I. Elasmobranchii,	Order Edestiformes
Cratoselachii, Cladoselachii,	
Ichthyotomi, Bradydonti	

(1958), Ørvig (1960, 1962), and Westoll (1962) have made a convincing case for this association. Ptyctodontid arthrodires and the Holocephali show many common features. A major problem, however, arises from the fact that the history of the Holocephali is very poorly known and the morphology of some of the supposed arthrodiran ancestors is only partly shown by a few well-preserved specimens.

The probable relationship of arthrodires and holocephalians, plus many additional characters that he has described and listed in detail, led Stensiö (1963a) to the classification that is quite different from Romer's (1966a) recent classification (see Table 8, I and II).

Of further interest is the suggestion of Jarvik (1968a and earlier) that the Dipnoi and Holocephali have some features in common, placing these supposed bony fishes somewhere in the area of the Elasmobranchiomorphi. C. Patterson (1965, 1968), however, held to the view, expressed by many earlier, that the Holocephali and Bradydonti are very close, and thus arrayed related orders and suborders in a very different relationship as shown in Table 8, III.

SUBDIVISIONS OF THE PLACODERMI

For the most part, recent efforts at classifications have recognized the placoderms as a coherent group, including the arthrodires and antiarchs as basic divisions. The antiarchs are a small group and pose few problems. The arrangements of the arthrodires, however, have not been stabilized. Some classifications have taken conditions and trends of limited parts of the skeletons as the basis for phyletic relationships. One of these, that of E.I. White (1952), was based on the concept that reduction of the shoulder was a basic trend. Reflections of this point of view are found in most recent classifications.

Stensiö, who has done monumental work on this group (see especially Stensiö, 1959, 1963a), has proposed a subdivision based on the character of the pectoral fin and the associated shoulder girdle. Three basic types of fins, two with subdivisions are recognized:

> Holosomactidial
> Euholosomactidial
> Apo-holosomactidial
> Merosomactidial
> Eumerosomactidial
> Apo-merosomactidial
> Monomesorhachic

The holosomactidial fin is presumed to be primitive. The first derived type, euholosomactidial, occurs in most of Stensiö's Pachyosteomorphi (see Table

7, IV). The second type has not been confirmed in the placoderms, although an indication of its development has been found in some of the pachyosteomorphs. It approximates the condition in some sharks. The merosomactidial fin characterizes the other arthrodires, with eumerosomactidial fins present in the majority and highly specialized in Acanthothoraci, Stensiöellida, Ptyctodontida, and possibly Macropetalichthyida. It is also found in some sharks and primitive actinopterygians. The monomesorhachic type is present only in antiarchs and is highly specialized.

The classification resulting from application of criteria of these fin types, along with other characters, is quite distinct from that proposed by Stensiö on other grounds and those used by other students. It is summarized in Table 7, IV. At the time of writing, Stensiö's studies of the arthrodires were continuing. Part of his detailed analysis of the arthrodire heads has been published (Stensiö, 1963a), and it suggests that additional modifications of the classification will be made as more information is obtained.

Comparisons of the classifications of Obruchev and Stensiö point up many of the problems that beset efforts to arrange such extinct groups in arrays that express their presumed phylogenies. Stensiö's two main units, under Euarthrodira, are based on criteria of fins and shoulder girdles. Obruchev has used a variety of characters, from various earlier studies including his own, and has stressed geological succession. The greatest differences are found in assignments to the "primitive" groups, Stensiö's Aspinothoracidi and Obruchev's Rhenanida. The orders of the Rhenanida of Obruchev are found scattered under the Spinothoracidi of Stensiö both as central and specialized groups. Stensiö's base group, Pachyosteomorphi, is an order of the Euarthrodira of Obruchev (order Pachyosteida).

Except for some major grouping differences and assignments of orders to the "primitive" group, there are resemblances in the classifications. The one other major difference relates to the grouping of "advanced" orders into two superorders by Obruchev and their inclusion under a single heading, on the basis of fin structure, by Stensiö.

PERSPECTIVES

Although the major structure of the Placodermi, as a group of extinct fishes, seems moderately secure, and the two major subdivisions, Arthrodira and Antiarcha, have firm foundations, the relationships of the placoderms to other groups of fishes and the internal organization of the arthrodires probably will continue to undergo modifications.

Recent years have seen a great increase in interest and study of the fishes of the Silurian and Devonian by students of many countries. Intensive field work has brought to light many new specimens of previously unknown morphology. Special studies of Mississippian fishes and of some from limited sites in the Pennsylvanian are adding to this fund of information. Coupled with the new materials has been a new conceptual freedom in the assessment of relationships, resulting in exploration of the potentials of relationships that could not have been accepted under the more restrictive concepts of earlier times. In addition, as well, divergences in the evaluation of criteria have arisen, especially with regard to the use of morphology as an exclusive basis for determining relationships, in contrast to use of stratigraphic and ecological information as well as morphology.

These factors suggest that a consensus on classification within the arthrodires and in relationships of the placoderms to other groups of fishes will not be reached in the near future. Widespread reorganizations involving placoderms, acanthodians, holocephalians, chondrichthyans, and even some of the bony fishes are to be expected in the next decade or so.

4 Class Acanthodii

POSITION IN THE SUBPHYLUM

Acanthodians form a readily recognizable group of extinct fishes. The most evident character is the presence of spines along the leading edges of both the paired and median fins, except the anal. Minute ganoid scales are found in all but the late members of the group. The acanthodians range from late Silurian into the Permian, and most of the detailed knowledge of the group has come from Permian representatives, *Acanthodes* in particular. Although the group itself is reasonably homogeneous, with few problems of assignment of well-preserved forms, its position relative to other major groups of fishes has posed problems and continues to do so. Some of these were taken up in the preceding chapter because acanthodians have been associated with the placoderms in the class Aphetohyoidea. They have also been placed with the Chondrichthyes (see, for example, Romer, 1933) and have been considered to represent the ancestors of Osteichthyes, having recently been tentatively placed in this class (Romer, 1966*a*).

Recognition of acanthodians as a separate class is to some extent an acknowledgment of lack of understanding of their actual relationships. The scope of radiation is much smaller than that of most classes of vertebrates as conceived by the majority of students, although in Berg's classification of 1955 classes are multiplied far beyond the numbers recognized by his contemporaries and several of his classes do have limited radiation patterns that do not differ greatly in scope from those of the acanthodians.

Fishes of this group were first described in the middle of the century and placed, on the basis of the scales, as ganoid fishes, thus being associated with the Osteichthyes as now understood. Later they were called elasmobranchs, although the presence of both dermal and skeletal bone technically rules them out of the Chondrichthyes as usually defined. They were sometimes called "bony sharks." The current tendency is to consider the acanthodians either as forming a separate class or to place them with the Osteichthyes, with the recognition that this may be a temporary position that could be altered as more information becomes available.

Active restudy of the nature and meaning of the visceral skeleton, following on Watson's earlier studies, has been under way for the last 20 years. The net effect has been to move the group closer to the Osteichthyes, as expressed in Romer's 1966 classification. It is important to understand, however, that much of the study pertains to the Permian genus *Acanthodes*, with only minor additions from earlier and presumably more primitive genera.

Nielsen (1949) made a strong case for resemblances of acanthodians and paleoniscoids, stressing in particular the nature of jaw suspension and jaw movement. In a series of publications Miles (1964, 1965, 1968) followed this work with a detailed analysis of structure and function. As he has shown, clear resemblances do exist between elasmobranchs and acanthodians, as had been stressed earlier on the basis of the endocranium, gills, and fins. However, he concluded that resemblances when viewed in detail were more superficial than had been recognized and that they were partly convergent and not phylogenetically significant. In particular he concluded that the articulations of the palatoquadrate in the otic region and with the basipterygoid process were typically paleoniscoid.

Recently, however, Nelson (1968) made a restudy of the gill arches of acanthodians, to reach a very different conclusion. He found little agreement between the acanthodians and primitive teleostomes and concluded that what resemblances there were expressed primitive features similar to those of the elasmobranchiomorphs. He considered the presence of gill rakers on both the dorsal and ventral portions of the hyoid in *Acanthodes* as particularly indicative of a primitive state. His conclusion was that the closest resemblances were with the primitive elasmobranchiomorphs.

It is thus not clear at present how these divers resemblances are to be evaluated with respect to phylogenetic position. Placement in a separate class, while avoiding commitment to one or another expression of relationships, tends to obscure the morphological resemblances to members of other classes and may seem, in the framework of classification, to offer a solution to problems for which there is in fact no current solution. Although it makes for desirable tidiness, the masking of problems of association seems a somewhat unjustified sacrifice. If it is recognized as an interim procedure, one that is subject to ready modification once there is greater insight into relationships, no harm is done. However, there is a very strong tendency, as witnessed by many cases in the past, for a classification to become fixed and by weight of its fixation, not intended by the original classifier, to become authoritative and to mold studies rather than merely being a vehicle of expression of thought as of a particular stage of development.

INTERNAL RELATIONSHIPS OF THE CLASS

Within the acanthodians there exists an array of some 25 to 50 or so genera, depending on the degree of credence given to many very incompletely preserved specimens. Fin spines, which preserve well, often suggest acanthodian affinities of their otherwise unknown possessors, but are not in themselves definitive of such relationship. It is these that have caused many of the problems. The better known genera differ in many ways and present a fairly evident evolutionary sequence viewed in the framework of the geological record. The degree of ossification is an important characteristic, with both internal and external ossification tending to become reduced in the course of evolution of the group. The number of dorsal fins, the number of fin spines intermediate between the pectoral and pelvic fins, and the scales serve as the principal characters for classification. Using such criteria, Berg (1955) recognized seven orders, Novitskaya and Obruchev (1964) recognized only four, and Romer (1966a), three. These classifications are summarized in Table 9.

They are in general similar, but differ in some important details. One point of importance is the weight given to the presence of one or two dorsal fins. Novitskaya and Obruchev and Romer group all genera with a single dorsal fin into one order, Acanthodida, or Acanthodiformes. Berg, however, regards the mesacanthids as more primitive, in spite of this feature. His classification puts more weight on the condition of the operculum. In both *Mesacanthus* and *Climatius* the operculum is not a fully developed gill cover and is thus considered primitive. The number of intermediate fin spines between the pectoral and pelvic fins enters into both classifications, with the greater number generally being considered more primitive. The degree of ossification of the endocranium, the extent and form of scales, and the form of elements of the upper and lower jaws all play roles. These are in large part used as characters for assignment without any more than superficial attention to their functional significance. The status as primitive or advanced is largely based on the time of appearance in the geological record and the associations of characters in the earliest genera.

TABLE 9   Classification of the Acanthodii

I. *Berg (1955)*
  Class Acanthodii
    Order Climatiiformes
    Order Mesacanthiformes
    Order Ischnacanthiformes
    Order Gyracanthiformes
    Order Diplacanthiformes
    Order Cheiracanthiformes
    Order Acanthodiformes

II. *Romer (1966a)*
  Class Osteichthyes
    ?Subclass Acanthodii
    Order Climatiiformes
      Family Climatiidae
      Family Diplacanthidae
      Family Gyracanthidae
    Order Ischnacanthiformes
      Family Ischnacanthidae
    Order Acanthodiformes
      Family Mesacanthidae
      Family Acanthodidae

III. *Novitskaya and Obruchev (1964)*
  Class Acanthodii
    Order Diplacanthida
      Family Climatiidae
      Family Diplacanthidae
    Order Ischnacanthida
      Family Ischnacanthidae
    Order Gyracanthida
      Family Gyracanthidae
    Order Acanthodida
      Family Mesacanthidae
      Family Cheiracanthidae
      Family Acanthodidae

Much of the difficulty in relating the acanthodians to other major groups of fishes comes from limitations on information on internal structures among the more primitive representatives. To some degree this has hampered efforts at internal classification as well, but for the most part the more superficial features, of little use in broad classification problems, have served to form a workable array that seems to express evolutionary pathways quite well.

5 Class Chondrichthyes

THE RECORD AND PROBLEMS OF CLASSIFICATION

The Chondrichthyes, or cartilaginous fishes, are abundant today and have an extensive fossil record. Thus they pose problems of classification that differ sharply from those considered to this point for extinct or essentially extinct groups. They compare more with the Osteichthyes, amphibians, reptiles, and mammals. However, the fossil record of the Chondrichthyes, though of great duration, is by and large unsatisfactory for an understanding of the phylogeny and classification of its constituents. The origin is essentially unknown, with possible but unidentified sources among placoderms and/or acanthodians. As a result of this isolation they form a coherent group whose composition is readily defined. The major problem that currently exists relates to the position of the chimeras, Holocephali, which have been variously associated with the Chondrichthyes proper and with the placoderms, as already discussed. The possible broadening of relationships by such associations has led to the use of a more comprehensive term, Elasmobranchiomorpha, to include shark and sharklike fishes, some of which do not have the chondrichthyan character of the absence of bone.

The problems of classification arise from inadequate preservation of the cartilaginous endoskeleton and the fact that the exoskeleton, with rare exceptions, consists only of dermal scales, or denticles, and spines. Cartilage seldom preserves well unless it has been calcified. Calcification in general, and particularly as it applies to the vertebrae, is predominantly a rather advanced feature among the Chondrichthyes, although the Paleozoic shark *Xenacanthus*

(*Pleuracanthus*) is heavily calcified. Occasionally, however, cartilaginous structures are well preserved, and in some instances the full body form, including impressions of the integument, is found. Even when preserved, cartilage does not lend itself readily to the usual methods of preparation and study. The use of X-rays has proven of great value for studying shark remains, but until now too little has been done to provide evidence over more than a limited range of specimens.

Much of the record until recently has consisted mainly of teeth, spines, and denticles, with little association. The specimens known from other remains are few (although some are excellent), and they provide a very spotty sampling of the total array of extinct cartilaginous fishes. Teeth, spines, and denticles, especially when not in association, are not generally reliable phylogenetic indicators. Chondrocrania, visceral cartilages (both in the gill apparatus and as jaws and jaw supports), axial structures, and fins are of greater value for determining relationships. Most of the characters used at higher categorical levels are from these parts of the skeleton. With respect to such features the sampling of the presumably numerous phyletic lines existing from the Devonian to the present is extremely irregular. It is inadequate even for recognition of particular lines for any long period of time and for the assessment of relationships between those lines that can be discerned.

The greatest hope of remedying what is currently a nearly chaotic condition in the classification of the Chondrichthyes seems to lie in the use of X-ray techniques for studying specimens encased in rock that cannot be removed without excessive damage or destruction. The work of Zangerl (1966) on *Edestus* shows the potential of this kind of work. Fortuitous finds, of course, will continue to provide spot information on one genus or another, but these cannot lead to anything approaching the coverage that phylogenetic synthesis and sound classification demand.

Classifications of the modern and the extinct Chondrichthyes, especially those of the Paleozoic, have proceeded along more or less independent courses, with only occasional efforts at integration, over much of the history of this group. Most of the efforts to establish relationships prior to the last 30 or 40 years were rather casual and not based on serious comprehensive studies of the whole array of cartilaginous fishes. From about 1930 on increasing attention has been paid to synthesis, but with generally rather indifferent success.

As a result rather different bases of classification have sprung up between students of modern and extinct sharks. This has contributed to difficulties in bringing the two together. In spite of recent efforts, there has been very little success in establishing a satisfactory overall classification of the Chondrichthyes. At present the ancient groups, mostly Paleozoic and very early Mesozoic, cannot be satisfactorily related to moderns either broadly by relationships based on general morphological trends or more narrowly by actual tracing of phyletic lines. All arrangements that take into account both fossil and recent sharks are based on less than sufficient evidence and rest on speculation for many parts of the structures. A number of such classifications are examined later in this section. First, classifications of modern Chondrichthyes, with special attention to the kinds of characters used in the classifications, are considered.

CLASSIFICATIONS OF MODERN CHONDRICHTHYES

Classification of present-day cartilaginous fishes has had a fair degree of stability over a long period of time, with differences pertaining to the placement and composition of some relatively small groups that are either primitive or aberrant. Below the highest levels of subdivision, the differences tend to be expressed by use of small, discrete suites of characters. The classifications are largely monothetic, with suitable keys, and with relatively little phylogenetic content.

One of the earliest of the "modern" classifications was developed by Müller and Henle (1841). The dorsal fins were used as a principal basis for familial association. Subsequently Hasse (1879) based a classification on vertebral structure, and this has continued to have an influence to the present time. Gill (1893) made use of the nature of the attachment of the palatoquadrate to the skull, and in this he contributed a feature that has played a prominent role, the nature of jaw suspension. Huber (1901) relied heavily on fin structure, especially the "stem joints," or myxopterygia.

Later classifications tended to be in large part based on several characters, mostly those that had been introduced by the earlier workers, but with differences in emphasis and introduction of some others. Regan (1906) produced an elaborate classification, with emphasis on the pectoral fins, skeleton, and rostral cartilages; Garman (1913) emphasized teeth, nasal structures, and the position of the last gills; Leigh-Sharp, in a series of papers from 1920 to 1926, treated groups singly with particular attention to the nature of the claspers in males. From this he developed a classification.

E. G. White (1937) reviewed the history of classification and developed a detailed classification based on what she considered phylogenetic characters, eliminating as physiological and thus less useful a number that had been used before. She attempted to detect the characters of broad adaptive value, which had implications over a wide range, as distinct from those that tended adaptively, to appear in different groups with somewhat similar life habits, Vertebral structures, the degree and pattern of calcification, rostral cartilages, pectoral-fin structures, and to some extent the nature of the intestinal spiral

TABLE 10 Classifications of Modern Chondrichthyes

I. *E. G. White (1937)*

Class Chondropterygia
  Subclass Plagiostomi
    Superorder Antaceae
      Order Hexanchaea
        Family Chlamydoselachidae
        Family Hexeptranchidae
      Order Galea
        Suborder Isurida
          Family Orectolobidae
          Family Charchariidae
          Family Rhineodontidae
          Family Scapanorhynchidae
          Family Vulpeculidae
          Family Isuridae
          Family Cetorhinidae
        Suborder Carcharhinida
          Family Catulidae
          Family Halaeluridae
          Family Atelomyeteridae
          Family Triakidae
          Family Geleorhinidae
          Family Sphyrinidae
      Order Heterodonta
      Order Squalida
        Suborder Squaloidea
          Family Squalidae
          Family Echinorhinidae
          Family Scymnorhinidae
          Family Pristiophoridae
          Family Rhinidae
    Superorder Platosomeae
      Order Narcobatea
      Order Batea
        Suborder Rhinobatoidea
          Family Rhinobatidae
          Family Pristidae
          Family Discobatidae
        Suborder Rajoidea
          Family Rajidae
        Suborder Dasybatoidea
          Family Dasybatidae
          Family Potomytrygonidae
          Family Myliobatidae
          Family Mobulidae
  Subclass Holocephali
    Order Chimaerea
      Family Callorhynchidae
      Family Chimaeridae
      Family Rhinochimaeridae

II. *Bigelow and Schroeder (1948–1953)*

Class Chondrichthyes
  Subclass Elasmobranchii
    Order Selachii
      Suborder Notidanoidea
        Family Hexanchidae
      Suborder Chlamydoselachoidea
      Suborder Heterodontoidea
      Suborder Galeoidea
        Family Carchariidae
        Family Scapanorhynchidae
        Family Allopiidae
        Family Orectolobidae
        Family Rhineodontidae
        Family Pseudotriakidae
        Family Triakidae
        Family Carcharhinidae
        Family Sphyrhinidae
        Family Scyliorhinidae
      Suborder Squaloidea
        Family Squalidae
        Family Dolatiidae
      Suborder Pristiophoroidea
      Suborder Squatinoidea
    Order Batoidea
      Suborder Pristoidea
      Suborder Rhinobatoidea
        Family Rhynchobatoidae
        Family Rhinobatidae
      Suborder Torpedinoidea
      Suborder Rajoidea
        Family Rajidae
        Family Acanthobatidae
      Suborder Myliobatoidea
        Family Dasybatidae
        Family Gymnuridae
        Family Urolophidae
        Family Myliobatidae
        Family Rhinopteridae,
        Family Mobulidae
  Subclass Holocephali
    Order Chimaerea
      Family Chimaeridae
      Family Rhinochimaeridae
      Family Callorhynchidae

valves, were used in formation of her major and intermediate groups. The resulting classification (Table 10, I) was elaborate in details of ranking and in terminology, both of which differed materially from those in earlier classifications.

Following E. G. White's classification and taking advantage of very extensive studies of new materials, Bigelow and Schroeder (1948–1953) arrived at a comprehensive classification of modern cartilaginous fishes (Table 10, II). The classifications of E. G. White and of Bigelow and Schroeder can be for the most part reconciled with relatively little difficulty. The subclass levels are the same. In general the suborders of Bigelow and Schroeder are comparable to the orders of White. Major differences are found in the association of the chlamydoselachoids and hexanchoids by White, the positions accorded the heterodontoids, and the great subdivision of White's squaloids by Bigelow and Schroeder. Differences of opinions on the positions of *Chlamydoselachus*, *Heterodontus*, and *Hexanchus* are found throughout classifications, both those concerned with moderns only and with moderns and ancients. The differences in the relationships of the sharks generally thought of as squaloids also are typical of a wider array of classifications.

The characters as used by Bigelow and Schroeder at different categorical levels present an insight into what has been considered important among moderns and the extent of taxonomic information that may be used in studies of relationships of modern cartilaginous fishes to extinct ones. They may be summarized as follows:

SUBCLASS LEVEL

Structure	*Elasmobranchii*	*Holocephali*
Gills	Five to seven pairs	Four pairs
Dorsal fin	Rigid	Erectile
Palatoquadrate	Not fused to cranium	Fused to cranium
Rostral cartilage	Fused to cranium	Not fused to cranium
Notochord	Constricted	Not constricted
Pelvis	Two halves fused	Halves separate
Cloaca	Present	Absent
Prepelvic and/or frontal tenacule	Absent	Present

ORDINAL LEVEL (Subclass Elasmobranchii Only)

Structure	*Selachii*	*Batoidea*
Gill apertures	Lateral	Ventral
Pectoral fin	Not attached to head or anterior to gills	Attached to head
Eyelid	Free	No free eyelid

SUBORDINAL LEVEL (Order Selachii) (in Key Form)
I. Anal fin present
  A. Six or seven gill apertures
    1. Margin of first gill does not cross throat; upper and lower teeth not alike toward center of mouth
        Notidanoidea
    2. Margin of first gill continuous across throat; teeth similar toward center of mouth
        Chlamydoselachoidea
  B. Five gills
    1. Dorsal fin preceded by spine; teeth unlike toward center of mouth
        Heterodontoidea
    2. No spine in front of dorsal fin; teeth in different parts of mouth not basically different
        Galeoidea
II. No anal fin
  A. Snout of moderate length, no lateral teeth on snout
    1. Trunk subcylindrical, anterior of pectoral not overlapping gills
        Squaloidea
    2. Trunk flat, eyes dorsal, fin overlaps gills.
        Squatinoidea
  B. Snout greatly elongated; lateral teeth
    1. Pristiophoroidea

The characters in the form outlined after Bigelow and Schroeder represent a typical example of key arrangement based in large part on a monothetic concept of classification. Most of the characters, being skeletal, have potential value for classification of extinct forms. Some are found well preserved in Cenozoic sharks, and in very occasional specimens from earlier in the record. The suite of characters used by Bigelow and Schroeder, plus some used by other students of modern sharks, are the principal ones used in classifications of extinct sharks as well. Calcification of vertebral centra, used by E. G. White but not found in the key of Bigelow and Schroeder, has importance. Of the problematic hexanchids and heterodontids, both of which have primitive features, the latter have well-calcified centra and a constricted notochord, presumably advanced features, and the former have weak calcification and little notochordal constriction, presumably primitive features.

MODERN AND EXTINCT FORMS

The primary problems in determination of relationships of the modern and ancient sharks do not depend on a need for use of different suites of characters, as is the case in some other groups. They stem simply from the fact, already stressed, that preservation of the fossils is not good over a wide enough range to permit judgment of the meaning and utility of these characters with respect to adaptive levels and phylogenetic lines. Not enough can be told of the probable phylogenetic relationships of moderns on the basis of the information that they provide, and much less can be determined concerning the relationships of extinct forms to each other and to moderns from the scant fossil record. The results are wide discrepancies between various classifications that have attempted to bring ancients and moderns into a single array.

Five classifications are shown in Table 11. These have been selected to cover the period from 1930 to 1964 and to represent the range of diversity and the types that have been proposed during that time. In addition, although not shown in any of the listed classifications, a grouping into Elasmobranchiomorpha, broadening the base of overall relationships, has come into some prominence. In each of the listed classifications, if only the moderns are considered, it is possible to see at least a reasonable resemblance to the classifications of E. G. White and Bigelow and Schroeder. The differences arise once again in the placement of *Heterodontus* and *Chlamydoselachus*, presumed primitive forms, and *Pristiophorus*. These classifications tend to keep the moderns and their immediate relatives, which go back into the Jurassic in most instances, distinct from the ancients, whose roots lie in the Paleozoic and many of which became extinct by the end of that time as far as the record shows. The distinctness holds in particular for the sharks proper, whereas a closer association is indicated in all but one of the classifications, that of Goodrich, between the Holocephali of the Mesozoic and Cenozoic, and the Bradyodonti of the Paleozoic. None of the listed classifications associates the Holocephali and placoderms, as taken up earlier on pages 215–216.

These classifications, especially in the ways that ancient lines are related to modern stocks, show the wide diversity of opinion that exists with reference to the class. As noted, the great majority of remains of fossil sharks consist of teeth, spines, and denticles. Teeth have been used at rather low levels of classification of moderns, but they have not proved particularly valuable for drawing broad lines. Yet, as stated by Moy-Thomas in the text accompanying his classification, if teeth are much of what is available, they are what must be used. In one classification of the elasmobranchs, that of Glickman, the histology of teeth is used as a basis for major subdivision. The results in terms of broad organization are evident in Table 11, V, and will be commented on specifically later in this section.

The Holocephali–Bradyodonti association illustrates the effects of use of teeth as a primary criterion. The Bradyodonti represent a large group of Paleozoic fish, presumably Chondrichthyes, which is based in large part on

TABLE 11 Classifications of Chondrichthyes

I. *Goodrich (1930)*
  Class Pisces
    Subgrade Chondrichthyes
      Subclass Elasmobranchii
        Order Selachii
          Group I. Notidani
          Group II. Division A:
            Suborder Heterodonti
            (including *Acrodus,*
            Cochliodus, Edestus,
            Helodus, Heterodontus,
            Hybodus)
          Division B:
            Suborder Scyllioidea
            Suborder Squaliformes
            Suborder Rajiformes
        Order Holocephali
          (including modern chimeras,
          Myriacanthus, Ptyctodus,
          Squaloraja)
        Order Pleuracanthodei
      Subclass Cladoselachii
      Subclass Acanthodii
      Subclass Coccosteomorphi

II. *Romer (1945a)*
  Class Chondrichthyes
    Subclass Elasmobranchii
      Order Cladoselachii
        Family Cladoselachidae
        Family Ctenacanthidae
      Order Selachii
        Suborder Hybodontoidea
          Family Coronodontidae
          Family Tristychidae
          Family Hybodontidae
          Family Edestidae
        Suborder Heterodontoidea
        Suborder Notidanoidea
        Suborder Galeoidea
        Suborder Squaloidea
      Order Batoidea
      Order Pleuracanthodii
    Subclass Holocephali
      Order Bradydonti
        Family Cochliodontidae
        Family Petalodontidae
        Family Copodontidae
        Family Psammodontidae
      Order Chondrenchelydi
        Family Chondrenchelydidae

III. *Moy-Thomas (1939b)*
  Class Chondrichthyes
    Subclass Elasmobranchii
      Division Pleuropterygii
        Order Pleuropterygii
          Suborder Cladoselachii
          Suborder Ctenacanthii
        Order Protoselachii
          Suborder Hybodonti
          Suborder Tristychii
        Order Euselachii
          Suborder Notidani
          Suborder Heterodonti
          Suborder Scyllioidei
          Suborder Raioidei
        Order Pleuracanthodei
      Division Bradydonti
        Order Eubradydonti
          Suborder Cochliodonti
          Suborder Holocephali
          Suborder Petalodonti
          Suborder Psammodonti
          Suborder Copodonti
          Suborder Edestidi
        Order Chondrenchelydi

IV. *Berg (1940, 1955)*
  Class Elasmobranchii
    Subclass Cladoselachii
      (Pleuropterygii)
        Order Cladoselachiformes
    Subclass Xenacanthini
      Order Xenacanthiformes
    Subclass Selachii (Euselachii,
      Plagiostomi)
        Superorder A: Selachioidei,
          Pleurotremata
          Order Heterodontiformes
            Suborder Ctenacanthoidei
            Suborder Heterodontoidei
          Order Hexanchiformes
            (Notidanoidei)
            Suborder Chlamydoselachoidei
            Suborder Hexanchoidei
            Suborder Lamnoidei (Isuridae)
            Suborder Scyliorhinoidei
            (Archarnida)
          Order Squaliformes
            (Tectospondyli)
            Suborder Squaloidei
            Suborder Squaltinoidei

TABLE II   (continued)

IV. *Berg (1940) (cont.)*

Order Pristiophoriformes
Superorder B: Skati, Batoideia,
Platysoma, Hypotremata
Order Rajiformes
Suborder Rhinobatoidei
Suborder Trygonidei
Order Torpediniformes
(Narcobatoidea)
Class Holocephali
Subclass Chondrenchelyes
Subclass Chimaerae
Order Chimaeriformes
Family Cochliodontidae
Family Menaspidae
Family Petalodontidae
Family Janassidae
Family Psammodontidae
Family Copodontidae
Family Myriacanthidae
Family Chimaeridae
Family Helicoprionidae
Family Rhinochimaeridon-
tidae

V. *Glickman (1964)*

Class Chondrichthyes
Subclass Elasmobranchii
Infraclass Orthodonti
Superorder Cladoselachii
(Pleuropterygii)
Order Cladoselachida
Order Cladodontida
Superorder Xenacanthini
(Ichthyotomi)
Order Xenacanthida
Superorder Polyacrodonti
Order Polyacrodontida
Superorder Chlamydoselachii
Order Chlamydoselachida
Superorder Carcharhini
Order Hexanchida
Suborder Hexanchoidea
Suborder Heterodontoidea
Order Squatinida
Suborder Echinorhinoidea
Suborder Squaloidea
Suborder Ginglyostoma-
toidea

V. *Glickman (1964) (cont.)*

Suborder Squatinoidea
Suborder Pristiophoroidea
Suborder Rajoidea
Superfamily Rhinobatoidea
Superfamily Pristioidea
Superfamily Torpedinoidea
Superfamily Rajoidea
?Superfamily Myliobatoidea
Order Carcharhinida
Family Palaeospinacidae
Family Scyliorhinidae
Family Triakidae
Family Carcharhinidae
Family Sphyrhinidae
Infraclass Osteodonti
Superorder Ctenacanthini
Order Ctenacanthida
Order Tristychiida
Superorder Hybodont
Order Hybodontida
Superorder Lamnae
Order Orthacodontida
Order Odontaspida
Superfamily Odontas-
pidoidea
Superfamily Isuroidea
Superfamily Scapano-
rhynchoidea
Superfamily Anacoracoidea
Subclass Holocephali
Order Bradydonti
Family Cochliodontidae
Family Menaspididae
Family Petalodontidae
Family Psammodontidae
Family Copodontidae
Family Orodontidae
Family Helicoprionidae
Family Edestidae
Family Pseudodontich-
thyidae
Order Chimaerida
Family Squalorajidae
Family Myriocanthidae
Family Edaphodontidae
Family Callorhynchidae
Family Rhinochimaeridae
Family Chimaeridae

teeth and relatively little else. These are crushing teeth with characteristic histology (see Radinsky, 1961). Just what most bradyodonts were, beyond their teeth, is not known, and it is possible that some of them were not cartilaginous fish at all. For a few animals part of the skeleton is known. In cochliodonts the cranial structure is preserved and indicates a possible relationship to the chimeras. Association of this one family with the Holocephali provides the vehicle that, in large part, has brought the others into this relationship. Similarly, however, the structure of the ptyctodont placoderms suggests affinities between *this* group and the Holocephali.

Recently much new information on the family Edestidae has come from Zangerl (1966). Edestids have been commonly placed with bradydonts, although not exclusively, and this was the determination reached by Nielsen. The conflicting, or possibly conflicting, evidence of Zangerl and other unpublished information from the fauna have brought him to the opinion that at present no conclusive statements on the broad relationships of the Paleozoic sharks can be made. There is some information beyond teeth for a few other genera—for example, *Janassa*, classified as a petalodontid by some; *Menaspis*, family Menaspidae; and the very odd creature *Chondrenchylys*, whose position is very uncertain. None of these gives clear information on placement. Added data may confirm the relationships of some bradydonts to holocephalans, as suggested by C. Patterson (1965), or they may throw one or more groups completely out of this range. This same general situation applies to many of the shark types of which only teeth are known.

ANALYSIS OF CLASSIFICATION

For a brief analysis of the nature of the different classifications listed in Table 11, it is convenient to look at the common features and the particular variants found in one or another arrangements. Most current classifications recognize the cartilaginous fishes as a coherent group. Acanthodians and placoderms are generally excluded, in contrast to older classifications (e.g., that of Goodrich in the table) and some recent ones in which ptyctodont-chimera relationships are included. The class is based on the absence of internal ossification—although this is not, of course, a completely diagnostic feature, being matched in some Agnatha, some Osteichthyes, and some Placodermi. The level of organization is different from that of the Agnatha, although this does not certainly apply in the case of the soft-bodied thelodonts. Differences from the bony fishes are immediately evident in comparisons of the exoskeletons. Sharks possess toothlike placodont scales, or denticles, in some cases with bone tissue at the base. In rare instances, as reported by Zangerl (1966), there may be fusion to form small dermal bone plates. There is no dermal bone comparable to that of the bony fishes and no ganoid scales such as are found in these fishes and in the acanthodians. Cases are rare in which there is any problem in the placement of reasonably well-known genera either as Chondrichthyes or non-Chondrichthyes.

Below the Chondrichthyes level, a class by most but a subgrade of the class Pisces by Goodrich, the usual practice is to recognize two major groups, usually subclasses, Elasmobranchii and Holocephali. In Table 11 only Goodrich's classification does not follow this pattern, and his work was cast in a framework that differed considerably from that of the classifications of the other four authorities. The basis of this subdivision is primarily that revealed in the morphology of moderns as specified, for example, by Bigelow and Schroeder. The resemblances of bradyodonts and chimeras, as discussed above, brings them into association, and the sharklike aspects of the other Paleozoic cartilaginous fishes is considered by many, but not all, to be a sound basis for their assignment to the elasmobranchs. The main differences in the classifications at this level are the categorical rank given the two major subdivisions. Berg, here as in most instances, places his divisions at a somewhat higher level than do the others.

The most troublesome elements are elasmobranchs. The simplest elasmobranch classifications are those of Romer and Moy-Thomas. They are similar in many respects and had common sources, with Romer's, which is later, depending in some major part on Moy-Thomas'. Romer's more recent classification of 1966 is little different. Berg's arrangements are basically similar, whereas Goodrich's older classification, made before the establishment of a number of relationships accepted by most after about 1935, differs in many respects.

The primitive nature of the cladoselachians is a feature of all of the classifications, and on this score there is little argument. Romer and Moy-Thomas agree in placing the cladoselachians and ctenacanths close together. Both are considered to be at the base-grade level, on the basis of features that pertain to paired and unpaired fins, jaw suspension, unconstricted notochord, lack of calcified centra, and the presence of the cladodont tooth. The teeth, which are abundant in the Paleozoic, have a high central cusp and lower lateral cusps and show these fishes to have been widespread and probably highly varied. The most important differences between the cladoselachians and ctenacanths relate to absence of an axis in the paired fins of the cladoselachians, the absence of spines in front of the median dorsal fins, and some aspects of jaw suspension. Berg places the two close to the heterodonts, but Glickman separates them widely.

A problem that is quite frequent in classification of fossils arises in the relationships of cladoselachians and ctenacanths. In both groups a few specimens are quite well known and give good indications of the structure.

Most, however, are very fragmentary, in this case predominantly teeth and spines. Assignments to one group or the other must be in large part based on these parts, which are far from satisfactory for separation or for assignment of forms to the respective categories. A result is that each category is likely to include arrays of forms that would not be associated were more known. Specifically, here it is clear that the ctenacanths are not a homogeneous group and even the genus *Ctenacanthus* includes specimens that are not at all closely related. *"Ctenacanthus" clarki* has many cladoselachian characters and clearly is at a cladoselachian level of organization. It has provided a considerable part of the information on which the concept of ctenacanths and their relationships to the cladoselachians are based. This "genus" and species are not properly *Ctenacanthus*, and the cladoselachian affinities of the group called ctenacanths, based on this shark, are highly misleading. This same problem will appear many times in discussions of other classifications. If *"Ctenacanthus" clarki* is removed, the other ctenacanths whose structure is well known appear to be closer to the hybodonts and to have only general resemblances to the more primitive cladoselachians. Woodward (1924, 1932) and Brough (1935) suggested hybodont affinities for the ctenacanths, and there seems little doubt that the Mississippian and later forms have attained a hybodont level, although this is not recognized in the older classifications.

The two groups are widely separated in Glickman's 1964 classification. He has recognized the histology of the teeth as the fundamental character for splitting the elasmobranchs. The result is that his arrangement differs widely from all others. One infraclass, Orthodonti, is characterized by teeth that Glickman describes as including a base of rhizodentine, a crown of orthodentine with an enamel cover, and a pulp cavity. The other infraclass, Osteodonti, has teeth composed of osteodentine, with irregular canals, no pulp cavity, peripheral dentine tubules, and an enamel outer layer. Figure 66 shows his interpretation of broad phylogeny based on this separation. The ctenacanths, hybodonts, and Lamnae represent a radiation long separated from all others. The advantage of classification on the basis of tooth structure is, of course, the fact that it is generally available and provides a criterion for placement in one of two major categories for most specimens. The disadvantage is that the use of a single feature employs only a very limited part of the information available. Unless the states of this character show a high correlation with states of the other features, so that they result in assignments commensurate with others, taken both one by one and together, the effectiveness as a basis for subdivision is open to question. If, as here, the structure involved is one that is likely to be subject to adaptive modification related to functional change, here change of diet and feeding, there is some a priori reason for questioning its effectiveness. Radinsky (1961) has given a detailed consideration of this broad problem.

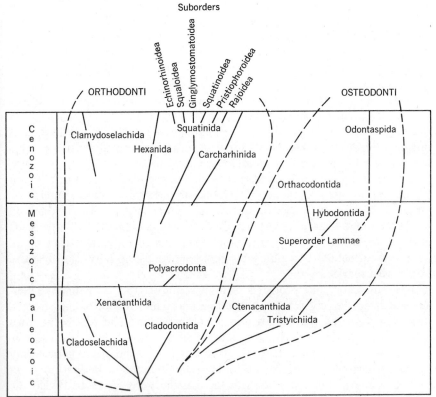

Figure 66. Suggested phylogeny for the Elasmobranchii, based on dental histology, as portrayed by Glickman. The envelopes enclose two major groups, Orthodonti and Osteodonti. To show the extent of radiation of the former suborders are entered under the Squatinida. No other order has such an extensive radiation. Glickman's classification and this phylogeny, which is based on it, are in sharp contrast with those of other students. (After Glickman, 1964.)

The associations and assignments in Glickman's classification do not correspond closely to those in others that are based on the use of as many characters as can be had from the materials. Although this does not indicate conclusively that it is a less natural classification, it does suggest that this possibility must be seriously entertained.

The position of another Paleozoic group, the xenacanths (or pleuracanths) is rather isolated in all of the classifications. These are specialized, predominantly freshwater sharks, abundant in the Carboniferous and the Permian, and unlike both their contemporaries and living sharks. Some xenacanth features—for example, jaw suspension and visceral structure—suggest a cladoselachian level of development. The teeth with high lateral cusps and a small central cuspule, the paired head fins, a high degree of calcification of the endoskeleton, and other special features, however, set them far apart. This generally is recognized in classifications, where the group is placed as a discrete order, usually as the last order of the elasmobranchs. Very little is known of the ancestry and they appear to have had no descendants.

Hybodonts, late Paleozoic and Mesozoic sharks, are based on the Triassic genus *Hybodus*. A categorical name derived from this genus figures prominently in most classifications, although in Berg's classification it is included in the heterodonts. Its usual position is intermediate between the ancient sharks, especially cladoselachians, and moderns. Moy-Thomas, for example, used the term "Protoselachii" for a group that includes this kind of shark, and Romer uses the term "Hybodontoidea" for a suborder under the Selachii. A fairly close kinship to the ctenacanths is generally suggested on one side and a relationship to heterodonts on the other. Both heterodonts and chlamydoselachians show some of the features considered to be hybodontoid, but neither has complete suites, tending also to have some features of modern sharks.

Here, as before, the problems of this group, or grade as it may be considered, stem largely from the fact that most information comes from a few well-preserved individuals and a host of teeth. The few well-known specimens are insufficient for the recognition of phyletic lines, because they cannot be clearly related to predecessors or to descendants. The necessity of hierarchical framework has led to the recognition of a hybodont category, but the patterns of organization in the classifications go beyond what the evidence justifies.

Recognition of the impossibility of making a realistic cladistic classification at present, prompted Schaeffer (1967c) to look at the elasmobranchs from the perspective of organizational grade levels. Since this throws the problems of classification of this group into perspective, a summary of his study will round out the comments on the classification of the elasmobranchs. He recognized three grades of organization: cladodont, hybodont, and modern. Each grade probably has several lines, separate phylogenetically, but following somewhat similar courses. These Schaeffer considered within the general framework of the De Beer concept of mosaic evolution, in which different parts of the organisms in different lines evolve at different rates, with respect to the organization levels. Thus any one line will have some features of older and more primitive, ancestral types, and some features of a more advanced type, those that typify its grade level. Grades, under this concept, do not, of course, have discrete definable upper or lower limits. Some genera may be recognized as just entering a grade, others as fully matured at the organizational level, and still others as retaining only some features of the grade as they move to a higher level of organization. Grade characters are broad adaptive features, and the particular members in a given grade may show a wide range of differences in narrower, more specialized adaptive features.

The levels, as recognized by Schaeffer, show some of the same features to be seen in the classifications that have been discussed. They permit more flexibility, since there is no intention to indicate specific taxonomic relationships. The broad adaptive features of the three grades may be summarized as follows:

CLADODONT GRADE

Examples: *"Ctenacanthus" clarki, Diademodus, Protacrodus, Denaea, Tamiobatis, Symmorium, Xenacanthus*

Broad adaptations:
1. Braincase with weak rostral region, strong subocular shelf, prominent postorbital process
2. Palatoquadrate with postorbital process articulated
3. Jaws with double articulation to chondrocranium
4. Prevalent "cladodont" dentition
5. Persistent, unconstricted notochord
6. Radials of pectoral and pelvic fins not divided, extend more or less to margin
7. Claspers present in conjunction with basal axis
8. Dorsal-fin spine absent (?secondarily) in some Dorsal fin(s) triangular, with broad base
9. Caudal fin long, unsegmented radials on hypochordal haemal rod, and epichordal radials
10. Body scales usually multiple (some single), each tooth or cusp with pulp cavity

From this base, it is thought, came a number of lines that passed through the hybodont grade. This is somewhat difficult to define, and, as Schaeffer points out, it would be possible to think of a primitive and advanced grade, with the hybodont merely a transition in which aspects of both are present. The grade, however, may be specified as follows:

HYBODONT GRADE

Examples: *Ctenacanthus* (excluding *"Ctenacanthus" clarki*), *Goodrichichthys, Tristychius, Sphenacanthus* (all late Paleozoic), *Lissodus, Hybodus* (Triassic)

Broad adaptations:[1]
1. Tribasal pectoral-fin skeleton, propterygia, mesopterygia, metapterygia

[1] They also retain many cladodont features.

2. Division and reduction of radial in paired fins
3. Possible first appearance of anal fin
4. Apparent loss of epichordal radials, division and reduction of hypochordal radials
5. ? Acquisition of haemal elements along the length of the notochord, and of ribs

The primary change from the primitive grade is in the fin structure, especially in the paired fins. This is found well developed in the Mississippian. In many features sharks at the hybodont level are close to the cladoselachian grade, as indicated in some of the classifications listed in Table 11.

MODERN GRADE

Representatives: The great majority of present-day sharks and their ancestors in the record extending back into the Jurassic; three general groups—the lamnoids, galeoids, and squaloids

Broad adaptations:
1. Hyostylic attachment of palatoquadrate to cranium complete, reduced postotic process
2. Shortened jaws, with protrusion mechanism
3. Calcified centra
4. Expanded neural and haemal elements
5. Fused pelvic plates
6. Single-cusped placoid scales

The origin of the modern grade is marked by the development of hyostylic jaw suspension, calcification of the axial skeleton, with acquisition of broad neural and haemal arches. A major feature that has been used variously in classifications of the sharks has been the nature of the attachment of the jaws to the chondrocranium. The change is from amphistylic to hyostylic in the elasmobranchs. This is brought out as a grade level feature by Schaeffer.

Heterodonts, *Hexanchus*, and *Chlamydoselachus* are intermediate in many respects between the hybodont and modern levels, each having some features of each. They are mosaics of moderns and intermediates and as such have posed problems of classification, as indicated in the systems in Table 11, and present in others as well.

The classification of the other major group, the Holocephali, relates the bradyodonts of the Paleozoic with the Holocephali, chimeras, of the Jurassic to Recent. The problems involved in this relationship have been noted before. They pertain primarily to the matter of relationships of Holocephali, bradyodonts, and placoderms.

In summary, it is clear that there is no satisfactory basis for classification of the Chondrichthyes and that none of the classifications can be considered anything more than temporary and highly provisional. All arrangements that take into account both fossil and living cartilaginous fish are beset with an insufficiency of vital information. Many parts of their structure remain largely conjectural.

6 Class Osteichthyes

THE CONCEPTS OF THE CLASS AND MAJOR DIVISIONS

Three large, coherent groups of fishes—actinopterygians, crossopterygians, and dipnoans—are generally grouped into a single class, Osteichthyes, or more casually associated as the bony fishes. This relationship was not clearly recognized in early stages of fish classification, and not all students agree with it at the present time; for example, Berg (1955) and Lehman (1966) have recognized the actinopterygians, crossopterygians, and dipnoans as separate classes. Even this procedure, however, does not necessarily deny the basic affinities that place these groups closer to each other than to other groups of fishes.

Most current assessments assume that the Osteichthyes form a coherent assemblage whose radiations stem from a common ancestral stock that had split off from other major radiations prior to the establishment of their basic adaptations. Although the source is unknown, it has been suggested, as discussed on pages 219–220, that it lay among the acanthodians. The evidence is not conclusive, and, as Jarvik (1968a) has emphasized in his contention that the dipnoans are remote from the others, the known acanthodians and the lungfishes have few fundamental features in common.

Wide differences existed between the major osteichthyan stocks when they first appeared in the record. This has led to the proposition that their beginnings were remote in time from the initial tangible records in the early and middle Devonian (see, for example, Jarvik, 1968b). Along with their fundamental and readily definable differences, however, the earliest members of the three groups do have some features in common. Schaeffer (1968), in a search for the features that are unique to Osteichthyes, noted as especially important the extensive dermal ossifications that cover or partially replace (functionally) the palatoquadrate in the jaw suspension and surround the primary mandible. The dermal parasphenoid may be unique—and the hypohyal, symplectic, and interhyal, or an element representing these, are not found elsewhere. The shoulder girdle is a structure unlike that of other fishes and pleural ribs are present only in Osteichthyes. Lepidotrichia are formed in a unique fashion in primitive forms.

These distinct features make it possible to define the class Osteichthyes on the basis of a few key characters. Fortunately they are structurally superficial and often preserved among fossil fishes. Little difficulty generally arises in assignment of well-preserved fossil specimens to one or another of the major groups or to the class, although this is not the case for isolated remains (e.g., scales). As for most groups, genera are usually placed in the major category, here the Osteichthyes, by association with some subgroup—for example, an order or family—that is itself considered osteichthyan. It is only in rare instances, involving early and primitive representatives, that the key characters come into play in the making of assignments.

Many other structural features are characteristically osteichthyan but not exclusive to the class. Both dermal bone and cartilage-replacement bone are primitively present, and the former is retained throughout. Dermal patterns of the skull found among the various Osteichthyes are not duplicated in other fish groups, but they are not consistent throughout the Osteichthyes. Scales, generally present, are either ganoid, cosmoid, or bony, the last formed by reduction of one of the other two types. They are primitively rhombohedral. Similar ganoid scales, however, are found among the acanthodians. Fins, both paired and median, are formed differently from those in other groups of fishes, but as in the case of the dermal patterns, they are not at all consistent throughout the different groups of Osteichthyes. Lungs or derived structures are consistently found among osteichthyans, but there is good evidence that lungs of similar nature were also present in the antiarch placoderms.

Similar circumstances exist for most of the features that can be considered characteristics of the bony fishes. Members of the class can be recognized by combinations of these characters, which are not found elsewhere, but also these combinations are not found throughout all of the Osteichthyes. The class, if it is accepted as such, is basically polythetic in its makeup, but the features noted by Schaeffer, while having a wide range of states, do make possible at least an operational monothetic characterization.

The general, although not universal, recognition that a radiation from a common base produced a reasonably coherent and recognizable category has not been matched by similar consensus on the relationships of the components.

That there are three distinct groups—lungfishes, lobe-finned fishes, and ray-finned fishes—is generally agreed. If, however, the technical names of Dipnoi (or Dipneustes), Crossopterygii, and Actinopterygii are applied to the three, as is often done, some problems of concept and nomenclature are introduced.

Before 1900 various dispositions were made of the little known lungfishes and lobe-finned fishes. Early in the current century the supposed remoteness of the lungfishes was indicated by Goodrich (1909), who abandoned Huxley's term "Crossopterygii" and divided the bony fishes into two groups, the Dipnoi and Teleostomi. The latter included both the Rhipidistia and Actinistia (Coelacanthini), the two groups of lobe-finned fishes, and the Actinopterygii. The term Teleostomi has persisted, with usage similar to that of Goodrich being found in Berg (1940). It also has been used as equivalent to Osteichthyes, as in the classification of Camp, Allison, and Nichols (1964). The term "Crossopterygii," earlier discarded by Goodrich, was used once again in his classification of 1930.

Opinion during the current century has swung partially over to the idea that lungfishes and lobe-finned fishes are rather closely related to each other, more so than is either to the ray-finned fishes. The early members are generally considered to be fairly close together. As the classification of Lehman (1966) shows (Table 12), not all students have agreed with this. The general concept of remoteness has been strongly supported by Jarvik. Lehman conceded that there are many common characters but felt that these were outweighed in importance by many that are distinct.

In recognition of a presumed relationship Romer (1937) proposed the term "Choanichthyes" to include crossopterygians, both rhipidistians and coelacanths, and the dipnoans. Later (1955) recognizing, as had been shown by Jarvik (1942), that some of the included fishes lacked choanae, Romer proposed the term "Sarcopterygia," referring to the fleshy nature of the fin lobe.

This led to a strong rebuttal, as a comment accompanying Romer's (1955) remarks, by Trewavas, White, Marshall, and Tucker, who proposed that the term "Crossopterygii" be retained in the sense proposed by Huxley with slight emendations, and as used by Regan (1929) and by Moy-Thomas (1939a) (see Table 12, III). The term so used includes both lungfishes and lobe-finned fishes and is equal in rank, as a subclass, to Actinopterygii.

Thus in recent years there have been many arrangements, some of which are listed in Table 12. Most are similar, differing mainly in terminology. In general, affinities between lungfishes and lobe-finned fishes have been recognized, and the somewhat separate position of the ray-finned fishes has been acknowledged. Usage thus becomes primarily a matter of preference, because there is no strong binding rule of priority at these taxonomic levels. Very recently, however, and not reflected in classifications, increasing doubts about the close relationships of the Dipnoi and Crossopterygii have emerged.

TABLE 12 Classifications of Osteichthyes

I. Regan (1929)

Class Pisces
Subclass Palaeopterygia
Order Archistia
(including paleoniscids,
platysomids, catopterids)
Order Belonorhynchia
Order Chondrostei
(including only modern forms)
Order Cladistia
(including Polypterus,
Calamoichthys)
Subclass Neopterygii
Order Platospondyli
(including semionotids,
macrospondylids, Amia)
Order Ginglymodii
(including Lepisosteus)
Order Halecostomi
(including Pholidophorus)
Order Isospondyli
and the other "teleost"
orders recognized by Regan
Subclass Crossopterygii
Order Rhipidistia
Order Actinistia
Order Dipneusti

II. Goodrich (1930)

Class Pisces
Subgrade Osteichthyes
Subclass Dipnoi
Subclass Teleostomi
Division Crossopterygii
Suborder Haplistia
Suborder Rhipidistia
Suborder Coelacanthini
Division Actinopterygii

III. Moy-Thomas (1939a)

Subclass Osteichthyes
Division Crossopterygii
Order Rhipidistia (Osteolepidoti)
Order Actinistia (Coelacanthini)
Division Actinopterygii

IV. Romer (1937, 1945a)

Class Osteichthyes
Subclass Actinopterygii
Subclass Choanichthyes
Order Crossopterygii
Suborder Rhipidistia
(Osteolepidoti)
Suborder Coelacanthini
(Actinistia)
Order Dipnoi

V. Romer (1955b)

Class Osteichthyes
Subclass Actinopterygii
Subclass Sarcopterygii
Order Crossopterygii
Suborder Rhipidistia
Suborder Coelacanthini
Order Dipnoi

VI. Berg (1955)

Class Dipnoi
Class Teleostomi
Subclass Crossopterygii
Superorder Osteolepides
Superorder Coelacanthi
Subclass Actinopterygii

**VII. Trewavas, White, Marshall, and Tucker
(1955) (in Romer, 1955b)[a]**

Subclass Crossopterygia
Order Rhipidistia
Suborder Osteolepidoti
Suborder Coelacanthini
Order Dipnoi

VIII. Vorobyeva and Obruchev (1964)

Class Osteichthyes
Subclass Sarcopterygia
Superorder Crossopterygii
Order Rhipidistia
Order Coelacanthini
Superorder Dipnoi
Subclass Actinopterygii

IX. Lehman (1966)[b]

Class Actinopterygii
Class Dipnoi
Class Crossopterygii
Subclass Actinistia
Order Coelacanthiformes
Subclass Rhipidistia

In addition to the position of Jarvik, noted above, Schaeffer (1968) has expressed the opinion that the Dipnoi, Crossopterygii, and Actinopterygii seem to be equally far apart, although they do form a coherent class. With these varied opinions as guidelines, it has seemed best to use the three subdivisions of the Osteichthyes—Dipnoi, Crossopterygii, and Actinopterygii—as the basis for analyzing the structure of the major subdivisions of the bony fishes. The use of the term "Crossopterygii" to include both lungfishes and lobe-finned fishes, in spite of statements to the contrary, is confusing and requires explanation when employed. If an association is to be expressed, to provide a group equal in rank to the Actinoptergii, then Romer's term "Sarcopterygii" has the advantage of being descriptive and not confusing by virtue of a long history of different usages. Both Rhipidistia and Coelacanthini are included under the Crossopterygii as used here, but the Dipnoi are excluded.

DIPNOI

The lungfishes—Dipnoi, or Dipneustes—have been used repeatedly in illustration of directional evolution. They show consistent trends in changes of body form and disposition of fins through time. The classification as generally used, provides a vehicle for this sort of analysis and is itself partly the result of this concept of change. The general changes of pattern are shown in Figure 67, after Schaeffer (1953).

Four classifications are shown in Table 13. Westoll (1949), in a detailed analysis of the evolutionary patterns of the Dipnoi, took the general arrangement as a basis for dealing with multiple characters and rates of change (see Table 14). He outlined an early period of rapid change in the later mid-Paleozoic, a slowing during the late Paleozoic, and a time of almost no change from the Triassic to the present. Although he did not present a formal classification in his work, he accepted the usual arrangement as a basis for representing the mode of progressive, directional change.

Except for the level of the categories, general consistency is found in the naming and composition of the major subdivisions in the classifications in Table 13. One partial departure is seen in the classifications of Berg and of Vorobyeva and Obruchev. In these an ordinal separation is accorded the Paleozoic Dipnoi, termed Dipterida, and the Mesozoic–Cenozoic forms Ceratodontida. *Gnathorhiza*, a Paleozoic genus with sectorial teeth, sometimes considered closely associated with the Lepidosirenidae of today, is placed with the Ctenodontidae and thus is not given a position ancestral to the modern forms with sectorial dentition.

Differences not evident in the major assignments appear in the placement of

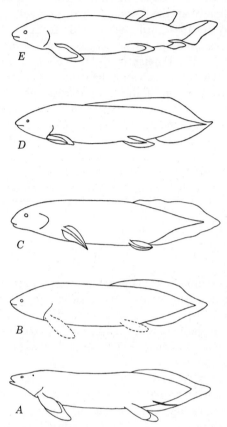

Figure 67.   Dipnoan genera shown in succession through geological time. If this sequence is taken as representing or approximating an evolutionary sequence, it can be considered as an example of directional evolution with changing rates (see text). *A, Dipterus,* from the Devonian; *B, Uronemus,* from the Carboniferous; *C, Conchopoma,* from the Permian; *D, Ceratodus,* from the Mississippian; *E, Neocera-todus,* one of the modern genera. (After Schaeffer, 1953.)

genera. For the most part, however, these are minor. The one relatively major difference relates to the genus *Gnathorhiza.* If this form is associated with Paleozoic forms, as a specialized offshoot, the modern forms *Protopterus* and *Lepidosiren* need not be considered as from an ancient lineage, but rather as recent departures from the ceratodontoid stock. If, on the other hand, as is sometimes done, *Gnathorhiza* is considered as directly related to the moderns, a very long lineage of a particular adaptive type from late Carboniferous to Recent is implied.

TABLE 13   Classifications of the Dipnoi

I. *Romer (1945a)*

Class Osteichthyes
 Subclass Choanichthyes
  Order Dipnoi
   Family Dipteridae
   Family Phaneropleuridae
   Family Ctenodontidae
   Family Conchopomidae
   Family Ceratodontidae
   Family Lepidosirenidae
    (?including *Gnathorhiza*)

II. *Romer (1966a)*

Class Osteichthyes
 Subclass Sarcopterygia
  Order Dipnoi
   Family Dipnorhynchidae
   Family Phaneropleuridae
   Family Ctenodontidae
   Family Sagenodontidae
   Family Ceratodontidae
   Family Lepidosirenidae
    (including *Gnathorhiza*)

III. *Berg (1955)*

Class Dipnoi
 Superorder Dipteri
  Order Dipteriformes
   Family Dipnorhynchidae
   Family Dipteridae
  Order Rhynchodipteriformes
  Order Phaneropleuriformes
   Family Phaneropleuridae
   Family Scaumanacidae
   Family Fleurantiidae
  Order Uronemiformes
   Family Uronemidae
   Family Conchopomidae
  Order Ctenodontiformes
   Family Ctenodontidae
    (including *Gnathorhiza*)
 Superorder Ceratodi
  Order Ceratodiformes
  Order Lepidosireniformes

IV. *Vorobyeva and Obruchev (1964)*

Class Osteichthyes
 Subclass Sarcopterygia
  Superorder Dipnoi
   Order Dipterida
    Suborder Dipteroidei
     Family Dipnorhynchidae
     Family Dipteridae
    Suborder Phaneropleuroidei
     Family Phaneropleuridae
     Family Scaumanaciidae
     Family Fleurantiidae
     Family Rhynchodipteridae
    Suborder Uronemoidei
     Family Uronemidae
     Family Conchopomatidae
    Suborder Ctenodontoidei
     Family Ctenodontidae
   Order Ceratodontida
    Suborder Ceratodontoidei
     Family Ceratodontidae
    Suborder Lepidosirenoidei
     Family Protopteridae
     Family Lepidosirenidae
      (?including *Gnathorhiza*)

V. *Lehman (1966)*

Class Dipneusta (Dipnoi)
 Order Dipteriformes
  Family Dipnorhynchidae
  Family Dipteridae
 Order Scaumanaciformes
  Family Scaumanacidae
  Family Fleurantiidae
  Family Rhynchodipteridae
 Order Phaneropleuriformes
  Family Phaneropleuriformidae
 Order Ctenodontiformes
  Family Uronemidae
   (including *Conchopoma*)
  Family Ctenodontidae
   (including *Gnathorhiza*)
 Order Ceratodontiformes
 Order Lepidosireniformes

TABLE 14   Summed Character Values after Westoll (21 Characters) Applied
to Dipnoi as Divided by Lehman (1966)

Order (Lehman)	Age	Range of Summed Characters in Order (or Suborder or Family)
Dipteriformes (*Dipterus, Dipnorhynchus*)	Middle Devonian	96 to 66.5
Scaumanaciformes (*Pentlandia, Scaumanacia, Fleurantia*)	Upper Devonian	62 to 44
Phaneropleuriformes (*Phaneropleuron*)	Uppermost Devonian	37.5 to 33.5
Ctenodontiformes (*Ctenodus, Uronemus, Sagenodus, Conchopoma*)	Upper Devonian to Permian	24.5 to 17.0
Ceratodontiformes (*Ceratodus, Epiceratodus*)	Triassic to Recent	12.0 to 1.0
Lepidosireniformes (*Lepidosiren, Protopterus*)	Tertiary to Recent	0

Recent evidence seems to favor the idea that the sectorial tooth pattern, such as found in these three genera, has been a common modification among the Dipnoi and that teeth are not reliable indicators of broad phylogenetic relationships (Carlson, 1968; Bertmar, 1968). Other features—skull pattern, roof pattern, burrowing and aestivation, and divergence in the developmental patterns—and nasal capsules in various living lungfishes may support the concept of long divergent evolutionary patterns (Thomson, 1965; H. Fox, 1965).

If the possibility of recent origin of the lepidosirenids is accepted, it opens a broader problem of similar nature, one that pervades the whole of lungfish evolution. The other major groups of vertebrates, as suggested by detailed studies, indicate that classifications based on levels of evolutionary advancement tend to be horizontal in nature. Representatives of several phyletic lines, each following a somewhat parallel course, may be included in a single taxon. This is notably true in the actinopterygians, taken up later in this section. The question thus arises as to whether this may not be the case for lungfishes and whether the picture of evolutionary change, reflected in most classifications, is not in fact somewhat misleading.

Westoll's study of 1949 illustrates what can be done under the usual framework, although he made no claim that he was showing more than rates of change expressed in trends. Using a total of 21 characters, he set up a scale of 7 to 0, with 7 the condition in the presumed ancestor of the Dipnoi and 0 the limit of development attained within the lungfishes. He then showed a succession of values based on the totality of these characters and also contrasted the head and postcranial structures. The successive changes, from group to group, were essentially in temporal series. Each character was found to have from three to six states. Decrease in the summed values of characters for each of the major subdivisions in classifications provided a graded scale that fits neatly into classifications predicated on progressive change. The pattern shown in Table 14 gives this general picture, based on the breakdown of major groups as used by Lehman.

The correspondence of geological age, the classification, and the values of characters is excellent. There are at best only minor problems in the classification. The question remains, however, of whether or not this is in fact a good expression of the phylogeny of the Dipnoi, or whether, as might be the case, it is largely an expression of grades of evolution. Was there essentially an order-by-order succession, with radiation in each, but a monophyletic origin of the succeeding order, or were there various lines passing through these grades and, in some cases, giving rise independently to various lines of the next grade?

The evidence at hand has not given a full answer to this question. The increasing number of cases in which this sort of situation is proving to be one of grades and polyphyletic origin of the next level suggests that this may be the case here. As matters now stand, however, the number of specimens available for analysis of critical structures is relatively small. The great mass of material consists of teeth, and, as noted, their significance as indicators of general relationships among the Dipnoi is open to serious question. That the sectorial type of tooth arose independently several times seems to be becoming well established. Only a relatively few specimens are well enough preserved to show cranial and postcranial characters in reasonably full array, and it is largely modern end members that provide anatomical and embryological details. From such limited data it is not possible to determine much more than general trends, certainly very little about participation of the genera in particular phyletic lines.

Current classifications of the Dipnoi are highly functional for assembling genera that resemble each other and for giving a broad pattern of change of adaptive levels, both their nature and their rates. They should not, however, be considered as final nor used without caution for presentation of a particular type of evolution in the vertebrates.

CROSSOPTERYGII

The fishes that form the Crossopterygii are distinctive, and the group presents few problems with respect to its coherency. These are the lobe-finned fishes, characterized by the presence of fleshy, scale-covered lobes in their fins, and roughly parallel rays reaching from the margins of the lobes to

TABLE 15   Classification of the Crossopterygii

I. *Moy-Thomas (1939a)*
  Division Crossopterygii
    Order Rhipidistia (Osteolepidoti)
      Family Osteolepidae
      Family Rhizodontidae
      Family Holoptychidae
    Order Actinistia (Coelacanthini)

II. *Berg (1955)*
  Subclass Crossopterygii
    Superorder Osteolepides
      (Rhipidistia)
      Order Osteolepiformes
        Family Osteolepidae
        Family Gyroptychidae
        Family Glyptomidae
        Family Parabatrachidae
          (Ectosteorhachidae)
      Order Holoptychiiformes
        (Porolepiformes)
        Family Porolepidae
        Family Holoptychidae
      Order Rhizodontiformes
        Family Rhizodontidae
        Family Eusthenopteridae
          (including *Incertae Sedis*
          Onychodus)
    Superorder Coelacanthi
      Order Coelacanthiformes
        Suborder Diplocercidoidei
        Suborder Coelacanthoidei
        Suborder Laugioidei

III. *Vorobyeva and Obruchev (1964)*
  Subclass Sarcopterygii
    Superorder Crossopterygii
      Order Rhipidistia

III. *Vorobyeva and Obruchev (1964)*—continued
        Suborder Holoptychoidei
          Family Porolepidae
          Family Holoptychidae
        Suborder Osteolepidotei
          Family Osteolepidae
          Family Eusthenopteridae
          Family Rhizodontidae
      Order Coelacanthini (Actinistia)
        Suborder Diplocercidoidei
          Family Diplocercididae
          Family Rhabdodermatidae
        Suborder Coelacanthoidei
          Family Coelacanthidae
          Family Latimeriidae
        Suborder Laugioidei

IV. *Lehman (1966)*
  Class Crossopterygii[a]
    Subclass Actinistia[a]
      Order Coelacanthiformes
        Suborder Diplocercidoidei
        Suborder Coelacanthoidei
          Family Coelacanthidae
          Family Latimeriidae
    Subclass Rhipidistia[a]
      Order Porolepiformes
        Family Porolepidae
        Family Holoptychiidae
      Order Osteolepiformes
        Family Osteolepidae
        Family Rhizodontidae
    Subclass Struniida[b]
      Order Struniiformes
        Family Onychodontidae

[a] Not specified by Lehman but implied in arrangement and text.
[b] Not used by Lehman but implied in separation.

the fin borders. The endocranium is divided into ethmosphenoid and oto-occipital portions, which are freely movable on each other. The dermal skull bones are disposed in patterns somewhat like those of tetrapods and with elements that possibly can be considered homologous with some in the skulls of actinopterygian fishes. Homologies between these elements and those of the lungfish do not appear to exist, and those with the actinopterygians may be in fact largely spurious. Teeth are characterized by moderately to strongly infolded enamel—that is, they are labyrinthine. Together these features are distinctive, but none is exclusive to the crossopterygians, and there is a wide diversity in their mode of expression within the group.

Several recent classifications are shown in Table 15. Each of these, as well as most others, recognizes two main subdivisions—Rhipidistia, or Osteo-lepida, and Coelacanthini, or Actinistia. Although members of these groups have common suites of crossopterygian characters, they differ in many respects, with differences to be found in almost all parts of the skeleton, as shown in the following comparison:

Rhipidistia	*Coelacanthini*
Endocranium well ossified, in two parts	Endocranium less well ossified, with various elements separate
Choanae present	Choanae absent
Large ectopterygoid	Ectopterygoid small or absent
Suboperculum present	Suboperculum small or absent
Vertebral centra partly ossified	Vertebral centra not ossified
Large lobes in paired fins, rays short	Small lobes in paired fins, rays long
Hyomandibular large	Hyomandibular small
Heavy, cycloid scales	Thin, cycloid scales
No ossified ribs	Ossified ribs
Interclavicle present	Interclavicle absent

These, and other similar characters, serve to separate the groups. There is, however, considerable developmental parallelism, and the primitive members of each have many features that are very similar. A few structures, such as choanae and the interclavicle, serve as absolute features as the stocks are now known. The suites of characters and differences in states makes separation easy, and there are no serious problems of assigning genera to one or the other group.

Coelacanthini

The coelacanths, or actinistians, form a coherent group of fishes with a long history from the Devonian to Recent. Their time of climax appears to have been in the Triassic. Three classifications listed in Table 15 make a three-fold division, with the categorical level a function of the way that higher categories were established in expressing the relationships of this group to other major groups of bony fishes. Each recognizes a primitive stock, Diplocercidoidei, or its equivalent. The range is Devonian and Carboniferous. Except for the late genus *Rhabdoderma*, the chondrocranium is well ossified, resembling that of the rhipidistians. In many other general features primitive members of the Coelacanthini resemble rhipidistians, and their general course of evolution is very generally parallel to that followed by the more conservative Dipnoi (see Figure 67).

The Coelacanthoidei, a second subdivision, includes lower Permian to Recent genera. *Latimeria*, a living genus, is characterized by the presence of "advanced" features, the end result of the trends implied in the list on page 251. Moderate expansion took place in the Triassic, and from then to the present there was only very modest change.

The third group, Laugioidea, is based on a Triassic genus from Greenland. The pelvic fin had migrated forward and developed an articulation with the shoulder girdle, expressing a trend also found among some groups of actinopterygian fishes. This is considered, along with other minor features, sufficient for separation from the other two groups. In most respects this subdivision has coelacanthoid features, but the braincase is somewhat better ossified than in most of its coelacanthoid contemporaries.

The Coelacanthini are known from relatively few genera. As far as these are concerned, there are no major problems in classification. Position of genera in the threefold division is based primarily on the degree of development of a series of characters that are expressions of progressive trends in development. Except for expansion in the Triassic, there is little evidence of a major radiation. Whether or not the trends were expressed in a complex of developing lines is uncertain from what is known. If the rather meager representation is a fair estimate of what actually existed, it may be that the progressive changes did take place in a reasonably restricted phylum and that the classification will remain adequate. If additional genera are eventually found, this pattern may prove to be erroneous; if it proves valid, problems of differentiation of levels may develop, as must occur when large arrays of types are known and classification is based on trends rather than discrete characters.

Rhipidistia

The Rhipidistia, like the Coelacanthini, form a coherent assemblage but, unlike the latter, do pose some interesting problems of classification. These are not immediately apparent in the classifications listed in Table 15, where only minor differences in categorical levels, related to the makeup of higher categories and to slight differences in family arrangements and assignments of genera to families appear. Berg, for example, shows a larger number of families than do the others, reflecting mainly a general characteristic of much of his classification of the fishes. More significant are some of the different bases on which similar groups have been established, as discussed later.

If the Rhipidistia are considered as a subordinal group, the families Osteolepidae, Rhizodontidae, and Holoptychidae (or Porolepidae) are generally considered to include most of the Rhipidistia. In some classifications a fourfold division results from separate listing of the Holoptychidae and Porolepidae, but generally they are taken to be closely allied. The three major groups may appear as orders (Berg) or as suborders. Also a twofold subordinal or ordinal array with Holoptychoidea (or Porolepiformes) and Osteolepoidea (Osteolepiformes) is often used. In such classifications the Osteolepidae and Rhizodontidae are included within the latter. *Onychodus* (see Berg) poses something of a problem, to which we shall return later.

Studies by Jarvik (1942 and later), devoted to an analysis of the snout of rhipidistians and the lower vertebrates in general, led him to the recognition of two distinct types of narial structures: one in the osteolepids, osteolepiform structure, and the other in the porolepids, porolepiform structure. This interpretation has been influential in subsequent formulations of classification in emphasizing the discreteness of the Porolepiformes and Osteolepiformes. The pattern of subdivision based on Jarvik's interpretation of narial features corresponds closely to that suggested by other characters. Thus no alteration of the formal pattern is involved. What is modified is the meaning given to some of the groups, something not expressible in the hierarchical arrangement.

Two important items are raised by a subdivision that gives special weight to the nasal structure. One, which is common to single-character assignments, is that the capsules are known in few representatives from the whole array of forms in which they are presumed to differ. Thus it must be assumed that there is a very high correlation between the states of nasal characters and the others that must serve to place genera in which nasal structures are not known in one or another major group.

The second, involving the nasal capsules here, but frequently entering into other classifications in one guise or another, is that the nature of classification or subdivision in a second major group, which is a presumed or known descendant of the one in question, may strongly influence the classification within the ancestral group. In this instance the separation of modern amphibians into urodeles and anurans, and the presumed derivation of the former from porolepiforms and the latter from osteolepiforms have exerted an influence on the conception of the internal structural relationships and hence the classification of the rhipidistians. The practice of recognition of complex relationships between organisms usually classified in widely separated categories, such as the porolepiform–urodele and osteolepiform–anura pairings, represents a revolt against monophyletic classification, the notion of monophyletic origin of major groups, and the use of grade levels as taxonomic units, longstanding practices in taxonomy. The polyphyletic aspects of many classifications of the mid-1900s are graphically emphasized by this treatment of rhipidistian–amphibian relationships. In very recent times, with some major exceptions, tendencies to return to some of the older practices are becoming increasingly evident.

There can be little doubt that the idea of marked discreteness of porolepiforms and osteolepiforms, and interpretations of their taxonomic positions stemmed in part from a knowledge of the structure of modern amphibians and the presumed relationships of these to fossil amphibian groups and to the rhipidistian ancestors. As a stimulus to examination of the fossil and recent forms such hypotheses about relationships play important roles in investigations of the vertebrates. As long as the hypotheses are treated as such, examined by objective testing, and rejected when found wanting, the cause of classification is advanced. Their history, however, has tended to be otherwise, with the hypothesis becoming the master and the factual information the handmaiden.

After some 20 years of detailed reexamination of the matter of nasal structure of the rhipidistians, including in particular additional studies by Jarvik and analyses by Vorobyeva (1960a,b), Kulcyzycki (1960), and Thomson (1962, 1964a,b), much evidence has accumulated that seems to show that the differences in the snouts are not truly fundamental. However, Jarvik (1965a, 1966, and 1967a,b) has strongly rejected some of the interpretations that point in this direction. What can be said about the relationship based on nasal structures remains up in the air. The problem as currently being considered, though indirectly taxonomic, is basically one of the origin of the amphibians, and the evolutionary implications are considered in more detail in Section VI, Chapter 4.

The classifications of the Rhipidistia today involve either a twofold or threefold division, and currently the tendency is to separate the rhizodontids and osteolepids, with each equivalent to the holoptychids. Thomson, in his current studies, indicates that this grouping probably expresses the relationships well, if taken without regard to descendants. Earlier Ørvig, on the basis of scales, recognized a twofold division, corresponding to the osteolepiforms and porolepiforms. Formation and function of jaws and dentitions and the generally great predatory capacities of rhizodontids as compared to osteolepids, suggests a radiation in which a threefold division was accomplished. In this case, as in many others of similar nature, the final decision comes down to the value that is accorded particular characters. Current classifications have had a gradual evolution and change is continuing, without any final agreement either on the value of different single characters or the ways in which they may best be associated.

What seems to be emerging is that most well-known representatives may best be grouped in three subdivisions (see, for example, Thomson, 1968), all of which are quite closely related. What bearing this has on the more complex problems of the relationships of each of the groups and the amphibians has not been considered in any detail. It is clear that in some features one subdivision is more amphibianlike, whereas in others another group is. None provides a good amphibian ancestor, either for all amphibians or for any particular group. Any attempt to sort out lines among the rhipidistians based on amphibian resemblances is at present completely unwarranted.

One other problem that has arisen from time to time is the placement of the Devonian genus *Onychodus*. This is a seemingly aberrant genus, with a strong whorl of symphyseal teeth. Berg (1955) has given it an ordinal *incertae sedis* status under Osteolepides (Rhipidistia). Lehman (1966), however, related it to *Strunius* from the base of the upper Devonian and concluded that these genera represent another major group, equivalent to the osteolepiforms and porolepiforms. Ørvig (1960), on the basis of scales, felt that relationship of *Onychodus* to osteolepiforms was reasonable. This was rejected by Lehman on the basis of the weight of various other characters of the jaws, trunk, operculum, and scapula. If this separation is taken to be valid, along with Berg's threefold division, as agreed to by Thomson, a fourfold pattern of radiation of the rhipidistians emerges.

ACTINOPTERYGII

As in the case of the Chondrichthyes, the early subdivision of the actinopterygians was based largely on living genera of fishes. Ways of separating the major types that were effective among moderns developed, but these had little phylogenetic content and proved to be inapplicable or at least extremely awkward, when applied to fossils as they came to be better known.

One early classification, which has had persistent influence, was that of Agassiz, based on scales. It included actinopterygians as well as some of the other bony fishes. He recognized four groups based on scales: placoids, ganoids, cycloids, and ctenoids. This is now known, of course, to cut across many lines recognized on the basis of characters that appear to be more reliable. Much more influential was the subdivision of the actinopterygians into three major groups by Müller in 1844. He recognized three basic types: Chondrostei (*Acipenser*, *Polyodon*), Holostei (*Lepisosteus*), and Teleostei. This scheme has been adopted by most later students, although Zittle, among others, continued use of Ganoidea (from Agassiz) in his classifications (Zittle, 1895, and later editions).

As fossils become better known, it became apparent that this threefold system was not phylogenetic but that its groups actually represented grades of development.

Because of the convenience of the system there has been great reluctance to abandon it, even though it has long been recognized that it is not a phylogenetic classification. Romer (1945a) retained this subdivision (see Table 16), with comments in the text to the effect that it was artificial but highly convenient, and he made no basic change in 1966. The terms "Chondrostei" and "Holostei," in spite of the problems they pose, remain firmly entrenched in most classifications. They are sometimes used only in a limited sense, applying primarily to modern fishes, in keeping with the original usage. Although it has long been realized that the chondrosteans and holosteans represent grades, this has been less true for the teleosts. These fishes exhibit a more coherent radiation pattern, and it was long agreed that they had a common base, expressed in the holosteanlike leptolepids of the Triassic and later Mesozoic. During the last 25 years the idea that teleosts may have stemmed from more than one source below the teleost level of organization has been developing steadily. Modern classifications are taking this into account (for examples see Greenwood et al., Table 17; Romer, 1966a; and C. Patterson, 1967).

Chondrosteans, Paleoniscoids, and Subholosteans

As fossils became well known, intermediates between the three initial divisions of Müller were found. Brough (1936) proposed the term "Subholostei" to indicate a stage between the Chondrostei, as then known, and the Holostei. These fishes were paleoniscoid in some persistent characters, but many holostean features had appeared or were incipient. Paleoniscoids (or Palaeonisciformes in some classifications), still generally classified as chondrosteans, had by then usurped the status of primitive actinopterygians originally accorded the living chondrosteans. The term "subholostean" represented an intergrade, but also a phase of radiation, in the sense of Brough. It found its way into many classifications and persists in usage, although in a less formal sense than was once accorded it (see, for example, Schaeffer, 1956).

It is now clear that beginning with the Devonian, the time of first appearance of the ray-finned fishes, first called Actinopterygii by Cope, there were many phyletic lines that passed independently through various grade levels at different rates and at different times. At any given time a "norm" of development can be detected, with some "relicts" and some "advanced" genera present. For any such time a horizontal or grade classification in which the majority of ray-finned fishes falls in a particular category is very useful. A few from earlier and later times can be associated with the grade. The majority of fishes from the middle and late Paleozoic, for example, can be included in Palaeoniscoidei, or "primitive" chondrosteans. During the Triassic most ray-finned fishes are "subholosteans," during the Jurassic, "holosteans," and today, "teleosteans." At each time some genera do not conform to the major group, and these can be placed in more primitive or more advanced categories, as appropriate. This makes for a highly functional classification, but not one that is strictly phyletic.

Most present-day classifications of the Actinopterygii represent a compromise between the old subdivision into Chondrostei, Holostei, and Teleostei and the recognition that these arrangements do not properly express phylogeny. Several classifications are presented in Table 16 for comparisons. Representative fishes are shown in Figures 120–129 in Section IV, Chapter 3, in which the evolutionary aspects of this group are taken up. The teleosts are omitted from some of these classifications because of the number of subdivisions and special problems they pose.

Berg (1955) did not employ a threefold major division, such as that used by Romer (1945a). In his text, however, he discussed the division and to some extent organized his orders under it. He also, however, made a primary breakdown in his discussion to conform to the twofold arrangement of Regan (1929), which recognizes Palaeopterygii (= roughly Chondrostei) and Neopterygii (= roughly Holostei and Teleostei) and gave consideration to Stensiö's recognition of gradation between categories. He thereafter listed orders, with only a grouping of the primitive forms into Division A, for Tarrasiiformes, and Division B for other primitive orders. The remaining orders were listed without grouping, but with a note that they pertain to Neopterygii.

Lehman (1966) also followed the plan of listing orders with a general discussion of their positions. Moy-Thomas (1939a) treated only Paleozoic fishes and presented a standard pattern for his time, one essentially like that used by Romer (1945a) (see Table 16). He did not use the term "Chondrostei," but in effect substituted Palaeoniscoidea for it, an appropriate use if only Paleozoic forms are considered. Like Romer, he used the term "Subholostei" in a formal sense.

Berg, Obruchev et al. (1964) made use of a fivefold division at the superordinal level. Three of these comprise more or less the original content of the Chondrostei, Holostei, and Teleostei, with many genera and families from the fossil record added as appropriate. The superorder Palaeonisci, essentially the Palaeonisciformes of Gardiner, includes the great majority of extinct families that are at a "chondrostean" or "subholostean" grade. The family Polypteridae, based on *Polypterus* and *Calamoichthys*, Eocene to Recent in range, is placed in a separate order. This classification resembles recent classifications by Gardiner (Table 16).

Each of these classifications, as well as others that are concerned with particular segments of the Actinopterygii, reflects the current instability of understanding in this group and of transition from essentially a typological to a phylogenetic scheme of classification. Much of this change has come from studies of the evolution of Actinopterygii, which have replaced efforts to assess relationship primarily on the basis of form. These are taken up in Section IV, and only matters directly pertinent to classification are considered at this time.

Current classifications all include, at one categorical level or another, a paleoniscoid assemblage, which comprises the vast array of primitive actinopterygians, predominant from the Devonian through the Permian, but persisting to the lower Cretaceous. In the various classifications the genera are arrayed into families, with ordering as the only indication of familial relationships. Lehman (1966) recognized two suborders, but Berg, Obruchev et al., and Gardiner place all genera in families, without grouping below the ordinal level. The major problems in making family assignments and in relating families have arisen from the fact that many of the characters available for analysis exist in a variety of states, with different genera containing different associations of the character states of common characters. The patterns of evolution thus display a classic example of mosaic evolution.

Lack of definitive descriptive information has been a blow to stable classification as well. A complete restudy of the families of the paleoniscoids has been undertaken to remedy this situation, but only parts of it had been completed at the time of writing (Gardiner 1963, 1967a).

Beyond the paleoniscoids, other chondrosteans have received very different treatments. Some of the arrangements of these orders of Chondrostei (or Palaeoniscida) are shown in Table 16. Extremes are those of Moy-Thomas (1935a) and Romer (1945a), who make complete commissions, and Lehman, who makes none, listing merely orders. None of the classifications can be considered as right or wrong, because each is appropriate within the framework within which it has been cast. In none is any pretense of full phylogenetic representation made.

Involved in these classifications is a vast array of largely late Paleozoic and Triassic fishes sometimes called subholosteans. These, in the words of Schaeffer (1956), were experimenting with a variety of adaptive modifications. They present a somewhat bewildering array of mosaics made up of both primitive and more advanced features. Along with modifications that appear to be of general adaptive value are more specialized adaptations to particular circumstances, reflected in such things as body form, fin distribution, and tooth arrangements and shapes. The parallel and convergent changes were taking place in a series of phyletic lines, but in general information has not been sufficient for an effective sorting of these lines.

TABLE 16   Classification of the Actinopterygii[a]

I. *Moy-Thomas (1939a)* (Paleozoic only)
  Division Actinopterygia
    Order Palaeonisciformes
      Suborder Palaeoniscoidei
        Cheirolepitormidae,
        Palaeoniscidae, Cryphiolepidae,
        Holuridae, Styracopteridae,
        Canobiidae, Eurylepidae,
        Trissolepidae
      Suborder Platysomida
        Cheirodidae, Platysomidae
      Suborder Tarrasiida
      Suborder Phanerorhynchida
      Suborder Urosthenida
      Suborder Dorypterida
      Suborder Polypterini
      Suborder Acipenseroidei
    Order Holostei
      Suborder Semionotida
      Suborder Teleostei
    Order Subholostei

II. *Romer (1945a)*
  Subclass Actinopterygii
    Superorder Chondrostei
      Order Palaeoniscoidea
        Cheirolepidae, Rhadinichthyidae,
        Canobiidae, Haplolepidae,
        Elonichthyidae, Palaeoniscidae,
        Pygopteridae, Acrolepidae,
        Amblypteridae, Aeduellidae,
        Cryphiolepidae, Scanilepidae,
        Coccolepidae, Dicellopygidae,
        Birgeriidae, Tarrasiidae,
        Boreolepidae, Trissolepidae,
        Tegeolepidae, Cocconiscidae,
        Cornuboniscidae,
        Styracopteridae, Amphicentridae,
        Platysomidae, Dorypteridae,
        Carbovelidae
      Order Polypterini (1)
      Order Acipenseroidei
        Chondrostei, Acipenseridae,
        Polyodontidae
      Order Subholostei
        Dictyopygidae, Platysiagidae,
        Cephaloxenidae, Peltopleuridae,
        Luganoiidae, Aetheodontidae,

        Ptycholepidae, Parasemionotidae,
        Pholidopleuridae, Saurichthyidae,
        Cleithrolepidae, Bobasatraniidae
    Superorder Holostei
      Order Semionotoidea
      Order Pycnodontoidea
      Order Aspidorhynchoidea
      Order Amioidea
      Order Pholidophoroidea
    Superorder Teleostei
      Order Isospondyli
        Suborder Clupeoidea
        Suborder Salmonoidea
        Suborder Opisthoproctoidea
        Suborder Osteoglossoidea
        Suborder Stomiatoidea
        Suborder Gonorhynchoidea
      Order Ostariophysi
        Suborder Cyprinoidea
        Suborder Siluroidea
      Order Apodes
      Order Heteromi
      Order Mesichthys
        Suborder Haplomi
        Suborder Iniomi
        Suborder Microcyprini
        Suborder Synentognathi
        Suborder Thoracostei
        Suborder Salmopercae
      Order Acanthopterygii
        Suborder Berycoidea
        Suborder Zeoidea
        Suborder Percoidea
        Suborder Carangoidea
        Suborder Scombroidea
        Suborder Trachinoidea
        Suborder Blennioidea
        Suborder Anacanthini
        Suborder Chaetodontoidea
        Suborder Plectognathi
        Suborder Heterosomata
        Suborder Scorpaenoidea
        Suborder Batrachoidea
        Suborder Pediculati
        Suborder Gobioidea
        Suborder Anabantoidea
        Suborder Mugiloidea
        Suborder Polynemoidea

TABLE 16   (continued)

II. *Romer (1945a)—continued*
    Suborder Ammodytoidea
    Suborder Echeneoidea
    Suborder Xenopterygii
    Suborder Allotriognathi
    Suborder Opisthomi
    Suborder Synbranchii

III. *Berg, Obruchev et al. (1964)*
    Subclass Actinopterygii
    Superorder Palaeonisci
        Order Palaeoniscida
            Cheirolepididae,
            Stegotrachelidae, Tegeolepidae,
            Palaeoniscidae,
            Cosmoptychidae,
            Centrolepidae,
            Canobiidae, Pygopteridae,
            Acrolepidae, Elonichthyidae,
            Commentryidae,
            Rhabdolepididae,
            Amblypteridae, Holuridae,
            Cornuboniscidae,
            Trissolepidae, Aeduellidae,
            Birgeriidae, Ptycholepidae,
            Coccolepididae (including
            Incertae sedis Urosthenidae)
        Order Tarrasiida (1)
        Order Haplolepidida (1)
        Order Platysomida
            (Platysomoidei, Dorypteriformes,
            Bobasatraniiformes of Berg)
            Platysomidae, Dorypteridae,
            Bobasatraniidae
        Order Phanerorhynchida (1)
        Order Perleidida
            Dictyopygidae,Cleithrolepidae,
            Colobodontidae,
            Aetheodontidae, Platysiagidae
        Order Luganoiida (1)
        Order Pholidopleurida (1)
    Superorder Chondrostei
        Order Sauricthyida (1)
        Order Errolichthyida (1)
        Order Acipenserida (3)
    Superorder Polypteri
    Superorder Holostei
        Order Ospiida (2)

Order Amiida (5)
Order Aspidorhynchida (1)
Order Pycnodontida (3)
Order Pachycormida (2)
Order Lepisosteida (1)
Order Pholidophorida (5)
Superorder Teleostei
    Order Clupeida (Isospondyli)
        Suborder Leptolepidoidei (1)
        Suborder Lycopteroidei (1)
        Suborder Clupeoidei (7)
        Suborder Ctenothrissoidei (1)
        Suborder Chirocentroidei (2)
        Suborder Chanoidei (1)
        Suborder Salmonoidei (2)
        Suborder Esocoidei (2)
        Suborder Stomiatoidei (4–5)
        Suborder Enchodontoidei (1)
        Suborder Gonorhynchoidei (1)
    Order Scopelida
    Order Cyprinida (Ostariophysi)
        Suborder Cyprinoidei (3)
        Suborder Siluroidei (4)
    Order Anguillida (Apodes)
        Suborder Anguillavoidei (1)
        Suborder Anguilloidei (4)
        Suborder Hemichthyoidei (1)
    Order Halosaurida (Lyopomi)
    Order Notacanthida (Heteromi) (1)
    Order Belonida (Synentognathi)
        Suborder Scomberescoidei (2)
    Order Gadida (Anacanthini pars)
        Suborder Gadoidei (3)
    Order Macrurida (1)
        (Anacanthini pars)
    Order Gasterosteida (2)
    Order Syngnathida
        Suborder Aulostomoidei (3)
        Suborder Syngnathoidei (3)
    Order Lampridida (2)
    Order Cyprinodontida
        Suborder Cyprinodontoidei (1)
    Order Percopsida (1)
    Order Berycida (5)
    Order Zeida (2)
    Order Mugilida
        Suborder Mugiloidei (1)
        Suborder Sphyraenoidei (1)

TABLE 16   (continued)

III. *Berg, Obruchev et al. (1964)—continued*
Order Percida (Acanthopterygii)
    Suborder Percoidei (15)
    Suborder Blennioidea (1)
    Suborder Ophidioidea (2)
    Suborder Ammodytoidei (1)
    Suborder Trichiaroidei (3)
    Suborder Scombroidei (2)
    Suborder Cottoidei (3)
Order Pleuronectida
    Suborder Pleuronectoidei (3)
Order Echeneidida (1)
Order Tetrodontida
    Suborder Balistoidei (2)
    Suborder Ostracioidei (1)
    Suborder Gobiesocidei (1)
Order Batrachoidida (1)
Order Lophiida (Pediculati)
    Suborder Lophioidei (1)
    Suborder Antennaroidei (1)

IV. *Lehman (1966)*

Order Tarrasiiformes
Order Paleonisciformes
    Suborder Palaeoniscoidei
        Cheirolepidae, Tegeolepidae,
        Palaeoniscidae, Birgeriidae,
        Aeduellidae
    Suborder Platysomoidei
        Amphicentridae, Platysomidae
Order Haplolepiformes
Order Dorypteriformes
Order Bobasatraniformes
Order Pholidopleuriformes
Order Peltopleuriformes
Order Platysiagiformes
Order Cephaloxeniformes
Order Aetheodontiformes
Order Perleidiformes
    Catopteridae, Perleididae
Order Luganoiiformes
Order Ptycholepiformes
Order Saurichthyiformes
Order Phanerorhynchiformes
Order Chondrosteiformes

IV. *Lehman (1966)—continued*
Order Acipenseriformes
    Acipenseridae, Polyodontidae
Order Parasemionotiformes
    Parasemionotidae
    Tungusichthyidae,
    Promecosominidae,
    ?Catervariolidae
Order Amiiformes
    Furidae, Amiidae, Macrosemiidae
Order Pachycormiformes
Order Semionotiformes
    Semionotidae, Dupediidae
Order Lepisosteiformes
Order Aspidorhynchiformes
Order Pycnodontiformes
    Gyrodontidae, Pycnodontidae,
    Coccodontidae
Order Pholidophoriformes
    Pholidophoridae,
    Archaeomaenidae,
    Oligopleuridae,
    Pleuripholidae, ?Majokiidae
Order Clupeiformes[b]

V. *Gardiner (1967a)[c]*

Class Actinopterygii
    Order Palaeonisciformes (39)
    Order Tarrasiiformes
    Order Haplolepiformes
    Order Saurichthyiformes
    Order Chondrosteiformes
    Order Acipenseriformes
    Order Polypteriformes
    Order Perleidiformes
    Order Luganoiiformes
    Order Peltopleuriformes
    Order Cephaloxeniformes
    Order Platysiagiformes
    Order Redfieldiiformes
    Order Pholidopleuriformes
    Order Ptycolepiformes
    Order Dorypteriformes
    Order Bobasatraniformes
    Order Parasemionotiformes

[a] Numbers in parentheses refer to the number of families.
[b] Other "teleosts" not in this classification.
[c] Chondrostei only.

Treatment has ranged from a horizontal grouping on the basis of grade level, as in Subholostei, as used by Romer (1945a) following Brough, to a classification listing only orders, arranged in sequence of relative advancement of members, as in the classification of Lehman, which includes in the single array the orders Parasemionotiformes, Amiiformes, and Semionotiformes, usually classed as holosteans. Berg, Obruchev et al. have grouped some of the families into orders, as Platysomida and Perleidida. The former is made up of deep-bodied fishes and may in some part be the result of parallel development of this feature. The second is an intergrade between paleoniscoids and holosteans.

Holosteans

The Holostei of Romer and of Berg, Obruchev et al. are generally similar and represent a stage of development in which most of the changes from primitive paleoniscoids have been completed. The most fundamental changes in this series were modifications of feeding mechanisms (Schaeffer and Rosen, 1961), which made possible the holostean radiations. The changes occurred, it would seem, in several lines and the holostean radiation did not come from a common base. The semionotids, pycnodontids, and the pholidophorids and amioids, the last two with their probable base among the parasemionotoids, are cases of independent origin from sources below the holostean level. Any classification retaining this general heading thus must be considered to express a grade level. When sufficient information is available either the term "Holostei" should be dropped or its usage greatly restricted. Currently, as in Gardiner's studies, two terms of equivalent rank, "Holostei" and "Halecostomi," are often used, the former including the majority of "holosteans," and the latter, the teleostlike "holosteans" (see Figure 116, Section IV, in which Gardiner's phylogeny shows this separation).

Teleosteans

The teleost grade also became a possibility once the feeding mechanisms found in fully developed holosteans were matured. The difference between the primitive halecostomes and members of other lines at the holostean level are minor but critical. Modifications basic to the teleost radiation, although dependent on earlier changes in feeding mechanisms, were immediately related to changes in locomotor adaptations, especially evident in the fins, girdles, and axial structures. The general consistency of the basic teleost structure has led to the concept of its members as forming a proper phylogenetic unit in classification. As noted, however, this has come more and more into question, and it now seems quite evident that the source and early deployment of fishes at the teleost grade is far from fully understood.

The majority of classifications of the teleosts (see examples in Table 17) have been based on the morphology of living fishes, with only moderate attention paid to the excellently preserved fossils of the late Mesozoic and Cenozoic. It has usually been more or less assumed that the early leptolepids, from the upper Triassic, were near the base of the radiation and that the later teleosts arose from them in a monophyletic manner. Questions about this conclusion have arisen in recent years. In 1942 Woodward suggested that the teleosts were polyphyletic. Although some of his proposed relationships have been shown to be doubtful (see, for example, C. Patterson, 1965a), the general concept of polyphyly has grown and is now widely accepted. It has been discussed in detail by Schaeffer and Rosen (1961) and by C. Patterson (1965a), among others. Thus the teleosts, like the chondrosteans and holosteans before them, are coming to be viewed more and more as a grade of evolution rather than a phyletic unit with a single preteleost ancestry.

This situation, if it is the true one, should be expressed in classifications if the objective of representation of phylogeny is to be met. The various classifications that have been considered more or less standard—for example, Regan (1929) and Berg (1940 and 1955)—in large part followed the patterns set in the late 1900s, with only relatively minor modifications. Bertin and Arambourg (1958) included some concepts of polyphyly, and earlier, in 1931, Garstang presented a phyletic classification that attempted to get away from the grade-level frame of reference. C. Patterson, working on the acanthopterygian fishes, gave emphasis to the idea of multiple sources of that group within the teleosts. For the most part, however, the classifications, as illustrated in some of these in Table 16, have maintained the groupings that appear to be based on levels of specialization rather than the phyletic lines involved in their actual development.

The first, and at the time of writing the only, full-scale reorganization of teleost classification with phyletic objectives was made by Greenwood, Rosen, Weitzman, and Myer (1966). This classification was adopted by Romer (1966a). An effort was made to sort out the major lineages, with separate origins among the holosteans, and, within these major groups, to detect the subradiations at superordinal, ordinal, and familial levels. The results are shown diagrammatically in Figure 68, from Greenwood et al. (1966). Comparisons with other classifications are to be seen in the tabular representation of this one along with others in Table 17. The general arrangement in this figure more or less speaks for itself. An understanding of the detailed basis of the divisions and the validity of the various assignments at different levels, however, require careful study of the very condensed and tightly written justifications in the cited publication. Since this publication was made many students have criticized various parts of it. This was

TABLE 17   Classification of the Teleostei[a]

I. *Berg (1940, 1955)*[b]

Order Clupeiformes (Isospondyli)
Suborder Lycopteroidei (2)
Suborder Clupeoidei
Superfamily Elopoidae (2)
Superfamily Alepocephaloidae (3)
Suborder Ctenothrissoidei (1)
Suborder Chirocentroidei (2)
Suborder Saurodontoidei (1)
Suborder Chanoidei (1)
Suborder Phractolaemoidei (1)
Suborder Cromerioidei (1)
Suborder Salmonoidei (1)
Suborder Esocoidei
Superfamily Dallioidae (1)
Superfamily Umbroidae (2)
Superfamily Esocoidae (1)
Suborder Stomiatoidei
Superfamily Gonostomoidae (2)
Superfamily Stomiatoidae (2)
Superfamily Astronethoidae (4)
Suborder Enchodontoidei (1)
Suborder Opisthoproctoidei (1)
Suborder Gonorhynchoidei (1)
Suborder Notopteroidei (2)
Suborder Osteoglossoidei
Superfamily Osteoglossoidae (3)
Suborder Pantodontoidei (1)
Suborder Anotopteroidei (1)
Order Bathyclupeiformes (1)
Order Galaxiiformes (1)
Order Scopeliformes (11)
Order Ateleopiformes (1)
Order Giganturiformes (1)
Order Saccopharyngiformes (2)
Order Mormyriformes
Suborder Gymnarchoidei (1)
Suborder Mormyroidei (1)
Order Cypriniformes
Suborder Characinoidei (6)
Suborder Gymnotoidei
Superfamily Sternarchoidae (2)
Superfamily Gymnotoidae
Suborder Cyprinoidei
Suborder Siluroidei (7)
Superfamily Siluroidae
Superfamily Diplomystoidae (1)
Superfamily Siluroidae (27)

Order Anguilliformes
Suborder Anguillavoidei (1)
Suborder Anguillioidei[c]
Suborder Nemichthyoidei[d]
Order Halosauriformes (1)
Order Notacanthiformes (2)
Order Beloniformes
Suborder Scomberesocoidei (2)
Suborder Exocoetoidei (2)
Order Gadiformes
Suborder Muraenolepidoidei (1)
Suborder Gadoidei (3)
Order Macruriformes (1)
Order Gasterosteiformes (3)
Order Syngnathiformes
Suborder Aulostomoidei
Superfamily Aulostotoidae (2)
Superfamily Centriscoidae (2)
Suborder Syngnathoidei (2)
Order Lampridiformes
Suborder Lampridoidei (1)
Suborder Veliferoidei (2)
Suborder Trachypteroidei (2)
Suborder Stylophoroidei (1)
Order Cyprinodontiformes
Suborder Amblyopsoidei (1)
Suborder Cyprinodontoidei
Order Phallostethiformes (2)
Order Percopsiformes
Suborder Percopsidoidei (1)
Suborder Aphredoderoidei (1)
Order Stephanoberyciformes (2)
Order Beryciformes (15)
Order Zeiformes (3)
Order Mugiliformes
Suborder Sphaenroidei (1)
Suborder Mugiloidei (2)
Order Polynemiformes (1)
Order Ophiocephaliformes (1)
Order Symbranchiformes
Suborder Alabetoidei (3)
Order Perciformes (Acanthopterygii)
Suborder Percoidei
Superfamily Percoidae (59)
Superfamily Cepoloidae (1)
Superfamily Embiotocoidae (1)
Superfamily Pomacentroidae (1)
Superfamily Labroidae (3)

TABLE 17   (continued)

I. *Berg (1940, 1955)*[b]—continued
    Superfamily Gadopsidae
    Superfamily Cirrhitoidae (5)
    Superfamily Trichodontoidae (1)
    Superfamily Trachinoidae (13)
    Superfamily Uranoscopoidae (3)
    Superfamily Champsodontoidae(1)
    Superfamily Chiasmodontoidae (1)
    Superfamily Notothenioidae (4)
  Suborder Blennioidei (18)
  Suborder Ophidioidei
    Superfamily Ophidioidae (2)
    Superfamily Fierasferoidae (2)
  Suborder Callionymoidei (3)
  Suborder Siganoidei
  Suborder Acanthuroidei (2)
  Suborder Trichiuroidei (2)
  Suborder Scombroidei
    Superfamily Scombroidae (2)
    Superfamily Xiphioidae (5)
  Suborder Luvarioidei (1)
  Suborder Tetragonuroidei (1)
  Suborder Stromateoidei (2)
  Suborder Anabantoidei (2)
  Suborder Kurtoidei (1)
  Suborder Rhamphosoidei (1)
  Suborder Gobioidei
    Superfamily Eleotroidae (1)
    Superfamily Gobioidae (2)

  Suborder Cottoidei
    Superfamily Scorpaenoidae (6)
    Superfamily Hexagrammidae (2)
    Superfamily Platycephaloidae (1)
    Superfamily Hoplichthyoidae (1)
    Superfamily Congiopodoidae (1)
    Superfamily Cottoidae (9)
Order Dactylopteriformes (1)
Order Thunniformes (1)
Order Pleuronectiformes
  Suborder Psettodoidei (2)
  Suborder Pleuronectoidei
    Superfamily Pleuronectoidae (4)
Order Icoteiformes (1)
Order Chaudhuriiformes (1)
Order Mastacembeliformes (1)
Order Echeneiformes (2)
Order Tetrodontiformes
  Suborder Belistoidei (5)
  Suborder Moloidei (1)
Order Gobiesociformes (1)
Order Batrachoidiformes (1)
Order Lophiiformes
  Suborder Lophioidei (1)
  Suborder Antennarioidei
    Superfamily Antennarioidae (3)
    Superfamily Oncocephaloidae (1)
Order Pegaciformes (1)

II. *Greenwood, Rosen, Weitzman, and Myers (1966)*[e]
Division I
Superorder Elopomorpha
  Order Elopiformes (Isospondyli in part, Clupeiformes in part)
    Suborder Elopoidei (two families, from division of one family)
    Suborder Albuloidei (one family, from two previous ones)
  Order Anguilliformes (Apodes, Lyomeri, Saccopharyngiformes,
    Monognathiformes, Anguillimorphi)
    Suborder Anguilloidea (23 + 22)
  Order Notacanthiformes (Lyopomi, Heteromi, Halosauriformes) (3)
Superorder Clupeomorpha
  Order Clupeiformes (Isospondyli in part)
    Suborder Denticipitoidei (1)
    Suborder Clupeoidei (3 + 6)

Division II
Superorder Osteoglossomorpha
  Order Osteoglossiformes (Isospondyli in part, Clupeiformes in part)
    Suborder Osteoglossoidei (2 + 3)

TABLE 17   (continued)

Division II—(continued)
  Suborder Notopteroidei (2)
  Order Mormyriformes (Isospondyli in part, Clupeiformes in part,
    Scyphophori) (2)

Division III
  Superorder Protacanthopterygii
    Order Salmoniformes (Isospondyli in part, Clupeiformes in part,
      Galaxiiformes, Haplomi, Xenomi, Iniomi,
      Scopeliformes, Myctophiformes)
      Suborder Salmonoidei (3 + 2)
      Suborder Argentinoidei (3 + 5)
      Suborder Galaxioidei (4 + 2)
      Suborder Esocoidei (2 + 2)
      Suborder Stomiatoidei (8 + 3)
      Suborder Alepocephaloidei (1 + 2)
      Suborder Bathylaconoidei (1)
      Suborder Myctophoidei (15 + 1)
    Order Cetomimiformes (Isospondyli in part, Clupeiformes in part,
      Ateleopiformes, Chondrobrachii, Giganturiformes)
      Suborder Cetomimoidei (3)
      Suborder Ateleopodoidei (1)
      Suborder Mirapinnatoidei (3)
      Suborder Giganturoidei (2)
    Order Ctenothrissiformes (1)
    Order Gonorhynchiformes (Isospondyli in part, Clupeiformes in part,
      Chanoiformes)
      Suborder Gonorhynchoidei (1)
      Suborder Chanoidei (3 + 2)
  Superorder Ostariophysi
    Order Cypriniformes (Plectospondyli in part, Heterognathi, Gymnonoti,
      Glanencheli, Evantognathi)
      Suborder Characoidei (16 + 10)
      Suborder Gymnotoidei (4 + 2)
      Suborder Cyprinoidei (31 + 25)
    Order Siluriformes (Plectospondyli in part, Cypriniformes in part,
      Nematognathi, Siluroidiformes)
  Superorder Paracanthopterygia
    Order Percopsiformes (Microcyprini in part, Cyprinodontes in part,
      Cyprinodontiformes in part, Amblyopsiformes, Salmopercae, Xenarchi,
      Percopsomorphi)
      Suborder Amblyopsoidei (1 + 2)
      Suborder Aphredoderoidei (1)
      Suborder Percopsoidei (1)
    Order Batrachoidiformes (Jugalares in part, Haplodoci, Perciformes in part,
      Pediculati in part (1 + 1)
    Order Gobiesociformes (Xenopterygii, Gobiesocomorpha, Perciformes in part)
      (1 + 1)

TABLE 17   (continued)

Division III—(continued)
  Order Lophiiformes (Pediculati in part)
    Suborder Lophioidei (1)
    Suborder Antennarioidei (4 + 4)
    Suborder Ceratioidei (10 + 4)
  Order Gadiformes (Anacanthini, Macruriformes, Gadomorphi, Perciformes in part)
    Suborder Muraenolepidoidei (1)
    Suborder Gadoidei (4 + 4)
    Suborder Ophidioidei (3 + 3)
    Suborder Zoarcoidei (1 + 3)
    Suborder Macrouroidei (1 + 1)
Superorder Atherinomorpha
  Order Atheriniformes (Synentognathi, Beloniformes, Gambustiformes, Microcyprini in part, Cyprinodontiformes in part, Percesoces in part, Mugiliformes in part, Mugilimorphi in part, Phallostethiformes, Perciformes in part)
    Suborder Exocoetoidei (3 + 7)
    Suborder Cyprinodontoidei (8 + 6)
    Suborder Atherinoidei (5 + 4)
Superorder Acanthopterygii
  Order Beryciformes (Xenoberyces, Berycomorphi, Berycoidei in part, Stephanoberyciformes in part)
    Suborder Stephanoberycoidei (3)
    Suborder Polymixioidei (1)
    Suborder Berycoidei (8 + 3)
  Order Zeiformes (7 + 6)
  Order Lampridiformes (Selenichthyes, Allotriognathi)
    Suborder Lampridoidei (1)
    Suborder Veliferoidei (1)
    Suborder Trachipteroidei (3)
    Suborder Styleophoroidei (1)
  Order Gasterosteiformes (Lophobranchii, Thoracostei, Aulostomi, Solenichthyes, Scleoparei in part, Syngnathiformes, Aulostomiformes, Rhamphosiformes)
    Suborder Gasterosteoidei (3 + 1)
    Suborder Aulostomoidei (4 + 2)
    Suborder Syngnathoidei (2 + 3)
  Order Channiformes (Labyrinthici in part, Ophiocephaliformes) (1 + 2)
  Order Synbranchiformes (Synbranchia, Synbranchii, Synbranchiformes, Alabetiformes)
    Suborder Alabetoidei (1 + 2)
    Suborder Synbranchoidei (2 + 2)
  Order Scorpaeniformes (Cataphracti in part, Scleroparei in part, Loricati, Sclerogeni, Cottomorphi, Perciformes in part)
    Suborder Scorpaenoidei (6 + 7 or 8)
    Suborder Hexagrammoidei (3 + 1)
    Suborder Platycephaloidei (1 + 3)
    Suborder Hoplichthyoidei (1 + 1)
    Suborder Congiopodoidaeei (1 + 1)
    Suborder Cottoidei (9 + 19)
  Order Dactylopteriformes (Craniomi in part, Seleropari, Cataphracti, and Perciformes) (1 + 1)

TABLE 17   (continued)

Division III—(continued)

Order Pegasiformes (Hypostomides, Perciformes in part) (1)
Order Perciformes (Percomorphi in part, Holcodoti, Labyrinthici in part, Chromides,
  Pharyngognathi, Gobioidea, Jugulares in part, Malacichthyes, Icosteiformes,
  Percesoces in part, Mugiliformes in part, Polynemiformes, Rhegnopteri, Bathy-
  clupeiformes, Xenoberyces in part, Berycoidei in part, Beryciformes in part,
  Thunniformes, Plecostei, Scombriformes, Echeneiformes, Discocephali, Mastacem-
  beliformes, Opisthomi, Chaudhuriiformes, Anabatiformes, Squamipenes, Embioto-
  comorphi, Gadopseiformes, Cryphaeniformes, Amphiprioniformes, Blenniformes,
  Trachiniformes, Gobiiformes, Carangiformes, Acanthuriformes)
  Suborder Percoidei (72 + 128)
  Suborder Mugiloidei (1)
  Suborder Sphyraenoidei (1)
  Suborder Polynemoidei (1)
  Suborder Labroidei (3 + 9)
  Suborder Trachinoidei (16 + 9)
  Suborder Notothenioidei (4 + 3)
  Suborder Blennioidei (15 + 19)
  Suborder Icosteioidei (1 + 1)
  Suborder Schindlerioidei (1)
  Suborder Ammodytoidei (2 + 1)
  Suborder Callionymoidei (1 + 1)
  Suborder Gobioidei (6 + 17)
  Suborder Kurtoidei (1)
  Suborder Acanthuroidei (2 + 9)
  Suborder Scombroidei (6 + 15)
  Suborder Stromateoidei (4 + 3)
  Suborder Anabanthoidei (4 + 2)
  Suborder Luciocephaloidei (1)
  Suborder Mastacembeloidei (2 + 1)
Order Pleuronectiformes (Heterosomata)
  Suborder Psettodoidei (1)
  Suborder Pleuronectoidei (4 + 6)
  Suborder Soleoidei (2 + 3)
Order Tetraodontiformes (Plectognathi, Diodontomorphi)
  Suborder Balistoidei (3 + 5)
  Suborder Tetraodontoidei (4 + 11)

---

[a] Numbers in parentheses refer to the number of families.
[b] Orders only, no higher subdivisions.
[c] Three groups: A, 1 family; B, 7 families; C, 10 families.
[d] Two groups: A, one family; B, three families.
[e] Numbers in parentheses refer to the number of families; numbers after the plus sign
refer to those earlier named families included in the families recognized in this classification.
Synonyms or new names due to changes of spelling are not included, only families or parts
of families brought together. In general this indicates the degree of condensation of the
classification at this level but does not indicate which of the families listed by number
include the earlier named families referred to the group of families under the suborder. The
number of included families is necessarily subject to differences in interpretation, but the
numbers give good order-of-magnitude statements.

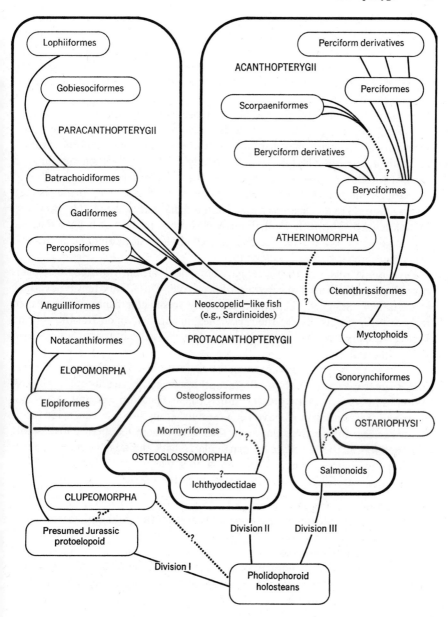

Figure 68.   Provisional phylogeny of the teleost fishes after Greenwood et al. (1966). Note especially the three main divisions, with III being much the largest, and the uncertain position of the clupeomorphs. Compare with Romer's phylogeny (Figure 117) and see representatives of the divisions in the next three figures, 69 through 71.

inevitable and expected by the authors, because their bold and sweeping reorganizations touched on many problems. No major consensus on the broad modifications and no comprehensive alternatives have as yet emerged. Thus this classification will be considered here as the most recent general statement of teleost classification. It makes a basic subdivision of the teleosts into three divisions.

Division I includes the fishes that are close to the holostean level, and the contained groups, except for the eels, each have primitive members. Isospondyli and Clupeiformes, usually given roles as coherent primitive groups, are split up in this classification between primitive orders spread through each of the three units. In division I the two superorders Elopomorpha and Clupeomorpha together include a wide variety of fishes. The former is very diverse, and there are relatively few common characters. The latter, however, is well defined. The authors of the new classification did not believe that the two major groups were very close and polyphyletic development of the two was considered a distinct possibility.

Division II includes the Osteoglossomorpha, primarily osteoglossiforms with the mormyriforms occupying a lesser role. The family Ichthyodectidae in tentatively assigned. These are primarily freshwater fishes, distinguished from others principally by their jaw mechanism, in which the bite occurs between the parasphenoid and tongue, and the presence of paired rods behind the second gill arch.

Division III is large, and contains the majority of teleosts arrayed in three major superordinal groupings: Protacanthopterygii, Paracanthopterygii, and Acanthopterygii, plus two lesser superorders, Ostariophysi and Atherinomorpha. The relationships have been established by use of a very wide range of characters. In this division, as in the others, there has been considerable modification at the family level, particularly with the assignment of several previous families to a single unit. The same is true at the ordinal level. Thus under the Ostariophysi, for example, the first few entries are as follows:

Superorder Ostariophysi
  Order Cypriniformes (Plectospondyli in part, Heterognathi, Gymnonoti, Glanancheli, Evantognathi)
    Suborder Characoidei
      Characidae (Characinidae, including Crenuchidae, Acestrorhynchidae, Serrasalmidae, Tetragonopteryidae, Creagrutidae, Glandulocaudidae)
      Erythrinidae
      Ctenoluciidae (Xiphostomatidae, including Hepsetidae in part)

Some of the characteristic fishes of each superorder are shown in Figures 69 through 71.

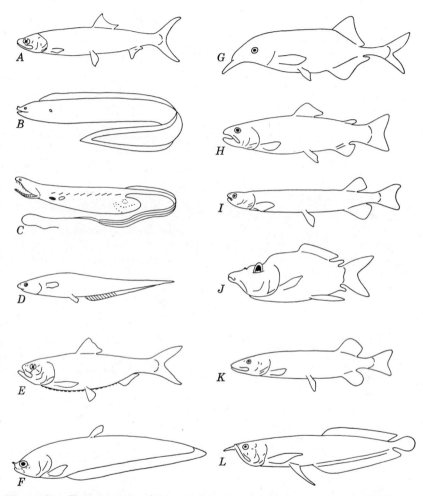

Figure 69.   Representative fishes of division I of Greenwood et al. (1966). Redrawn from figures in this reference. *A*, Elopidae; *B*, Muranidae; *C*, Saceopharyngidae; *D*, Notacanthidae; *E*, Clupeidae; *F*, Notopteridae; *G*, Mormyridae; *H*, Salmonidae; *I*, Galaxiidae; *J*, Opisthoproctidae; *K*, Esocidae; *L*, Osteoglossidae.

Figure 70.   Representative fishes of division II of Greenwood et al. (1966). Redrawn from figures in this reference. *A*, Melanostomatidae; *B*, Chlorophthalmidae; *C*, Chanidae; *D*, Cynodontidae; *E*, Electrophidae; *F*, Cyprinidae; *G*, Siluridae; *H*, Callichthyidae; *I*, Melanocetidae; *J*, Exocetidae; *K*, Cyprinodontidae; *L*, Ateleopodidae.

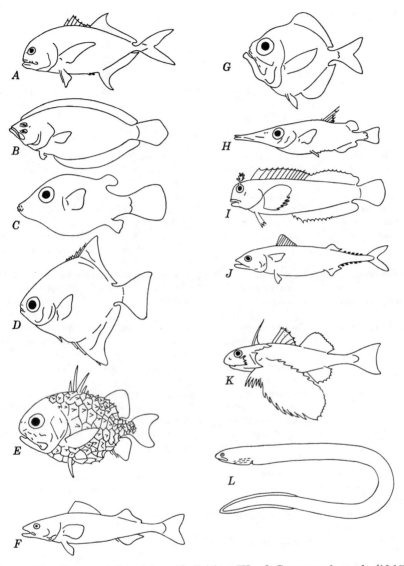

Figure 71. Representative fishes of division III of Greenwood et al. (1966). Redrawn from figures in this reference. *A*, Carangidae; *B*, Citharidae; *C*, Tetraodontidae; *D*, Monodactylidae; *E*, Monocentridae; *F*, Scombridae; *G*, Diretmidae; *H*, Macorhamphosidae; *I*, Blenniidae; *J*, Anoplopomatidae; *K*, Dactylopteridae; *L*, Synbranchidae.

TABLE 18   Characters with Different States in the Three Divisions of the
Teleostei of Greenwood et al. (1966)

Character	Division I	Division II	Division III
Maxilla	Maxillary teeth, seldom excluded from gape	Simple, toothed maxilla generally contributing to gape, partially excluded in a few genera	Maxilla partly or fully excluded from gape by a large, mobile premaxilla
Teeth on palate bones	Parasphenoid and pterygoid teeth developed	Parasphenoid, glossalhyal, pterygoid teeth developed	Loss of pterygoid and parasphenoid teeth
Cranial nerves and vessel passages in basicranium	Numerous interosseous passages for parts of each	Loss of many of interosseous passages in prootic for nerves V, VII, and major vessels in some species	Common passage (trigeminofacialis chamber) in basicranium for nerves V, VII, and head vein
Intermuscular bones	Development of full complement	Upper intermuscular bones only	Reduction of intermuscular bones
Ductus pneumaticus	Functional	Functional	Disappearance of ductus pneumaticus
Otophysic connection	Otophysic connection developed, not involving intercalation of bony elements	As in division I in young or adults, except Osteoglossoidei with no connection	Otophysic connection involving intercalation of bony elements
Lateral lines	Ethmoidal commissure present, confluence preopercular and infraorbital, formation of recessus lateralis	Separated preopercular and infraorbital canals; suprapreopercular bone in one genus	Separated preopercular and infraorbital canals; frequent occurrence of suprapreopercular bone

Other features:
Division II has features not in I, and III has features not in II. Thus in I there are 11 characters listed; in II, 14; and in III, 24.
Characteristic distinctive features are the following:
Division II:
  1. Reduction of size, or loss, of subopercular bone.
  2. Development of somatic electric organs in 1 order.
Division III:
  1. Lowering of center of gravity.
  2. Loss of supraorbital bone.
  3. Covering of posterior part of dorsicranium by epaxial muscles.
  4. Reduction in the number of pectoral radials.
  5. Reduction in the number of vertebrae.

The full merit of this vast revision of former classifications will be judged only after a great deal of study by many specialists in limited fields. The authors considered it as preliminary and expressed their awareness of the many problems that it raises. Basically the arrangements depend on the morphology, largely skeletal, of modern fishes. Many of the characters can be studied in fossils, but little of this was done. Within the very wide range of characters used, some occur through much of the classification, being represented by different states. A great many, however, although used to distinguish one group from another, are not necessarily definitive of one particular rank, being found as well in one or more other groups of equal rank. Basically the groupings must be considered polythetic, based on a wide variety of characters weighted heavily by phylogenetic interpretations.

In Table 18 are listed characters of the three divisions that occur in different states in each. These must be considered as the characters that best express the general adaptive levels. It will be noted that these, and other characters, are largely expressed as trends and are not entirely similar in their degree of development throughout the superorders and lesser categories in the three divisions.

Another point of importance is that each successive division, from I to III, has a larger number of characters specified than did the one preceding it. Thus division I has listed 10 characters; division II, 14; and division III, 24. The added features in division III, not noted in I or II, are in large part new developments, either additions or losses. Although discrete in the listings, they actually represent plus or minus character states.

Within the superorders of a division the numbers of characters that occur in all in different states are much reduced. Some few features are listed for each, but in large part no mention is made of the particular state of a character listed for one order in the other included orders. Table 19 presents a typical example.

This classification represents a new step in treatment of the teleosts. Full reading of the report is necessary for appreciation of the fine points, the strengths and the weaknesses, and the reservations of the authors. The classification is based on the concept of polyphyletic origin of the teleosts from relatively limited sources and of polyphyletic development within the teleosts and the subgroups of this grade. The characters are morphological and mostly expressed as trends. Relatively little dependence on the fossil record accompanied the analysis, but where they were available, syntheses, such as that of C. Patterson (1965a) were taken into account.

The classification provides a new perspective in which the fossil evidence can be examined. As in the case of other classifications based on moderns, a weakness may come from consideration of end members of radiations as the basis for phylogenies, rather than a study of the developing lines and

TABLE 19 Examples of Character States in the Superorders of the Third
Division of the Teleostei, after Greenwood et al. (1966)

Character	Protacanthopterygia	Ostariophysi	Paracanthopterygii	Atherinomorpha	Acanthopterygii
Upper jaw	Trend toward exclusion of maxilla from gape by premaxilla; development of premaxillary process; jaw slightly protrusile in few	Upper jaw protrusile in numerous species	Ascending process of premaxilla is often joined to premaxilla by cartilage or absent; upper jaw not protractile	Upper jaw protractile in many species, with no true ascending process	Upper jaw protractile in many species, premaxilla with ascending process
Photophores	In oceanic representatives	Absent, not noted in listing	Virtual loss of photophores	Not noted, absent	Photophores very common
Hypurals	On one to three centra, basic acanthopterygian caudal skeleton in some, paracanthopterygian in others	Hypurals on one centrum	When caudal skeleton present, two large hypural plates on terminal half-centrum, never more than four, two broad and fan-shaped	Two large hypural plates on terminal half-centrum; no more than four, two broad and fan-shaped	Virtually always from single centrum; when on two, no more than six in number, in no case forming plates

TABLE 19 (continued)

Character	Protacanthopterygia	Ostariophysi	Paracanthopterygii	Atherinomorpha	Acanthopterygii
Branchiostegia	Very numerous in many, reduced to two or three in some cases	Generally few, as many as 15 in some species	Not exceeding six	Four to 15	Epihyal with four bladelike branchiostegals, hairlike branchiostegals, when present, on ceratohyal
Number of vertebrae[a]	Commonly more than 24; precaudal elements commonly 15 or more			High in most species, precaudal number modally 20	Vertebrae commonly numbering 24, normally equal numbers caudal and precaudal, some exceptions
Baudelot's ligament	To first vertebra		To first vertebra, or to basicranium if first vertebra fused to basioccipital	To basicranium	Usually to basicranium, rarely to first vertebra

[a] This is an example of characters that are present in different states and are noted in some, but not all, of the suborders.

recognition of the end members as their products. The tendency in studies of fossils, with this classification in existence, will likely be to fit what is found into the classification, rather than to build a classification on fossils, independently, and examine how the two relate to each other. To the extent that fossils were considered, and in some cases they were given attention, these problems are reduced.

Finally, this new classification is set in the framework of polyphyletic origins for major groups, which enjoys popularity today. How much of this is actually deserved is certainly a matter open to debate. Romer (1965) recently has argued strongly against it. It has been carried to extremes for the whole of the vertebrate world by some authors, to produce classifications with lines passing separately far back in time for all major categories (Jarvik, 1968b). A major point at issue is the reliability and completeness of the fossil record. Polyphyletic origin of the mammals, the reptiles and amphibians, and even the birds has been suggested, and similarly polyphyletic origins within each of these groups have been indicated. The evidence today seems to point in this general direction, yet there can be no question that current trends in classifications, such as this one, are to some extent affected by a theoretical base that outruns the tangible evidence. Reassessments from a more conservative basis may be expected to result in modifications.

7 Class Amphibia

DEFINITIONS AND ORIGINS

The amphibians are currently conceived to form a coherent group of class rank, characterized by possession of tetrapod limbs and the absence of an amnion in developmental stages. This pair of features is definitive except, of course, in those instances in which the tetrapod limbs are absent. It is assumed for such cases that the animals involved had a tetrapod ancestry and that the limbs were lost in the course of evolution.

Most authorities agree that the amphibians arose from a restricted group of fishes among the rhipidistian crossopterygians. Jarvik (1942 and later) has championed the concept of dual origins, with one branch from the osteolepiform and one from the porolepiform crossopterygians. Säve-Söderbergh (1934), following a rather undefined earlier suggestion of Wintrebret, proposed a dual origin that involved lungfishes as ancestral to one branch and crossopterygians as ancestral to another.

Holmgren (1949 and earlier) also strongly supported the concept that one branch of the amphibians arose from the Dipnoi. Of the various suggestions of multiple origins, only that of Säve-Söderbergh and some of his followers strongly modifies the internal classification of the amphibians, which is the concern of this chapter.

Reptiles arose from the amphibians, and many ideas concerning their source or sources have been advanced over the years. Some creatures, now usually classified as anthracosaurs or batrachosaurs (see Table 20) possess many features commonly attributed to amphibians in association with others

TABLE 20   General Amphibian Classifications

I. *Zittle (1895)*

Class Amphibia
  Order Stegocephalia
    Suborder Phyllospondyli
    Suborder Lepospondyli
    Suborder Temnospondyli
    Suborder Stereospondyli
  Order Coecilia
  Order Urodela
  Order Anura

II. *Watson (1919, 1926, 1940)*

Class Amphibia
  Order Labyrinthodontia
    Grade Embolomeri
      Superfamily Anthracosauroidea
      Superfamily Loxommoidea
    Grade Rhachitomi
    Grade Stereospondyli
  Order Phyllospondyli
  Order Lepospondyli
  Order Adelospondyli

III. *Romer (1933)*

Class Amphibia
  Order Labyrinthodontia
    Suborder Embolomeri
      Family Otocratiidae
      Family Palaeogyrinidae
      Family Cricotidae
      Family Pholidogasteridae
      Family Loxommidae
    Suborder Rhachitomi
      Family Eryopsidae
      Family Rhinesuchidae
      Family Dissorophidae
      Family Trematopsidae
      Family Archegosauridae
      Family Trimerorhachidae
      Family Micropholidae
      Family Cochleosauridae
    Suborder Stereospondyli
      Family Capitosauridae
      Family Trematosauridae
      Family Metoposauridae
      Family Brachyopidae
  Order Lepospondyli

IV. *Romer (1945a)*

Class Amphibia
  Subclass Apsidospondyli
    Superorder Labyrinthodontia
      Order Ichthyostegalia
      Order Rhachitomi
      Order Stereospondyli
      Order Embolomeri
      Order Seymouriamorpha
    Superorder Salientia
      Order Eoanura
      Order Proanura
      Order Anura
  Subclass Lepospondyli
    Order Aistopoda
    Order Nectridea
    Order Microsauria
    Order Urodela
    Order Apoda

V. *Tatarinov and Konzhukova (1964)*

Class Amphibia
  Subclass Apsidospondyli
    Superorder Labyrinthodontia
      Order Temnospondyli
        Suborder Ichthyostegalia
        Suborder Rhachitomi
        Suborder Phyllospondyli
        Suborder Stereospondyli
      Order Plesiopoda
    Superorder Salientia
      Order Proanura
      Order Anura
  Subclass Batrachosauria
    Order Anthracosauria
      Suborder Embolomeri
      Suborder Seymouriamorpha
  Subclass Lepospondyli
    Order Nectridea
    Order Aistopoda
    Order Lysorophia
    Order Urodela
    Order Apoda
  Amphibia *incertae sedis*
    Order Microsauria

TABLE 20   (continued)

VI. *Romer (1966a)*

    Class Amphibia
      Subclass Labyrinthodontia
        Order Ichthyostegalia
        Order Temnospondyli
          Suborder Rhachitomi
          Suborder Stereospondyli
          Suborder Plagiosauria
        Order Anthracosauria
          Suborder Schizomeri
          Suborder Diplomeri
          Suborder Embolomeri
          Suborder Seymouriamorpha
      Subclass Lepospondyli
        Order Nectridea
        Order Aistopoda
        Order Microsauria
      Subclass Lissamphibia
        Order Proanura
        Order Anura
        Order Urodela
        Order Apoda

VII. *Säve-Söderbergh (1934, 1935)*[a]

    Class Amphibia
      Superorder Batrachomorpha
        Order Ichthyostegalia
        Order Labyrinthodontia
          Suborder Loxommoidea

      Suborder Capitosauroidea
        Superfamily Eryopoideae
          Eryopidae
          Zatrachyidae
          Rhinesuchidae
          ?Acanthostomatidae
          ?Achelomidae
          ?Melosauridae
          ?Trematopsidae
        Superfamily Capitosauroideae
          Wetlugasauridae
          Capitosauridae
          Mastodonsauridae
        Superfamily
          Trematosauroideae
          Trematosauridae
          Peltostegidae
          ?Archegosauridae
        Suborder Brachyopoidea
          Dvinosauridae
          Brachyopidae
      Labyrinthodontia *incertae sedis:*
        Actinodontidae
        Lydekkerinidae
        Rhytidosteidae
      Order Phyllospondyli
        (includes ?order Anura)
    Superorder Reptiliomorpha
      Order Anthracosauria
      Order Seymouriamorpha

[a] Also includes Reptilia, Aves, and Mammalia as orders.

that are usually found in reptiles. These animals appear to have been near the amphibian–reptilian boundary and have variously been classed as reptiles and as amphibians. The one aspect of the origin of the reptiles that has materially affected the classification of the amphibians relates to this group, and in particular to the included seymouriamorphs and diadectomorphs. The arrangement of major categories of the Amphibia is somewhat altered as these groups are included or excluded.

THE MAJOR GROUPS OF AMPHIBIANS

The term "Amphibia" was introduced by Linnaeus in 1758. It has been used in many senses since that time. Linnaeus's Amphibia included some of the animals now called amphibians, but also others now considered to be reptiles and fishes. Recognition of what are now called "modern amphibians," or Lissamphibia, the Urodela, Anura, and Apoda (Gymnophiona), and definition of these as tetrapod anamniotes, and thus amphibians, posed little problem. As fossils came to be known, however, the concept of amphibian became less concrete, and the sense in which it is now used was slow to emerge and stabilize. The key feature of the amphibians among the tetrapods—the nature of reproduction—could not, of course, be determined for most fossil groups. Osteological features provided the principal criteria and the lack of correspondence between many of these features in ancients and moderns has been a constant source of difficulty. Many genera now considered as Amphibia were first placed among the Reptilia. The labyrinthodonts, as grouped in early classifications by Cope, are a case in point.

An early and influential classification was that used by Zittle in his *Textbook of Paleontology* (1895):[1]

Class Amphibia
   Order Stegocephalia
      Suborder Phyllospondyli
      Suborder Lepospondyli
      Suborder Temnospondyli
      Suborder Stereospondyli
   Order Coecilia (Apoda, Gymnophiona)
   Order Urodela
   Order Anura

Three things stand out in this classification. First, all of the extinct amphibians, following the pattern set by Cope, are placed in a single order, the Stegocephalia, distinct from any of the modern orders. This broad gap has never been fully bridged. Second, the moderns are placed in three distinct orders without indications of relationships between them and without indication of relationships to any particular groups among the fossils. The practice of treating the three modern groups separately has tended to persist in herpetology. In recent years, however, interest in the relationships of fossil and modern orders has spurred new studies of the interrelationships of the moderns. The idea that they may be grouped in a single order, Lissamphibia, has gained headway. Third, and very important in Zittle's classification is the recurrence of the root "spondyl" in the terms used in classifications of fossils. This is an indication of the weight given to vertebral characters in the formation of the major groups. This weighting has persisted and has also entered into efforts to relate modern and fossil genera into common groups. The central structure of the vertebrae is a key in all of the

[1] See also classification I in Table 20.

major classifications of the Amphibia made to the present time. Once again, however, its meaning and value in classification have recently been challenged. Several classifications of the last decades are listed in short form in Table 20. They fall into three general groups.

1. Watson's classification and those related, such as Romer (1933).

2. Romer's classifications (1945*a*, 1947, 1966*a*) and those that are related—for example, Tatarinov and Konzhukova (1964) and Lehman and Piveteau (1955).

3. Säve-Söderbergh (1934).

Each of the classifications suggests some integration of the ancient and modern orders, but the closeness suggested by the tabular classifications has not been met by an equal understanding of the actual relationships. Currently there exist many areas of uncertainty and disagreement.

The discreteness of the major groups extends beyond the schism of the ancients and the moderns, and occurs equally between the fossils called labyrinthodonts, or apsidospondyls, and those called lepospondyls. Each has been treated in large part in isolation, except in occasional studies purposefully designed to draw relationships between them or among all of the major groups of Amphibia. The history of classification of amphibians has for the most part been one in which the major groups have been treated separately. Thus we can most meaningfully analyze the classifications by treatment of each of the groups, rather than by stressing interrelationships. This is the format of this section, and interrelationships are dealt with in context of the major groups as appropriate.

The labyrinthodonts and lepospondyls are primarily distinguished by characters of the vertebral centra (Figure 13). The centra of the former consist of an intercentrum (hypocentrum), alone in some cases, and a centrum (pleurocentrum) that presumably had cartilaginous embryological precursors. The centra of lepospondyls are spool shaped and were presumably formed by ossification in the perichordal sheath of the notochord. This method of formation also occurs in the urodeles, apodans, and with some modifications, among the anurans. The similarities of the vertebrae of the modern and fossil amphibians have been used as evidence of possible relationships. Resemblances of the adults, however, do not necessarily imply common embryological development.

Studies of the vertebral characteristics of the two major living groups by E. Williams (1959) have raised questions about the efficacy of vertebrae as primary indicators of relationships. He has questioned not so much the existence of differences or the use of vertebral form as a basis for major subdivision, as the transfer of criteria founded on embryological studies of a limited group across class boundaries and their application to adults. This is what has been done and largely accepted in the application of Gadow's arcualial determinations to interpretation of the nature of labyrinthodont vertebrae and the consequent explanation and definition of differences between this group and lepospondyls. The difficulties in classification, resulting from the practices questioned by Williams, pertain mostly to the degree of association of the anthracosaurs and temnospondyls and of the relationships of these with modern orders.

Williams' study noted that there was a well-formed sclerocoel in most tetrapods, which, at an early stage, separated the sclerotome into a cranial and caudal half. The Anura and Urodela provided exceptions. Early embryological formation of part of the perichordal tube in the region of the sclerocoel gives rise to an intervertebral disk in each segment, whereas the remainder of the perichordal tube forms in the region of the future centrum. Chondrification and subsequent ossification occur in the perichordal tube, except in apodans and urodeles, in which the ossifications have no cartilaginous precursors.

The single centrum is considered in Williams' study to be homologous in most living tetrapods, being formed intersegmentally in the fused adjacent sclerotome halves. The intercentrum, when present, has its source in the intervertebral disk. Panchen (1967b), in an analysis of the relationship of myocommata to the parts of the centrum, concluded that the whole centrum of the labyrinthodonts is homologous with that of amniotes—that is, with the pleurocentrum. From this he envisaged a twofold evolution, one of the temnospondyls and the other the anthracosaurs and reptiles. As in the case of Williams' study, no major reorganization of classification of the amphibians results from these interpretations.

Schaeffer (1967a) has given evidence, based on Osteichthyes, that there may be considerably greater diversity in the formation of vertebral centra than has generally been thought. He found that modifications in development and in adult design may reflect many changes in the generally uniform morphogenetic pattern of the Osteichthyes. No sclerotomal segmentation occurs in the Osteichthyes, suggesting that this feature may have originated in the rhipidistian–tetrapod transition. He also suggested that the so-called pleurocentra of the rhipidistians (see Figure 72) may be intercalaries, raising further question about the identity of the elements in these fishes and amphibians (also Thomson and Vaughn, 1968).

As in the other cases, these considerations of the variability in development of the centra of vertebrates need not have major effects on classifications that have long been in use. They do, however, raise serious questions about use of adult central structures as a basis for interpreting the embryological course of their development. This becomes especially critical in analyses that use these sorts of data in consideration of the relationships of the lepospondyls

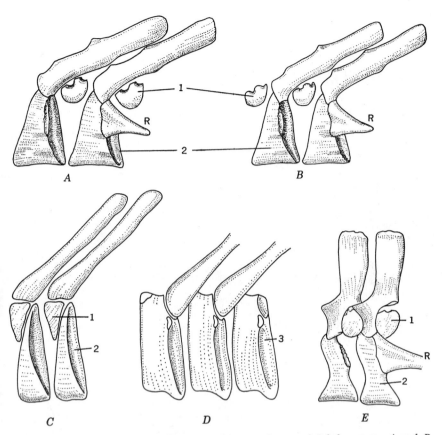

Figure 72.   Vertebrae of rhipidistian crossopterygians and *Ichthyostega*. *A* and *B*
Eusthenopteron. *A*, the usual interpretation (after Jarvik 1955*a*) in which the
element labeled "1" is considered to be a posterior element in the vertebra in which
it occurs, usually considered to be equivalent to the pleurocentrum of tetrapods.
E, Ichthyostega, a labyrinthodont amphibian showing this same interpretation: *B*,
Eusthenopteron as interpreted by Thomson and Vaughn (1968), in which element
"1" is considered to be an intercalary element rather than a pleurocentrum, lying
in the anterior part of the vertebra, having origin from the dorsal anlage. *C*,
Lohsania, a rhipidistian with structure somewhat similar to that of *Eusthenopteron*
(after Thomson and Vaughn, 1968) and treated in the same fashion as this genus by
these authors. *D, Ectosteorhachis*, a Permian rhipidistian, with a single central
element, containing, as interpreted by Thomson and Vaughn, the elements of both
of the central units of *A*, *B*, *C*, and *E*. Thomson and Vaughn concluded that the
patterns in the various rhipidistians and *Ichthyostega* bear no direct relationship to
that seen in the main lines of tetrapod evolution. 1, the element in dispute as
pleurocentrum or intercalary; 2, the principal central element, intercentrum of most
authors; 3, central element with source possibly in several embryological anlagen;
R, rib.

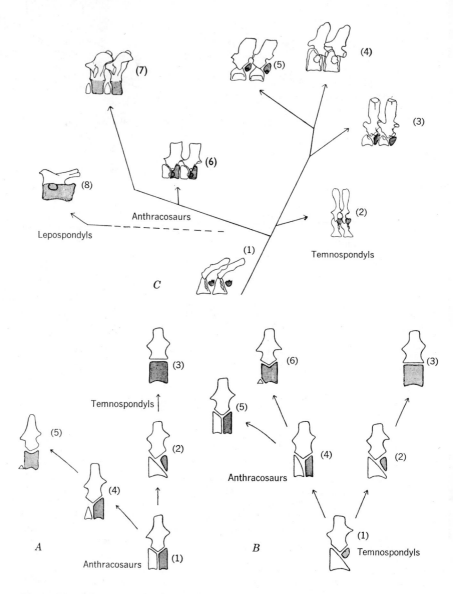

Figure 73. Alternative phylogenetic schemes for vertebral evolution. The general differentiation into anthracosaurs and temnospondyls has been used in each, although it was not so used, except in *C*, in the original presentations. *A*, Watson's scheme (1919, 1926, from Romer, 1964*a*): 1, embolomere; 2, rhachitome; 3, stereospondyl; 4, embolomere with reduced intercentrum; 5, seymouriamorph reptile. *B*, Romer's scheme (1964*a*); a somewhat more elaborate arrangement was

and "modern" orders. Without firsthand knowledge of embryological development, at best very difficult to obtain for fossils, it is not clear how the adult
structures can be interpreted in taxonomic studies.

Many of the features of recent classifications, in which fossils have played
a dominant role, stemmed from Watson's classical studies of 1919 and 1926.
He dealt initially in large part with the extinct labyrinthodonts, but implications of the work went far beyond these limits. Within the labyrinthodonts,
as Watson demonstrated, there are many common trends in skull and
postcranial structures. Irrespective of the particular specializations of one or
another group, many of these trends—such as flattening of the skulls, loss of
the kinetic palatal-basicranial articulation, reduction of braincase ossification,
and enlargement of the interpterygoidal vacuities—were repeatedly realized in
successively more recent representatives. In addition, many similar developmental patterns, more mature when they first appeared, are found among the
lepospondyls.

Judging that he had substantial evidence that these trends could be keyed
to certain kinds of vertebral development (embolomerous, rhachitomous, and
stereospondylous) (Figure 73), Watson proposed a three-grade classification:

> Class Amphibia
> Order Labyrinthodontia
> Grade Embolomeri
> Grade Rhachitomi
> Grade Stereospondyli

To this, of course, was added the Lepospondyli. Later Watson added the
Phyllospondyli (creatures lacking vertebral centra), eventually developing a
full classification as given in Table 20, II. Romer (1933) essentially followed
this classification (see Table 20, III), with one very important difference. He
gave subordinal rank—that is, a formal categorical status—to the grades of
Watson. Technically this raises the problem of polyphyletic origins of the
taxa, a question that does not arise with the use of the informal term "grade."

---

given by Romer (1966a): 1, basal pattern, rhachitomous; 2, rhachitome; 3, stereospondyl; 4, schizomerous form; 5, diplomerous form (= Watson's 1); 6, seymouriamorph reptile. C, Williams' (1959) scheme: 1, basal form, crossopterygian; 2,
ichthyostegid, rhachitomous form; 3, rhachitome; 4, stereospondyl; 5, trematosaur,
rhachitomous form; 6, embolomere; 7, seymouriamorph reptile; 8, lepospondyl,
not given in other diagrams. The element usually called the pleurocentrum (or true
centrum) is marked by the stippled pattern, that termed hypocentrum (or intercentrum) is without pattern.

Apsidospondyli

Continuing study by Romer led to a revision of the basic threefold sequence of labyrinthodonts by Watson. The new arrangement was presented in his 1945 and 1947 classifications (Table 20, IV, and Table 21, I). In the course of these studies he established a new term, "Apsidospondyli." This was necessary to provide a broad category in which both Labyrinthodontia and Salientia, anurans and their ancestors, could be included to show their supposed relationships. In these studies he concluded that the general rhachitomous form of vertebral centrum was primitive among the labyrinthodonts. Information had become available to suggest that this type of centrum was present in the rhipidistians and the very earliest of the amphibians, ichthyostegals, from the upper Devonian. Although Romer later dropped the term "Apsidospondyli," it gained wide acceptance in classifications that were published between 1945 and 1965.

Some of Romer's modification of Watson's general classification was the result of new information (e.g., the addition of Ichthyostegalia from the Devonian of Greenland). Other changes represent judgments on degrees of relationships—for example, the distinctness of the Trematosauria and Stereospondyli. The fundamental change, however, was the treatment of the embolomeres as one element of the anthracosaurs, a group considered equivalent to temnospondyls, in which rhachitomes and stereospondyls were included. Not only was the embolomerous vertebra no longer considered primitive, it was thought to have been derived from the labyrinthodont type. Completing this pattern was the classification of the seymouriamorphs, often considered as reptiles, among the amphibians as part of the parallel phylum. Inclusion of the anurans essentially followed the work of Watson (1940) in which he assigned some Paleozoic forms to the Eoanura. On the basis of his own studies in which he concluded that the branchiosaurs, the mainstay of the Phyllospondyli, were larval labyrinthodonts, Romer eliminated the Phyllospondyli from his classification.

The phylogeny of vertebral structure as envisaged by Romer is shown in Figure 73, along with that of Watson (1919, 1926) and of E. Williams (1959). A threefold array in which all moderns are considered as closely related is shown in Williams' phylogeny. He did not present a formal classification, but implications are clear in the diagram.

To date most formal classifications since 1947 have followed, with some modifications, the scheme of Romer. Examples are given in Table 21. These are all based on the concept that there is a distinctive vertebral type, apsidospondylous, which is found in the labyrinthodonts and Anura or Salientia. A modification of the sort that was suggested by Williams, but not based on his work directly, is seen in the Tatarinov–Konzhukova classification.

TABLE 21   Classifications of the Apsidospondyli and Labyrinthodontia

I. *Romer (1947a)*

Subclass Apsidospondyli
  Superorder Labyrinthodontia
    Order Temnospondyli
      Suborder Ichthyostegalia
        Elpistostegidae,
        Ichthyostegidae,
        Otocratiidae, Colosteidae
      Suborder Rhachitomi
        Superfamily Loxommoidea
          Loxommidae
        Superfamily Edopsoidea
          Edopsidae, Dendrerpetontidae
          Cochleosauridae
        Superfamily Trimerorhachoidea
          Peliontidae,
          Trimerorhachidae,
          Dvinosauridae
        Superfamily Micropholoidea
          Lysipterygiidae,
          Micropholidae,
          Chenoprosopidae,
          Archegosauridae
        Superfamily Eryopsoidea
          Eryopsidae, Trematopsidae,
          Dissorophidae, Zatrachydidae
      Suborder Trematosaurea
        Trematosauridae
      Suborder Stereospondyli
        Superfamily Rhinesuchoidea
          Rhinesuchidae,
          Lydekkerinidae,
          Uranocentrodontidae,
          ?Sclerothoracidae
        Superfamily Capitosauroidea
          Benthosuchidae,
          Capitosauridae
        Superfamily Brachyopoidea
          Brachyopidae,
          Plagiosauridae,
          Metoposauridae
    Order Anthracosauria
      Suborder Embolomeri
        Anthracosauridae,
        Palaeogyrinidae,
        Pholidogasteridae, Cricotidae

      Suborder Seymouriamorpha
        Diplovertebrontidae,
        Discosauriscidae,
        Seymouriidae, Kotlassiidae
  Superorder Salientia
    Order Eoanura
    Order Proanura
    Order Anura

II. *Tatarinov and Konzhukova (1964)*

Class Amphibia
  Subclass Apsidospondyli
    Superorder Labyrinthodontia
      Order Temnospondyli
        Suborder Ichthyostegalia
          Ichthyostegidae,
          Acanthostegidae
        Suborder Rhachitomi
          Superfamily Colosteoidea
            Otocratidae, Colosteidae
          Superfamily Loxommoidea
            Loxommidae
          Superfamily
          Cochleosauroidea
            Edopsidae,
            Dendrerpetontidae,
            Cochleosauridae
          Superfamily
          Trimerorhachoidea
            Trimerorhachidae
          Superfamily Micropholoidea
            Lysipterygidae,
            Micropholidae
          Superfamily
          Archegosauroidea
            Archegosauridae,
            Chenoprosopidae,
            Platyopidae,
            Melosauridae
          Superfamily Eryopsoidea
            Eryopsidae,
            Trematopsidae,
            Intasuchidae,
            Dissorophidae,
            Zatrachydidae

TABLE 12—continued

II. *Tatarinov and Konzhukova (1964)*—Cont.

Suborder Phyllospondyli
  Peliontidae,
  Branchiosauridae
Suborder Stereospondyli
  Superfamily Capitosauroidea
    Rhinesuchidae,
    Benthosuchidae,
    Capitosauridae,
    Mastodonsauridae,
    Cyclotosauridae,
    Yarengiidae
  Superfamily
  Trematosauroidea
    Trematosauridae,
    Rhytidosteidae,
    Peltostegidae
  Superfamily Brachyopoidea
    Brachyopidae, Tupilako-
    sauridae, Metoposauridae
  Superfamily Plagiosauroidea
    Plagiosauridae
Suborder Plesiopoda
  Hesperoherpetontidae
Superorder Salientia
Order Proanura
Order Anura
Subclass Batrachosauria
Order Anthracosauria
  Suborder Embolomeri
    Anthracosauridae,
    Palaeogyrinidae,
    Pholidogasteridae,
    Cricotidae
  Suborder Seymouriamorpha
    Gephyrostegidae,
    Discosauriscidae,
    Seymouriidae,
    Kotlassiidae,
    Bystrowianidae,
    Chroniosuchidae,
    Lanthanosuchidae,
    Waggoneriidae

III. *Lehman, Jarvik, and Pivateau (1958)*[a]

Series Apsidospondyli
  Labyrinthodontia
    Order Temnospondyli

Suborder Ichthyostegalia
  Ichthyostegidae,
  Acanthostegidae
Suborder Rhachitomi
  Primitive:
    Superfamily ?Otocratiidea
    Superfamily ?Colosteidea
    Superfamily ?Loxommoidea
    Superfamily Edopsoidea
    Superfamily
    Trimerorhachoidea
  Evolved:
    Superfamily Micropholoidea
    Superfamily Eryopsoidea
Order Anthracosauria
Order Phyllospondyli
  Eugyrinidae,
  ?Peliontidae,
  Branchiosauridae
Anura
Order Proanura
Order Euanura

IV. *Romer (1966a)*

Class Amphibia
  Subclass Labyrinthodontia
    Order Ichthyostegalia
      Elpistostegidae,
      Ichthyostegidae,
      Otocratiidae
    Order Temnospondyli
      Suborder Rhachitomi
        Superfamily Loxommatoidea
        Superfamily Edopoidea
        Superfamily Trimerorhachoidea
        Superfamily Eryopoidea
        Superfamily Trematosauroidea
      Suborder Stereospondyli
        Superfamily Rhinesuchoidea
        Superfamily Capitosauroidea
        Superfamily Brachyopoidea
        Superfamily Metoposauroidea
      Suborder Plagiosauria
    Order Anthracosauria
      Suborder Schizomeri
      Suborder Diplomeri
      Suborder Embolomeri
      Suborder Seymouriamorpha[b]

---

[a] Mostly the same as Romer (1947a). Only the parts that are different are listed here.
[b] Six to eight families, including Diadectidae.

Although the Salientia are retained as apsidospondyls, the Batrachosauria (Anthracosauria) are given subclass recognition. This followed mostly from earlier work of Efremov (1946) which established the basis for such a scheme. Romer (1966a) concluded that there was no solid evidence that the anurans had apsidospondylous vertebrae and revised his classification to a threefold pattern, dropping Apsidospondyli and including as equal in rank Labyrinthodontia, Lepospondyli, and Lissamphibia.

Labyrinthodontia

The term "Labyrinthodontia" was introduced into classifications very early, dropped as noted above, and then reestablished. Irrespective of the ideas of phylogeny and the efficacy of application of embryonic distinctions to classifications of adults, the nature of the vertebral centrum in classifications of extinct amphibians remains dominant. The ease of separation of basic types by vertebral features and the close correlation of changes in vertebrae and other osteological structures have resulted in its persistence.

Within the labyrinthodonts, Romer's use of the ordinal term "Temnospondyli" has become quite general, with the subordinal designations of Rhachitomi and Stereospondyli broadly recognized. The inclusion of Ichthyostegalia as a suborder is common, but Romer (1966a) has given it an ordinal rank. Separation of the Trematosauria as a distinct suborder has been followed, but not as consistently. The term "Temnospondyli" was introduced by Zittel, and Romer's usage did not modify its original intent seriously. Below the ordinal level, in the formation of the suborders, there is a partial departure from the strict use of vertebral characters. The distinction between the rhachitomes and stereospondyls is basically vertebral, but skull characters play an important role both with respect to these two and also to the separation of Ichthyostegalia and Trematosauria. In addition temporal considerations, as in the case of the Ichthyostegalia, and adaptations (e.g., the aquatic adaptations of the Trematosauria) are important.

An elaborate classification that depended basically on skull characters was given by Säve-Söderbergh (1934, 1935). A number of the characters of the skull that he used were those used by Watson before him, and they have come over into more recent classifications. Importance is accorded the presence or absence of the intertemporal bone, the contact or lack of contact of various dermal skull elements, the relative length of the posterior platform of the skull, the nature of the otic notch and the presence or absence of a tabular extension, the position of the jaw articulation relative to the occipital condyle, the nature of the occipital condyle, the kinetic or nonkinetic character of the palatobasicranial joint, the expansion of the parasphenoid, and the size of interpterygoid vacuities.

Most of these characters are expressed as trends, but those of the dermal elements, especially presence or absence and particular contacts, are prone to be absolute and are used as strictly definitive at various levels. Thus skulls with the flat, semitriangular pattern characteristic of *Trimerorhachis* are considered to pertain to the Trimerorhachoidea if an intertemporal element is present and Colosteoidea if this element is absent.

The problem of the Phyllospondyli still persists. Romer eliminated the group as comprising only larval labyrinthodonts. Certainly some of the included forms were just this. Other students—Watson, Lehman, and Tatarinov and Konzhukova—have disagreed and have retained the group as valid. The membership, however, varies between the various classifications (see Table 21).

A feature of recent classifications is the inclusion of a suborder Plesiopoda under the temnospondyls. This is based on a Pennsylvanian amphibian given subordinal position after a detailed study by Eaton and Stewart (1960). It appears to have been a very aberrant amphibian, as compared with others known from the same general time. The placement must be tentative until more is known, for the only materials are very badly preserved.

Aside from rather minor groups, such as the one just mentioned, the classification of the Temnospondyli, or the labyrinthodonts as a whole, approaches the ideal of a phylogenetic arrangement more closely than do most classifications. The group is self-contained, and within it the sequences, as expressed in classifications, seem to represent coherent evolutionary successions.

Anthracosauria: Batrachosauria

These two terms may be considered synonymous or, as treated by Tatarinov and Konzhukova, Batrachosauria may be used to include the Anthracosauria. In Watson's classification the embolomeres, one group of Romer's (1947a) Anthracosauria, represented the most primitive labyrinthodonts—and the seymouriamorphs, the other subdivision of Romer's (1947a) Anthracosauria, were considered to be reptiles. Romer (1966a) made a more extensive subdivision on the basis of vertebral structures, as shown in Table 21, III, and Figure 73.

Here, as earlier, Romer considered seymouriamorphs as amphibians, as had Watson in his later classifications. Much of the support for this assignment has come from the evidence of presence of external gills in immature members of the family Discosauriscidae (Spinar, 1952) and discovery of lateral-line canals in *Seymouria* (T. E. White, 1939). In 1964 Romer suggested that the diadectids, long considered reptiles, also be included within the amphibians.

Primary to recognition of the Anthracosauria (or Batrachosauria) as a coherent group is the structure of the vertebrae. The centrum ranges from schizomerous through embolomerous and to typically reptilian form in the seymouriamorphs and diadectids. The last two also have swollen neural arches, a character shared with many captorhinomorphs and procolophons. In addition to the vertebral characters, the states of various skull features, similar to those found in some of the temnospondyls, are significant. It is important, however, that most trends of evolutionary development seen among the temnospondyls, and also in lepospondyls, were absent. This alone provides a strong basis for considering that the anthracosaurs diverged widely from other labyrinthodont amphibians. Anthracosaurs, in general, retain or emphasize these characters of primitive amphibians that persist among the early reptiles.

The anthracosaurian assemblage has raised some interesting problems of classification, problems not restricted merely to this particular group. Some were noted briefly in the introduction to this section. Morphological features of the adult, particularly among the seymouriamorphs and diadectids, are those usually identified partly with amphibians and partly with reptiles. Since the fundamental differentiation of the two classes hinges on reproduction, in particular on the presence or absence of an amnion, the morphological structures in question are of a secondary sort; that is, they consist of features usually found in association with the definitive reproductive characters and by virtue of the association given special meaning. The associations must be based largely on modern representatives, where the reproduction can be known. On the basis of morphological characters alone, *Diadectes* could only be classed as a reptile, and this is what was done before Romer's recent suggestion. *Seymouria*, the best known seymouriamorph, has both reptilian and amphibian features, but on the basis of balance between them it was long classified as a reptile. A strictly numerical evaluation of the adult features indicates such a placement. *Kotlassia*, the well-known genus from Russia, gives much the same answer.

T. E. White (1939), in his detailed study of *Seymouria*, attempted to find adult characters that would relate closely to reproductive processes. He found a trace of lateral-line canals, suggesting an aquatic larval stage, but he also found what appeared to be a sexual dimorphism in the haemal arches, much like that found in recent lizards, which lay large eggs. Torn, he opted in favor of a reptilian assignment. There is, of course, nothing sacred about the categories of reptile and amphibian, and it may well be that such forms as these do not fit either category. A suggestion acknowledging the insolubility of assignment to categories that must develop artificiality at the intersects was the recognition of a separate class, Batrachosauria, for this limited radiation (Olson, 1962).

Among the seymouriamorphs, however, the discosauriscids, as noted, do have a well-defined, gilled immature stage, and by definition members of this family are amphibians. Thus it can be thought that all similar animals—that is, all seymouriamorphs—are amphibians, although this can be questioned. Then the next step is to consider *Diadectes*, which has some otic and cranial features that are similar to those of *Seymouria* and swollen vertebral arches, to have been an amphibian, although here the ontogeny is totally unknown and most of the definitive morphological features are those generally called reptilian (Romer, 1964c; Olson, 1965b).

Classifications based on such tenuous lines of reasoning are certain to be controversial, and this has been the story of the classification of the embolomeres, seymouriamorphs, and diadectomorphs. The resolutions of the problem of assignment bear immediately on the constitution of the amphibians and the reptiles. They have far-reaching implications concerning the origin of reptiles as will be discussed in the next chapter. No complete solution is at hand at present, and none seems likely to be forthcoming unless new evidence of a fairly direct kind is unearthed. It is evident that the anthracosaurs had an extensive history of radiation, that this was quite distinct from that of other amphibians, and that in the course of this radiation reptilelike animals developed. Very likely from within it, by means not known, the reptiles proper came to be. This course was not, however, through such end members as diadectids or known seymouriamorphs. If any of these attained a reptilian mode of reproduction, it must have been independent of that in the main reptilian line or lines.

Within the subgroups of anthracosaurs, irrespective of their major assignment, there are problems of classification. The embolomeres appear to have carried out much of their radiation prior to the time for which a good record is available. Thus knowledge of them, except for a few genera (*Pholidogaster*, *Diplovertebron* (?*Gephyrostegus*)) and late members, such as *Archeria*, is relatively slight. Two of the suborders in Romer's 1966 classification contain only one certainly defined genus. One of the big problems has been the assignment of various genera that actually belong elsewhere to this group. It was this difficulty that led in part to Watson's conclusion that almost all of the early amphibians were embolomeres. Vertebrae were found associated with a few, so comparable genera, for which vertebrae were unknown, were similarly assigned. Some of these problems have been reduced by new finds and restudy of old materials, but placement of some genera, for which vertebrae still are not known, remains uncertain.

Increase in knowledge of the seymouriamorphs since the time of Romer's classification in 1947 has enlarged the number of families and given a picture of a rather extensive adaptive radiation (see Tatarinov and Konzhukova, Table 21; Olson, 1965b). Most of the eight families now usually included are seymouriamorphs, but questions have been raised about the placement of the Lanthanosuchidae and Waggoneriidae (Romer, 1966*b*). The diadectids form a very limited group, known mainly from *Diadectes* and closely allied genera. The radiation of the family appears to have been limited. Whether placed among the amphibians or the reptiles, the diadectids introduce difficulties. Watson (1954) considered them to be the earliest known representatives of the Sauropsida reptilian radiation. Olson (1947) considered them close to the Chelonia and the base of a separate reptilian radiation termed Parareptilia, a position reversed later (Olson, 1965*b*). No intermediates between diadectids and other groups have been found, and the family goes well back into the Pennsylvanian. Recently, however, Vaughn (1964*a*) has described a genus *Tseajaia* from the lower Permian, and this has some features that resemble those of *Diadectes* and others that seem somewhat seymouriamorph. It may be a persistent member of an intermediate stock.

Lepospondyls

This subclass is one of the most remote and self-contained of any among the tetrapods. Its origins are unknown and, in spite of frequent proposals of associations with the urodeles and apodans, positive evidence of close relationships to any other amphibians is very slight.

Key characters are the spool-shaped centrum and the usual absence of any "intercentral element". Recent studies have revealed the presence of intercentra in some of the gymnarthrid microsaurs and in *Lysorophus* (Brough and Brough, 1967). The structure of the centrum does not provide a clear basis for separation from the modern orders of amphibians.

A suborder Lepospondyli was erected by Zittel in 1890 to include the microsaurs of Dawson (1860) plus some known nectrideans and aïstopods. The three groups are distinctly separate, but they have enough in common to justify placement in the same subclass. There are no known links between them. The association is generally accepted by modern students, although Tatarinov and Konzhukova (1964) place the microsaurs in Amphibia, *incertae sedis*. Two recent classifications by Dechaseaux (1955) and Tatarinov and Konzhukova (1964) illustrate some of the differences that occur (see Table 22). These revolve almost entirely around the microsaurs.

Classification of the microsaurs has had a long and rather hectic history. As the name suggests some features of the animals so classed are suggestive of reptiles. Figure 74 shows comparisons of the skull pattern of microsaurs and captorhinomorphs. The problems of classification were aggravated by the fact that some animals classified as microsaurs were in fact reptiles (see J. T. Gregory, 1948). Once the reptiles are removed, many of the characters of the

TABLE 22   Classifications of the Lepospondyli

I. *Dechaseaux (1955)*	II. *Tatarinov and Konzhukova (1964)*
Series Urodeliformes	Subclass Leposondyli
Subclass Lepospondyli	Order Nectridea
Order Aïstopoda	Urocordylidae
Ophiderpetontidae	Keraterpetontidae
Dolichosomidae	Order Aïstopoda
Order Nectridea	Ophiderpetontidae
Urocordylidae	Phlegethontiidae
Keraterpetontidae	Order Lysorophia
Lepterpetontidae	Lysorophidae
Order Microsauria	Lysorocephalidae
Adelogyrinidae	(Plus orders Urodela, Proteida,
Dolichopareiidae	Apoda)
Lysorophidae	Amphibia—*incertae sedis*
Microbrachidae	Order Microsauria
Gymnarthridae	Adelogyrinidae
Ostodolepidae	(including Dolichopareiidae)
Pantylidae	Microbrachidae
(Plus orders Urodela, Apoda)	Gymnarthridae
	Pantylidae
	Ostodolepididae

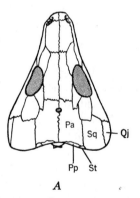
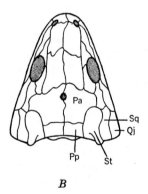

A   B

Figure 74.   A captorhinomorph skull and a lepospondyl (microsaur) skull compared in dorsal aspect. Note especially the differences in the temporal region and the close resemblances of much of the rest of the skull. *A, Captorhinus; B, Hyloplesion.* Pa, parietal; Pp, postparietal; Qj, quadratojugal; Sq, squamosal; St, supratemporal. (After Romer, 1950.)

palate, temporal region, occiput, and postcranium clearly show the non-reptilian nature of the microsaurs. Resemblances, on the other hand, have led students to seek a source of the reptiles within this group (Westoll, 1938; Vaughn, 1962). Complicating things still further was Watson's recognition of a separate group, Adelogyrinoidea, on the basis of the separation of the centrum and the neural arch by a suture. Romer (1950) reviewed the whole problem of the microsaurs, concluding that they formed a coherent group, that they had nothing to do with the reptiles, and that there were no particular indications of relationships to any other tetrapods beyond the lepospondyls. J. T. Gregory (1965) restudied the constitution of the group and arrived at similar conclusions, in particular with respect to the lack of relationships to the reptiles.

Aïstopods and nectrideans do not pose internal problems, because they are readily definable and known from so few different kinds that phylogenetic problems rarely arise. The aïstopods form an isolated, consistent group of highly specialized, limbless, aquatic amphibians, first known in the lower Mississippian and persisting into the Permian. There is no knowledge of their source and they have no recognized descendents. The "horned" members of the Nectridea form a highly distinctive and successful group, best known from *Diplocaulus* and *Diploceraspis* of the Permo–Carboniferous.

Lysorophus and its few known relatives are placed either with the microsaurs or in a separate group of equivalent rank. They were highly distinctive, elongated animals, with a rather stout body and small limbs. Evidence that *Lysorophus* was an aestivator sets it well apart from most early amphibians. Relationships to the apodans and also to amphisbaenid lacertilians have been suggested, but there is little basis for either proposition. Like the other lepospondyls they are highly specialized but possess a vertebral centrum that suggests a common origin.

The Modern Orders: Lissamphibia

The problems of classification of the modern orders fall into three general categories:

1. The internal relationships within each order.
2. The relationships of each of the orders to other moderns.
3. The relationships of each to the various extinct orders.

To some degree the last two are interdependent and are treated together, because neither can be studied without some consideration of the other. Gadow (1896) proposed the term "Lissamphibia" for the modern orders, indicating a concept of their general relationship to each other. This rarely found its way into formal classification until rather recently, when it has come into frequent use. The term was used formally, for example, by Romer (1966a). If a derivation of the modern orders from different ancient stocks is thought to be correct, whether this be from separate amphibian lines or from different groups of fishes, no such term of association is needed or proper.

Recent studies, especially those of Parsons and Williams (1962, 1963), have given evidence for close relationships of the Apoda, Anura, and Urodela, the last two in particular, and have had a considerable effect on classification during the last few years.

One of the major features presumed to show the unity of the modern groups is the possession of pedicellate teeth in which the crowns and bases are partially separate. This had been known elsewhere only among some of the teleost fishes. Very recently, however, Bolt (1968) has found good evidence of the existence of such teeth in a new Permian genus *Doleserpeton*, a dissorophid temnospondyl. This obviously has bearing on the nature of modern groups, but its significance has not yet been clearly determined.

Internal Relationships of the Orders

The classifications of Noble (1931) of the Caudata (or Urodela) and Salientia (or Anura) set patterns that have been used as points of departure since their establishment (see Table 23, I and II). The very restricted group of Apoda (Gymnophiona or Coecilia) has to date received little attention above the genus and species level.

URODELA (CAUDATA). Noble's classification of the Urodela is based on a complex of characters that gives a polythetic aspect to the arrays. It was constructed on the basis of modern animals and for the most part fossil urodeles have played little part in the formulation of any classifications. Much of the fossil evidence now available has come to light rather recently, except for sporadic finds to which little attention was paid. The accumulation of new data is gradually setting the stage for a sweeping modification of the classi-fication of Noble. No general rearrangement has been made as yet, although there have been reorganizations of particular groups and many studies showing that problems exist in various generally accepted relationships. Estes (1965) summarized the current situation, indicating in particular that changes are to be made in the Salamandroides and the Meantes. He did not, however, propose a comprehensive new classification.

Detailed studies of sound-transmitting systems, musculature, vertebrae, and biochemistry are leading the way to a better comprehension of moderns. To date the fossil evidence has had only minor impact on the classification. That its role will increase is suggested by some recent and problematical finds. Vaughn (1963) described vertebrae from the lower Permian that Kuhn (1964, 1965) considered to be urodele. This is based on very tenuous evidence.

TABLE 23   Classifications of the Anura and Urodela

I. *Noble (1931)*
Order Caudata
  Suborder Cryptobranchoidea
    Family Hynobiidae
    Family Cryptobranchidae
  Suborder Ambystomoidea
    Family Ambystomidae
  Suborder Salamandroidea
    Family Salamandridae
    Family Amphiumidae
    Family Plethodontidae
  Suborder Proteida
    Family Proteidae
  Suborder Meantes
    Family Sirenidae

II. *Noble (1931)*
Order Salientia (Anura)
  Suborder Amphicoela
    Family Liopelmidae
  Suborder Opisthocoela
    Family Discoglossidae
    Family Pipidae
  Suborder Anomocoela
    Family Pelobatidae

  Suborder Procoela
    Family Palaeobatrachidae
    Family Bufonidae
    Family Brachycephalidae
    Family Hylidae
  Suborder Diplasiocoela
    Family Ranidae
    Family Polypedatidae
    Family Brevicipitidae

III. *Tatarinov and Konzhukova (1964)*
Order Urodela
  Suborder Cryptobranchoidei
    Family Cryptobranchidae
  Suborder Meantes
    Family Sirenidae
  Suborder Ambystomatoidei
    Family Ambystomatidae
  Suborder Salamandroidei
    Family Salamandridae
    Family Amphiumidae
    Family Batrachosauroididae
  Suborder Proteida
    Family Proteidae

More recently Tatarinov (1968*b*) has described a maxillary from the upper Permian of the Soviet Union (Zone II) as a representative of a caudate amphibian. This new genus and species, *Permotriturus herrei* Tatarinov, does show dental and osteological features similar to those in some salamanders. Yet it is very fragmentary and must be viewed with some skepticism pending more complete materials. If more such finds are made and Mesozoic genera that are more primitive than those known to date are found, fossils may eventually provide the necessary evidence for understanding the broad relationships.

ANURA (SALIENTIA).   Noble's classification (Table 23, II) of the Anura has been closely followed by subsequent workers, with a few notable exceptions. It is clear that it has inadequacies and alternative classifications such as that of Reig (1958) have been proposed. Reig united the Procoela and Diplasio-coela into Neobatrachia (an order) and the Opisthocoela and Anomocoela into the Paleobatrachia. The modifications have not been widely accepted. The basis of Noble's classification, as evident from the names of the major groups, is the condition of the centra of the vertebrae, including the nature of ribs and the coccygeal condyle. Tihen (1965) gave a summary of the problems of classification. He recognized levels of organization and attempted to evaluate various of Noble's groups with regard to them. Some of the diffi-culties of Noble's classification are pointed out, but no overall modification is proposed. Fossil evidence for the Anura is more extensive than that for the Urodela. As far back as the Jurassic, however, there is little to indicate the paths that classification should take. Prior to that there is but a single speci-men. Gradually studies of fossil frogs are beginning to have influence, but to date there has been little concrete contribution to an understanding of the general relationship among moderns.

APODA (GYMNOPHIONA, COECILIA). These odd, nearly limbless am-phibians are virtually unknown as fossils, having been reported only from very problematical remains. The vertebral centra are of the same general type as those of other modern orders, and in this there is a possible basis for assuming relationship. Association with the urodeles has generally been loosely made, but without any very firm basis. The ancestry of the apodans has been sought among the lysorophids, but once again, this is based more on general re-semblances than on sound morphological evidence. The problems of classi-fication include this major one, the placement of apodans within the Amphibia, for which there are no satisfactory answers, and the problems of the genus and species composition and relationships within the restricted group.

Relationships between Modern and Ancient Orders

Most recent classifications place the anurans under Romer's (1947a) Apsidospondyli and the urodeles with the Lepospondyli (see Table 20). Romer (1966a), as noted, abandoned this arrangement, dropping the concept of Apsidospondyli and accepting the evidence of close relationships of modern orders by placing them in the subclass Lissamphibia.

The association of the Anura and Urodela respectively with Apsido-spondyli and Lepospondyli, which are separate as far back as the record goes, implies of course a very ancient separation, either within the amphibians or even among the fishes. This relationship was based mostly on vertebrae and some of the problems in this approach have been discussed earlier. Among other things that are pertinent, E. Williams (1959) noted that among the anurans alone there are two rather distinct ways of formation of the centrum, one perichordal and the other epichordal. The very rapid modification of vertebral structure at the time of metamorphism also undoubtedly clouds this issue.

Much of the evidence for relationships of anurans and labyrinthodonts, prior to recent work such as that of Bolt on *Doleserpeton*, was embodied in Watson's (1940) study of *Miobatrachus-Amphibamus* and analyses of *Triadobatrachus* (*Protobatrachus*), a specimen from the lower Triassic of Madagascar. Watson's study gave the basis for Eobatrachia, included under Salientia or Anura in various classifications (see, for example, Romer, 1947a). J. T. Gregory (1950) reexamined this problem and indicated that the specimens involved appeared to be rhachitomes and that, although they were plausible ancestors for anurans, they did not show anything sufficiently definitive to argue for placement in a subdivision of the Salientia (i.e., the Eoanura). The closest affinities were thought to be with the Dissorophidae. The anurans, under such an interpretation, still fall within the apsidospondyls, but Eoanura is deleted.

With such a deletion two orders remain, Proanura, based on the Madagascar specimen, and Euanura (or Anura). In many respects the Triassic fossil resembles the frogs and it is generally considered to be a primitive relative. Hecht (1962), however, has questioned the evidence for the relationship. All remaining anurans, from the Jurassic to the present, fall well within the limits of the Euanura. At present the evidence for association of the Anura with the labyrinthodonts is not strong, although there is nothing that seriously contradicts this assignment.

The supposed relationship of Urodela and Apoda to the lepospondyls is on much more tenuous grounds. The vertebral central elements of the two are spool shaped and do have general resemblances. Resemblances of body form between some lepospondyls and moderns, microsaurs and some salamanders or lysorophids and apodans, for example, are suggestive, but detailed comparisons show little fundamental morphological similarity. Both E. Williams (1959) and Eaton (1959) questioned the validity of the criterion of the vertebrae. The latter made a thorough review of available structures and found little positive to suggest relationships.

Evidence for close association of the modern orders, from the work of Parsons and Williams (1962, 1963), supported by Szarski (1962), Reig (1964), and to some extent Estes (1965), suggests a common origin of these groups from some single ancestral group. Tendencies in the most recent work have been to seek this source among the dissorophid temnospondyls. Reig has argued for a branchiosaur origin, raising the problem of just what these amphibians are and how they are to be interpreted. Such an origin is not necessarily incommensurate with a temnospondylous source. Bolt's (1968) discovery of pedicellate teeth in a dissorophid strengthens the concept that the origin of moderns may lie within this general group, as does the supposed caudate amphibian described from the Soviet Union by Tatarinov (1968*b*). Through all of these considerations the Apoda remain somewhat of an enigma, without clear relationships to other modern orders and without any information about possible ancestors.

PROSPECTUS

Reevaluation of the problems of amphibian classification at all levels is continuing. Settlement of many of the major problems on the basis of the information now available seems out of the question. New finds among fossils are being made and detailed studies of modern groups—especially their embryology, morphology, and biochemistry—are being intensively carried out.

Romer's (1966*a*) classification furnishes a solid base from which new work may proceed. Currently the most stable group among the extinct amphibians appears to be the labyrinthodonts. The lepospondyls may remain little changed, but the associations are far from clear and new discoveries may result in drastic alterations of the concept of lepospondyl or of the internal organization of the group. The greatest remaining problems, however, are those of the relationships of the extinct and living orders. The perspectives have changed rapidly over the last several years and solutions appear closer. Without considerable additional information from fossils from critical stages, however, problems of relationships will remain unsolved.

8 Class Reptilia

The class Reptilia as currently conceived includes all cold-blooded, strictly poikilothermic, amniotes. These two characters are sufficient for decisions concerning the reptilian or nonreptilian nature of all vertebrates in which they are available. The class so defined has monothetic properties. Direct evidence on neither of these properties is available for extinct animals and related characters, largely those of the skeleton, must be used for determination. This runs into difficulties of the sorts outlined in the general comments on problems of classification, pages 169–178, and discussed for the amphibians.

For the great majority of extinct vertebrates that appear to be reptiles, the inferred association of osteological features and key morphological and physiological properties of the reptiles lacks confirmation. In a few instances —for example, in the preservation of eggs of the Asiatic dinosaur *Protoceratops*, some other dinosaurs, and turtles—the resemblances of the eggs to those of known amniotes lend strong support. In others—for example, the ichthyosaurs—where young are known to have been born at a reasonably advanced, nonlarval stage, evidence of an amnion is very strong. In another direction, absence of metamorphism, evident in the lack of a larval stage or persistence of larval characters, is suggestive. In this regard, however, it is noteworthy that there also is no evidence of metamorphism among the lepospondylous amphibians.

In large part, in spite of these theoretical difficulties, skeletal evidence produces classifications that are not seriously contested and appear to be sound in view of their phylogenetic consistency. Most students have been willing to accept features of the skeleton as evidence for the existence of an amnion in ontogeny.

With regard to physiology, reflected in blood temperature, the case is somewhat different. All modern reptiles are essentially poikilothermic, but the birds and mammals, descended from reptiles, are homoiotherms. These arose from two different stocks, and it is not clear in either case whether warm-bloodedness developed within the reptiles or after the general reptilian grade had been passed. It may also be, of course, that at least partial warm-bloodedness developed in other reptilian lines as well. Flight, for example, would seem to require maintenance of a high rate of metabolism, and this may have been developed in the pterosaurs. Warm-bloodedness has also been suggested for some dinosaurs. Such speculations are interesting, but they play no role in the determinations of major relationships with which we are concerned in this discussion.

Seymouria and *Diadectes*, as discussed under the Amphibia, show that osteological features alone are not sufficient for differentiation of reptiles and amphibians. Similarly no definitive osteological features occur in the transition from reptiles to mammals. The existence of only the dentary in the lower jaw, often used for this purpose, has proven to be less and less useful as new information has come to light.

It is natural to suppose that were reproductive and physiological criteria available an end could be put to the confusion. This is almost surely not true. Such features must have had transitions and, like osteological structures, probably developed at different rates in different groups. Technically, thus, the difficulties of assignments of vertebrates as reptiles or nonreptiles are serious, but actually in only a few cases do any practical problems arise. The class Reptilia is a large, coherent assemblage in which the osteology is sufficiently standard to be generally adequate as a basis for definition of the class. Both theoretically and on the basis of the available evidence, however, the class must be defined polythetically.

MAJOR SUBDIVISIONS

An excellent brief review of the history of the classification of reptiles was given by Romer in 1956. There have been few fundamental changes at or above the ordinal level since that time. For sake of completeness a brief summary of the history is included here, one that can be supplemented in detail by reference to Romer's review. In the reptiles, as in other major groups with good modern representation, early classifications were based largely on moderns. As fossils were found, they were fitted into the already designated groups as far as this was at all possible. Crocodilia, Chelonia, Ophidia, and Lacertilia were the commonly recognized modern groups, and all but a few very distinctive fossils, such as the ichthyosaurs, were associated with them. With the discovery of the tuatara, *Sphenodon*, and the establishment of the Rhynchocephalia, a fifth group was added. This was considered to be representative of very primitive reptilian conditions, and a strong tendency to place extinct forms within the Rhynchocephalia developed.

Proliferation of kinds of reptiles that were found in the record produced a complex situation, with a large number of groups. A beginning of a resolution of this difficulty was made by Osborn (1903), who recognized two primary subdivisions based on the temporal region of the skull (Table 24, I). One of these he termed Synapsida, with a single temporal arch, and the other Diapsida, with two arches. Although this classification associated many reptiles that now are placed in separate categories and did not entirely follow its own criteria (e.g., placing anapsids as synapsids), it did establish firmly the use of temporal structure as a basis for defining major subdivisions of the reptiles. Most later classifications have followed this lead.

The structure of the temporal region is closely related to the adductor jaw musculature and tends to alter with adaptations to new diets. As a mobile region it has advantages in classification, but in being responsive to dietary shifts its changes are not necessarily closely associated with the development of new levels of organization that mark the differences between major taxonomic units. Considerable parallelism in the development of temporal fenestrae appears to have occurred, and classifications that take only this area into consideration have not proven satisfactory.

A strong tendency in the study of the temporal region has been to relate its modifications largely or completely to changes in the action of the jaws. The fenestrae often have been considered to act mainly to accommodate expansion of jaw muscles during adduction. This was basic to Adam's (1919) analysis of the evolution of jaw musculature. W. K. Gregory's concept of the relationship of jaw musculature and the temporal region was more sophisticated. In contrast to Osborn, who emphasized the arches, he placed major emphasis on the openings.

Analyses that relate temporal structure to a single functional system are vastly oversimplified. Jaws and mastication are important, but there are also geographic and functional relationship to the middle ear, the braincase, the occiput, and through it to the axial musculature. In spite of extensive recent work with fairly sophisticated methods, the temporal region has so far defied complete mechanical analysis. Its role in the complex of systems is evident, and its relative stability, in spite of its relationship to the highly adaptive

TABLE 24   General Reptilian Classifications

I. *Orborn (1903)*

  Subclass Synapsida
    I. Cotylosauria
    II. Anomodontia
    III. Testudinata
    IV. Sauropterygia
  Subclass Diapsida
    I. Diaptosauria (superorder)
    II. Phytosauria
    III. Ichthyosauria
    IV. Crocodilia
    V. Dinosauria (superorder)
    VI. Squamata (superorder)
    VII. Pterosauria

II. *Williston (1925)*

  Class Reptilia
    Subclass Anapsida
    Subclass Synapsida
    Subclass Synaptosauria
    Subclass Parapsida
    Subclass Diapsida

III. *Romer (1956a)*

  Class Reptilia
    Subclass Anapsida
      Order Cotylosauria
      Order Chelonia
    Subclass Lepidosauria
      Order Eosuchia
      Order Rhynchocephalia
      Order Squamata
    Subclass Archosauria
      Order Thecodontia
      Order Saurischia
      Order Ornithischia
      Order Pterosauria
    Subclass uncertain
      Order Mesosauria
    Subclass Ichthyopterygia
      Order Ichthyosauria
    Subclass Euryapsida
      Order Protorosauria
      Order Sauropterygia
    Subclass Synapsida
      Order Pelycosauria
      Order Therapsida

(Sauropsid orders; Theropsid orders — bracketed groupings)

IV. *Säve-Söderbergh and Van Huene*[a]

  Ramus Reptiliomorphoidea
    Order Microsauria
    Order Anthrembolomeri
    Order Seymouriamorpha
    Order Diadectomorpha
    Order Procolophonia
    Order Captorhinomorpha
    Order Testudinata
    Order Ichthyosauria
  Ramos Theromorphoidea
    Order Mesosauria
    Order Pelycosauria
    Order Therapsida
    Order Placodontia
    Order Synaptosauria
    Order Protorosauria
  Ramus Sauromorphoidea
    Order Eosuchia
    Order Rhynchocephalia
    Order Squamata
    Order Thecodontia
    Order Saurischia
    Order Ornithischia
    Order Crocodilia
    Order Pterosauria

V. *Olson (1947)*

  Class Reptilia
    Subclass Parareptilia
      Diadectomorpha
      (Seymouriamorpha)
      Procolophonia
      Pareiasauria
      Chelonia
    Subclass Eureptilia
      Captorhinomorpha
      Synapsida
      Ichthyopterygia
      Ichthyopterygia
      Synaptosauria
      Lepidosauria
      Archosauria

[a] As modified by Romer (1956a).

structures and functions of mastication, appears to be a function of these complex relationships, which lower the probabilities of radical change except when accompanied by a major reorganization of the several systems.

Williston (1925) brought together many modifications that had followed Osborn's proposal, incorporating studies of his own as well as those of Watson, Case, Von Huene, Broom, and many others (see Table 24, II). Stemming from his work, with additional subsequent modifications, was the classification presented by Romer in 1956 in a much expanded and altered version of Williston's study (Table 24, III).

In Williston's scheme Osborn's Synapsida and Diapsida were retained, but much modified in concept. The Anapsida were recognized as a separate, coherent group, and the Sauropterygia, with a single upper opening, also were removed from the Synapsida. They were combined with the Placodontia in a group called Synaptosauria. Parapsida, animals with a high, single temporal opening, lying above the junction of the postorbital and squamosal, were separated from the Diapsida. This group included the Mesosauria, Ichthyosauria, Protorosauria, and Squamata. Within the subclass Diapsida two major subdivisions were recognized: the Diaptosauria (Rhynchocephalia) and Archosauria (crocodiles, saurischians, ornithischians, pterosaurs, and primitive relatives).

Romer's 1956 classification, which has been followed by most students in its broad outlines, included some fundamental changes, but shows clearly its Willistonian heritage. Anapsida and Synapsida were retained without substantial modification. The two diapsid groups of Williston were raised to subclass rank, Lepidosauria and Archosauria, and, very importantly, the diapsid heritage of the Squamata was recognized and the lizards and snakes were placed in a single category with the eosuchians and rhynchocephalians. This resulted in partial reduction of the Parapsida. Mesosaurs, considered as Parapsida by Williston, were removed and not given a subordinal position, and protorosaurs, the parapsids of Williston's classification, were associated with the Sauropterygia in a group termed Euryapsida (= Synaptosauria). In 1966 Romer placed the mesosaurs with the Cotylosauria. Only the ichthyosaurs remained of the Parapsida, and Romer assigned these to a separate subclass Ichthyopterygia. This new classification brought together changes that had been made in a series of steps after 1925, both by Romer and many other students.

In Romer's 1956 classification the subclasses were grouped informally by brackets as Theropsida and Sauropsida. These terms came into being through developments in classification not closely related to temporal fenestrae, but to circulatory systems. In his 1966 classification Romer dropped this grouping. Many features of the synapsids had indicated that this group was long separate from all others. In 1916 a subdivision of the reptiles recognizing this

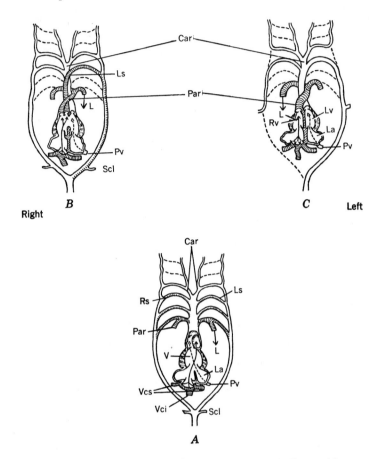

Figure 75. Goodrich's differentiation of Sauropsida and Theropsida among the reptiles on the basis of the circulatory system. *A*, amphibian; *B*, sauropsid (lizard); *C*, theropsid (mammal). Car, carotid artery; L, to lung; La, left auricle; Ls, left systemic arch; Lv, left ventricle; Par, pulmonary artery; Pv, pulmonary vein; Rs, right systemic arch; Rv, right ventricle; Scl, subclavian artery; V, ventricle; Vci, inferior vena cava; Vcs, superior vena cava. (After Goodrich, 1916.)

long separation and dividing the reptiles into two main groups was made by Goodrich. This was done primarily on the basis of the major differences of the aortic structures of contemporary reptiles, birds, and mammals (Figure 75). He applied the term "Sauropsida" to those reptiles that had the "reptilian" circulation and "Theropsida" to those with "mammalian" circulation. The groupings noted in Romer's (1956a) classification stem from this breakdown. Obviously there is no good basis for direct determination of the circulation

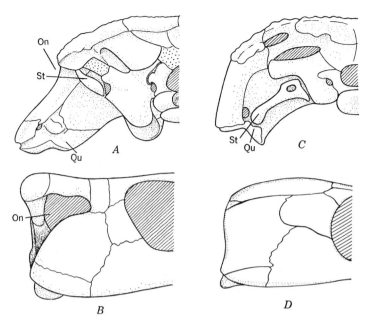

Figure 76.   D. M. S. Watson's differentiation of the Sauropsida and Theropsida on the basis of the temporal and otic regions of the skull. *A* and *B*, *Seymouria*, posterior and lateral, respectively, showing the high position of the stapes and the otic notch that carried a tympanic membrane, typical of the supposed sauropsid condition; *C* and *D*, a captorhinomorph, with no otic notch and a stapes that passes lateroventrally, illustrating the supposed theropsid condition. On, otic notch; Qu, quadrate; St, stapes. (Mostly after D. M. S. Watson, 1954.)

among extinct types. If it be granted, as seems reasonable, that the "mammalian" pattern was present somewhere within the mammal-like reptiles and the reptilian pattern among the ancestors of modern reptiles and birds, there is still no way in which the point or points of origins of the two can be determined. Goodrich found a hooked fifth metatarsal associated with the reptilian type and tried to use this as an associated character. It does not hold up in fossil groups. Furthermore, the discreteness of the supposed "sauropsid-aortic" pattern is open to serious question.

Watson (1954) followed this scheme and felt that the sauropsid condition was associated with one sort of middle ear and suspensorium and the theropsid condition with another (Figure 76). Anapsids and synapsids were theropsids, and other reptiles, some by definition and some more or less by default, were sauropsids. The implication of Watson's work was that there was a subdivision of the two groups at a prereptilian, amphibian stage. Romer's (1956a) usage is somewhat different. Following some very shaky leads, but the best available, he grouped together in the theropsids (informally) Ichthyopterygia, Euryapsida, and Synapsida. The cotylosaurs, which in 1956 included the seymouriamorphs, diadectomorphs, and captorhinomorphs, were not placed in either category, implying that both arrays came from this, and not from a prereptilian, stock.

A threefold division, following Säve-Söderbergh and Von Huene, and modified by Romer to fit his terminology of 1956, is given in Table 24, IV. The bracketed groupings from Romer show a fair degree of conformability to his arrangement, with, however, the big difference that the three do not have a common ancestry in the reptiles.

In 1947 (Olson, 1947) I suggested still another arrangement with two major groups Parareptilia and Eureptilia, with once again separate origin among the amphibians (see Table 24, V). This was based on resemblances of diadectids and chelonians and the presumed relationship of diadectids to procolophons and pareiasaurs. A seymouriamorph source for this branch, the Parareptilia, was postulated. The Eureptilia included the others, both theropsids and sauropsids, for which separation at a lesser categorical level was indicated.

In each of the various classifications characters of three regions have been important: the temporal fenestrae and arches, the middle ear and suspensorium, and the basic arterial circulation. Of these, the temporal fenestration has held a dominant position, with its indications being challenged only in cases in which other characters clearly showed associations to be false.

Some important points still have not been resolved. The question of diphyletic or polyphyletic origin of reptiles, which had considerable vogue in the 1940s and 1950s, has many less followers now. Identification of more than a single reptilian source among the amphibians, although suggested by various persons (Vaughn, 1962; J. T. Gregory, 1948; Westoll, 1942a,b; Olson, 1947; Watson, 1954) has not been very successful. Romer (1965) has argued strongly against it. The primitive structure of the middle ear found in the anapsids and synapsids now appears to have been widespread among various groups of "primitive" reptiles and probably is the one from which both the sauropsid and theropsid types arose (Olson, 1965a).

THE SUBCLASSES

The major subdivisions of the reptiles center around the subclass arrangement of Williston as modified over the years to take the form shown in Romer (1956a and Table 24, III). The subclass Anapsida as now conceived includes captorhinomorphs, procolophons (both procolophonids and pareiasaurids) and chelonians. The basis for the association is the roofed temporal region. This is clearly a primitive feature and within the reptiles is an absolute criterion of anapsid affinity. It is highly modified in the Chelonia. Only in a recently named *Acleistorhinus* and in *Lanthanosuchus*, if these are considered procolophon reptiles, is a temporal fenestra present in creatures that seem best associated with the anapsids.

The Cotylosauria for a long time formed a well-recognized unit of the Anapsida. The name was given by Cope because of the supposed existence of a long basioccipital which formed a stemmed condyle in *Diadectes*. This appearance was actually misleading, since it resulted from the loss of the exoccipitals during preservation. Gradually the name, meaning "stem-reptile," took on the sense of the stem of the reptilian stocks, which was most appropriate. The term "Cotylosauria" became the name of a group including Seymouriamorpha, Diadectomorpha, and Captorhinomorpha. The sey-mouriamorphs now are generally considered amphibians and must be dropped from the Cotylosauria. Recently as well (Romer, 1964*c*; Olson, 1965*b*, 1966*b*) a tendency to place *Diadectes* among the amphibians has emerged. Even if this is not strictly followed, there still is strong evidence that this genus is not at all close to the captorhinomorphs. Serious question has arisen on the propriety of association of *Diadectes* with procolophons and pareiasaurs, and thus the Diadectomorpha disappears. This leaves the Cotylosauria with only Captorhinomorpha and Procolophonia as members. Romer (1968) has argued for the retention of the name. This presupposes a fairly close relationship between Captorhinomorpha and Procolophonia, which may exist but is far from clear. An alternative is to recognize the Captorhinomorpha, Procolophonia, and Chelonia as three separate orders of the subclass Anapsida.

The subclass Synapsida is not definable on the basis of a single character or on discrete suites of characters. It must be conceived in terms of a central, primitive pattern that is definable and a series of related evolutionary trends that in their realization modify the primitive pattern materially. Two sub-orders are generally recognized: Pelycosauria, the more primitive, and Therapsida, the more advanced. The existence of the two groups resulted in part from accidents of preservation, with the primitive and advanced members generally geographically separated, and from the consequent course of study in which the two were long kept somewhat separate.

A "lower" temporal fenestra, below the junction of the postorbital and squamosal (Figure 77), is characteristic of the primitive synapsids and is found throughout the pelycosaurs. It occurs elsewhere only in *Lanthanosuchus*, in several primitive reptiles of uncertain affinities, and among some of the primitive therapsids. This feature is the only general characteristic that differentiates primitive pelycosaurs and some of the captorhinomorphs.

In the course of evolution of the therapsids the temporal condition was modified as the fenestra migrated or expanded dorsally, reaching the parietal

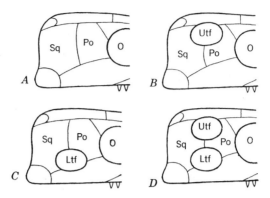

Figure 77.   Diagrams of four types of temporal regions in reptiles, showing the nature of the structures that are critical in some classifications of the reptiles. *A*, anapsid, no temporal fenestra; *B*, synapsid, a lower temporal fenestra only; *C*, parapsid or euryapsid, with the single temporal fenestra above the junction of the postorbital and squamosal; *D*, diapsid, with both an upper and a lower temporal fenestra. Ltf, lower temporal fenestra; O, orbit; Po, postorbital bone; Sq, squamosal bone; Utf, upper temporal fenestra.

with the loss of the postorbital-squamosal contact. Many lines of therapsids became extinct, but others were part of the gradual transition between the reptiles and the mammals. Although the criterion of a single element, the dentary, in the lower jaw and the existence of a squamosal-dentary contact in the jaw articulation have been used to indicate mammalian status, no hard and fast limits of the subclass Synapsida can be drawn on these or any other osteological features.

The subclass Ichthyopterygia includes only the ichthyosaurs. The origin and early history are unknown, and in the absence of transition forms, there is no problem insofar as definition of the group is concerned. Suites of characters—such as body shape, relatively large orbits, the structure of the finlike paired appendages, elongated snout, and vertebral centra formed of checkerlike disks—serve for monothetic definition. Although none of these characters is in itself unique, all members have them, and the combination is not to be found in any other group of vertebrates. The relationship of ichthyosaurs to ancestral types is not understood. There is an upper temporal opening. It had long been supposed that it lay above the junction of the postfrontal and squamosal, but Romer (1969) has shown that the more usual postorbital–squamosal array forms the ventral border. Many of the characters are primitive and suggest relationships with captorhinomorphs or pelycosaurs. This in turn has suggested theropsid affinities, but this is most tenuous.

The subclass Euryapsida, or Synaptosauria, is characterized in large part by aquatic animals. Like the Ichthyopterygia, they have only an upper temporal opening, margined below by the junction of the postorbital and squamosal. In addition to the marine forms a number of terrestrial or semi-terrestrial animals are sometimes included, grouped together as Protorosauria or Araeoscelida. These have the high temporal fenestra and a large squamosal. Wide differences of opinion exist on what belongs in this group and whether in fact such a group is valid. Similarly there is no agreement on how closely various reptiles placed by some students as protorosaurs or araeoscelids are related to the members of the other main group of Euryapsida. Romer (1956*a*) recognized two major divisions: Protorosauria and Sauropterygia. The latter includes three groups: Nothosauria, Plesiosauria, and Placodontia. The homogeneity of the Sauropterygia as far nothosaurs and plesiosaurs are concerned is beyond question, but the position of the placodonts is somewhat uncertain.

Archosauria and Lepidosauria are two groups, subclasses of Romer, many of whose members have diapsid temporal regions; that is, they have two fenestrae separated by the junction of the postorbital and squamosal bones. Together they include the great majority of terrestrial reptiles, being rivaled in only a rather minor way by the synapsids.

Archosaurs are characterized by trends that are mostly associated with the development of bipedalism. Such features, rather than any unique, specifiable characters, provide a basis for the recognition of a homogeneous unit. There are a few problems of classification at the level of the major subdivisions, in some major part as a result of the fact that in truly primitive forms transitions from ancestral stocks are poorly known. Possibly this group stemmed from the Lepidosauria, but this is not known from direct evidence in the form of ancestral types.

The Lepidosauria comprise rhynchocephalians, lizards and snakes, the squamata, and the eosuchians. Basically they are all diapsids, but both the snakes and lizards have lost part or all of this distinctive feature of the temporal region. It was the absence of a lower temporal arcade that induced Williston to place lizards with the Parapsida, with a single upper opening, and assume a relationship to the protorosaurs. Lepidosaurian characters are difficult to specify. Lepidosaurs are diapsids (or descendants of diapsids) that have not developed the special features of the Archosauria. This produces a peculiarly negative definition, meaningless without comparison, but one that is commonly used. There are no precise defining characters.

A few genera of reptiles that may be transitional between the captorhinomorphs and the primitive lepidosaurs, the eosuchians, have been found. These are grouped in the family Millerettidae. They were placed with the captorhinomorphs by Broom, their original describer, but have been considered lepidosaurian or an equivalent by some others, including Romer and Watson. The latter (Watson, 1954) argued strongly that they did not represent a transition from captorhinomorphs, whereas Romer, as well as Parrington, has made a case for this position.

As in all major groups, there are a few isolated reptilian genera that are difficult to place in any of the major subdivisions. The most important of these are *Mesosaurus* and *Stereosternum*, Permo-Carboniferous aquatic reptiles that have been variously placed with the Ichthyopterygia, Synapsida, and Parapsida. Romer (1956*a*) gave the mesosaurs no subclass position, noting that they might have arisen independently from the captorhinomorphs. In 1966 he placed them with the cotylosaurs. *Eunotosaurus* from the upper Permian of Africa is another highly problematical form. It has been associated with the Chelonia, but on relatively minor evidence. *Petrolacosaurus* from the upper Pennsylvanian of Kansas has been controversial, being placed in the Eosuchia by Peabody but among the edaphosaurian pelycosaurs by Romer. In each of the problematical cases the temporal region is not well known, attesting to its strong influence in classifications even today.

PROBLEMS OF CLASSIFICATION BELOW THE SUBCLASS LEVEL

Subclass Anapsida

The Captorhinomorpha and Procolophonia are grouped as Cotylosauria by Romer (1966*a*). This order is taken as equivalent in rank to Mesosauria and Chelonia. Because of the extensive modification of the meaning and content of the Cotylosauria confusion may arise, and this can be avoided by eliminating the term from formal classification, as suggested above. It is, however, very firmly entrenched and, for all of the confusion that it carries, likely to persist. In this section Captorhinomorpha, Procolophonia, and Chelonia are used without further grouping below the subclass level.

Captorhinomorpha

This is a coherent assemblage that is quite limited in its morphology but has an extensive stratigraphic range. It appears in the record near the base of the Pennsylvanian and persists through the upper Permian. Three families are usually recognized: Limnoscelidae, Romeriidae and Captorhinidae. Of these only the last is known to have had an extensive radiation, but the Romeriidae are of great importance, lying near the base of the pelycosaurs, and the Limnoscelidae are the most primitive members and have various "amphibian" features. The radiation of the captorhinids was carried out for the most part as an omnivorous and herbivorous unit associated with pelycosaurs.

When Broom first described the millerosaurs from South Africa, he indicated captorhinomorph affinities. Although it has been more common practice recently (see, for example, Romer, 1956) to associate these animals with the Lepidosauria, there is little in their structure, except the incipient to well-developed temporal fenestra, to exclude them from the Captorhino-morpha. They have been considered as intermediate between the capto-rhinomorphs and eosuchians by some writers (e.g., Parrington, 1958a) but as totally unrelated to the captorhinomorphs under the general sauropsid–theropsid division by Watson (1954, 1957). This basic difference does not, however, show up in formal classifications, because inclusion of the millero-saurs with Lepidosauria tends to mask it.

Procolophonia

The procolophons, pareiasaurs, and diadectids for many years were re-ferred to the Diadectomorpha. Removal of the diadectids from this group as suggested by Romer (1964c, 1966a), and by the writer (1965b) requires considerable modification not only of the concepts but of the formal classi-fication as well. The resulting group, Procolophonia, included the procolo-phons and the pareiasaurs, for which there is no close ancestral base unless it be somewhere among the captorhinomorphs. Very recently, however, a skull from the lower Permian, *Acleistorhinus*, has been assigned to the Procolophonia (Daly, 1969). This will undoubtedly be controversial, but if accepted, it extends the range of the group to lower Permian and seems to suggest an ancestry among the very primitive, limnoscelid captorhinomorphs.

Association of the procolophonids and pareiasaurids in a single group, order Procolophonia, forms a useful array (Romer, 1966a). It depends, however, on the interpretation that procolophons and pareiasaurians are in fact closely related. Both lack a temporal fenestra and have some generally similar features of the palate and postcranium.

The main link between procolophons and pareiasaurs is provided by *Rhipaeosaurus*, a genus from the Russian Permian. This genus is rather similar to a presumably primitive procolophon, *Nycteroleter*, also from the Russian Permian. *Rhipaeosaurus* has a dentition that appears somewhat like that to be expected in an ancestral pareiasauran, with three longitudinal cuspules on the cheek teeth. The skull, which is badly preserved, has a some-what pareiasaurian and somewhat procolophonian look. It could well be intermediate. The association may be valid, but it is based on slender evidence.

Since the time that the teeth of *Rhipaeosaurus* were first used as one of the impelling bases of association, it has been discovered that this type of tooth is widespread among early reptiles. It occurs in caseid pelycosaurs and in some dinocephalians (*Estemmenosuchus*) and also, of course, in some dino-saurs and lizards. There is need for much additional study of relationships at this general level of development. This is true for this particular relationship, for the relationships of diadectids and these reptiles, and for determination of whether such genera as *Nycteroleter* and *Lanthanosuchus* belong to the procolophonids or to the seymouriamorphs. Current confusion makes any classification highly tentative.

Chelonia (*Testudinata*)

The Chelonia, or turtles, form an extremely coherent group. The only genus that poses any problem of inclusion or exclusion is the aforementioned *Eunotosaurus*. Origins of the Chelonia are unknown, and at the first appearance in the fossil record, barring *Eunotosaurus*, characters are fully expressed. The skull has many characters not found in other reptiles, particularly those centering around the suspensory apparatus of the jaw and the otic region. The carapace and plastron and the inclusion of limb girdles within the rib cage are unique features of the postcranium. Thus there is no difficulty in arriving at a monothetic definition as things stand.

Chelonia range from the Triassic to Recent and throughout that time have a good record. Bergounioux (1938) described *Archaeochelys* from the Permian of France, but this has been generally discounted as being based on inorganic remains. The genus appears in his classification in the French *Traité* in 1955 (Bergounioux, 1955).

Three classifications are given in Table 25. The first is quite standard and has been taken from Romer (1956*a*). A very similar classification is to be found in Sukhanov (1964) and Camp, Allison, and Nichols (1964). As was true for other groups of animals with good modern representation, the early classification of the Chelonia was based in large part on moderns. The distinction between the two major modern groups—the side-necked turtles (pleurodires) and joint-necked turtles (cryptodires)—goes back to J. E. Gray (1844). The name "Amphychelydia" was proposed for the fossil pleurosternids and baenids by Lyddeker. The fossil groups were associated and maintained as distinct from moderns. From a standpoint of pure convenience this is effective, but it neglects the phylogenetic aspects of classification. Nevertheless, the "standard" present-day classifications in large part depend on this base.

Hay's classification (Table 25, III) is in many ways very different. It owes its major divisions to Cope (1871), who recognized the leatherback turtles as very primitive and distinct from all others. Since they lack a well-defined shell, leatherbacks were called Atheca (Athèques, Dollo). The shelled turtles then were named Thecophora (Thecophores, Dollo). Under the latter were placed Trionychia, Cryptodira, Pleurodira, and Amphychelydia. In Hay's classification (1902) the Trionychia were classified as a separate major unit. This general classification was carried by Hay into his 1930 bibliography and,

TABLE 25   Classifications of the Chelonia

I. *Romer (1956)*
Order Chelonia (Testudinata)
Suborder Amphichelydia
Superfamily
Proganochelyoidea
Superfamily
Pleurosternoidea
Superfamily Baenoidea
Suborder Cryptodira
Superfamily Testudinoidea
Superfamily Chelonioidea
Superfamily
Dermochelyoidea
Superfamily
Carettochelyoidea
Superfamily Trionychoidea
Suborder Pleurodira
?Suborder Eunotosauria

II. *Bergounioux (1955)*
Chéloniens
Thécophores
Sous-orde Gymnodermes
Super-famille
Trionychoïdes
Famille Trionychidés
Famille
Carettochélyidés
Sous-ordre Lépidodermés
Super-famille
Cryptodirés
Groupe
Protocryptodirés
Famille
Archaeochelyidés
Famille
Triassochélydés
Famille Baenidés
Famille
Pleurosternidés
Famille Miolaniidés

Groupe Cryptoridés
Famille Testudinidés
Famille Emydidés
Famille Chélydridés
Famille
Dermateamyidés
Famille Chéloniidés
Famille Protostegidés
Famille Toxochelyidés
Super-famille Pleurodirés
Groupe
Protopleurodirés
Famille
Proterochersidés
Famille
Kallokibotiidés
Groupe Pleurodirés
Famille
Thalassemyidés
Famille
Plesiochelyidés
Famille Pelomedusidés
Famille Bothremydés
Famille Chélyidés
Famille Eusarkiidés
Athèques
Famille
Dermochélyidés

III. *Hay (1902, 1930)*
Order Testudines
Suborder Athecae
Family Dermochelyidae
Suborder Thecophora
Superfamily Pleurodira
Superfamily Cryptodira
Suborder Trionychia

with the modifications of Bergounioux (1955), is given as the third classification in Table 25.

In Hay's arrangement the Amphichelydia are not singled out but distributed among the cryptodires and pleurodires as protocryptodires and protopleurodires. Technically this appears a reasonable procedure, because it seems probable that the two modern groups have their roots well back in turtle phylogeny. Practically there are problems of assignment. The protocryptodires, after Bergounioux, contain the Triassochelyidae (Proganochelydiae), Baenidae, Pleurosternidae, and Miolaniidae, or all of the groups for which the suborder Amphychelydia was established. Only the Proterochersidés, with one fairly-well-preserved genus, a second less well-preserved, and two questionable, and the Kallokibotiidés, with a single genus, form the protopleurodires. Assignment is based in large part on the pelvic structure.

The primary differences between the two types of classifications rest on (a) the proposition that the leatherback turtles are primitive and not close to the others and (b) that the Amphychelydia represent a grade level. The degree of separation accorded the Trionychia, which differs in the classifications, is largely a matter of judgment of the distinctness of this line, not one of fundamental difference of opinion on the nature of radiation of the Chelonia.

Most modern students have come to doubt that the leatherbacks, the dermochelyids, are in fact primitive. The fossil record is poor, but the structures are thought to have come from the general chelonid line in the course of adaptation to marine conditions of life. It seems highly probable that lumping of all fossil turtles of the Mesozoic that do not show strong affinities to modern groups into the single group Amphychelydia does produce a horizontal or grade classification. The problem of determining discrete lines in part accounts for retention of this group. The separation of Bergounioux, with which many do not agree, shows some of the problems involved in making a classification that attempts to avoid this horizontality.

Subclass Synapsida

The problem of the identity of some advanced members of this subclass as reptiles or mammals, already noted, has only a minor bearing on the internal composition of the Synapsida. Various proposals have been made to include some of the synapsids as mammals or some of the mammals as therapsids (e.g., Reed, 1960). Irrespective of the logical justification for any such arrangements, they have not until now met with any general acceptance.

Two major groups, usually designated as orders, are widely recognized as including all or almost all of the synapsids. These are the Pelycosauria (Theromorpha) and Therapsida (Table 26). In some instances the Ictidosauria have been given ordinal status, but this has usually been the case when

TABLE 26   General Classifications of the Synapsida

I. *Williston (1925)*
  Subclass Synapsida
    Order Theromorpha
      Suborder Pelycosimia
      Suborder Edaphosauria
      Suborder Poliosauria
      Suborder Caseosauria
    Order Therapsida
      Suborder Dinocephalia
      Suborder Dromosauria
      Suborder Anomodontia
      Suborder Theriodontia

II. *Romer (1933)*
  Subclass Synapsida
    Order Pelycosauria (families only)
    Order Therapsida

      Suborder Anningiamorpha
      Suborder Dromasauria
      Suborder Dinocephalia
      Suborder Dicynodontia
      Suborder Theriodontia
      Suborder Ictidosauria
      Suborder Protodonta

III. *Romer (1956a)*[a]
  Subclass Synapsida
    Order Pelycosauria
      Suborder Ophiacodontia
      Suborder Sphenacodontia
      Suborder Edaphosauria
    Order Therapsida
      Suborder Theriodontia
      Suborder Anomodontia

[a] Mostly after Romer and Price, 1940.

this array has been used to include some rather diverse forms including tritylodonts, diarthrognathids, and haramyids.

Early studies of the two major groups were carried out in North America, where pelycosaurs are well represented, and in South Africa, where therapsids are dominant. In addition, some early studies were made in Russia, but they reached the mainstream of analysis rather late in the history of the study of synapsids. The result of the geographic separation of members of the two groups is that they tended to be considered as much more widely separated than was the actual case. Studies initiated in the 1940s began to close the gap by leading to the discovery of essentially contemporary vertebrate-bearing deposits in North America and the Soviet Union (Chudinov, 1960; Olson, 1962). Gradually the sharp distinctions between the pelycosaurs and the therapsids have begun to diminish. To date not enough evidence has been obtained to allow a revision, but a suggestion that this may be eventually required and the general lines that it may follow was made by the writer (Olson, 1962).

At present, except for this possible need for an eventual thorough overhaul of the structure of classification, the principal problems of classification lie within the orders and are relatively minor.

Pelycosauria (Theromorpha)

Romer and Price (1940) made a definitive study of this group and produced a classification that has been little altered in subsequent work. This study stabilized the nomenclature for the major subdivisions, selecting terms to be found in earlier classifications by Cope, Case, Watson, Williston, Broom, and others who were students of the pelycosaurs. The name "Pelycosauria" was used in preference to "Theromorpha," partly on the basis of priority and partly on the basis of clarity. The classification proposed also added various new names at the family level, to accommodate new genera and genera earlier assigned to families, to which, in the opinion of the authors, they did not pertain. The classification of Romer and Price given in 1940 is as follows:

> Order Pelycosauria
> Suborder Ophiacodontia
> Family Ophiacodontidae
> Family Eothyrididae
> Suborder Sphenacodontia
> Family Varanopsidae
> Family Sphenacodontidae
> Suborder Edaphosauria
> Family Edaphosauridae
> Family Lupeosauridae
> Family Nitosauridae
> Family Caseidae

As conceived in this relationship, the Pelycosauria showed three major radiations, here at the subordinal rank. The most primitive forms are included in the Ophiacodontia. The major radiation of this group was toward an aquatic, fish-eating existence. This trend is not, however, reflected in the family Eothyrididae, erected to include a series of small, primitive pelycosaurs, with short faces and rather generalized dentition. This was designed as a classic "wastebasket" group, with the recognition that it probably contained genera that were not closely related and that may have represented several phyletic lines.

The sphenacodont radiation produced primarily terrestrial carnivores. There is good reason to suppose that they fed in large part on aquatic animals (Olson, 1961*a*, 1966*b*), but most of them did not become specially adapted for swimming. Associated with them in Romer and Price's classification, however, are the varanopsids, which did develop aquatic, fish-eating characteristics.

A principal departure of Romer and Price's classification from earlier ones is found in the composition of the Edaphosauria, which represents the third radiation. In addition to the first three families, two of which played little role in earlier classifications, containing largely new genera, is the fourth family, Caseidae. In most earlier studies the caseids had been considered to be widely separated from other pelycosaurs, even being placed in another order in some cases. Williston used the term "Caseosauria" for them, as a suborder of the Pelycosauria. Romer and Price concluded, on the basis of considerable morphological evidence, that caseids and edaphosaurids were sufficiently close to be included in a single suborder. Langston (1965b) revived the term "Caseosauria," feeling that this group was distinctive, and earlier (1962) I had in a somewhat similar way suggested that subordinal separation had some merit. A more recent study (Olson, 1968) has shown that the caseids form an extremely coherent family and underwent a radiation that is very well expressed at the familial level.

The caseids represent an early, terrestrial radiation of herbivores. In some respects they resemble some of the primitive herbivores among the therapsids. The possibility that they may have been ancestral, at least generally, to these therapsids is considered under the classification of that group.

Therapsida

The main outlines of therapsid classification go far back into the history of the study of fossil vertebrates, since specimens of this group were found very early. In large part the early students of these reptiles were more interested in morphology and in evolution viewed typologically than in classification and phyletics. They made only relatively general and sporadic efforts to arrive at a coherent classification of the therapsids as a whole. Watson (1917), Williston (1925), and some others progressed in this direction, but the most stabilizing effort was that of Broom (1933) in his comprehensive work on the mammal-like reptiles and the origin of mammals. Although his book did not contain a systematic section, it provided a framework in its sequential treatment of major groups. This has been followed, but with considerable modification, by most subsequent students.

Broom recognized two groups of dinocephalians: herbivores (tapinocephalids) and carnivores (titanosuchids); early and late therocephalians; gorgonopsians; anomodonts; cynodonts; ictidosaurs; and two small groups, Dromasauria and Burnetiamorpha. Romer's 1933 classification (Table 26, II) was similar, adding formal grouping and including also a suborder Protodontia to accommodate two problematic Triassic American genera *Dromatherium* and *Microconodon*. In his 1945 classification (Romer, 1945) he made several changes. Anningiamorpha was dropped and included in the pelycosaurs; dromasaurs, tritylodonts, and a few genera known only from teeth were placed in the order Ictidosauria; and Protodontia were considered as questionably reptilian.

Haughton and Brink (1954) presented a comprehensive classification, along

TABLE 27   Classifications of the Therapsida

I. *Romer (1945a)*[a]
Order Therapsida
Suborder Dinocephalia
Infraorder Titanosuchia
Infraorder Tapinocephalia
Suborder Dicynodontia
Suborder Theriodontia
Infraorder Gorgonopsia
Infraorder Cynodontia
Infraorder Therocephalia
Order Ictidosauria
(including Dromatheriidae, Trithelo-
dontidae, Microcleptidae, Tritylo-
dontidae)

II. *Romer (1956a)*
Order Therapsida
Suborder Theriodontia
Infraorder Titanosuchia
Brithopodidae (Titanophoneidae)
Anteosauridae, Jonkeriidae
Infraorder Gorgonopsia
Phthinosuchidae[b]
Infraorder Cynodontia
Infraorder Ictidosauria
"Ictidosaurs," Tritylodontidae,[c]
Microcleptidae
Infraorder Therocephalia
Infraorder Bauriamorpha
Suborder Anomodontia
Infraorder Dinocephalia
Infraorder Venjukoviamorpha
Infraorder Dromasauria
Infraorder Dicynodontia

III. *Olson (1962)*[d]
Order Therapsida
Suborder Theriodontia
Infraorder Eotheriodontia
Phthinosuchidae (including
Eotitanosuchus), Biarmosuchi-
dae, Brithopodidae (Titano-
phoneidae)
Infraorder Eutheriodontia
(including gorgonopsians,
therocephalians, cynodonts, etc.)
Suborder Dinocephalia
Infraorder Eodinocephalia
(including Deuterosauridae)
Infraorder Eudinocephalia
Tapinocephalidae

Suborder Anomodontia
Infraorder Venjukovioidea
Infraorder Dicynodontia

IV. *Boonstra (1963a,b)*[e]
Order Therapsida
Suborder Eotitanosuchia
Eotitanosuchidae
Phthinosuchidae
Suborder Anomodontia
Infraorder Dinocephalia
Brithopodidae, Anteosauridae,
Estemmenosuchidae, Titanosuchi-
dae, Styracocephalidae,
Tapinocephalidae
Infraorder Dicynodontia
Otsheriidae, Endothiodontidae,
Dicynodontidae
Infraorder Dromasauria
Suborder Theriodontia
(gorgonopsians, therocephalians,
cynodonts, etc.)

V. *Romer (1966a)*
Order Therapsida
Suborder Phthinosuchia (Eotitano-
suchia)
Phthinosuchus, Biarmosuchus,
Biarmosaurus
Suborder Theriodontia
Infraorder Gorgonopsia
Infraorder Cynodontia
Infraorder Tritylodontoidea
Infraorder Therocephalia
Infraorder Bauriamorpha
Infraorder Ictidosauria
Diarthrognathidae,
?Haramyidae
(Microcleptidae)
Suborder Anomodontia
Infraorder Dinocephalia
Superfamily Titanosuchoidea
Brithopodidae, Estemmeno-
suchidae, Anteosauridae,
Titanosuchidae
Superfamily Tapinocephaloidea
Deuterosauridae[f]
Infraorder Venjukoviamorpha
Infraorder Dromasauria
Infraorder Dicynodontia

more or less classical lines, but included tritylodonts as a family of cynodonts. Romer and Watson (1956) and Romer (1956a) (Table 27, II) made a somewhat radical departure in recognizing the two main groups, the carnivorous Theriodontia and the herbivorous Anomodontia. The dinocephalians, often considered as basal therapsids, were divided between the two groups. Ictidosaurs were reduced to an infraorder, but still included the divergent tritylodonts and Broom's "ictidosaurs" later to be named *Diarthrognathus*.

New finds made after this time and restudies of some of the older Russian materials began to change the picture. The later classifications in Table 27 (III, IV, and V) show some of the effects. Classification III in Table 27 (Olson, 1962) represented an effort to integrate new Russian and North American materials into a grade system. Starting from the pelycosaurs as ancestors it arrived at a pattern quite different from that of Boonstra (Table 27, IV), who viewed the new materials in the perspectives of his South African background (Boonstra, 1963). Romer (1966a) shows still another arrangement (Table 27, V), one in which the earlier separation of dinocephalians is somewhat modified.

Other recent classifications (e.g., Tatarinov, Vjuschkov, and Chudinov, 1964) primarily followed Watson and Romer (1956), with some modifications. The treatment of the ictidosaurs, however, differs because the studies of Crompton (1958), Kühne (1956), C. C. Young (1947), Watson (1942), Parrington (1958a), and Hopson (1964a) have shown that the tritylodonts are a very distinct group, related to the gomphodont cynodonts.

Although there is internal stability in the classifications of the therapsids, problems arise at the transition between the pelycosaurs and therapsids, and the therapsids and the mammals. The evolutionary aspects of these transitions are treated in Section IV. The classifications have not been seriously altered by new discoveries, for the mammal-like forms have largely been called mammals, and, at the lower limit, it has been possible to include the new genera in the therapsids without seriously disrupting the structure.

The fundamental problem in the relationships of the pelycosaurs and therapsids is whether the latter stemmed from a single stock of pelycosaurs or whether they may have come from more than one source. Generally it has been supposed that the former is true. A few cases will show the general nature of the problem as it affects classification.

---

[a] Representative of patterns of that time.
[b] Also many other families.
[c] Later named *Diarthrognathus*.
[d] Refers to early therapsids only.
[e] Composite of listings from South Africa and the Soviet Union.
[f] Plus some ?North American families.

The Phthinosuchidae, named by Efremov from a Russian genus, are generally accepted as very close to the base of the gorgonopsians and often classified with them. *Eotitanosuchus* is very close to *Phthinosuchus* but has been placed with the titanosuchid dinocephalians by Chudinov (1960) and as a phthinosuchid by Olson (1962). In many respects it is almost ideally intermediate between advanced sphenacodonts and gorgonopsians. Another pair of genera, *Biarmosuchus* and *Biarmosaurus*, are somewhat similar, but on the basis of their temporal regions seem to fall very close to the base of the titanosuchids. They too have many sphenacodont features, but the temporal structure is not that encountered in any known sphenacodonts. Different placements of these genera within the therapsids produce only minor modifications in the formal structure of classifications. However, as evaluation of their relationships to pelycosaurs differs, these modifications can become of major proportions with reference to the nature of the origin of therapsids.

A problem of comparable scope is found in the origin of the anomodonts. *Otscheria*, described by Chudinov (1960), appears to lie in the ancestry of the anomodonts, as does the genus *Venjukovia*. Both are from the lower part of the upper Permian of Russia. In the San Angelo deposits of North America occurs *Dimacrodon*, known from very poor material but seemingly vaguely related to venjukovioids. *Otscheria* and *Dimacrodon*, although contemporaries of the phthinosuchids, are very different in structure from the sphenacodonts. The principal argument for origin of all therapsids from sphenacodonts is the presence in both of an inflected process of the angular bone of the lower jaw (Figure 78). At the moment much of the problem of single or multiple origins of therapsids hangs on the question of whether this flange originated once or whether it appeared in more than one line, arising separately among the

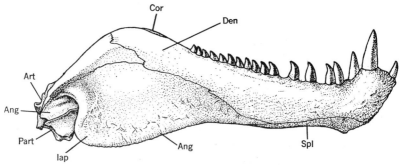

Figure 78.   Right lower jaw of *Dimetrodon*, a sphenacodont pelycosaur, in lateral aspect, showing the inflected process of the angular bone in an initial (evolutionary) stage of development. Ang, angular bone; Art, articulan bone; Cor, coronoid bone; Den, dentary bone; Iap, inflected process of the angular bone; Part, prearticular bone; Spl, splenial bone. (After Romer and Price, 1940.)

primitive anomodonts since it is not present in pelycosaurs other than the sphenacodonts. If it did arise but once, it shows that the record has very major gaps.

There are no current solutions to these problems. The present classifications that recognize by their framework a twofold radiation of the synapsids, one pelycosaurian and one therapsid, cannot express the uncertainty of the relationships, but rather give an impression of a meaningful disjunction and an orderly transition through a single line to a distinct therapsid level. This is almost certainly an oversimplification. Once more adequate materials have come to light, as they slowly are doing, this problem will be opened to what may be a major revision.

Subclass Ichthyopterygia

As currently conceived, the subclass Ichthyopterygia includes only the ichthyosaurs—very distinctive, highly specialized marine reptiles of the Mesozoic era. Even the most primitive ichthyosaurs are thoroughly adapted to marine life and show no close morphological resemblances to any other group of reptiles. This isolation, while posing baffling problems as to origins and relations, does at the same time reduce the major problems of classification to a minimum. There are no well-known genera that pose the problem of inclusion or exclusion with respect to the Ichthyopterygia.

The remoteness from all other reptiles, on the other hand, has resulted in wide ranging pronouncements concerning the ancestry of the ichthyosaurs. These appear in classifications in two ways. One is in the ordering of the subclasses, designed to place the Ichthyopterygia in sequence after their progenitors, and the other is inclusion in one of two or more major divisions where a dichotomy—such as that of Goodrich and Watson (sauropsids and theropsids) or of Olson (Parareptilia and Eureptilia)—or the trichotomy of Säve-Söderbergh (Reptiliomorphoidea, Theromorphoidea, and Sauromorphoidea) is used.

Ichthyosaurs were among the earliest of fossil vertebrates to arouse scientific interest, being known early in the eighteenth century (Baier, 1708), and they have received a great deal of attention from many paleontologists through the years. During the current century their origins have been sought among the mesosaurs (Von Huene, 1922) as well as among various other groups. Williston (1925) bracketed the ichthyosaurs and lizards as parapsids. Appleby (1961) indicated in a review of concepts of origin and on the basis of a mistaken interpretation of a footnote by the writer (Olson, 1936b) that I had suggested a relationship to Eosuchia. This was deduced from the suggestion that the meniscus pterygoideus might be present in both. Sources have been sought among amphibians; embolomeres (Von Huene, 1937), loxommids (Von Huene, 1944a), and among ancestors to trematosaurs (Nielsen, 1954). Romer (1948) argued for an origin in a preophiacodont synapsid stock. The diadectomorph–microsaur branch of the reptiliomorphs was suggested as a source by Von Huene (1952b). Most recently Appleby (1961) has suggested an anapsid origin, one close to the source of the Chelonia in a chelonian–ichthyosaurian stock that arose from the procolophon-pareiasaur complex near the beginning of the upper Permian. He presented a review of evidence of morphology that he feels leaves little doubt of the chelonian affinities of ichthyosaurs. The criteria are many, based on mostly relatively minor details, not on one or two major resemblances.

This is where the matter stands. Most of the determinations have been based on very slender evidence, as they must be in view of the major morphological gap between all ichthyosaurs and other reptiles. The use of different morphological criteria yields very different results, and it would seem that no definitive conclusion can be reached until some intermediate ancestors are discovered.

Classifications within the subclass, except as nonichthyosaurs have sometimes been included, for example mesosaurs, have been fairly stable. The most important single character that has been used is the width and length of the paddle. Length of the rostrum of the skull, length of the maxilla, size and extent of the temporal fenestra, relative size of the orbit, the nature of the teeth, and general body proportions all have entered in. Representative classifications are given in Table 28.

At the time of their first appearance in the Triassic, the ichthyosaurs had already undergone considerable diversification. Narrow-finned and broad-finned types were present as well as heavy-toothed, apparently mollusk-feeding genera, the omphalosaurs. Most classifications used in recent times recognize a twofold division in the Triassic. This is reflected in descendant lines from the Jurassic and Cretaceous.

Romer's (1956) classification is one of the simplest and least committed with reference to phyletic lines and recognizes five families without additional grouping. Tatarinov (1964a) made a primary separation of the omphalosaurs from the rest, grouping the remainder into two superfamilies, based mostly on broad and narrow fins. This is similar in most respects to the classification of Camp, Allison, and Nichols (1964), except that in the latter the omphalosaurs are considered merely a family of the latipinnates. The difference lies in the weight given to the adaptive dietary shift from active predation to a molluscivorous habit.

There are thus few major problems recognized in the internal classifications of the Ichthyopterygia, differences existing mostly in terminology, the degree to which the evidence is considered valid for the detection of phyletic lines, and the importance accorded the dental specializations of the omphalosaurs.

TABLE 28   Classifications of the Ichthyopterygia

I. *Romer (1956)*
Subclass Ichthyopterygia
Order Ichthyosauria
Family Mixosauridae
Family Omphalosauridae
Family Shastasauridae
(including *Californosaurus,*
Cymbospondylus, and
Shastasaurus)
Family Ichthyosauridae
Family Stenopterygiidae
(= Longipinnati)

II. *Tatarinov (1964)*
Subclass Ichthyopterygia
Order Ichthyosauria
Suborder Omphalosauroidei
Family
Omphalosauridae
Suborder Ichthyosauroidei
Superfamily
Ichthyosauroidea
(= Latipinnata)

Family Mixosauridae
Family Ichthyosauridae
Superfamily
Shastasauroidea
(= Longipinnata)
Family
Cybmospondylidae
Family Shastasauridae
Family Stenopterygiidae

III. *Camp, Allison, and Nichols (1964)*
Subclass Ichthyopterygia
Order Ichthyosauria
Suborder Latipinnati
Family Mixosauridae
Family Macropterygiidae
Suborder Longipinnati
Family Californosauridae
Family Cymbospondylidae
Family Omphalosauridae
Family Shastasauridae
Family Stenopterygiidae

Subclass Lepidosauria

Most current classifications of reptiles divide the Diapsida of Osborn (1903) into two major groups, Lepidosauria and Archosauria, following the lead of Romer (1933), and dismiss any close relationship of either to the Ichthyosauria. Both groups are fundamentally diapsid, but there are various departures from this condition in each, by closure of one or both fenestrae or by loss of the arches that bound them. If the Millerettiformes (= Millerosauria) are included, as is sometimes done, then the diapsid condition is not fully expressed in the most primitive stages.

Within the limits of present knowledge the lepidosaurians and archosaurs do not approach each other closely in morphology. The latter, even in their most primitive form, have specializations not found in the lepidosaurs. As a result of the general lines that the course of study and classification has followed, a rather curious sort of definition of the lepidosaurians has come into common use, as noted earlier in this section. They are frequently defined as diapsids that lack the specializations found among archosaurs, which, of course, presupposes the existence of the archosaurs as a basis for definition. They can be defined on a positive basis, but this requires statement of a primitive, or central condition, and of the trends or departures from these conditions found among derived lines. This is both more difficult and cumbersome.

Three primary subdivisions are generally recognized within the Lepidosauria: Eosuchia, Rhynchocephalia, and Squamata (Table 29). Romer (1956a) Tatarinov et al. (1964) give these three subdivisions ordinal rank. Hoffstetter (1955) has used a similar grouping of Eosuchiéns, Rhynchocéphalés, and Squamatés. These are not given specified categorical ranking, but under the Eosuchiéns are four orders, established in recognition of the difficulties of evaluating relationships of a number of early and poorly understood kinds of lepidosaurs. In addition to the ordinal breakdown Hoffstetter made a differentiation of Triassic and later forms among the rhynchocephalians and among the squamatans. This kind of classification cannot be fitted well into the formal categorical framework, but it nicely expresses differences in grade of probably related animals and also allows expressions of doubts that tend to be engulfed in the formalities of assignment in a more rigid system. In 1962 Hoffstetter modified this somewhat, reducing the snakes, lizards, and amphisbaenians to subordinal rank.

Underwood (1957a), following Schmidt's (1950) plea that the Ophidia be considered as a separate order, proposed a twofold division into the superorders Holapsida, including eosuchians and rhynchocephalians, and Squamata, with orders Sauria (= Lacertilia) and Serpentes (= Ophidia).

Various classifications in current use are shown in Table 29. That of Tatarinov et al. (1964) erects a separate order for the Millerosauria, places the forms usually considered as eosuchian as well as rhynchocephalian under the Eosuchia, and assigns the Lacertilia and Ophidia to separate orders. Romer (1968) indicated that his use of subordinal rank for lizards and snakes in 1966 may have been overly conservative. The distinctions between these different classifications reflect judgments rather than fundamental differences in concepts of relationships. The overall concept of the Lepidosauria and the general relationships of its major units are little affected.

The problems and their reflections in the practices of classification of the more primitive Lepidosauria, grouped as eosuchians and rhynchocephalians, are for the most part those that are common to groups with essentially only fossil representatives. In this case only the genus *Sphenodon* is still extant. Problems arise from the fact that only a few genera are well known, along with many fragmentary remains of other genera. Breaks in the record occur, and sampling from deposits from many parts of the world has not provided an

TABLE 29   Classifications of the Lepidosauria

I. *Williston (1925)*
Subclass Parapsida
  Order Proganosauria
    (Mesosauria)
  Order Ichthyosauria
  Order Protorosauria
  Order Squamata
Subclass Diapsida
  ?Order Proterosuchia
  ?Order Eosuchia
  Superorder Diaptosauria
  Order Rhynchocephalia

II. *Romer (1956a)*
Subclass Lepidosauria
  Order Eosuchia
    Suborder Millerettiformes
    Suborder Younginiformes
      Family Younginidae
      Family Tangasauridae
    Suborder Choristodera
      Family Champsosauridae
    Suborder Thallatosauria
      Family Thallatosauridae
  Order Rhynchocephalia
      Family Sphenodontidae
      Family Rhynchosauridae
      ?Family Sapheosauridae
      ?Family Claraziidae
      ?Family Pleurosauridae
  Order Squamata
    Suborder Lacertilia
    Infraorder Iguania (Pachyglossa
      and Rhiptoglossa)
      Family Iguanidae
      Family Agamidae
      Family Chameleontidae
    Infraorder Nyctisauria (Gekkota)
      Family Gekkonidae
      Family Pygopodidae
    Infraorder Leptoglossa
      (Scincomorpha)
      ?Family Ardeosauridae
      Family Xantusiidae
      Family Teiidae
      Family Scincidae
      Family Lacertidae
      Family Cordylidae
      ?Family Dibamidae

Infraorder Diploglossa
  Superfamily Anguoidea
    Family Anguidae
    Family Anniellidae
  Superfamily Varanoidea
    (Platynota)
    Family Helodermatidae
    Family Varanidae
    Family Aigialosauridae
    Family Mosasauridae
    (Pythonomorpha)
Infraorder Annulata
  (Amphisbaenia)
  Family Amphisbaenidae
Suborder Ophidia (Serpentes)
  Superfamily Typhlopoidea
    (Scoleophida)
    Family Typhlopidae
    ?Family Leptotyphlopidae
  Superfamily Booidea
    Family Dinilysiidae
    Family Aniliidae (Ilysiidae)
    Family Boidae
  Superfamily Colubroidea
    Family Colubridae
    Family Elapidae
    Family Hydrophiidae
    Family Viperidae

III. *Hoffstetter (1955)*
Subclass Lepidosauria
  Order Eosuchia
    Family Younginidae
    Family Prolacertidae
    Family Tangasauridae
  Order Rhynchocephalia
    Suborder Sphenodontia
      Family Sphenodontidae
      Family Sapheosauridae
      Family Monjurosuchidae
    Suborder Rhynchosauria
  Order Choristodera
  Order Thallatosauria
  Order Pleurosauria
  "Squamates du Trias"
    Tanystropheus
    Macrocnemus,
    Askeptosaurus

TABLE 29—continued

III. *Hoffstetter (1955)*—continued
"Squamates de type moderne"
Order Sauria
  Suborder Ascalabota
    Infraorder Gekkota
    Infraorder Iguania
    Infraorder Rhiptoglossa
  Suborder Autarchoglossa
    Infraorder Scincomorpha
      Superfamily Xantusioidea
      Superfamily Scincoidea
      Superfamily Lacertoidea
    Infraorder Anguinomorpha
      Superfamily Anguioidea
      Superfamily Varanoidea
Order Amphisbaenia
Order Serpentes
  "Suborder" Cholophidia
  "Suborder" Scolecophidia
  "Suborder" Alethinophidia
  a. Henophidia
    Family Boidae
    Family Palaeophidae
    Family Aniliidae
    Family Uropeltidae
    Family
      Xenopeltidae
    Family
      Coniophidae
  b. Caenophidia
    Family Colubridae
    Family Elapidae
    Family
      Hydrophidae
    Family Viperidae

IV. *Underwood (1957a and 1967)*
Subclass Lepidosauria (1957)
  Superorder Holapsida
    Order Eosuchia
    Order Rhynchocephalia
  Superorder Squamata
    Order Sauria
      Suborder Ascalabota
        Infraorder Gekkota
          Superfamily Gekkonoidea
          Superfamily Pygopoidea
        Infraorder?

Family Xantusiidae
Infraorder Iguania
Infraorder Rhiptoglossa
  Suborder Autarchoglossa
    Infraorder Scincomorpha
      Superfamily Scincoidea
      Superfamily Lacertoidea
    Infraorder Anguinomorpha
      Superfamily Anguinoidea
      Superfamily Varanoidea
    Infraorder Amphisbaenia
Order Serpentes (1967)
  Infraorder Scolecophidia
    Family Typhlopidae
    Family Leptotyphlopidae
  Infraorder Henophidia
    Family Aniliidae
    Family Uropeltidae
    Family Xenopeltidae
    Family Acrocordidae
    Family Boidae
  Infraorder Caenophidia
    Family Dipsadidae
    Family Viperidae
    Family Elapidae
    Family Homalophidae
    Family Natricidae
    Family Colubridae

V. *Tatarinov et al. (1964)*
Subclass Lepidosauria
  Order Millerosauria
    Family Millerettidae
    Family Mesenosauridae
  Order Eosuchia
    Suborder Younginiformes
    Suborder Rhynchocephalia
      Family Sphenodontidae
      Family Sauronodontidae
      Family Claraziidae
    Suborder Rhynchosauria
      Family Mesosuchidae
      Family Rhynchosauridae
    Suborder Choristodera
      Family Champsosauridae
  Order Lacertilia (= Sauria)
    Suborder Prolacertilia
      Family Protorosauridae
      Family Prolacertidae

TABLE 29—continued

V. *Tatarinov et al.* (*1964*)—continued
  Family Tangasauridae
Suborder Thallatosauria
  Family Askeptosauridae
  Family Thallatosauridae
Suborder Tanystrachelia
  Family Tanystropheidae
Suborder Iguania
Suborder Chameleonia
Suborder Gekkota
Suborder Scincomorpha
  Superfamily Xantusioidea
  Superfamily Scincoidea
  Superfamily Lacertoidea
Suborder Anguinomorpha
  Superfamily Anguinoidea
  Superfamily Helodermatoidea
  Superfamily Varanoidea
Suborder Cholophidea
Suborder Amphisbaenia
Order Ophidia
  Suborder Typhlopidia
  Suborder Anilidia
  Suborder Alethinophidia
    Superfamily Booidea (Henophidia)
    Superfamily Colubroidea
      (Caenophidia)

VI. *Romer* (*1966a*)

Subclass Lepidosauria
  Order Eosuchia
    Suborder Younginiformes
    Suborder Choristodera
    Suborder Thallatosauria
    Suborder Prolacertiformes
  Order Squamata
    Suborder Lacertilia
      Infraorder Eolacertilia
        Family Kuehneosauridae
      Infraorder Iguania
        ?Family Bavarisauridae
        ?Family Euposauridae
        Family Iguanidae
        Family Agamidae
        Family Chameleontidae
      Infraorder Nyctisauria (Gekkota)
        Family Ardeosauridae
        Family Broilisauridae
        Family Gekkonidae

        Family Pygopodidae
      Infraorder Leptoglossa
        (Scincomorpha)
        Family Xantusiidae
        Family Teiidae
        Family Scincidae
        Family Lacertidae
        Family Cordylidae
          (Gerrhosauridae,
          Zonuridae)
        ?Family Dibamidae
      Infraorder Annulata
        (Amphisbaenia)
      Infraorder Diploglossa
        Superfamily Anguoidea
        Family Anguidae
        Family Anniellidae
        Family Xenosauridae
        Superfamily Varanoidea
          (Platynota)
        ?Family Necrosauridae
        Family Helodermatidae
        Family Varanidae
        Family Lanthanotidae
        Family Aigialosauridae
        Family Dolichosauridae
        Family Mosasauridae
        Family Palaeophidae
          (Cholophidae, Cholophia)
        Family Simoliophidae
    Suborder Ophidia (Serpentes)
      Superfamily Typhlopoidea
        (Scolecophidia)
        Family Typhlopidae
        Family Leptotyphlopidae
      Superfamily Booidea
        (Henophidia)
        Family Dinilysiidae
        Family Aniliidae
          (Ilysiidae)
        Family Boidae
      Superfamily Colubroidea
        (Caenophidea)
        ?Family Archaeophidae
        Family Colubridae
        Family Elapidae
        Family Hydrophidae
        Family Viperidae

adequate basis for understanding temporal relationships or geographical interconnections. The result, stemming in some major part from the insufficiency of information, tends to be a fairly consistent major breakdown of the groups, marred, however, by wide differences of opinion on the placement of many of the fragmentary remains within the general framework. A strong tendency to adopt a monothetic classification, with the use of one or a few key characters, follows this sort of record. In this instance strong reliance tends to be placed on the condition of the temporal region.

The Squamata pose many difficult problems, but their classification is based on a very different sort of evidence. By far the greatest number of genera are known as living organisms, and these have provided much of the basis of classifications. As in similar cases, problems have arisen from the difficulties of reconciling the fossil record with the established classification. Phylogenies, as they have emerged as fossils became better known, have not always coincided with those established primarily from extant genera. In the case of the Squamata close relatives of the living families and genera can be found in the record back to the early Tertiary, earlier in some cases, but as this record dims in the older record, an array of more primitive kinds of animals is found and connections with the moderns are most difficult to establish. This situation is reflected, for example, in Hoffstetter's recognition of "Triassic Squamata" or Romer's Eolacertilia and is evident in the confusion of assignment of genera between Eosuchia and Squamata, and Protorosauria and Squamata, with the Protorosauria, if the group exists at all, being in dispute as members of Synaptosauria and Lepidosauria.

The simplest framework for consideration of the nature of some of these problems is one that includes Eosuchia, Rhynchocephalia, and Squamata, but with the recognition that these three categories are convenient but not inviolate.

Eosuchia

Eosuchia is recognized as a category that includes primitive Lepidosauria by essentially all students of the reptiles. It is a category that has received a wide array of primitive and often poorly known, generally lizardlike creatures. Discussion of the problems of the group demand treatment in detail, since placement of particular genera have much to do with the concept of the Eosuchia.

Most of the eosuchians are diapsids, but if this is applied as an absolute criterion, some genera, as included, for example, by Romer (1956a), are excluded (as he did in 1966a). The definition by Tatarinov et al. (1964) recognizes the diapsid condition as fundamental and thereby excludes millerettiforms, prolacertids, thallatosaurs, and tanystrachelians. The definition, on the other hand, brings in the rhynchocephalians, placed in a separate order by most students. Only relatively minor characters separate the primitive rhynchocephalians from such central eosuchians as *Youngina*, so that separation or inclusion hinges mainly on preference, that is, whether or not the radiation of the rhynchocephalians is sufficient to warrant an ordinal rank. Removal of some problematical rhynchocephalians, the Choristodera and Thallatosauria, from this group, reduces the extent of the radiation and thus the reason for considering rhynchocephalians as a separate order. This is the course followed by Tatarinov.

Of major importance to general concepts of the course of reptilian evolution and of particular relevance to the problems of the lepidosaurs is the group known as millerosaurs or millerettiforms. A few small specimens form this array. They were defined as Cotylosauria by Broom in his initial descriptions. Hoffstetter has followed this practice, whereas Romer (1956*a*) and Tatarinov (1964*a*) place them within the Lepidosauria, as a suborder of Eosuchia and as an order, respectively. Romer (1966*a*), however, placed them as a superfamily of the procolophonian Cotylosauria. The most extensive work on this group was by Watson (1957), who had earlier (1954) used some members in the framework of his theropsid–sauropsid dichotomy, as primitive representatives of the sauropsids.

Millerosaurs are not diapsids. One genus, *Millerosaurus*, has a distinct lower temporal fenestra, synapsid in character, and an unusual development of the suspensorium. *Milleretta* is described as having only an incipient temporal fenestra. *Mesenosaurus* from the Russian Permian has a synapsid-like temporal region. Basic dermal skull patterns are very much like those of captorhinomorphs.

Of the classifications listed only Tatarinov's (1964*a*) and Romer's (1966*a*) were prepared after publication of Watson's study. *Mesenosaurus* was first described as a pelycosaur by Efremov (1940*a*). It is possible that these animals came from primitive pelycosaurs or from captorhinomorphs. If the several genera are in fact closely related and this is subject to some doubt, a captorhinomorph ancestry seems most probable. The proximity to primitive pelycosaurs, which seems evident, casts very strong doubt on the concept of theropsid–sauropsid dichotomy as envisaged by Watson. It may be noted also that Reig (1967) has suggested that the archosaurs came from a branch of the pelycosaurs as well. If this were in fact the case, it would strengthen the picture of a basically monophyletic origin of most of the major reptilian types. The role of the millerosaurs, of course, is also dependent on the credence given relationships of the group to the eosuchians proper. Much more study of this whole area, hopefully with more adequate specimens, is necessary.

Eosuchians, such as *Youngina*, probably did arise from the millerosaur stock, although this has not been demonstrated beyond question. Under this assumption inclusion of the millerosaurs in the Eosuchia is an acceptable solution, but recognition of them as a separate order under the Lepidosauria is a better expression of the uncertainty of close ties.

The central core of the Eosuchia is a well-defined group, consisting of small, lizardlike diapsids represented by the *Youngina* complex. From this kind of core presumably arose rhynchocephalians and squamatans. In addition, however, there exists a series of generally poorly known genera that are variously placed among the Lepidosauria and seem to lie some-where in the vicinity of the Eosuchia. These pose some of the typical problems of the fossil record met with in early phases of radiation before a few stable lines have been "sorted out." The following illustrate this kind of problem as expressed within the later Permian–Triassic complex of lepidosaurian and lepidosaurianlike reptiles.

TANGASAURIDAE.   This family is placed in the Younginiformes of the Eosuchia by Romer (1956a), as a separate family of Eosuchia by Hoffstetter (1955), and under the Lacertilia as a family of the Prolacertilia by Tatarinov et al. (1964). The differences of opinion are actually slight, because in his text Tatarinov notes that this may be an Eosuchian group. The necessity of placement in a categorical scheme emphasizes differences unduly. The problem of position arises from two considerations. First is the matter of the criterion to be used in the separation of Eosuchia from Lacertilia. No hard-and-fast criterion applies, but assignment generally settles around the nature of the temporal region, with two complete arcades in eosuchians and loss of the lower one in lacertilians. If this is applied in this case, the problem resolves itself to the inadequacy of materials, because in *Tangasaurus* as in many of the delicate skulls of lepidosaurianlike creatures, the temporal is not well preserved.

THALLATOSAURIA.   In the case of the Thallatosauria assignment has been made to the Eosuchia, as a suborder (Romer (1956a), as a separate order of the Rhynchocephalés (Hoffstetter 1955), and as a suborder of the Lacertilia (Tatarinov et al. 1964). Kuhn-Schnyder (1952), who described *Askeptosaurus*, a thallatosaurianlike reptile, considered it, on reasonable grounds, to be an aquatic lacertilian. Romer (1966a, 1968) concluded that thallatosaurs rep-resented an offshoot of the eosuchians. The age is mid-Triassic. There is no question that there was a rather extensive radiation of lizardlike creatures during the Triassic, with even the development of a gliding lizard in the upper Triassic, (Robinson, 1962; Colbert, 1966). The primary question, as raised by Romer, is whether this radiation stemmed from an already realized lacer-tilian condition, from which it is assumed later lacertilians came, or whether a similar grade of organization was attained within the radiation that had its roots independent of the lacertilians proper. With respect to *Askeptosaurus*, it must be noted that Peyer and Kuhn-Schnyder (1955b) did not place it in the thallatosaurs but in their "Squamates du Trias," implying an early, separate lacertilian radiation.

Three other families—Sapheosauridae, Claraziidae, and Pleurosauridae—pose somewhat similar problems. The first two are difficult to place because of the possession of somewhat indeterminate general structure, essentially primitive lepidosaurian, along with the possession of acrodont teeth, found mainly among the rhynchocephalians. The pleurosaurs pose a problem because of the apparent absence of the lateral temporal opening, but possession of otherwise lepidosaurian features. On the basis this family is sometimes placed with the Protorosauria among the Synaptosauria, or Euryapsida.

Tanystropheus and *Macrocnemus* are representative of a number of Triassic reptiles that have posed problems of placement. The principal difference has been whether they are to be placed with the Lepidosauria, generally among the lizards, or as Protorosauria (Araeoscelida). If, as is usually done, the Protorosauria are considered members of the Euryapsida, then this decision determines to which of one of two subclasses they belong and is of major consequence. Arguments hinge primarily on the temporal region, its structure in the various genera, and the weight that is to be given to this structure.

Among these are a few rather primitive animals that have only upper temporal openings (see pages 311–312). There is general, if somewhat uneasy agreement that they form a group related somehow to the sauropterygian euryapsids. In addition, such animals as *Tanystropheus*, *Macrocnemus*, and others somewhat less well known, although highly specialized, have many features that appear to be lacertilian but for which there is varying information on the nature of the temporal region. In some at least it appears that the lower temporal fossa was absent. *Macrocnemus* now can be shown to be lepidosaurian, but *Tanystropheus* is still a problem.

Additional aspects of this problem are taken up under the consideration of euryapsids. It arises mainly from the lack of adequate morphological information about the genera in question and less than sufficient material to establish the early radiation patterns and understand the phyletic lines of vertebrates in the Triassic.

Rhynchocephalia

Most of the problems of classification of the Rhynchocephalia pertain to the assignment of the genera and families just discussed under the Eosuchia. If these are removed, the remaining animals generally placed within this category form a coherent assemblage consisting of rather generalized sphenodontids and more specialized rhynchosaurids. Some genera pose problems of assignment, mostly as the result of less than adequate information.

Squamata

Under this heading we consider the problems of classification of the Lacertilia and Ophidia. At the highest level, as discussed on page 328, the question has been raised as to whether they should be considered as separate orders or a subdivisions of a single order. Beyond the matter of mere preference or convention, this question has real meaning, since it involves concepts of the degree of discreteness of origin of the two and of the extent of their independent radiations.

The question of assignment of some of the primitive, lizardlike materials to the Lacertilia or to the Eolacertilian "Triassic Lacertilia" by Hoffstetter has been considered above. Beyond these general questions, the classifications of both the Lacertilia and the Ophidia have been based in large part on modern materials, and the problems have been more those of herpetology than of paleontology. Synthesis of the modern and fossil evidence will eventually enter more importantly into the classifications, but as yet the latter has had relatively little effect. Snakes are unknown in the Jurassic. Lizards have a record that shows considerable diversification to have occurred by that time (Hoffstetter, 1964) but does not give a broad picture of the evolution during this period.

LACERTILIA. The classifications of lizards generally used today are in large part based on Boulenger (1884) and Camp (1923). Earlier (1871) Cope had first used skeletal characters in classification, bringing a stability to a host of previous classifications that had been based largely on external features. Boulenger followed closely on this (1884). Camp's work introduced the use of a much wider variety of characters, building on the studies of Cope and Boulenger, and paid particular attention to what he called "adaptive" characters. Romer's (1956a, 1966a) classifications depended heavily on Camp and also incorporated the results of McDowell and Bogert (1954) on the relationships of the anguinomorphs. Underwood (1957b) and also Tatarinov et al. (1964) raised serious objections to some of the conclusions of McDowell and Bogert, and made alterations in classifications as a result. This sort of reshuffling and restudy is still going on at the time of writing.

The classifications listed in Table 29 all show similarities, stemming from a common base. Romer's 1966 classification is the most recent. It includes a new infraorder Eolacertilia, based on rather recently discovered upper Triassic lizards including the gliding forms Kuehneosaurus and Icarosaurus (L. Robinson, 1962; Colbert, 1966). The general agreement, however, is not actually indicative of a real consensus among workers.

Only a few points in the classifications in Table 29 need mention. All recognize the amphisbaenids (Annulata) as very distinct, but only Hoffstetter represents them as an order. Certainly on the basis of the morphology (Zangerl, 1944, 1945) there is a basis for such a separation.

Hoffstetter (1955) and also Underwood (1957a) used the categories Ascalabota and Autarchoglossa, old names from before 1850, as subordinal designations to differentiate the geckos, iguanas and chameleons from the scincomorphs and anguinomorphs. Other classifications in the ordering of the categories indicate much the same relationships, but without use of the larger, inclusive units.

Both Romer (1966a, 1966a) and Tatarinov et al. (1964) regarded the obscure, extinct Cholophidae as lacertilians, whereas Hoffstetter (1955) considered them ophidians. These are Cretaceous and Eocene materials, ophidian in general appearance but suggestive of varanid affinities in more detailed characters.

These and other similar problems that arise in this area of reptilian classification are primarily the results of the poor fossil record and the necessity of relying on modern materials for the primary classifications. Parallelism and convergence are difficult to recognize in end members of phyletic lines, and the lines are thus difficult to sort out. Most fossils that are well known can readily be placed within modern families and do little to aid in understanding the early deployment of the lines. Early and more primitive animals —particularly those of the Triassic, when the radiations seem to have been commencing—are known from mostly fragmentary remains. Relationships thus are so vague that assignment even to suborders is often problematical.

OPHIDIA (SERPENTES). Classifications of Ophidia from recent sources are given in Table 29. These differ in some respects but for the most part are basically alike. Outside of the question of ordinal or subordinal rank, noted before, the major difference relates to the inclusion or exclusion of the Cholophidia. This group brings together several families on the basis of very incomplete remains from the Cretaceous and earlier Tertiary. Their affinities are very uncertain. Romer (1956a, 1966a) has placed them with the varanopsid lizards, but considers *Archaeophis* of the Eocene as a forerunner of the colubroids (cenophoids).

Ophidian classifications have been erected largely on living materials because the fossil record is scant and difficult to interpret. The commonest remains of fossil snakes are vertebrae, and these are not entirely satisfactory as taxonomic indicators.

The origin of the snakes has long been a subject of discussion and argument, and this can be an important item in the development of the classification. Aquatic, terrestrial limbless, and fossorial lizards have all been proposed as ancestors. A varanoid (platynotan) ancestry has received strong support from many osteological resemblances between the two groups. Walls (1942) has put forth a case for fossorial origin on the basis of eye structure. Bellairs and Underwood (1951) gave additional evidence that the source was from more or less subterranean creatures. Somewhat different was the view of Underwood (1957a) that nocturnal lizards, in particular members of the pygopodid group, may have been in the ancestry of snakes. Aquatic origins were championed in particular by Nopsca. The problem has not been solved and perhaps will be most difficult to solve, for it would appear to depend in particular on fossil evidence, which is at present not to be had. The strongest case appears to be for a subterranean source.

Present classifications, since they are based on morphological features for the most part, have not felt any important effect of this discussion. In fact the contrary has occurred. The primitiveness of certain groups has been determined primarily on a morphological basis and the habits of these primitive snakes have formed a basis for inferring the habits of the ancestral types. Some of the presumed primitive features of the vertebrae and skulls may, however, be adaptations to burrowing. Whether the two families Leptotyphlopidae and Typhlopidae, sometimes grouped under Typhlopoidea, are closely related is open to question. In addition, the extent of relationship of each to other ophidian groups is uncertain.

Characters of the jaws, the suspensory apparatus, and arterial systems are the basic items that separate the booids (Henophida) from the colubroids (Caenophida). Burrowers do occur among the booids, but they show little resemblance to the typhlopoids. By far the most successful snakes are the colubroids. Taking advantage of the highly modified and mobile skulls and jaws, this group has undergone a successful and diversified radiation. Within it there have been many questions of relationships at a familial and lesser level, some of which are evident in the classifications. These are problems which at present must be settled largely through modern herpetology and which can be little aided by paleozoology until more adequate materials are discovered and studied.

Subclass Synaptosauria (Euryapsida)

Some of the most difficult problems of reptilian phylogeny and classification cluster around the concept of the Synaptosauria, or Euryapsida. Some of these have come up in the discussion of the Lepidosauria, inasmuch as they involve the assignment of some genera and families to one or the other subclass. The difficulties revolve largely around the group of Paleozoic and early Mesozoic reptiles considered as possible lepidosaurs or protorosaurs in the last section.

The temporal fenestrae have played an important role in studies of the

TABLE 30   Classifications of Protorosauria–Araeoscelidia[a]

I. *Romer (1945a)*	II. *Romer (1956a)* continued
Subclass Synaptosauria	Suborder Trilophosauria
Order Protorosauria	Family Trilophosauridae
Family Araeoscelidae	
Family Protorosauridae	III. *Romer (1966a)*
Family Tanystropheidae	Subclass Euryapsida
Family Trilophosauridae	(Synaptosauria)
Family Weigeltisauridae	Order Araeoscelidia
	(Protorosauria)
II. *Romer (1956a)*	Family Araeoscelidae
Subclass Euryapsida	?Family Protorosauridae
(Synaptosauria)	?Family Tanystropheidae
Order Protorosauria	?Family Weigeltosauridae
Suborder Araeoscelidia	Family Trilophosauridae
Family Araeoscelidae	
Family Protorosauridae	
Family Tanystropheidae	
Family Weigeltisauridae	

[a] Three classifications after Romer to show changes in the concept of this problematical group. Some families included here are placed under Lepidosauria (Squamata) by other writers.

position of these rather poorly known reptiles, but recently a tendency to place more reliance on other features has arisen, partly because of the absence of data on temporal openings and partly because information from other structures seems to indicate that the temporal fenestrae may be inadequate for decisions on assignment. The work of Vaughn (1955) on *Araeoscelis*, in which he found it necessary to downgrade the evidence of the temporal region, is a case in point.

In addition to the problematical protorosaurs, or areoscelidans (Table 30) Nothosauria, Plesiosauria, and Placodontia (Table 31) are usually placed in the Euryapsida, grouped together as Sauropterygia. The first two are closely related, but the affinities of the placodonts are uncertain. All groups have an upper temporal fenestra bordered below by the postorbital and squamosal bones. The openings tend to be widely spaced, and the squamosal tends to be a large element. Both of the features may be reduced or absent in some of the advanced types.

The ancestry of these aquatic synaptosaurs is uncertain because there are no known reptiles that can be considered directly on the line to even the most

TABLE 31   Classifications of the Sauropterygia

I. *Von Heune (1952b)*
Order Sauropterygia
  Suborder Nothosauroidea
    Family Proneusticosauridae
    Family Pachypleurosauridae
    Family Nothosauridae
  Suborder Plesiosauroidea
    Family Cymatosauridae
    Family Pistosauridae
    Family Plesiosauridae
    Family Elasmosauridae
    Family Pliosauridae

II. *Saint-Seine (1955)*
Order Sauropterygia
  Suborder Nothosauria
    Family Pachypleurosauridae
    Family Nothosauridae
  Suborder Plesiosauroidea
    Infraorder Pistosauroidea
      Family Pistosauridae
    Infraorder Pliosauroidea
      (Brachydeira)
      Family Pliosauridae
      Family Polycotilidae
    Infraorder Plesiosauroidea
      (Dolichodeira)
      Family Plesiosauridae
      Family Elasmosauridae
Order Placodontia
  Suborder Helveticosauroidea
    Family Helveticosauridae
  Suborder Placodontoidea
    Family Paraplacodontidae
    Family Placodontidae
  Suborder Cyamodontoidea
    Family Cyamodontidae
    Family Placochelydae
    Family Henodontidae

III. *Tatarinov and Novoshilov (1964)*
Order Sauropterygia
  Suborder Nothosauria

Family Lariosauridae
Family Pachypleurosauridae
Family Simosauridae
Family Nothosauridae
  Suborder Plesiosauria
    Infraorder Pistosauroidea
      Family Cymatosauridae
      Family Pistosauridae
    Infraorder Plesiosauroidea
      Family Plesiosauridae
      Family Cryptocleididae
      Family Brancasauridae
      Family Elasmosauridae
    Infraorder Pliosauroidea
      Family Leptocleididae
      Family Polycotylidae
      Family Trinacromeriidae
      Family Pliosauridae
      Family Branchaucheniidae
Order Placodontia
  Suborder Placodontoidei
    Family Helveticosauridae
    Family Placodontidae
  Suborder Cyamodontoidei
    Family Cyamodontidae[a]
    Family Placochelyidae
  Suborder Henodontoidei
    Family Henodontidae

IV. *Romer (1956a, 1966a)*
Order Sauropterygia
  Suborder Nothosauria
    Family Nothosauridae
    Family Cymatosauridae[a]
    Family Pachypleurosauridae
    Family Simosauridae
  Suborder Plesiosauria
    ?Superfamily Pistosauroidea
      Family Pistosauridae
    Superfamily Plesiosauroidea
      (Dolichodeira)

TABLE 31—continued

IV. Romer (1956a), (1966a)— continued	Family Polycotylidae
Family Plesiosauridae	Family Leptocleididae
Family Thaumatosauridae	Suborder Placodontia
Family Elasmosauridae	Family Helveticosauridae
Superfamily Pliosauroidea	Family Placodontidae
(Brachydeira)	Family Placochelyidae
Family Pliosauridae	(Cyamodontidae)
	Family Henodontidae

ª Placed under Pistosauria (1966).

primitive, the Nothosauria. The protorosaurs may lie somewhere in the general line of ancestry, but this is far from certain. Only about 20 genera fall within the protorosaurs even if all possible members are accepted. Most, however, are of uncertain taxonomic position. To add to the confusion is the question of the propriety of using the name "Protorosauria" for the group. The name is based on the reptile *Protorosaurus*, whose position is not clear. It has been considered a lepidosaur (see Camp, 1945), but Romer (1947b) argued strongly for retention of the term, whatever the position of this reptile might be. An alternative, followed by Romer and others in more recent classifications, has been to use the term "Araeoscelidia," based on the Permian genus *Araeoscelis*. This genus has an upper temporal opening, and the lower part of the temporal region is formed by a large, imperforate squamosal. *Kadaliosaurus* is similar, and a few other poorly known fossils may be fairly close to *Araeoscelis*. *Dictybolos*, a recently described aquatic reptile from the lower Permian (Olson, 1969) has tentatively been assigned to this group. If all problematical associates are avoided, however, *Araeoscelis* and the Triassic reptile *Trilophosaurus* are the source of most of the detailed information on the Araeoscelidia.

SAUROPTERYGIA. The classifications in Table 31 represent a fair sampling of recent opinions on the constitution of the Sauropterygia and indicate that, in contrast to the Protorosauria, a broad consensus with respect to the major groupings exists. No real doubt about the associations of the Nothosauria and Plesiosauria is found, and this has been the case almost since the first discoveries of fossils of the two groups, before 1850.

Tatarinov and Novoshilov (1964) have recognized the placodonts as a separate order, equivalent to the order Sauropterygia, but grouped with it under the Euryapsida. Romer in his text (1956a) indicates somewhat similar doubts as to the proximity of the placodonts to the nothosaurs and plesiosaurs, but in the formal classification he retains them as a suborder of Sauropterygia.

NOTHOSAURIA AND PLESIOSAURIA. The immediate source of the nothosaurs is not known, for they appear in the marine beds of middle Triassic age
in Europe without any earlier history. The lack of evidence results in a clear
separation from any other reptiles and reduces problems of classification and
definition of the group. Within the nothosaurs, as well, problems of classification, except those relating to placement of very poorly preserved materials,
relate largely to familial groupings. These are expressions of the opinions of
the various students on the structure of the adaptive radiation. The morphology of nothosaurs shows many consistent features throughout the
group. The geological range is very short, Triassic only. Added to this is the
problem that some genera are known only from skulls and others only from
postcranial elements. These factors have made it difficult to sort out the
evolutionary lines.

Skull shape—relative length, width, depth, and regularity of outline—
and the size and shape of temporal fenestra have entered into the formulations of classifications of the nothosaurs. In addition the size, relationships,
and disposition of the nasal bones are considered important. Reduction of
nasal bones represents a plesiosaurlike trend, because these bones were absent
among plesiosaurs.

Two general types of nothosaurs have been recognized, the pachypleuro-
saurids with small temporal openings and nothosaurids with large temporal
openings. The latter also have generally heavier skeletons and limbs. Von
Huene (1952a) has used these differences as the basis for his taxonomy,
although he also maintained as separate the primitive lower Triassic Pro-
neusticosaurus. Saint-Seine (1955) made a similar division, but Tatarinov and
Novoshilov (1964) recognized four lines of descent. Romer (1956a) included
the cymatosaurids in the nothosaurs but shifted them to the pistosaurs in
1966.

Plesiosaurs presumably were derived from nothosaurs during the late
Triassic. Von Heune has suggested a dual origin and this may well have been
the case. The evidence for or against this interpretation is far from sufficient for
a decision.

A point of some general interest arises in the assessment of the nothosaurs
as ancestral to the plesiosaurs. This relates to the concept of irreversibility of
evolution, a principle that is indispensible to the interpretation of phyletic
histories but which may be readily misapplied. Nothosaurs lack an interpterygoidal vacuity, but plesiosaurs have one. Without doubt there was such
a vacuity in the ancestors of the nothosaurs. Hence, if the known nothosaurs
are considered to include the ancestors of the plesiosaurs, the opening, once
lost, must have reappeared, technically violating the "law" of irreversibility.
The result has been that the known nothosaurs often have been excluded from
the ancestry of plesiosaurs, with the thought that some other nothosaurs,
which retained this vacuity, were in the ancestry.

It is certainly true that the strong directional tendencies of evolution militates against reversals, especially when some major features are lost. Nothing currently known about evolutionary mechanisms, however, indicates that such a change as the loss and reappearance of a vacuity is out of the question. In this case it may or may not have happened, but a presumption of impossibility cannot be taken as a bar to direct nothosaur ancestry to plesiosaurs if the weight of other evidence strongly supports this position.

Some interesting side effects appear from the interpretations of the meaning of the interpterygoidal vacuity. One of the major characters cited by Williston (1925) for separation of nothosaurs and plesiosaurs is the closed-versus-open palate. The two are monothetically distinct on the basis of this character under his designation. Thus arises the curious condition that by definition no nothosaur can have an open palate; hence, since a closed palate cannot be ancestral to an open palate, no nothosaur can be ancestral to plesiosaurs, which, again by definition, have open palates.

In this particular case a genus from the mid-Triassic, *Pistosaurus*, is directly involved. This genus has often been considered to be a nothosaur, but it does have interpterygoidal vacuities and could be an excellent nothosaur antecedent of the plesiosaurs. Now, however, it is usually referred to the plesiosaurs, partly because of its palatal condition and partly because some bones in deposits with it, not certainly part of it, are rather plesiosaurlike. In many respects the skull has nothosaurian features. It can be placed either with the nothosaurs or the plesiosaurs, but such placement in either event does violence to the usually accepted definitions. It has, for example, nasal bones, not known in other plesiosaurs, but, as indicated, it has the palatal vacuities, not known in other nothosaurs.

Tatarinov and Novoshilov (1964) and Romer (1966a) have gone somewhat farther, bringing *Cymatosaurus* along with *Pistosaurus* into the plesiosaurs. What the evidence shows is that an intermediate between the two groups did exist in the mid-Triassic and that the nothosaurian habitus was ancestral to the plesiosaurs. The placement of the genus *Pistosaurus* is important only in that it may be taken as a source of information on relationships, information that is more definitive than the knowledge of the situation warrants.

Classification within the plesiosaurs poses problems very much like those noted for the nothosaurs. Recent reviews have been presented by Pehrson (1963), Tarlo (1960), and Welles (1962). Basically two groups, longnecked and shortnecked, are recognized and these represent two adaptive types. Sometimes the terms "Dolichodeira" (Plesiosauroidea) and "Brachydeira" (Pliosauroidea) are used to express this aspect of the adaptive pattern. As the classifications of Table 31 show, however, there are differences in familial groupings under this broad differentiation, or in some instances the differentiation is not taken as a basis for familial grouping. Differences arise largely from interpretations of the meaning of skull features, particularly proportions, teeth, limb proportions, and the number of elements in the column and extremities.

PLACODONTS. Placodonts include a rather small number of highly diversified aquatic reptiles, characterized by highly specialized skulls and crushing teeth. They appear to have undergone a very rapid radiation. Diversification is expressed in classifications either at a familial level or, in the case of Tatarinov and Novoshilov (1964), at a subordinal level. The major problems in classification of the placodonts revolve around the relationships of the group as a whole rather than internal relationships. During their history, once reptilian affinities had been established, for they were first classified as fish, they have been referred either to the Sauropterygia, as relatives of nothosaurs and plesiosaurs, or to the Theromorpha, in particular the Therapsida.

Most students have followed the sauropterygian assignment as the most probable. Romer (1968) stated that he had reached the conclusion that the placodonts should be placed in a separate order, but with sauropterygian affinities. Kuhn-Schnyder (1961, 1963) has made extensive studies of the group and has indicated that they are very remote from all other reptiles and stemmed independently from the amphibians. The evidence for this is highly problematical. The skulls, even in the most primitive forms, are specialized, but *Helveticosaurus* (Peyer, 1955) lacks the typical crushing dentition. The upper temporal opening suggests euryapsid affinities, but the remainder gives little in the way of clues. The postcranium in primitive forms is very nothosaurlike. Such features are perhaps sufficient for assignment to Sauropterygia or at least to the Euryapsida, but the chance of convergence cannot be ruled out in view of the complex and unclear nature of the reptilian radiation of the very early Mesozoic. Origin certainly was from a stage more primitive than that represented by any known sauropterygian. The problem of relationships, it would seem, must await more information on the source of the sauropterygians; the nature of protorosaurs, or areoscelids; and the affinities of the early Triassic reptiles with such Permian predecessors as the captorhinomorphs, synapsids, and araeoscelids.

Subclass Archosauria

Romer's (1933) use of the term "Archosauria" to designate a subclass encompassing the thecodonts, crocodiles, saurischians, ornithischians, and pterosaurs has been adopted in most subsequent classifications. In large measure this subclass is equivalent to the superorder Archosauria of Williston (1925), but the increase of the categorical rank recognizes a greater distinction between the Archosauria and other diapsids. Camp, Allison, and Nichols

(1964) have retained the subclass Diapsida, with the archosaurs and lepido-saurs arrayed in two infraclasses. Williston's (1925) classification employed Osborn's term "Diapsida" as follows:

> Subclass Diapsida
> ?Order Proterosuchia
> ?Order Eosuchia
> Superorder Diaptosauria (Rhynchocephalia)
> Superorder Archosauria

Williston placed the Squamata, as noted in a preceding section, in a remote position, with the Parapsida rather than with the Lepidosauria, as is now commonly done. His orders and the superorder Diaptosauria, with the Squamata added, form Romer's (1933) Lepidosauria. Except for the removal of the Proterosuchia to the Archosauria, later classifications have in general followed the concepts of Archosauria and Lepidosauria as defined by Romer in his modification of Williston's classification.

The major outlines of the classification of diapsids have thus remained stable for a period of over 30 years, and there seems no reason at present to expect any major reorganization in the near future. The Archosauria, like the Lepidosauria, as noted in a preceding section, are undefinable on the basis of an evident set of common characters. The group must be considered in a polythetic context. Quite generalized members can be recognized among the primitive stocks. These are diapsids with antorbital fenestrae and strong laterosphenoid walls of the braincase. Radiation of the diapsids from this central, primitive condition is accompanied by expression of a number of trends, common in part to all phyla. Many of these have generally been thought to be related to the development of bipedalism. In recent years, however, this concept has been challenged. The opinion that archosaurs may have originated from semiaquatic ancestors and that the relatively elongated hind limbs may have arisen as an adaptation to swimming has gained some acceptance. In some lines accentuation of this feature led to bipedalism, but in others, which are quadrupeds, it has merely been retained. These quadrupeds, contrary to the usual interpretation, are thought not to have passed through a bipedal stage prior to development of a quadrupedal gait.

This is certainly worthy of consideration for at least some of the groups. Whatever circumstances may have led to the various patterns, there are common skeletal properties among the archosaurs, expressed in trends of development rather than as stable, key characters that provide a basis for monothetic classification.

The major subdivisions of the Archosauria are much the same in current classifications. Romer's 1956 and 1966 classifications list the orders as follows:

> Subclass Archosauria
> Order Thecodontia
> Order Crocodilia
> Order Saurischia
> Order Ornithischia
> Order Pterosauria[1]

The same arrangement is found in most present-day classifications, and even Williston's 1925 classification differs very slightly, appearing as follows:

> Superorder Archosauria
> Order Parasuchia (= Thecodontia in large part)
> Order Crocodilia
> Order Saurischia
> Order Ornithischia
> Order Pterosauria

Camp, Allison, and Nichols (1964), in addition to the use of Archosauria as an infraclass, include other minor differences, which are apparent in the following arrangement:

> Infraclass Archosauria
> Superorder Thecodontia
> Superorder Crocodilia
> Superorder Dinosauria
> Order Saurischia
> Order Ornithischia
> Superorder Pterosauria

The stability of classification reflects two principal things. There is a real consistency in the major groups and recognizable structural continuity between the various component units. Also little is known in general about the ancestry of each of the groups or of their interrelationships with each other. Consequently overlaps that result in taxonomic problems do not exist. Presumably ancestors of the more advanced groups occur among the thecodonts, but even in cases in which one or another of these gives some indication of relationship it is generally not close enough to cause confusion. In one instance, involving the thecodonts and saurischians, in which confusion developed, the contributing component, *Ornithosuchus*, has been shifted from the thecodonts to the saurischians by some students, as discussed later in this section.

[1] Placed after Crocodilia (1966).

As the history of the lower and middle Triassic is more fully elucidated, relationships will presumably become more evident and classification more complex and difficult. Until such time, which is approaching very slowly, the stability of the classification will tend to persist.

Order Thecodontia

The classifications in Table 32 from Williston (1925) through Romer (1966a) give a fair summary of the recent history of opinions about the constitution and arrangement of this group. Rather than "Thecodontia," Williston used the term "Parasuchia," which is often applied to the Phytosauria. As in later classifications, Pseudosuchia was recognized as central among thecodonts by Williston. Within the pseudosuchians is an array of primitive archosaurian reptiles that presumably includes the ancestry of most of the more progressive archosaurs. It fits well into the "horizontal" pattern of classification that is often imposed on a "basal complex" in which lines are difficult to sort out. With increased knowledge, this sorting may take place, and if fully successful, the primitive complex may disappear from classifications.

Sorting of this sort has taken place in the Pseudosuchia, but it has only progressed to a stage in which lines are recognized somewhat more clearly within the assemblage. The process is well illustrated in the table of classifications. Williston recognized three subgroups of thecodonts as distinct from the central pseudosuchian complex: Pelycosimia for a single very primitive form, Phytosauria, and Desmatosuchia.

Von Huene expressed much the same pattern in 1936, but used only two distinct suborders, Phytosauria and Pseudosuchia. Within the latter his classification has an odd structure, with a single superfamily Pelycosimioidea, including three families, one of which, Stagonolepidae, is comparable to Williston's suborder Desmatosuchia. Six other families are listed and are numerically equivalent to the superfamily as inspection of the classification will show. The meaning is clear: he recognized one family group as a separate radiation and listed the others without specifying grouping, thus making no commitments on interrelationships.

Both Hoffstetter (1955) and Rozhdestvenski (1964) recognized two suborders, as did Von Huene. There are minor differences in familial arrangements and content. Romer (1956a) separated out the Proterosuchia, equivalent to Williston's Pelycosimia. Part of Von Huene's Pelycosimioidea is included.

Continuing the process of differentiation, Romer (1966a) retained his earlier subdivisions but sorted out the Aetosauria (= Stagonolepidea) to include a series of armored forms that he considered to be related. This includes the old suborder Desmatosuchia of Williston and is more or less equivalent in content to the family Stagonolepidae of Von Huene.

TABLE 32   Classifications of the Thecodontia

I. *Williston (1925)*

Superorder Archosauria
Order Parasuchia
Suborder Pseudosuchia
Family Aetosauridae
Family Ornithosuchidae
Family Scleromochlidae
Suborder Pelycosimia
(for *Erythrosuchus*, Broom)
Suborder Phytosauria
Family Phytosauridae
Family Stagonolepidae
Suborder Desmatosuchia
(for *Desmatosuchus*)

II. *Von Huene (1936)*

Order Thecodontia
Suborder Phytosauria
Suborder Pseudosuchia
a. Superfamily Pelycosimioidea[a]
a.1. Family Proterosuchidae
a.2. Family Erythrosuchidae
a.3. Family Stagonolepidae
a.3.a. Subfamily Stagonolepinae
a.3.b. Subfamily Desmatosuchinae
a.3.c. Subfamily Episcoposaurinae
a.3.d. Subfamily Rauisuchinae
a.3.e. Subfamily (*Procerosuchus*)
b. Family Euparkeriidae
c. Family Aetosauridae
d. ?Family (*Erpetosuchus*)
e. Family Ornithosuchidae
f. Family Stegomosuchidae
g. Family Scleromochlidae

III. *Romer (1956a)*

Subclass Archosauria
Order Thecodontia
Suborder Proterosuchia
(Pelycosimia)
Suborder Pseudosuchia
Family Euparkeriidae
Family Ornithosuchidae
Family Sphenosuchidae
Family Scleromochlidae
Family Aetosauridae
Family Stagonolepidae
Suborder Parasuchia (Phytosauria)
Family Phytosauridae

IV. *Hoffstetter (1955)*

Subclass Archosauria
Order Thecodontia
Suborder Pseudosuchia
Superfamily Proterosuchoidea
Superfamily Elachistosuchoidea
Superfamily Stagonolepoidea
Superfamily Ornithosuchoidea
Superfamily Sphenosuchoidea
Suborder Phytosauria (Parasuchia)

V. *Rozhdestvenski (1964)*

Subclass Archosauria
Order Thecodontia
Suborder Pseudosuchia
Family Proterosuchidae
Family Elachistosuchidae
Family Ornithosuchidae
Family Scleromochlidae
Family Sphenosuchidae
Family Aetosauridae
Suborder Phytosauria

VI. *Romer (1966a)*

Subclass Archosauria
Order Thecodontia
Suborder Proterosuchia
Family Chasmatosauridae
Family Erythrosuchidae
Suborder Pseudosuchia
Family Euparkeriidae
Family Erpetosuchidae
Family Elachistosuchidae
Family Prestosuchidae
Suborder Aetosauria
Family Aetosauridae
(Stagonolepidae)
Suborder Phytosauria

[a] Three Families

As currently conceived, the Thecodontia include a very primitive array, Proterosuchia; a central array, Pseudosuchia; and two somewhat specialized, sterile side branches, Phytosauria and Aetosauria. Although primitive, the Proterosuchia, with such genera as *Chasmatosaurus* and *Erythrosuchus* (see Section IV for their evolutionary position), lower Triassic, are all in some respects too specialized to include the actual ancestry of later archosaurs.

The Pseudosuchia retains a highly varied array of small, partially to fully bipedal reptiles characterized by basic archosaurian features and few specializations. Within this assemblage are ancestors of later forms or at least close relatives to these ancestors. No more than general statements on relationships are now possible, but as more information is obtained and new analyses are made, it may well be that these currently obscure relationships will be clarified and some of the members of this complex will find their way to the base of phyletic lines now best known from end members. Either differentiation within the Pseudosuchia or assignment of such genera to the appropriate more advanced categories will follow. *Ornithosuchus*, long considered to be a thecodont, was placed in the saurischian carnosaurs by Romer (1966a) following A. D. Walker (1964) but with some misgivings.

Neither the Phytosauria nor the Aetosauria appear to have left any descendants. Phytosaurs very successfully filled the general ecological zone later occupied by some of the crocodiles, but the similarities of structure between the two are clearly a matter of parallelism and not of genetic origin.

Order Crocodilia

Unlike most archosaurian reptiles, the crocodiles have flourished since the end of the Mesozoic and, with a good contemporary representation, played a fundamental role in the early history of the classification of reptiles, being one of the initial groups recognized. Modern representatives are, however, sufficiently limited in their taxonomic spread for their history to be rather readily traced in the fossil record, and there is less than the usual confusion in drawing relationships between moderns and ancients.

Current classifications (Table 33) are very similar. Five suborders were recognized, except by Romer (1956a, Table 33, II), who retained as separate a suborder Thallatosuchia for the Metriorhynchidae, following Williston (1925). Even at the familial level differences are minor. In his 1966 classification, however, Romer made some changes, adding the suborder Archaeosuchia and dropping Thallatosuchia.

In general the most primitive genera are placed in the Protosuchia, intermediate genera in the Mesosuchia, and advanced genera, modern forms, and their fossil relatives in the Eusuchia. Two somewhat aberrant genera first known from the Cretaceous and early Tertiary of South America are the

TABLE 33 Classifications of the Crocodilia

I. *Williston (1925)*
Order Crocodilia (Loricata)
Suborder Eusuchia
Family Teleosauridae
Family Pholidosauridae
Family Atoposauridae
Family Goniopholidae
Family Dryosauridae
Family Hylaeochampsidae
Family Gavialidae
Family Tomistomidae
Family Crocodylidae
Suborder Thallatosuchia
Family Metriorhynchidae

II. *Romer (1956a)*
Order Crocodilia
Suborder Protosuchia
Family Notochampsidae
Suborder Mesosuchia
Family Teleosauridae
Family Pholidosauridae
Family Atoposauridae
Family Goniopholidae
Family Notosuchidae
Family Dryosauridae
Suborder Eusuchia
Family Hylaeochampsidae
Family Stomatosuchidae
Family Crocodylidae
Subfamily Crocodylinae
Subfamily Alligatorinae
Subfamily Gavialinae
Suborder Sebecosuchia
Family Baurusuchidae
Family Sebecidae
Suborder Thallatosuchia
Family Metriorhynchidae

Kalin (1955)
Order Crocodilia
Suborder Protosuchia
Family Notochampsidae
Family Protosuchidae
Suborder Mesosuchia
Family Teleosauridae
Family Metriorhynchidae
Family Pholidosauridae
Family Notosuchidae
Family Theriosuchidae
Family Goniopholidae
Family Atoposauridae
Family Libycosuchidae
Suborder Sebecosuchia
Family Sebecidae
Family Baurusuchidae

Suborder Eusuchia
Family Hylaeochampsidae
Family Bernissartidae
Family Stomatosuchidae
Family Gavialidae
Family Crocodilidae
Subfamily Alligatorinae
Subfamily Crocodilinae
Subfamily Tomistominae

IV. *Konzhukova (1964)*
Order Crocodilia
Suborder Protosuchia
Suborder Mesosuchia
Family Teleosauridae
Family Metriorhynchidae
Family Pholidosauridae
Family Notosuchidae
Family Goniolophidae
Family Atoposauridae
Family Libycosuchidae
Family Paralligatoridae
Suborder Sebecosuchia
Family Sebecidae
Family Baurusuchidae
Suborder Eusuchia
Family Hylaeochampsidae
Family Bernissartidae
Family Stomatosuchidae
Family Gavialidae
Family Crocodylidae

V. *Romer (1966a)*
Order Crocodilia
Suborder Protosuchia
Suborder Archaeosuchia
Family Notochampsidae
Family Proterochampsidae
Suborder Mesosuchia
Family Teleosauridae
Family Pholidosauridae
Family Atoposauridae
Family Goniopholidae
Family Notosuchidae
Family Metriorhynchidae
Suborder Sebecosuchia
Suborder Eusuchia
Family Hylaeochampsidae
Family Stomatosuchidae
Family Gavialidae
Family Crocodylidae

basis of the Sebecosuchia. Recent finds suggest that this odd group may have had a wider distribution (Langston, 1965a).

Early classification recognized only a single suborder (as Eusuchia of Von Zittel, 1890), although the term "Mesosuchia" had been proposed earlier by Huxley. Recognition of two groups, Mesosuchia and Eusuchia began in the 1930s, and at about the same time the term "Protosuchia" was used in classification (Mook, 1934). The suborder Thallatosuchia, for the metriorhynchids, was used by Williston (1925) and continued by Romer as late as 1956.

Separation of the Mesosuchia and Eusuchia is made primarily on a single feature, the inclusion or exclusion of the pterygoid bone from the secondary palate. Protosuchians, of course, lack the pterygoid in the secondary palate and have many primitive features in addition.

Three different modes of classification are found in present-day classifications of the Crocodilia. First, there is a simple monothetic separation of the Mesosuchia and Eusuchia on the basis of the palate. In addition, however, there is a much more comprehensive use of multiple characters. Kalin (1955) has prepared a chart of 26 characters and listed three states for each—primitive, intermediate, and advanced. This produces a mosaic of characters with some characters in each of the three states present in many of the genera.

On this basis Protosuchians, taken as primitive, are characterized by an array of primitive states of characters. All well-known mesosuchians, except *Metriorhynchus*, show a greater number of primitive characters than advanced, usually an appreciable number. In Eusuchians the number of advanced characters equals or exceeds the number of primitive ones, with the gavials showing the greatest ratio, 16:7, with two intermediate and one not entered. Under this scheme *Metriorhynchus* appears at a grade that is much in advance of other mesosuchians, with a 14:10 ratio of advanced to primitive characters, with two being intermediate.

Superimposed on these two types of classification are special features, presumed to indicate phylogenetic relationships. These are not clearly reflected in any of the formal classifications. That Kalin's phylogeny is precisely correct, or meant to be so interpreted, is questionable. Among other things, it appears to include the highly marine forms in the general ancestry of modern crocodiles. It does, however, bring out a number of interesting points.

First, a multiple origin of the mesosuchians is recognized. That this was true seems highly likely. The group is quite surely representative of a grade of evolution. Parallelism of adaptive types is also indicated. Important in understanding of the classification of crocodiles is recognition of the distinctions between the phylogeny as envisaged and the actual classification, which shows only a very general outline of it. Were the phylogeny actually depicted in a classification, the arrangement would be much more complex than any now in use. The classification of crocodiles now in general use is a reasonably functional compromise between a grade and a phyletic classi-fication. In view of the evidence now available any attempt to present a fully phyletic classification would be premature, because much that it would portray would be highly controversial.

Dinosaurs: Orders Saurischia and Ornithischia

Most classifications do not use the term "Dinosauria" in a formal sense, although as first proposed by Owen (1841) it was intended to cover a single array of Mesozoic reptiles. Increasing knowledge has long since shown that the animals placed together actually fall into two very distinct groups, with common origin presumably near the most primitive level of the Archosauria.

The saurischian theropods are close to the small, bipedal thecodonts, and there is question as to whether some of the major divisions of the Saurischia might not have arisen separately from the pseudosuchians. The Ornithischia, however, show no close relationships to any known primitive archosaurs.

The two major groups are separated, as is well known, by the pelvic structure, and within the archosaurs the ornithischians are precisely character-ized by this structure. In addition, of course, there are other notable dif-ferences between saurischians and ornithischians. These are evident in comparisons of a particular type of saurischian with a particular kind of ornithischian, but it is difficult to arrive at broad generalizations that are totally applicable over the orders as a whole. Except in cases where remains are extremely fragmentary, difficulty in placing appropriate forms as saurischians or ornithischians rarely arises.

Typical current classifications of the dinosaurs are listed in Table 34. It will be seen that the classification of Rozhdestvenski does employ a super-order Dinosauria, in contrast to the others. A similar practice was followed by Camp, Allison, and Nichols (1964), but with the superordinal designation following from the use of subclass Diapsida and infraclass Archosauria. Superorders are used as well for Crocodilia, Thecodontia, and Pterosauria.

Perhaps the most striking single feature of the reptiles generally called dinosaurs is the consistency of the classifications at the high categorical levels both by contemporary students and by those of the earlier parts of this century. That differences do exist is shown in Table 34, but these concern only minor reshuffling resulting from shades of opinions on relationships and on origins.

Consistency at a major level generally reflects the existence of sharp discontinuities between categories, due in large part to an incompleteness of the record. The dinosaurs are no exception, both with regard to the two orders and the suborders of the two categories. Below the subordinal level there are many problems of classification, stemming largely from the difficulties of establishing phylogenies with the materials available. The practices in each

TABLE 34   Classifications of the Saurischia and Ornithischia

I. *De Lapparent (1955)*
Ordre des Sauripelviéns
Sous-ordre des Théropodés
Super-famille des Coelurosauroïdés
Podokesauridés, Compsognathidés,
Segisauridés, Coeluridés,
Ornithomimidés
Super-famille des Carnosauroidés
Teratouridés, Megalosauridés,
Tyrannosauridés, Spinosauridés
Super-famille des Prosauropoidés
Platéosauridés, Anchisauridés
Sous-ordre des Sauropodés
Cetiosauridés, Brachiosauridés,
Camarasauridés, Astrodontidés,
Diplodocidés, Titanosauridés
Ordre des Avipelviéns
Sous-ordre des Orthopodés
Super-famille des Iguanodontoïdés
(= Ornithopodés)
Iguanodontidés,
Hypsilophodontidés,
Pachycéphalosauridés
Super-famille des Stégosauroïdés
Stégosaurides, Syrmosauridés,
Nodosauridés, Acanthopholidés
Super-famille des Cératopsoidés
Pachyrhinosauridés, Cératopsidés,
Protocératopsidés

II. *Romer (1956a)*
Order Saurischia
Suborder Theropoda
Infraorder Coelurosauria
Ammosauridae, Hallopodidae,
Podokesauridae, Segisauridae,
Coeluridae, Ornithomimidae
Infraorder Carnosauria
Palaeosauridae, Teratosauridae
(Zanclodontidae),
Megalosauridae,
Tyrannosauridae
(Deinodontidae)
Infraorder Prosauropoda
Thecodontosauridae,
Plateosauridae,
Melanorosauridae

Suborder Sauropoda
(Opisthocoela, Cetiosauria)
Brachiosauridae, Titanosauridae
(several subfamilies)
Order Ornithischia
Suborder Ornithopoda
Hypsilophodontidae,
Iguanodontidae,
Hadrosauridae,
Psittacosauridae,
Pachycephalosauridae
Suborder Stegosauria
Stegosauridae
Suborder Ankylosauria
Acanthopholidae, Nodosauridae
Suborder Ceratopsia
Protoceratopsidae,
Ceratopsidae,
Pachyrhinosauridae

III. *Rozhdestvenski, (1964)*
Superorder Dinosauria
Order Saurischia
Suborder Theropoda
Superfamily Coeluroidea
Ammosauridae,
Procompsognathidae,
Podokesauridae,
Segisauridae,
Compsognathidae,
Coeluridae,
Ornithomimidae
Superfamily Deinodontoidea
(Carnosauria)
Palaeosauridae, Teratouridae,
Megalosauridae,
Ceratosauridae,
Spinosauridae,
Deinodontidae
Suborder Prosauropoda
Thecodontosauridae,
Plateosauridae,
Melanorosauridae,
Plateosauravidae
Suborder Sauropoda
Cetiosauridae,   Brachiosaur-
idae

TABLE 34—continued

III. *Rozhdestvenski, (1964)*—continued
   Order Ornithischia
     Suborder Ornithopoda
       Hypsilophodontidae,
       Laosauridae,
       Psittacosauridae,
       Iguanodontidae,
       Hadrosauridae,
       Thescelosauridae
     Suborder Stegosauria
       Scelidosauridae, Stegosauridae
     Suborder Ankylosauria
       Acanthopholidae,
       Nodosauridae,
       Syrmosauridae
     Suborder Ceratopsia
       Protoceratopsidae,
       Ceratopsidae,
       Pachyrhinosauridae
     Incertae Sedis suborder
       Pachycephalosauridae

IV. *Romer (1966a)*
   Order Saurischia
     Suborder Theropoda
       Infraorder Coelurosauria
         Procompsognathidae
         ?Segisauridae
         Coeluridae (Coelurosauridae,
         Compsognathidae)
         Ornithomimidae
         ?Caenagnathidae
       Infraorder Carnosauria

         Ornithosuchidae
         ?Poposauridae
         Megalosauridae
         Spinosauridae
         Tyrannosauridae (Deinodontidae)
     Suborder Sauropodomorpha
       Infraorder Prosauropoda
         Thecodontosauridae
         (Gryponychidae)
         Plateosauridae
         Melanorosauridae
       Infraorder Sauropoda
         Brachiosauridae
         Titanosauridae
   Order Ornithischia
     Suborder Ornithopoda
       Heterodontosauridae
       Hypsilophodontidae
       Iguanodontidae
       Hadrosauridae
       Psittacosauridae
       Pachycephalosauridae
       (Troödontidae)
     Suborder Stegosauria
       Scelidosauridae
       Stegosauridae
     Suborder Ankylosauria
       Acanthopholidae
       Nodosauridae
     Suborder Ceratopsia
       Protoceratopsidae
       Ceratopsidae
       Pachyrhinosauridae

instance are necessarily tailored to the materials available and the temporal and spatial distributions, and add little to the general consideration of different approaches to vertebrate classification that are central in this discussion.

ORDER SAURISCHIA. Two or three major subdivisions of the saurischians are generally recognized. Morphologically two principal types existed, the bipeds, or theropods, and the quadrupeds, or sauropods. Some genera among the very primitive saurischians and among animals often classed as prosauropods were intermediate in structures related to gait, being only partially quadrupedal, and had some features found both in the sauropods and the theropods. This group has been placed differently by different authors. It was considered as an infraorder of the suborder Theropoda by Romer (1956a) and later as an infraorder under the suborder Sauropodomorpha by Romer (1966a). The genera are placed in a separate suborder by the classification of Rozhdestvenski (1964). The evolutionary position is reasonably clear, and the different treatments represent shades of differences in estimating degrees of relationships and differences in opinion as to how best to show these in formal classification.

Primitive theropods, Coelurosauria (Coelurosauroidea, Coeluroidea), include genera very close to their pseudosuchian ancestors. They were small, lightly built bipeds or semibipeds and are generally rather poorly known. Evidence of the foot structures has suggested that different families may have come from different pseudosuchian ancestry. As so often at such transition levels, the problem of possible polyphyletic origin arises. The usual solution, as made here, is to group on a more or less horizontal basis, drawing a line between the source and issue at some place where a morphological and/or temporal gap exists. General agreement on the familial constitution of the coelurosaurs has been attained, but there exist some problems of placement of genera in particular families.

The consensus shown in recent classifications in recognition of a group of theropods termed Carnosauria represents a practical compromise. Early representatives of this group are structurally close to early prosauropods and are difficult to distinguish. Later genera may or may not represent parallel development from coelurosaur ancestors. The familial array, such as that of Romer (1956a), is in part stratigraphic, with the very primitive Palaeosauridae and Teratouridae Triassic, the Megalosauridae mostly Jurassic, and the Tyrannosauridae Cretaceous. In Romer's later classification (1966a) important changes appear, with the Ornithosuchidae, earlier classed as pseudosuchians, referred to the Carnosauria and the Poposauridae tentatively included. These changes reflect the uncertainty of the boundary between the pseudosuchians and theropods and indicate a position of instability that does not appear on the examination of a single classification. The effect of the change in one important aspect is to reduce the polyphyletic nature of the saurischians, by the simple expedient of including supposedly ancestral forms in the taxon. This is much the same technique of bringing high categories into conformity with the concept of monophyletic origin as that suggested by Reed (1960) for the mammals (see page 375).

Three major factors contribute to the problem of the base of the Theropoda and in turn, of course, of the Saurischia as a whole. One of these is the relatively poor preservation of Triassic remains. A second is the fact that the record has tapped the time of gradation from pseudosuchians to theropods. A third is the pattern of the stratigraphic record, which, although it has tapped this transition, has done so in a temporally selective fashion. The well-known genera pertinent to this change are known mostly from the upper Triassic, with very limited occurrences earlier. The early and middle Triassic representatives give indications of the existence of genera important in the transition, but they give only fleeting glances of their morphology and little idea of phyletic continuity.

The geologic record of the Mesozoic as a whole has contributed to similar difficulties in assessing other saurischian relationships. Deposits formed during the upper Triassic, upper Jurassic, and upper Cretaceous have yielded the most genera. The lower and middle parts of these periods still are beset with frustrating gaps as far as vertebrate-producing terrestrial deposits are concerned (see Figure 11).

Below the family level a great deal of confusion exists in classifications of the saurischians. Solutions of problems of assignment at this level could in the future have a great effect on the constitution of the higher categories. The nature of much of the dinosaur material, its sheer size being far from the least important factor, is such that definitive treatments are very difficult and at best slow progress in modification of classifications is to be expected.

ORDER ORNITHISCHIA. The four classifications in Table 34 reveal only one major difference—namely, the inclusion of stegosaurs and ankylosaurs (nodosaurs) in a superfamily by De Lapparent and their separation in the other three. The fourfold classification has been standard for many years (e.g., Romer, 1933). De Lapparent's abandonment of it was on morphological and stratigraphic bases, primarily on supposed resemblances of pelvic structures of *Syrmosaurus* from Mongolia to the pelvic structures of some of the stegosaurs. This arrangement has not been followed by Maleev (see Rozhdestvenski, et al. 1964), the describer of *Syrmosaurus*, after a complete consideration of the morphology.

The four groups are associated in a single order primarily on the basis of structure of the pelvis, and, of course, overall resemblances. All were herbivores. Except for the most primitive group, the ornithopods, all were tetrapods, and within the ornithopods some groups appear to have trended in this direction. There are common features of the skulls and jaws, such as the presence of a predentary bone, a strong trend toward reduction and loss of anterior teeth, and various postcranial resemblances, many stemming, it seems, from superposition on a common structural heritage of trends accompanying the development of quadrupedal, herbivorous habits. Attainment of dermal armor is a common feature, in contrast to the saurischians.

The early record is extremely poor, although improved by recent finds (Crompton and Charig, 1962; Simmons, 1965). A few specimens are known from the uppermost Triassic, and these, as described to the present, are mostly fragmentary with but a single skull. No evident ancestor is found among the pseudosuchians, although presumably this is the source of the group. Genera transitional between the major groups, with the possible exception noted by De Lapparent, are not known. Even within the groups there is less than satisfactory evidence of evolutionary patterns from primitive to advanced. The result is that classification into major types is relatively easy, and rarely do problems of assignment arise. On the other hand, internal classifications, in view of the lack of clear phyletic intragroup patterns, is difficult and has been rather unstable.

The Ornithopoda comprise a highly varied series of bipedal and semi-bipedal herbivores. One of the principal bases for association is the bipedal gait and accompanying structure, because clearly a number of evolving lines are included and their common ancestry cannot be adequately demonstrated. By far the largest number of genera are from the upper Cretaceous, and these include a variety of types: fast-running, seemingly primitive animals; highly specialized, toothless, medium-sized bipeds; and the crested or duckbilled dinosaurs, the hadrosaurs. Records of primitive ornithopods from early beds, even Triassic, are few.

Stegosaurs, nodosaurs (ankylosaurs), and ceratopsians are all distinctive. In these, as in the hadrosaurs among the ornithopods, some knowledge of intragroup radiation has aided in the construction of tentative phyletic classifications. These must, however, be based on well-matured and end members of the radiations. Morphology and stratigraphic position are the principal criteria.

The geological record of the ornithischians, like that of the saurischians, is irregular. Most well-known genera are from the upper Jurassic (or upper Jurassic to lowest Cretaceous) and the upper Cretaceous. Major discontinuities occur, contributing, as in the saurischians, to ease in erecting major groups and difficulties in classification within these groups.

Contributing to problems of classification, over and above the problems of distribution and preservation, is the great size of the animals. Collecting is difficult and expensive, and collection and preparation of samples adequate for analysis of variation is essentially impossible. Studies necessarily must be cast at a level quite different from those that can deal with large suites of specimens representing genera and species. The total effect of factors such as these cannot be satisfactorily estimated, but there can be no doubt that they are of considerable magnitude, not only with reference to the dinosaurs, but in many cases where the problems of size become a controlling factor in collecting and analysis.

ORDER PTEROSAURIA. The flying reptiles, Pterosauria, form an isolated group of archosaurs known from rocks of Jurassic and Cretaceous age.

Their ancestry presumably lay among the pseudosuchian thecodonts, but there are no solid clues to the actual ancestral stock. Since no intermediates between pterosaurs and any other reptiles are known, a simple monothetic separation from all other reptiles is possible, either on such a single feature as the wing structure or on the basis of a complex of characters.

There are two distinct types of pterosaurs: rhamphorhynchoids and pterodactyloids. These are sharply separated on morphological grounds, and there are no intermediates. Length of tail alone is a sufficient character, with the tail being long in rhamphorhynchoids and short in pterydactyloids, and various other features of the skeleton serve equally well. There are thus no problems of classification at this, the subordinal level. Within the suborders representation is generally not sufficient for establishment of dependable phylogenetic lines. Classifications tend to rely on simple morphological features—for example, presence or absence of teeth—and groups are for the most part typological assemblages, which may have some (largely undemonstrated) phylogenetic validity.

The pterosaurs are of extreme interest from many points of view, and they pose many interesting problems concerning the modes of life and adaptation to flight. As far as classification is concerned, problems are largely nonexistent, except at subfamilial levels, due to the lack of knowledge of ancestors and the probability that the known pterosaurs represent but a very meager part of the group as it once existed.

9 Class Aves

In 1867 T. Huxley presented a classification of the birds in which three major groups were designated: Sauriurae, including only the extinct *Archae-opteryx;* Ratitae, the running birds with flat, unkeeled sternums; and the Carinatae, the birds with well-developed, keeled sternums. Drawing on the work of his predecessors in the nineteenth century, he formulated a classification and formalized terminology that has been influential ever since. The terms "Ratitae" and "Carinatae" were proposed much earlier, by Merrem in 1813, but the fossil birds had been described in detail only a short time before this classification was formulated.

Until the beginning of the nineteenth century only the carinate birds had entered into classifications, because the running birds had not come to the attention of naturalists and the Jurassic fossils, of course, had not been discovered. The singing birds, however, had long attracted the attention of both casual and serious naturalists, and such early treatises as Pierre Balon's *Histoire de la nature des oyseaux* (1555) revealed a broad knowledge of living birds and also of many of the details of their anatomy.

The sport of falconry, or hawking, the abundance of birds, and their esthetic appeal generated much interest in studies of their habits, and early classifications show strong effects of humanistic elements in their structures; for example, in his introduction to the section on Aves of the 10th edition of his *Systema Naturae* Linnaeus begins: "This beautiful and cheerful portion of

359

created nature. . . ." Throughout the classifications of the birds, even from very early times, morphology along with behavior has provided the principal criteria on which the major categories have been based. The features used first were primarily superficial, but even before the nineteenth century some internal morphological features had found their way into classification procedures. Lesser categories tended to be based more on particular habits and ways of life as they applied rather directly to man's relationships to birds.

One of the earliest classifications, which has had a lasting influence, was that of Willughby and Ray (Ray, 1678). This is shown in its entirety in the frontispiece as an example of the flow-diagram type of classification of that time and is summarized in Table 35. It shows the influence of the ways in which nature, and birds in particular, was viewed at the time and the descriptive designation of the categories generally used. Only carinate birds were known, and, although a few fossils had been noted in the literature as early as the thirteenth century, they played no role in the classification. The criteria employed are clear from the flow diagrams, showing the special attention paid to the nature of the life environment and the ways of occupancy. At a secondary level evident external morphological features were used. Characteristics of the bills and feet were taken to be of special importance, and, as seen in the classification of Linnaeus (Table 35, II), these criteria were carried over into more formal arrangements. The effects of falconry, reflecting the behavioral characteristics of the categories, are to be seen in the classifications of the hawks.

Linnaeus' (1758) classification in the 10th edition of the *Systema Naturae* recognized six orders. These too included only modern, carinate birds. The major divisions were based mainly on beak and foot characteristics, as were secondary categories as shown in their descriptive designations.

Late in the eighteenth century explorers began to make known to the ornithologists birds from the Southern Hemisphere, including the ratites, which stirred special interest. The large, relatively massive bones of the running birds are more subject to preservation than those of smaller, flying birds and, when exposed in the rocks in which they have been buried, more likely to catch the eye of casual students. Sub-Recent and Pleistocene remains were discovered early enough for them to play some role in the formulation of ideas about this group. Even the extinct genera, however, are so similar to moderns that they added only slightly to a general understanding of the biology of the ratites.

In one form or another the separation of living birds into ratites and carinates as proposed by Merrem has persisted in most later classifications. The names in their original sense have been used in the recent classification of Brodkorb (1963, 1964, 1967). A problem of assignment of the living *Tinamus* and its Pliocene and Pleistocene relatives has existed since this clumsy flier

TABLE 35   Some Classifications Showing the Arrangements
of Major Groups of Aves

I. *Willughby (Ray, 1678)*

Land Fowl
  Crooked beaks and talons
    Carnivorous (Rapacious)
      Diurnal
        Greater
        Lesser
      Nocturnal
    Frugivorous
  More straight beaks and claws.
    Greatest kind
    Middle size kind
    Least kind
Water Fowl
  Frequent watery places
    Greater
    Lesser
  Swim in water
    Whole footed
    Cloven footed

II. *Linné (Linnaeus) (1758) (Latin)*

Class Aves
  Order Accipitres
    A. Feet formed for perching
    B. Feet formed for climbing
    C. Feet formed for walking
  Order Anseres
    A. Bill toothed
    B. Bill without teeth
  Order Grallae
    A. Feet four-toed
    B. Feet three-toed
  Order Gallinae
  Order Passeres
    A. Bill thick
    B. Upper mandible somewhat
       hooked at point
    C. Upper mandible straight near
       end
    D. Bill strong, simple, tapering

III. *Haeckel (1866)*

Series Amnioten
  Monocondylien
    Class II. Aves
      Subclass Saurirae
      Subclass Ornithurae
        Legion Autophage
        Legion Paedotrophe

IV. *Huxley (1867)*

Order I. Saururae Haeckel
    Archaeopteryx
Order II. Ratitae Merrem
    Struthio—group
    Rhea—group
    Casuarius-Dromaeus—group
Order III. Carinatae Merrem
    Suborder Dromaeognathae
    One Family, Tinamidae
    Suborder Schizognathae
    Six families
  Suborder Desmognathae
    Seven families
  Suborder Aegithognathae
    Six families and two other groups

V. *Cope (1898)*

Class Monocondyla
  Subclass Reptilia
  Subclass Aves
    Superorder Saururae
      Order Ornithopappi Stejneger
    Superorder Eurhipidurae Gill
      Tribe Ratitae
        Eight orders, including
          Tinamus
      Tribe Odontolcae Marsh
        Dromaeopappi Stejneger
      Tribe Odontotormae Marsh
        Pteropappi Stejneger
      Tribe Euornithes Stejneger
        Desmognathae
        Schizognathae
        Aegithognathae
      Tribe Impennes

TABLE 35—continued

VI. *Hay (1902)*
  Class Aves
    Subclass Saururae
      Order Ornithopappi
        Archaeopterygidae
    Subclass Eurhipidura
      Superorder Odontotormae
        Order Pteropappi
          Ichthyornithidae
          Apatornithidae
      Superorder Odontolcae
        Order Dromaeopappi
          Hesperornithidae
      Superorder Dromaeognathae
        Order Gastronithes (*Diatryma*)
      Superorder Euornithes
        Eight orders (following Huxley,
        1867; Bedden, 1897; Cope,
        1898; and Blanchard, 1859)

VII. *Abel (1919)*
  Class Aves
    Subclass Archaeornithes
    Subclass Ornithurae
      Twenty-two orders without
        grouping, including;
      Odontolcae (*Hesperornis*)
      Odontormae (*Ichthyornis*)
      Ratites, tinamiforms, sphenisci-
      forms, and orders of carinates

VIII. *Romer (1933)*
  Class Aves
    Subclass Palaeornithes
      Order Archaeopterygia
    Subclass Neornithes
      Superorder Odontognathae
        Order Ichthyornithiformes
        Order Hesperornithiformes
      Superorder Palaeognathae
      Superorder Neognathae
        Order Sphenisciformes and
        other carinate orders

IX. *Wetmore (1960)*
  Class Aves
    Subclass Archaeornithes
      Order Archaeopterygiformes
    Subclass Neornithes
      Superorder Odontognathae
      Superorder Ichthyornithes
      Superorder Impennes
      Superorder Neognathae
        Orders including ratites, tinami-
        forms, and carniates

X. *Brodkorb (1963)*
  Class Aves
    Subclass Sauriurae
      Archaeopterygiformes
    Subclass Odontoholcae
      Hesperornithiformes
    Subclass Ornithurae
      Infraclass Dromaeognathae
        Tinamiformes
      Infraclass Ratitae
      Infraclass Carinatae

XI. *Romer (1966a)*
  Class Aves
    Subclass Archaeornithes (Sauriurae)
      Order Archaeopterygiformes
    Subclass Neornithes
      Superorder Odontognathae
        (Odontoholcae)
        Order Hesperornithiformes
      Superorder Palaeognathae
        Order Tinamiformes plus orders
        of ratites
      Superorder Neognathae
        Twenty-one orders, including
        Ichthyornithiformes and
        Sphenisciformes

was found. It has a keeled sternum but a palatal structure very like that of the ratites.

Fossil birds were sporadically noted in the literature before Huxley's classification, but up to the discovery and description of *Archaeopteryx* (Von Meyer, 1861; Owen, 1863) had had no effect on bird classification. The record has remained poor, mostly from quite recent beds, and sporadic in both geography and time. Within the last few decades, however, a considerable increase in the number of fossils and in taxonomic spread has been realized. The fossil occurrences known until about 1930 have been summarized by Lambrecht (1933), and later ones can be found in the studies of Wetmore (1960 and earlier) and Brodkorb (1963, 1964, 1967).

Before the formulation of Huxley's classification, Haeckel (1866) had taken cognizance of *Archaeopteryx* and proposed the subclass Sauriurae for its reception (see Table 35, III). This term was adopted by Huxley (as Saururae) and is found in some later classifications, including Brodkorb's in 1963. Many, however, substituted the synonymous term "Archaeornithes," as coordinate with "Neornithes," which usually designated a subclass including both carinates and ratites.

Huxley's major subdivisions were based primarily on osteological criteria, especially those of the shoulder girdle and the palate. This has in large part been followed in later classifications and stemmed from practices that had grown up in the preceding decades. His four subdivisions of the carinates, based primarily on palatal and associated skull structures, have not in general been followed. Problems arise particularly among his Aegithognathae in which there is a considerable variation in basic palatal structure and questionable association of the six families and two other groups on the basis of its structures.

Shortly after 1867 the Cretaceous "toothed" birds from North America were described by Marsh, in a preliminary fashion in 1872 (*Ichthyornis*, Marsh, 1872*a*; *Hesperornis*, Marsh, 1872*b*) and in much more detail in 1880 (Marsh, 1880*a*). These birds have entered into all later classifications that cover the major groups of birds and have been treated in a variety of ways (see Table 35, IV–XI). Stejneger's (1885) terms "Ornithopappi" (under Saururae), "Pteropappi" (an order containing *Ichthyornis*), and "Dromaeopappi" (an order containing *Hesperornis*) were used by Cope (1898) and Hay (1902) and have appeared sporadically in later classifications. Other terms from this period, some of which have remained in usage whereas others have not, appear in the classifications in Table 35. The source of the various terms is indicated in the first listed classifications in which it appears. Except for the special problems considered below, the somewhat confusing differences in the various classifications are more terminological than conceptual.

SPECIAL PROBLEMS

Archaeopteryx

Two partially preserved skeletons, both with feather impressions, are the principal bases for the very extensive studies of this extinct Jurassic bird. A third specimen and a few isolated feathers have not had an important role. The placement of the genus in bird classification has posed little problem, because it is so distinct that it has always been widely separated from all other known birds. Placement in a separate subclass under the class Aves is generally accepted, but it has also been suggested that it might better be placed with the reptiles, a position taken quite recently, for example, by Lowe (1944b). The closeness to reptiles is implied by Haeckel's term for the subclass, Sauriurae, a designation recently revived by Brodkorb (1963), in place of the more commonly used Archaeornithes.

The two well-known skeletons of this bird have been placed in separate genera, *Archaeopteryx* and *Archaeornis* (see, for example, Heilmann, 1926), but a restudy of the materials by De Beer (1954) and evidence from other students as well have shown quite conclusively that a single generic designation is proper. With this problem fairly well settled, the remaining questions concern the relationships of this creature to the other birds and its source among the reptiles. An archosaurian derivation is clear, but beyond that little can be said. Whether *Archaeopteryx* represents a sideline, not on the course to birds proper, or is close to the base of the ancestral stock of all birds cannot be settled from what is known. Opinion has tended to favor the former interpretation. Finally the proposition that the birds are polyphyletic has almost inevitably arisen. This has been explored, with a generally favorable interpretation by Lowe (1944b). The great gap, both temporally and morphologically, between *Archaeopteryx* and other birds and the increasing realization that much of the differentiating radiation of the birds took place prior to the time for which there is any adequate record must leave this question completely up in the air.

Hesperornis and Ichthyornis

The extinct, aquatic, flightless, toothed bird called *Hesperornis* lies far from all others and has not been integrated into classifications based on modern birds. It and *Ichthyornis* were for many years taken as evidence of a primitive, toothed stage of birds dominant in the late Mesozoic and preceding the major radiation of the birds, which was presumed to have been mainly Cenozoic. Although this is a thoroughly exploded myth, *Hesperornis* remains a puzzling creature. It was toothed, and in some respects its dental and jaw apparatus showed remarkable parallelism with that of the mosasaurs (J. T. Gregory, 1951, 1952). Like *Archaeopteryx*, it is so distinct that it has little effect on bird classifications beyond requiring an erection of a separate major category.

The majority of recent classifications (see, for example, Wetmore and Romer, Table 35, IX and XI, respectively) place this genus under a superorder Odontognathae (Odontoholcae). Unless some names usually thought to be synonyms are accepted as valid, it is the lone genus. The superorder is generally placed within the subclass Neornithes. Brodkorb (1963), however, has placed *Hesperornis* in a separate subclass Odontoholcae, thus emphasizing its distinctness and lack of direct relationship to living birds.

Ichthyornis until recently was often accorded a position as separate as that of *Hesperornis*, and the two were placed together in the superorder Odontognathae (see, for example, Romer (1933), Table 35, VI), which followed for the most part Wetmore (1930). It was also placed in similarly high categories by earlier students, such as Cope (1898) and Hay (1902), following the usual practices that developed after Marsh's early descriptions and evaluations of the position of *Ichthyornis* and *Hesperornis*. The supposed teeth formed a primary criterion. Their existence was based on a lower jaw, originally given a separate generic name and later associated with the type of *Ichthyornis* by Marsh. J. T. Gregory's (1952) study has cast serious doubts on the association, and he suggested that the jaw might be that of a young mosasaur. Although some doubt must remain on the assignment of the jaw, the weight of evidence has led to the assignment of this genus to an order of the superorder Neognathae and suggestions of relationships to the Procellariiformes and Colymbiformes.

Ratitae, Palaeognathae, and Tinamidae

Since they first became known to the zoologists of the Western World in about 1800, the ratites have occupied a fairly firm position in the classifications of birds. They formed one of the three primary groups of Huxley, a tribe in Cope (1898), a superorder (Dromaeognathae) in Hay (1902), and a superorder under Neornithes in many later classifications. There are some exceptions. Wetmore (1960) has included the orders of ratites without further grouping under the Neognathae, giving them equal rank with orders of carinates and not distinguished from them. Brodkorb (1963), having used the subclass Ornithurae, recognized the Ratitae as an infraclass under it.

As flightless birds the ratites have many features that distinguish them from the majority of other birds, although some of these are found as well in non-fliers among the carinates and also in *Hesperornis*. The structure of the palate

is distinctive and not readily explicable as an adaptation to cursorial existence. It is this more than any other feature that indicates that the ratites form
a coherent group that stemmed from a common ancestor with this type of
palate. Many other features, such as the platelike sternum, can reasonably be
thought to have arisen in adaptation to terrestrial habits, and along with the
isolated occurrences of the different ratite groups—in South America,
Africa and Arabia, and the Australian realm—have suggested that the running habit may have developed in parallel in several places. Although this
may well have happened, the common palatal features strongly indicate that
there was a common base distinct from that of the carinates.

It has been urged (see, for example, Lowe, 1944a) that the ratites did not
descend from flying birds at all, but either from nonflying avian ancestors or
directly from the reptiles separately from the other birds. The basic morphology of all members of this group argues strongly to the contrary, while
supporting the taxonomic integrity of the array.

Fossil ratites are well known from the Pleistocene and late Tertiary, and
representatives of one group go back into the Eocene. Although various
extinct genera form part of this record, the fossils all conform closely to the
modern representatives in overall structure and offer little evidence on the
origin or phylogeny that is not available from moderns.

The principal problem of classification relates to the position of *Tinamus*
and its fossil relatives from the Pliocene and Pleistocene. This bird is a clumsy
flyer and, as noted earlier, has a palaeognathus palate, like the ratites, and a
keeled sternum and coracoids of the type found in carinates. This contradiction of definitive characters has resulted in its placement with the ratites
in some classifications, with the carinates in others, and in a separate category
in a few. Huxley used it as the basis for his suborder Dromaeognathae, and
recently Brodkorb followed a somewhat similar plan, with Tinamiformes
being placed in the infraclass Dromaeognathae under the subclass Ornithurae.
Cope placed the tinamou with ratites, as did Romer (1966a), but Wetmore
(1960) relegated it to a separate order under the Neognathae, equivalent to
others of the ratites and carinates.

The problem, of course, revolves around whether or not members of this
group can be considered as related to the ratites on the basis of their palatal
structure. If so, they may perhaps be considered as a persistent stock of the
early flying ancestors of the ratites proper. If the palatal structure of the
ratites is used as an indication of the homogeneity of this group, it follows
that the palate of tinamiforms should be given similar weight. If this is
accepted, the alternatives of viewing this bird as a stage in the loss of flight or
as one that developed flight from flightless bird ancestors arise. Unfortunately
the fossil record does almost nothing to clarify the problem, so that the actual
history can only be inferred from living and recently extinct birds.

Impennes (Spheniseiformes)

The fossil record of the penguins—the Impennes or Spheniseiformes of various classifications—more than that of most groups of living birds has contributed something to an understanding of their position in the major classification of birds and a comprehension of some aspects of the radiation of the group itself. Extended treatment is to be found in the works of Simpson (1946) and Marples (1952), and many other detailed studies of both recent and extinct penguins have contributed to an extensive literature. A number of positions on the origins and relationships of the penguins have emerged from these various studies, and these are reflected in classifications.

The recent tendency has been to include the penguins in an order of the Neognathae. Wetmore (1960), however, has maintained the distinct superorder Impennes. Simpson (1946) has given a full account of the history of arguments about the nature and origins of penguins, which can be briefly summarized as follows. Three distinctive positions, each with some variants, were current at the time of his review. One, supported in particular by Fürbringer (1888), maintained that penguins were related to modern birds, in particular to the Procellariformes. This position, which also brings in tangentially possible relationships to Colymbiformes and Pelecaniformes, is the one supported by most students of the present time. The second point of view, held by M. Watson (1883), was that the penguins were ultimately derived from flying birds but that the source was not relatable to moderns because of the remoteness of relationships. The third point of view was a part of a more general one that from the very beginning various groups of birds arose from the reptiles independently and that flight was not developed in the ancestry of either the penguins or the ratites. This view, as far as the penguins are concerned, was held by Menzbier (1887).

Adherence to one or another of these positions is shown by the degree of independence given the penguins in classifications. In recent years Lowe (1933, 1939) has been the principal proponent of the third view, the extreme concept that the penguins arose independently from the proavians. This conclusion was part of a broader conviction that birds originated more than once, noted earlier in connection with *Archaeopteryx* and ratites. It has received little support from other recent students.

The position of the penguins in the classifications of Romer (1966a) and Brodkorb (1963) listed in Table 35 represents the current opinions of the vast majority on these questions. The Impennes, or Spheniseiformes, are considered as an order of the Neognathae (Carinatae of Brodkorb) and in their ordinal listings following the Podicipediformes (Colymbiformes) and Procellariformes (Tubinares) and preceding the Pelecaniformes suggest general relationships to these groups.

Neognathae (Carinates)

The composition and primary subdivisions of this major group, except for terminology, have been fairly stable in recent years. The principal modern birds are arrayed in orders that are for the most part distinct and readily separable. No intermediate grouping between the ordinal and higher level, Neognathae, is made to indicate radiation patterns. The order of listing in general displays opinions on possible relationships. Both the penguins and ichthyornithiforms are usually included among these orders. The latter and the Diatrymaformes from the early Tertiary are the only orders that do not have living representatives. In the listing of Romer (1936a) only one order, Coliformes, has no fossil representatives, but several suborders are known only from recent representatives.

The number and variety of known fossil birds have increased rapidly over the last several decades. The poorest representation generally is among the passerines. Lambrecht (1933) listed 691 described fossil birds, of which about 620 pertained to the Neognathae. These were spread through most orders, but the preponderance was in the Pelecaniformes, Ciconiiformes, Falconiformes, Galliformes, and Telmatoformes (using his ordinal designations). By far the greatest number was known from Europe and North America. Recent work has greatly increased the number of known and described fossil birds and has broadened both the geographic and taxonomic spread. The 14 or 15 extinct families now known, their increasing number of genera, and extinct genera of modern families have made a substantial contribution to the knowledge of the extent of the radiation of the birds.

At and below the familial level, however, taxonomic difficulties plague efforts to coordinate recent and fossil birds because the criteria used in the ultrafine subdivisions of modern birds are largely based on features that do not allow development of complementary taxonomies among the fossils.

The record, although continually improving, is still far from satisfactory. It is now clear that the radiation seen among Tertiary birds and culminating in those living today is the matured expression of an expansion of the class that had occurred well before the Tertiary began. The concept of a primitive Cretaceous avifauna and rapid Tertiary expansion can no longer be held. Without question the initial radiations of birds took place in the middle and late Mesozoic, and if the fossil record is to become meaningful as a means of understanding the phylogeny of birds, it must include records from this time.

10 Class Mammalia

GENERAL

Mammals constitute only one of the several classes of vertebrates, but the literature concerning their classification is immense. Excellent syntheses accompanied by examinations of the principles of mammalian classification greatly aid in the study of this vast literature. Two of the most important are by W. K. Gregory (1910) and Simpson (1945). Individual orders have been extensively studied, and, of course, there exist innumerable investigations of subordinal and lesser categories. The problems of classification within orders have depended largely on the specific composition of the subgroups and the particular interests that they have generated. Many are not pertinent to our broad aims of analysis of the general classification of vertebrates. In some instances, however, where principles of classification are involved or where there are broad areas of disagreement, discussions are extended to these lower levels.

Major problems in the classification of mammals are the definition and establishment of the limits of the class, the nature of the major subdivisions, the kinds of characters used, and how they have been applied in arriving at classifications. As for all major taxa that have abundant living and fossil representatives problems of coordination have arisen. The excellent records of mammals in strata of Cenozoic age have led to a good understanding of phylogenetic relationships between many living and extinct groups, but this is much less the case for the very earliest Cenozoic mammals and their Mesozoic predecessors.

369

The pre-Tertiary record of mammals is much less complete, and the samples are from deposits widely scattered in space and time. The roots of several of the modern groups reached into the late Cretaceous, but they are somewhat obscure. Earlier Mesozoic records reveal the existence of a large variety of mammals that have no modern representatives and are, with one exception, confined to the Mesozoic. They range in age back into the uppermost Triassic, the Rhaetic, but the record is far from continuous and lineages are difficult to trace. The results have been confusion and broad differences in interpretation of relationships and classifications of these "Mesozoic mammals."

DEFINITION AND BASES OF CLASSIFICATION OF MAMMALS

The class Mammalia was originally defined on the basis of modern animals. These provided characters that made possible a strictly monothetic definition. The presence of mammary glands is a sufficient criterion: all species possessing them are mammals, and all lacking them are nonmammals. The glands, of course, are related to reproductive structures and functions, several of which also are diagnostic. In a different category hair is a definitive character.

Most such characters are not available among fossils, but associated with them are osteological features that, technically at least, can be found among fossils. Some of these are present in all living mammals and are completely diagnostic—for example, the structure of the lateral walls of the braincase and the ossicles of the middle ear. Others are diagnostic only in combinations. Among these are the single element of the lower jaw, the dentary, and the double occipital condyle, which are especially useful because they are readily accessible for study.

The osteological systems throughout the class Mammalia are more uniform than those of other classes of vertebrates excepting the Aves. The greatest change in constituent elements has occurred in the distal parts of the extremities, where loss of elements is frequently encountered. Partly as a result of this stability, modifications of the skeleton have been less important in the development of classifications of mammals than in the other classes of vertebrates. Also contributing strongly to this tendency is the fact that mammalian dentitions are excellent taxonomic indicators and are well preserved in the fossil record.

All of the kinds of characters that can be used for vertebrates have been applied in the development of classifications of mammals, including those of hard and soft anatomy, physiology, behavior, biochemistry, cytology, and zoogeography. However, as far as fossils are concerned, studies of dentitions predominate, and both the phylogenies and taxonomies depend very heavily on information from this source. In this respect mammalian classifications tend to differ from those of other classes of vertebrate.

The crown patterns of the molar teeth of marsupials and placentals have stemmed from a single type, called tribosphenic (= tuberculosectorial). Mammals with this sort of dentition and some of their immediate predecessors are usually called therian mammals and placed in a subclass Theria. Mammalogists have generally paid little attention to the dentitions of extant mammals, being more concerned with external features, physiology, habits, and distributions. Thus there has been something of a hiatus between studies of modern and extinct groups. It is, for example, only quite recently that definitive investigations into the meaning of dental variation relative to genetic and environmental effects, possible only with living animals, have been undertaken (Bader, 1965a–d).

The monotremes and many of the extinct animals from the Mesozoic have much fewer clear-cut dental affinities because they did not originate from any known common base and relationships between their crown patterns are far from fully understood. At least some of the dental patterns of primitive non-therian and pretherian mammals were present among their predecessors, the therapsids. Until recently, however, little attention was paid to the dentitions in these reptiles, because the usual practice was to concentrate on osteology. Current studies are beginning to reveal marked similarities between the dentitions of some of the therapsids and primitive mammals, and clarification of relationships is certain to follow from work along these lines.

CONSTITUTION OF THE SUBDIVISIONS OF THE CLASS MAMMALIA

Among living mammals there are three distinct groups: monotremes, marsupials, and placentals. The separation is based primarily on reproductive structures and function. These groups provide a basis for a simple subdivision of mammals into subclasses:

> Subclass Prototheria (monotremes)
> Subclass Metatheria (Didelphia, marsupials)
> Subclass Eutheria (Monodelphia, placentals)

This arrangement goes fairly well back into the history of the classification of mammals. The placentals and marsupials, however, resemble each other closely and often are placed in the single subclass Theria (Eutheria).

As fossils became better known, the classification proved satisfactory as far as Cenozoic forms were concerned. All known Cenozoic fossils, except for the multituberculates, can be placed with either the marsupials or the placentals. The multituberculates represent a "holdover" from an ancient lineage that is

TABLE 36   General Classifications of the Mammalia

I. *Simpson (1945)*
Class Mammalia
  Subclass Prototheria
  Subclass Allotheria
  Mammalia of uncertain subclass or order
    Family Microcleptidae
  Mammalia of uncertain subclass
    Order Triconodonta
  Subclass Theria
    Infraclass Pantotheria
      Order Pantotheria
        (including Docodontidae)
      Order Symmetrodonta
    Infraclass Metatheria
    Infraclass Eutheria

II. *Piveteau (1961)*
Class Mammalia
  Subclass Prototheria
  Subclass Allotheria
  Subclass Theria
    Infraclass Pantotheria
      Order Symmetrodonta
      Order Eupantotheria
    Infraclass Metatheria
    Infraclass Eutheria
  Mammalia *incertae Sedis*
    Order Triconodonta
    Order Docodonta

III. *Camp, Allison, and Nichols (1964)*
Class Mammalia
  Incertae sedis family Microcleptidae
  Subclass Eotheria
    Order Docodonta
  Subclass Prototheria
  Subclass Allotheria
  Subclass Theria
    Infraclass Trituberculata
      Order Triconodonta
      Order Pantotheria
      Order Symmetrodonta
    Infraclass Metatheria
    Infraclass Eutheria

IV. *Gromova, et al. (1962)*
Class Mammalia
  Subclass Prototheria
  Subclass Allotheria
  Subclass Triconodonta
  Subclass Pantotheria
    Order Symmetrodonta
    Order Trituberculata
      (Eupantotheria)
  Subclass Metatheria
  Subclass Eutheria

V. *Romer (1966a)*
Class Mammalia
  Subclass Prototheria
  Subclass uncertain
    Order Docodonta
    Order Triconodonta
  Subclass Allotheria
  Subclass Theria
    Infraclass Trituberculata
      Order Symmetrodonta
      Order Pantotheria
    Infraclass Metatheria
    Infraclass Eutheria

VI. *B. Patterson (1956)*
Class Mammalia
  Subclass Prototheria
  Subclass Allotheria
  Mammalia of uncertain subclass
    Order Triconodonta
    Order Docodonta
  Subclass Theria
    Infraclass Pantotheria
      Order Symmetrodonta
      Order Pantotheria
    Infraclass Metatheria
    Infraclass Eutheria

VII. *D. Kermack, K. A. Kermack, and Mussett (1968)*[a]
  Infraclass Pantotheria
    Order Eupantotheria
      Suborder Amphitheria
      Suborder Dryolestoidea
      Suborder Symmetrodonta

[a] Pantotheres and symmetrodonts only.

predominantly Mesozoic and usually considered with the "Mesozoic mammals." They form an isolated assemblage placed in a separate mammalian subclass Allotheria, first proposed by Marsh (1880b).

The remaining fossils of animals generally considered as mammals have come from rocks of Mesozoic age and raise many problems in the formulation of classifications that attempt to express phylogenetic relationships. The questions are of two sorts: (a) are these animals to be placed in the class Mammalia? (see also page 318 in discussion of therapsids); (b) if so, under what major subdivisions, if any, are they to be placed and how are these subdivisions related to each other?

Table 36 includes a number of classifications and indicates a basic consistency in recent treatments. The allocations of the triconodonts and docodonts introduce the major differences. Both have long been a source of difficulty, but the triconodonts in particular have been subject to widely different interpretations.

Triconodonts are mainly Jurassic but broadly conceived range from latest Triassic into the early Cretaceous. Their characters, shown in Figure 175 in Section IV, do not clearly define a relationship to any other mammalian group. As a rule they have not been assigned to any subclass (Simpson, 1945; Romer, 1966a; Piveteau, 1961), but some students have considered them as ancestral to pantotheres and related to symmetrodonts, with the three forming an infraclass of the therians termed Trituberculata (see Camp, Allison, and Nichols, 1964, Table 36, III).

Differences stem in part from the impact of the history of study of these small Mesozoic mammals, because they have played a role in one concept of the development of the tribosphenic dental pattern. Part of the well-known Cope–Osborn tritubercular theory of the origin of tuberculosectorial (tribosphenic) molar patterns in placentals and marsupials involved a sequence passing from a single reptilian cusp, through a stage with three cusps in line (triconodont), then by rotation of cusps to a triangular stage with upper and lower molars, mirror images (symmetrodont), and finally to a tribosphenic stage (Figure 79; also see Figure 175, Section IV). Classification of the groups in accordance with this scheme was natural, but it has now been demonstrated (see B. Patterson, 1956, for an excellent summary) that this was not precisely the course followed.

The role and position of triconodonts have been further complicated by the group known as docodonts. The Jurassic docodonts were placed among the pantotheres by Simpson (1929, 1945), which were considered to be ancestral to therians. B. Patterson (1956), however, suggested that there was no such relationship and that the Rhaetic forms known as morganucodonts (eozostrodonts) were docodonts. He and others considered these to be related only distantly to other mammals. Kühne (1958) followed Patterson's assignment of

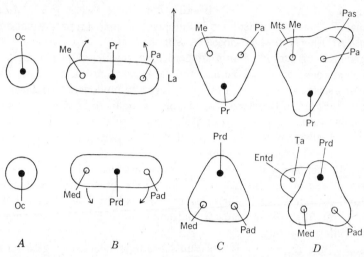

Figure 79.   Diagrammatic representation of the Cope–Osborn tritubercular theory of the origin of the cusp pattern of therian mammals. Upper row, upper molars; lower row, lower molars; labial to top of page; anterior (proximal) to the right. *A*, reptilian stage with a single cone (black circle), with equivalent cusps in following diagrams so indicated; *B*, triconodont stage, lateral cusps added, one anterior and one posterior, with directions of subsequent, presumed rotation indicated by arrows; *C*, symmetrodont stage, after rotation, producing the fundamental triangle; *D*, therian (tuberculosectorial, or tribosphenic) stage, in which the talonid is added to the posterior margin of the lower molar. The cusp names that are indicated have been retained, although there have arisen serious doubts on the order and position that the names imply. Entd, entoconid; La, labial; Me, metacone; Med, metaconid; Mts, metastyle; Oc, original cone; Pa, paracone; Pad, paraconid; Pas, parastyle; Pr, protocone; Prd, protoconid; Ta, talonid (heel).

the morganucodonts to the docodonts, but considered them as triconodonts, which he placed ancestral to pantotheres and thus on the line to therian mammals. This was somewhat like the classification of Camp, Allison and Nichols, (Table 36, III), but they placed the docodonts apart, as a separate subclass, Eotheria. Finally the classification of Gromova et al. (1962) avoids any commitment by placing the problematical forms in a separate subclass.

Confusion has arisen partly because classifications must be based on dentitions, especially on the form and crown patterns of the cheek teeth. Although the table of classifications seems to indicate a consensus, only the Prototheria, known from two living genera, the Allotheria, a distinctive extinct group, and the marsupials and placentals rest on solid ground. Future work may well alter opinions on other groups markedly.

The problems of classification sketched above depend on the supposition that the animals involved are all mammals. There is no positive evidence that the "Mesozoic mammals" possessed the reproductive features, mammary glands, hair, or homoiothermy that we associate with modern mammals. Within recent years suggested limits of the class Mammalia have been placed at the base of the therapsid reptiles at one extreme (Reed, 1960) and at the base of the Theria at the other (MacIntyre, 1967). Various intermediate positions have been taken as well (Figure 80). Simpson (1961b) has suggested that without the influence of the history of classification an impartial assessment might include all synapsid reptiles and mammals in a single class, under some term other than "Mammalia." The Theria might then form one subclass, possibly under the term "Mammalia," and the remainder might be placed in another subclass. This has some resemblance to the arguments of Von Huene and Kuhn noted in consideration of the synapsids on page 194. Classifications are, as Simpson notes, the product of history and in general radical departures so reduce their capacity to communicate that they have a hard time gaining acceptance.

The problem of limits of a class relate, as Reed stressed, to the general concept of class. Evolution shows a nodality, with times of origin of new levels of organization followed by adaptive exploitations of these new levels. The attainment of such a level, under the usual taxonomic concepts, should occur in a single or very limited source so that the new class is monophyletic. Classes, however, represent major radiations within their phyla, and each radiation takes place after initiation of a major reorganization. The fundamental organizational change, in at least some instances, was neither rapid nor definitive at any one level, and was not confined to a single phyletic line. Only in the broad sense that their source(s) lay in a lesser category can major groups, such as classes, be conceived to be monophyletic. If this is coupled with the apparently universal mosaic pattern of attainment of new levels, it becomes impossible to define the precise limits of a class.

A class, like any other category, must be practical and convenient. It represents a compromise of many aims. It is the effort to satisfy these different aims that gives rise to confusion, as one person considers one aspect, say monophyly, most important, whereas another considers ease of assignment, and still another is concerned with the closeness of fit to a major radiation pattern.

There is, of course, no correct answer to the problem of mammalian limits, because the category Mammalia is technically an artifact, although the concept of the category has vast biological significance. The state called mammalian was attained gradually, beginning well back among forms generally recognized as reptiles. Short of abandoning classification, which has been seriously suggested by Kühne (1968), the solution must be in a compromise, expressing the phylogeny as well as possible and allowing ready assignment of the majority of animals as mammals or nonmammals.

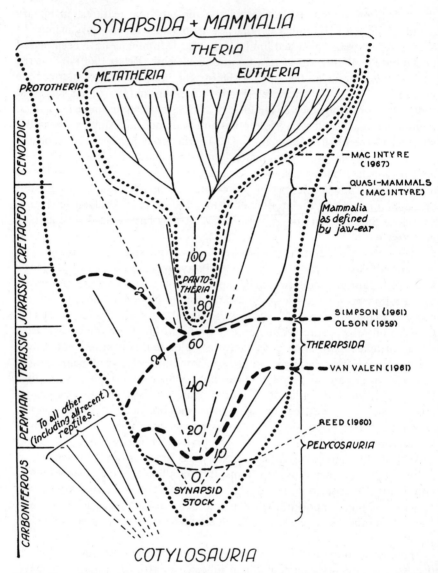

Figure 80.   A diagrammatic representation of the phylogeny of the synapsid reptiles and their mammalian descendants, illustrating several different concepts about the assignment of these groups to mammals. The basic diagram was taken from Simpson (1961*b*) and modified by addition of other information. Reed (1960) included most of the pelycosaurs, sphenacodonts specifically, and all therapsids in the Mammalia. Simpson and Olson essentially have agreed on placing the division between reptiles and mammals in such a way that all "Mesozoic mammals" (triconodonts, morganucodontids, symmetrodonts, and pantotheres) are included with

Most students today use arrangements that include both the Cenozoic and Mesozoic groups of "mammal-like" creatures as mammals. This is not necessarily the best arrangement, and it reflects the natural reluctance to change even in the face of information that suggests other organizations might be superior. The current scheme, however, is useful and, except to those most familiar with the matter, seems to present a straightforward arrangement. The evolutionary aspects of the problem, which though distinct in one way may sometime serve as the basis for a complete reorganization, are treated in detail in Section IV, Chapter 6.

NONTHERIAN MAMMALS

The subclasses Prototheria and Allotheria and the diarthrognathids and haramyids, if the last be considered mammals at all, are clearly nontherian mammals. If the term "Theria" is restricted to mammals with fully expressed tribosphenic molar dentitions, only the marsupials and placentals, plus some very poorly known lower Cretaceous animals, are included. This could be a satisfactory statement of the present knowledge and a satisfactory arrangement, but it is not one that is commonly followed.

The tendency is to include members of presumed pretherian lineages in the Theria (see Table 36). But this poses a problem, because it depends on interpretations of what is actually pretherian, and on this there is no consensus. Most agree that the pantotheres were ancestral to the mammals with tribosphenic molars, but there agreement ends. The symmetrodonts are considered as a likely source of pantotheres. If this is accepted, the symmetrodonts should follow the pantotheres and be classed as therians. This is often done. When it is, the question of the possible relationship of symmetrodonts and triconodonts arises, and this may involve the role of the morganucodonts and docodonts. We pass from a fairly certain relationship of pantotheres and primitive forms with tribosphenic dentitions to very questionable relationships, with no well-defined position for a break.

Efforts to make a phylogenetic classification at present cannot portray a stable and agreed-on series of relationships, and the various attempts must

---

the Mammalia, with the therapsids, as usually defined, considered Reptilia. MacIntyre (1967) included only the therians (pantotheres, marsupials, and placentals) in the Mammalia, but placed the "Mesozoic mammals" as "quasi-mammals," not in formal classification. Various other authors have proposed modifications of these schemes, but these show the major kinds of subdivisions that have been given serious consideration.

be considered what they actually are—expressions of opinions to indicate a current but perhaps fleeting assessment.

The breakdown used in this chapter is based on the current practices of association of symmetrodonts and pantotheres with the therians, but with the recognition that this introduces problems that cannot be solved on the basis of the rather scant current evidence. The following, thus, comprise the nontherian mammals.

The subclass Prototheria, the monotremes, is based on two living genera and is known only from sub-Recent, poorly preserved fossil remains. The two genera, *Tachyglossus*, the echidna, and *Ornithorhynchus*, the platypus, are adaptively very different and linked mainly by such fundamental features as reproduction by means of large eggs and the presence of rudimentary mammary glands, along with similarities of parts of the hard and soft anatomy. Presumably they are survivors of a very primitive stock that originated early and retained many features of its reptilian ancestry. This interpretation, though logical, lacks support of a fossil record, and W. K. Gregory (1947), along with some others, has suggested that they have in fact come from a marsupial-like stage and that the seemingly primitive features are regressive. This point of view has not been widely accepted, but it is obvious that it would require extensive modifications in mammalian classification were it demonstrated to be true.

The subclass Allotheria, or multituberculates, appeared in the Jurassic and persisted into the Eocene, thus having a very long life span. During this time they had a modest adaptive radiation in which three suborders were developed. All are placed within a single order. The members show superficial resemblances to the tritylodonts, but there is no basis whatsoever for presuming a close relationship. It is generally thought that the multituberculates form an isolated group whose source is unknown and gave rise to no descendants beyond its own subclass. Bohlin (1945) and Butler (1939), among others, have suggested a triconodont origin, and a morganucodontid origin is not out of the question.

Farther afield have been proposals that the multituberculates lay at the base of some or all of the stocks of eutherian mammals. One suggested link has been with the marsupials through the South American caenolestids. This and similar suggestions deny the origin of therian cheek teeth from a tribosphenic base, a theory rooted in so much evidence that any denial is unlikely to be valid.

The order Docodonta was placed with the Pantotheria by Simpson (1929) and was associated with morganucodontids by B. Patterson (1956). Later students have generally followed Patterson's lead, but their interpretations vary with respect to the placement of the morganucodontids, as noted on page 373. This creates a more complex array than was recognized by Simpson and raises possibilities that the docodont assemblage may have been related to the triconodonts.

The order Triconodonta in a strict sense includes Jurassic and lower Cretaceous genera. These are of two rather distinct types: the amphilestids and the more typical triconodontids. Various specimens from the Rhaetic have been associated with the triconodonts, either broadly or specifically. Among these are morganucodontids, which have been considered triconodonts by Kühne and others. *Sinoconodon*, from Yunnan, China (B. Patterson and Olson, 1961) was placed in the triconodontids. This assignment has been questioned with suggestions that this genus may be a cynodont, a mammal but not a triconodont, and that part of what was described as *Sinoconodon* might be actually *Morganucodon*. This is the sort of confusion that has accompanied many of the finds of this age.

Highly problematical are some of the small teeth of the Haramyidae or Microcleptidae from the late Triassic. They may be reptiles or mammals, and there is little basis for choosing. The structure of the diarthrognathids is much better known, and the question as to whether they are nontherian mammals or not hinges strictly on what criteria are used for association of borderline fossil genera with the mammals.

THERIAN MAMMALS

Infraclass Pantotheria (Trituberculata)

The majority of recent classifications place the symmetrodonts and pantotheres together, generally in an infraorder, either as Pantotheria or Trituberculata (see Figure 175, Section IV, for representative jaws). Trituberculata has a terminological advantage, avoiding confusion in reference, but otherwise there is no basic difference in the procedures. Differences in grouping range about as shown in Table 36. Confidence that the relationship of symmetrodonts and pantotheres shown in the classifications actually exists is much less than the hierarchical association might suggest.

Simpson's summary statement of the problems of relationships (1961) expresses simply and clearly the primary difficulty, the absence of evidence needed for decisions. Patterson's (1956) interpretation of cusp homologies (Figure 81), which gives a logical course of origin of the tribosphenic pattern depends on the assumption of symmetrodont–pantothere relationship. The transition as he visualized it was made before a very primitive pantothere or pantotherelike symmetrodont, *Kuehneotherium* from the Rhaetic, was known. Although the cusp homologies depend to some extent on the understanding of relationships, it is also true that the understanding of relationships in this

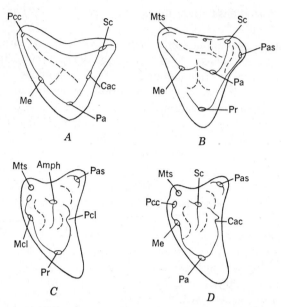

Figure 81.   Diagrams of the pattern of cusp origin of the upper molars from the symmetrodont stage as proposed by B. Patterson (1956), in contrast to the "orthodox" cusp pattern generally followed before that time. Note that this is based on a "two-triangle concept" rather than one in which the fundamental triangle persisted. *A*, symmetrodont, with inner cusp designated as paracone (see Figure 79C) at the apex of a paracone–metacone–stylocone triangle; *B*, upper molar of a therian (based on the lower Cretaceous Forestburg therian) as interpreted by Patterson; note that the protocone is the base of the added triangle; *C*, the pantothere *Melanodon*, with the "fundamental" triangle consisting of protocone, paracone, and metacone, under the "orthodox" interpretation; *D*, same, but as interpreted under Patterson's system. Amph, amphicone; Cac, cusp on anterior crest; Mcl, metaconule; Pcc, postero-external cingular cusp; Pcl, protoconule; Sc, stylocone. Other abbreviations as in Figure 79.

instance is dependent largely on dentition interpretations of cusps and wear facets (Crompton and Jenkins, 1967, 1968).

The structure of lower jaws and interpretations of jaw dynamics also have played roles in assessments of the relationships of these primitive mammals. The lower jaws of the symmetrodonts lack an angular process. The Rhaetic–Liassic forms possess a Meckelian sulcus and internal or splenial groove. In later forms of symmetrodonts only the internal groove persists. Pantotheres have an angular process (see Figure 175, Section IV) and maintain the internal groove. Thus passage from the symmetrodont stage to the pantothere stage, provided it did occur, involved a marked reorganization in jaw and dental mechanics, presumed to be intimately related to the modifications of the cheek teeth.

The changes are plausible and can be "explained," as B. Patterson has done in the development of his hypothesis of origin of tribosphenic dentitions. The hypothesis cannot be denied on the basis of the evidence available, but neither can it be confirmed. The sparseness of data is not altered by the feeling of confidence that comes from the fact that these data do fall nicely into place in a plausible model.

A few specimens from the Trinity formation of the lower Cretaceous are important in this problem. These consist of fragments of therians that are definitely not pantotheres. B. Patterson, who described the first specimens, has suggested that they were ancestral to both metatheres and eutheres. Thus the tribosphenic dentition occurred in a stock that was ancestral to both. Mills (1964) expressed the opinion that the common ancestry lay among the amphitheriid pantotheres of the late Jurassic. Slaughter (1968) has identified some of the teeth from the Trinity as marsupial, as follows:

Subclass Theria
Order Marsupialia
Family Didelphidae
Clemensia texana Slaughter

On the basis of this determination the two lines had already diverged, although slightly. The only older tooth of this grade, *Aegealodon* (K. A. Kermack, Lees, and Mussett, 1964) is indeterminate in position. The only characters available for study are those of the crown patterns, and necessarily any interpretations are to be viewed with caution.

Most comprehensive classifications have not found a place for the therians of the lower Cretaceous or for the one genus from the middle Cretaceous of Asia, which is of a similar nature. Romer has tentatively included them as Insectivora, under the Eotheria. The subdivision suggested by Slaughter, however, alters this to the extent that some of the included specimens are removed to the marsupials.

Infraclass Metatheria

General Considerations

Most classifications recognize but a single order, Marsupialia, under the infraclass Metatheria, and this term is frequently used in references to the group rather than Metatheria. Today marsupials have a limited distribution, with the greatest variety in Australia and two groups, opossums and caenolestoids, in the New World, with the latter confined to South America. The fossil record is likewise limited. A moderate variety of didelphoid, opossumlike creatures occurs in the late Cretaceous and early Tertiary of North America, and in the early and mid-Tertiary of South America and Europe. The only earlier fossils are those from the lower Cretaceous assigned to this group by Slaughter.

TABLE 37   Classifications of the Marsupialia

I. *Jones (1923–1925)*
  Subclass Didelphia (living, Australia only)
    Order Marsupialia
      Suborder Didactyla
        Family Didelphidae
        Family Dasyuridae
        Family Thylacinidae
        Family Myrmecobiidae
        Family Notoryctidae
      Suborder Syndactyla
        Section Syndactyla Polyproto-
          donta
        Family Peramelidae
        Section Syndactyla Diproto-
          donta
        Family Phalangeridae
        Family Macropodidae

II. *Romer (1933)*
  Subclass Didelphia (Metatheria)
    Order Marsupialia
      Suborder Polyprotodonta
        Family Didelphidae
        Family Borhyaenidae
        Family Dasyuridae
        Family Notoryctidae
        Family Peramelidae
      Suborder Caenolestoidea
        Family Caenolestidae
        Family Polydolopidae
      Suborder Diprotodonta
        Family Phalangeridae
        Family Phascolomidae
        Family Macropodidae
        Family Diprotodontidae

III. *Simpson (1945)*
  Subclass Theria
    Infraclass Metatheria
      Order Marsupialia
        Superfamily Didelphoidea
          Family Didelphidae
          Family Carloameghiniidae
        Superfamily Borhyaenoidea
          Family Borhyaenidae
        Superfamily Dasyuroidea
          Family Dasyuridae
          Family Notoryctidae
        Superfamily Perameloidea
          Family Peramelidae
        Superfamily Caenolestoidea
          Family Caenolestidae
          Family Polydolopidae

Superfamily Phalangeroidea
  Family Phalangeridae
  Family Thylacoleonidae
  Family Phascolomidae
  Family Macropodidae
  Family Diprotodontidae

IV. *Romer (1966a)*
  Subclass Theria
    Infraclass Metatheria
      Order Marsupialia
        Suborder Polyprotodonta
          Family Didelphidae
          Family Carloameghiniidae
          Family Borhyaenidae
          Family Necrolestidae
          ?Family Microtragulidae
          Family Dasyuridae
          Family Notoryctidae
        Suborder Peramelida
          Family Peramelidae
        Suborder Caenolestoidea
          Family Caenolestidae
          Family Groeberiidae
          Family Polydolopidae
        Suborder Diprotodonta
          Family Phalangeridae
          Family Thylacoleonidae
          Family Phascolomidae
          Family Macropodidae

V. *Ride (1964)*
  Class Mammalia
    Infraclass Metatheria
      Superorder Marsupialia
        Order Marsupicarnivora
          Superfamily Didelphoidea
          Superfamily Borhyaenoidea
          Superfamily Dasyuroidea
        Order Paucituberculata
          Family Caenolestidae
          Family Polydolopidae
        Order Peramelida
          Family Diprotodonta
        Order Diprotodonta
          Family Phalangeridae
          Family Wynyardiidae
          Family Vombatidae
          Family Diprotodontidae
          Family Macropodidae
        Marsupialia *incertae sedis*
          Family Notoryctidae

The South American Tertiary in addition has a fair record of marsupials, including the Borhyaenoidea and the family Polydolopidae, aberrant caenolestoids. The fossil record in the Australian region ranges from Oligocene to Recent, but, except for the late Pleistocene and sub-Recent, it is very sparse. Slowly diligent work by many students is unearthing some traces of the history of the Australian radiation.

At one time or another most of the Mesozoic mammals have been placed among the marsupials. This is partly in accord with the long-held concept of a monotreme, marsupial, placental course of mammalian evolution, now thought to be incorrect. Partly, however, the relationship has been based on supposed resemblances of dentitions. In particular the caenolestoids and multituberculates have been brought together because of some rather vague dental resemblances. Origins and classifications have become rather curiously intertwined with theories of origin of marsupial and placental cheek dentitions. If the general proposition of the Cope–Osborn tritubercular theory that tribosphenic patterns form the base of all metatherian and eutherian dental patterns is followed, none of the so-called Mesozoic mammals—triconodonts, docodonts, symmetrodonts, pantotheres, and multituberculates—can be classified as marsupials. Other theories of origin, and many have been proposed, do permit this practice. The evidence for the tribosphenic ancestry, however, is so strong that the other theories have gained little headway. Summaries of these may be found in various places, particularly in "A Half-Century of Trituberculy" (W. K. Gregory, 1934).

The caenolestoids have posed special problems in classification because they are isolated, small, and very different from other marsupials. They became known rather late in the studies of marsupials and did not play an important role in the formulation of classifications. Their special case is taken up at the end of this section.

The present distribution and the nature of the fossil record have influenced classifications of marsupials. Although the principal criteria have been morphological, the effects of these other factors are significant. One peculiarity of the classifications presented in Table 37, except for that of Ride, is that but a single order is recognized, in contrast to the Eutheria in which 25 or more are recognized. This is partly the result of the somewhat limited radiations of marsupials. Adaptive diversity, however, is greater than that found in any of the placental orders. Cain (1959) made a point of this in a discussion of general matters related to bases of taxonomy. He suggested that the relatively late discovery of the marsupials, resulting from their geographic isolation and the history of human civilization, cast them in the perspective of the norm of placentals, as abnormal with respect to reproduction. This was sufficient to result in their placement in a single order characterized by this "abnormal" condition. Once established, even after the initial reason for establishment had been abandoned, the disposition tended to persist tenaciously. Ride (1964) took particular note of this aspect of the history in making his proposal of four orders in his classification shown in Table 37, V.

Absence of a good fossil record in Australia and the presence of one in South America have played important roles in the later stages of development or marsupial classification. Inability to trace phylogenies of the highly diversified Australian marsupials has limited recognition of the degree of diversity. The duration of the separation of different stocks, often important in assessing relationships and classifications, cannot be determined. Recent evidence, although scant, suggests that major subdivisions go far back in the Tertiary (Stirton, Tedford, and Miller, 1961; Ride, 1964). For many years after their initial discoveries neither the North nor the South American marsupials were placed within this order. Once the South American borhyaenoids were recognized as maruspials, controversies concerning their relationships to Australian forms arose, bringing up problems of zoogeography, including concepts of continental movements and intercontinental isthmian links. These too assumed importance in classifications. As the widespread nature of the early didelphoids became better understood and their position among marsupials was properly assessed, the current concepts of origin of marsupial groups from a common didelphoid base emerged and classifications of the sorts used today developed.

Originally, of course, the marsupials were defined on the basis of living animals, which provided many features of the soft anatomy that were distinct from those found in any other known mammals. Early, because of the nature of reproduction, it was thought that marsupials were closer to monotremes than to placentals, and the classifications reflect this understanding, but this has long since been dismissed. Characters of the marsupial female reproductive system (Figure 24) are distinctive along with other morphological and physiological features related to reproduction. Associated with these definitive structures are osteological features, which, at least in the absence of knowledge of intermediate ancestors, serve readily to separate the marsupials and placentals. Formation of the auditory bulla by the alisphenoid, rather than tympanic elements as in placentals, is one such feature. The existence of the marsupial bone anterior to the pubis represents another feature, one shared, however, with monotremes.

Dental features are also distinctive, in particular the primitive dental formulas. Primitive placentals have but three incisors above and below. Primitive marsupials have more. The basic marsupial formula in the cheek region includes three premolars (a replaced or partly replaced series of teeth) and four molars. Placentals have four premolars and three molars. Such features have made it possible to differentiate almost all fossil mammals from the upper Cretaceous to the present as marsupials or placentals. Among fossil forms for which even very modest data are available, only two or three pose any problem of placement. Currently, then, monothetic classification is applied at the infraclass level.

Major Subdivisions

The classifications in Table 37 show the major subdivisions of the marsupials that have been made over the last several decades. Australian and non-Australian marsupials were linked early in the nineteenth century, although before that the marsupial nature of *Didelphis*, the one known non-Australian genus, had not been recognized. The term "Marsupialia," commonly used as the major unit below the infraclass level, was first used formally by Illiger (1811), whereas the term "Didelphia," also used for this major unit, was first formally applied by De Blainville in 1816. The former thus has priority and is now generally accepted. De Blainville's Didelphia included both Australian and non-Australian didelphoid marsupials, combining them for the first time, and also monotremes. The early classifications dealt only with living marsupials. It was not until 1906 (Sinclair, 1906), for example, that the South American borhyaenids were placed with the marsupials, and fossil didelphoids played no role in early considerations.

A fundamental division of the living marsupials on the basis of incisor dentition into Polyprotodontia and Diprotodontia (Polyprodonta and Diprotodonta) was made by Owen in 1866. This was a phenetic rather than a phyletic subdivision because Owen was not an evolutionist. It served well to separate all marsupials into two easily distinguished groups. This basis of classification has persisted, as witnessed by Romer's (1933) classification in Table 37 (II). In addition in this classification Romer included a third major group, Caenolestoidea, not known at the time of Owen's initial breakdown.

That there were two types of pes in the marsupials had long been known and recognized as a basis for differentiation when Jones (1923) used it as a formal basis for separation of marsupials into two major groups, Didactyla and Syndactyla. He again was working only with moderns and restricted his classification to Australian marsupials. Jones argued that his classification was phyletic, in contrast to Owen's, which, though useful for separating collections into two groups, was phenetic.

As could be expected, the two ways of subdivision, each based on one feature, did not entirely coincide, and thus in Jones' classification is found the awkward use of two sections of the Syndactyla, one polyprotodont and the other diprotodont. If the caenolestoids are added, there is also a didactylous-diprotodont. Romer's (1966a) classification resolves the awkwardness by the use of four suborders—Polyprotodonta, Diprotodonta, Caenolestoidea (using a term proposed by Osborn), and Perameloidea—thus distinguishing the peramelids, syndactylous polyprotodonts, from the others. This resolution uses paired characters to maintain the concept of monothetic separation.

Simpson (1930, 1945) argued that there were six natural groups of marsupials and erected a classification (Table 37, III) in which each group constituted the basis for one superfamily. These six superfamilies were not grouped under any higher category below the level of order Marsupialia. This subdivision has been widely followed. It was based on the understanding that there was insufficient evidence of actual phylogenies to detect any grouping that might exist. Each of the possible relationships was examined in detail by Simpson in making the decisions.

Romer (1966a) did not follow this practice; rather he recognized the four major groups noted above, bringing together the first three of Simpson's superfamilies Didelphoidea, Borhyaenoidea, and Dasyuroidea under the suborder Polyprotodonta.

Simpson (1941) summarized the differences of opinion that existed at that time or had been voiced earlier. The differences are relatively subtle but have importance with regard to zoogeographic interpretations. The Thylacinidae, a family of Dasyuroidea, occurs only in the Australian realm today. Members of this family show strong resemblances to some of the Miocene borhyaenids, and this has led to the conclusion that the borhyaenids are thylacinids (see, for example, Sinclair, 1906; H. E. Wood, 1924). A somewhat different point of view is that the borhyaenoids and dasyuroids had a common base above the didelphoid level (Cabrera, 1927). This implies the existence of such a stock at a time and place that it could have given rise to both the South American and Australian members of the group, but no such stock at the suggested level of development is known.

The third position, that of Simpson (1941, 1945) and also of Ameghino (1906) and Matthew (1915), was that the borhyaenoids and dasyuroids arose separately from a didelphoid ancestry and that the close resemblances of some of the borhyaenids and thylacinids, as in Cabrera's interpretation, were the result of strong parallelism. This is basic to the concept of three superfamilies. Romer's classification does not deny this contention but requires that his other three major groups had their origin within the first, assuming that the ancestry of all lay among the didelphoids, as Romer clearly does. In this sense they are more or less equivalent to the borhyaenoids and dasyuroids of the Polyprotodonta. The two positions are thus similar but expressed very differently in the classifications.

These problems of relationship and the ways of handling interpretations in formulation of classifications are intimately related to geographic distribution, concepts of continental stability and interconnections, to dating of beds in widely separated areas, and to ideas on phylogeny based on very scant data. They emphasize the many facets of study that enter into the erection of a classification and the intricate role of the history of study of a group in current arrangements.

Ride's 1964 classification was established on the basis of a review of the evidence that led him to upgrade the taxonomic levels within the marsupials. This was based primarily on the conclusion that the radiation was sufficiently extensive for four major groups to rate ordinal ranking. He believed that the lines represented had been long separated temporally, which seems to be supported by the fossil evidence. His classification, except for the use of orders rather than suborders and of some different names, resembles that used by Romer (1966*a*), which of course was formulated with awareness of Ride's study. The Polyprotodonta of Romer are equivalent to the Marsupicarnivora of Ride, and the Caenolestoidea are equivalent to the Paucituberculata of Ride, with differences of categorical levels. Romer retained the Notoryctidae in his first major subdivision, whereas Ride placed it as Marsupialia *incertae sedis*.

The mere change in rank made by Ride might be of little importance, but in his discussion he shows that he means something quite different in terms of radiation from what is implied by the usual single order and series of suborders or superfamilies. He states, for example, that his order Marsupicarnivora is equal to the placental order Carnivora in scope. Actually, although radiation is clearly somewhat less among the marsupial carnivores than it is among placentals, his group in a sense is more inclusive because it also brings in what must be considered equivalent to the placental Insectivora. This, of course, rounds the circle and raises the question of association of didelphoids and the borhyaenoids and dasyuroids in a single group, here at the ordinal level. Ride also feels, certainly with some justification, that the degree of radiation of phalangeroids and perameloids warrants ordinal recognition.

CAENOLESTOIDS.   Among the marsupials the caenolestoids occupy a rather special position, and they have not for the most part entered into considerations of classifications in a very important way. Suggestions of relationship with both phalangers and peramelids have been made (e.g., Osgood, 1921) as well as of relationships with various polyprotodonts. Simpson (1930, 1945) concluded that they were independently derived from the didelphoids, and this has been the most commonly adopted position.

An odd, small group of fossils from the Paleocene and Eocene of South America has been at the heart of some of the problems. These are the polydolopids. When first known, they were called multituberculates and as such, of course, provided a bridge between the multituberculates and marsupials once their affinities to caenolestids were recognized.

Currently there seems to be little disagreement that the caenolestoids and polydolopids, along with the more recently described Groeberiidae, represent a coherent group of marsupials that had an independent origin in the didelphoids. All recent classifications are drawn up with this general concept in mind.

EUTHERIAN MAMMALS

The eutherian, or placental, mammals were first classified as distinct from other mammals by De Blainville (1816). Because of their abundance and dominance over all other mammals in the regions where most zoologists lived and worked during the eighteenth century, the placentals had played a primary role in classifications of mammals even before their recognition as a discrete group. Marsupials and of course monotremes had very little effect on the overall formulations, being lost, so to speak, among the placentals.

Initially only the divisions of Monodelphia (Eutheria) and Didelphia (Metatheria) were recognized by De Blainville, both the marsupials and the monotremes being assigned to the latter. Later, in 1834 and 1839, De Blainville proposed a threefold division, separating Ornithodelphia and Didelphia, and thus establishing the arrangement of modern mammals generally accepted today. Until about 1900 there was somewhat sporadic acceptance of this division, but since that time it is found in most major classifications.

The history of mammalian classification up to about 1905 was outlined in considerable detail by W. K. Gregory (1910) in his "Orders of Mammals" (see Table 38). In this lucid account he not only gave the course of development but also analyzed the principles of classifications and the nature and use of characters at the various stages. Portions of what follows in this section are to be found in more detail in his study. Some parts of the work of W. K. Gregory were summarized and important additions were made by Simpson (1945) in his critical work on mammalian classifications.

Four steps in particular mark the progress in mammalian classification within the "scientific" era, which began during the seventeenth century. The first is the classification proposed by Ray (1693), which initiated many of the modern practices. His was a dichotomous classification, a keylike form similar to that of the Aves (see frontispiece) now rarely used. It was, however, constructed on the basis of morphological and functional features and downgraded the importance of locus and medium, which were in frequent use at the time. His major groups of animals that we now call mammals do not show this particularly well, since they were designated as aquatic (cetacean) and terrestrial (quadrupedal). The major divisions of the classification as a whole, which dealt with all vertebrates, and the lesser groups among the mammals were in large part established with morphology as the major criterion.

TABLE 38   Major Subdivisions of the Placental Mammals[a]

I. *Linnaeus (1758–1766)*[b]
Modern Orders 1910

Class Mammalia
Unguiculata
Primates   Primates + Dermoptera + Chiroptera
Bruta   Proboscidea + Sirenia + Xenarthra (part) + Pholidota
Ferae   Carnivora
Bestiae   Suilline artiodactyls (part) + Xenarthra (part) +
           Insectivora + Polyprotodonta (part)
Glires   Perissodactyla (part) + Rodentia
Ungulata
Pecora   Artiodactyla (minus *Sus* and *Hippopotamus*)
Belluae   Perissodactyla (part) + Suilline Artiodactyla (part)
Mutica   Mystacoceti + Odontoceti

II. *Storr (1780)*
Class Mammalia[c]
Phalanx I. Pedetorum
Cohors I. Unguiculatorum
Order Primates
Order Rosores
(rodents, rabbits)
Order Mutici
(sloths and anteaters)
Cohors II. Ungulatorum
Order Jumenta (*Equus*)
Order Pecora
Order Belluae
Phalanx II. Pinnepedium
Phalanx III. Pinnatorum

III. *Lacépédé (1799)*
Division I. Quadrupeds
Division II. Flying Mammals
Division III. Marine Mammals

IV. *De Blainville (1816)*
Mammifères
Sous-classe I. Monodelphes
Quadrumanes

Carnassiers (?)
Edentates
Ronguers (?)
Gravigrades
Sous-classe II. Didelphes
Normaux
Anormaux (egg-layers)

V. *De Blainville (1834)*
Sous-classe I. Ornithodelphes
Sous-classe II. Didelphes
Sous-classe III. Mondelphes

VI. *Gill (1870)*
Class Mammalia
Subclass Monodelphia
I. Primate Series
Order Primates
II. Feral Series
Order Ferae
Order Cete
III. Insectivore Series
Order Insectivora
Order Chiroptera

TABLE 38—continued

VI. *Gill (1870)*—continued
  IV. Ungulate Series
    Order Ungulata
    Order Hyracoidea
    Order Proboscidea
    Order Sirenia
  V. Rodent Series
    Order Glires
  VI. Edentate Series
    Order Bruta (or Edentata)

VII. *Cope (1891–1898)*
  Class Mammalia
    Subclass Monodelphia
      Cohort Mutilata[d]
      Cohort Unguiculata
      Cohort Ungulata

VIII. *Osborn (1910)*
  Class Mammalia
    Subclass Monodelphia
      Cohort Unguiculata
      Cohort Primates
      Cohort Ungulata
      Cohort Cetacea

IX. *W. K. Gregory (1910)*
  Class Mammalia
    Subclass Monodelphia
      Superorder Therictoidea
      Superorder Archonta
      Superorder Rodentia
      Superorder Edentata
      Superorder Paraxonia
      Superorder Ungulata
      Superorder Cetacea

X. *Simpson (1945)*
  Class Mammalia
    Intraclass Eutheria
      Cohort Unguiculata
      Cohort Glires
      Cohort Mutica
      Cohort Ferungulata
        Superorder Ferae
        Superorder
          Protoungulata
        Superorder
          Paenungulata
        Superorder Mesaxonia
        Superorder Paraxonia

[a] Mostly after W. K. Gregory (1910).
[b] Includes all mammals.
[c] Moderns only.
[d] After Owen (1868).

The classifications of Linnaeus (1758 to 1766) (Table 38, I) were important in the development of mammalian classifications, just as they were in biological classification as a whole. These represent the second step, setting forth patterns and procedures that have since become dominant. The use of the binomial system is of course important in general. With special reference to the mammals, Linnaeus' use of several characters rather than one in formulation of categories, in contrast to the usual practice of the time, was of special significance. He coined the term "Mammalia" in recognition of the importance of mammae as a criterion in taxonomy. His methods also led to the placement of man among the primates, a truly significant advance in unifying concepts of the vertebrate world. For his orders Linnaeus used the nature of the front teeth ("laniaform" teeth), limb structure, and the nature and ways of procurement of sustenance as primary criteria. His classifications naturally drew heavily on what had gone before, and it was his systematization of knowledge that was outstanding for his time.

The third signal step was that taken by De Blainville (Table 38, IV, V), who, as already noted, recognized the importance of the reproductive system in making fundamental divisions of modern mammals into Monodelphia, Didelphia, and Ornithodelphia (1816, 1834, 1839).

The fourth step, which set the stage for current classifications and firmly established many of the principles generally in use, is the classification and comments on its formulation set forth by Gill (1870, 1872) (Table 38, VI). The Darwinian epoch was well under way, and this classification involved phylogenetic concepts. Many classifications of this general period were once again wedded to a single criterion as the basis for major subdivisions. Owen, for example, placed reliance on features of the brain, whereas Haeckel made primary use of the nature of placentation. Gill, like Cuvier and Linnaeus before him, made use of a system of multiple characters and in discussing processes of classification carried this process to its logical end, stressing that morphology should be the basis of classifications and that relationships could be properly assessed only by the sum of agreements of morphological characters. In his classification Gill used the three major subdivisions of De Blainville. Within the placentals he recognized six series (equal to superorders or cohorts). In the principles employed his was a truly "modern" classification.

None of the four advances, of course, was isolated from the influence of the work of others being carried on at the same time or completed earlier. Each owes many debts to others, but in the particular synthesis that it accomplished stands out as unique. The names of T. S. Huxley, Cope, Weber, Flower, Osborn, W. K. Gregory, Matthew, and Simpson loom large among the hundreds who concerned themselves with mammals in the latter part of the nineteenth and in the twentieth century. Phyletic considerations have been fully integrated into procedures of classification, and zoogeography has become increasingly important. Increased understanding of the fossil record and the many new forms that it has added to the list of known mammals have made this recorded history a dominant aspect of classification. Morphology has remained a primary consideration in both living and extinct groups, and among the former newer methods of analysis involving cytogenetic and biochemical studies have assumed an increasingly important role. Numerical taxonomic procedures have long been used in classification in one form or another, but only recently have more formal methods become important in their direct formulation.

Principles of classifications have not been modified drastically since the time of Gill. There is continued interplay between phyletic and phenetic methods. Although some of the major subdivisions go far back in history (such as the three proposed by De Blainville), most currently used major categories, even though they may retain older names, differ considerably in content from those of 75 or 100 years ago.

Four classifications are listed in Table 39: two from 1910, showing the status of classification of eutherians at that time; the classification of Simpson (1945), which has been immensely influential; and the classification of Romer (1966a), which, among other things, embodies much information not available at the time Simpson erected his classification. Many others exist, of course, and these differ in various ways from the four in Table 39. Some deal primarily with modern mammals, whereas others, like those of Simpson and Romer, lean heavily on the fossil record. The range of difference at the high levels is not much greater than the differences between the classifications of Simpson and Romer. Most differences lie below the ordinal level.

Superordinal Subdivisions of the Eutherian Mammals

A marked difference between the classifications of Simpson and Romer in Table 39 is in their treatment of major subdivisions of the infraclass Eutheria. Simpson used cohorts, with superorders in one cohort, whereas Romer uses orders as his highest subdivision. The cohort (or a similar division) as currently used is designed to indicate relationships of the included categories in the sense that they are all parts of a major radiation. Presumably each had a common basal stock, distinct from the ancestral stock, from which all lines in the cohort originated. Such relationships are less evident in the sequential position of orders, such as that in Romer's classification, because the breakpoints cannot be specified without some device for separation. A full cohort system, on the other hand, requires that all orders be placed in one cohort or another, and in this process knowledge of relationships may be stretched far beyond the available information.

Most students of mammalian classification have employed some sort of superordinal grouping of the eutherian mammals, but some (e.g., Flower, 1883, and Weber, 1904) have not done so. Representative examples of ways of major subdivision are included in the selected classifications in Table 39.

In Linnaeus' classification the nature of appendages occupies a primary role in the formulation of his major groups: Unguiculata (clawed mammals), Ungulata (hoofed mammals), and Mutica (finned mammals). These include all mammals, not just placentals, but the noneutherians play a minor role.

Storr (1780) used two divisions between class and order—phalanx and cohort—in a system that is based primarily on single characters (see Table

TABLE 39   Four general Classifications of Monodelphia or
Placentalia

I. Osborn (1910)	II. W. K. Gregory (1910)
Subclass Eutheria	Subclass Theria
Infraclass Monodelphia	Infraclass Eutheria
Cohort Unguiculata	Superoder Therictoidea
Order Trituberculata	Order Insectivora
Order Insectivora	Order Ferae
Order Tillodontia	Superorder Archonta
Order Dermoptera	Order Menotyphla
Order Chiroptera	Order Dermoptera
Order Carnivora	Order Chiroptera
Order Rodentia	Order Primates
Order Taeniodonta	Superorder Rodentia
Order Edentata	Order Glires
Order Pholidata	Suborder
Order Tubulidentata	Duplicidentata
Cohort Primates	Suborder
Order Primates	Simplicidentata
Cohort Ungulata	?Superorder Edentata
Order Condylarthra	?Order Taeniodonta
Order Amblypoda	?Order Tubulidentata
Order Artiodactyla	?Order Pholidota
Order Perissodactyla	Order Xenarthra
Order Ancylopoda	Superorder Paraxonia
Order Proboscidea	Order Artiodactyla
Order Barytheria	Superorder Ungulata
Order Sirenia	Order Proungulata
Order Hyracoidea	Order Amblypoda
Order Embrithopoda	Order Barytheria
Superorder Notoungulata	Order Sirenia
Order Toxodontia	Order Proboscidea
Order Litopterna	Order Hyraces
Order Pyrotheria	Order Embrithopoda
Cohort Cetacea	Order Notoungulata
Order Zeuglodontia	Suborder
Order Odontoceti	Homalodotheria
Order Mystacoceti	Suborder Astrapotheria
	Suborder Toxodontia
	?Suborder Pyrotheria

TABLE 39—continued

II. *W. K. Gregory (1910)*—continued
 Suborder Litopterna
 Order Mesaxonia
 Superorder Cetacea
  Order Zeuglodontia
  Order Odontoceti
  Order Mystacoceti

III. *Simpson (1945)*
 Infraclass Eutheria
  Cohort Unguiculata
   Order Insectivora
   Order Dermoptera
   Order Chiroptera
   Order Primates
   Order Tillodonta
   Order Taeniodontia
   Order Edentata
   Order Pholidota
  Cohort Glires
   Order Lagomorpha
   Order Rodentia
  Cohort Mutica
   Order Cetacea
  Cohort Ferungulata
   Superorder Ferae
    Order Carnivora
   Superorder Protungulata
    Order Condylarthra
    Order Litopterna
    Order Notoungulata
    Order Astrapotheria
    Order Tubulidentata
   Superorder Paenungulata
    Order Pantodonta
    Order Dinocerata
    Order Pyrotheria
    Order Proboscidea

 Order Embrithopoda
 Order Hyracoidea
 Order Sirenia
 Superorder Mesaxonia
  Order Perissodactyla
 Superorder Paraxonia
  Order Artiodactyla

IV. *Romer (1966a)*
 Infraclass Eutheria
  Order Insectivora
  Order Tillodontia
  Order Taeniodontia
  Order Chiroptera
  Order Primates
  Order Creodonta
  Order Carnivora
  Order Condylarthra
  Order Amblypoda
   (including Pantodonta, Dino-
   cerata, Pyrotheria)
  Order Proboscidea
  Order Sirenia
  Order Desmostylia
  Order Hyracoidea
  Order Embrithopoda
  Order Notoungulata
  Order Astrapotheria
  Order Litopterna
  Order Perissodactyla
  Order Artiodactyla
  Order Edentata
  Order Pholidota
  Order Tubulidentata
  Order Cetacea
  Order Rodentia
  Order Lagomorpha

38, II). Foot structure is the primary criterion. Blumenbach (1779) (not in the table) largely followed Storr and led to the general pattern to be found in the work of Cuvier.

Lacépédé (1799) was somewhat off of the main line of development in his threefold division of mammals into quadrupeds, flying mammals, and marine mammals (see Table 38, III). His major divisions were related to ways of life, and his subgroups were based, in a rather intricate system, on the numbers and kinds of teeth.

Cope (1891–1898), Osborn (1910), and Simpson (1945) have each, with considerable modification, followed the general cohort plan of Linnaeus. Gill (1872), however, used a sixfold (series) division, with the six elements being approximately at a superordinal level. In this arrangement there is less commitment to relationships than in the others noted. The 1910 classification of Osborn (Table 39, I), as he noted in his book, was prepared under his direction by W. K. Gregory. In "The Orders of Mammals," however, published the same year, Gregory (Table 39, II) used a superordinal arrangement of a rather different sort, making some subdivisions and associations that have since proven to be unfounded, in particular his wide separation of the Paraxonia from other ungulates.

Simpson's cohort arrangement, although it finds its roots in Linnaeus, differs in many particulars. Most sweeping is the inclusion of a large number of orders, drawn from Linnaeus' Unguiculata and Ungulata in the cohort Ferungulata. A wide variety of animals was considered to have taken origin in common in a creodont–condylarth-like stock. As shown in Table 39, III, 5 superorders and 15 orders are assigned to this cohort. The cohort Mutica is unchanged from Linnaeus, whereas the cohort Unguiculata is greatly reduced. From this reduction, in which some elements are transferred to the cohort Ferungulata, comes the cohort Glires, encompassing the rodents and rabbits. This array, unlike the others of Simpson, is highly questionable, because evidence at hand today strongly suggests that rabbits and rodents, although of somewhat uncertain origins, did not arise from a common stock above the order Insectivora.

Although the formal roots of Simpson's classification can be found in the Linnean arrangement, the fundamental basis is different and resemblances exist only to the extent that Linnaeus' strictly morphological system corresponds to Simpson's phylogenetic arrangement. The extent to which parallelism and convergence have contributed to morphological similarity gives some measure of the degree of difference to be expected in the two approaches.

The fact that radiations of the eutherian mammals are relatively well understood makes a phyletic grouping such as that of Simpson both possible and extremely useful. Among his cohorts the Ferungulata is the most extensive and in many respects the most coherent. The main questions of authenticity and consistency occur among the most primitive members of the order Carnivora under the superorder Ferae and the order Condylarthra under the superorder Protungulata. Assignment of some genera to these orders raises questions of cohort position, since they might equally be assigned to the Insectivora under the cohort Unguiculata.

The cohort Unguiculata is somewhat less coherent. The first four orders—Insectivora, Dermoptera, Chiroptera, and Primates—have a high degree of phyletic consistency, although the position of primitive genera in one or another is open to question in a number of cases. The Tillodontia and Taeniodontia presumably came from insectivores and may be associated with the four other orders on the basis of this presumed relationship. Neither had an extensive radiation. The same more or less applies to the Edentata as far as associations are concerned, but the radiation is much greater. Pholidota have an uncertain history. The association of none of these four groups is particularly close to the first four orders as far as is shown by known intermediates.

The cohort Mutica appears to be consistent, but it is known only from animals that have departed far from a common ancestral stock so that the base of the radiation is not known. Glires, as noted, includes two groups that probably are not closely related. The rodents stand very much alone, although they presumably did come from insectivorous placentals, thus finding a place in the Unguiculata, or they may have some relationship to Ungulata, possibly relating somehow to the Condylarthra.

The vast radiation of the ferungulates has prompted Simpson to use a superordinal classification within this cohort. The groupings in general conform to well-accepted phylogenies, although for each some problems can be raised, as Simpson has done in his text. The superorders, just as the cohorts, have the merit of showing probable groupings, with phyletic bases, and the danger of inclusions in a common superorder of forms that are not in fact related.

Orders of Eutherian Mammals

Marked changes in ordinal subdivisions of the eutherians have followed the great increase in knowledge and understanding of the infraclass since the time of Linnaeus. He recognized eight orders. These include units distributed through 25 to 30 orders in the current century. This proliferation has resulted partly from additions of newly discovered groups of mammals, especially from the fossil record, but also involved is the less tangible difference in the comprehensiveness of the term "order." Some multiplication accompanied introduction of phyletic concepts, particularly influenced by the recognition of zoogeographic restrictions on possibilities of relationships.

Early in the era of scientific classification, however, there were wide differences in the numbers of orders recognized by the various authorities. Differences over the years have been as follows: Geoffroy and Cuvier (1795) recognized 14; Lacépédé (1799), 22; De Blainville (1834), 12; Gill (1870), 13; Flower (1883), 9; Cope (1898), 13; Weber (1904), 22; Osborn (1910), 30; W. K. Gregory (1910), 24; Simpson (1945), 26; Romer (1966a), 25. Important differences exist in the contents of the orders (see, for example, Linnaeus, Table 38), but within the current century differences mainly represent differences in judgment on relationships rather than any fundamental distinctions in methods or concepts.

The differences between the four classifications in Table 39 are slight as far as ordinal composition is concerned. Gregory's 1910 classification is the most distinctive, primarily, however, in areas for which he later modified his interpretations. Below the level of the order there is much less conformity, both in the genera assigned to the orders and in arrangements of the included groups. Differences here also stem in large part from judgments and information available at a given time, rather than from any real changes in methods and use of data.

Many of the orders currently found in major classifications need little consideration because they are widely accepted and their contents are moderately standard. This does not imply that there are few or no internal problems, but rather that modifications are not likely to reshape the ordinal classification. Some orders are in a greater state of flux and marked by major uncertainties. Others for one reason or another have complex internal problems that can contribute to an understanding of how the mammalian classification has come to be, and, in a few, methods of classification are of special interest. These will form the basis of the discussions that follow.

Primitive Placental Mammals

The best available evidence indicates that all placentals came from small Cretaceous mammals that morphologically were little differentiated from their common ancestors with the marsupials. Their cheek dentitions, it would seem, were tribosphenic, and the general habitus was that associated with the descriptive term "insectivore." Very little is known of this presumed ancestral group. *Pappotherium*, the Forestburg therian of the lower Cretaceous Trinity formation of Texas, and the middle Cretaceous *Endotherium* from Asia may be actual representatives of such a group, but this is not at all certain. Otherwise, no representatives are known before the late Cretaceous.

Finds of placental mammals in the late Cretaceous, though relatively unusual, do show that radiation of this group had commenced by this time. In beds of the early Tertiary, Paleocene and Eocene, many lines can be recognized. Some of these can be tied neatly to well-known phyla of later times, but others cannot.

Thus in the late Mesozoic and very early Cenozoic there existed a complex array of evolving lines, generally poorly known, with constituent genera that differed from each other only moderately at least in the portions of their skulls, jaws, and dentitions that were preserved. These pose interesting but difficult problems of classification.

Several still living orders can be traced far back into the record and can be seen to merge with this complex; the same is true for some groups that died out in the Tertiary. The general tendency in classification is to attempt to assign newly found genera and species to already existing groups; thus the orders that were erected on moderns and well-known fossils generally became the receptacles for the less-well-known early placentals. Naturally, in view of the nature of the early record, awkwardnesses have developed and classifications have been unstable, reflecting marked differences of opinion. Adding to the instability in recent years has been the addition of a great deal of new material, resulting from intensive exploration of early mammal-bearing beds and use of more effective methods of recovery of small specimens that occur in low concentrations.

The conventional orders that include members of the primitive complex noted above and are involved in the problems that it poses are as follows:

Order Insectivora
Order Dermoptera
Order Primates
Order Carnivora
Order Condylarthra

In the case of the Insectivora, which was initially based on living animals, a great expansion of the order took place as the early Cenozoic and late Cretaceous placentals were discovered. Many were assigned to this order. Closeness of the relationships between modern and ancient genera, the latter often known from relatively meager remains, cannot always be accurately assessed, and not only were the limits of the order Insectivora extended but the order became more or less indefinable. This is but one facet of the larger problem—namely, that the same genera assigned by one or another student to one of the five orders can often, with equal logic, be assigned to one or more of the other orders in the list. This, of course, is not unique to these five orders.

The classifications of Insectivora by Romer (1966a) and by Simpson (1945) give concrete examples of the kinds of differences that exist (see Table 44). Some have resulted from finds made after Simpson's study, such as *Pappotherium* and *Endotherium* and the family Ptolemaiidae, but this is not all that is involved. One order, Dermoptera, of Simpson has been reduced to a suborder by Romer, and the family Mixodectidae, placed in a separate superfamily by Simpson, has been placed under the suborder Dermoptera by Romer. This sort of change, the reduction of the level of Dermoptera, is largely a matter of judgment, because there has been no question that Dermoptera were close to the Insectivora.

In addition, however, fundamental regroupings, inclusions, and exclusions occur, and these represent significant major differences. For example, the suborder Proteutheria (Insectivora) of Romer includes some groups not similarly associated under Insectivora by Simpson and also two families that Simpson placed with the Primates. Some differences also appear in the treatment of the "modern" insectivore groups, but these are not profound.

An important distinction is found in exclusion of the superfamily Deltatheridoidea of Simpson from the Insectivora and its placement within the order Creodonta by Romer. This related to Romer's elevation of the suborder Creodonta (under the order Carnivora) of Simpson to an order. The Creodonta, so considered and reduced in content as noted below, correspond closely in content to the order Deltatheridia proposed by Van Valen (1966) (Table 40, III). The remaining Creodonts, in the sense of Simpson, are placed in the order Condylarthra by Romer. Both Simpson and Romer agree, as do others today, in placement of the Hyopsodontidae in the order Condylarthra, but earlier (e.g., Osborn, 1910) the family had been placed with the Insectivora and relationship with the Primates had also been suggested.

Simpson classified Tupaioidea and Apatamyidae with the Primates. The position of these forms has long been in question, and in the classification of Romer they are found under his large group of primitive insectivores, the Leptictoidea.

This comparison is used not with the thought that one or the other of these two classifications is "right" and the other then "wrong," but to indicate the extent of the problems of assignments at this evolutionary stage in the eutherians. Other similar problems exist in other major groups of vertebrates, but it is only where there are relatively well-known materials that the outlines and difficulties become so clear. This sort of instability can be expected to continue as more materials come to light. When relative stability will be attained cannot be predicted. It will depend in part, of course, on new materials, but also on intangibles, such as attainment of an arrangement that meets with general approval of the students in the field at that time. Accepted classifications, which are generally compromises, may endure for a long time, lasting until their inadequacies become intolerable. It probably will not be impossible to eliminate subjectivity in classifying materials at this

TABLE 40   Classifications of the Insectivora and Some Related Groups

I. *Simpson (1945)*

Cohort Unguiculata
Order Insectivora
Superfamily Deltatheridoidea
Superfamily Tenrecoidea
Superfamily Chryso-chloroidea
Superfamily Erinaceoidea
Superfamily Macroscelidoidea
Superfamily Soricoidea
Superfamily Pantolestoidea
Superfamily Mixodectoidea
Order Dermoptera
Order Primates
Suborder Prosimii
Infraorder Lemuriformes
Superfamily Tupaioidea
Infraorder uncertain
Family Apatamyidae
Cohort Ferungulata
Superorder Ferae
Order Carnivora
Suborder Creodonta
Superfamily Arctocyonoidea
Superfamily Mesonychoidea
Superfamily Oxyaenoidea (including Hyaenodontidae, Oxyaenidae)

II. *Romer (1966a)*

Order Insectivora
Suborder Proteutheria
?incertae sedis ?Endotheriidae, ?Pappotheriidae
Superfamily Leptictoidea (seven families, including Tupaiidae, Pantolestidae)
Suborder Macroscelida
Suborder Dermoptera
Suborder Liptotyphla
Superfamily Erinaceoidea
Superfamily Soricoidea
?Suborder Zalambdodonta
Superfamily Tenrecoidea
Superfamily Chrysochloroidea
Order Creodonta
Suborder Deltatheridia (including Deltatheriidae, (Palaeoryctidae)
Suborder Hyaenodontia
Order Condylarthra (including Arctocyonidae, Mesonychidae)

III. *Van Valen (1966)*

Order Deltatheridia
Superfamily Palaeoryctoidea
Family Palaeoryctidae
Family Didymoconidae
Superfamily Oxyaenoidea
Family Hyaenodontidae
Family Oxyaenidae

level of radiation, but gradually phyletic classifications are coming to replace the less satisfactory grade, or horizontal classifications.

Carnivores

Two aspects of the classification of carnivores are of general interest. One relates to the composition of the order Carnivora as a whole and the other to the methods of subdivision within the Fissipedia.

Constitution of the Order Carnivora

Recognition of the flesh-eating mammals as a coherent group goes far back in time, beyond the beginning of scientific classification and is found in the earliest of these classifications. It has, for the most part, persisted to the present time. Association of the flesh eaters at about the level of an ordinal rank has been the usual practice. Fossil carnivores were discovered quite early, during the first part of the eighteenth century, and thus have played a role in initial conception of the group. Main interest centered around some of the creodonts and their possible relationships to the marsupials on the one hand and the fissipedes on the other. Considerations of this matter, involving widely different opinions, continued well into the current century. It was discussed, and largely disposed of, by W. K. Gregory as late as 1910.

In recent years, in fact until very recently, the constitution of the order Carnivora has been quite stable. It has been considered to contain all the animals usually termed creodonts, fissipedes, and pinnipedes. Marsupial carnivores have been excluded. Some cracks in this bulwark have started to appear. Van Valen, and following him Romer (see Table 40), have eliminated all "creodonts" from the Carnivora, except for the Miacoidea (Eucreodi of Matthew), which most authorities have placed with the Fissipedia in recent years. Arctocyonoid "creodonts" are placed at the base of the Condylarthra, and the mesenychoids are considered a side branch of this basal stock. Matthews' Pseudocreodi, Oxyaenoidea (Hyenodontoidea) are referred to the Deltatheridia by Van Valen (= Creodonta, an order by Romer). Such changes result from continuing efforts to arrive at a phyletic rather than phenetic classification.

Classification of various animals as carnivores has depended on a wide range of criteria: general habits in feeding, other behavioral patterns, general morphological resemblances to other flesh eaters, recognized sets of skeletal features, phyletic relationship to "known" carnivores, and, more than anything else, dental characteristics. Within the Carnivora, as brought out more in detail below, characters of the auditory, glenoid, and basicranial regions have figured in classifications. In Van Valen's discussions of his reasons for his modifications many of these features were considered, but primary emphasis was on dentition. His materials, of course, were fossils, and limitations, particularly in view of the fragmentary nature of most specimens, were rather severe. His conclusions depended largely on morphology and stratigraphic relationships, cast in a phylogenetic context.

It cannot be judged how acceptable this and other modifications will prove to be. Such conclusions are necessarily expressions of opinions based on limited evidence and will continually be subjected to contradiction by others whose opinions, based on similar criteria, differ for one reason or another. Not the least of such reasons is the not unhealthy reluctance to abandon a functional and well-established system until it has been shown beyond any doubt to be false. Less meritorious, but also a strong reason for rejection of new systems, is an emotional response to change, which, so to speak, requires the abandonment of one's favorite prejudices.

Subdivisions of the Carnivora

If the Creodonta are eliminated from the Carnivora, only the Fissipedia and Pinnipedia remain. Beyond the question of inclusion of the Creodonta there are a few matters of internal classification that are of special significance. The Pinnipedia, while having their own problems of origin and relationships (e.g., the suggestion that the group is synthetic and arose from more than one source), do not add much to a consideration of the methods of classification among the mammals. The opposite is true, however, for the Fissipedia. Interesting problems have arisen from discrepancies in the methods of formulation of major groups, based largely on living fissipedes, in the methods of assignment of extinct forms to these groups, and in ways of reconstruction of evolutionary history from the fossil record.

Familial assignments have been based for the most part on dentition and related masticatory structures, especially among extinct groups. This appears to have produced a reasonably satisfactory structure, with some exceptions noted at the end of this section. Larger superfamilial groupings, however, have been based primarily on the moderns and have utilized the nature of the auditory region and basicranium, including the carotid circulation.

Flower (1869) proposed a threefold division based on these characters and somewhat set the stage for later arguments involving these features. At one extreme in his classification were his Aeluroidea (Viverridae, Felidae, Hyaenidae) and at the other were the Arctoidea (Mustelidae, Procyonidae, Ursidae). Somewhat intermediate were the Cyonoidea (Canidae). Carotid circulation and characters of the glenoid region appeared to offer support for the basic data of the auditory region.

Later studies, especially that of Winge (1895), questioning the homologies of the bullar septum accepted by Flower, suggested that the cyonoids were close to the arctoids and that there were but two groups, aeluroids (feloids of some authors) and arctoids (canoids of some authors). Evidence from various sources, especially from dental studies of extinct carnivores, lent support, and this general plan is found in most major classifications today.

Hough (1948, 1953), however, investigated the cranial characters of some of the fossil carnivores and determined that the cranial characters of these two groups did not occur in all of the fossils that were assigned to them, mostly on the basis of teeth. Dental criteria and auditory and basicranial criteria produced conflicting results. Most of the early carnivores approached the arctoid cranial patterns, and it appeared that the specializations were developed in the groups after divergence and in various ways.

Hough, in part as a result of her earlier work, adopted the position that auditory and related structures were more fundamental phyletic indicators than were teeth and masticatory structures. To eliminate the confusion of dual and conflicting criteria, she applied this conclusion to form a new classification, as follows:

Hough (*1953*)	*Romer* (*1966a*)
Superfamily Machairodontoidea	Infraorder Aeluroidea
Hoplophoneidae	Viverridae
Eusmilidae	Hyaenidae
Machairodontidae	Felidae
Superfamily Aeluroidea	Infraorder Arctoidea
(Herpestoidea)	Mustelidae
Daphoenidae	Canidae
Vivervidae	Procyonidae
?Hyaenidae	Ursidae
Superfamily Cyono-feloidea	
Canidae	
Felidae	
Nimravinae	
Felinae	
Superfamily Arctoidea	
Procyonidae	
Ursidae	
?Mustelidae	

The comparison with Romer's 1966 classification shows the great contrast in approaches. Hough's tentative system has not been accepted, so that the dual system of Winge still is followed in most classifications. Although Winge's system is useful and provides a "home" for the families of fissipedes, it does not take into account the objections raised when its primary criteria are applied to fossils, and, persistent as it has been and may continue to be, it involves a basic conflict of criteria that restricts meaningful analysis of evolutionary developments of the fissipedes as they are conceived within its limits.

The problems of the use of teeth as principal criteria for assigning carnivore genera to families were brought out clearly by Hough (1948) in the work that eventually led to the more sweeping modification of fissipede classification. In the course of study of auditory regions in procyonids and canids it became apparent that some "procyonids," on dental bases, had canid auditory regions. *Phlaocyon*, *Cynarctos*, and *Cynarctoides* were involved. *Zodiolestes*, whose dentition is distinctly canid, proved to have a procyonid auditory structure. Accepting the auditory structure as the more fundamental, Hough made assignments on the basis of the canid or procynonid nature of this region. Assignment of genera proposed by Hough on the basis of this procedure has been generally accepted, in contrast to the less enthusiastic response to her broader reorganization of the fissipedes.

Subungulate and Ungulate Mammals

Condylarths appear to have been the source of most, probably all, of the groups of placental "ungulate" mammals. Two clearly recognizable stocks, perissodactyls and artiodactyls, occur first in the early Eocene and, being quite coherent, pose very few problems at the ordinal level. A few points are discussed concerning them in short sections that follow this one.

A wide array of placentals that did not become fully ungulate also stemmed from the same general condylarth stock along with some, which, though being neither artiodactyl nor perissodactyl, attained highly ungulate features. Some points of special interest with respect to modes and methods of placental classification have arisen in the course of studies of some of these groups. The animals concerned are conventionally grouped, much as was done by Simpson in 1945, into the following orders:

Order Pantodonta	Order Embrithopoda
Order Dinocerata	Order Proboscidea
Order Pyrotheria	Order Hyracoidea
Order Litopterna	Order Sirenia (including
Order Notoungulata	Desmostylia)
Order Astrapotheria	Order Tubulidentata

It may be that each of these orders stemmed independently from the Condylarthra, as implied by classifications that make a simple listing of the orders without grouping at higher levels or without combining some of the listed orders into a single order. Most classifications, however, do employ some sort of grouping to indicate probable common origins above the basic condylarth habitus. Several possible ways of grouping are available, and, combining zoogeography, time of occurrence, and morphology, various students have come up with somewhat divergent answers. In large part the differences relate to different concepts of the role of taxonomy, different evaluation of the roles of parallelism and convergence, different estimations of the reliability of the fossil record, and different interpretations of morphology.

By the use of chiefly morphological criteria, Simpson (1945) grouped seven of the orders under a superorder Paenungulata, a subdivision of the cohort Ferungulata. The orders Pantodonta, Dinocerata, Pyrotheria, Proboscidea, Embrithopoda, Hyracoidea, and Sirenia are his Paenungulata. Within this superfamily the order of listing carries implications on relationships, not, however, confidently interpretable from this listing alone.

These orders occur in the early Tertiary in one or more of three regions: North America, South America, and North Africa. Migrations quickly carried some to other areas, especially to the Eurasian region. All of the animals involved tended toward large size, attained early, with the Hyracoidea expressing this tendency the least. All tended to develop cross-lophs on the crowns of the molars (especially the uppers), to develop some sort of anterior tusklike teeth, to be mesaxonic, and to have nails rather than hoofs. Claws occur in some instances.

The array of similar trends suggests the existence of common features, or the potential for development of common features, in the ancestry, and thus suggests a single ancestry separate from that of other Ferungulata. Whether such an ancestral group, if such existed, could or should be associated with the Condylarthra, perhaps as a distinct branch, is purely a theoretical question, because no such group has been identified. Grouping of these orders into a superorder implies the existence of a widespread ancestral stock, separate from the ancestral stocks of other placentals, in the earliest Tertiary or possibly latest Cretaceous.

Equally reasonable, however, in view of the zoogeographic problems posed by the very early separation of North and South America and the presumed early isolation of Africa is a system that accepts all, or nearly all, as ungrouped orders, whose resemblances result from convergence and parallelism after they arose from a condylarth stock that was itself primitive and did not have either an expression or specific potential of the features that are common to these orders.

Two, sometimes three or four, of these orders have variously been grouped together under a single order, termed Amblypoda. This term has most recently been used by Romer, as noted in the list below, but over the years it has had many usages and many meanings, so that without specification of exactly what is meant, interpretation is impossible. Romer (1966a) used the term as follows:

Order Amblypoda
   Suborder Pantodonta
   Suborder Dinocerata
   Suborder Xenungulata
   Suborder Pyrotheria

The order includes mammals that have attained large size rapidly and had a tendency to develop cross-lophed upper molars. Initial association of the Pantodonta and Dinocerata was based in part upon resemblances of the molar teeth in late members. It was, however, pointed out by H. E. Wood (1934) that the two attained this somewhat similar structure in very different ways. Romer, in making his association, however, felt that the general resemblances were sufficient for inclusion in a single array.

The amblypods, so constituted, developed in the Western Hemisphere, with migrants to Asia in the Eocene. The presumption in the grouping appears to be that these suborders had a common ancestry, different from that of other placentals, among the condylarths or above the condylarth level. This stock must have been in existence before the isolation of North and South America and was spread over the continents but was not extended, at least as far as the records show, beyond the Western Hemisphere. It is, however, difficult to specify what the common and distinctive characters of the ancestry might have been if these are to separate them from other subungulate ancestors, because most of the common features of the amblypods appear to have developed within the known stocks.

In contradistinction to the Amblypoda, but not formally specified by Romer, are the subungulates.[1] As used by Romer (1966a), this comprises the creatures that are found in the Eocene of North Africa and their descendants—Proboscidea, Embrithopoda, Hyracoidea, and Sirenia. They have similar structures that suggest common ancestry, and the sites of the early finds, combined with morphology, suggest a common ecology in swampy inlets, close to the sea.

Until the postcranial structures of the desmostylids became known, it was usually assumed that these were marine mammals related to the sirenians. They are now known, however, to have been quadrupeds and are usually placed in a separate order. Although the morphology suggests that they may have originated along with other subungulates, their distribution around the Pacific basin poses a zoogeographic puzzle.

The breakdown of paenungulates into amblypods and subungulates is first geographic and secondarily morphological, with the morphological affinities looming greater among the subungulates than among the amblypods. In addition, the morphological resemblances of the pyrotheres and the proboscideans seem to bring them close together, closer than the pyrotheres and any of the other amblypods of Romer. The seemingly insurmountable zoogeographic obstacles have convinced most students that a pyrothere–proboscidean association, often made in the past, is incorrect.

In addition to the orders discussed to this point, the South American ungulates come into the picture as Paenungulata. The three groups—litopterns, notoungulates, and astrapotheres—currently are treated as separate and coherent orders. It is generally thought that they originated separately from condylarths, but that this base lay in a particular group of "South American" condylarths. A curious situation, however, pertains in the distribution of one order, Notoungulata. A key structure in this group is the presence of an isolated entoconid in the lower molars. *Palaeostylops*, a Paleocene genus from Asia, has teeth that resemble those of notoungulates, even to

---

[1] The term "subungulate" has been used both formally and informally and has a complex history in systematics. By some it is essentially equivalent to "Paenungulata," but in great contrast, earlier it was, in an extreme case, affixed to the caviid rodents.

the extent of having the isolated entoconid. On this basis it has been placed with the notoungulates in spite of the zoogeographic difficulties that this poses.

The early classification of the South American ungulates is complex and interesting. Much of the initial collecting and study was by Carlos and Florentino Ameghino. During their productive careers they performed an immense task of assembling and describing the fauna of the Tertiary of Argentina. Estimating the ages of the fossils as greater than they were, Florentino Ameghino saw in them the ancestors of placental stocks in other parts of the world and so classified the materials. Egged on by patriotism, the arrogance of his colleagues, and by increasing scientific isolation, he added to objective classification an intense subjectivity.

Few major schemes of classification can be considered to be free of somewhat similar subjective elements. They are little discussed and often mainly the property of those "in the know." But phyletic classification is inevitably in part subjective, depending on value judgments at its best and intense emotionalism at its worst, and this factor must always be reckoned with.

Roth (1903) began to straighten matters out, particularly in his recognition of the Notoungulata, the members of which Ameghino distributed among eight orders. Astrapotheria were included within this order but were later eliminated. The classification of today, in which the three orders are considered as coherent and separate, stands with little challenge.

The order Tubulidentata is a very isolated group, with a poor fossil history. It was associated with the edentates, in the broad sense of this group, on the basis of its habits and its odd tooth structure. This association has long since been abandoned by most students and it has been recognized that the Tubulidentata form a distinctive group that probably stemmed from the condylarths independently of others. Decision is based largely on morphology and on lack of information on the early history. Colbert (1941) has made a good case for the interpretation, viewing the tubulidentates as essentially highly specialized persistent condylarths.

Most of the orders that have been discussed do not raise any notable problems in their subordinal structure, although of course there are the ever present problems of phyletic relationships involving constitutions of particular groups. The methods of study and ways of seeking solutions to the problems do not for the most part introduce any aspects of classification that have not been fully explored for other groups. One order, however—the Proboscidea—provides a striking exception, because in its analysis a philosophy of classification quite different from that usually applied has been brought to fruition.

Order Proboscidea

Elephants have long attracted the attention of vertebrate zoologists and paleontologists, and remains of extinct proboscideans from the relatively

TABLE 41   Classifications of the Proboscidea

I. Osborn (1936)

Order Proboscidea
  Superfamily Moeritherioidea
    Family Moeritheriidae
      Subfamily Moeritheriinae
  Superfamily Deinotherioidea
    Family Curtognathidae
      Subfamily Deinotheriinae
  Superfamily Mastodontoidea
    Family Mastodontidae
      Subfamily Palaeomastodontinae
      Subfamily Mastodontinae
      Subfamily Zygolophodontinae
      Subfamily Stegolophodontinae
    Family Bunomastodontidae
      Subfamily Rhynchorostrinae
      Subfamily Longirostrinae
      Subfamily Gnathabelodontinae
      Subfamily Amebelodontinae
      Subfamily Tetralophodontinae
      Subfamily Notorostrinae
      Subfamily Brevirostrinae
    Family Humboltidae
      Subfamily Humboltinae
    Family Serridentidae
      Subfamily Serridentinae
      Subfamily Platybelodontinae
      Subfamily Notomastodontinae
  Superfamily Stegodontoidea
    Family Stegodontidae
      Subfamily Stegodontinae
  Superfamily Elephantoidea
    Family Elephantidae
      Subfamily Loxodontinae
      Subfamily Mammontinae
      Subfamily Elephantinae

II. Simpson (1945)

Order Proboscidea
  Suborder Moeritherioidea
    Family Moeritheriidae
  Suborder Elephantoidea
    Family Gomphotheriidae
    (Bunomastodontoidae)
      Subfamily Gomphotheriinae
      (Longirostrinae, including
      Palaeomastodontinae,
      Serridentinae)
      Subfamily Anancinae
      (Brevirostrinae)
      Subfamily Cuvierioninae
      (Notorostrinae)
      Subfamily Rhynchotheriinae
      (Rhynchorostrinae)
      Subfamily Platybelodontinae
    Family Mammutidae
    Family Elephantidae
      Subfamily Stegodontinae
      (including Stegolophodontinae)
      Subfamily Elephantinae
      (including Loxodontinae,
      Mammontinae)
  Suborder Deinotherioidea
    Family Deinotheriidae
  Suborder Barytherioidea
    Family Barytheriidae

III. Romer (1966a)

Order Proboscidea
  Suborder Moeritheroidea
  Suborder Euclephautoidea
    Family Gomphotheriidae
    Family Mastodontidae
    Family Elephantidae
  Suborder Deinotherioidea
  ?Suborder Barytherioidea

recent past were among the first of the mammalian fossils to be studied with some scientific thoroughness. For many years classifications followed the general trends of their particular times, and current classifications (e.g., Simpson, 1945; and Romer, 1966a) are within the framework of the usual traditional systems (see Table 41).

By far the most extensive work on Proboscidea was performed by Osborn in the course of some 40 years of study. Results were published posthumously in two volumes of a great monograph (Osborn, 1936). This massive effort brought together and organized the wealth of information accumulated in the work of Osborn and many others. In the course of his studies Osborn developed a system of classification which is unique and which, although not followed in most classifications, has had a major effect on systematics, particularly in the ways it has influenced perspectives on evolutionary patterns.

The system is explained in the first part of Volume I of the monograph by Osborn and has been carefully analyzed by Simpson (1945). It is of sufficient interest, however, that some of its most pertinent features must be repeated in any treatment of the nature of classification of the vertebrates. The system is what Osborn has called his *phylogenetic classification*. He drew a distinction with the Linnean "typological" classification and would not, offhand, seem to be departing far from what most post-Darwinian systematists had done. Most of their classifications are phylogenetic. Even as seen in tabular form, Osborn's classification does not appear to be very much different from others. In actual fact, as clearly outlined by Simpson (1945), it is based on very different concepts of relationships than those expressed in usual phyletic classifications.

The organization is by phyla, with the subfamilial phyletic unit as the basis of the classification. Each subfamily phylum is composed of genera, usually successive in time, with rare divergence to form two contemporary genera and occasional persistence of an old genus to overlap its successor in time. These phyla trace back to origins, mostly unknown, in the Eocene. The elephants proper, whose early history is considered to be absent from the record, however, are thought to have differentiated from all others in the Oligocene. The general pattern is that shown in Figure 82.

No subfamily split to produce two subfamilies that can be grouped into a family. The larger units consist of bundles of discrete phyla that persist far back to an initial differentiation. Osborn recognized four major subdivisions of the Proboscidea: Moeritherioidea, Deinotherioidea, Mastodontoidea, and Elephantoidea. Under these were families and the basic subfamilial units with the genera in phyletic lines and finally species and mutants. In his system he did not follow the rules of nomenclature but named taxa as he felt was convenient and most meaningful.

The characters used were dental, and these were formally organized into types significant at particular levels:

Order: profound adaptation of incisor teeth.
Suborder: profound modifications of incisors.
Family: profound adaptation of the grinding teeth.
Subfamily: secondary modifications of the grinding teeth.

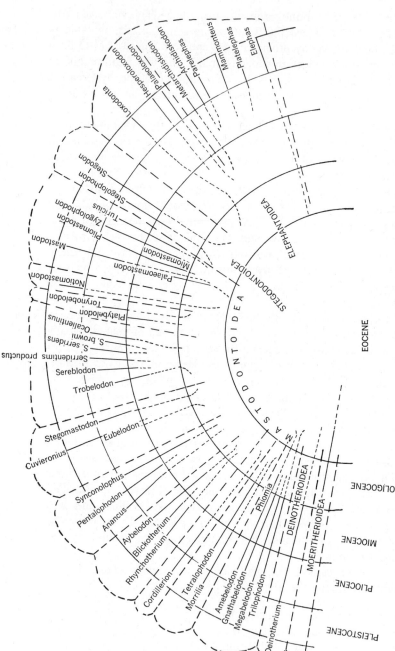

Figure 82. The "bush" type of phylogeny as exemplified in its most extensive application in Osborn's (1936) interpretation of the Proboscidea. Other phylogenies of the more orthodox "tree" type have been more usually applied to this order. The phyla, outlined by heavy dashed lines, closed at the top, are at the subfamilial level and consist of successions of species and genera, with some branching at the generic level, but none to form other subfamilies. The basic differentiation took place in the Eocene or earlier, but this is not known in the record. Note the five suborders that encompass the subfamilial phyla. Their primary differentiation also was before the beginning of the record. Quinn (1955) used a somewhat similar explanation of the evolution

Genus: successive stages in the evolution of the secondary and compensatory adaptations of the teeth (with much parallelism, convergence, and homoplasy).

The Osbornian system thus is formal. It is workable if Osborn's general evolutionary philosophy is followed. As in most formal systems, there are some exceptions when it is applied practically. Generic rather than subfamilial phyla are necessary in some cases. Use of such a system requires commitment to one and only one pattern of evolution. The system thus becomes a ruling pattern and one that tends to inhibit study of anything that may fall outside its framework.

The effectiveness of any classification must be sought in analysis of how well it expresses the relationships it is intended to express. Osborn's classification performed this function extremely well for his interpretation. The existence of some 21 lines of independent descent was what he saw. Earlier he had introduced the general ideas of this classification in monographic treatment of the titanotheres. Once again discrete phyla and trends were what he saw.

Long-lived phyla of the Osbornian type most certainly do occur. Many generic lines seem to have integrity from the Miocene to the present. Some seem to show some generic branching, others do not. Quinn (1955) (see pages 413–414) recognized this sort of pattern in the Miocene to Recent Equidae. The problem is not that this sort of "bush" evolution may not exist, but that adherence to a system that admits only this kind of evolution restricts all studies to the frame of reference that it imposes.

In Simpson's classification moeritheres, deinotheres, and barytheres are placed with the Proboscidea. This was done with recognition that the moeritheres and barytheres were questionably assigned. Romer (1966a) followed much the same practice, indicating by a question mark the tentative nature of assignment of the barytheres. In his comments on the classification (1968) Romer pointed out that each of the three groups may have stemmed independently from the subungulate base and that the mastodons and mammoths may constitute the full proboscidean radiation. If this is followed, the subungulate evolution was more or less of the "bush" type, with somewhat similarly adapted animals arising from a common stock but in divergent phyletic lines.

Order Perissodactyla

RECOGNITION OF THE ORDER.   The unity of the odd-toed ungulates was recognized by De Blainville (1816) in his classification in which he separated all ungulates on the basis of foot structure:

De Blainville (*1816*) (*After W. K. Gregory, 1910*)

Ongulograd[es]
Normaux, doigts
Impairs
Pachydermes
Solipèdes
Pairs
Non Ruminans ou Brutes
Ruminans

Although this distinction was adopted rather slowly, it has become standard and occurs in all recent classifications and in most presented since the middle of the nineteenth century.

Whether the artiodactyls and perissodactyls were to be considered as suborders within a single order of ungulates or as separate orders remained an open question for some time, and whether other placentals were to be included in any such common order was similarly debated. Some examples of the various classifications which were proposed and which show the general nature of the spread of opinions are presented in Table 42. One of the most marked and persistent divergences related to the position of the ancylopods, or chalicotheres. In many classifications they have been placed in an order of their own or separated at lesser levels from the perissodactyls.

SUBDIVISIONS OF THE ORDER. The basic subdivision of perissodactyls in current classifications is that to be seen in Simpson's (1945) classification, as shown in Table 42. This has essential similarities to the subdivision made by De Blainville (1816) in his Pachydermes and Solipedes. H. E. Wood (1934) recognized this subdivision and extended it to fossil forms, suggesting the use of two terms: "Solidungula" (a horse–palaeothere–titanothere–chalicothere complex) and "Tridactyla" (a tapir–rhinoceros complex), using the terms of Blumenbach (1779) and Latreille (1825), respectively. Later H. E. Wood (1937) substituted the terms "Hippomorpha" and "Ceratomorpha" for the two groups, respectively. These have been used by Simpson (1945) and most subsequent students of perissodactyls. Romer (1966a) has maintained the Ancylopoda as a separate suborder, on morphological and phyletic grounds (Table 42, VIII).

The initial perissodactyl grouping was based on extant animals. In contrast to what occurred in many groups, development of the fossil record strengthened rather than weakened this classification. The end members of the phyla in this instance gave good indication of the actual nature of the phyletic separation.

Many problems of relationships and evolutionary patterns within the perissodactyls still exist and will continue to do so. Even for the very-well-known horses there are problems at every level. Rhinoceroses, with their

TABLE 42   Some Classifications of Ungulates with Special Reference to the Position of the Perissodactyla

I. *Gill (1870)*
Ungulate Series
Order Ungulata
Suborder Artiodactyla
Suborder Perissodactyla
Order Hyracoidea
Order Proboscidea
Order Sirenia

II. *Flower (1883)*
Order Ungulata
Suborder Artiodactyla
Suborder Perissodactyla
Suborder Hyracoidea
Suborder Proboscidea

III. *Cope (1891–1898)*
Ungulata
Order Taxeopoda
Order Toxodontoidea
Order Proboscidea
Order Diplarthra
Suborder Perissodactyla
Suborder Artiodactyla

IV. *Weber (1904)*
Subclass Monodelphia
Order Perissodactyla
Order Artiodactyla
Order Ancylopoda
Order Hyracoidea
Order Proboscidea
Order Sirenia

V. *Osborn (1910)*
Cohort Ungulata
Order Artiodactyla
Order Hyracoidea
Order Proboscidea
Order Perissodactyla
Order Ancylopoda

VI. *Romer (1945a)*
Order Perissodactyla
Suborder Hippomorpha
Superfamily Equoidea
Superfamily
Titanotherioidea
(including
Chalicotheriidae)
Suborder Tapiromorpha
Superfamily Tapiroidea
Superfamily
Rhinocerotoidea

VII. *Simpson (1945)*
Order Perissodactyla
Suborder Hippomorpha
Superfamily Equoidea
Superfamily
Brontotherioidea
Superfamily
Chalicotherioidea
Suborder Ceratomorpha
Superfamily Tapiroidea
Superfamily
Rhinoceratoidea

VIII. *Romer (1966a)*
Order Perissodactyla
Suborder Hippomorpha
Superfamily Equoidea
Superfamily
Brontotherioidea
Suborder Ancylopoda
Suborder Ceratomorpha
Superfamily Tapiroidea
Superfamily
Rhinocerotoidea

extremely complex phylogeny, have defied definitive phyletic analysis. Recently important advances in the understanding of tapirs have come about, in particular as a result of the studies of Radinsky (1963–1966). Classifications of titanotheres are far from satisfactory at detailed levels. These problems, however, are the standard problems of paleontology, approached and solved by usual morphological, geographic, and stratigraphic studies.

Quinn (1955), working with horses from the Gulf Coastal Plain, arrived at a pattern of equid evolution from the Miocene to the present that is basically different from that of most other students. His new perspectives arose partly from morphological analysis and interpretation of resemblances in terms of his analyses, partly from study of new materials, and in some important measure from zoogeographic considerations in which he drew contrast between the evolution in the Gulf Coast and in the midwestern interior.

His picture is one of the development of the basic horse types in the Miocene, with a rapid burst in early Miocene, and the continuation of phyla, generic in constitution, to the present. This contrasts with the usual branching pattern in the *Merychippus* stage from which were sorted out the Pliocene to Recent lines. Much as in the contrast of Osborn's system of Proboscidea to the usual one, Quinn adopted, as a result of his interpretation of his evidence, a system that with its persistent phyla contrasted with the usual view. This pattern has not been widely adopted, but it is an example of the increasing recognition of the great persistence of little-branched lines that is being found in many groups of vertebrates. The form of tabular classification does not clearly differentiate this strictly "phyletic" classification from the more usual branching type; hence explanations are generally necessary if the underlying philosophy is to be definitely established.

Order Artiodactyla

The coherence of the even-toed ungulates has been recognized since the time of De Blainville's early classification (1816), and in general a dichotomy of major types as ruminants and nonruminants has been in use in one form or another. Within this dichotomous framework, however, the positions of the camels, Tylopoda, and the oreodonts, Merycoidodontoidea, have been widely disputed. The constitution of the primitive artiodactyl group or groups has likewise been controversial. Some of the treatments are shown in Table 43.

The problems have stemmed from several sources. First of all the dentitions of artiodactyls are in general less definitive and less easy to study than those of perissodactyls. It is clear that today there are ruminants and nonruminants and that their dentitions are very different, selenodont and bunodont, respectively. But this general split between selenodont and bunodont patterns does not necessarily hold as an indication of ruminant and nonruminant

TABLE 43   Some Classifications of the Artiodactyla

I. *Flower (1883)*
  Order Ungulata
    Suborder Artiodactyla
      Suina
      Tragulina
      Tylopoda
      Pecora

II. *Weber (1904)*
  Order Artiodactyla
    Non ruminantia (Suoidea)
    Tylopoda
    Pecora
    Traguloidea
    Dichobunoidea
    Anthracotheroidea

III. *Osborn (1910)*
  Order Artiodactyla
    Section primitive artiodactyls
    Section Suina
    Section Oreodonta
    Section Tylopoda
    Section Tragulina
    Section Pecora

IV. *Romer (1933)*
  Order Artiodactyla
    Suborder Ruminantia
      Infraorder
        Protoselenodontia
      Infraorder Tylopoda
      Infraorder Pecora
    Suborder Suina

V. *Simpson (1945)*
  Order Artiodactyla
    Suborder Suiformes
      Infraorder Palaeodonta
      Infraorder Suina
      Infraorder Ancodonta
      Infraorder Oreodonta
    Suborder Tylopoda
    Suborder Ruminantia
      Infraorder Tragulina
      Infraorder Pecora

VI. *Romer (1966a)*
  Order Artiodactyla
    Suborder Palaeodonta
    Suborder Suina
      Superfamily
        Entelodontoidea
      Superfamily Suoidea
      Superfamily
        Hippopotamoidea
    Suborder Ruminantia
      Infraorder Tylopoda
        Superfamily
          Cainotheroidea
        Superfamily
          Anoplotheroidea
        Superfamily
          Merycoidodontoidea
        Superfamily Cameloidea
      Infraorder Pecora
        Superfamily Traguloidea
        Superfamily Cervoidea
        Superfamily Bovoidea

affinities in the record. Oreodonts have selenodont teeth but appear to be nonruminants. On other bases there are possible relationships to the camels, which also have selenodont teeth but are not strictly ruminants. Nondental structures are more definitive in many instances but, of course, are much less generally available. Also they tend to provide better evidence of within-group affinities and less good evidence of between-group relationships.

A second series of problems arises from the geographic distribution of the artiodactyls. Some major groups appear to have evolved primarily in North America, whereas others are characteristically Old World. For example, oreodonts and camels are characteristically North American, whereas anthracotheres are Old World. Some groups, such as the cervids, present a very complex record in which the geographic center or centers from which they radiated are very uncertain. In addition the artiodactyls have been extremely mobile and have had complex patterns of immigration and emigration, so that relationships between Old and New World representatives are highly complex.

Added to the problems of morphology and zoogeography are those of times of occurrence in the geological record. In general the Eocene record of artiodactyls is much less good than that of the perissodactyls, and this is particularly true in the United States. This has seriously complicated an understanding of the early deployment of the order and of the nature of the primitive assemblages. By late Eocene the artiodactyl radiation was well under way. It was rapid in North America, and, it would seem, somewhat less so in the Old World. Oreodonts and camels reach their peaks in the Miocene. However, bovoids, which form the largest and most complex branch of the artiodactyls, were late in radiation and present a welter of forms from the Miocene to the Recent. Much of their radiation was in the Old World and extremely complex. In the New World is the confusing array of prongbuck-like creatures that developed in isolation of the main Old World radiation.

Each of these circumstances has contributed to the difficulties of developing a satisfactory classification of the artiodactyls. Although there is currently fairly general agreement on broad relationships, most matters are far from settled, even at the higher levels. The approaches to the study of this group, however, are in no way out of the ordinary, and no unusual ways of using criteria or interpreting evolutionary patterns are involved. The great number of existing genera and species, spread over a rather wide range of the higher categorical levels, offers an opportunity for extensive application of techniques not available for fossils—for example, serological or cytological studies. To date these have been rather spotty and have lacked the systematic basis necessary for application to some of the more general problems of artiodactyl classification.

Order Rodentia

RECOGNITION OF THE ORDER.   Rodents, by virtue of their distinctive dental structure and habits, have long been recognized as a distinct group. In the past, however, various animals, not now generally considered to be rodents, have been associated with them, primarily on the basis of tendencies toward persistently growing front teeth. Most of the large, primitive Paleocene and

Eocene animals for which such suggestions have been made (e.g., the tillo-donts and the taeniodonts) have long since been eliminated from any such considerations. The most persistent association has been with the lagomorphs, with the rodents and lagomorphs being considered as two suborders under a single order, as follows:

Order Rodentia
Suborder Simplicidentata
Suborder Duplicidentata

Such an assignment, for example, is found in Romer (1933). The term "Simplicidentata" persists as an ordinal rank in some classifications today, such as that of Schaub (1958) (Table 44, IV).

The origin of neither the lagomorphs nor the rodents is known, but it is now generally conceded that they have little to do with each other and that they should be placed in separate orders. As currently viewed the order Rodentia appears to be strongly isolated from all other placentals.

The scant evidence of their early history in the late Paleocene and early Eocene suggests that the rodents form a monophyletic order. The source is not certain, but presumably it lay among the insectivores.

SUBDIVISIONS OF THE ORDER.   In a study of extant rodents Brandt (1855) suggested a threefold division based on the masseteric musculature. The three groups were the Sciuromorpha, Myomorpha, and Hystricomorpha, generally considered as suborders when ordinal rank is given the rodents. This classification has had a remarkable persistence. Most fossils, as they were discovered, were fitted into it, except for the most primitive, which in many instances were placed in a separate fourth group, more or less a waste-basket for the rodents whose masseter patterns did not fit one of the three basic types of Brandt. Among moderns only the Aplodontia fall within this category.

The nature of the masseteric musculature is generally determinable from the skulls and lower jaws; hence it has been possible to determine it for most extinct rodent groups. Since the time of Brandt a great many students have been concerned with the classification of rodents and have come up with a wide variety of interpretations. Yet in most of these at least some aspects of the Brandt classification are still evident.

Simpson (1945), as for so many placental groups, has given a rather full summary of the history prior to 1940. A. E. Wood (1962) has treated the same period and added later stages to the history. Because of the different attitudes of these two students toward classification in general and classi-fication of the rodents in particular, the history as interpreted is somewhat different and both references form important guides to the development of important concepts.

TABLE 44   Classifications of the Rodentia

I. *Romer (1933)*
  Order Rodentia
    Suborder Simplicidentata
      Infraorder Sciuromorpha
        Superfamily Aplodontoidea
        Superfamily Castoroidea
        Superfamily Sciuroidea
        Superfamily Geomyoidea
        Superfamily Anomaluroidea
      Infraorder Myomorpha
        Superfamily Myoxoidea
        Superfamily Dipodoidea
        Superfamily Myoidea
      Infraorder Hystricomorpha
        Superfamily Bathyergoidea
        Superfamily Hystricoidea
      Infraorder uncertain:
        Family Eurymylidae
    Suborder Duplicidentata

II. *Romer (1945a)*
  Order Rodentia
    Suborder Sciuromorpha
      Superfamily Aplodontoidea
      Superfamily Sciuroidea
      Superfamily Geomyoidea
      Superfamily Bathyergoidea
    Suborder Hystricomorpha
      Superfamily Anomaluroidea
      Superfamily Hystricoidea
        (including Old and New World
        families)
    Suborder Myomorpha
      Superfamily Dipodoidea
      Superfamily Muroidea

III. *Simpson (1945)*
  Cohort Glires
    Order Lagomorpha
    Order Rodentia
      Suborder Sciuromorpha
        Superfamily Aplodontoidea
        Superfamily Sciuroidea
        Superfamily Geomyoidea
        Superfamily Castoroidea
        Sciuromorpha *incertae sedis*
        Superfamily Anomaluroidea

      Suborder Myomorpha
        Superfamily Muroidea
        Superfamily Gliroidea
        Superfamily Dipodoidea
      Suborder Hystricomorpha
        Superfamily Hystricoidea
        Superfamily Erethizontoidea
        Superfamily Cavioidea
        Superfamily Chinchilloidea
        Superfamily Octodontoidea
        ?Hystricomorpha *incertae sedis*
        Superfamily Bathyergoidea
        ?Hystricomorpha or
        ?Myomorpha *incertae sedis*
        Superfamily Ctenodactyloi-
          dea

IV. *Schaub (1958)*
  Order Simplicidentata (Rodentia)
    Suborder Pentalophodonta
      Infraorder Palaeotrogomorpha
        Superfamily Theridomyoidea
        Superfamily Hystricoidea
        Superfamily Castoroidea
        Incertae sedis
          Family Thryonomyidae
          Family Eupetauridae
          Family Petromyidae
          Family Bathyergidae
          Family Spalacidae
          Family Rhizomyidae
          Family Pellegriniidae
      Infraorder Nototrogomorpha
        Superfamily Erethizontoidea
        Superfamily Cavioidea
        Superfamily Dinomyoidea
        Superfamily Octodontoidea
        Incertae sedis
          Family Dasyproctidae
          Family Cephalomyidae
    Suborder Non-Pentalophodonta
      Superfamily Aplodontoidea
      Superfamily Sciuroidea
      Superfamily Geomyoidea
      Superfamily Eomyoidea
      Superfamily Gliroidea
      Superfamily Ctenodactyloidea

TABLE 44—continued

IV. *Schaub (1958)*—continued

  Non-Petalophodonta *incertae*
    sedis
    Family Pedetidae
    Family Mellissiodontidae
    Family Iomyidae
    Family Eutypomyidae
    Family Diamautomyidae
  Infraorder Myodonta
    Superfamily Dipodoidea
    Superfamily Muroidea

V. *A. E. Wood (1958)*[a]

Order Rodentia
  Suborder Protrogomorpha[b]
  Suborder Caviomorpha[b]
  Suborder Myomorpha[b]
  Perhaps new suborder, perhaps
    Myomorpha (superfamily Gliroidea)
  Groups not allocated to suborders:
    Family Sciuridae
    Superfamily Castoroidea
      Castoridae
      Eutypomyidae
    Superfamily Theridomyoidea
      Theridomyidae
      Pseudosciuridae
    Family Ctenodactylidae
    Family Anomoluridae
    Family Petetidae
    Family Hystrididae
    Superfamily Thryonomyoidea
      Theryonomyidae
      Petromuridae
    Family Bathyergidae

VI. *A. E. Wood (1958)*[c]

Order Rodentia
  Suborder Protrogomorpha
    Superfamily Ischyromyoidea
    Superfamily Aplodontoidea
  Suborder Caviomorpha
    Superfamily Octodontoidea
    Superfamily Chinchilloidea
    Superfamily Cavioidea
    Superfamily Erethizontoidea

  Suborder Myomorpha
    Superfamily Muroidea
    Superfamily Spalacoidea
    Superfamily Geomyoidea
    Superfamily Dipodoidea
  Perhaps a new suborder, perhaps
    Myomorpha
    Superfamily Gliroidea
  Suborder Sciuromorpha
    Sciuridae
  Suborder Castorimorpha
    Castoridae
    Eutypomyidae
  Suborder Theridomorpha
    Theridomyidae
    Pseudosciuridae
  New suborder
    Ctenodactylidae
  New suborder
    Anomaluridae
    Pedetidae
  Suborder Hystricomorpha
    Superfamily Hystricoidea
    Superfamily Thryonomyoidea
  Suborder Bathyergomorpha
    Bathyergidae

VII. *Romer (1966a)*

Order Rodentia
  Suborder Sciuromorpha
    (Protrogomorpha)
    Superfamily Ischyromyoidea
    Superfamily Aplodontoidea
    Superfamily Sciuroidea
  Suborder Caviomorpha
    Superfamily Octodontoidea
    Superfamily Chincilloidea
    Superfamily Cavioidea
    Superfamily Erethizontoi-
      dea
  Suborder Myomorpha
    Superfamily Muroidea
    Superfamily Dipodoidea
    Superfamily Geomyoidea
    Superfamily Gliroidea
    Superfamily Spalacoidea

TABLE 44—continued

VII. *Romer (1966a)—continued*	*Not assigned to superfamiles:*
Not assigned to suborders:	Family Hystricidae
Superfamily Castoroidea	Family Ctenodactylidae
Superfamily Theridomyoidea	Family Anomaluridae
Superfamily Thryonomyoidea	Family Pedetidae

[a] Suggested classification in which higher taxonomic units are used only when reasonably well authenticated.

[b] Lesser categories as in classification VI.

[c] A possible classification with all families in suborders.

Among the studies of the last several decades that stand out in classification of the rodents are Tullberg (1899), Miller and Gidley (1918), Winge (1924), Ellerman (1940–1941), Schaub (1953a,b, 1958), A. E. Wood (1954, 1955, 1958, 1962). Simpson (1945, 1959a), Lavocat (1951, 1955, 1956), A. E. Wood and Patterson (1959), Stehlin and Schaub (1951), and Viret (1955). There are others, but these men and their work provide a good cross section of the various ideas that have been advanced.

Simpson pointed out that in about 1940 there were three general sorts of classification:

1. Explosive, with little effort of grouping (e.g., Winge).

2. Wastebasket, with subdivisions such that all forms were included one way or another (e.g., Miller and Gidley).

3. Orthodox, with three or four groups, using the three of Brandt and one for more primitive forms. All rodents distributed among these groups as well as possible (e.g., Simpson, 1945; Romer, 1945a).

In addition Simpson noted a developing system, which he called eclectic, in which well-recognized members of related major groups are placed together, whereas superfamilies or families, for which relationships are obscure, are not associated at higher levels, below the ordinal level. This is the system that was being developed by Wood and has attained prominence since that time.

Examples of classifications of these various sorts are to be seen in Table 44. The strong effect of Brandt's initial system is evident in many of them. As Simpson predicted in his review (1945), activity in the field of rodent classification has continued apace. Matters still remain in a state of flux. Romer's 1966 classification (Table 44, VII) indicates this when compared with his 1945 classification (Table 44, II). In his unassigned families and superfamilies Romer reflects in general A. E. Wood's point of view. This is the position that only those rodents whose major relationships are well understood should be placed in suborders. Any additional grouping, above the family or superfamily level as the case may be, is considered to serve, to paraphrase Wood, only to conceal the lack of knowledge.

This point of view, however, is not universal. Simpson (1959a) and Schaub (1958, Table 44, IV) have felt it better procedure to include all rodents under suborders if suborders are to be used. Simpson's argument was basically that either all rodents should be put in suborders or that none should. A. E. Wood (1962), on the contrary stated the position that those for which subordinal relationship is evident should be so grouped and that others should be put in such categories only as information makes this possible. In making this argument he notes that the Caviomorpha, Protrogomorpha, and Myomorpha have a high probability of being "good" groups.

From about 1950 to the present time various trends of thinking have come to the fore, all resulting from a general dissatisfaction with Brandt's classification, largely because it does not take into account the great amount of parallelism that appears to be involved in the evolution of the rodents.

Schaub's 1958 classification, including modifications by Viret (1955) of Schaub (1953b), shows a basic concern with dentition as the primary criterion for separation of major groups. All rodents were placed under one of two suborders, one characterized by the presence, and the other by the absence, five-lophed teeth. Some superfamilies were used, assigned to larger groups in some cases and not in others, and there are many incertae sedis families.

Romer's (1966a) classification, though reflecting A. E. Wood's philosophy, used the term "Sciuromorpha" (= Protrogomorpha) and includes the family Sciuridae, excluded by A. E. Wood and B. Patterson (1959) and A. E. Wood (1962). He thus makes a fundamentally different interpretation of this group. In his 1968 notes and comments on the classification Romer speaks of this as only a mild modification, on the basis that the Sciuridae arose from the primitive group and led to no other.

Two classifications of A. E. Wood (1958) are shown in Table 44 (V and VI). The first, the eclectic pattern of Simpson's terminology, is the one that he felt to be superior. The second was included to show what, in his opinion, would be necessary were all the known rodents to be assigned to suborders. Obviously it is something of a mess.

Each of the classifications has abandoned the Brandtian subdivision of Hystricomorpha. This was based on the resemblances of jaw musculature of the South American rodents, which first appeared in the Deseaden Oligocene, and the various Old World rodents. Geographic considerations strongly suggested that these were not the result of common descent, thus overruling the morphological evidence that they were derived from a common ancestor with the "hystricomorph" jaw musculature. The New World representatives are included in the very homogeneous group of Caviomorpha. The Old World forms find no relatives among other rodents of the Eastern Hemisphere, and the Hystricidae tend to be treated as an unattached family. Recently A. E. Wood (1968) described the Oligocene rodents from the Fayum. He has pointed out many similarities between them and the early caviomorphs of South America. The great increase in knowledge of these early African rodents does nothing to alter the interpretation that development of similar genera in the Old and New World was the result of parallelism.

The difference in classifications and the continuing flux of ideas relate to various factors, already noted or implicit in the discussions, as follows:

1. The basic philosophy of use or nonuse of major categories to include all known forms, here a matter of placing all in suborders. Related is a predilection for a sense of symmetry in classification by those who wish to include all subgroups in intermediate high categories.

2. Differences in the evaluation of the meaning of different suites of characters—for example, teeth by Schaub and Viret, mostly following Stehlin and Schaub (1951), myology by Brandt and his many followers, and multiple characters by A. E. Wood and B. Patterson.

One other very important point is fundamental to differences in classifications. A. E. Wood, in general agreement with Lavocat and with Simpson, believed that the evidence indicates a very rapid radiation of the rodents in the Eocene, with most of the major lines coming into being at that time. If this is the case, at least the possibility of arriving eventually at a phyletic classification exists. The records indicate that sufficient collecting and sufficient study can produce the materials and interpretations necessary.

Schaub, on the other hand, has felt that there was a long pre-Eocene history, going well back into the Paleocene and possibly into the Cretaceous. Deployment into the various lines was thus thought to have taken place well before the time in which there is any trace of rodent remains. From the fossil record to date it seems most unlikely that an array of fossil rodents from the Paleocene or earlier will ever be found. If this is the case, the chances of arriving at a reasonable phyletic classification seem remote and the best solution might be to erect one that comes as close as possible but meets the needs of practicality, with less than full regard for phyletic niceties.

Clearly, even if the first suggestion on the time of radiation is correct, an immense amount remains to be done before a classification without many loose ends can be developed. Until such evidence is at hand, continued differences of opinion and ways of approaching rodent classification will undoubtedly persist.

Order Primates

The problems of primate classification fall into three main categories. Perhaps the most fundamental and difficult problems relate to the classification of the very early Tertiary and late Cretaceous mammals, which bear resemblances to primates and to insectivores. Among these presumably are creatures that lie close to the origin of the primates, but the systematic position of many is most uncertain. This problem has been already treated (see pages 398–399 and Table 40). A second suite of problems relates to the ways of subdivision of the order into major categories, a matter on which there is no strong consensus at the present time. The third matter, which has been of particularly lively interest lately, is the classification of manlike creatures and man.

Of these three, the second requires the most attention in this section, because the first has been considered and the third, interesting as it is, falls below the level of generality with which this brief survey is concerned. One observation, however, is pertinent with respect to methods of classification. In efforts to know and portray the ancestry of man and to arrange the participating stocks in some sort of order many students have tended to depart widely from normal taxonomic procedures and, often, to substitute strongly subjective judgments for dispassionate assessment of available data. Throughout the history of the classification of primates, with of course notable exceptions, the direct involvement of man in the system that he studies has been a hindrance. Simpson (1945) has made a strong point of this:

The peculiar fascination of the primates and their publicity value have almost taken the order out of the hands of sober and conservative mammalogists and have kept, and do keep, its taxonomy in a turmoil.

Concepts of the internal organization of the order Primates have fallen roughly into two general types: one which considers a dual arrangement, usually with two suborders, and one in which a greater number of equal categories is used (see Table 45). Very early in the history of classification the question of whether or not man was to be considered related to other animals now called primates was an important issue. This is, of course, now more or less beside the point, because man's inclusion has long been accepted.

The term "Quadrimanes," found in many early classifications (e.g., Cuvier, 1800), was used by Blumenbach (1779) in distinguishing between Bimanes (*Homo*) and Quadrimanes (other Primates). More generally, however, two major divisions came to be recognized under the Quadrimanes, essentially simian and nonsimian. Within this dual pattern, in some instances, animals now not considered as related to the Primates were included in the Quadrimanes. In De Blainville (1816), for example, taligrades and *Galeopithecus* were so treated.

TABLE 45　Classifications of the Primates

I. *Simpson (1945)*	II. *Romer (1966a)*
Order Primates	Order Primates
Suborder Prosimii	?Suborder Plesiadapoidea
Infraorder Lemuriformes	Family Phenacolemuridae
Superfamily Tupaioidea	Family Carpolestidae
Superfamily Lemuroidea	Family Plesiadapidae
Family Plesiadapidae	Suborder Lemuroidea
Family Adapidae	Family Adapidae
Family Lemuridae	Family Lemuridae
Family Indridae	Family Daubentoniidae
Superfamily Daubentonioidea	(Cheiromyidae)
Infraorder Lorisiformes	Family Lorisidae
Infraorder Tarsiiformes	Suborder Tarsiodea
Family Anaptomorphidae	Family Anaptomorphidae
Family Tarsiidae	Family Omomyidae
Prosimii—infraorder uncertain	?Family Microsyopsidae
Family Apatamyidae	Suborder Platyrrhini
Family Carpolestidae	Family Callithricidae
Suborder Anthropoidea	Family Cebidae
Superfamily Ceboidea	Suborder Catarrhini
(Platyrrhina)	Superfamily Parapithecoidea
Family Cebidae	Superfamily Cercopithecoidea
Family Callithricidae	Superfamily Hominoidea
Superfamily Cercopithecoidea	
(Catarrhina)	
Superfamily Hominoidea	

Tendencies for multiple groupings also occur early, as seen in a classification by Illiger (1811):

> Order Pollicata
>   Family Quadrumana
>   Family Prosimii
>   Family Macrotarsi (tarsiers)
>   Family Leptodactyla (*Chiromys*)
>   Family Marsupialia

Flower (1883) and Gill (1870) used two divisions—suborders Anthropoidea and Lemuroidea—and in Weber (1904) the terms "Prosimii" and "Simii," common in many current classifications, were introduced together in the current sense. Simpson's (1945) classification (Table 45, I), with its two major groups—suborder Prosimii (Illiger) and Anthropoidea (Mivart)—follows this pattern. His subgroupings are essentially those of Gregory (1915b), with some modifications resulting from subsequent work. His three super-families of the Anthropoidea include the New and Old World monkeys and the apes, respectively.

Romer's (1966a) classification (Table 45, II) represents the other point of view—basically that inclusion of all primitive primates in a single suborder does not properly express their evolutionary patterns. His problematical suborder Plesiadapoidea tentatively brings together many of the early presumably primate types. Most of these do not appear to be near the ancestry of other primates.

Romer's distinction of the Old and New World monkeys, indicated by placing them in separate suborders, was based on the belief that they stemmed independently from a primitive, perhaps lemuroid stock. This is backed by strong geographic evidence but requires that allowance be made for a great deal of parallelism between New and Old World forms. The source of the South American monkeys is not known, because they appear first in the very lowest Miocene (?latest Oligocene) and, in the course of this epoch, attain essentially their modern dispersal as far as morphology is concerned. Currently the most reasonable hypothesis seems to be that they were derived from primitive primates independently, and this is well expressed by Romer's treatment.

The coherency of the catarrhine primates seems to be unquestioned. The fossil record, though vastly improved in recent years, is still not sufficient for the construction of a detailed phylogeny. Apparently each of the major lines goes well back into the Tertiary. Recent finds in the Fayum region of Africa, by Simons (1960–1968) in particular, are gradually clarifying the picture, but it still is not clear how the various lines are related and how they are best to be expressed. This, of course, extends into the hominoids and particularly into the relationship of the hominids to other members of the group. New finds of "fossil men" have excited great interest and have brought forth a wide-ranging series of systematic speculations, based for the most part on evidence that would be looked on as far from sufficient for any sound conclusions in other groups of mammals.

Order Edentata

A variety of animals, both living and extinct, have developed somewhat similar characters of dentition, skulls, jaws, and extremities and have been placed both informally and formally in an array of mammals termed edentates. Reduction of front teeth, development of peglike, enamelless cheek teeth, various extreme modifications of the usual placental dental formulas, some-what tubular braincases, extended secondary palates, and large claws are common features to be found within this array. In some instances, perhaps initially in all, these features were adaptations to somewhat similar dietary habits. Many are feeders on small invertebrates and some specifically feed on ants and termites. Others, however, are primarily herbivores, and it is not clear that they came through a path of invertebrate feeding.

During the development of mammalian classifications a more or less classic concept of "edentate" developed in which hairy anteaters, pangolins, aardvarks, sloths, and armadillos were related. The term "edentate" was applied, although only a few are in fact edentulous and some have many more than the primitive number of teeth in the placentals.

Such a grouping was not made in early classifications. In Linnaeus' (1758) classification, for example, most "edentates" were placed with the Bruta, in association with proboscideans, and the armadillos were placed in the Bestiae, with the swine. Monotremes have been included with the edentates as well by various students. One of the early names given to the group was Mutici (Storr, 1780), but this long ago dropped out of use. De Blainville (1816, 1834), in his generally advanced classifications, included the Cetece under the Edentata. Several classifications are listed in Table 46.

Flower (1883) and Weber (1904) followed the more or less classical approach. One of the last to suggest a possible relationship between the Old and New World forms of edentates was W. K. Gregory (1910), who tentatively included the full suite on his superorder Edentata. The inclusion of Taeniodonta was on the basis of a similar allocation by Wortman, and the subordinal terms of the Xenarthra were adopted from Ameghino.

In most early classifications attention was paid to similarities, many of which were recognized later as convergent. Once emphasis was shifted to dissimilarities, both in morphology and in geographic distribution, division of the New and Old World "edentates" became evident. Virtually all classifications of recent decades recognize the New World forms as separate, termed either Edentata or Xenarthra. In addition to their geographic isolation, for they developed almost entirely in South America, xenarthrans have many skeletal features rarely found elsewhere among the mammals. The presence of extra-articular processes on the vertebrae, the basis for the name "Xenarthra," is an example, and there are many others throughout the skull and skeleton. These features are so distinctive that they have prompted various students to consider xenarthrans as very distinct from all other mammals, a position accorded recognition by assignments to a separate subclass or superorder.

Early in the 1900s, in North America, a small Eocene animal that appeared to have edentate affinities was discovered. Later other genera came to light, one from the Eocene and one from the Oligocene. These two formed the basis of a group termed Palaeanodonta by Matthew in 1918. The edentate affinities were recognized shortly after the discovery of the first specimen. In 1942 another genus was added (Colbert, 1942).

Scott (1937) included the Palaeanodonta in the Edentata, but as a separate

TABLE 46   Classifications of the Edentata

I. *Flower (1880)*
    Order Edentata
        Suborder Pilosa
        Suborder Loricata
        Suborder Squamata
        Suborder Tubulidentata
II. *Weber (1904)*
    Order Edentata
        Tubulidentata
        Pholidota
        Xenarthra (New World edentates)
III. *W. K. Gregory (1910)*
    Superorder Edentata
        ?Order Taeniodonta
        ?Order Tubulidentata
        ?Order Pholidota
        Order Xenarthra
            Suborder Anicanodonta (Pilosa)
            Suborder Hicanodonta (Loricata)
IV. *Osborn (1910)*
    Order Edentata
        Suborder Pilosa (Anicanodonta)
            Section Gravigrada
            Section Vermilingua
            Section Tardigrada
        Suborder Loricata (Hicanodonta)
            Section Dasypoda
            Section Glyptodonta
V. *Scott (1937)*
    Superorder Edentata
        Order Xenarthra
            Series A. Pilosa
            Series B. Loricata
        Order Palaeanodonta

VI. *Simpson (1945)*
    Order Edentata
        Suborder Palaeanodonta
        Suborder Xenarthra
            Infraorder Pilosa
                Superfamily Megalonychoidea
                    (Gravigrades)
                Superfamily Myrmecophagoidea
                    (Vermilingua)
                Superfamily Bradypodoidea
                    (Tardigrada)
            Infraorder Cingulata
                (Loricata)
                Superfamily Dasypodoidea
                Superfamily Glyptodontoidea
VII. *Romer (1966a)*
    Order Edentata
        Suborder Palaeanodonta
        Suborder Xenarthra
            Infraorder Loricata
                (Cingulata)
                Superfamily Dasypodoidea
                Superfamily Palaeopeltoidea
                Superfamily Glyptodontoidea
            Infraorder Pilosa
                Superfamily Megalonychoidea
                Superfamily Mylodontoidea
            Infraorder Vermilingua
    Order Pholidota
    Order Tubulidentata

order, thus being an advocate of the superordinal nature of the group, but with the South American stocks in a single order (Table 46, V). Simpson's (1945) and Romer's (1966a), classifications (Table 46, VI and VII, respectively), which are very similar, stem from well back into the present century, being much like that of Osborn (1910) except for the subordinal position given the Palaeanodonta.

The history of the classification of the Edentata illustrates many of the factors that have entered into mammalian classification in particular and most vertebrate classifications in general. The initial classifications were formulated largely on modern animals, and groupings were based on morphological resemblances, with some influence of patterns of behavior and conditions of life. Fossils, as found, were fitted into the classifications, modifying them little at the beginning. This was generally during a period of groping for major affinities and before the establishment of the practices of phyletic classification. With the introduction of concepts of phyletic relationships and wider knowledge of distribution of animals, it became apparent that zoogeography was an important aspect of classification. Reexamination of morphology followed, and on the basis of geographic separation, requiring separate genesis from very primitive placentals, and differences in morphology, the old assemblage of classical edentates was separated into several orders.

Later fossil discoveries, as searches were intensified and more attention was paid to small animals, revealed a totally extinct group. Initially this was integrated into a known array, but later it was separated as distinct. In the case of the edentates this distinction has a strong geographic and morphological basis. The more general and common case, such as noted for the primates, is that the primitive stocks, usually recognized rather late in the development of classifications, include within them some animals that may represent the base of phyletic lines from which the readily distinguished members were derived. This area remains the most difficult one in all vertebrate classifications, because it tends to defy expression in a strictly phyletic manner.

section IV
Patterns of evolution

The challenge of life in water, I

WATER AS A MEDIUM FOR LIFE

Water has been the primary abode of life for all but about the last 10 percent of the fossil record. The extent of primitive plant life on land before the time of the first definitive record is uncertain, but no major radiations have been recorded before the Devonian period. To the best of present-day knowledge water was the medium in which life arose and in which subdivision of the primitive stocks to form all of the phyla took place. Only with the coming of a suitable atmosphere and of organisms sufficiently preadapted to push into what must have been a hostile environment did massive population of the continents begin. The first terrestrial organisms probably were plants. The time of first occupancy is uncertain, but by the very latest Silurian primitive vascular plants were making inroads on the continents, and during the Devonian a forest cover had developed in places. After the plants came several waves of invasion of land by animals, and, following successes on land, came reinvasions of the aquatic realm. Plants were much less successful in this second phase.

Today, as at all times for which there is a record of life, the greatest abundance and variety of animals are found in the oceans, estuaries, lakes and streams. Plants similarly are abundant and varied in water, but the marine floras are composed in large part of nonvascular plants and in this respect are more limited than the floras of the land and continental fresh waters.

431

The unique properties of water make it an ideal medium for the development and existence of organisms with the carbon base and molecular constitution common to all life forms that we know. Required is a physically and chemically stable medium that can act to ensure a continuing source of energy and the materials necessary to its utilization, and the removal of the products of metabolic activity. Life, of course, having originated in water, was necessarily adapted to it, at least in its initial stages. To say then that water is the medium particularly suited to life is in a sense redundant. On the other hand, no other medium has properties necessary to the origin and early evolution of life as we know it, and to date we have had no evidence of any other chemical systems that exhibit the properties that are called "living."

The fitness of water as an abode for life has been discussed by many biologists, from a variety of points of view. In his introduction to "Adaptation to Water" in the *Handbook of Physiology* Wilber (1964) has summarized the principal points in a concise way. The important features are as follows. Water has the highest heat capacity of any solid or liquid except ammonia. Heat transfer thus makes for thermal stability of the medium, both external to the organism and within it, in relationship to physiological processes. Coupled with this property is the unusual nature of its thermal expansion, with maximum density reached at 4°C, well above the freezing point of pure water. Circulation in both large and small bodies of standing water is essential to the maintenance of temperature distributions and relative uniformity of distributions of dissolved solids and gases as well as transport of suspended materials. Density relationships of water at different temperatures, which result from the properties of thermal expansion, are critical in the development of patterns of circulation that are necessary to this mixing.

Radiant energy penetrates fairly readily through water with differential screening of wavelengths. Visible light penetrates to considerable depths in clear water and provides the source of energy for basic photosynthetic processes. Water is often called the universal solvent. It has the capacity to dissolve many of the ingredients that are essential to the roles of metabolism, making these materials available to the organism in its external environment and providing a medium for transfer and reactions within the organism. Adding to its importance in physiology is the fact that water has a very high surface tension, the highest of any liquid.

The specific gravity of water, which is generally taken as a standard of comparison, is neither extremely high nor low relative to other liquids concerned. It is, however, low relative to most solids so that they sink to the bottom of bodies of water and permanent suspension of introduced materials does not occur.

All of these properties are essential to life as we know it today. Seawater, while varying in physical properties in its areal and vertical distribution over the earth, does not exceed rather restricted limits in many of its features. Temperatures range from 0 to about 30°C, with a general decrease poleward and with depth. Local variations over the year rarely exceed 10°C and as a rule are lower. Both oxygen and carbon dioxide show moderate variations. Except for unusual circumstances, both gases are dissolved in seawater in amounts that are within the tolerance ranges of many living organisms. The salt content of the seas is quite stable, although subject to local variations for a variety of reasons. Pressure, of course, is depth dependent.

All of these variables act to control the distribution of life and play important roles in the course of adaptations witnessed in the course of evolution. Few circumstances that they produce either singly or in combination have resulted in aquatic conditions to which organisms have found it impossible to adjust.

What has been described, of course, relates to conditions of the sea today. The best information available suggests that conditions not greatly different with respect to these variables have existed during the Paleozoic and probably for a considerably longer time. It does not appear that even under the most extreme conditions have the critical variables exceeded by any appreciable amount the upper and lower limits known today. Distributions, of course, have been very different, especially with respect to temperature and temperature-dependent features of the waters.

The properties of bodies of waters in earlier, pre-Cambrian, times are less well known, and of course it is far from clear what they were during the period of gestation and appearance of the first life. If, as is believed by many students, the atmosphere in these early stages of life contained no free oxygen but was reducing in nature, rich in such gases as carbon dioxide, methane, ammonia, and water vapor, the properties of water in relationship to dissolved gases would have been greatly different. Its salt content too may have departed appreciably from that known today. In part it is the necessity of these differences for the accumulation of organic materials and the origin of life that has led to their postulation.

If this is a generally correct picture, then during the early phases of the development and evolution of life this evolution was accompanied by an evolution of the environment, as the atmosphere and waters were modified to become eventually the fluids that exist today. The evolution of more advanced forms of life with metabolisms in which free oxygen was an end product is considered as a critical event under this hypothesis, one that led to the atmospheric changes that in turn made it possible for the evolution of life of the sort that is known during the Phanerozoic.

As far as vertebrates proper are concerned, their full evolution took place during a time of a moderately stable, oxygen-bearing atmosphere, with conditions in the waters of the seas and continents that seem not to have differed sharply from those we know today. It is not known what steps led to protochordate ancestors and to the chordates proper and where these took place. That the origins of these ancestral types occurred in marine waters seems highly probable, although not subject to demonstration. Whether the early stages occurred during the continuing evolution of the properties of the atmosphere and the seas or after relative stability had been attained is equally uncertain.

The environment of the origin of the vertebrates themselves has been a matter of continued controversy. Possible origin in fresh water has received some support from physiological considerations related to osmoregulation, and it does seem quite possible that some groups of fishes, Osteichthyes and perhaps Chondrichthyes, did originate there. Arguments supporting the hypothesis that the first vertebrates originated in saline waters—based on the habitats of the early vertebrates, the probable marine source of proto-chordates, and, somewhat paradoxically, on physiological and structural aspects of osmoregulation—seem very strong.

It seems reasonably safe to assume that vertebrate ancestors did inhabit saline waters, perhaps coastal waters or estuaries, and that the general environmental circumstances were at least somewhat like those found in these environments today. It was only a considerable time after the first record that vertebrate remains are known from unequivocally freshwater deposits, and even longer before they are found in "normal" marine sediments laid down in open waters away from coastal and estuarine conditions.

Very early in their history, then, vertebrates were adapted to saline waters as far as their physiology was concerned. Passage into the open marine waters of the seas probably involved adaptations in locomotive, feeding, and perhaps reproductive activities. On the contrary, passage into nonsaline waters, perhaps gradually through the upper reaches of estuaries, would have involved modifications in osmoregulation, very likely with the development of adaptation of special structures of the kidneys to meet the new problems occasioned by entrance of an excess of water into the system.

Little of what went on can be known with any assurance. Many of the kinds of physiological changes that presumably took place did not leave a clear morphological record, even when detail of the sort available for moderns can be studied. It is probably too much to hope that the relatively crude and limited morphological analyses possible for the ancient fishes and fishlike creatures can give many hints as to the nature of such adaptive changes.

VERTEBRATES AS AQUATIC ANIMALS

Scope and Limits

The two sources of aquatic vertebrates, one with its full history in the water and the other with an intervening terrestrial phase, have produced a vast and varied array of animals. These have occupied most of the kinds of environment available in view of the limitations imposed by their basic forms and functions. The upper extremes of water temperatures have been a bar to vertebrate habitation, as they have been for all but a very few organisms. Likewise, very high salinities found only in limited environments have acted as a block. Water pressures encountered at abyssal depths have proved no barrier to occupancy of those regions. Both fish and tetrapods have adapted to life in waters that range from violently mobile to almost completely quiescent.

All vertebrates are reasonably active, mobile creatures. None has adopted a completely sessile way of life. They have not developed into animals, such as those found in various other phyla, that are totally dependent on movements of waters and of materials carried by the waters for substances of life, such as oxygen and food. They are not, strictly speaking, elements of the infaunal part of benthic communities or entirely passive members of planktonic assemblages, depending on currents alone for movement.

A second limitation to radiation, common to most metazoans, is size. Vertebrates are all relatively large in the full scale of organisms. Even the smallest far surpass in size animals grouped roughly as microorganisms. Their system organization and life processes appear to preclude development of creatures sufficiently small to invade this immense realm of animals and plants. This fact, of course, has vastly reduced the potential of sheer numbers of vertebrates that can be supported in possible environments. On the other hand, restrictions of size at the other end of the scale have been less stringent than for the great majority of other kinds of animals. Only a few, such as the giant squid, have approached the proportions of some of the large cetaceans. Many other aquatic and terrestrial vertebrates, although less extreme, have attained sizes far above those found in most other groups of animals.

A third limitation, usual among animals, is the presence of an oxygen-based metabolism. Occupancy of anaerobic circumstances for any appreciable period of time is impossible. The demands of constant supplies of the vital gas vary widely within the vertebrates, in part as a function of the overall metabolic rates of major groups, but also within different subdivisions of the major groups. Adaptations to conditions that vary widely in oxygen supply have taken place among fishes, and marine tetrapods have developed capacities to adjust to long periods without a supply of oxygen.

Many aspects of vertebrate evolution have been related to the attainment and utilization of oxygen. The epithelium as a whole, special parts of it, and highly specialized structures, the lung and gills, take part in the procurement of oxygen. The basic problems are to move the oxygen-bearing medium over surfaces so constituted that oxygen is infused into the system and to transfer the oxygen to the tissues of the body. Major portions of the vertebrate body have been modified in special adaptation to the requirements of different environments and various ways of life.

In their radiations in the water, both from the bases of aquatic and of terrestrial ancestors, the vertebrates have penetrated into a vast number of circumstances stretching almost to the ultimate the potentials of the environments as they are limited by the basic vertebrate properties. Wherever mobile, macroscopic, oxygen-based aquatic animals can exist, vertebrates will be found if the history of the site has been such that they have had physical access to it.

ADAPTATIONS TO LIFE IN WATER

General Considerations

Although every animal is an adapted unit, a product of its genetic and physiological responses to internal and external environments, anything beyond a general "natural history" description requires that the activities and systems involved be somewhat isolated for analysis. The activities of vertebrates on land and in water are of the same general kinds, but the problems posed by the properties of the media are different and adaptive responses are expressed in the sharp differences in structure and function.

Adaptations to special aquatic circumstances led to the development of vertebrates able to make a transition to land; others, on land, permitted a return to water. The ways in which the common problems of aquatic life have been met by vertebrates with different adaptive heritages is one of the most interesting stories of vertebrate history and offers one of the clearest avenues to an understanding of the nature of adaptive evolution.

It is convenient for a study of these problems of adaptation to water to subdivide the activities into the following categories:

Movement
Feeding
Sensing
Respiration
Reproduction
Osmoregulation

The fossil record gives information very directly applicable to the first item in the list and considerable information on the second. For each of the other activities some direct information is available for special aspects and for some groups of extinct animals. The least interpretable are the last two, reproduction and osmoregulation. Both are critical aspects of adaptation to water, particularly related to major adaptive shifts between waters of different salinities and between water and land.

Many of the basic data on the processes involved in these activities must come from moderns. Within recent years, as new techniques became available, an immense amount of study has been devoted to the life of vertebrates in water. A vast literature has developed, and a short treatment must leave out much that is significant. To remedy this, references that give greater coverage of special areas are cited in the text. Precise data have been gathered as bases for critical physical and chemical treatments of many of the phenomena. One of the best examples is the application of concepts of hydrodynamics and fluid mechanics to problems of propulsion. Another is the study of sound production and reception in cetaceans; a third is the use of radioactive tracers in physiological analyses. Concurrently rising interest in the oceans has increased many times over observational studies of the life and interrelationships of a vast variety of marine vertebrates. Studies are still going on apace and with no signs of any diminution, but rather of continuing increase.

The problem of the paleozoologist is first to be aware of this vast volume of work and to assimilate it sufficiently to use it in his interpretations. This alone is no mean task. He must then be able to understand his own materials in the detail necessary for application and adopt a position that neither ignores what is known nor goes overboard in totally unwarranted extrapolation. The remainder of this chapter and the next are applied to the special problems of vertebrates in water. The data have been drawn from many sources, the most important of which are cited in the text.

Movement through Water

Propulsion through water is critical to all aquatic vertebrates, since none is truly sedentary. Movements of the body and fins provide the principal forces, but hydrostatic adjustments and jet actions also are significant. The problems of propulsion relate on the one hand to the properties of the medium, its viscosity and specific gravity in particular, and on the other to the form and the drive mechanisms of the organisms. Most studies of propulsion have held the critical parameters of the medium constant and centered attention on the drive mechanisms.

Body shape, the form and position of appendages, their mobility, and the nature of the body surface are the external features that are critical in the transmission of the forces of the organism to the medium and its progress through that medium. The source of the power is the muscle system. Within broad limits it acts in propulsion in one of two ways. Rhythmic differential contractions of axial muscles induce body and fin flexures; muscles of the

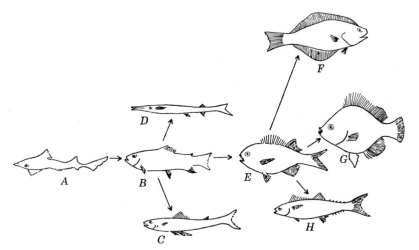

Figure 83. Various types of fishes, shown in flow-sheet form (not a phylogeny) in which the various ways of origin of different types of swimming are indicated. *A*, a somewhat flattened form, with propulsion by body flexure and tail fin, and with but small, mobile paired fins. *B*, normal fusiform body, with pelvic fin well posterior from the pectoral and locomotion induced primarily by body flexure and caudal fin. *C*, elongated body form, with pelvic fin forward in a more effective position for controlling movement. *D*, slender body with pelvic fin posterior. *E*, showing a common trend toward a deep body in which emphasis in locomotion comes to lie in movements of the posterior most part of the body and the caudal, pectoral, and pelvic fins become increasingly involved in small movements and in braking. *F*, dorsal and ventral fins elongated, giving the potential, realized in some fishes, for their use in locomotion in conjunction with the caudal fin. *G*, deep-bodied fish carrying the conditions noted in *E* to a more complete development. *H*, a streamlined, pedunculate fish, a strong and fast swimmer, with the caudal fin and movement of peduncle the principal propulsive mechanisms. (Redrawn from J. E. Harris, 1953.)

paired appendages induce paddling, rowing, or undulating motions in the paired appendages. Minor appendicular motions are important in positioning, maneuvering, and slow swimming, but they are mostly supplemental to the primary sources of thrust. Departure from the two principal modes do of course occur, but most of the patterns of movement found among the vertebrates fall under one or the other of the main types.

In a survey of fin structure and function J. E. Harris (1953) presented a general morphological evolutionary series of fin types and uses in the fishes. Figure 83, from J. E. Harris, gives an excellent pictorial summary of many of the items taken up in this section with reference to caudal fins, body movements, and uses of the pectoral fins. Figure 84, portraying a more specialized group, illustrates in some greater detail modifications that delegate propulsion, in various ways, to one or another part of the whole body-fin propulsive system.

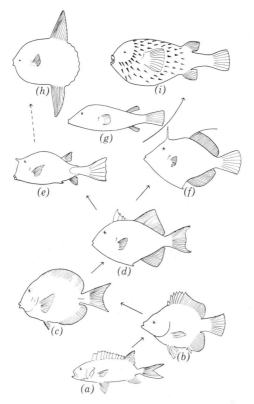

Figure 84.   Fin and body patterns among the advanced teleosts, illustrating the extent of modifications that occur and the general relationships between body change and feeding structures. *A*, a holocentrid; *B*, a carpoid, *C*, an acanthurid; *D*, *E*, and *F*, balastids; *G*, a tetraodontid; *H*, a molid; *I*, a diodontid. (After J. E. Harris, 1953.)

The fact that externally visible features include the important variables of the gross aspects of propulsion has made study of locomotion a ready subject for analysis among moderns and has made fossils a good source of information on the actions of extinct creatures. With some notable exceptions, the analytical work with recent animals, however, has paid only passing attention to the skeleton and musculature and has been focused more on the observations of activities of aquatic animals and physical interpretations of the interactions of their bodies and the surrounding water medium. This has resulted in some difficulty in coordinating the information from moderns and fossils. In spite of this, experimental studies have provided a dynamic framework into which the structures of the skeleton of both modern and extinct animals can be fitted for interpretation.

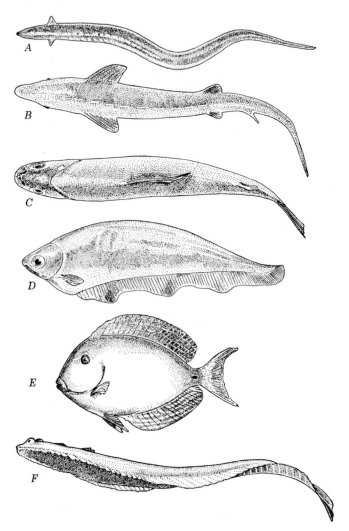

Figure 85.   Several types of swimming patterns found among fishes, to illustrate variation in the kinds of body movements and their relationships to fin structures. *A*, an eel, *Anguilla*, the basis for the concept of anguilliform motion, with body oscillation extending the length of the trunk. *B*, a sturgeon, somewhat reduced anguilliform motion, with body oscillation largely confined to the posterior part of the trunk. *C*, a trout, somewhat further reduced body flexure, but swimming primarily by the movement as it involves the caudal fin. *D*, a knife fish, with a ribbon-like ventral fin and propulsion depending on traveling waves along this fin. *E*, a surgeonfish, with elongated dorsal and ventral fins that move laterally only with body action acting to stabilize when the fish is still or moving slowly and are folded down when the fish is in rapid motion, with propulsive force furnished by the caudal fin. *F*, a sole, which, like similar flatfishes, moves by undulation of the paired fins, either in conjunction with body oscillations as in the sole, or by traveling waves developed in the fins themselves, as in some skates and rays. (Redrawn from photographs in *Structure, Form, Movement* by H. Hertel, 1966, published with permission of Krauskopf-Verlag, Mainz, West Germany.)

The problem of "how fish swim" has been of interest almost from the beginning of man's study of nature. As other swimming animals became known, the additional aspects of motion in the water that they revealed were incorporated into the studies. Analytical approaches to the problems were made sporadically during the last half of the nineteenth century and the early part of the twentieth. Abel (1912) made a survey of these studies and added much of his own. Breder (1926) published the results of his classic experimental and analytical studies on fish locomotion, noting that, in spite of all that had been done previously, very few critical probes had been made into mechanisms. J. Gray, beginning studies in the 1930s, pioneered many facets of locomotion and set a pattern that many students have since followed.

New types of equipment, particularly high-speed movies, have greatly broadened studies, and engineering analyses have been prompted by man's search for solutions to his own problems of high-speed, efficient locomotion of his vehicles through fluid media. Many of the analytical studies have aspects that are directly applicable to extinct organisms, and they establish a basis for interpretations that should dispel older misconceptions that have been made without such a background.

Drive Mechanisms

Propulsion, turning, control of yaw and pitch, stopping, and all similar adjustments of swimming animals require the interaction of all parts of the locomotor apparatus. Most structures function to some extent in each of the actions. To keep considerations within bounds, however, it is necessary to treat the systems of propulsion and their components separately.

BODY AND MEDIAN FIN MOTIONS. The majority of swimming animals move through the water as a result of motions of the trunk and the median fins. Those that move primarily by rowing and undulatory actions of the fins, by jet action, and by hydrostatic adjustments are relatively few in number but include very important adaptive types, which are treated in a subsequent section.

Figure 85 illustrates the body and fin forms that exemplify some of the principal varieties of propulsion among fishes. In these propulsion results from flexures of the system that produce full or partial sine waves that pass from anterior to posterior along the body. An accelerating posterior motion of the water is induced by these waves, and forward motion of the animal, relative to the water, occurs. Relations between the forces propelling a fish and forces exerted by the lateral undulations, after J. Gray, are shown to illustrate this general process (Figure 86).

In most analytical studies efforts have been made to isolate types of movement, to provide standards for comparison. Breder developed a system based primarily on the extent of development of the sine wave. At one extreme was

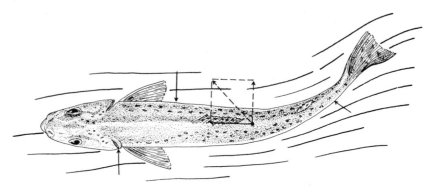

Figure 86.   The swimming action of the trout, showing the primary components of force. Heavy solid arrows on body show forward motion with components in dashed arrows. Lighter solid arrows represent opposing forces of the medium against which the body acts. Redrawn from *How Fishes Swim* by J. Gray. Copyright ©, 1957, Scientific American Inc.

the eel-like fish and at the other fishes that showed no trunk flexure. The one he called *anguilliform*, with more than half a sine wave developed, the other *ostraciform*, with virtually no body wave. Intermediate, with less than half a sine wave, is the *carangiform* type, much the most common. Hertel (1966) summarized studies of "slender swimmers," the extreme anguilliforms, of Breder, and powerful swimmers. The very rapid swimmers with carangiform motions and fast-starting fish, lunging carangiforms with special flexibility, fall within the realm of fast and powerful swimmers.

The spectrum over the range of actions is continuous. Most swimming animals fall somewhere in the middle carangiform range and vary within the general limits between flexures of one-half to about one-sixth of a sine wave. Fishes within this general range are mostly those with moderately fusiform bodies and discrete caudal fins, not coalesced with the dorsal or the anal. An extremely effective development of the propulsive system within this group is found in the fast and powerful swimmers—tarpons, barracudas, and some of the sharks among fishes and ichthyosaurs and cetaceans among tetrapods. Their capacities for rapid movement in the water have made them special targets for analytical studies.

Fast Swimmers.   Two aspects of the speed of swimming are important in adaptations toward very rapid movement. One is the absolute speed relative to the water, and the other is the ratio of the speed to the size of the animal. A great many measurements, under many different conditions, have been made of the speed of fishes, and some have been made for tetrapods (Table 47), as given by J. Gray (1957).

TABLE 47   Swimming Rates of Some Fishes and Mammals[a]

Animal	Length (ft)	Maximum Observed Speed (mph)	Maximum Observed Speed/Length
Trout	0.656	3.1	8.5
	0.957	6.5	11.0
Dace	0.301	3.6	17.8
	0.594	3.8	9.0
	0.656	5.5	13.5
Pike	0.529	4.7	13.0
	0.656	3.3	7.5
Goldfish	0.229	1.5	10.3
	0.427	3.8	13.0
Rudd	0.730	2.9	6.0
Barracuda	3.937	27.3	10.0
Dolphin	3.529	22.4	5.0
Whale	90.000	20.0	0.33

[a] From "How Fishes Swim" by J. Gray. Copyright © 1957, Scientific American Inc.

Of the fast and powerful swimmers, the barracuda appears to have reached a maximum completely authenticated speed, with an absolute rate of 27 miles per hour. Others, both fish and mammals, have come fairly close to this speed and perhaps exceed it considerably, and reports of speeds up to 60 miles an hour for sailfish and swordfishes have been reported. Flying fishes near takeoff have been reported to reach 35 miles per hour, and rates up to 40 miles per hour have been cited for porpoises. These rates are not in general well confirmed. Figure 87 shows diagrams of some of the fast swimmers to bring out the common characters that they display. The dorsal and anal fins lie relatively far forward and are not involved in the force developed for propulsion. The caudal fin is symmetrical, isobatic in action, and somewhat bifurcated. The ratio of body depth to length is high, ranging from 21 to 28 percent in the forms shown. However, there is considerable variation, with the barracuda having a ratio of only 16 percent and some of the fast predacious pike even less.

The deep, relatively broad body provides the necessary powerful musculature, the power source, and also tends to approach fairly closely a "laminar" body, the ideal shape in which the flow of water at the boundary layer remains laminar for the length of the body. Drag by turbulence does not occur in the ideal case and is low in forms that approach it. The deepest part of the body lies in the vicinity of the longitudinal midpoint, generally departing less than 10 percent plus or minus.

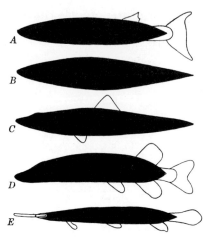

Figure 87.   The body form in some powerful swimmers, shown in lateral silhouette, to portray the position of greatest body depth and the depth-to-length ratio. In each the position of the deepest portion, DP, expressed as percentage of distance from the front, and the depth-to-length ratio D/L, as a percentage, differ. *A, Barracuda:* DP = 40 percent, D/L = 16 percent; *B,* shark: DP = 44 percent, D/L = 18 percent; *C,* smooth dogfish: DP = 45 percent, D/L = 14 percent; *D,* pike: DP = 55 percent, D/L = 18 percent; *E,* alligator gar: DP = 70 percent, D/L = 11 percent. (Redrawn and modified from *Structure, Form, Movement* by H. Hertel. Published with permission of Krauskopf-Verlag, Mainz, West Germany.)

Drag by slowing of the flow behind the moving body, in the wake, is also kept very low, but the mechanism in these active swimmers is less evident. The problem can be approached only by analysis of the simultaneous operation of all of the factors during fast swimming, while the body is propelling itself through the medium. Just how the body form, its motions, and the motions of the fins act to keep this component of drag low and prevent a forward creep of turbulence along the laminar body layer has not been fully analyzed.

The body surfaces of all of the fast swimmers evidently are well adapted to maintenance of laminar flow in the boundary layers. The smooth-scaled, slimy surfaces of fishes appear to produce an effective structure for this purpose (Rosen, 1959). Perturbing structures—the eyes, mouth, and gills—lie far forward, in positions appropriate to their main functions and also little disruptive to flow. Little can be said about the surfaces of extinct fast-swimming reptiles. The skin of the ichthyosaur was smooth, but whether it approached the firm, smooth base of fishes or was more like those of swimming mammals is uncertain. From what is known of living reptiles, it probably was more like the body surface of some fishes, perhaps sharks, than like its adaptive counterparts the cetacean among mammals. Some

Figure 88.   Drawing of a dolphin in rapid motion, showing the portion of the body below water. The characteristic "high-speed" skin folds related to reduction of drag are shown. (Drawn from a photograph in Hoerner, 1958.)

fishes—for example, the tunny—have fatty layers under the skin and may show responses in swimming somewhat like those of the porpoise. It is by no means impossible that a similar adaptation occurred in the ichthyosaurs.

Much attention has been devoted to the problem of surface drag in swimming mammals. A very high level of adaptation has been attained. The skin in cetaceans tends to be deformed at high speeds, and this appears to suppress the tendency for turbulence to form in the boundary layer (Figure 88). Many hypotheses to account for the low turbulence and high laminarity have been advanced, but none has received very broad acceptance as a full explanation. A summary of the problems is to be found in a paper by Land (Norris, 1966) on the hydrodynamics of cetacean performance. The problem of the relationships between power, drag and speed, which resulted in what has been called Gray's paradox (i.e., that the power observed is far less than necessary for the speed attained), has been analyzed by many students. Bainbridge (1963) has shown the relationships in a series of graphs, here given in Figure 89. What emerges from his studies is that there are no serious problems in small and in large fishes. The power and speed are related in terms of levels of drag that are reasonable from observations. In the middle groups, however, some problems still exist, although they may be resolved by assuming the power to have been estimated as somewhat less than it in fact is. The case of the barracuda (Figure 90) remains a puzzle, because unless speed

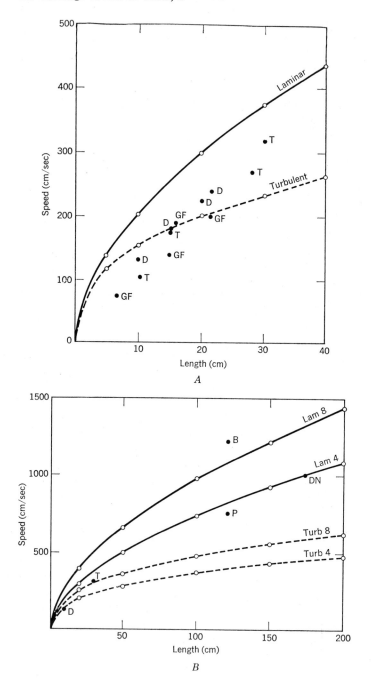

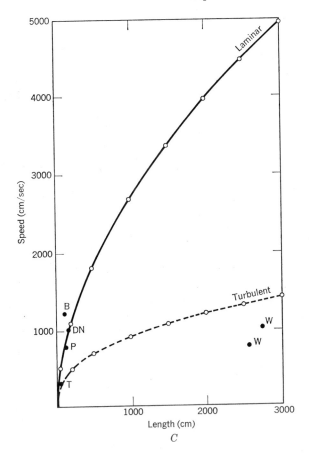

Figure 89. Three graphs based on experimental studies by Bainbridge (1963). These show the calculated curves for fishes relating speed and size, expressed as length. The plots of points show the correspondence of experimental results to the calculated curves. Lam, laminar flow; Turb, turbulent flow, the number indicates the power factor used in the calculations. A, power factor of 4×10^5 ergs per second per gram of muscle; the solid line is completely laminar flow, the dashed line is completely turbulent flow. The relationships shown in A extended in B to specimens up to 200 cm long. The heavy lines (Lam 4 and Turb 4) are derived by using a power factor of 4×10^5, the finer lines (Lam 8 and Turb 8) are derived by using a power factor of 8×10^5. In C the same relationships are extended to specimens up to 3000 cm long, by using a power factor of 4×10^5. The points indicate measured values for D, dace; T, trout; GF, goldfish; B, barracuda; P, porpoise; DN, dolphin.

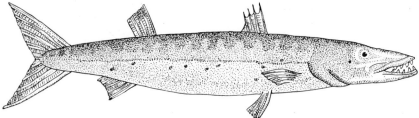

Figure 90.   The barracuda, considered by many, on the basis of repeated accurate measurements, to be the fastest swimming fish. The figure emphasizes the strong, muscular body, the large bifurcated caudal fin, the small median and paired fins, with pectorals and pelvics far forward. Redrawn from *How Fishes Swim* by J. Gray. Copyright ©, 1957 by Scientific American Inc.

estimates have been badly made, no adequate explanation of how it attains such speed has been found. Hertel (1966) states:

Gray's paradox is solved by demonstrating that the net power requirement is much less than computed previously. The correctness of the solution is seen in the fact that behind fast swimming fishes we find no velocity difference of the nature of $V_S > V_0$ and $V_W < V_0$, where V_0 is the initial velocity of the stream, V_S is the velocity resulting from a thrust (S) normal to the direction of flow (increase in velocity), and V_W is the velocity resulting from a drag (W) normal to the direction of flow.

Variations in Fast Swimmers.   Marked departures from the fusiform body in swimming animals indicate swimming to be less rapid than in the fishes and tetrapods treated to this point. Even among the fast swimmers in

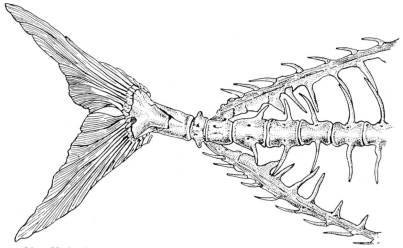

Figure 91.   Skeletal structure of the caudal fin in a powerful swimmer, the scombroid fish *Luvarus imperialis*. Note especially the well-developed peduncle and its articulation with the more anterior part of the vertebral column. (After W. K. Gregory and Conrad, 1942–1943.)

which the body shape is not drastically modified from the "ideal" form, however, wide variations do occur. These are associated with special adaptations within the group and very often are interpretable from fossil remains. One of the most common and easily recognized series of modifications is in the slenderness and structure of the posterior part of the trunk and the caudal fin. If only the external structure is available in the fossil, much can be determined, and if the bony apparatus is preserved, a great deal can be learned about the mechanics of the propulsive movements.

Fast-swimming fishes, in particular scombroid fishes, have a slender region, just anterior to the caudal fin, termed the peduncle (Figure 91). This tends to act more or less independently of the rest of the trunk, and its independence is sometimes emphasized by the development of a joint in the vertebral column at the anterior end of the fin. In very slender peduncles, as in *Caranx* and *Sarda*, the internal structure may be complex and set with a ventral keel or keels and intricate muscle-and-tendon systems. A somewhat similar structure is found in cetaceans, and the motions, although basically vertical, resemble those of the horizontal fin in fishes (Figure 92).

Some scombroid fishes, in addition, develop small supplemental fins along the peduncle. It has been suggested that these act like slots in an airfoil to increase the thrust of the tail and to decrease the turbulence in the lateral motions of the peduncle (Nursall, 1958). G. E. Gadd (in J. Gray, 1949), however, has proposed that they act as flow separators.

The function of the caudal fins, which are isobatic in character, in fast- and moderately-fast-swimming fishes is to produce a forward drive. A number of fairly readily determined characteristics of the fins are related to the effectiveness of this mechanism and to the speed of the fish bearing the fin. Bainbridge (1958) has shown that, except at very low frequencies, the speed of swimming in a species is a function of the frequency of the stroke of the caudal fin. Amplitude is important at low frequencies, but above these levels it remains fairly constant. This simple relationship is of considerable importance in providing a stable base from which other variables may be assessed. It is not, of course, a feature that can be determined from fossils.

Nursall, in considering the caudal fin as a hydrofoil, applied the concept of *aspect ratio* to the fins of various fishes. The aspect ratio (AR) of the fin as he used it is the span^2/area. High efficiency is reached at ratios of 8 to 10. The ARs for various fishes are given in Table 48, and it is evident that, although some approach these ratios, the majority lie far below them. Those with high ratios are fast swimmers, with tails that are somewhat rigid and noncollapsible. Other variables, of course, enter into the picture. Most of the fast swimmers also have forked caudal fins, possibly related to the action of the tips in reducing flow turbulence. The fins in this fast-swimming group also

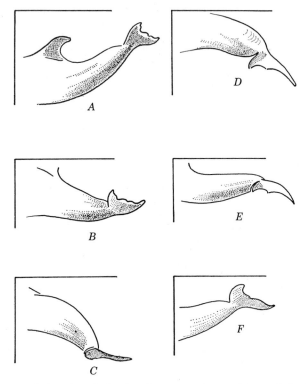

Figure 92.   The fluke of a whale, showing successive positions of the fins and the related torsion of the posterior part of the body through one cycle during loco- motion. *A* through *C*, the downstroke; *D* through *E*, the return; *F*, the upstroke. (Modified after Slijper, 1962.)

have relatively small amplitudes, and motions are restricted largely to the peduncular part of the body.

The fishes in the list with lower ARs are less rapid swimmers but are capable of fair speeds and also tend to be highly maneuverable, able to start and turn rapidly. The caudal fins are more mobile, and parts can be moved differen- tially. Both the structure and the hydrofoil produced by this differential flexure are complex. It appears that it is primarily in the fast swimmers with re- stricted motions that the AR is important, and this distinction makes it possible to estimate with some success the speeds of fishes for which tests are not possible.

Highest levels of proficiency in fast swimming have been attained by some of the teleosts among the bony fishes, some of the recent sharks, and advanced cetaceans. Many of the specializations that they show, however, appear well back in the fossil record, in other lines that moved in similar directions but

TABLE 48   Aspect-Ratio Measurements in Some Fishes as
Presented by Nursall (1958)

Species	Aspect Ratio
Bowfin (*Amia calva*)	1.0 (fin extended)
Lake trout (*Salvelinus namaycush*)	1.0 (fin relaxed)
Lake trout	2.0–2.5 (fin extended)
Smallmouth bass (*Micropterus dolomieu*)	2.0 (fin extended)
Striped pargo (*Hoplopagrus guntheri*)	1.5 (fin relaxed)
Striped pargo	2.9 (fin extended)
Yellow perch (*Perca flavescens*)	3.2–3.8 (fin extended)
Green jack (*Caranx caballus*)	3.8
Sierra mackerel (*Scomberomorus sierra*)	4.1
Black skipjack (*Euthynnus lineatus*)	5.0
Little tuna (*Euthynnus alletteratus*)	5.0
Yellowfin tuna (*Neothunnus macropterus*)	5.2
White marlin (*Makaira albida*)	6.1

did not carry the full suite of adaptations to such high levels. Porpoises and ichthyosaurs, modern sharks and cladoselachians, the pike and rhipidistians (e.g., *Eusthenopteron* or *Holoptychius*) represent pairs of living and extinct creatures that show such resemblances. Similar structural modifications have appeared repeatedly in successive groups as they have become adapted to rapid swimming. Parallel evolution has been extensive. The differences between successive groups are mostly increases in refinements, of greater integration of the various systems involved, and of accommodations to the special characteristics of the groups in which the specializations occur.

Swimming animals with other than fusiform bodies can in large part be grouped into three more or less distinct categories: elongated eel-like forms, deep-bodied compressed forms, and flat-bodied or depressed forms. Some body shapes, particularly among the teleost fishes (e.g., sea horses or mud-skippers), cannot be placed in any of these categories.

For the most part a particular body shape is matched by a characteristic form and position of the median fins. The caudal fin, however, is somewhat special in this regard, because it is related not only to propulsion but also to the specific gravity of fishes and, in this role, assumes an important phylogenetic position. Some of its aspects thus are best treated separately from considerations of general body form.

Special Aspects of the Caudal Fin. The primary function of the caudal fin, as has been emphasized, is propulsion, and in fast-swimming fishes the principal component of its thrust is forward. In many fishes, as well as some aquatic tetrapods, however, it also serves to raise or lower the position of the posterior end of the body while propelling the creatures forward. Caudal fins may be defined on the basis of three types of lift effect: *isobatic*, in which there is only forward motion; *epibatic*, in which the thrust is partly upward, and *hypobatic*, in which it is partly downward. These terms have long been in use, having been extensively employed by, for example, Abel in 1912.

Caudal fins of vertebrates are generally classified on a morphological basis, with the curvature and position of the axial column relative to the fin lobes the crucial items. Various kinds of caudal fins have been recognized, among them such familiar types as heterocercal, homocercal, diphycercal, and hypocercal (Figure 14). Adaptively it is the function of these types that is important, but they also have proven very useful in systematics quite apart from their functions.

To resolve conflicting statements about the action of caudal fins, in particular that of the heterocercal fin, Grove and Newell (1936) performed a series of experiments, using models of different types of fins. The results have very clearly demonstrated that the actions of the fins in their "lift effect" were directly related to the relative extent of the epichordal and hypochordal lobes, rather than to the disposition of the column itself. Each type, of course, also has its own characteristics with respect to the forward component of thrust, with only the isobatic tail expending its full energy in this direction.

Of primary interest among the fishes is the epibatic tail, which is generally morphologically heterocercal. This appears to be the primitive caudal fin in all major groups of fishes. In each group, by one path or another, tendencies toward an isobatic condition have developed. The studies of Grove and Newell indicate that the epibatic caudal fin tends to be found in fishes whose specific gravity is relatively high in comparison to the medium in which they live.

No truly fast-swimming fishes of today have the epibatic tail, but some of the relatively rapid swimmers of the past did have this feature. It would appear that such fishes—for example, some of the sharks, crossopterygians, and arthrodires—although the dominant fast fishes of their times, did not approach the proficiency in speed developed in more recent fishes. Except for the sharks, most of the early fishes, throughout much of the Paleozoic, were equipped with heavy armor of scales and bony plates. It is likely that their specific gravity was high and that the epibatic caudal fin was adaptively important. Many, of course, lived in fresh water, which emphasized the need for mechanical lift. Sharks, lacking lungs or a gas bladder, may have been relatively heavy as well.

Modifications toward an isobatic condition came about repeatedly, as they became adaptively feasible. Specific gravity, of course, is a function of several variables, few of which can be assessed directly in fossil fishes, and no very close correspondence between times of detectable weight reduction and modification of the caudal fin is to be expected.

Other types of caudal appendages encountered among fishes and tetrapods are somewhat more difficult to interpret. The hypobatic fin is present in anaspids and ichthyosaurs and appears to have depressed the caudal region. This may have related in some way to the structures of the anterior paired fins, but interpretations that have been made of the adaptive significances seem very problematical. In some fishes and tetrapods an effective caudal organ of propulsion does not exist. Seals utilize the hind limbs to some extent as a substitute. Some fishes and amphibians employ undulations of the other median fins, either in conjunction with body movements or alone. Deepening of the tail, as in various swimming reptiles, produces a flat surface that to some extent acts similarly to the posterior part of the body-fin system in creatures with well-defined caudal appendages. Many avenues to similar ends are encountered, but much the most common one, developed in many different lines, is evolution of a caudal fin as the main source of thrust.

Slender Swimmers.   Elongated animals with flexible bodies have developed in several groups of fishes and in amphibians and reptiles. In the latter aquatic amphibians may have developed this form in the water, but reptiles very clearly developed it on land, as a locomotor adaptation, and carried it over to the water (Figure 93). The swimming of the elongated, slender forms falls into the anguilliform category of Breder, typified by the eel, *Anguilla*. He accepted within this group, however, many swimmers in which flexure is quite restricted (about half a sine wave). In truly elongated animals flexibility is an important characteristic, and in those with land ancestry persistent evidence of once important vertical axial support is seen in special features of the vertebrae, such as presence of extra articular processes.

No sharp differentiation between the slender swimmers and moderately elongated fishes and tetrapods can be made. Elongated fishes—such as chlamydoselachian sharks, rhipidistian crossopterygians, many lungfishes, and a wide variety of actinopterygians—approach this level but do not show its extreme specializations. Among the reptiles and some extinct aquatic amphibians a somewhat special type of subanguilliform body action has developed. In their aquatic locomotion the body flexure was largely confined to the tail. Some of the placodonts, nothosaurs, primitive ichthyosaurs, and crocodiles fall within this group. The ancestry, of course, was very different from that of fishes, but the actions in propulsion converged into a very similar pattern. In many of these creatures and some others less aquatic on the whole

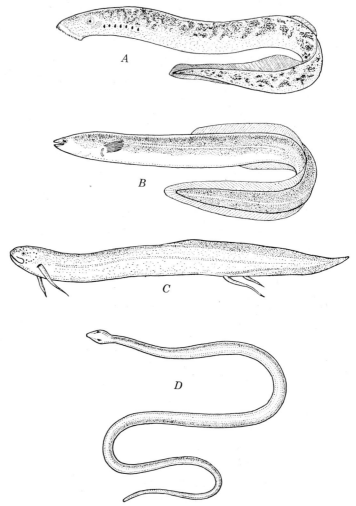

Figure 93.   Examples of elongated or slender swimmers (see also Figures 85*A* and 95). *A*, lamprey; *B*, eel; *C*, *Lepidosiren*, an elongated lungfish; *D*, an aïstopodan amphibian, *Phlegethontia*, extinct, representative of the form in some aquatic snakes as well. In each swimming is accomplished in some part by throwing the body into sine curves, usually with changing amplitude along the length of the body. Each, to some extent, has its own particular supplemental methods of locomotion.

the dual functions of land locomotion and swimming are reflected in distinctive specializations of the trunk region for land locomotion and the caudal region for water propulsion.

The swimming of animals in which the eel-like or snakelike form is very highly expressed has attracted much attention among students who have

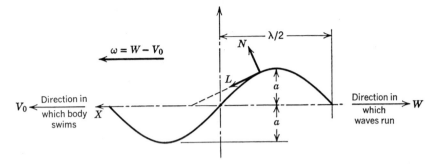

Figure 94.   The oscillation of the body in a slender swimmer, shown on the basis of a simplified diagram of a single body wave (sine curve). It assumes that the amplitude is constant along the wave and that there is but a single wave. a = amplitude of wave $(\lambda/6)$, λ = wavelength.

	Speed	Direction
Body wave relative to head	W	$-x$
Head forward motion relative to water	v_0	$+x$
Body wave relative to water	$w = W - v_0$	$-x$

sought to make hydrodynamic analyses of locomotion in water. Gray has led the way in such studies, in particular by making motion pictures of actions and analyzing some of their principal components. Taylor (1952), Gero (1952), Lighthill (1960), and Hertel (1966 and earlier) have approached the problems from a strictly analytical point of view, setting them in hydrodynamic and engineering frameworks. The relative simplicity of form and motion is attractive, but complete analyses have proved extremely difficult. Some of the general results are of considerable interest for paleozoology, but the more precise and complex analyses can, at present at least, have only very indirect applications.

Oscillation of the body of the slender swimmer, abstracted as a single sine wave, is shown in Figure 94. In this diagram the wave is of consistent amplitude, but in many of the slender swimmers, such as *Natrix*, which has been extensively studied, the amplitude increases posteriorly and is not constant even in this increase as the snake throws its body into successive waves.

The pattern of snake locomotion probably was approached by some of the slender amphibians, such as *Phlegethontia* (Figure 93 *D*), but other amphibians and slender fishes also have median fins that play important roles in propulsion. If the fins are continuous and the dorsal, caudal, and anal fins are coalesced and more or less symmetrically disposed, as they are in eels and a wide variety of other elongated fishes, the qualitative aspects of swimming are not greatly different from those of finless creatures that have been studied in more detail. A large number of anguilliform and subanguilliform creatures, however, have not attained this extreme condition, but rather have the caudal, dorsal, and anal fins closely spaced but separate. They are thus brought into full play at different stages of body flexure, producing a complex, cyclic situation. This, the more usual circumstance, is much more difficult to analyze and interpret in terms of contributions of the various different components to locomotion. In still another pattern of locomotion of slender swimmers the undulation of median fins is an important source of thrust. This may be supplemental, as in the electric eel and the bowfin, *Amia*, or the sole source of locomotion, as in some of the knife fishes (Figure 85*D*). Although in some ways very inefficient, this type of propulsion allows ready reversal of movement and movement in narrow passages in which thrust cannot come from body flexure. Similar fin actions are also present in some of the deep-bodied fishes.

The extremely slender swimmers among the fishes and some amphibians are predominantly bottom dwellers, often living in contact with the substrate. Larvae of some of the eels, and leeches and nerian worms among the invertebrates, however, are free swimmers of this type. Slender swimmers among the tetrapods tend, on the other hand, to be pelagic swimmers, whether swimmers of the snake type, utilizing much of the body in propulsion, or of the more conservative "tail swimming" type. The issue here becomes complicated by the fact that there is a gradation between solely aquatic vertebrates and those that are both aquatic and terrestrial. In the latter the anterior part of the body, the precaudal region, maintains the structure necessary for life on land and the caudal portion becomes somewhat modified for swimming by anguilliform motion.

Among fishes the less extreme anguilliforms range widely in habitats and in habits. Many—for example, *Neoceratodus, Polypterus*, sturgeons, and a number of the sharks—are basically bottom dwellers, but not confined to this level. Others range widely, and slender-bodied forms with somewhat greater stiffness to their bodies, such as pike, pass over from the anguilliform to the carangiform way of swimming, although the body itself might suggest the former. No hard and fast rule can be set for estimating the nature of swimming and the probable habitats from body and fin form alone. It is, on the other hand, possible to limit considerably the possibilities on the basis of just these features, which are often available for fossils, and within these limits to reduce them, since other particular features of the animals and their environments may be available.

Deep-bodied Fishes.   The deep-bodied form of fish is found fully expressed only among the actinopterygians. It occurs in partial development in coelacanths, such as *Latimeria*, and some of the ichthyosaurs tended toward fairly deep, compressed bodies. The presence of a well-developed gas bladder appears to be a requisite of the very-deep-bodied fishes, related to the delicate relationship of balance and center of gravity. No simple characterization of the adaptive significance of the deep-bodied form in fishes is possible, since this feature has developed in many lines in adaptation to a large suite of circumstances. Some limitations have occurred. Deep-bodied fishes are not bottom dwellers and do not belong in the ranks of the very fast swimmers. This body form is not characteristic of abyssal fishes.

As in the other groups of fishes based on form and propulsion patterns discussed to this point, all gradations from fusiform to deep-bodied, compressed fishes occur, and within these grades there is vast specialization of different parts of the structures, including the median fins. In a sense the deep-bodied fishes as a whole have evolved in a way that gives them the best of the two worlds of speed and of maneuverability, with the compromise reducing speed somewhat in favor of maneuverability. Paired fins are important in slow propulsion, in quick starts, and in maneuvering, as taken up specifically in a subsequent section. Attention here is focused on the propulsive mechanisms of the body and the median fins.

In most deep-bodied fish it is the caudal, dorsal, and anal fins that provide the primary thrust. As for other types of fishes, the functions are most evident in the extreme development of the mode, in the disk-shaped fishes (Figure 85E). In these the caudal fin is seen to be columnar, with its span about equal to that of the body. The angle of the leading edge with the body axis is high, up to 60 degrees, and the area of the fin is relatively large. Thus, in spite of the span, the aspect ratio, though fairly large, is not as high as it is in some of the rapid swimmers.

The caudal fin is mobile and intricately controlled by muscles. It is collapsible in many deep-bodied fishes. The sweep is much as in stiff-bodied fast swimmers, with high frequency and rather low amplitude. Only the very posterior part of the body is involved, and the trunk itself develops little of the thrust. The condition is subcarangiform. Caudal propulsion thus is effective, and some of the deep-bodied fishes are quite fast swimmers.

The dorsal and anal fins of disk-shaped fishes are essentially extensions of the posterior rims of the disk and do not engage the water current with their leading edges. They do not function importantly to provide thrust in conjunction with body movements. On the other hand, undulatory movement is a common feature, and this mechanism of thrust, in conjunction with that of the caudal fin, opens possibilities for intricate movements, forward and backward and upward and downward. These actions are coordinated into the full system that involves as well the highly mobile, forwardly placed paired fins.

Mobility is a critical aspect of the deep-bodied fish and occurs in all of them, whether their specialization is that of the disk-shaped fish or more moderate. Locomotor agility, however, is not considered the single dominant aspect of their habitus. It is a necessary adjunct to the many ways of life that they pursue, but alone it does not indicate any one particular mode. Similar shapes and propulsion mechanisms have developed in adaptation to various ways of feeding, to life in fresh and in salt water, to life in areas of heavy aquatic vegetation, to rocky habitats, to life in coral reefs, and to life in the open waters of seas and lakes. Any circumstances in which delicate balance, moderate speed, high mobility, and maneuverability are important are a target for adaptation for fishes with the capacity to develop the deep-bodied form of body.

Flat-bodied, or Depressed, Fishes. Depressed body form in fishes is an adaptation to bottom living. This was true as well, it seems, for many of the very-flat bodied amphibians developed during the late Paleozoic and early Mesozoic. Depressed bodies are not characteristic of aquatic reptiles or mammals, and when they do occur, as in some partially aquatic turtles, they are not related to bottom life but to other special aspects of existence. Among fishes extreme depression of the body is found primarily among Chondrichthyes, as far as living groups are concerned, and in extinct agnathans and arthrodires. Osteichthyes show an unusual approach to this condition in the flounders and similar fishes, which attain the equivalent of a depressed form by turning over. The lateral expanses of fairly-deep-bodied creatures become "dorsal" and "ventral" in orientation and function.

Living sharks include intermediates between slightly depressed body forms and extremes, such as the manta ray. In all but the most extreme the posterior part of the body and median fin remain important in propulsion. No new patterns are introduced in the actions of these structures, but their actions are supplemented by motions of the paired fins, as taken up in the next section. In the flounder type of modification body undulations and actions of the median fins are little different from those in deep-bodied fishes in which the vertical axial plane has maintained its position. The body tends to be more flexible but otherwise little altered for locomotion.

Even the most highly developed depressed fishes of ancient groups used their caudal structures in locomotion, with paired appendages supplemental when present. Even in the agnathans, which had no proper paired fins, the propulsion was by movement of the caudal apparatus much as in modern carangiforms. Innovations in the flat-bodied fishes thus pertained to the pectoral fins, not to axial motions.

PAIRED APPENDAGES.   *General Matters.*   The most primitive and some of the highly specialized vertebrates lack paired appendages, some acanthodians have more than two pairs, and a small number of vertebrates have only one pair. For the most part, however, vertebrates carry two pairs of lateral appendages: one the pectoral and the other the pelvic. When present, the paired appendages are involved in one way or another in locomotion. Among aquatic vertebrates they are the sole source of propulsive thrust in some groups — chelonians and aquatic birds—whereas in others they function almost entirely to confer stability during swimming and to alter it with directional change. In the majority of aquatic vertebrates the paired appendages play multiple roles with respect to locomotion. In addition to these primary functions paired appendages also serve as part of the thermoregulatory system in cetaceans and as supplemental respiratory surfaces in some of the teleost fishes that spend part of their time out of water.

Thrust by paired appendages is developed by movement of their external portions, which produces simple to very complex motions in water, with resultant movement of the animal. It may be developed either by flapping or sculling motions or by undulation of the fins in a horizontal plane. The skeletal structure, nature and extent of the musculature, disposition of the paired appendages, and the characteristics of their surfaces all bear rather direct relationships to the functions, and thus they can, within limits, be reasonably ascertained for extinct vertebrates.

Active propulsion by paired appendages contrasts with the more passive roles, in which they act primarily as hydroplanes, producing stability during motion or reducing it selectively as the angles of attack of the appendages to the water are adjusted to depart from the stabilizing positions. This rather limited range of function is characteristic of most primitive fishes with paired appendages and has been secondarily developed in a number of the most highly aquatic groups of tetrapods—ichthyosaurs, mosasaurs, and cetaceans. In such roles the paired appendages act more or less as do the elevators and ailerons of airplanes.

The principal functions of paired appendages can be arrayed into the following categories:

I. Paired appendages not the principal organs of locomotion
   A. Functions with animals in active motion in water
      1. Acting with median appendages: establishment of equilibrium or stability with respect to pitch, yaw, and roll during horizontal motion
      2. Disruption of stability, as in function 1 by inducing pitch, yaw, and roll, usually in combination
      3. Slowing or stopping forward motion, without directional change
      4. Development of system for gliding in water and in the air
   B. Functions with animals motionless in water
      1. Induction of "small" motions by sculling, with resultant minor changes in position in any direction

2. Maintenance of "still" position by small motions countering gentle water currents and actions of other parts of the body

II. Paired appendages the principal organs of locomotion
   A. Motions induced by horizontal undulation of paired fins
   B. Motions induced by flapping, rowing, or paddling actions
   C. Motions induced by undulations in vertical plane (as in seals)

Fewer mechanical analyses have been made for the actions of paired appendages than for those of the body and median fins. Much of the information has come from observations, some under experimental conditions and with models, and much from observations under natural conditions. Observation has varied from critical study to gather data for analyses to casual observations and statements that range widely in reliability. The simple cases, in which paired appendages are involved primarily in stability relationships during locomotion, are mechanically straightforward and susceptible to analysis. Complexity is added by the flexibility of the appendages, both that developed in passive reaction to the water and that induced by the musculature. As these more complex motions of the paired appendages relative to the water come into play, especially with initiation of locomotor functions, the situation becames increasingly complex and has been little analyzed.

The bases of interpretation have come from the work of many persons, among whom the following may be noted in particular. Breder (1926) pioneered in efforts to introduce rigor into descriptions of the actions of paired fins in fishes. The studies of J. Gray covered many aspects of functions of paired fins, as they have done for other structures involved in animal locomotion. J. E. Harris (1936–1938) carried out and reported interesting and definitive experiments on the actions of paired fins in fishes. There have been many engineering approaches to fish locomotion involving to some extent systems interrelationships exemplified by the studies of Taylor (1952), Gero (1952), and Ohlmer (1964). Extensive analyses of the paired fins of cetaceans have been made by many workers, of whom Slijper (1961) and Felts (1966) may be noted in particular.

Observations and speculations on the functions of the appendages in marine reptiles of the past have been many. Among those that have been most influential are works by Dollo (1904), Williston (1914), D. M. S. Watson (1924, 1951), and Camp (1942). Oehmichen (1938) undertook a study of Jurassic ichthyosaurs from the standpoint of mechanics, in one of the most detailed attempts to understand extinct-reptile locomotion. Others, too many to mention specifically, have dealt with problems of many of the aquatic groups. These cover the full range from terrestrial animals that spend some time in active swimming through swimmers that are occasionally on land, to

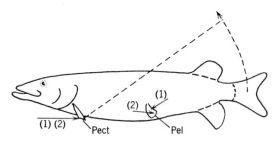

Figure 95.   A fusiform-bodied fish, *Esox*, whose pectoral fins are well separated from the pelvics. In normal swimming maneuvers the pelvic fins so placed take a secondary role. When braking by the pectoral fin alone occurs (1), there is a tendency for the caudal portion to elevate, as indicated by the dashed lines. In this event the caudal fin may act to counteract this motion, as indicated by the arrow pointing roughly to 8 o'clock. For rapid braking both fins may be brought into play, as shown by arrows (2). (After Breder, 1926.)

animals that are totally or nearly totally confined to water. It is with the latter that we are most concerned in this section.

Paired Appendages not Primarily Involved in Propulsion. With Limited Mobility. The paired appendages grouped under this general heading function primarily to maintain stability during forward motion, to disrupt stability in change of direction and orientation, and to form gliding surfaces in the water. These functions are, of course, also performed by more mobile appendages, but only as a part of the complete role.

Some cephalaspids, acanthodians, many selachians, and most primitive osteichthyans have a paired-fin system of this type. It is generally found in close association with an epibatic caudal fin. The pectoral fin in its more or less horizontal orientation and angle of attack in swimming has a resultant upward force, countering the negative pitch induced by the epibatic tail and also, to some degree, by the pelvic paired fin. Among tetrapods, relatively immobile paired appendages are found among ichthyosaurs, mosasaurs, and cetaceans. In each of these however, mobility is somewhat in excess of the most simple expression of this mode, as found, for example, in cladoselachian sharks (Figure 15*B*).

The appendages are characteristically broad based and set well apart, far down on the flanks of the body. The anterior, or pectoral, appendage is generally larger than the pelvic one and plays the major role in the various functions of the system (J. E. Harris, 1936, 1938). The smaller pelvic appendage has less important and well-defined roles, and in experiments its removal has been found to have little serious effect on the locomotion of the animals. It produces a slight negative pitch and appears to play a small role in body equilibrium during swimming activities. The relatively immobile

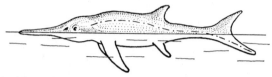

Figure 96.   Position of a floating ichthyosaur as analyzed by Oehmichen (1938). Based on *Eurypterygius*. The dashed line shows the approximate position of the vertebral column. The floating position was determined by the requirements of the position of the nostrils, the paired fins, and the "reversed" heterocercal caudal fin, of which only the ventral part is under water in this position. (From Oehmichen, 1938.)

fins do not form effective brakes but do operate in this way to a minor extent. In braking the pectoral and pelvic fins act together (Figure 95). Figure 15 shows examples of typical appendages of this type.

Characteristically the internal structure of the external part of the fin consists of more or less parallel rays, developed from various sources but of generally similar patterns. Greater mobility is attained with development of a skeletal axis, with lateral rays in some sharks and crossopterygians and dipnoans, and by a basal restriction of the fin and reorganization of the internal parts in actinopterygians. Only in the latter group did the mobility necessary for delicate adjustments of body position by fin movement develop to a high level.

The course of changes in tetrapods, passing from land to water existence, was marked by a trend opposite to that toward greater mobility found in the fishes. The axial elements tended to become shorter, and the effective external portions of the appendages became the metapodials and phalanges. As a rule these were augmented by addition of phalangeal elements, producing hyperphalangy. Motions in general became restricted.

In his analysis of the ichthyosaur Oehmichen (1938) found evidence in the bones of the anterior appendage that flexibility increased distally and from anterior to posterior. The principal motions were deduced to have been vertical abduction and adduction, thereby giving an upward component of force. This he found to be necessary to maintain the floating position indicated as necessary by various body structures (Figure 96). This motion of the fin was, of course, in addition to the stabilizing and maneuvering functions employed during active swimming. Oehmichen interpreted the reverse heterocercal tail as an adaptation to surface swimming in which the main muscular force was below the water level. The effect of the tail when fully submerged would have been hypobatic, but in the surface position indicated in Figure 96 this would not have applied. Hyperphalangy increased the mobility of the appendages and presumably was related to particular swimming habits but it is not at all clear what this relationship may have been.

Mosasaurs generally have a somewhat similar but less modified condition. The appendages seem to have been primarily related to control while swimming and not to propulsion, although this clearly would have been possible at slow speeds. These animals still were capable, it would seem, of rather awkward progress on land. Differences in the degree of development and of ossification of the appendages have been related, along with other features, to different habits. Poorly ossified appendages have been supposed to be indicators of diving habits, but it is not evident from this alone that this was the case. Association of this type of appendage with strongly ossified and protected middle-ear parts perhaps lends some strength to the argument.

Cetaceans, many of whose functional features are best known from porpoises, possess only an anterior flipper. The elbow is at the body contour; there is hyperphalangy and spongy bone, resulting from retarded ossification; and lack of a marrow cavity is characteristic. The weight of the bone is reduced without, however, sacrifice of strength in the direction of major stress (Felts, 1966). Functionally the fin is an adjustable hydroplane that moves by abduction and adduction and has added rotation around its axis. The flippers act together, in rather complex and varied patterns, in stabilization and induction of the swimming patterns in conjunction with the dorsal fin, body motions, and the fluke. There is no evidence that they function in any way as a hydrofoil.

Mobile Appendages. Whereas tetrapods had generally mobile appendages when they first entered water and have specialized by reduction of the mobility, fishes have tended in the other direction. The extent of mobility, the way that it has been attained, and the functions of the more mobile fins differ widely. Form, the nature of the internal skeleton, and the cover of the fins are a general guide in estimating the mobility and functions of fins in extinct fishes. Unless care is used, however, such estimation may, it would seem, be wide of the mark. For example, the paired fins of *Polypterus* (Figures 15 and 63), being short, covered by rather heavy scales, and with a relatively broad base reflected in the internal skeleton, might be judged to be fairly rigid, whereas in fact they are remarkably mobile and used for "bottom walking." Much the same applies in the case of *Latimeria*, whose rather massive-appearing fins are mobile and highly flexible.

Mobility has developed in several ways. Among the selachians, where it has rarely been carried to a very high level, the tendency has been for the formation of a central axis to which are attached strong radiating rays. A good example of this development is seen in the extinct shark *Xenacanthus* (Figure 15). A fin capable of some rotation and of considerable distal flexure has developed. Both Dipnoi and Crossopterygii have undergone strong intensification of the central axis of the paired fins, which was to some degree developed even in the primitive forms. Among the Dipnoi the extremes are reached in *Protopterus* and *Lepidosiren* (Figure 93) in which much of the functional aspect of the fins in locomotion has been lost. In the less extreme condition, as in *Neoceratodus* (Figure 67), the mobile, slender fin is of some use in bottom walking.

Various fin types are found among the early crossopterygians, all trending toward a structure centered around a median axis. Undoubtedly these were mobile fins, and probably they were of use not only in swimming but also in bottom walking. Coelacanths, as they passed from fresh to marine waters, trended in a somewhat different direction, developing fins more nearly like the general type found in actinopterygians, with some suppression of the fleshy lobe and greater emphasis on the ray structures.

A very high level of mobility has been attained by the actinopterygians and is carried to its highest level in some teleosts. Delicate maneuvering while maintaining position in the water, moving in all directions, and stopping and starting all lie within the functional province of the paired fins and in the coordination of their actions with those of the median fins. Very commonly pectoral fins are the dominant source of thrust in slow swimming. Gliding, a somewhat special process both in and out of water, is accomplished by these fins. Mobility is required so that the large, gliding fins may be retracted during active rapid swimming but expanded and held fairly immobile during gliding activities.

As in the case of less mobile fins, the pectoral appendages play the dominant roles, but here the pelvics tend to become somewhat more important in locomotion. In rather general teleosts, like the cypriniforms, the disposition of the fins is somewhat like that in sharks and the pelvics tend to parallel the actions in that group, providing a minor lift force. This function occurs throughout most teleosts, but with increased mobility and modification of position other functions are acquired.

A common modification seen in teleosts, and partially attained in some paleoniscoids and holosteans (Figures 83 and 84) is an upward displacement of the pectoral fin and a forward movement of the pelvic one. The two come to lie more or less in the same vertical plane and together, since both are highly mobile, they form a complementary functional system (Figure 97). An important contribution of the fins disposed in this manner is their braking action. The larger, more dorsally placed pectoral fin forms an effective brake when extended, but an upward increment of force is developed. This is countered by the lesser pelvic fin. Applications of these forces differentially on the two sides of the animal as it moves through water results in turning, rolling, and vertical displacement, which depend on the precise angles of attack of each of the fins to the water. Flexures of each of the fins may change this angle along each fin, adding to the delicacy of adjustment,

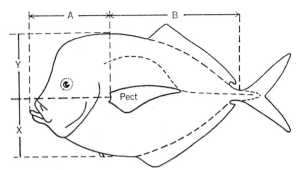

Figure 97.   A deep-bodied fish, *Vomer*, whose pectoral fin is centrally located, as indicated in the ratios of X:Y and A:B, which approximate 1. The fin lies about at the center of gravity. Thus in braking, as well as in small movements induced by the fin, the position of the body is maintained without compensation by other structures. (After Breder, 1926.)

and sculling actions of the fins, coordinated with their passive attack on water further the possibilities of rapid and delicate maneuvering.

Pectoral and pelvic fins tend to be in constant motion when teleost fishes are still in water, performing intricate patterns of back-and-forth sculling in adjustment to minor currents and body actions. The body at rest is in equilibrium, and the motions of the fins, in contrast to their functions in locomotion, are not related to the maintenance of equilibrium. Breder has considered counteraction of the jet action of exhaled water as one important aspect of the action of fins in maintenance of the still position of the fish relative to water. Just as in swimming, the fin system in standing fishes is in constant operation counteracting all of the forces that would result in movement.

The extremely varied expression of body form and development of appendages in teleosts make any attempt to generalize or systematize their basic swimming patterns rather futile. Outside of the main types there are many unclassifiable extremes. One particularly interesting modification in which the fin is an adjunct to locomotion is the development of gliding. Gliding occurs both in water and in air. To some extent it is developed in fishes with rather immobile fins—for example, some of the bottom-living sharks—and it probably was even present, as a function of the very flat body, in finless forms such as the heterostracan *Tremataspis*. Rays that swim by paired fins alone also use them in gliding.

Within active swimmers, propelled largely by the caudal fin and body undulations, and with mobile paired appendages, gliding has been highly developed in a number of groups and has been carefully studied both by observation and by theoretical analyses. Hertel (1966) has summarized

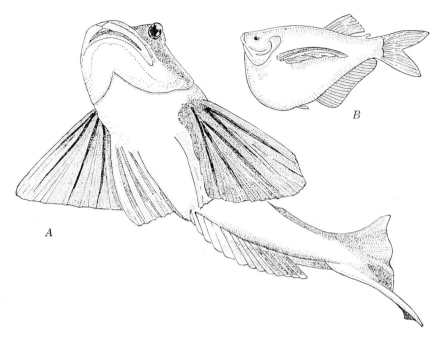

Figure 98.   *A*, a gurnard, *Trigula hirundo*, with very large pectoral fins used for braking and for gliding in the water, as well as for movements on the bottom. This fish, lacking a swim bladder, has high specific gravity, and the tendency to sink provides the primary motion necessary for water gliding. *B*, the hatchetfish *Gastropelecus strigatus*, with strong, massive muscles in the pectoral region. In "flight" these fishes skim the water, not leaving it, and the pectorals beat in a winglike fashion. (Drawn from photographs in *Structure, Form, Movement* by H. Hertel, 1966, published with permission of Krauskopf-Verlag, Mainz, West Germany.)

various aspects of this specialization, drawing on his own studies and many outside sources. His work provides a good résumé for those interested in the engineering aspects of this particular adapation.

The gurnards are heavy fishes, lacking air bladders, that have attained a high level of adaptation to gliding. During rapid swimming the large, winglike pectoral fins are collapsed against the body, but for gliding they are expanded and held rigidly in position. *Trigula hirundo* (Figure 98) is primarily, perhaps totally, a water glider, which, when it slows its rapid swimming, glides on its pectorals to the bottom. The fins are double and the smaller, more ventral portion is used to aid locomotion on the bottom. The flying gurnard, *Dactypoterus volitans*, is both a water and an air glider. In the air its fins flutter, but this is a passive action, not one induced by the musculature of dynamic flight.

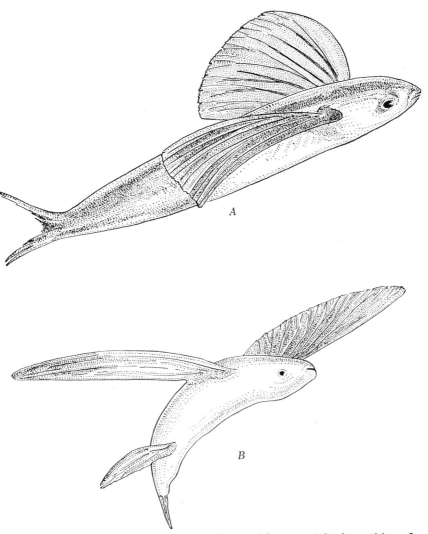

Figure 99.   The Catalina flying fish, *Cypelurus californicus*. *A*, in the position of a taxiing glide, with the caudal fin still in the water, before free glide. *B*, braking action during the free glide, with lowering of the caudal fin, before entering the water, acting as a brake. (Drawn from photographs by Edgerton and Breder, 1941.)

True "flying fishes," such as the Catalina flying fish (*Cypelurus californicus*) come to the surface of the water at high speed, immediately spread their fins, make a taxiing run, increasing speed, and then launch into gliding flight (Figure 99, Edgerton and Breder, 1941). By lowering the caudal portion of the body, they slow progress and return to the water, either entering it or undertaking another taxiing run.

Some three dozen species of this family, Exocetidae, have developed, with various differences. Some have two wings, others four. The loading of the wings has been estimated to be about 1.3 g/cm², a very light loading. The wings, however, are not suitable to beating, since they are too large and too soft to create thrust. The musculature of the fins is only about one-thirty second of the total body weight, whereas in birds it averages about one-fourth. In *Gasteropelecus* and some other "surface flying fishes" the pectoral musculature (Figure 98) is heavy and there is a "wing" flapping action that appears to support the heavy-bodied fishes. Some of these do attain very short flight.

Paired Fins as Principal Organs of Locomotion. Within the actinopterygian fishes remarkably varied patterns of paired fins have developed, and these enter into locomotion in many ways, either acting alone or in conjunction with other sources of thrust. As a rule, where the paired fins do act alone, they do so in special circumstances, whereas under other conditions other parts of the body either take over or supplement the thrust of the paired appendages. Pectoral fins, for example, provide the thrust in many of the fishes that move by short, rapid, darting movements. A wide variety, including of course the darters themselves, show this feature. They are moved up and down, somewhat as are wings, in the slow swimming of many teleosts. In other instances paired fins act against the substrate, either under water or on land. Some "bottom walkers" were mentioned in the preceding section. More spectacular both in their actions and in the fin modifications related to them are the mudskippers, such as *Periophthalmus* (Figure 63D). These fishes have come to land for feeding and have become well adapted to moving about on it, with heavily muscled fins, as shown in the figure. Fins used for underwater locomotion along with pectoral suctorial discs mark another high degree of specialization. Within the teleosts, in fact, the modifications are so many and so varied that no proper summary is possible.

Most of these modifications are consequent on the increasing mobility of the paired fins cited in the last section. Three groups of aquatic animals, however, have reached essentially full dependency on paired fins in locomotion through quite other channels. One group, the rays, has come through successive development of bottom-living adaptations; the other two, chelonians and plesiosaurs, have come from land-living tetrapod stocks.

Some of the rays have carried the development of pectoral fins as propulsive agents to the extreme, with the caudal part of the body completely losing any role in this function. Progress, as in the less specialized forms, depends on undulatory motion of the broad, flexible fins. The motion is essentially the anguilliform type, with the dynamics being much the same as those of the body and fin undulations in the vertical plane, but depending on coordinated action of paired structures, increasing considerably the potential range of activities. This development, like that of the highly specialized use of paired appendages of the actinopterygians, represents the extreme of a long line of evolution, one toward which various groups of fishes at various times in the past have trended.

Chelonians and plesiosaurs come from land ancestors and have passed through stages in which the limbs were used both on land and in water. Many modern chelonians, of course, are in this stage, and it is only the marine turtles that have nearly lost their capacity to move on land. Plesiosaurs presumably came through a nothosaur stage, in which land locomotion was certainly possible, but those that are known appear to have been completely aquatic.

The sea turtles have modified the forelimbs as propulsive organs, whereas the posterior limbs have become merely steering organs. The pectoral flipper is much altered from the terrestrial condition, being rather inflexible and largely confined to vertical movements. In swimming the turtle in effect flies through the water. Zangerl (1953) has made a careful analysis of the osteological changes of the flipper and its resultant functions. The structure has become suited only for propulsion in swimming and can be used only very awkwardly on land. The hind limb is almost equally modified, but in a different manner, so that it serves as a rather stiff, but somewhat mobile rudder.

Freshwater turtles are excellent swimmers and are also dependent on the paired appendages for locomotion. In sharp contrast to the marine turtles, however, it is the hind limbs that are the principal locomotor organs and the forelimbs, though also functioning in this capacity, are less powerful. Both can be adducted into the body, under the carapace, in contrast to marine forms. Modifications are much less, and the limbs act by a more or less fore-aft paddling, not the flying action of the marine turtles. Somewhat intermediate in structure are members of the family Toxochelidae of the Cretaceous. In these the forelimb is modified to form a flipper and to act, it would appear, much as in some of the other marine turtles. The hind limb, however, has maintained many of the characteristics of the hind limb of the river turtle. It is about the same size as the forelimb, in contrast to other marine turtles, and it has not undergone specializations to become primarily a steering organ.

Marine chelonians appear to have developed independently from terrestrial turtles several times in the past. Each stock has its own peculiar features, but each has developed a swimming mechanism that is very much like that of other groups.

Plesiosaurs present a somewhat more varied group in which there appear to be two general lines of specialization for aquatic life. In both of these the pectoral and pelvic limbs become oarlike and participate in locomotion. The limbs are characterized by elongated proximal phalangeal elements. They were quite inflexible, and undoubtedly the whole "fin" was a single structure, without separate digits externally. Pectoral and pelvic girdles supported a very strong musculature, which produced the rowing actions of the paddles.

One line of plesiosaurs during its evolutionary history developed increasingly long necks and small skulls. Along with these two changes went modifications of the locomotor system by which the paddles became somewhat smaller and the increments of musculature that controlled forward and backward movements became more even. In the early forms emphasis was on the muscles from the coracoids and the belly ribs to the humerus, but in later forms (D. M. S. Watson, 1924) the structure of the scapula and clavicle indicates very strong muscles between them and the humerus. D. M. S. Watson's interpretation of these changes is that they made possible rapid rotations of the body, swinging the long neck rapidly over a broad arc in the quest for prey. Control of the inertia would be supplied by the high degree of maneuverability created by the capacity of the limbs to act effectively in both the fore and aft directions.

The other line changed less, becoming somewhat more streamlined and probably somewhat faster. Great speed, in view of the rowing type of locomotion, seems improbable. This group retained short necks and had long heads. In it, as well, there was some reduction in flipper size in later forms, but the changes are not striking. D. M. S. Watson concluded that such forms used their rowing propulsion to catch their prey, without the turning and twisting that the presumably slow, long-necked forms had adopted.

Rowing, as a subsidiary source of movement, is very widespread, but only occasionally has it been a successful solution to the primary problems of progress through water among the vertebrates. Two of the three families of seals, Odobenidae and Otariidae, depend to a major extent on the sculling action of the paired appendages for propulsion. The walruses receive the greatest thrust from the hind limbs, using the forelimbs for paddling and steering. Both are somewhat modified, with the external expression of free digits lost. In otariids the principal sculling is with the forelimbs, and a side-to-side movement of the hind flippers adds to the force. In both groups, although the limbs are rather strongly modified as organs of water propulsion, they are usable for locomotion on land. In the slow swimming of the manatees and dugongs, in which only the pectoral appendage remains, the flipper plays an important role, but this is augmented by the flattened tail, which is more whalelike in action. Except at very slow speeds, this is a mixed action.

The phocids, among the seals, have developed fast swimming, with speeds up to 10 miles per hour. This has been done by an adaptation not closely approached by other aquatic animals, although somewhat like that in aquatic birds and flying reptiles. The hind limbs are turned backward, are short shanked, and have the phalangeal portion formed into a large fish-tail-like organ. The broad surfaces of these limbs are used in propulsion. The body is flexed during the action, in the sacral region. As it is flexed to the right, the left limb is extended and the right is flexed. When the flexure is altered to the left, the right limb is extended and the left is flexed. In slow swimming the two may both be extended, and in fast starts both may be flexed and then extended. The limbs are sharply modified in connection with this action. The forelimbs, also somewhat modified, are flippers that act in steering and are used in sculling during this process but do not provide any material thrust in locomotion.

The pattern of alternating thrusts by the hind limbs, of course, is the common one in aquatic birds. In some, such as the extinct *Hesperornis*, the structure suggests an action somewhat like that of the phocid seals, although the structure is considerably different. The hind limbs of pterosaurs were also turned back and probably were somewhat analogous in their actions. A striking adaptation is seen in the penguins, whose wings have become modified as propulsive oars for swimming, acting together with the hind limbs. These birds have taken up the rather rare use of both pairs of limbs in swimming, in a way analogous to that of freshwater chelonians and plesiosaurs. This has been attained, however, by modification of organs previously adapted for flight. In both penguins and turtles, of course, this is not the sole locomotor function of the paired appendages, because one or both pairs are also used for moving over land. Multiple uses necessarily have accompanied all invasions of aquatic habitats by tetrapods, except as the appendages may have been lost prior to this move.

HYDROSTATIC ADJUSTMENT.   Animals that lack gas bladders or lungs tend to be somewhat heavier than the water in which they live. When at rest, they tend to drop toward the bottom, either vertically or with a gliding motion when the appropriate paired fins or similar gliding surfaces are present. In such fishes adaptations of the drive mechanisms, such as the epibatic tail, serve to maintain their position in the water while swimming. Tetrapods, whose lungs are of considerable volume, tend to be lighter than water. In various of these among the reptiles (e.g., *Mesosaurus*, placodonts), birds (e.g., penguins), and mammals (e.g., dugongs and manatees) some of the bones are enlarged and very densely ossified, a condition termed pachyostosis This is generally considered to function to reduce the buoyancy for creatures living in rather shallow waters and feeding underwater. It tends not to be developed in the more actively swimming reptiles and mammals.

It is largely within the actinopterygian fishes that adjustments of position by hydrostatic changes are such that they can be considered as having a role in propulsion. In most such fishes the gas bladder has the capacity to alter the specific gravity of the organism, resulting in a rise or drop in the water. This function is but one of several of the gas bladder, which also may participate in respiration and in sound reception and production. The hydrostatic function, though apparently important, is at best a very minor part of the total drive mechanism.

JET PROPULSION. The importance of jet propulsion in the fishes as a contributor to locomotion has been a matter of debate. To even its strongest advocates it is nothing more than an occasional and subsidiary action. Breder (1926) gave a rather full account of his interpretation of the importance of jet action developed by the expulsion of water from the gills. He noted that it could have considerable force and that in some fishes it appeared to act rather effectively. This action was interpreted to be in part that of producing forward motion and more importantly in the reduction of drag forces during locomotion. Few students of fish locomotion have considered that this source of thrust and its subsidiary actions are of any great significance in fish locomotion. In no other groups of aquatic vertebrates can it be considered a factor at all. In invertebrates, of course, quite the opposite is true and jet propulsion is a principal source of thrust in various groups, the best known of which are cephalopods and some of the pelecypods.

2 The challenge of life in water, II

General Matters

Vertebrates are active animals and generally require relatively large amounts of food. A major part of their active life is devoted to obtaining sustenance, and many of their adaptations are closely related to food gathering and associated patterns of behavior. The majority of vertebrates, and aquatic vertebrates in particular, are rather unselective in feeding, although many have a preferred food that they will select if it is available. Seasonal differences in feeding as well as changes in feeding practices with age also occur in various groups of vertebrates.

All body systems are involved to some extent in feeding activities, but those most directly associated relate to locomotion and to acts of procuring, eating, and digesting foodstuffs. The hard structures of the locomotor systems and of the mouth and pharynx are the main bases for interpretation of food habits of extinct animals, but general body form and occasional indications of the food itself, in the digestive tracts and in coprolites, also are of some importance.

Both locomotor and feeding mechanisms often appear to be very highly specialized to perform particular acts of food procurement and ingestion. It would seem at first glance, that they would provide an excellent basis for interpretation of the feeding habits of extinct animals. Many such interpretations have been made and some of them undoubtedly have been correct, at least in general terms. Studies of living aquatic animals, however, cast severe doubts on the validity of specific interpretations and show the chances of major errors to be quite high if structure is taken as the basis for any but the very broadest conclusions. Structure and function, in this area of vertebrate behavior, as in some others, are often not closely related.

Structures do, on the other hand, give rather precise evidence on the limits of possible behavioral patterns. These may be evident in two somewhat distinct types of features: those that merely restrict the foodstuffs that can be used, and those that relate especially to one particular kind of food. The gape of the mouth, limiting the size of food, or the speed of the animal, limiting what can be caught, are of the first sort. The holding and cutting teeth of the upper and lower jaws of many predacious sharks, which serve to cut out pieces of flesh, the pavement teeth of various aquatic vertebrates, suctorial mouths, and highly developed gill rakers are examples of the second type, indicating special types of food within the general potential that the animal can handle. The ability is not necessarily an indication that the potential was realized. Taken together, the permissive and the special features show the potential and suggest possible preferred foodstuffs and ways of feeding. Only the generally limiting features are definitive, because animals can and often do eat a very wide range of foods within their general capabilities.

In some instances the features of the feeding system are so tightly meshed into an integrated complex that the potential limits of feeding can be rather clearly ascertained from them. In the case of the baleen whale the large gape, the absence of teeth, the restricted gullet, and the whalebone filter system seem to point directly to an odd sort of filter feeding. Whether or not this would be deduced were this animal extinct, however, is difficult to know. Comparable interpretations of other dietary habits have been made for forms known only from the fossil record, but it cannot be known whether or not they actually fall wide of the mark. For example, the apparently logical interpretation that pavement-toothed rays feed on hard-shelled mollusks is widely made. This is suggested for ancient pavement-toothed sharks as well. Rays do eat mollusks, but they also eat many other kinds of food, and mollusks may be a relatively small part of the diet, depending, it would seem, on the abundance and accessibility of various foodstuffs, the time of year, and very likely on the individual.

In any event it is not the specialized cases but the great majority of aquatic animals, especially the fishes, that do not show special features—such as tooth plates, sensitive probing rostra, or highly developed gill rakers—that cause the major difficulties in assessing diets from feeding mechanisms. These are, for the most part, the animals that are on major evolutionary lines, not the specialized offshoots that frequently reach dead ends. If all of the related parts, including the anatomy of the digestive system, are known, judgments can be narrowed, but only moderately. For fossils interpretations must be very broadly drawn. It thus becomes very difficult or impossible to place particular aquatic creatures in their proper places in food webs and to reasonably evaluate the significance of their structure in the evolution of fish groups.

The latter is a primary goal of paleozoology, and much of the consideration of the evolution of aquatic animals, and terrestrial as well, has been based on analyses of the locomotor and feeding systems. Usually interpretations of major thresholds that are thought to have led to new levels of organization have involved these two systems. With respect to feeding, in view of what is known of form and diet in many fish, this poses a serious problem.

Schaeffer and Rosen (1961) point out quite clearly that the teleost radiation followed the acquisition of a particular pattern of the mechanics of the skulls and jaws, along with associated muscles. Greater mobility of feeding mechanisms and greater variety of activities became possible (see Chapter 3 in this section). The potentials of feeding were thus increased, and these became very effective in view of a shift in locomotor potentials attained within the holostean grade level.

Now it seems evident that the capacity for rapid and mobile locomotion and development of a large gape of mouth, coordinated with greatly increased mobility of parts and strength of bite, would be advantageous in feeding and could, as it apparently did, set the stage for the teleost radiation. Yet the development of special features requires selective processes of a specific kind, and here, at least so it seems, it is not possible to point to any specific adaptation. What appears to be involved is a totally integrated system that in general is more effective, but not for any one particular mode of behavior.

Such evolution, if such is the case, can be brought into the mutation-selection fold only by a very broad concept of system integration, perhaps with some particular modification, which itself may be under some selective pressure, as a center around which the many changes are organized. As far as can be told, however, although such a major shift as that shown by Schaeffer and Rosen can set the stage for a new radiation in which diet is important, it cannot be related to a specific adaptation either on the evidence of the structures or in light of the lack of specificity in feeding of most fishes.

This is one important aspect of the feeding problem. A second aspect of the problem of adaptation of feeding mechanisms and evolution of aquatic animals relates to the apparent specificity of structures to a way of life. A large number of fishes fall into a general category of predatory feeders, and this is a common mode of the tetrapod aquatic animals. Within this range it is only in rather unusual cases that there is any high selectivity. Foodstuffs of creatures that are quite different are frequently much the same; conversely, those of creatures that seem adaptively quite similar may be quite different. One creature may feed on a variety of foodstuffs either at the same time or at different times.

Yet there do exist recognizable types of feeding mechanisms, and many different major groups of aquatic animals have developed very similar types. The problem is to relate these to principal modes of feeding, and moderns suggest that this cannot be done with any assurance. It may be thought, perhaps rightly, that the capacity to feed in one way or another, even though the mode is not even a principal way of feeding, has a selective advantage. Even if this is the case, and it is difficult to substantiate, it is of relatively little help in understanding the feeding behavior of a particular animal from its structure alone. As Springer (1961) has pointed out, in at least some instances and probably in many, the foodstuffs taken by modes other than those suggested by the particular specializations may be enough to be critical to survival.

Because of these problems, both in interpretation of habits of feeding and in explanations of their evolutionary meanings, analyses of adaptations to water life must be made with caution. On the other hand, they are of great importance in the total picture of adaptation to water.

Feeding Habits and Adaptation

No clear-cut classification of feeding in water comparable to that generally used for terrestrial animals is possible, largely because of the catholic feeding habits of aquatic vertebrates noted in the preceding section. It is possible to recognize a number of principal modes of feeding and to single out the sorts of structures that are generally associated with them. Vertebrates tend to group rather broadly into these categories, but almost all cross over to some extent between two or more. A breakdown on the basis of place, kinds of animals, and mode of obtaining food gives the following utilitarian list:

Bottom feeders, on animals, plants, and organic matter.
Feeders on small, nonplanktonic animals (vertebrates and invertebrates).
Predatory feeders on (relatively) large prey (vertebrates and invertebrates).
Plankton feeders (phytoplankton and animals).
Browsing feeders (plants, corals, etc.).
Parasitic feeders.

Inevitably there is overlap, as between the second and third categories, but in general the behavior patterns are sufficiently different that distinction can be made. Even though one animal may cross between categories, its behavior in pursuing each tends to be somewhat different.

Bottom Feeders

A wide variety of foodstuffs exists in the benthic realm, and many vertebrates have become adapted to life on the bottom and to feeding on the nutrients that it supplies. A particular form of body structure is a key to this adaptation, related, however, to locomotion more than to feeding. Most bottom feeders are somewhat depressed, slow swimmers. Food consists of organic materials in the mud, of invertebrates in the infauna, and of animals and plants of the epifauna. Mollusks, worms of various sorts, and arthropods are important food elements. Among these are the hard-shelled mollusks to which many of the bottom feeders appear to be specially adapted. To the extent that their structures permit, bottom feeders tend to range widely over this array of foodstuffs, with relatively little discrimination.

MUD FEEDERS.   Some fishes, much like angleworms, take in mud and pass it through their systems, retaining the nutrients. Suctorial mouths and sensory devices for testing the mud, as found in the various sturgeons, are characteristic specializations. How common this habit has been in the past is uncertain. Mud feeders existed, and at least some of these, as indicated by the antiarchs in which mud-filled intestines have been found, did pass the mud through their systems. For the most part this cannot be determined.

Many mud feeders, however, are fully or partially filter feeders, in that the mud is taken in and expelled through the gills, which, by the use of gill rakers and mucous nets, take out the organic portions. The combination of a suctorial mouth or one with very little biting capacity and gills with rakers is suggestive of this habit. Filter feeding on both mud and food particles suspended in water probably played an important role in the origin of vertebrates. In various nonvertebrate chordates, including tunicates, water is passed over a gill system by the use of cilia and the food is extracted. The primary function of the gills is food gathering, not respiration. This likely was the feeding mechanism in the protochordates.

Among the vertebrates today the ammocete larva of the lamprey has a filter feeding system that may somewhat resemble that of the vertebrate ancestors. In it, however, a muscular pump is present for water movement rather than cilia. Many forms of filter feeding have been developed in other vertebrates, but these have taken on a variety of forms and many are not associated with bottom living.

Among the agnathans filter feeding has been suggested as the mode of feeding of each of the major groups. The cephalaspids appear to have had suctorial mouths (Figure 46) and gills that were appropriate for filter feeding,

Figure 100.   Lateral view of the skull of the extinct aquatic nectridean amphibian *Diploceraspis*. The apparent great elongation results in large part from the very extensive "tabular horn," which passes far back of the occiput. The short jaws and simple teeth, characteristic of this type of amphibian, have suggested bottom feeding, possibly mud grubbing or picking up small particles from the bottom. (After Beerbower, 1963.)

although the presence of a sievelike system is not known. Denison (1961) and Heintz (1958) have argued convincingly for this feeding habit. If this is correct, it strengthens the concept of filter feeding as a primitive way of feeding in the vertebrates. Heintz and others have also suggested this as a mechanism in the anaspids, but this has not been widely accepted. The hypobatic tail and fusiform body do not argue strongly that this was a bottom-feeding animal.

Heterostraci, in cases in which the structure is well known, had somewhat better developed, more mobile, and somewhat denticulated tooth plates. The mouth does not appear to be strictly suctorial, but rather suited for nibbling and picking up small bits of organic material from the bottom. Presumably, if this were done, considerable mud would have been taken in and some filtering action may have occurred. The gills, however, open into an orifice with a single outlet, suggesting some sort of specialization that may not have been conducive to mud filtering.

Many bottom feeders probably were to some extent mud feeders. Flat arthrodires, such as *Arctolepis* and the antiarchs, as mentioned, fall in this category. In all major groups of fishes bottom feeders that existed partly as scavengers developed, and these were partial mudfeeders. Various bottom-living sharks, skates, and extinct pavement-toothed forms, and Osteichthyes, including some lungfish, catfish, carp, and flounders, are examples. Very probably some of the extinct amphibians, such as the nectrideans *Diploceraspis* (Figure 100) and *Diplocaulus*, fit the picture. Their jaws are short and scooplike, the teeth small, usually little worn, and the bite was probably not strong.

Among other tetrapods many of the pavement-toothed forms, usually considered mollusk feeders, may have ingested considerable amount of mud, and a few have special feeding patterns involving the intake of sediments. Of these, *Eschrichtius gibbosus*, the gray whale, is of special interest. It stirs up the bottom with its snout, in shallow water, and feeds on the amphipods, with of course anything else that comes along. Feeding is not carried out

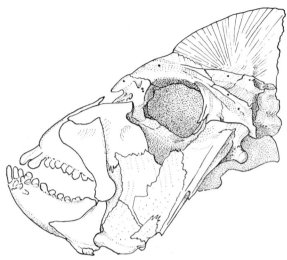

Figure 101.   The skull of the teleost sheeps head, showing extreme modifications related to prying and crushing actions of the jaws and teeth. These fishes are capable of tearing mollusks from their moorings and of crushing coral. (After W. K. Gregory, 1933.)

during migration and breeding. This raises an interesting point, pertinent to many cases in which feeding and breeding grounds are different. Undoubtedly fossilization of such creatures can occur in places other than their feeding grounds. They may thus become associated with other animals and with plants in what might appear to be a relationship interpretable in terms of a food web. This would, of course, be completely erroneous, but with little or no information available to indicate the actual circumstances.

"MOLLUSK FEEDING." Many large living vertebrates, both fish and tetrapods, having crushing dentitions accompanying very strong jaws. They tend to be bottom living, but active pelagic forms also show this development. Some elasmobranchs, skates, rays, ptyctodont arthrodires, lungfishes, reptiles (e.g., the placodonts), and mammals, walruses, manatees, and dugongs) are examples. In addition some other types—for example, such teleosts as the sheepshead (Figure 101), some ichthyosaurs, and some mosasaurs among reptiles—have developed heavy, blunt teeth and very strong jaws. Anterior teeth in many are equipped to tear shellfish from the objects to which they are attached. All such forms have the potential of feeding on hard-shelled mollusks, pelecypods in particular, and other shellfish. The existence of such dentition and jaws is usually taken as presumptive evidence that these animals did in fact feed on mollusks. No doubt this is in part the case.

There are, however, problems. Dugongs and manatees are almost exclusively vegetarians. Rays often turn up with their stomachs filled completely with soft-bodied food, such as worms. The beluga porpoise, on the other hand, when young feeds mostly on bottom-living crustaceans but shows no evidence of this habit in its structures. The capability to eat mollusks, then, does not prove that mollusks were in fact eaten or that they made up more than a small portion of the diet if they were eaten. On the other hand, the lack of dentition suited to crush mollusks is not a firm basis for concluding that they were not consumed.

Many forms with crushing dentition undoubtedly did feed on hard-shelled invertebrates. If moderns are a guide, this was not an exclusive diet and may not have been the principal food source. Yet, it would seem, this capacity must have been of value in survival because it has been developed repeatedly in many widely separated groups. Just how this took place cannot be determined for extinct animals and, if determinable at all among moderns, would require extended studies of populations over long periods of time, something almost beyond practical reach.

Feeders on Small Nektonic, Pelagic Animals

The great majority of aquatic vertebrates find much of their food in this general part of the aquatic biota. Many kinds of small invertebrates and small fishes are the rather indiscriminate bill of fare. No sharp line between bottom-living creatures on the one hand and planktonic ones on the other can be drawn, and, of course, both bottom feeders and plankton feeders generally get part of their food from this somewhat intermediate zone. Moderate mobility is necessary, to move with the food supply. The gape of the mouth may be small to large, and teeth need not be well developed. The great majority of such feeders tend to have fusiform to moderately deep bodies. Many of them appear to have special features of the feeding and locomotor systems, suitable more for one particular source of food, and some select this food if it is available.

Small fishes and mobile arthropods, crustaceans, and insects in particular are a main source of food. Insects in various stages of development are most important in fresh waters, and crustaceans predominate in marine waters. Food is generally swallowed whole.

Except for a generally permissive habitus, there is little in the structure of the animals that feed on these small creatures to indicate their food habits. The presence of sharp, cutting teeth does not necessarily rule a creature out of this mode, even when it is the dominant mode. Springer (1961) has noted that the sandbar shark, *Eulmania milberti,* feeds on small fish, swallowed whole and not cut, although it has strong teeth, and the silky shark feeds predominantly on fingerlings.

Predators on Moderate to Large Animals

This mode of feeding was adopted by many fishes, including large arthrodires, various elasmobranchs, large crossopterygians, several kinds of actinopterygians, possibly some amphibians, aquatic reptiles, and, among the mammals, otters, seals, and cetaceans. Although the mode is dominant, most also obtain food from other sources. Food is taken in two ways: (a) by grasping, positioning, and swallowing whole and (b) by cutting or tearing pieces from the prey. The latter may be done by small fish, such as the piraña, but the former is predominantly a way of feeding of relatively large creatures. Gradation of predation into the feeding on small animals, discussed in the last section, necessarily occurs, and no hard-and-fast limits can be drawn.

Over the whole range of animals in this feeding category, certain similarities of structure occur. Most, with the exceptions noted above, are fairly large, a foot or more in length. Size, of course, varies in relationship to the size of the prey. The body form and locomotor system are suitable for moderate to rapid swimming, and most of the fast swimmers belong in this category. An exception is found among the plesiosaurs, which, as already discussed, evolved long-necked forms with very special locomotion thought to be related to predation on fish and cephalopods. Jaws tend to be strong, and teeth are strong and commonly slanted somewhat forward in the mouth. The gape is large, often greatly increased by the development of long, tooth-bearing beaks. Teeth are sharp, suitable for grasping and holding prey and, in some, for cutting.

Many variations on these general features occur, but as a rule evidence of a predatory mode of life is shown fairly clearly even in extinct aquatic animals. Not all creatures with the "distinctive" features are predators that grasp their prey, as noted for the sandbar shark in the preceding section. The foodstuffs known to be taken by modern predators vary widely to include fish, cephalopods, birds, and mammals as well as a wide range of smaller animals that are swallowed whole in the manner of feeding discussed in the preceding section. The killer whale gives an example of the variety of foods; stomach contents have yielded seals, dolphins, porpoises, other cetaceans, diving birds, shellfish, large bony fish, and squids. Porpoises variously feed on cuttlefish, prawns, and fish. River dolphins feed on freshwater shrimp, eel-like catfish, herring, and other fish, depending largely on what is available.

Predacious sharks of the present conform to this general pattern, in particular to the cutting and tearing mode. The jaw support is hyostylic, with the jaws supported by the hyomandibular. Teeth generally include a cutting array in the upper or lower jaw and a holding array in the other. After the bite, one set serves to hold while the other, accompanied by twisting of the body utilizing the pectoral fins, acts to saw through flesh. More primitive sharks, mostly with amphistylic jaw suspension involving articulation of the

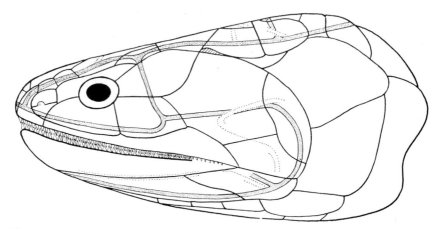

Figure 102.   The skull and jaws of a rhipidistian fish, *Eusthenopteron*, illustrating the structures in crossopterygians related to their presumed carnivorous habits. In addition to the lateral teeth, strong, tusklike labyrinthine teeth are carried on the palate and on the inner margins of the lower jaws. (After Jarvik, 1944.)

jaw and braincase, appear to have conformed to the pattern of holding and swallowing whole. In the Paleozoic such sharks as *Cladoselache, Xenacanthus,* and *Ctenacanthus* had teeth with high, sharp cusps, suitable for holding but not for cutting. Fins were less mobile. Hybodonts of the Mesozoic had a mixed dentition, like that of the Port Jackson shark, with the anterior teeth suitable for holding and the posterior ones for crushing. They could have been both predators and shellfish eaters as the Port Jackson shark of today actually is.

Rhipidistian crossopterygians, although only moderate in size, have dentitions and jaws (Figure 102) that suggest predatory habits, with fish as the probable main prey. Among actinopterygians many of the large active fishes are predators, but again, some capable of this habit do not follow it. Even within small groups, such as that of the tunny, there are marked feeding differences between species that are not revealed by morphological differences.

The long jaws, strong teeth, and probably rapid locomotion of some of the Triassic trematosaur amphibians strongly suggest this way of life (Figure 103). Earlier long-snouted amphibians, cricotids and rhachitomes, such as archegosaurs, though not completely aquatic, probably were fish predators. Among aquatic and semiaquatic reptiles predation appears to be a primary mode, expressed very early in *Mesosaurus* and later in a wide variety of reptiles, including some of the crocodiles, phytosaurs, ichthyosaurs, plesiosaurs, mosasaurs, snakes, and at least some of the nothosaurs. Few of these were adapted to cutting and tearing, although this is found in some of the partially aquatic crocodiles. Most, it would seem, grasped their prey and swallowed it

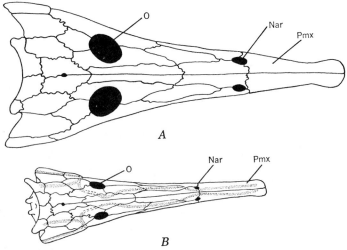

Figure 103. Skulls of two amphibians with very elongated snouts and nares set well posterior. The elongation and simple, sharp, somewhat forward-directed marginal teeth suggest a habit of fish eating. The posterior position of the nares appears to result from the elongation of the premaxilla. *A*, *Platyops*, a Permian rhachitomous amphibian; *B*, *Aphaneramma*, a lower Triassic trematosaur amphibian. Not to scale. Nar, external naris; O, orbit; Pmx, premaxilla. (Mainly after Romer, 1947*a*.)

whole. There is not much evidence on the extent of variability of feeding among the fossil forms. Stomach contents of ichthyosaurs indicate belemnite cephalopods and fish as important elements. Plesiosaurs appear to have had well-developed gastroliths, which may have given them the capacity to feed on a very wide range of foods, including both vertebrates and hard-shelled invertebrates.

Many types of predators occur among the aquatic mammals, ranging through otters and seals, basically fish eaters, to the many cetaceans. In most of these, as in reptiles, the primary habit was to take the prey whole and swallow it. Otters, however, like some of their close terrestrial relatives, do use cutting and tearing as a common means of reducing food to suitable size.

Plankton Feeders

The plankton of the seas and lakes today provides a rich source of food to which vertebrates of a number of classes have turned. The phytoplankton of the seas accounts for at least half and probably a much larger fraction of food production from inorganic sources at the present time. At other times in the past, when continents were less high and covered by vast stretches of epicontinental seas, it probably accounted for a very large part. The diatoms of the present seas form the base of the oceanic food web, along with other phytoplankton, such as algae, and shallow-water plants, which make a relatively small contribution. Many animals that feed in the water depend directly on the phytoplankton, but vertebrates, with some important exceptions, do not. The vertebrate plankton feeders subsist in large part on the zooplankton, with phytoplankton forming a somewhat indeterminate supplement.

Essentially no generalizations can be made about the morphological characteristics of vertebrate plankton feeders. Filter feeding is one of the mechanisms. In fishes it tends to be accompanied by small teeth or no teeth and gill rakers. But these structures alone, although they may suggest filter feeding, do not indicate plankton feeding. In the absence of definite specializations that suggest bottom feeding, filter feeders may well be plankton feeders. *Acanthodes*, a somewhat fusiform-bodied acanthodian, lacks teeth and has excellent gill rakers. Very likely it was a planktonic filter feeder. The modern shark *Rincodon typus*, as mentioned, is a plankton feeder with poorly developed teeth and a good filter system. Among actinopterygians, the mackerel is an example of a common plankton feeder, but there is little in its structure that would lead to such a conclusion were it preserved as a fossil fish. The plankton feeding of the baleen whale, of course, is well known.

Plankton feeding may have been a common way of life among aquatic vertebrates, especially among the fishes. The structures in general are not sufficient to determine which of the fossil fishes may have had this habit and which may not among those that seem to have the potential.

Herbivores and Browsers

Vertebrates that feed on phytoplankton directly are herbivores, but on the basis of their adaptations they are better placed with other plankton feeders. There is much resemblance between some of the browsing plant eaters and browsing coral and reef feeders among the fishes. They may be considered under the same heading as far as feeding habits and adaptations are concerned. On the whole there are relatively few strict herbivores among aquatic animals, but many, particularly fish, make some use of vegetable foods.

Carp are herbivorous fishes, but they eat a great variety of materials, weeds, organic mud, and a wide range of animal foods. They have been called the pigs of the fish world. Lack of teeth and a leathery mouth, such as they have, are often associated with plant eating in fishes not highly specialized for one or another type of herbivorous diet. These fishes are usually, but not always, rather slow moving, sluggish, and with some tendencies toward bottom living. Examples are roach, dace, carp, and mullet. Such fish, mainly cyprinids, are characterized by prominent pharyngeal teeth, used for crushing food and swallowing.

As among land animals, it is the special forms of herbivores that are easily recognized among aquatic animals. "Nibbling" fishes in which the teeth form platelike structures with various kinds of cutting edges may be plant or coral feeders. This is a form that has developed repeatedly among actinopterygians, being found at all organization levels. Among them it is a tooth pattern that is commonly associated with deep bodies and high mobility.

Among the tetrapods the development of aquatic herbivores has been rather unusual. Some of the sea-going turtles, with their horny, toothless beaks, very strong jaws, and rather slow swimming, feed largely on seaweed, but they also eat animal matter as it may be available. Among the therapsids an aquatic form, *Lystrosaurus*, of the Triassic (Figure 104) is marked by a strong, toothless skull, except for tusks in some (males?), and bone structure indicating a horny beak. It seems logical to assume that it was a herbivore, quite possibly an algal feeder. Among mammals the manatee and dugong represent highly developed herbivores. Both have strongly modified skulls (Figure 104), with the jaws heavy, the anterior parts specialized, and the teeth strong and flattened. It is, of course, known that they feed on plants—dugongs on algae and manatees on freshwater plants. Were these known only as fossils, it might well be concluded that they were mollusk feeders. The desmostylids—very odd, partially aquatic relatives—have somewhat similar modifications and also probably were plant feeders (Figure 104). Without the modern analogs, however, it would once again be difficult to reach a sound conclusion on the habits of these creatures.

Parasitic Feeders

Few fish and no tetrapods are strictly parasitic. Among moderns only the lamprey is a parasite. It is highly specialized for this mode of feeding, with a round, suctorial mouth, rasping tongue, and separate gill chamber into which water is taken and then expelled, without passing through the mouth. The hagfish, sometimes called a parasite, is more strictly a scavenger.

Stensiö (1958) has made a strong argument that the anaspids were parasitic feeders. As noted under other categories, these little fish have been placed in essentially all the feeding categories that their size permits. The mouth of the anaspid is circular and apparently suctorial. The gill chamber somewhat resembles that of the lamprey and could have been similar in function. There is sufficient space for a rasping tongue, but no direct evidence that one was present. The suggestion is then not out of the question, but, like the others, it cannot be backed up by solid evidence.

Beyond these two forms—the lamprey and the ancient ancestor, the anaspid, if indeed it was parasitic—there is little in the vertebrates to suggest that parasitic feeding was a common habit. It has been suggested that the little-known fish *Palaeospondylus* was a parasite, but this rests on very slender evidence.

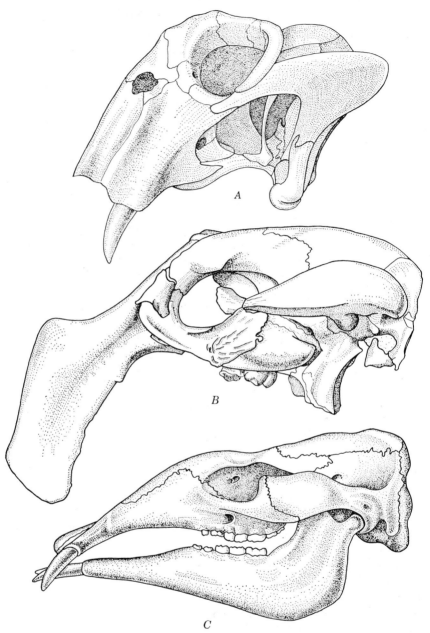

Figure 104. Skulls of three aquatic (or semiaquatic) tetrapods, showing similarities attained by convergent adaptation to water. *A, Lystrosaurus*, a therapsid reptile; *B, Halitherium*, an extinct sirenian; *C, Desmostylus*, a marine, ungulatelike animal of uncertain affinities. (*A* after Broom, 1933; *B* after Reinhart, 1959; *C* after Romer, 1966*a*, after Gregory, 1951.)

SENSING THE ENVIRONMENT

Fishes living today display the full range of types of sensory perception that occur throughout the vertebrates. In the course of evolution few things that are truly new have developed, but some of the sensory functions have been modified and the receptor systems have undergone commensurate changes. These are related directly to modifications of the soft tissues, particularly to parts of the nervous receptors and brain, but in part they are recorded in the surrounding hard structures. The principal organs of special sense—eyes, ears, and olfactory organs—are housed in capsules, which are usually formed of cartilage and bone. Thus some record of their general features may exist in fossilized animals. The lateral-line system similarly is associated with bone and scales in many organisms. The other sensory receptors, such as those involving touch, taste, and temperature, for the most part have not left interpretable records.

All of the organs of special senses were present in the most primitive known vertebrates, although their functions probably were less developed than in many more advanced forms. Evolution of the organs in the fishes appears to have been toward greater effectiveness in orientation based on sound and in sensing the chemical aspects of the water. As far as can be told sensitivity to light was modified principally in accommodation to particular aquatic circumstances rather than in any strictly progressive way.

The transition to land was marked by notable modifications, especially in the organs associated with the acoustico–facialis nerve complex. Some modification of olfaction took place as well as adjustments in the eyes. Beyond the relatively minor modifications necessitated by the different properties of the gaseous and aquatic media, the nasal and optic structures show their principal changes during the evolution of tetrapods as they invaded new environments.

One of the environments invaded by tetrapods, of course, was the water. Adaptations to water necessarily were based on a morphological and behavioral background that was very different from those of fishes, whose ancestry was directly aquatic. In addition the backgrounds were different in the major groups that returned to water—the amphibians, reptiles, birds, and mammals. In all but the amphidians the principal problems related to the loss of the lateral-line system and the presence of a sound-conducting middle ear. The structures related to reception of sound vibrations and complementary structures related to sound transmission underwent very significant modifications in the adaptation of tetrapods to aquatic life, but these were realized only in the most highly adapted aquatic tetrapods.

The sensory systems of aquatic vertebrates to some extent form an inter-related functional array of organs involved in the coordination of behavior. In the various adaptations to water, however, one or another of the receptor systems tends to predominate, with the others playing somewhat subsidiary roles. It is convenient to treat them singly. Most amenable to analysis in the fossil record are the structures related to acoustics and balance, and they are emphasized in the discussions that follow.

The Hearing-and-Balance System

The labyrinth of the inner ear, the lateral-line system, the gas bladder, the middle ear, the ampullae of Lorenzini, the sound-producing structures, and the nerves that supply them, principally those of the acoustico–facialis system, are the primary components of this system. Of these, only the labyrinth is present in all aquatic vertebrates. The lateral-line system is present in all fishes, including agnathans, and in amphibians. The gas bladder is present in many Osteichthyes. In primitive bony fishes this structure was represented by its forerunner, a lung, and this persisted in dipnoans and some cros-sopterygians. Lungs are known to have occurred very early, but little is known of their possible relationships to acoustics and balance in the nonosteichthyan fishes. Ampullae of Lorenzini occur in Chondrichthyes only. Middle-ear structures developed with the advent of land-living tetrapods and occur only in these and in their immediate predecessors, the rhipidistians. Sound pro-duction takes place in many osteichthyan fishes and in aquatic mammals. Presumably this was true for strictly aquatic reptiles as well, for it is in par-tially aquatic forms, but there is no direct evidence on the matter.

Primitive Vertebrates

The most primitive known vertebrates possess a labyrinth. In hagfishes, with the simplest known condition, it is rudimentary, with a single semicircu-lar canal and a more or less undifferentiated ventral sac. Lampreys have two semicircular canals but lack a horizontal canal. The sacculus–utriculus complex is somewhat better defined. Extinct agnathans, both the osteo-stracans and heterostracans, have at least two semicircular canals. In the latter no circumlabyrinthine ossification took place and there is a possibility that an unrecorded third, horizontal, canal was present.

Throughout the living aquatic vertebrates the labyrinth operates as a sensor of gravity and of angular acceleration and as a regulator of muscle tone. In most, possibly all, it is a sensor of sound waves. The lateral-line system receives low-frequency vibrations. Very extensive studies of bio-acoustics in recent years have crystallized an important distinction between the reception of the lateral-line system and the inner ear. G. G. Harris (1964), G. G. Harris and Van Bergeijk (1962), Dijkgraaf (1963), and Van Bergeijk

(1966) have shown that the lateral-line system responds to *water motion*, not to pressure waves—that is, sound waves in the conventional sense. A sound source in water produces *water motions* in its "near field" and *pressure waves*, which alone reach its "far field." The near field extends about one-sixth of a wavelength from the sound source and thus, especially in view of the greater wavelength of waves in water at a given frequency as compared with those in air, covers a range significant to the receiver. The low frequencies of the "near field" are received by the lateral-line system directly, recorded by movement of its hairs, and are directional. The far-field waves are converted into "water motions" by various body structures and are recorded by the special sensory hairs of the sacculoutricular parts of the inner ear.

The extinct agnathans probably represent a stage through which all other vertebrates passed. The functions of the labyrinth were much the same as in other vertebrates, except that they probably did not hear far-field sounds. If transmission of such sound waves did occur, it was through the body tissues without any special organs for reception or conduction. Fishes, on the whole, appear to be "transparent" to sound waves, but the evidence that some sharks do respond to far-field sound indicates that this is not completely the case and that some degree of interruption, probably by the skeleton, does occur.

The lateral-line system in the Osteostraci is associated with the pore-canal system (Figures 20 and 40), as pointed out in detail by Denison (1947). Heterostraci appear to have this, although less well developed, and it occurs as well in some acanthodians, lungfishes, and crossopterygians. Denison (1947, 1966) has suggested that this pore-canal system may represent the primitive receptor system of the acoustico–facialis nerve complex and that the lateral-line system and the inner ear may have been derived from it. Both the lateral-line system and the inner ear arise embryologically from the ecto-dermal dorsal–lateral placodes. Both are characterized by similar sensory hairs. The pore-canal system is structurally closely akin to the lateral-line system, and the two, in cases where both occur, appear to form a single system. Differentiation, if this is the actual course of events, had already taken place in the most primitive known vertebrates. The lateral lines are derived from the lateral placodes and supplied by nerves VII, IX, and X. Denison has noted that the pore-canal system, since it is dispersed in the ectoderm of the whole body, might be presumed to have derived from the neural crest. This poses a problem of just what the courses of evolutionary derivation may have been.

Very likely the lateral-line and pore-canal systems were the most effective and dominant receptors in the agnathans. The eyes were small, especially in the heterostracans. The olfactory organs opened into the mouth in hetero-stracans; in the osteostracans, as in lampreys, they were housed in a single, dorsal capsule. A pineal eye was present, but it was small. Hearing in agnathans probably was restricted to water motions in the near-field range. Denison has suggested that possibly the motions of the bodies of these creatures set up low-frequency disturbances that were reflected back from nearby objects and received, thus producing a very crude echolocation system.

Cephalaspids, in addition to pore canals and lateral lines, possessed two or three areas on the head shield that were covered by small plates. They appear to have been genetically related to the lateral-line and pore-canal systems and were supplied by very large nerves from the acoustico–facialis complex (Figure 43). Their function is not clearly understood. It has been suggested (Stensiö) that they were electrical organs, somewhat like those of electric eels, but Westoll (1941a) has made a case for their being special sensory receptors. Nothing like them exists among living animals, so this question must remain up in the air.

The functions of the lateral-line systems and labyrinths outlined for agnathans occur throughout the fishes and the amphibians. How much evolutionary change there may have been in their sensory structure cannot be told. The great resemblances of the receptor mechanisms in the lateral-line system and the inner ear throughout living forms, however, suggest that they have not been materially modified. Elasmobranchs, in addition, have the ampullae of Lorenzini, as noted. These lie in association with the lateral lines (Figure 40) and have similar innervation. Various suggestions have been made concerning their functions. They appear to be sensitive to temperature changes, because the rate of transmission of pulses has been shown to drop as temperature is increased (Sand, 1938). Pressure responses have also been detected. The structures do not occur, as far as known, in other fishes, nor have they been detected in the fossil record.

Evolution of Hearing in Water

Hearing, as considered in this section, is restricted to reception of pressure waves and does not include the near-field hearing discussed in the preceding section. Water is an excellent transmitter of pressure waves, which have a velocity exceeding 5000 feet per second in seawater. Wavelengths are about five times as long as for the same frequencies in air. More input is required to produce the waves, but the sound goes farther and faster. Modern fishes, teleosts, generally hear in a range centering around 500 to 600 hertz but are sensitive well below this range and up to 1500 to 2000 hertz.

For hearing to occur some sort of acoustical interruption must take place, transforming the pressure waves to "water motion" to which the sensory receptors of the inner ear can respond. The vibrations so induced must reach the inner ear. Primitively, it would seem, the inner ear was not an organ related to hearing, and just when it developed this capacity is not

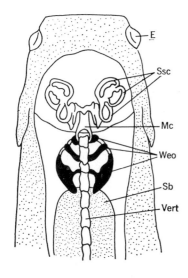

Figure 105. Diagram of "Weberian ossicles," characteristic of Ostariophysi, shown in relationship to the swim bladder, vertebral column, and inner ear. E, eye; Mc, central cavity; Sb, swim bladder; Ssc, semicircular canals and other structures of the inner ear; Vert, vertebrae; Weo, "Weberian ossicles." (After Pincher, 1948.)

known. Bone or other skeletal interruption transmission could have occurred very early, but that it did cannot be demonstrated.

Van Bergeijk (1964, 1966) has stressed the "air-bubble" concept of a receptor and developed a theory of evolution of hearing based on it. The gas bladder or its lung predecessor and the middle ear, in tetrapods, are considered the primary converters of pressure waves into water motion. With the development of this structure, the gas-bladder "bubble," in fishes effective hearing became possible. As has been noted before, however, hearing of sound waves does appear to occur in elasmobranchs, in which the gas bladder is absent. In various ways the gas bladder and inner ear have come into proximity (Figure 105), and in the Ostariophysi Weberian ossicles, formed from elements of the vertebral column, provided an osseous avenue from the bladder to the fluids of the labyrinth. In effect the gas bladder forms a middle ear.

Only in the actinopterygians is hearing known to be well developed. Experiments indicate that it is most effective in the Ostariophysi, in which the Weberian ossicles appear to increase the effective frequency range of sensitivity and perhaps have some effect in equalizing frequencies and acoustical filtering. Even in the most efficient "hearers," however, the system is nondirectional and the range of frequencies is relatively restricted.

Many fishes produce sounds. This is done by the grating and clicking of teeth, especially pharyngeal teeth, by fin movements and by bone scraping (stridulent sounds), by vibration of the gas bladder and by emission of air bubbles from the bladder, and by rapid movements and changes of patterns of movement (hydrodynamic sounds). A summary of types of sounds and mechanisms of production can be found in Tavolga (1965). The gas bladder is vibrated by extrinsic muscles in various cases. Within the teleosts for which sound production has been studied little taxonomic distribution of the various ways of production occurs. Some of the most effective noise production comes from resonation of the stridulent sounds by means of the air bladder. Among the fishes that are noted as "noise makers" are grunts, drumfishes, groupers, marine catfishes, toadfishes, and squirrelfishes.

The sounds are mostly within the frequency range over which fishes are sensitive, and there is considerable evidence that behavior is affected in various ways by the sounds—for example, spawning practices, schooling, feeding (including attraction to food sources made by crunching noises), predator–prey relationships, and territorial defense. Much more needs to be learned about patterns of transmission and sensitivities before this field of study can become precise. So far no clear cases of echolocation have been found among the fishes, although this has been suggested on very limited data.

The transition from hearing in water to hearing in air that accompanied the development of tetrapods has received considerable attention and is taken up in some detail in Chapter 4 of this section. Amphibians, by and large, developed a capacity to hear in the air and lost the capacity to hear effectively in water. Some aquatic forms, such as the extinct nectrideans *Diploceraspis* and *Diplocaulus*, may have been unable to hear far-field sounds, but they retained lateral lines and probably relied on near-field hearing (Beerbower, 1963).

On return to water by reptiles and mammals, considerable modification of the auditory mechanisms took place in a few forms. Total reliance on the far-field waves was necessary, since the lateral-line system no longer existed. No new mechanism developed in its stead. The great majority of reptiles and mammals that became partially to almost entirely aquatic underwent at most only moderate ear modification. The most common event was a development of a facility to close the ear so that water could not reach the tympanic membrane, as in crocodiles and many mammals. Marine turtles retained the ear elements, but the tympanic membrane was covered by heavy dermal elements. Hearing in most of these animals remained essentially adapted to air vibrations, with hearing in water being an unimportant part of the behavioral pattern.

Among extinct reptiles that were highly adapted to aquatic existence some give indications of moderate to extensive structural modification that probably was related to hearing in water. For one large group, the plesiosaurs, there is almost no knowledge of the nature of the middle ear. Williston (1907) described a possible stapes in *Brachauchenia*. This is a small stout bone lying

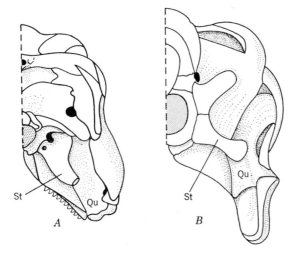

Figure 106.   Occiput and stapes of *A*, *Dimetrodon*, a sphenacodont pelycosaur and *B*, *Baptanodon*, an ichthyosaur, showing the general resemblances in gross morphology. Qu, quadrate; St, stapes. (*A* after Romer and Price, 1940; *B*, after Gilmore, 1905.)

in an appropriate position. However, if it were the full stapes, the tympanic membrane would have lain very deep below the surface. Except for this no indication of the middle ear has been found and no conclusions are possible. The middle ear is known in ichthyosaurs from a very heavy stapes that is solidly articulated with the otic region of the skull. No fenestra ovalis seems to have been present. The element passes toward the large quadrate, and in many respects the structure resembles that of a sphenacodont pelycosaur, such as *Dimetrodon* (Figure 106). It is far from certain that a tympanic membrane was present. The evolutionary antecedents of the ichthyosaurs are not known, so the history of this stapedial condition is uncertain. It may represent an accommodation to hearing in water, or it may merely represent the continuation of a primitive condition. In any event, if there was hearing at all, the mechanisms by which it was accomplished are completely uncertain.

Only in the mosasaurs is there any detailed information on the middle ear. In one species, *Platecarpus* (Figure 107), the stapes was a simple, slender, lizardlike element, presumably carrying an unossified extrastapedial. The quadrate is hollowed out and emarginate to produce what appears to be a large tympanic cavity. The structure probably was well adapted to hearing in air or possibly in shallow water. No indications of its suitability for hearing

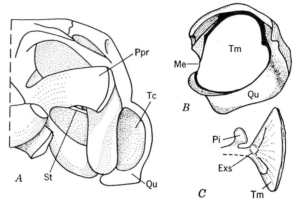

Figure 107.   The otic region and stapes in mosasaurs, showing possible adaptations to aquatic life. *A, Platecarpus*, with a large, open tympanic cavity, excavation for the tympanic membrane, and a slender stapes. *B* and *C, Plioplatecarpus; B*, ossified membrane in lateral view and *C*, posterior view of heavy stapes with ossified membrane. Exs, extrastapedial; Me, meatus; Pi, internal process of stapes; Ppr, parocippital process; St, stapes; Qu, quadrate; Tc, tympanic cavity; Tm, tympanic membrane. (*A* after Williston, 1898; *B* and *C* after Dollo, 1905.)

at any depth are present. *Plioplatecarpus*, on the other hand, had a very robust stapes with a strongly calcified extrastapes. The latter forms a broad plate that fits closely into the recess of the quadrate and has been identified either as a tympanic membrane (calcified) or extrastapedial. The external auditory meatus is small. Such a middle-ear arrangement could indicate a high degree of specialization and even possible accommodation to maintenance of an air-filled middle ear at considerable depth. This interpretation has been made, and *Plioplatecarpus* has been thought to be a deep-diving form. Devillers' (1943) analysis of the cranial circulation, however, has cast some doubt on the diving capacities of this creature.

Among the mammals only the cetaceans exhibit auditory systems that are highly adapted to water hearing, in particular to hearing at depths. This is found most highly developed in the odontocetes. Modifications of the mysticetes are less, and among the archaic whales, the archeocetes, only a moderate change appears to have taken place. In the odontocetes both sound reception and transmission have been developed to a remarkable degree, in particular with the development of echo sounding, which can be carried out at great depths. Many of the modifications are expressed in the bones of the skull and jaw and can be detected in fossil representatives of this group.

The odontocetes emit two types of sounds—clicks related to echolocation and whistles. The former issue at very high energy levels and are as short as

0.01 second in duration. They range up to 100 kilohertz in the ultrasonic range. They appear to be made in the nasal passages leading to the blowhole in a complex structure of plugs and air sacs. They are reflected forward from the skull (Figure 108*A*) and focused by an auricular lens, the oily mass termed the spermatocete organ, or melon. Whistles, apparently produced by the larynx, range in frequency from about 1 to 10 kilohertz. In the baleen whales sounds are made only in the lower ranges and pass to very low frequencies, around 400 hertz, but down to as low as 50 hertz.

A great deal of work has been done on the bioacoustics of the whales, especially on the bottle-nosed dolphin (*Tursiops truncatus*). In spite of this there has remained considerable disagreement on sound reception. The inner ear of the odontocete whale is enclosed in dense bone that is virtually suspended (in pendulumlike fashion) in a cavity filled with mucous foam. In this way the two ears are separated, and each is free from reception of vibrations of the skull bones. Separation is critical if the sound-receiving structures are to be accurate directional indicators, as they must be in echolocation.

The middle ear is similarly highly specialized and adjusted to the processes of hearing underwater. It appears to be so constructed that air spaces around the ossicles are maintained even under the pressures of very great depth. Pterygoid air sacs or sinuses are present and serve as a principal mechanism in this process (Figure 108*B*).

The pathway of sound to the middle ear has received much attention, without full agreement. The external auditory meatus is very slender and filled with wax. Fraser and Purves (1960*a,b*) and Purves (1966) have advanced evidence that the sound passes through this meatus. Reysenbach de Haan (1957, 1966) has argued strongly to the contrary, saying that it is the skin, muscle, and blubber around the meatus that form the channel of communication. Norris (1964) pointed to the oil-filled canal of the lower jaw as the sound conduit, noting in particular its remarkable adjustment for echo reception relative to the mechanism of projection (Figure 108*A*).

These whales have one of the most remarkable adaptations of sensing structures to be found among vertebrates, in particular in the development of capacities to perform echolocation at great depths. The odontocete whales go back in the geological record to the upper Eocene. Only very moderate specialization of the skulls had occurred, but by the late Oligocene and early Miocene skull modifications had proceeded quite far. Little study has been devoted to the development of the structures associated with echolocation, but the possibilities for tracing its evolution seem reasonably good.

The archeocetes probably represent a primitive type of whale. A study of the braincast and the eighth nerve by Edinger (1955*a*) shows that they were little specialized as compared with later whales. In the acoustic nerve (nerve VIII)

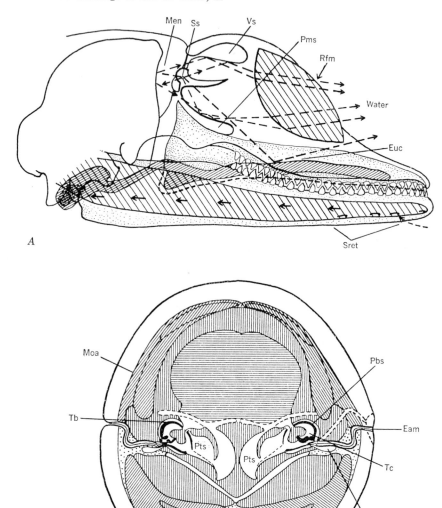

Figure 108. Two diagrammatic drawings of structures related to hearing in whales. *A*, a composite drawing of the skull and jaws, with data from several species. The arrows and lines indicate possible pathways of vibrations in emission and reception of sounds, as tentatively proposed by Norris (1964). The sound source is thought to be the tip of the nasal plug at its insertion into the tubular sac slit. The various reflecting and refracting positions are indicated. Euc, pneumaticized portion of the Eustachian canal; Men, mesethmonasal region, a reflecting surface; Pms, premaxillary sac; Rfm, refractile substance, the lenslike "fatty melon"; Sret, path of

the vestibular ramus was the largest, not the cochlear, in contrast to the odontocetes, in which the opposite occurs. Pompeckj (1922) studied the ossicles of the otic region and concluded that the heavy bones of the middle ear could not have conducted airborne waves and that their rigidity suggested transmission by bone conduction and not movement of the middle-ear elements. Hearing thus may have been somewhat specialized for sound reception in water but not, it would seem, for the highly specialized sort of behavior found among the modern odontocetes. Mysticetes have been much less studied, although their sound emissions are well known. They too are much less specialized for sound reception than odontocetes and, with the possible exception of *Eschrichtius glaucus*, the gray whale, are not known to produce high-frequency clicks. Their use of sound is not well understood. They go back in the fossil record to the Oligocene, with possible relatives transitional from the archeocetes in the middle of the epoch. Miocene genera had approached fairly close to the present level or organization and probably had acquired about the same level of hearing specialization.

Smell, Taste, and Touch

Much less is known about the functions of the organs of smell, taste, and touch than is the case for the sense of hearing. They have been less studied and are much less amenable to investigation in the fossil record. The sense cells of the three are somewhat similar, terminating in sensory hairs. Only if there is an intimate association with parts of the hard anatomy do the organs leave any fossil record.

Organs of the sense of touch, responsive to solid substances in aquatic animals while in water, are widely scattered through the skin. Special tactile organs and structures with highly sensitive areas occur in various vertebrates. The elongated, slender, paired appendages of lepidosirenian lungfishes and the barbels that are developed around the mouth and snout of various fishes are examples. Occasionally such structures are found in the fossil record, and their interpretations are made in accord with their modern

---

returning vibrations along the lower jaw; Ss, sound source; Vs, vestibular sac (inflated). (Redrawn after Norris, 1964.) *B*, a composite cross section of the air-sinus system bringing the various parts into a single plane to show the relationships of the sinuses to the middle ear and their effects in isolating the ear structures from other hard structures of the skull. Eam, external auditory meatus; Moa, occipital-auricularis muscle; Ms, stapedial muscle; Pbs, peribullary air space; Pts, pterygoid air sacs; Tb, tympanic bulla; Tc, tympanic cavity. (After Purves, 1966.)

counterparts. Vibrissae are very important sensory organs in seals and walruses. They receive heavy nerve fibers. Enlarged areas of innervation such as those of the vibrissae or other sensitive parts of the skin, are accompanied by increased innervation. The course of nerves may be evident in the skulls, giving some hint of this condition among fossils. In addition appropriate areas of the brain may be enlarged, and this too may be detected in endo-cranial casts in some instances (Radinsky, 1968). Little is known about the adaptive evolution of the tactile sense in general, but it appears to have been well developed in the most primitive of vertebrates and to have changed principally by specializations in parts of the body related to adaptations to specific ways of life.

The sense of taste falls in much the same category. In tetrapods, except for some of the amphibians, receptor organs are in the area of the mouth. In fishes, however, they are more widely dispersed over the body, with special concentration in barbels, probing organs, fins, and so on. For the most part there is no record of the condition of taste sensation among fossils. It has been suggested that barbels may have been present in cyathaspid hetero-stracans and, if so, these probably had a taste function. The elongated snouts of sturgeons are sensitive to chemical stimuli and, when present in fossils, probably may be so interpreted. As for touch, however, at present there is little information that is useful in assessing general evolution of taste.

Fishes react to electrical stimulation and orient themselves with respect to electrical fields. The receptors appear to be nerves in the muscles. Like the other senses, just noted, little can be said of the evolution of this, and the fossil record gives no evidence of it.

The situation with regard to the sense of smell is somewhat more favorable, although much of what is known from fossils has come to light in con-junction with studies of changes in respiratory systems. The olfactory bulbs of the brain are housed in nasal capsules, usually cartilaginous but in some cases partially ossified. The sensory nerves pass from these into the associated epithelium. Both the capsules and supporting connective tissue of the epithelium may be preserved in the fossil record. In addition, the nares, which provide for passage of water or air to the sense organs, open through dermal bone in vertebrates in which this is developed, either onto the surface of the head only or also into the buccal cavity. These various associations, im-pressions of the olfactory bulbs, traces of olfactory tracts when the bulbs are spatially separated from the rest of the brain, and structures of the nares and passages of the olfactory nerves give some morphological evidence of the development of the sense of smell. It is necessarily very general, and few generalizations on olfactory evolution are possible.

Living cyclostomes possess olfactory structures that are rather different from those of the other living vertebrates but quite similar to those of some of the extinct agnathans. The nasal capsule and narial opening are single structures. The olfactory organ, however, is paired, and in at least the lampreys there is a partial division of the capsule by a median, vertical septum. The single opening is a nasopharyngeal structure that passes back to the olfactory organ and thence in a hypophyseal canal to the region of the first branchial arch in lampreys, ending in a closed sac. The canal passes to the pharynx in hagfishes, where it opens into the pharyngeal cavity. The nasohypophyseal opening is terminal in hagfishes but dorsal in lampreys.

In both of these creatures the structure and position of the nasohypophyseal opening and internal passages appear to be adapted to the particular ways of life of the animals. In lampreys they appear to form a system by which water is brought in and then expelled, by manipulation of the sac. In hagfish it forms an alternate route of water when the head is partially buried during feeding. The single nasohypophyseal opening in cephalaspids and anaspids, however, is very similar to that of the lamprey, and probably there was functional similarity. It has been suggested that anaspids were parasitic, but this is far from clear and certainly does not apply to the cephalaspids, probably mud feeders with suctorial mouths. Thus a specialized structure like that of the lamprey, was, it would appear, present in primitive fishes with other habits, and its particular function may have been appropriate in various situations, not just parasitic life.

Heterostraci appear to be different. Efforts have been made to demonstrate similarity to hagfishes. Problems arise because of the lack of internal ossification, restricting the evidence to impressions of organs on the dermal skeleton. The nasal capsules were impressed on the under surface of the dorsal shield. They appear to have been paired and probably there were two narial openings (Denison, 1967b). Stensiö, however, has restored this area to resemble that of the hagfish. Tentative identification of lateral paired nares have been made, but these are uncertain, and the openings to the olfactory organs may have been within the buccal cavity. If these interpretations are correct, the structures of the heterostracans are in many ways similar to those of the gnathostomes.

All other vertebrates have paired capsules and nares, except as the septum between the external nares may break down in some reptiles and in mammals. Elasmobranchs, and probably placoderms and acanthodians, possess a single nasal opening into each narial chamber. Water passes in and out of this one opening. In some elasmobranchs, the adults possess a naso-buccal groove covered by a flap of skin (Figure 109). This structure occurs in the embryos of various gnathostomes and appears to have been a common heritage of the group. In sharks it appears to be an aid to flow of water in

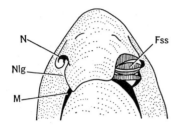

Figure 109. Diagram of the ventral surface of the head region of a shark, showing the nostril, nasolabial groove, and the folds carrying the sensory apparatus, with the flap of skin dissected away. Fss, folds of sensory organs, M, mouth; N, naris; Nlg, nasolabial groove. (After Pincher, 1948.)

olfaction, which is a highly developed sense in some of the sharks. It has been speculated that this may have also been its function in other, extinct, fishes, in which it may have persisted as an adult structure. Evidence for the groove or canal among fossils is very slight, although suggestions of it have been seen in detailed studies of some of the rhipidistians. The nature and development of the olfactory structures in the fishes leading to choanate fishes and tetrapods are part of the story of the invasion of land, and many of the details are deferred until the treatment of this subject in Chapter 4 of this section.

Osteichthyes in general have a pair of nasal canals; an incurrent canal through which water is carried to the olfactory organ and an external one through which it is ejected. In Dipnoi the excurrent canal empties into the buccal cavity but does not appear to be involved in air breathing. In actinopterygians the two openings are on the outer surface of the head and internally a system of sacs has developed, supplying mucus to the narial cavity. No nasobuccal groove is present in the adults, but in some it is present embryologically.

The situation in rhipidistians, from which the tetrapods arose, has been studied in great detail and has been the subject of a great deal of debate (see Panchen, 1967a, for summary). The discussion hinges on the origin of the choanae through which air is passed into the mouth in tetrapods. Some, osteolepiforms, appear to have a single external orifice and others, porolepiforms, have two (Jarvik, 1942, 1967b). This has been part of the basis for arguments about a dual origin of amphibians. The nasobuccal groove likely was the source of the choana, and its initial function probably was olfaction rather than breathing. A lacrimal tube is presumed to have developed from the position of the external nasal tube (see Figure 143B). Air breathing was developed in the course of the changes, as discussed in detail in Chapter 4 of this section.

Similar trends toward the development of passages for water into the mouth have developed in parallel fashion in the teleosts. Astriscopus (Atz, 1952a,b) has approached the choanate condition closely, and other members of its family (Uranoscopidae) have trended in this direction. Other teleosts of the families Echelidae and Ophichthidae have external incurrent and internal excurrent nares, with water being passed into the mouth (Szarski, 1962).

Changes in the fishes, such as those outlined, clearly do relate in part to modifications of the systems of olfaction. Some greater efficiency appears to have been attained, in different ways in different lines. Relatively little experimental work has been done to determine the functional and behavioral aspects of the changes so that little useful information on the adaptive nature of the changes is available.

Aquatic tetrapods have not, as far as is known, developed olfaction in the water. The tendency rather has been for general reduction of the capacity to smell. In mammals, and apparently in reptiles, the first modification, carried out in various ways, is the development of structures to close the nares. In most mammals closure is a voluntary action, but in whales in particular opening is voluntary and closing involuntary. Changes in nasal structure, which can be seen in fossils (e.g., the backward movement of the nares), are connected with respiration and not directly with olfaction.

Loss of the sense of smell has been carried to an extreme in some of the whales. The Mysticeti still have a diminutive bulb and pedicle, but it is nonfunctional, at least as an olfactory organ. Archaeocetes had well-developed nasal structures, although it cannot be known how much they were used. Seals, today, have excellent organs of smell, but the use must be extremely limited. The embryos of odontocetes have rudimentary olfactory organs, but in the course of ontogeny these are lost and no organs of smell occur in postembryonic stages.

Vision

The orbits, sclerotic plates, and the pineal, or parietal, openings are the principal sources of information on the structures related to vision, or light sensitivity, in the vertebrates. From these it is clear that paired eyes and a median pineal organ were present in the very primitive vertebrates, as seen in living cyclostomes and in the agnathans from the late Silurian. Eyes in hagfishes are poorly developed, somewhat rudimentary, but this may well represent degeneration rather than a primitive state, as has been suggested.

Investigations of the paired eyes among the major groups of living vertebrates has revealed a marked stability of fundamental structures. The differences that do exist (see Walls, 1942, for a comprehensive treatment of all aspects of the organs of vision) relate in most instances to mechanisms of focusing images on the retina. Among tetrapods the greatest differences, except for cases of degeneration of eyes, are between the snakes and other tetrapods. Superimposed on the basic structural differences are specializations to particular ways of life.

Problems of sight in the water, in air, and between the media have been responsible for most of the gross modifications that have occurred in evolution. Other changes have related to specializations in one or the other medium, with development of high resolution of images at great distances in some hunting birds, of binocular vision, of night vision and vision at depths in the sea, and similar adaptations. These involve the soft anatomy—lens musculature, cornea, retina, and nervous connections to the brain—features not often reflected in the skeleton and not found in the fossil record.

Relative size and proportion of the orbits and their positions in the head give information on the status of vision among some of the extinct aquatic vertebrates. Among the primitive forms, agnathans, the orbits differed considerably between groups. They were small and widely spaced in heterostracans. Except in very flat animals, they were directed somewhat laterally and slightly anteriorly. The size suggests that they were not of prime importance in sensing the environments, and they may not have been highly developed. It is possible, of course, that they were similar to the eyes of hagfishes, but there is no real basis for the comparison other than size. In anaspids the eyes were larger, relative to the size of the body. Members of both of these groups appear to have been fairly active swimmers. In the latter the eyes may have been important in this activity. The cephalaspids have orbits that are closely spaced and face dorsally. In this way they exhibit a pattern common to bottom-living animals.

For other aquatic vertebrates the position of the orbits in the head, indicating the direction of sight, the raising of eyes above the level of the head, the size of orbits, the presence of sclerotic plates, and similar features all may give some insight into habits of the animals but at best tell only a little about the nature of vision. The eyeballs of ichthyosaurs were truly enormous, but what this implies functionally remains uncertain.

Sclerotic plates occur widely among aquatic vertebrates and may be specializations related to the strengthening of the eyeball against the impact of water. They are present as well in birds, in some mammals, and in various nonaquatic reptiles. No specific adaptive significance can be attached to the presence of these plates, but they do, as in the case of the ichthyosaurs, aid in determining the size of the eyeball, which is not necessarily completely correlated with that of the orbit.

The most intriguing part of the visual apparatus is the pineal or parapineal eye. Two structures, the pineal and parapineal, or epiphysis and paraphysis, are present in lampreys, and there is some evidence to suggest that these may, at an even more primitive stage, have been laterally paired. In adult gnathostomes there is no more than a single structure, either the epiphysis or paraphysis, but embryologically two can be identified. In some fishes and in various amphibians and reptiles the pineal organ lies in or beneath an opening in the parietal bone on the roof of the skull. This opening is relatively small in most animals in which it occurs, but it is large in some primitive reptiles. In lampreys and hagfishes the pineal structures have a retina with pigmented cells and a lenslike structure. In the young, at least, they are light sensitive, and exposure stimulates the migrations of chromatophores. In some of the vertically migrating teleosts, such catfishes as *Arius* and *Madrones*, pigmented pineal spots are present.

The pineal opening is found in agnathans, small but well developed, and it occurs here and there in some other groups of fishes, in particular in flat, bottom-dwelling types. Among placoderms it is present in the very flat *Arctolepis* and in the antiachs. Among the bony fishes, only the rhipidistians have an open pineal, although thinning of the skull roof in the region of the pineal body is not uncommon. It may be supposed that the presence of a parietal opening implies that the underlying organ had a function relative to external stimulation. Most reasonably this may be assumed to be light sensitivity. Whether this was of a low order or whether image formation may have taken place cannot be told. In some way, however, the organ must have been of value in sensing changes in light coming from above the organism.

Most flat, bottom-living amphibians and many aquatic reptiles show the presence of a parietal opening. Some—for example, the plesiosaurs—do not. The relatively largest parietal foramina are found not in aquatic animals but among land-living creatures such as *Diadectes*, caseids, and some of the primitive therapsids. A well-defined but modest opening occurs in ichthyosaurs, nothosaurs, placodonts, and mosasaurs. Turtles lack one. The distribution of the parietal opening among amphibians and reptiles does not provide a key to the specific functions of the pineal. In modern lizards it acts in sensing of temperatures as related to insolation. That this was the function in swimming reptiles is not, of course, impossible, but it does not seem likely. It may well be that the basic property of light sensitivity has been adapted to various shades of use in different vertebrates. Unfortunately few of the possessors have survived, so that the functions must remain speculative.

RESPIRATION

The Fishes

The earliest vertebrates, since they were aquatic and arose from aquatic ancestors, necessarily were adapted to respiration in water. Evidence from the most primitive living vertebrates, cyclostomes, and from the ancient agnathans shows with little question that the first vertebrates breathed by means of gills. This mode has been retained by the great majority of fishes, modified in particulars but fundamentally unchanged.

The sequence of changes in gills, circulation, gill supports, and the mechanics of water exchange shown by cyclostomes, elasmobranchs, and the grades of actinopterygian fishes (see Chapter 3 in Section IV) represent roughly the modifications that took place in the evolution of fishes from their agnathan base. Many special adaptations developed along the way, but on the whole these changes represent progress related to the assumption of increasingly active life in water.

It is mostly the evidence of the branchial arches and gill covers that relates the morphological series known from modern fishes to the evolutionary stages shown by the fossils. The striking resemblances of the gill baskets of cyclostomes to those of agnathans, along with the remarkably close resemblances of structures in *Petromyzon*, cephalaspids, and anaspids, argue convincingly that, except for specializations related to particular habits, the living cyclostomes do represent the respiratory habits of the early vertebrates very well. Theirs was a sluggish life, and the respiratory needs of more active fishes with higher metabolic rates were not yet evident.

Only at one stage has the fossil record given a glimpse of the gills of ancient fishes, this in the remarkable *Bothriolepis* from the upper Devonian studied by Denison (1941) (Figure 25). Enclosed gills with a single external outlet are indicated. Gill positions appear to have been similar in other placoderms, as indicated by adjacent circulatory structures impressed on the bones and evident in some very-well-preserved individuals. Acanthodians show some features indicating a level found in primitive elasmobranchs, but the presence of an operculum on each branchial arch and from the mandible is not shark-like. *Polypterus* may find its counterpart among the primitive paleoniscoids, and the earliest dipnoans and crossopterygians were at this general level as well. Fast swimming and sustained activity had not yet developed.

Within the actinopterygians the successive grades of locomotor adaptive alterations and respiratory modifications developed more or less simultaneously to produce the active swimming capacities found in modern teleosts. Grossly these are the adaptive steps in the evolution of gills. Specializations at all levels took place, often with marked departures from the general direction of evolutionary advances. These are seen among present-day fishes and are used in interpretation of the probable modifications of soft anatomy in extinct counterparts.

A second major respiratory organ appeared early in fish evolution— namely, the lung, developed as a diverticulum from the esophagus. Probably it was preceded by initiation of buccopharyngeal breathing as a supplement to gill breathing in situations in which available oxygen in the water was low. The first actual evidence of a lung, as was noted earlier, has been found in *Bothriolepis*. It appears to have been a bilobed, ventral structure with a duct to the anterior part of the gut. Like many air-breathing fishes, *Bothriolepis* lived a rather sluggish existence, probably in quiet waters.

Existence of lungs at this stage of development suggests that this structure may have been widely dispersed among primitive fishes, and this may have been the case. It does not necessarily follow, however, since lunglike structures have developed repeatedly in different lines of fish evolution, from various parts of the buccopharyngeal cavity. The only impelling argument from the lung of *Bothriolepis* is that it does seem to have originated in much the same area as the lungs of other primitive fishes.

That some descendants of all of the early groups of Osteichthyes possessed lungs suggests that they were a common feature in the primitive phases of evolution of this group. It is generally held by anatomists that the gas bladder of actinopterygians has been derived from lungs and is homologous with them. Homologies are, however, uncertain, and it is not entirely clear that all gas bladders have had the same origin. Goodrich (1930) has considered some of the contrary evidence. In any event lungs as breathing organs did not maintain a prominent position in respiration in the development of the active swimming fishes developed by the actinopterygians.

Air-breathing structures, rather, were mostly adaptations to very special conditions, where gill breathing was ineffective, and eventually in occupancy of land by various groups of fishes. Very likely the initial development was in response to the freshwater circumstance in which the actinopterygians may have taken origin. Low oxygen content over at least part of the year may have been a characteristic of such environments.

Somewhat paradoxically, the conditions that probably were fundamental to the initiation of this group were those that it soon largely abandoned, either by adaptation to more active life in the waters on the one hand or by occupancy of land, leading to amphibians, on the other. In more recent times as well several lines of teleost fishes have become once more adapted to life in very sluggish waters or on land, among them the mudskippers (*Periophthalmus* and *Boleophthalmus*), climbing perch (*Anabas*), eels (*Anguilla*), and various catfishes. These, however, are basically adaptations to land, or very sluggish water, not to usual aquatic conditions. They are taken up in more appropriate context in Chapter 4 of this section.

Tetrapod Aquatic Animals

Amphibians

Amphibians present a special case of aquatic animals among the tetrapods and must be considered separately from the amniotes. They are the most primitive tetrapods and are aquatic animals in their larval, ontogenetic stages, adapted to water life by the presence of gills. If one may generalize from the larval stages of living amphibians and the occasional larvae known from the Permo-Carboniferous, the organs of respiration in the larvae of the first amphibians were external gills. Such gills are present in larval stages of various fishes, including dipnoans, and probably were present in early onto-genetic stages of rhipidistians. If this were the case, this aspect of aquatic adaptation of amphibians was directly inherited from the fish ancestors.

Panchen (1967a) has proposed that the critical step in amphibian evolution in the transition to land was the loss of gill breathing, involving closure of the external opening of the branchial chamber. This closure relates to internal gills and hence probably involved a process of metamorphism, at which stage it would seem that pathways of development were either toward internal gills suited to aquatic respiration of the fishes or to the terrestrial, air-breathing mode of the amphibians.

Only among adult amphibians thus could there have been adaptation to strictly water conditions analogous to that found in other tetrapods and involving preceding adaptation to land. Even this was mostly a process that at most emphasized one side of the usually amphibious existence. This minor shift, back to totally aquatic life, appears to have occurred early and re-peatedly, and it has been suggested that some lines—for example, aïstopods—never abandoned aquatic existence while taking on the guise of amphibians. Even if this was the case, it must be assumed that air breathing was a con-dition that was attained as a dominant form of respiration, whether later lost or not.

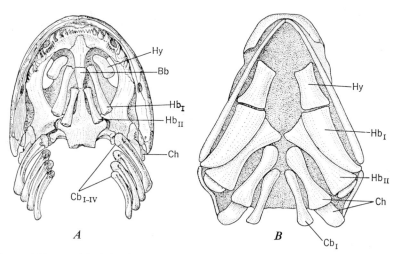

Figure 110. Ossified hyoid apparatus in two late Paleozoic amphibians. *A*, *Dvinosaurus*, a labyrinthodont, usually interpreted as neotonic; *B*, *Lysorophus*, an aquatic aestivating lepospondyl. Bb, basibranchial; Cb_{I-IV}, ceratobranchials as numbered; Ch, ceratohyal; $Hb_{I,II}$, hypobranchials as numbered; Hy, hypohyal. Original length of specimen of *Dvinosaurus* about 20 cm, specimen of *Lysorophus* about 1 cm (*A* after Bystrov, 1938.)

Adult amphibians that are strictly aquatic, at the present time, have adapted to this mode of life by retention of the larval mechanisms of respiration by external gills. This may be supplemented by skin and buccopharyngeal breathing, but not by appreciable air breathing.

Some of the strictly aquatic ancient amphibians became adapted to water life by this same means (Figure 110), among them *Dvinosaurus*, a rhachitome, and *Gerrothorax*, a stereospondyl, as shown by persistent branchial arches. Neotony clearly is a basic aspect of their developmental patterns. Adults of *Lysorophus* and some of its earlier relatives (lepospondyls) also have well-formed branchial arches that presumably bore gills. These were aestivating amphibians whose mode of life may not have been greatly different from that of contemporary gnathorhizan lungfishes. Their active life was entirely aquatic, and in this phase external gills probably provided the means of respiration. Air breathing may have functioned during aestivation.

Nothing is known about respiration in the very early ontogeny of other groups of lepospondylous amphibians since no larval stages have been identified. In *Diplocaulus*, in which skull lengths reach 150 mm, small skulls no more than 10 mm long show no clear evidence of larval features. Rhachitomous amphibians and discosauriscid seymouriamorphs had larval stages with larvae adapted to aquatic respiration by possession of external gills. This probably was true throughout the labyrinthodont amphibians.

Many adult extinct amphibians appear to have been adapted to aquatic life not by retention of larval gills but in conjunction with air breathing. Many of these were sluggish, shallow-water creatures—lepospondyls, such as *Diplocaulus*, *Eryops*-like rhachitomes, and many stereospondyls. Others became active predators, probably fish feeders that caught their prey in active pursuit. *Archeria*, among embolomeres, the rhachitome *Platyops*, and trematosaurs, such as *Aphaneramma*, are examples (Figure 103). In the last two the narial opening lies well back from the tip of the elongated rostrum. Although this superficially resembles the condition found in many marine reptiles, where it is a respiratory adaptation to aquatic life, the condition in amphibians appears to have resulted merely from an elongation of the premaxillaries. Where a long rostrum, suitable for fish catching, has developed without elongation of the premaxilla, it appears that the nares have remained essentially terminal.

Amniotes

All aquatic amniotes breathe by lungs, and none has developed any new organs or modified any already existing ones in response to the problems of aquatic respiration. Aquatic reptiles of today—turtles, sea snakes, and crocodiles—are essentially poikilotherms, and it may be presumed that this was the case for extinct groups as well. This being the case, the intermittent breathing demanded by aquatic existence poses less serious problems than those encountered by birds and mammals, homoiotherms with continuing high metabolic rates. Some of the responses found highly developed in these warm-blooded creatures are, however, to some extent developed among cold-blooded animals.

Adjustments to the physiological problems of intermittent breathing in aquatic animals are found widespread among mammals and birds. Extensive investigations have found a common phenomenon of slowing of the heart rate (bradycardia) at the initiation of periods of nonbreathing (apnea). Postapneal recovery is a highly specialized process in some aquatic mammals, such as seals and cetaceans. The rapid onset of the reaction when apnea begins led to studies that have shown it to be the result of an unconditioned reflex, but a reaction that can be improved by voluntary control in some animals.

Accompanying bradycardia is a selective vasoconstriction by which blood supplies to the muscles are limited but that to the brain maintained. Irving (1966) has presented an excellent summary of the physiological aspects of circulatory regulation in diving animals. Bradycardia, as he indicates, occurs in alligators, essentially cold-blooded creatures. It is widespread in mammals, even those but slightly aquatic in habits. An interesting sidelight is that the phenomenon appears in the air-breathing mudskipper, *Periophthalmus*, when it is submerged, and that some gill-breathing fishes show a similar reaction when removed from their source of oxygen.

As far as fossil, aquatic amniotes are concerned, the evidences of their respiratory adaptations to life in the water are slight, relating mostly to modifications in the narial structures as seen in the bones. In many aquatic reptiles the narial openings are well back from the snout, and their positions relative to skull elements are such that this cannot be due primarily to elongation of the premaxillae. Air was introduced far back in the buccopharyngeal cavity, and presumably the mouth could have been opened during breathing. Nostrils as a rule were also rather high on the head. Ichthyosaurs, nothosaurs, placodonts, plesiosaurs, and mosasaurs all show this tendency. Turtles and crocodiles do not. In the latter the well-developed osseous secondary palate carries air far back, and in the former the same is accomplished by a shorter soft palate. Many partially aquatic reptiles show the tendency for nostrils to be posterior. Phytosaurs, generally crocodilian in their form and presumably behavior, lacked a secondary palate, and the external nares lay far back in a crater on the dorsal surface of the skull.

Most aquatic mammals have no marked specializations of the narial openings related specifically to respiration. As in the crocodiles, there is an efficient secondary palate. Even moderately aquatic mammals have developed an effective mechanism for closing the nostrils while swimming, but this is not well recorded in the fossil record. Cetacean skulls, of course, are extremely modified, and with this modification has come the posterior position of the external naris and its marked asymmetry in odontocetes. The posterior position relates to the total reorganization of the skull and probably to various of the functional modifications that have taken place. Partly it may be an adaptation to venting after long periods of apnea. Clearly it is related to sound production in echolocation. Neither the posterior position nor the asymmetry can be fully explained as a respiratory adaptation, and the reasons for these features are still the subject of lively debate.

REPRODUCTION

All vertebrates reproduce sexually and have basically similar reproductive processes. Some differences do exist in cleavage patterns in the eggs and in later developmental processes. No systematic relationship of these kinds of differences has emerged from embryological studies. Structural modifications of the reproductive systems, which are of course of major scope, and other related modifications of structure and behavior are adaptations that aid in ensuring that fertilization of the ovum takes place and that a suitable environment is provided for development. The significance of such modifications is evident in the fact that reproductive criteria are important in definitions of some of the major categories of vertebrates, anamniotes versus amniotes, or amphibians versus reptiles and mammals. Major radiations followed the basic changes.

The evidence from the fossil record is such that it is only these adaptive modifications that have left some record of their occurrences. Even these contributions are necessarily general and quite restricted. Reproduction in the earliest vertebrates was, of course, adapted to the aquatic existence of the animals. Initially, it would appear, reproduction was carried on in relatively shallow waters, perhaps in estuarine environments, the presumed habitat of the first vertebrates. Adaptations among the anamniotes were toward the exploitation of the many other types of environments that exist in the water. The variety of such adaptations is extremely broad and involves all aspects of the reproductive mechanisms and the behavioral patterns associated with sexual activities and development and care of the young. Among these many adaptations were those leading to successful reproduction on land. At various stages of the ramifications of reproduction among land forms new adaptations to aquatic life occurred, with modifications of the reproductive structures and behavior to the extent necessary for success in this medium. The amniotes, judged by the marine forms alive today, did not undergo drastic changes in their reproductive systems as they reinvaded the waters.

Fishes today exhibit an extremely wide array of reproductive patterns, which seem to have arisen as new environments were invaded successively by the different groups of fish. Many fishes are oviparous, with the number of eggs varying from a few tens to many millions per female. Eggs range from very small to relatively large and have many specializations that aid in survival in special circumstances. They may be merely deposited free in open waters, but specialized ways of protection, by nests and by parental brood pouches, have developed. Fertilization is commonly external, but internal fertilization has developed independently in many different lines. It is usually associated with such identifiable structures as claspers or intromittent organs of males, some of which preserve readily as fossils. Parental care is non-existent in many fishes, but others have developed it to a high level. Patterns of social behavior are widespread and varied. In a number of different groups of fishes there are some that are viviparous. Some fishes display distinct larval stages in their ontogeny, others do not.

Full understanding of extinct species of fishes requires that their reproductive habits be understood. This is, of course, never fully attainable, but very often morphological features, such as secondary sexual characters that are evident in osteology, existence of larval stages, nature of the eggs, associations of young and adults, evidences of schooling, and similar data, can give some pertinent information.

Reproductive adaptations within the fish are for the most part not closely related to taxonomic boundaries of the major groups. Thus little can be done to expand information from a few species over the broader taxa that they represent. Individual instances in which information is available do indicate the extent to which some features known among moderns were characteristic of the group to which they belong. Thus the existence of claspers on the pelvic fins of sharks from the Paleozoic suggests that this feature, and hence internal fertilization, was widespread within the sharks and not something developed in recent forms only.

The existence of larval stages among the extinct labyrinthodonts similarly shows the extent of this way of development, with its implications on the nature of reproduction. It is evident that the method of reproduction found among modern amphibians was not something developed late in their history but retention of a fundamentally primitive feature. In a somewhat different vein, the larval stages known for some of the seymouriamorphs indicate that these creatures, sometimes considered reptiles, were reproductively amphibians and thus, by definition, are to be placed with that class.

Among the amniotes very little evidence on the nature of adaptations of the reproductive system to aquatic life has come from the fossil record. Studies of modern aquatic mammals, especially the cetaceans, show the extent to which adaptation has taken place. Adjustments to the feeding of young in water have occurred, but on the whole modifications have been modest.

None of them would show up in the fossil record. The only evidence on adaptation of reproduction to aquatic life among fossil reptiles comes from a few specimens of ichthyosaurs that suggest that the young were born alive. This is based on the presence of young in the body cavity of an adult, and in one case, in cloacal position. Information on the reproductive practices and life histories of other marine reptiles would be of considerable value in understanding the course of their evolution and possible reasons for their extinction.

OSMOREGULATION

General Matters

All vertebrates are faced with the necessity of maintaining moderately constant ionic concentrations in their body fluids. The problem that aquatic animals face in attaining osmotic homeostasis arises from the fact that the base of the fluid in which they live, water, is the same as that within the body. Active exchange of both water and dissolved salts thus tends to take place across all permeable membranes. The uptake and loss of water and salts through such membranes are relatively slight in amniotes, and osmotic adjustments ordinarily must eliminate as excess only the salt obtained in feeding and drinking. Fishes and amphibians, however, tend to gain excessive water in freshwater environments, to which the body fluids are strongly hypertonic, and to lose water in the sea, to which, in most cases, the body fluids are hypotonic. Amphibians have not in general met the second challenge successfully.

Gills are the most important of the permeable surfaces and, along with oral membranes, the most active sites of water and ion exchange. The skin is highly permeable in many amphibians, but in most fishes it has been rendered relatively impervious by scales, mucus, and similar devices. Gills act both to secure and to eliminate salt, depending on the environment. Kidneys are the other organs most directly involved in osmoregulation, with their structures and functions adapted to very different roles in the major groups of organisms, and, within these groups, to different environmental stresses, ranging from water with almost no salt to that with concentrations higher than in normal seawater. Some animals, of course, face such ranges in the course of their lives.

Few of the organs intimately involved in osmoregulation are closely associated with the skeleton. Exceptions are the salt-secreting glands, which in some birds are impressed on the bones of the skull. The result is that relatively little can be done directly with study of this important adaptation to aquatic environments among fossil vertebrates. A possible avenue, which has been little explored, is the use of bone histology as an indicator of some properties of the blood. Clear evidence of deposition and resorption of calcium can be seen in thin sections of fossilized bone, but as yet no clear evidence of any other ionic changes of the blood has been found.

The role of the kidneys as possible indicators of the environment of origin of the vertebrates and some of the subgroups of fishes has brought the problem of osmoregulation to bear on a major area of vertebrate phylogeny. It is only by the use of modern relatives and analogs of the creatures involved in the actual history that any statements about the processes and structures can be made. It is true that among moderns there is a fairly close correspondence between osmoregulatory physiology and structure and the major taxonomic boundaries. This undoubtedly does reflect ancestral conditions.

Several things, however, argue against interpretations that take this as a basis. Most important is the fact that there is only a loose correspondence between structure and function in osmoregulation, as shown by the wide range of strictly physiological adaptations in modern aquatic animals, particularly in diandromous fishes. The moderns, also, are several hundred million years remote in time from the critical ancestors and undoubtedly have evolved in fundamental features during that long period of time. Convergences in adaptations to common circumstances are hard to sort out. The result of these difficulties, coupled with the extremely small and biased samples available, is that only very small bits of evidence are available, and these represent the whole in such a way that they may be fitted to almost any pattern of explanation. Any one of several hypotheses predicated on other evidences, those of environments in which fossils have been found, for example, may both be supported and falsified by the same evidence from moderns. Thus, with regard to the glomerulus of the kidney, Romer (1955a) and Robertson (1957) have found support of two opposite hypotheses on the place of origin of the vertebrates.

The Major Groups of Anamniotes

Cyclostomes

Myxinoids, the hagfishes, are marine, and the petromyzontoids live in both marine and fresh water. Relatively little is known of the marine cyclostomes, but the freshwater phases of the lampreys have been studied in some detail. The blood of the hagfishes is isotonic to slightly hypertonic to seawater and thus rather like that of many marine invertebrates. Blood concentrations undergo rapid changes in response to external conditions, but little is known about the causes or about functions in general. The kidneys are quite primitive in structure.

Lampreys have pronephritic kidneys in the larval stage, but in the adults the mesonephritic kidneys take over. The ammocete larval stage is passed in fresh water. A glomerulus is present in the kidney, active in water transfer in fresh water. In both fresh and salt water the blood concentrations are close to those of teleost fishes in the same medium.

From an evolutionary standpoint the importance of the structures and physiologies of these primitive vertebrates is that they show no major differences from higher fishes, even from the teleosts. It is, of course, impossible to be sure that resemblances are not in part the result of convergence, but the great similarities of the cyclostomes and their ancient agnathan relatives in other regards argue that there probably were also no major differences in the processes of osmoregulation.

Some of the ancient agnathans appear to have been marine (heterostracans), some largely nonmarine (cephalaspids), and some probably euryhaline (anaspids). If this was the case, then even in the first known vertebrates a wide range of adaptation in osmoregulation had occurred.

Elasmobranchs and Holocephali

Chondrichthyan fishes are on the whole relatively high in urea content and thus maintain osmotic concentrations at least slightly higher than those of the environment, with tendencies for water to enter across the permeable membranes. Kidneys possess glomeruli, and the tubules have the unusual property of reabsorbing urea. Gills are almost completely impervious to the passage of urea and trimethylamine oxide, both of which are critical to the high osmotic level. Both gills and kidneys actively excrete nitrogenous wastes.

Most of the living Chondrichthyes are marine but, a few have taken up life in fresh waters. The earliest known sharks are from marine deposits of the middle Devonian, but during the Paleozoic the xenacanths came to inhabit fresh water, first appearing in the late Devonian. The first appearance of sharks in marine waters is abrupt, and no predecessors are known. It has been suggested that the Chondrichthyes may have originated in fresh waters. Support for this assumption has been sought in the structures and processes related to osmoregulation, with the result that positive statements on freshwater origin have been made (e.g., Carter, 1967). The argument compares the concentrations of salts in the blood of elasmobranchs and of actinopterygians, fishes that perhaps did have a freshwater origin. It is noted that they are somewhat the same in these two and very different from those of other animals with marine origins, in which the blood is about isotonic with seawater. Glomeruli in kidneys may also be cited to advance the argument.

Conditions, however, provided little basis for any such statements, unless very special cases are considered. Table 49, which is based on studies by Nicol, shows some examples of a range of concentrations, based on freezing points, of vertebrate pairs that inhabit marine and fresh waters. It is the marine invertebrates and the hagfishes to which the marine shark shows the greatest

TABLE 49   Osmotic Pressure of the Body Fluids
of Various Marine and Nonmarine Pairs of
Vertebrates[a]

Animals	Osmotic Pressure
Aquatic mammal	0.78
Terrestrial mammal	0.60
Marine bird	0.72
Terrestrial bird	0.60
Marine turtle	0.79
Freshwater turtle	0.46
Marine cod	0.79
Freshwater carp	0.53
Marine dogfish	1.90
Freshwater sawfish	1.02
Marine hagfish	1.90
Freshwater lamprey	0.60
Marine water (normal)	1.85

[a] Data from Lagler, Bardach, and Miller (1962) after Nicol.

resemblance. Freshwater sharks tend to have concentrations considerably higher than those of freshwater teleosts. Any attempts to sort out origins from data of this sort are at best hazardous, and as additional animals and variations with changes in concentrations of the medium are added the task becomes essentially futile.

Osteichthyes

Almost all of the information on osmoregulation in osteichthyans concerns the actinopterygians and, within this group, largely the teleosts. Data on Dipnoi are chiefly from their aestivating phases, which are not pertinent to this discussion, and almost nothing is known of the osmoregulation of the one surviving crossopterygian.

Osmotic concentrations in freshwater teleosts average about 300 milliosmoles per liter. In marine teleosts they are around 400 milliosmoles per liter (Gordon, 1964). Thus freshwater fishes have concentrations about 10 times that of the medium, and marine fishes have about 0.33 times. Physiologically they are thus very different, but this does not show up in the hard anatomy. Glomerular kidneys function to remove copious amounts of water in freshwater fishes, but they may be reduced or even absent in marine fishes. Different groups of teleosts vary greatly in their tolerances of salinity changes, ranging from stenohaline forms that cannot tolerate changes of more than a few parts per thousand, to euryhaline fishes which can pass from fresh to salt water and some of which can tolerate concentrations as high as three times that of seawater.

Adaptations to various special conditions occur within the Osteichthyes. Aestivating dipnoans, for example, pass no urine during this phase and undergo a strong rise in osmotic concentrations. Once water returns, rapid dissipation takes place. Diandromous fishes undergo marked physiological changes as they pass into waters of different salt concentrations. Changes in kidney action and permeability of surfaces are under hormonal control and have rather tightly controlled limits that appear to be genetic in origin. Ease of adjustment depends, to some extent, on such factors as the ratio of gill-surface area to body-surface area, a strictly morphological character.

Very striking modifications in osmoregulation are found in some of the "terrestrial" teleosts, such as the mudskippers. These may possibly be somewhat like modifications that accompanied the development of land animals in the Devonian, but the initial osmotic base may have been so different that comparisons would be misleading. This matter is explored further in Chapter 4.

None of the morphological features associated with differences in osmoregulation in the teleosts is directly reflected among fossils. If the different major groups of fishes are considered, differences in the ratio of gill-surface area to body-surface area might be deduced, but it is dubious that this sort of difference would outweigh the between-group differences in other aspects, such as relative permeability.

Amphibians

The skins of most modern amphibians are permeable, and this seems to have been an important determinant in the pathways of their evolution. Some have evolved to reduce this permeability, largely in overcoming the needs of large amounts of fresh water for living and for reproduction.

Blood ionic concentrations approximate those of freshwater teleost fishes. Salt tends to be lost to surrounding water by the skin, the urine, and probably by skin glands as well. Much less study has been made for amphibians in different environments than for some of the fishes, and the rates of loss of salts and sources of uptake are not as well understood. Food is the probable source of much of the salt, but the skin and possibly the gills may be involved in aquatic forms.

Most present-day amphibians are strongly stenohaline and unable to tolerate changes of more than about 9 percent in salt concentration. Gordon (1964) has summarized some of the data on the best known euryhalines, *Rana cancrivera*, a frog, and *Bufo viridis*, a toad. Both seem to attain tolerances to about 29 parts per thousand salt concentration. The toad has somewhat reduced skin permeability, but this is not the case for the frog. High ionic blood content is tolerated, thus inducing loss of water. This high concentration depends principally on high urea levels, higher even than those in the sharks. The urine is not hypertonic to blood.

Mostly on the basis of the evolution of the conditions in modern amphibians, Gordon suggests that amphibian evolution began in humid conditions, probably in tropical fresh waters. Although this is not unreasonable, it does run contrary to the interpretations of others who have focused more on the nature of some of the ancient amphibians. Unlike moderns, many, perhaps all of them, had scale-covered skins, which perhaps were relatively impervious to water and salt transfer much as those of comparable fishes. Scales are well developed in some of the gymnarthrid lepospondyls and various laby-rinthodonts—for example, *Trimerorhachis*. Thus desiccation on land may not have been a critical problem, and, in the absence of gills, water and salt exchange in the water may have more closely approached that of amniotes than of modern fishes and amphibians.

The situation with regard to eggs and larvae, however, appears to have been quite different, with a freshwater habitat being mandatory. As far as osmo-regulation is concerned, many of the ancient amphibians probably were more strictly amphibious, with respect to larval and adult stages, than are many of the moderns that by various modifications have aborted one or the other aspect of this duality.

Freshwater habitats appear to have been primarily critical for reproduction and for larval stages. Most amphibians did not attain capacities for existence in salt water, probably for this reason. Occasionally remains are found in marine deposits, but explanations on the basis of transportation by rafting from freshwater habitats seem generally the most satisfactory. The tremato-saurs of Spitzbergen may be an exception. These were highly aquatic am-phibians (Figure 103) and are known only from marine deposits, where they occur along with marine fishes. Very possibly the adults lived in marine waters, perhaps even developing salt-secreting glands, as have various reptiles and birds. If so, they must either have reproduced in fresh waters or have developed some capacity to obviate the need of freshwater larval stages. That such marine amphibians could have developed, given the problems of osmoregulation that they would have faced, seems quite possible. How these problems may have in actual fact been solved is almost completely speculative.

Amniotes

Marine reptiles, birds, and mammals that live primarily in saltwater obtain their food from this medium and take in salt in water as they drink and feed. Fish, one of the main items of diet, do not have a high salt content, but

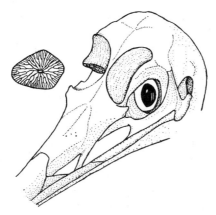

Figure 111.   Sketch of the skull of a marine bird, showing the impression of the salt gland in the orbital region and a section of the gland. (After Schmidt-Nielsen and Fänge, 1958.)

seaweeds and invertebrates, which also figure prominently, do. Mammalian kidneys, in general, are structured so that they can dispose of the additional salt. The loops of Henle, which produce a countercurrent multiplier system, seem to be the principal structures that make this possible. Kidneys of some mammals, such as those of man, are incapable of handling concentrations as high as those of seawater, but this is not generally the case. Seals and whales, and apparently various other aquatic mammals, produce highly concentrated urine, and their efficient kidneys cope with the problems of osmoregulation without difficulty.

The kidneys of birds and reptiles are incapable of maintaining osmotic concentrations of the body fluids in this way. Birds do have some development of the loops of Henle, but they are insufficient for the needed salt disposal. Marine birds have developed glands in the orbital region that secrete solutions with concentrations of salt several times that of their urine. The glands vary in size and increase in individuals under stress conditions. They have a distinctive histology and form. The glands in some cases leave a record in the bone (Figure 111), and this could, given good materials, be detected in fossils.

Modern marine reptiles are far from representative of those that existed in the past and may not give a very clear picture of the osmoregulation in the extinct forms. Several genera of turtles, sea snakes (Hydrophidae), a crocodile (*C. porosus*), and the Galapagos lizard, *Amblyrhynchus cristatus*, are the best modern examples. Few of these are completely marine. They have not been studied in nearly the detail accorded some of the marine birds. All of these reptiles have large orbitonarial glands, with histologies rather like those of some of the salt-secreting glands of birds. Turtles have been shown to secrete salty fluid, in nature and under stimulation by saline injections. Turtle "tears" have been considered to show this function in operation in nature (Schmidt-Nielsen, 1958, 1960). *Amblyrhynchus*, which feeds on seaweed, also excretes a salty fluid from its gland. This was not confirmed for the sea snakes and marine crocodiles by very limited studies by Schmidt-Nielsen. In none of these creatures are the positions of these large glands distinctly on the bone so that their presence could be detected unmistakably among fossils.

Certainly the strictly marine reptiles of the past faced the problems of salt concentation by feeding and drinking. It is likely, on the basis of modern birds and reptiles, that they met this by development of salt-excreting glands. Skulls do not, however, give clear evidence of the presence of such glands, although the orbits and narial chambers are sufficiently large to have accommodated them. It is possible, but only speculation, that in their extensive evolution some or all of these reptiles underwent kidney modifications functionally comparable to those of marine mammals. It is also possible that they developed salt glands in other parts of the body. It would only be the most extraordinarily favorable circumstances of preservation that could give any answers on the correctness of any of these suggestions.

3 Actinopterygian evolution: radiations in the aquatic medium

THE RECORD

Geological and Geographical Aspects

Information on the radiation of the actinopterygian fishes has come from sedimentary deposits covering the time span from the middle Devonian to the Pleistocene and from the great array of ray-finned fishes that populate the marine and fresh waters of the present time. This is one of the longest and in many respects best documented records to be found for any large subdivision of the vertebrates. Some scales that probably pertain to the group have been found in lower Devonian beds, but these yield so little information that to all intents and purposes the interpretable record begins in the middle Devonian. Devonian specimens come from several areas in Europe, North America, and also from Australia, both from freshwater and marine deposits. The relatively few sites and their scattering, along with the high diversity of Devonian actinopterygians, suggest that sampling of what actually was present has been rather incomplete.

A large array of actinopterygians has come from deposits of Carboniferous age. Many of these are from freshwater beds, associated with coal measures, but excellent remains have been found in marine beds as well. The situation is much the same through the early and middle Permian. During these times the best known assemblages have come from the Northern Hemisphere, from

519

North America, Europe, including the British Isles, and Greenland, but scattered finds have also been made in southern continents. By the late Permian actinopterygian-bearing deposits are widespread, with both marine and nonmarine beds providing highly varied arrays of specimens.

Many of the actinopterygians of the early Triassic have come from marine beds, and from this time on these fishes are common fossils in many kinds of marine sediments. The freshwater actinopterygians of the Mesozoic and later times appear to have been derived from marine ancestors, with frequent invasions having resulted in excellent records throughout much of the Mesozoic and Cenozoic wherever and whenever such beds were deposited and preserved.

The Major Subdivisions and Taxonomic Problems

The record, viewed very broadly through geological time, shows a succession of levels or grades of organization of the actinopterygians. Each of these has living representatives, with the most advanced level widely represented by the teleost fishes and the other two, holostean and chondrostean, by a few survivors of earlier radiations. Dominating the record before the Triassic are fishes that form a somewhat coherent, central group, the paleoniscoids. Along with some of their more modified descendants, they fall into the organizational level termed chondrostean. From the mid-Triassic to the late Jurassic the dominant group is formed by the grade generally termed holostean, including holosteans proper and a second group variously called Pholidophoriformes or Halecostomi. Thereafter the third level, teleosteans, dominate the record. During the early Triassic many fishes somewhat intermediate between the central chondrosteans and the typical holosteans dominate the record. These, following an original suggestion by Brough (1935), are often called subholosteans. Although generally discarded from formal use, this term remains very convenient to describe this intermediate developmental level.

Early in the history of classification of the actinopterygians, as discussed on pages 255–256, the three main grades formed the basis for the major taxonomic groups, usually placed at the ordinal level. Subholostei was added to these three. The effectiveness and convenience of this classification have resulted in its persistence in the face of compelling evidence that the groups it designates had multiple origins and that the levels they represent are organization grades attained independently by distinct evolutionary lines. Strictly, in accord with generally recognized taxonomic practices, these categories are not proper taxonomic groups, but practically they are extremely useful. The tendency in recent classifications has been to drop the term "subholostean" but to retain the others (see pages 258–262).

The grade levels express the extent of departure from the common morphological base and in this sense may seem to reveal linear trends in evolution. Were but a few representatives of the various levels known, as was true in earlier phases of study, a reasonable morphotypic phylogeny might be developed and the grades then neatly equated to taxonomic groups on an assumption that they were monophyletic in origin. The records, however, show this not to be the case.

The pattern that emerges, if the complexities are ignored, or that would emerge were they unknown, is reminiscent of those that have developed at various stages of studies in other groups of vertebrates—for example, in the temnospondylous amphibians, the anthracosaur–primitive reptile–synapsid array, or in the therapsid–mammalian evolutionary sequence. It is suggested, although less well documented, by the rhipidistian crossopterygians and by the heterostracan agnathans. In each of these, if successive types are treated as a linear series with evolutionary content, directional and seemingly "inevitable" changes occur. In some for which information is limited, there have been tendencies for the apparent patterns to be considered as representing the actual pattern, and oversimplified interpretations of the nature of the evolution and monophyletic origin of successive levels may follow.

That all of the patterns were similar to that shown by the actinopterygian fishes is by no means certain and cannot be deduced from the records. The resemblances do suggest the possibility of general similarities and at least argue for recognition of this possibility in making interpretations.

Interpretability of the Record

The nature of the morphology and preservation of the actinopterygians make the actinopterygian record especially suitable for studies of evolutionary change. The full body and head are frequently preserved so that the outlines and the external bones are well shown. The body is commonly scale covered in those fishes in which scales are heavy, but in some the internal ossifications may be evident. Usually fishes are preserved resting on their sides and compressed, although there are many exceptions in position of preservation, and occasionally fishes are preserved "in the round." In the lateral aspect many of the major features of the locomotor and feeding systems tend to be well shown. Both, as noted in the preceding chapters, are subject to ready interpretation; the locomotor adaptations on the basis of body form, fin form and structure, and scales; the feeding adaptations on the basis of the skull and jaw structures and the teeth. These two systems figure importantly in the adaptive evolution of actinopterygian fishes. Other major modifications in the development of the new levels of organization—for example, changes in the lung–gas bladder complex, the gill respiratory system, the circulatory system, and physiological processes—must be determined largely from modern representatives, from changes that modifications of the life environment may suggest, and from inferences based on alterations of the locomotor and feeding systems.

Knowledge of the life environment, also critical to an understanding of evolution, poses difficult problems, especially for the earlier fishes. On the whole, for later groups, interpretations can be made with some confidence. The primary environmental distinction is between nonmarine and marine environments. As a rule, if sediments and their organic contents are studied in detail, deposits can be specified as marine or nonmarine. It is generally more difficult to determine that the creatures in question lived in the environment of deposition indicated by the sediments. The majority of fishes live above the substrate on which deposition is taking place and may inhabit a variety of environments and yet be deposited under the same conditions. In addition, of course, are the problems occasioned by transportation, which may carry fishes far from their habitats. Sometimes such problems are subject to interpretations, but there are few generalizations that apply to solutions.

Phylogeny and Evolution

The assessment of the utility of the fossil representatives of a particular category of organisms in understanding its evolution and phylogeny raises some interesting and difficult problems. A meaningful study of the evolution of any group requires an understanding of the phylogenetic relationships of the constitutents, but this necessary understanding may be cast at very different levels, each with some significance, and the suitability of the record is strongly dependent on the nature and extent of preservation as pertinent to the level being attacked. If the roles of variation and selection are to be studied directly, relationships at low taxonomic levels, the species level or lower, must be known. It is rare in paleozoological studies for any appreciable part of an evolutionary sequence to be known at this level. Even were such complete information available over a long time span, it would be so overwhelming in detail that meaningful generalizations pertinent to the evolution of the group as a whole could hardly emerge from analyses. The transition from the population level to studies of modifications as seen in higher categories, however, is a somewhat complex process.

Studies of paleontological materials commonly utilize generic or higher categorical levels as their basic units. At each successively higher level an abstraction of common features of the subsets formed by the next lowest (acknowledged) category is made. From an evolutionary point of view such an abstraction is useful if it includes the properties that have resulted from processes of natural selection. These maintain a link to the evolutionary processes, but this link tends to be increasingly tenuous at higher categorical levels.

A familial characteristic may be used as a pertinent example. It may cover a wide spectrum of character states of the specified property, as evident in the frequent necessity of stating a character as a trend rather than as a little-varied property. The selective and adaptive aspects of the character may exist only at the level of a subfamilial (or lower) radiation and be expressed by the character states distributed among the units of this radiation. Familial characters, of course, may have single states, and this may be true even for characteristics definitive of higher categorical levels. If only such characters are used, however, much of the richness of the evolutionary content is lost in the categorical definition. When a familial or higher category is used as the unit in phylogeny and multiple-state characters are used in the conception of the properties of the category, statements that can be made about "evolution" may differ sharply from those that can be made at, say, species or genus levels. The higher the category, the wider the breach becomes.

The interrelationships of the various levels are shown diagrammatically in Figure 112. In these diagrams it is assumed that all groups are strictly monophyletic—that is, that they arose from one species each. If polyphyly of any sort is admitted, circumstances become much more complex and the capacities for interpretation are much reduced. Another complicating factor, shown in Figure 113, arises with the use of different categorical levels in a single phylogeny. Although the intent may be quite evident, irregularities in the levels of the included units cannot but lead to at least technical difficulties in interpretation of patterns and modes of evolution.

Each of the types of phylogeny illustrated in Figures 112 and 113 has been used to depict actinopterygian relationships and evolution, as in the examples in Figures 114 through 117. For the most part these have been taken from the more primitive grades since these are less influenced by "horizontal" classifications based mostly on moderns. These phylogenies form the framework within which many of the concepts of the evolution of actinopterygians have been expressed and from which others have been deduced. Clearly, statements about evolution that can be derived from Figure 114 have a very different content from those based on the familial relationships in Figure 115. Both are as complete as the pertinent known materials permit. Only rather small segments of the total portrayed in Figure 115 are sufficiently complete to be illustrated at the level of Figure 114. Some parts of the familial pattern, which follows Gardiner, are known to be more accurate than others, and of course at no level is a phylogeny better than the accuracy of the relationships that it portrays.

Figure 116 shows still another sort of phylogeny in which the lines of descent are shown by using selected families and higher categories to indicate gross relationships. Here, in a different way, an abstraction of all of the available information has been made, relegating what has been considered less

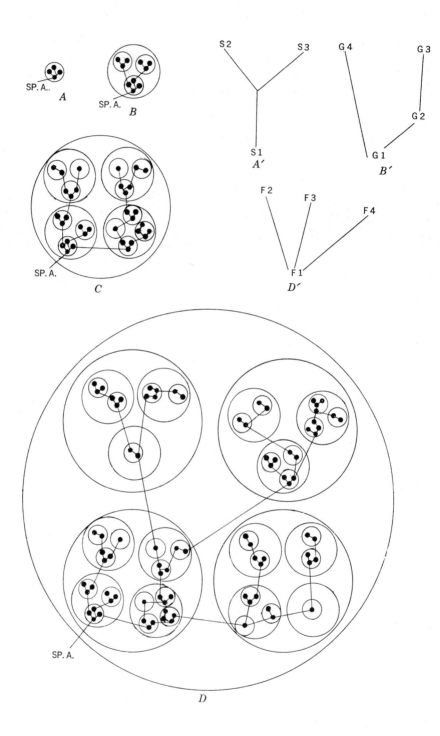

SP. A.

A

SP. A.

B

SP. A.

C

SP. A.

D

S 2   S 3

S 1

A'

G 4   G 3

G 2

G 1

B'

F 2   F 3

F 4

F 1

D'

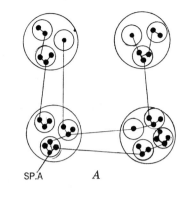

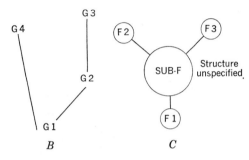

SP.A   *A*

B   *C*

Figure 113. Diagrams of evolution to supplement those shown in Figure 112. *A*, polyphyly at the species–genus level, with genera coming from more than one species within the ancestral genus. As shown in *B*, this phenomenon is not evident in the conventional diagram, with single or multiple origins undifferentiated. This of course applies at each higher level, fitting the concept that polyphyly is not recognized as existing as long as the ancestral group is at an equivalent or lower categorical level. From an evolutionary point of view, however, the difference may be significant. *C*, diagram of a common way of depicting evolutionary sequences, using different categorical levels, here F_1 (family 1) leading to F_2 and F_3 (families 2 and 3) through a subfamilial level. This is the type of sequence shown in Figure 116, where, for example, parasemionotiforms are shown leading to pholidophorids and this family in turn to aspidorhynchiforms and several higher categorical levels. Although clear and convenient, this sort of diagram poses serious problems in interpretation of the nature and modes of evolution involved.

Figure 112. (Left) Diagrammatic representation of various levels of evolution that are depicted in charts illustrating phylogenies. The small black circles represent subspecies, and the increasingly large circles are species, genera, families, and suborders, shown respectively in *B*, *C*, and *D*. In each case, *B* through *D*, the evolution through species and subspecies, as shown, is implied in all such charts, but as shown in *B'* (S = species), *C'* (G = genus), and *D'* (F = family), this is not evident in the usual diagrams, with the result that interpretations must depart to some degree from the precise course of development and become to some extent typological.

Rarely is the level in *A* detectable in the fossil record, although see Westoll's interpretation in Figure 114. *B* is the sort of pattern primarily depicted in Figure 114. *C* is the sort of pattern that is often used, and to the extent that many of the families are monogeneric, as far as known, is effectively that of part of Figure 115. *C*, the familial level of evolution, is used in Figure 115 and in parts of Figure 116. Essentially ordinal evolution is the basis of Figure 117.

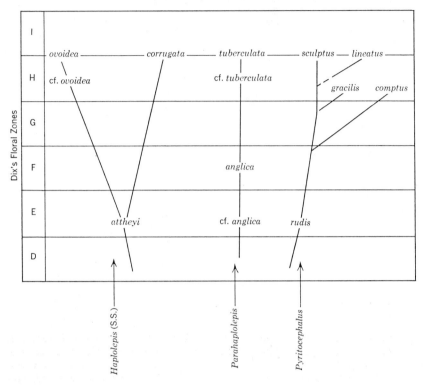

Figure 114. Relationships between genera and species of haplolepids as suggested by Westoll (1944). Compare Figure 112*B*.

important to a position where it does not enter into the main interpretation. The complexities of the use of different categorical levels and the problems posed in reading out evolutionary information from this sort of classification are self-evident.

At this level, along with the introduction of selective use of information, portrayals begin to assume a strongly typological character and the units assume a somewhat morphotypic role. Deductions about adaptive trends often made at such levels, though perhaps valid, are necessarily based on subjective simplifications in which intracategorical variations at each lower level, leading eventually to the evolutionary processes of variation and selection in a causal scheme, are necessarily lost.

The ordinal phylogeny within the teleosts (Figure 117) is at a still higher categorical level. It is at approximately this level that many of the major disagreements on the relationships between actinopterygian groups exist. Ultimately answers from any level trace back to the materials and the collation of morphological and distributional characteristics that have emerged in the course of their evolutionary developments. The greatest understanding of the evolution of a group of organisms requires materials that allow investigation of the gross patterns of change that portray in intelligible form the full radiation in terms of events at successively lower levels, so that both the history of descent and the causal aspects of change can be revealed and integrated. No major groups of animals are represented in the fossil record in such a way that these aims are fully met. Among the vertebrates the actinopterygian fishes and the mammals approach the ideal most closely. Both have contributed importantly to understanding of patterns of evolution. The advantages of the actinopterygians have come largely from the fact that the early parts of their radiation, before the establishment of the major diverging lines, are reasonably well recorded, whereas in the mammals the early phases are known largely from very limited parts of the organisms, mostly the teeth and the jaws.

MAJOR ADAPTIVE CHANGES

From the central and primitive chondrostean level, best represented by paleoniscids, have come many evolving lines of fishes. These have followed somewhat divergent courses but have undergone extensive parallelism, which has led repeatedly to similar thresholds of morphological development both in the feeding and locomotor systems. Following the attainment of these thresholds, new adaptive radiations have taken place. The modifications are readily recognized in single structures, such as the caudal fin or the cheek region, but each of the individual changes is to some degree correlated with other changes both within its own functional complex and in other complexes. In various combinations these have produced fishes adapted to particular life circumstances and, in some instances, patterns of organization that have allowed new, broad adaptive radiations. Only a small percentage of the adaptive experiments led to such radiations, because most of them were evolutionarily abortive.

Many of the problems of development of a satisfactory classification and phylogeny of the actinopterygians, over and above those posed by limits and difficulties of preservation, have arisen from the fact that changes of different parts of the functional systems did not arise at the same time relative to each other in separate lines and did not proceed at constant rates once introduced. Thus different lines show various combinations of "primitive" and "advanced" characters, producing the mosaic patterns of De Beer. A study of "subholosteans" by Schaeffer (1956d) has nicely illustrated this point (see Table 50). As he has expressed it, there is a broad spectrum of adaptive experimentation with regard to combinations of characters. The carryover from

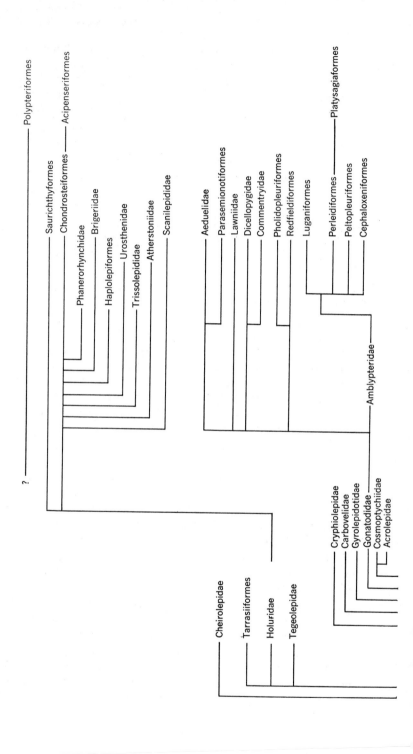

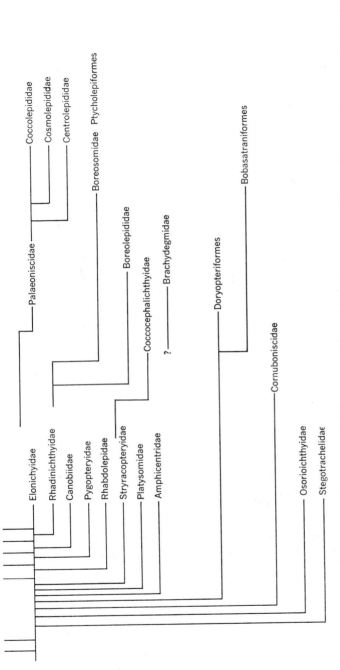

Figure 115. Gardiner's suggestions on probable relationships of chondrosteans, with special attention to paleoniscoids, based on his continuing analyses of the families and genera. Compare Figures 112*C*, *C'*, and 113*C*. (After Gardiner, 1967*a*.)

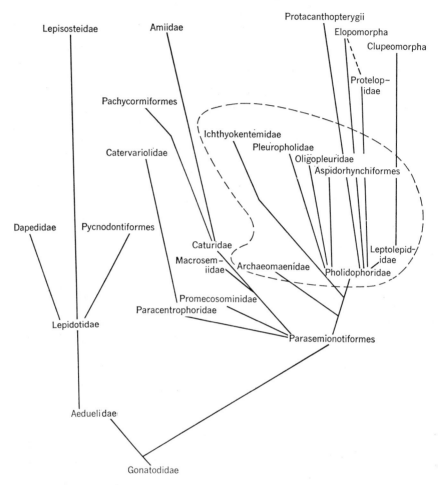

Figure 116. Evolutionary tree of principal holosteans, halecostomes, and relatives of some teleosts as determined by Gardiner, using primarily the evidence of the preopercular element. Compare with Figures 112D and 113C. (After Gardiner, 1967b.)

this general concept to other specific cases in order to follow the process in action is of course extremely difficult and for the most part impossible among available fossils. One of the closest approaches among the primitive actinopterygians is the study of the Haplolepidae by Westoll (1944) in which he attempted an analysis at the species level, at which much of the causal explanation must be sought (see Figures 114 and 128).

The realization of successful combinations, producing a new level of organization, can be recognized by the subsequent development of a new

Figure 117. Suggested evolutionary tree of the teleost fishes somewhat modified from Greenwood et al. (1966) by Romer (1966a). Compare Figures 112D and 68.

adaptive radiation. The first major attainment of a new level among the actinopterygians was the holostean grade. This level, it would appear, was fully realized by two distinct lines: the holosteans per se and the pholido-phorids (halecostomes), as shown in Figure 116. It was very nearly realized a number of times within the subholosteans. The fully accomplished holostean

TABLE 50   Summary of Results of Chi-Square Tests of Association on Comparable Measurements of Structures of Samples of the Paleoniscoid *Pteronisculus* and the Holostean *Chondrosteus*[a]

	Body axis in caudal fin reduced	Dorsal and anal fin rays or = radials	Transverse segmentation of dorsal and anal fins	Transverse segmentation of pelvic fin	Transverse segmentation of pectoral fin	Interopercular present or absent	Suspensorium inclined or vertical	Coronoid process present or absent	Maxillary free or attached	Anorbital present or absent
Postrostral present or absent		+	+			X		X		
Antorbital present or absent										
Maxillary attached or free		+				X	x	X		
Coronoid process present or absent		+				X	x			
Suspensorium inclined or vertical	X	X								
Interopercular present or absent										
Transverse segmentation of pectoral fin				X						
Transverse segmentation of pelvic fin										
Transverse segmentation of dorsal and anal fins		X								
Dorsal and anal fin rays or = radials										

[a] As presented by Schaeffer (1956) in a study of the mosaic pattern of evolution of subholosteans. Probabilities based on 2×2 tables: X, $P < .01$; x, $P > .02$, $< .05$; +, $P > .05$, $< .10$.

grade represents a somewhat stable association of morphological features, which, along with associated functional and physiological properties, tended to persist through various avenues of adaptive radiation.

The pholidophorid version of the holostean level included some features and combinations that differed from those of the more typical holosteans. These were morphologically minor, but in retrospect, in view of the subsequent teleost radiation, they can be seen to have been of the greatest importance. The significance, as in most instances, is not recognizable in the incipient stages, and the importance is not intrinsic to the structure but in its potential *and* the fact that this potential was realized. The potential possibly could be recognized, but its realization, considering the many factors involved, certainly could not have been predicted.

Historically the study of the evolution of the actinopterygians has been slowly assembled from many bits of information that finally have led to the current understanding. Early, especially on the basis of modern fishes, the basic grades were recognized as taxonomic units. To understand the intricacies of the actual record, it is important to be aware of the morphology exhibited at these developmental levels, since they provide the background against which the intricacies of parallelism and the importance of environment and adaptation can be studied and assessed. We treat the two major systems for which there is a good record within this framework, first taking up the conditions at the different levels, without regard for the precise sources of origin, and then considering some of the details of analyses that have led to the interpretations of the overall patterns of evolution.

The Locomotor System

Scales, body form, size, fin structure, and internal ossifications all relate to locomotion. In part these structures act in the production of propulsive forces, but they also play roles in positioning and to some extent in the hydrostatic adjustments of the organisms. Specific gravity, in relationship to the water in which particular fishes lived, is important in the assessments of locomotion, and the heavy scales and thick external bones of some of the early fish strongly suggest a high specific gravity, which must have had important effects, especially in freshwater environments.

All of the actinopterygians, except for some specialized teleosts, appear to have had some facility for hydrostatic adjustment by use of their lung–gas bladder systems. That modifications of the hard structures were accompanied by changes in hydrostatic systems can hardly be doubted, but the nature of the latter can be known directly only in moderns. Consistent changes in the locomotor systems, as outlined below, probably were closely allied to increased efficiency of the structures related to hydrostatic modifications, but the correlation of these presumed associations is largely speculative.

The primary data about fossils come from the fins and body structures. Although these show many concurrent changes, brought out in consideration of particular patterns of change later in this section, the most important points are best clarified by individual treatment.

Caudal Fin

The most evident change in body structure and one of the most interpretable is that shown by the caudal fin. The fundamental forms are shown in Figure 14, but there are many variations. The primitive caudal fin was clearly heterocercal, an epibatic fin that lacked or nearly lacked an epichordal lobe. Recognition of this condition in actinopterygians is facilitated by the presence of the scale-covered lobe of the body that extends to the termination of the fin along its dorsal margin. The large hypochordal lobe was deeply cleft, and the two portions tended to be somewhat different in size.

The action of such fins, as discussed on pages 451–453, was to produce a forward thrust and to cause the tail to rise and the body to pitch downward anteriorly. This action was countered by the pectoral fins and to some extent by the rostrum, both of which tended to raise the anterior end of the fish.

Evolution in the actinopterygians proceeded through successive stages to a reduced heterocercal tail, found in subholosteans and some holosteans, and to a strictly homocercal condition in which the tail fin is symmetrically disposed, as in many teleosts. Functionally the fin so produced is isobatic, with the thrust being entirely forward.

With the evolution of the homocercal tail the opportunity for many specializations was opened. Modifications, such as those described on pages 448–451, produced highly efficient arrangements of bones and tendons (Figure 91), trending particularly toward isolation of the motions of the caudal fin from other body motions in fast-swimming fishes. Some fishes at the holostean level showed similar trends, but relatively few of them reached the high level of specialization and efficiency found among the teleosts. Within the teleost level, starting from a stage in which the heritage of the asymmetrical caudal fin is clear, modifications of the fins went in many directions, and although few generalizations are possible, each of these depended basically on the attainment of the initial homocercal condition.

The Dorsal and Anal Fins

Primitively both the dorsal and anal fin were triangular and had rather elongated bases. The dorsal fin was supported by two rows of radials, and the anal by a single row. Fin rays, lepidotrichia, were numerous, exceeding the number of radials, bifurcated distally, and set close together. The rays were thick and heavy, of the same ganoid composition as the scales. The result was a strong, but relatively inflexible, fin. The primary function appears to have been one of stabilization (see pages 440–442). The single dorsal fin tended to oppose the anal fin fully or partially, in some instances lying somewhat in front of it. Even in the array of very early paleoniscoids there is some variation in positions and relationships of these fins, but only in the very aberrant *Tarrasius* (see Figure 123*G*), in which there seems to be a continuous fin that incorporates both, did they enter significantly into production of locomotor thrust. In all primitive actinopterygians the dorsal and anal fins, as well as the caudal and paired fins, tended to be set with well-developed fulcral scales.

Both the dorsal and anal fins underwent many modifications in the course of actinopterygian evolution, but the trends are less consistent than those for the caudal fin. The rays became fewer, lighter, and more flexible, coming to match the radials in number. Distal bifurcation was lost. The bases tended to become more restricted, and mobility and flexibility were increased. This produced the holostean condition of the fin structure when fully developed, but the trends in many lines, though evident, did not carry fully to this stage. Continued modifications in the same general directions produced the fin pattern of the primitive teleosts, and within this group developed the spiny fins of the acanthopterygians, with the soft rays of the primitive teleosts being replaced by stiff, pointed spines, especially in the dorsal fin.

At all levels special adaptive modifications of the dorsal and anal fins took place. With the development of eel-shaped, elongated bodies, the fins, either separately or merged with the caudal fin, became part of the mechanism of locomotor thrust. Anguilliform swimming was developed at all grade levels. At each level as well, elongated dorsals and anals developed into propulsive organs that depended on fin undulations as their primary mechanism (see pages 453–456). Deep-bodied types arose in many lines, and in these the dorsal and anal fins were modified to occupy the steeply sloping posterior margins of the disklike body and to function in locomotion and positioning in conjunction with the caudal fin. At each of the levels these modifications were superimposed on the basic fin structure that characterized the particular stage of development. The progressive evolution, especially the increasing mobility of the fins among some of the teleosts, permitted development of very highly specialized dorsal fins, such, for example, as the dorsal fins of the sea horses or of the knife fishes.

Paired Fins

The pectoral and pelvic fins have undergone two principal types of modification: one of structure and the other of position. The fundamental significance of the changes has been explored to some extent on pages 458–471. The primitive actinopterygian paired fins were more or less triangular, with a very broad base (Figure 15). Even in the Devonian, however, there was

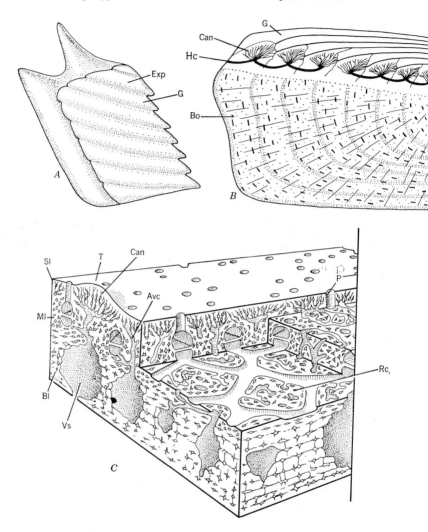

Figure 118.   The ganoid scale of a paleoniscoid, showing the full scale, *A*, and a section, *B*. *C* shows the cosmoid structure, common in some scales, but here based on section from the exoskeleton of *Tremataspis* to show maximum detail. Avc, ascending vascular canal; Bl, basal layer, Bo, bone lamellae; Can, canaliculli; Exp, exposed surface; G, ganoine; Hc, horizontal canal; Ml, middle layer; P, pore of sensory canal system; Rc, radiating canal; Sl, superficial layer; T, tubercle; Vs, vascular space, A, B, after Gross (1966); C, after Denison (1947).

considerable variation in this basic plan. The pelvic fin was small and lay well anterior to the anal. Its function at this stage is not particularly clear (see pages 460–462). The pectoral was moderate to large, in proportion to body size, and was placed near the ventral margin of the lateral surface of the fish. Its function clearly was to provide a lift of the anterior part of the body, counteracting the pitch induced by the epibatic caudal fin.

Both fins, like the medians, had numerous, closely spaced, heavy rays. Internal radials articulated with the girdles, the pectorals forming a somewhat fanlike structure, and the pelvics converging to a small area of articulation on the small pelvic element. These fins had little mobility and probably were restricted mostly to vertical movements. They could have had little function in propelling or braking actions. Fleshy lobes were in general little developed but were present in *Cheirolepis* (Figure 121) and are, of course, well developed in the living *Polypterus* and *Calamoichthys*.

Evolution of the paired fins, as far as shape and structure are concerned, saw the narrowing of the bases and reduction in the number of rays, accompanied by wider spacing and the loss of distal bifurcations. Spines developed within the teleosts, in the acanthopterygians.

Structural changes were accompanied by position changes. As in other evolutionary modifications, these proceeded at different rates in different lines, and rather similar modifications occurred repeatedly and at different adaptive levels. A common change was the forward movement of the pelvic fin, which brought it to the level of the pectoral or, in some cases, anterior to the pectoral. Accompanying this change was the tendency for the pectoral fin to lie higher on the lateral body wall. This modification took place in a number of lines of deep-bodied fishes in which the function of the two fins appears to relate to intricate maneuvers involving rapid stopping, starting, and turning. The pectoral fins in particular assume an important role in propulsion in many groups, especially among teleosts. Once flexible, narrow-based fins had developed, the way was open for a wide variety of adaptive uses, including even gliding in air, and these were repeatedly developed by different lines under differing adaptive circumstances.

Scales

The primitive scale was of the ganoid type (Figure 118) in which active growth took place both on the outer and inner surfaces, with the former producing a thin enameloid, or ganoid, layer. Large, rhomboid scales were characteristic of the primitive actinopterygians, although one very early genus, *Cheirolepis*, had very small scales, reminiscent of those of the acanthodians. Scales, like most other structures, underwent successive modifications in parallel in various lines. These changes witnessed reduction and eventual loss of the ganoid layer, preceded by reduction of the vascular "cosmoid," middle layer of bone. Eventually only the inner layer of bone remained. The rhomboidal shape gave way to a cycloid condition. Roughly a chondrostean, holostean, and teleostean level can be defined, but the trends are very general. Holosteans, and some subholosteans, lack the cosmoid layer, and teleosts lack both the ganoid and cosmoid layers.

Actually there has been a great deal of variation in scale structure at all levels. Aldinger's (1937) extensive studies, for example, have demonstrated great variation among the paleoniscoids, sufficient for scales to be used to some extent for familial taxonomic assignments. Detailed information on the scales of many of the chondrosteans, however, is not available, so that the extent and possible significance of scale modifications related to adaptive changes remain uncertain. Reduction of the scales clearly tends to lessen the specific gravity of the fishes, and it has been suggested that this is related to transfer from fresh to marine waters. The evidence, however, is far from conclusive. Greater body agility has accompanied scale reduction, and in some lines at all levels there has been a tendency for almost complete loss of the scale cover.

Body Form

The primitive actinopterygian body was fusiform, and the main lines of evolution to more advanced levels on the whole appear to have been carried out by fishes that have maintained this shape. At all levels of adaptive advance, however, there has been a strong tendency for some lines to depart from this basic form into others. The functions of different body forms, the eel-like form, deep-bodied form, and others have been already discussed (pages 442–458). Repeatedly these types have arisen, successively and with greater perfection at higher levels. Both the chondrosteans and holosteans were moderately conservative in body form, with adaptations predominantly toward elongation or deepening bodies. Some lines of teleosts, however, departed markedly from these common shapes, assuming the many bizarre patterns seen in the fishes that populate the seas at the present time.

Skulls, Jaws, and Feeding Mechanisms

The Patterns of Change

Schaeffer and Rosen (1961), Westoll (1944), and Gardiner (1963, 1967a,b) have done much to bring together and clarify the work of many ichthyologists who have made very extensive studies of both living and fossil fishes. They have concentrated especially on lower levels of the actinopterygians, leading up to the teleost level. The basic data have come from a wide array of descriptions and analyses, in particular those by Stensiö, Lehman, Moy-Thomas, Traquair, Smith, Woodward, Gross, Rayner, Aldinger, and

Brough, as cited in the references. Figures 119 through 130 illustrate representatives.

In the transition from the most primitive of the actinopterygians to the multitudes of specialized types found among the teleosts today, several adaptive levels of the skull, jaws, and related visceral elements have come into being. Modifications have involved both feeding and movements related to respiration, but the latter have in general remained fairly stable, whereas the former have undergone important modifications. The adaptive levels are those related primarily to the feeding mechanisms. As is true for the postcranium, three more or less distinctive stages can be delimited. The first is the central, paleoniscoid stage; the second is the holostean stage; and the third, in contrast to the third stage in the postcranial structure, is found developed mainly at the acanthopterygian level *within* the teleosts.

Intermediate between the paleoniscoid and holostean stages is the subholostean level, attained independently in several lines. Between the general holostean level, developed from a special expression of it, and the acanthopterygian stage, is the primitive teleost stage, which differs in small but extremely important ways from the more usual holostean conditions.

The evidence of these adaptive levels comes in part from modern fishes, in which the full structure of the cranial anatomy can be studied, and in part from fossils. The former provide some of the needed information on the relationships of hard and soft anatomy and a key to the functional relationships of the different structures. The latter supply critical data on variations of the general patterns of change and on intermediate steps. Understanding of the evolutionary history, that the several radiations have followed in the attainment of the new levels, comes from the fossil record. Without it there could be little understanding of the immense amount of "experimentation" that occurred in the course of development of the somewhat stable bases from which new radiations arose.

A multitude of skull and jaw modifications took place during the development of the adaptive levels. These were not confined to single lines but arose independently under varied environmental conditions. An immense amount of parallelism was involved, and although there were strong tendencies for suites of features to vary together, the differences in rates of change and the emphases on one or another part of the whole adaptive system were commonplace. The evidence of mosaics of characters changing at different rates and in different combinations is nowhere more evident that in these patterns of modification.

Detection and understanding of the patterns of change have come from a long series of intricate and detailed series of fishes ranging in age from Devonian to Recent. Only when this wealth of information became available could syntheses such as those taken up in this section be made.

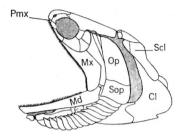

Figure 119. Jaw opening and adduction in a primitive actinopterygian, *Pteronisculus.* Cl, cleithrum; Md, mandible; Mx, maxilla; Op, operculum; Pmx, premaxilla; Scl, supracleithrum; Sop, suboperculum. (After Schaeffer and Rosen, 1961.)

The Basic Structure

The early paleonisciforms, although showing considerable diversity in the structure of their skulls, jaws, and associated visceral elements, point toward the generalized condition basic to the actinopterygians as a whole. Figures 119 and 120A show a pattern that appears to lie more or less in this area, although the genus is from the Eotriassic. The suspensorium of the jaw is strongly oblique, and the maxilla is long and firmly joined to the preopercular and infraorbital. The mandible is long and the gape quite wide. Branchiostegal rays are numerous, closely spaced, relatively small, and all similar. The cheek region is completely covered with a large opercular and somewhat smaller subopercular, two suborbitals, a dermosphenotic, and an infraorbital. The suborbitals and dermosphenotic, however, are quite variable, being absent in some of the "primitive" forms.

Two narial openings are present on each side of the snout. They lie opposite one another on each side of a bone that is conveniently called the nasal, without implying homologies to nasals in other classes. The snout has an overhanging rostrum, and the orbits are small to medium and rather far forward.

The palatoquadrate is a long element, with its articulation with the palate rather far forward. The chamber of the adductor mandibulae muscle is closed in most early paleonisciforms, flanked laterally by the cheek bones, dorsally by a flange of the preopercular to the palatoquadrate and internally by the palatoquadrate. The muscle appears to have been rather simple in structure, to have taken origin along the length of the palatoquadrate, and to have inserted in the Meckelian fossa just anterior to the jaw articulation. No coronoid process was developed, and the lower jaw functions as a straight lever. In some of the early paleonisciforms the chamber of the adductor muscle was open dorsally, and it is from groups with this feature that some of the later orders of chondrosteans appear to have arisen. Gardiner (1967A) has indicated such sources within four families—Holuridae, Gonatodidae, Acrolepidae, and Amphicentridae—partly on the basis of this structure. In these families the suspensorium is somewhat more vertically disposed than in the most primitive of the paleonisciforms.

Figure 55 shows the reconstructed adductor system of a primitive actinopterygian, illustrating the small muscle and its origin and insertion. The general nature of adduction of the jaw is shown in Figure 119, the pattern as envisaged by Schaeffer and Rosen (1961). The shoulder girdle has been drawn back and down by contraction of muscles inserting on the cleithrum. By this process the orobranchial chamber is expanded, but the expansions appear to have been limited by restricted movement of the oblique hyomandibular and slight mobility of the ceratohyal. The neurocranium is raised in the process of adduction. Adduction brings the adductor mandibulae into play, producing a biting action of rather limited force. Small sharp teeth are characteristic of the primitive paleonisciform level, and these fish probably were predators, using the bite and teeth largely to hold and position food.

This general structure and functioning of the feeding mechanism were common among the early paleonisciforms and persisted in the paleoniscids. Among even the early paleonisciforms, however, there were many modifications, many of which trended in the directions displayed by the lines that eventually culminated in the holostean level. Loss of the rostrum, assumption of a more vertical suspensorium, changes in proportions of the palate, and various patterns of "fragmentation" of the cheek elements were characteristic modifications.

The Subholostean–Holostean Stage

Figure 120 shows the transitional stages to the holostean levels through the subholostean condition as envisaged by Schaeffer and Rosen (1961). *Amia*, the modern bowfin, is a more or less central holostean in this regard, although it represents but one aspect of the full radiation and is not close to the stock from which the teleost pattern arose. In the course of transition the suspensorium became more nearly vertical and mobility of the hyomandibular was increased. The adductor mandibulae muscle increased considerably in size and complexity, expanding its origin onto the hyomandibular, the braincase, and the anterior border of the preopercular. The maxillary-palatoquadrate chamber was eliminated, with various modifications of the cheek region, as the maxilla became separated from the preopercular and the infraorbital bones. In some instances a proliferation of anamestic suborbital elements occurred, as in *Aeduella* and *Acentrophorus* (see Figure 124F and J).

The adductor mandibulae, freed from many restraints, became enlarged and subdivided. During abduction the maxilla was swung forward (Figure 120). Although this bone was freed of its major contacts with cheek elements, it still retained teeth and acted as a functional part of the jaw margin. A coronoid process developed, with the result that part of the jaw acted as a

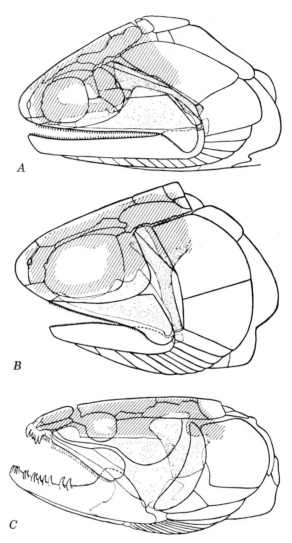

Figure 120.   A series of actinopterygian skulls, showing changes in the transition from the paleoniscoid to the teleost state. *A* paleoniscoid, *Pteronisculus; B*, sub-holostean, *Boreosomus; C*, holostean, *Amia* whose musculature is shown in *D; E*, acanthopterygian teleost, *Epinephelus* whose musculature is shown in *F*. Note especially the change in the angle of jaw suspension and the modifications of the maxilla and premaxilla. (After Schaeffer and Rosen, 1961.)

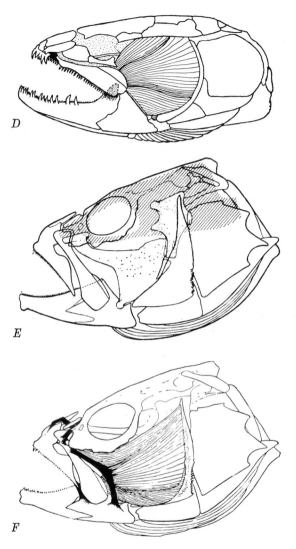

Figure 120 (continued)

bent lever, increasing the effective force of the bite. The number of branchio-stegal rays was reduced, and the most dorsal element was enlarged to form the interopercular. This entered into a new system of abduction of the mandible, induced by action of an interopercular-mandibular ligament. Contraction of the levator opercularis muscle, acting on this ligament, focused force on the end of the mandible and depressed the jaw. A somewhat greater capacity for expansion of the orobranchial chamber accompanied these changes, along with a modification of the serial relationship of the symplectic and interhyal which brought the latter into direct attachment with the hyomandibular.

These changes increased the general effectiveness of the feeding apparatus, mainly by increasing the efficiency of abduction and the strength of the bite. They opened the way for a rather modest radiation in feeding habits, which, along with locomotor modifications, is evident in the patterns of radiation at the holostean grade level.

The Teleostean Level

Teleostean fishes appear to have arisen from Parasemionotiformes by way of the pholidophorid holosteans, which with the leptolepiform teleosts are often called halecostomes, as shown in Figure 116. One line, including the leptolepids and their probable derivatives, the clupeids, was somewhat off the main course and, although prominent within the teleosts, did not participate in many of the changes of feeding mechanisms found in the line from which the acanthopterygians arose. The latter appear to have stemmed from the elopids, which preserved most of the feeding structures of the holosteans but had a few significant modifications. It was within the teleosts rather than during their origins that the major feeding modifications took place. The initial changes leading to teleosts seem to have been related to locomotion rather than feeding. One important change in the feeding mechanism found in the pholidophorids is the reduction of the dorsal process of the premaxilla. This eventually set the stage for a freeing of this bone and its development as the primary tooth-bearing element of the upper jaw, an element that eventually became protrusible in many teleosts.

Primitive teleosts, such as elopids, exhibit for the first time a ball-and-socket joint between the maxilla and palatine (Figure 120E). The premaxilla is the prominent tooth-bearing element, and in the course of evolution the tooth-bearing ramus was extended and the maxilla was completely excluded from the jaw. This change was accompanied by a series of other modifications (Figure 120). All are keyed to the freeing of the maxilla from the cheek elements, which permitted it to be pulled down and forward by the lower jaw as the mouth was opened. Schaeffer and Rosen (1961) have listed the following five modifications as significant:

1. A forward shift of the palatoquadrate and a forward movement of the lower jaw.

2. Freeing and elongation of the premaxilla and exclusion of the maxilla from the gape of the jaw.

3. Consolidation and simplification of the subdivisions of the adductor mandibulae.

4. Insertion of part of the mandibular musculature on a tendon to the maxilla.

5. Origin of a new ligament from the maxilla to the ethmoid and/or palatine bone and from the upper end of the premaxilla to the palatine bone.

Through these changes the teleosts arrived at the basic acanthopterygian pattern. Other teleosts underwent important radiations, but it was with the attainment of this development of the feeding mechanism that the stage was set for a new and incredibly complex burst of radiation for which the locomotor apparatus had provided the groundwork and the jaw modifications the basis for exploitation of an immensely diverse array of environments.

EVOLUTION AND ORIGINS

Patterns of Change

At some undetermined time during the early mid-Paleozoic the actinopterygian fishes came into existence. This probably followed soon after osteichthyan fishes had diverged from an ancestral complex that also included the forerunners of the acanthodians. The time was not later than the latest Silurian, because the organizational pattern that appears to have been common to all osteichthyans must have been in existence by that time. From this base the particular habitus that characterizes the fundamental paleoniscoid condition arose.

Neither the nature of the evolutionary shift nor the pattern of change that accompanied the divergence of acanthodians and osteichthyans, if in fact they are closely related, is known. Possibly the attainment of a lunglike hydrostatic system, found throughout the Osteichthyes, may have provided an innovation to which other changes were related. A somewhat similar lunglike structure is found among freshwater placoderms (antiarchs), but there is no positive evidence that it was present in acanthodians and no indication of it in the chondrichthyans. Thus it may be that this feature was attained independently in various lines. The various air-breathing structures in some of the teleosts indicate the relative ease of adaptation of vascular tissues to the ends of air breathing.

If there was a key change, presaging the initiation of the osteichthyan organization, it likely related to the respiratory system. Whether there was a single line from a general preosteichthyan ancestry, or whether the different groups of Osteichthyes came from somewhat divergent lines within such a complex is not known. Similarly, it is not clear whether all actinopterygians came from some single, central stock, or whether they developed from several related but somewhat divergent lines with bases below the organizational level recognizable as actinopterygian.

The usual concept of the evolution of the early actinopterygians, exemplified by the interpretations of Gardiner (1967a), recognizes the existence of a central, ancestral actinopterygian, the primitive paleoniscoid, from which later lines diverged. This may well express the actual situation, but the evidence is not definitive. If there were such an early ancestral type, the pattern of evolution would appear to have been one that first produced an array of species which differed only moderately in form and adaptations and from which then developed a more varied complex of adaptive types increasing the range of the central group but not altering materially the basic morphological patterns.

It is this basic paleoniscoid condition which is recognized in the Mississippian assemblage of central paleonisciforms by Gardiner and which persists without severe modification into the early Cretaceous. From it, at various times and with various combinations of characters, arose many lines from which the more divergent paleonisciforms and other major groups of chondrosteans developed.

Given a beginning in a coherent, central type, an early, modest radiation, and the persistence of a central type, a definable pattern of successive, multiple adaptive radiations culminating in new levels of organization emerges. From the Mississippian time on there is considerable documentation of an evolution that involves the repetitive acquisition of similar characters in many different evolutionary lines. A wide variety of mosaics of primitive and advanced features was attained by different lines in course of this adaptive experimentation, and some particularly successful combinations were sorted out from the many to produce new radiations, including those that attained the holostean level of organization. This is the general pattern of change that seems to emerge from the Paleozoic and early Mesozoic record. Both the record and the interpretations of the causes of the patterns of change are of considerable interest in regard to the actinopterygians, and they reveal modes of evolution that appear to have occurred in many other groups.

The Record

Devonian

Six well-documented genera of paleonisciforms are known from the middle and upper Devonian, but only scales, of problematical affinities, are known from the lower beds of the period. The six genera fall into four distinct groups, at about the family level. None of the genera can be considered as directly ancestral to the "central" Mississippian stock, and at least two of the families represent lines that left no known descendants. Middle and upper Devonian specimens have come from both marine and nonmarine rocks, and no evident adaptive differences separate the fishes obtained from these two environments of deposition. With the scant record, various interpretations are possible: that the fish were all freshwater, with some washed into marine environments; that some were marine and some freshwater; or that members of this group were euryhaline. In addition, it may be argued either that the origin of the actinopterygians was in fresh water or that it was in salt water. The geographic distribution is extensive and has been used as a basis for assuming an originally marine habitat. The data, however, are simply insufficient for any reliable conclusions, and interpretations of later evolutionary patterns based on the information from early environments must be suspect.

Closest to the generalized Mississippian types are the Devonian members of the family Stegotrachelidae, marine and nonmarine fishes known from both the middle and upper Devonian (Figure 121). Unlike the central Mississippian forms, however, they possess a well-developed pineal opening as well as other unique morphological features. The pineal opening is found in a Mississippian descendant, *Kentuckia*. *Osorioichthys*, of the upper Devonian, shows some resemblances in its opercular apparatus to later Rhabdolepidae, but there is no clear evidence of close relationships. *Tegeolepis*, representing a different family, is a large form, a meter long, which shows some rather striking resemblances to the Chondrosteiformes of the Mesozoic and suggests that this pattern may have already emerged in the Devonian. The lepidotrichia, scales, including absence of fulcral scales, and features of the palate and suspensorium are suggestive of relationships, but in view of the repeated acquisition of similar suites of characters in many different lines, the long time gap must make this proposed relationship highly tentative.

Perhaps the most widely known Devonian genus, from both the middle and upper beds of the period, is *Cheirolepis*. It seems to have left no descendants in spite of its wide geographic range and considerable duration in the Devonian. The scales, unlike those of most paleonisciforms, are small and resemble those of acanthodians quite closely, providing one of the strong arguments for relationship between these two groups.

The Devonian record offers no particular support for the concept of early development of a central paleonisciform complex of the sort outlined in the preceding section, but it does not argue strongly to the contrary. Much depends on interpretations of how complete the record of Devonian paleoniscoids actually is. Clearly it is far from complete, but it is difficult to tell how biased the sampling may have been, whether what is known is representative or whether it has tapped only a few lines that were divergent from a little-known main stock.

If there were a fairly abundant central stock, with many species of generally similar nature, then the record must be extremely faulty. If, however, a hypothesis of independent divergence of a number of lines from one or

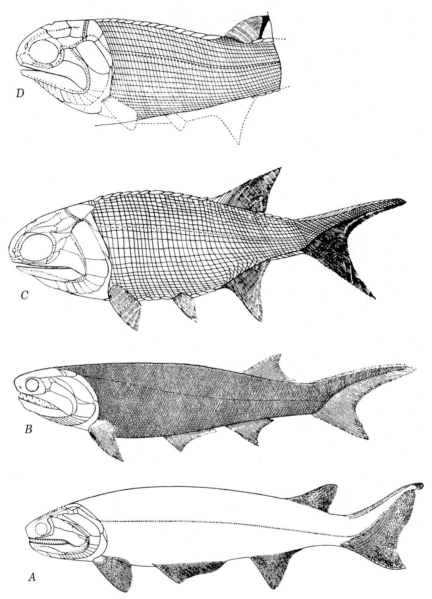

Figure 121.   Some Devonian paleoniscoids. *A* and *B*, two examples of *Cheirolepis*—
A, New World and *B*, Old World—showing the great resemblances of these fishes
from the two continents at this time. Note the small scales in *B*, characteristic of this
genus but not of paleoniscoids in general. *C*, *Stegotrachelus*, a composite; *D*,
Moythomasia. (*A* after Woodward, 1898; *B* after Lehman, 1947; *C* after Lehman,
1966; *D* after Gardiner, 1963.)

several bases be adopted, then the record may be considered to be more representative. In this event one of these lines, perhaps represented by the Stegotrachelidae, might be thought of as one that radiated markedly, rather late in the Devonian, to produce the Mississippian central complex.

Any such hypothesis leaves the source of some of the more divergent Mississippian forms unknown and requires rather rapid deployment. Furthermore, even the closest Devonian family, the Stegotrachelidae, does not appear to contain the actual ancestors of any of the central Mississippian types. Such a hypothesis likely is invalid, but on the basis of the Devonian record alone no hypothesis has more than a very tenuous basis in fact.

The Mississippian

A major part of what is known of the Mississippian actinopterygian faunas comes from the Glencartholm in Eskdale of the British Isles. From a thin belt of shaly sediments has come a wide variety of fishes. From their structural differences it is clear that they were adaptively diverse, but there is little information on the environment that gives clues as to their ways of life or the exact habitats in which they lived. In general they seem to have been deposited in marginally marine to nonmarine beds, perhaps associated with a very-low-lying delta. They variously have been interpreted as marine, estuarine, and freshwater. Marine invertebrates occur in the sequence, but it has been noted by Peach and Horne (1903) that the fishes usually occur in a separate, thin bed, which might be of freshwater origin, with the fishes and crustaceans. Only the most meticulous study could produce data necessary for fully reliable interpretations, and this has not been made.

Over 40 genera have been named from the Mississippian, and these have been arrayed in 15 or more families, the number depending largely on the personal judgment of the student making the classification. All genera except the very aberrant *Tarrasius*, whose position has been much debated, have been placed by Gardiner (1967a) among the paleonisciforms without further grouping above the family level. They have, however, been grouped into three suborders—Palaeoniscoidea, Platysomoidea, and Tarrasioidea—under the single order Palaeonisciformes by Romer (1966a), following in general Moy-Thomas. His order includes all major groups of chondrosteans, except those with living representatives, the acipenseriforms and polypteriforms. Gardiner has recognized 18 orders. In spite of these seemingly great differences in classification, the position and evolutionary significance of the Mississippian genera are reasonably similar in the two treatments.

Of the wide variety of paleonisciforms developed at this time, a number of families form a central core group that is presumed to reveal the basic paleoniscoid condition, in the sense noted in discussions of the general pattern of evolution. It is presumed as well that this is a continuation of a

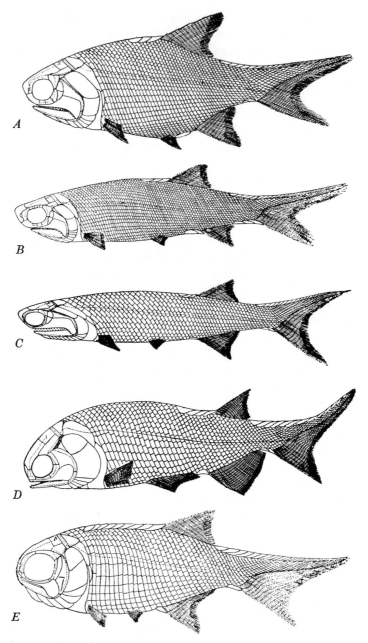

Figure 122. "Central" paleoniscoids of the Carboniferous. *A, Elonichthys pulcherius; B, Elonichthyes serratus; C, Cycloptychius; D, Whiteichthys; E, Canobius; F, Aetheretmon; G, Gonatodus; H, Rhadinichthyes; I, Carboveles.* (*A, B, C, E,* and *H* after Moy-Thomas and Dyne, 1938; *D* after Moy-Thomas, 1942; *F* and *I* after E. I. White, 1927; *G* after Gardiner, 1967a.)

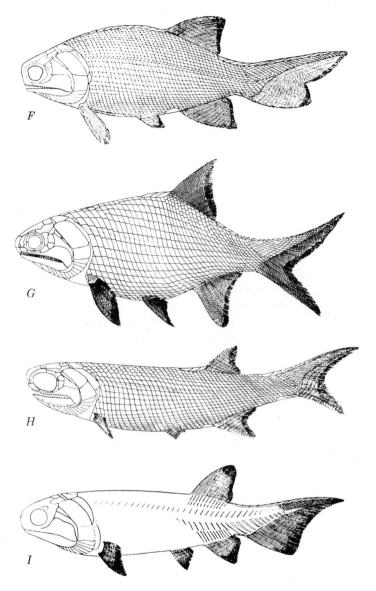

Figure 122 (continued)

similar but little-known central group of Devonian age. Gardiner has recognized 12 families as constituting this array. Many of these have but one or two genera, and Gardiner's recent analysis has increased the number of families over those recognized by most earlier students.

Some members of this central group are illustrated in Figure 122. The Elonichthyidae and Rhadinichthyidae fall close to the basic paleoniscoid condition, but even within these families and their genera, as seen in Figure 122A, B, E, and F, there is considerable variation both in body shape and in details of the structure of the fins and feeding structures. Gardiner's phylogeny (Figure 115) shows the constitution of this group as he conceived it and its derivation from a common base somewhat earlier. Diverging even before this time, from the common stem, were some of the distinct Devonian and Mississippian types.

A large number of Mississippian paleonisciforms, though hewing in general to the central paleoniscoid pattern, depart from it sufficiently to be hardly considered as part of the central complex (see examples in Figure 123). Among these, and initially very close to the common base, are the deep-bodied fishes. These have been placed together in the Platysomoidea by Romer (1966a), but there is at least some evidence in their divergent structures that they developed somewhat independently (see Figure 123B, E, F, and J). Fusiform variants are found in the heavily scaled and sharp-snouted *Phanerorhynchus* (Figure 123C). Although the structure of the head resembles that in some later chondrosteans, relationships are at best remote. The same is true of *Cornuboniscus* (Figure 123D), whose pectoral fin resembles superficially that found in *Polypterus* of the present day. *Holurus* (Figure 123A) has elongated dorsal and anal fins, foreshadowing developments in later forms to which it may have some relationship. It is one of several fishes of the Mississippian complex in which the chamber of the adductor mandibulae is not completely closed dorsally, a feature that Gardiner believes to be essential for any fish that was ancestral to more advanced chondrostean stocks.

Very far off the main line, but with generally paleoniscoid features of the skull and jaws, is *Tarrasius* (Figure 123G). The place of this strange fish and its meaning in the interpretation of paleoniscoid evolution are something of a puzzle. It appears to have developed swimming by means of undulatory motions of the combined dorsal, anal, and caudal fins, much like that found in fishes at considerably higher levels of organization. That this was accomplished so early in the history of the actinopterygians suggests either a long period of independence from other lines or else a capacity for very rapid adaptive modification.

The Mississippian complex as a whole is sufficiently diversified for all or almost all later stocks to have descended from it. This includes both lines that were ultimately successful, in that they persisted for a considerable

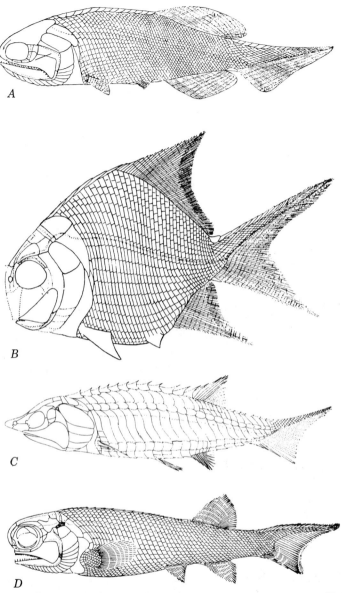

Figure 123. Paleozoic paleoniscoids that have diverged from "central" pattern both in fundamental organizational features and in special adaptations. *A, Holurus,* Mississippian; *B, Paramesolepis,* Mississippian; *C, Phanerorhynchus,* Pennsylvanian; *D, Cornuboniscus,* Mississippian; *E, Chirodus,* Carboniferous; *F, Eurynothus,* Carboniferous; *G, Tarrasius,* Mississippian; *H, Dorypterus,* Permian; *I, Cheirodopsis,* Mississippian. (*A, B,* and *I* after Moy-Thomas and Dyne, 1938; *C* after Gardiner, 1967a; *D* and *F* after Lehman, 1966, with *D* from E. I. White and *F* from Moy-Thomas; *G* after Moy-Thomas, 1934; *H* after Westoll, 1941b.)

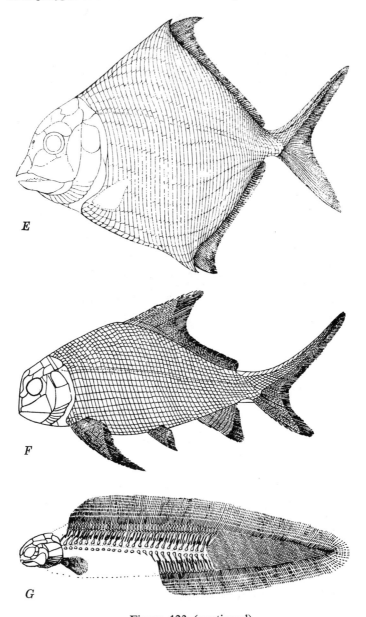

Figure 123 (continued)

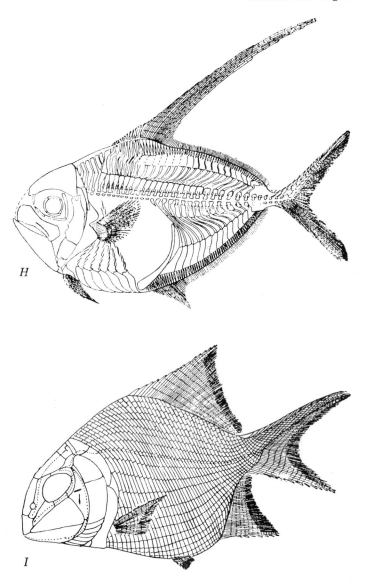

Figure 123 (continued)

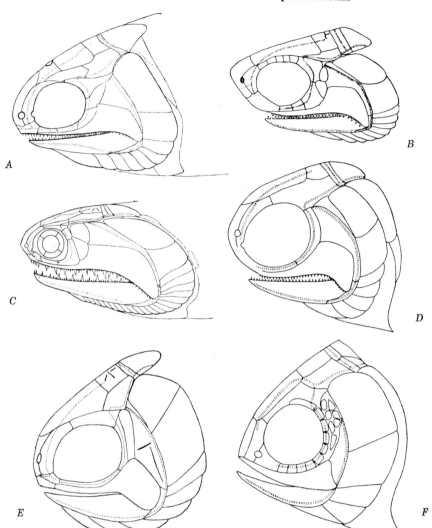

Figure 124.   Skulls and jaws of a variety of paleoniscoids and primitive holosteans, illustrating the basic modifications that took place during evolution, especially in structures related to feeding. *A, Stegotrachelus,* Devonian; *B, Eloniththys,* Carboniferous; *C, Nematoptychius,* Carboniferous; *D, Mesopoma,* Carboniferous; *E, Canobius,* Carboniferous; *F, Aeduella,* lower Permian; *G, Holurus,* Carboniferous; *H, Cheirodopsis,* Carboniferous; *I, Paramesolepis,* Carboniferous; *J, Acentrophorus,*

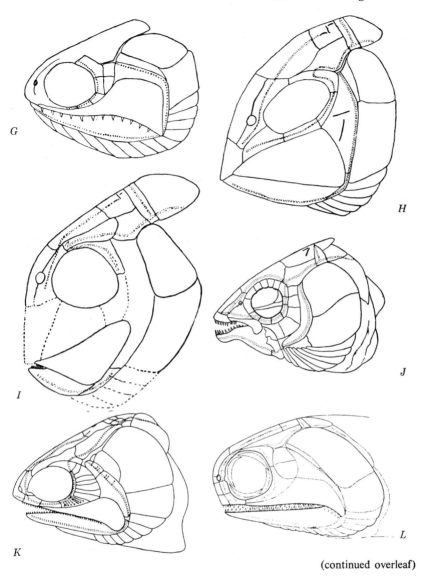

(continued overleaf)

Figure 124 (continued)—upper Permian; *K*, *Boreosomus*, lower Triassic; *L*, *Meso-nichthys*, Carboniferous; *M*, *Sakamenichthys*, lower Triassic. (*A* and *C* after Gardiner, 1963; *B*, *E*, *G*, *H* and *I* after Moy-Thomas and Dyne, 1938; *D* and *F* after Gardiner, 1967b, from Moy-Thomas; *J* after Lehman, 1952; *M* after Lehman et al., 1958.)

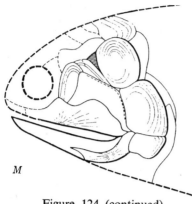

M

Figure 124 (continued)

period, and lines that were short lived, in some instances highly successful for a time but then disappearing from the record. Within the variants of the complex are found incipient, and sometimes fully developed, many of the features that, when found in full association, define the holostean condition. In the early complex these characters occur singly or in various small suites in many different combinations. Examination of Figures 122 and 123 with these features in mind will make clear some of the patterns, both those that result in some particular adaptive type and others that appear to be of quite general adaptive significance. Departures from the central pattern involve reduction of heavy scales, reduction in the number of fin rays, loss of distal bifurcation of the rays, narrowing of the bases of the paired fins, differences in the cleft of the caudal fin, and dorsal migration of the pectoral fins. Various modifications of body shape, trending toward deep and elongated body forms, have taken place. The skulls and jaws show reductions of the rostrum, countered by development of beaks in a few cases, trends toward vertical suspensoria, various "fragmentations" of the bones in the postorbital area of the cheek region, and modifications of the nares and of the lateral-line system as impressed on the dermal bones. Some of these modifications are shown in detail in Figure 124.

In the presence of all of these modifications, however, the caudal fin remained heterocercal, except in the extreme case of *Tarrasius*, and the maxilla remained firmly attached to the cheek elements. These key characters held inviolate.

The relationships of the members of the Mississippian complex to fishes of later times are, of course, somewhat speculative and open to various interpretations. Very little opportunity of following lines through closely graded stages exists. The phylogeny of Gardiner (1967a) is one of the most recent (see Figure 115). In his discussion he indicates that he considers some parts of it well supported and others quite speculative, and other students have found reasons to question some of the parts he believes to be well documented. In spite of these problems, however, the existence of a central paleoniscoid complex along with some highly divergent and aberrant lines, and the presence of the sources of later fishes partly within the complex and partly within more divergent lines, outside it, are well documented. The evolutionary pattern appears to be reasonably clear, although details are often obscure. What happened after the Mississippian and how the Mississippian complex related to it is much clearer than what preceded this time and how the earlier faunas stood in relationship to those of the Mississippian.

Post-Mississippian Paleozoic Radiations

Evolution of the chondrosteans during the Pennsylvanian and Permian has several distinctive features, presumably more or less typical of a major radiation that follows the origin and early deployment of a stock. Most of the representatives found in these times can be traced back to antecedents in the Mississippian. The central, conservative pattern, represented by several families in the Mississippian, is carried on by the paleoniscids proper through the Permian and by very similar descendants all the way to the base of the Cretaceous (see representatives in Figure 125). A number of families persisted from the Mississippian with only modest modification involving departures from the central complex of the sorts already initiated. In addition new directions of evolution are found in a number of genera, sufficiently distinct for the genera to have been placed in new families. *Aeduella, Lawnia, Sphaerolepis, Paramblypterus,* and *Haplolepis* (Figure 126) are examples of some of these new trends.

Gardiner (1967a,b) concluded that all of the later orders of chondrosteans, 18 in his classification, arose from but 4 Mississippian families: Holuridae, Gonatodidae, Acrolepidae, and Amphicentridae (Figure 115). Representatives of these are shown in Figures 122, 123, and 124: *Holurus* in Figures 123A and 124G, *Gonatodus* in Figure 122G, *Cheirodopsis* (amphicentrid) in Figures 123I and 124H, and *Mesonichthys* (acrolepid) in Figure 124L. Members of these families had an open adductor mandibulae chamber and a more or less vertical suspensorium. Other special features of the scales, fins, and skulls suggest relationships with one or another of the later lines. Although members of some other families had open adductor chambers, they do not appear to have led to successful radiations, and none of the higher orders appears to have stemmed from fishes with closed chambers.

The links between the Mississippian genera and the Pennsylvanian and Permian genera are necessarily somewhat problematical. Regardless of this, however, two aspects of the evolutionary patterns stand out. First, a number of separate, developing evolutionary lines were sorted out from the earlier complex, and, along with some more persistently conservative stocks, these made up an important part of the fish faunas of these times. Second, many members of the various independent lines developed different arrays of holosteanlike characters.

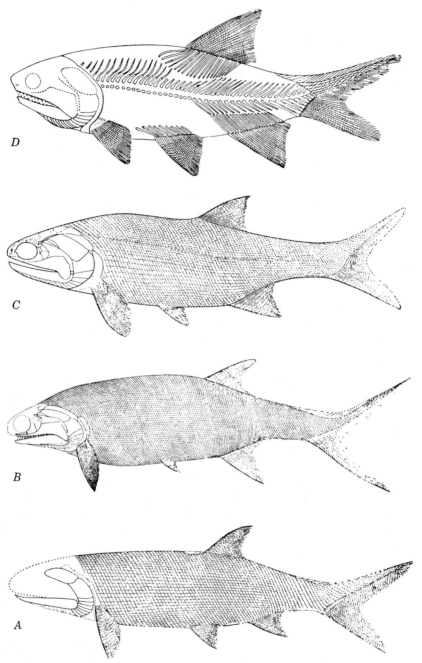

Figure 125. Late Paleozoic and Mesozoid paleoniscoids. *A*, *Pygopterus*, middle
Permian to lower Triassic; *B*, *Palaeoniscus*, Permian to Triassic; *C*, *Pteronisculus*
lower Triassic; *D*, *Coccolepis*, lower Cretaceous. (*A* and *B* after Aldinger, 1937; *C*
after Lehman, 1952; *D* after Lehman, 1966, from Traquair.)

Gonatodids seem to lie close to the base of the aeduellids, and these in turn are not far from the ancestry of the parasemionotiforms (Figure 127*F*), a very important chondrostean group that appears to have close ties to the holosteans. The upright suspensorium and modifications of the maxilla and of the preopercular suggest these relationships, although they are far from conclusive. Also probably stemming from this stock are members of the Amblypteridae (Figure 126*B*), which possess an upright suspensorium, long-based paired fins, a single tooth row, and a large opercular. The body is deeply fusiform, paralleling the platysomids and amphicentrids, which seem to have stemmed from another, common, base. During the Permian the spectacular fish *Platysomus* (Figure 126*E*) was developed, and seemingly related were *Dorypterus* (Figure 123*H*) and the Triassic bobasatraniiforms (Figure 127*D*).

The most conservative lines appear to be closely related to the acrolepids of the Mississippian. Holurids, already well separated from the central stock in the Mississippian (Figure 123*A*), appear to lie close to the base of several lines, some of which were successful and others, abortive (Figure 115). Among possible derivatives are members of the family Haplolepidae, which form a short-lived but interesting group treated at length by Westoll (1944) in a study that contributed importantly to concepts of chondrostean evolution.

Haplolepis (Figure 126*F*) shows the general characteristics of this family, the phylogeny of which is shown in Figure 114. The members of the family lived, for the most part, in coal swamps of the Pennsylvanian, and their evolutionary patterns have been attributed in part to the oxygen-deficient environment of these waters. During their rather short history these fishes made a number of modifications suggestive of holostean morphology. The skulls reveal rather striking convergence with those of later bony fishes, and the postcranium, with its narrow, somewhat mobile fins, with few rays that lacked distal bifurcation, went in a similar direction. Branchiostegals were reduced, the opercular apparatus was very small, the rostrum absent, the suspensorium vertical, and the cheek region considerably fragmented.

The various characters developed within the haplolepids are also found in other paleonisciforms, but, as Westoll stressed, the particular combinations are not found elsewhere. The mosaics of patterns of the skulls within this group are illustrated in Figure 128, which shows three types, in comparison with conditions later realized in two chondrosteans, *Peltopleurus* and *Perleidus*, and the primitive halecostome, or teleost, *Leptolepis*.

Late Permian and Mesozoic Radiations: Subholosteans–Holosteans–Halecostomes

Both the marine and fresh waters of the late Permian and the Triassic were dominated by a wide variety of chondrostean fishes. With the coming of the

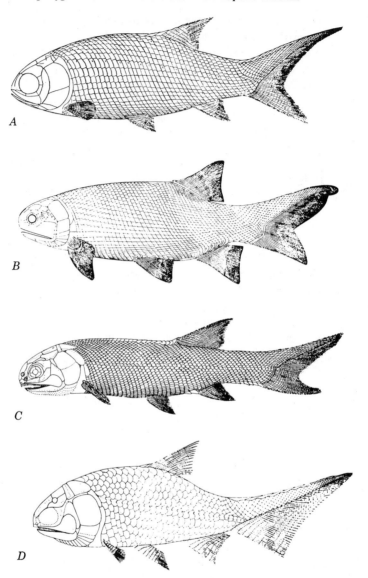

Figure 126.   Moderately to highly specialized Carboniferous and Permian paleoni-
sciforms. *A*, *Aeduella*, lower Permian; *B*, *Paramblypterus*, Upper Carboniferous
(Pennsylvanian); *C*, *Lawnia*, lower Permian; *D*, *Sphaerolepis*, lower Permian; *E*,
Platysomus, Carboniferous and Permian; *F*, *Haplolepis*, Upper Carboniferous
(Pennsylvanian). (*A* after Westoll, 1937*b*; *B* after Blot, 1966; *C* after Wilson, 1953;
D after Gardiner, 1967*a*; *E* after Moy-Thomas and Dyne, 1938; *F* after Westoll,
1944.)

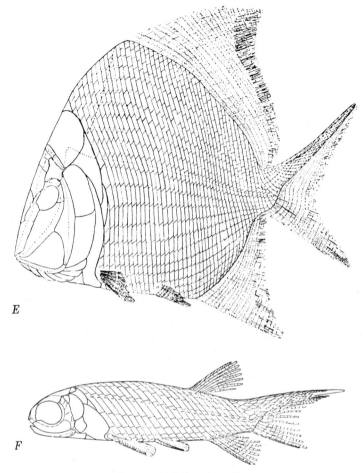

E

F

Figure 126 (continued)

Jurassic the numbers and variety reduced as they were replaced by fishes of the holostean grade, which arose from within their ranks.

The radiation complex of the chondrosteans, in many different lines with parallel and convergent adaptive types, witnessed the independent attainment of various holostean features, in various sequences and combinations. From this bewildering array it has been possible to sort out, tentatively, the major aspects of this evolution, although some loose ends still exist. The most recent work on these lines has been by Gardiner (1967a), and prior to his work Romer (1966a) made a comprehensive synthesis. Although the two do not agree in all details and have somewhat different classifications, the major aspects are treated in comparable ways.

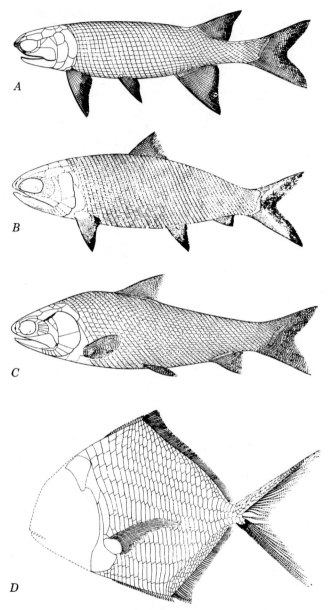

Figure 127. "Subholosteans," showing the range of adaptive types during the early Mesozoic. *A, Cionichthys,* upper Triassic, *B, Ptycholepis,* middle to upper Triassic; *C, Boreosomus,* Eotriassic; *D, Bobasatrania,* Eotriassic; *E, Saurichthys,* Eotriassic; *F, Parasemionotus,* Eotriassic; *G, Cleithrolepis,* Triassic; *H, Luganoia,* middle to upper Triassic; *I, Peltopleurus,* Triassic; *J, Atopocephala,* Triassic; *K, Perleidus,* Triassic. (*A* after Schaeffer, 1967*b*; *B* after Wenz, 1960; *C* and *K* after Lehman, 1952; *D* after Stensiö, 1932*b*; *E* after Stensiö, 1926; *F, H* and *J* after Lehman, 1966, from Brough, 1939; *G* after Brough, 1939.)

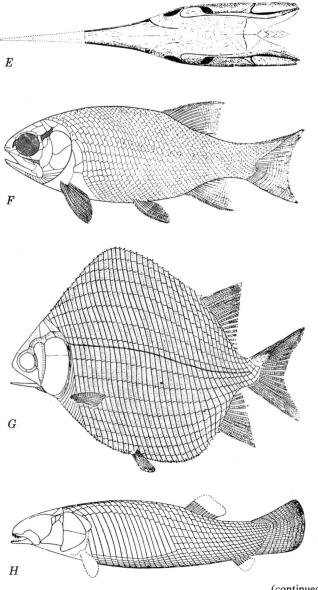

E

F

G

H

(continued overleaf)
Figure 127 (continued)

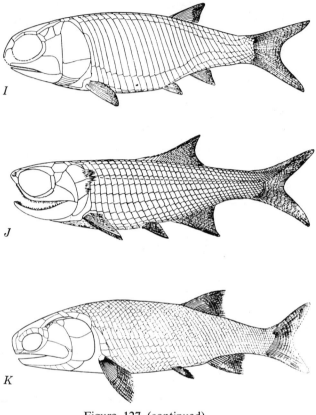

Figure 127 (continued)

The two major problems in assessing the patterns of evolution relate to the phylogenetic sources of the major groups and the relationships between members of rather well-defined groups. In addition the inevitable loose ends, genera that have no evident close relationships to other chondrosteans, make the picture somewhat unclear. Gardiner has made an effort to suggest at least tentative affiliations for most of the chondrosteans, whereas Romer has left a number of genera unassigned at a lower level. Although some of these problematical fishes may eventually fill important niches in the phylogenetic scheme, it appears unlikely that they will cause major alterations in what now has been proposed.

Several phases of evolution can be recognized in the late Permian and Mesozoic. They involve lines that have survived with little change. These in some instances have merely persisted without becoming prominent, but in

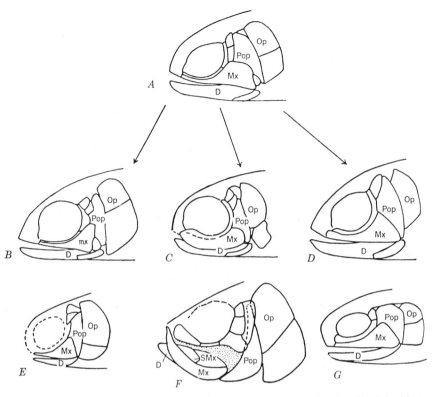

Figure 128.   Development of three types of skull pattern in the Haplolepidae, showing convergence with patterns developed as well in later and more advanced actinopterygian fishes. *A*, unnamed central type; *B*, *Haplolepis* (*Parahaplolepis*) *tuberculata*, convergent with *E*, *Peltopleurus lissocephalus*, an upper Triassic chondrostean; *C*, *Haplolepis* (*Haplolepis*) *ovoidea*, showing convergence with *F*, a halecostome *Leptolepis* of the upper Jurassic and close to teleosts; *D*, *Haplolepis* (*Haplolepis*) *corrugata*, showing convergence with *G*, *Perleidus madagascarensis*, a lower Triassic chondrostean. *D*, dentary; Mx, maxilla; Op, operculum; Pop, pre-operculum. (Somewhat modified from Westoll, 1944.)

others they have undergone considerable adaptive proliferation, without great change in the basic characteristics. Other lines have persisted, without marked radiation, but have become more progressive within one or another partially set adaptive mode. Still other lines became progressive, assuming many features suggestive of holosteans. These tended to undergo notable radiations and to provide a great many of the fishes during this time of major chondrostean evolution. From at least two of these lines stemmed fishes that became holostean in structure.

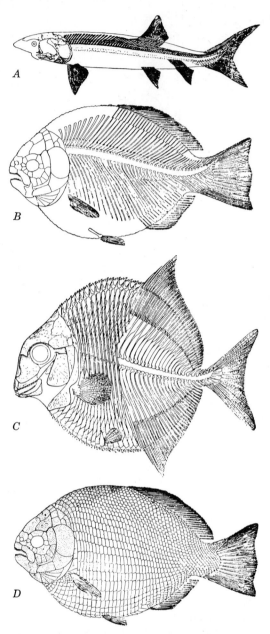

Figure 129.   An advanced chondrostean and various holosteans and halecostomes from the Jurassic and lower Cretaceous. *A*, *Chondrosteus*, a chondrostean trending in structure toward *Acipenser; B* and *D*, *Dapedius*, a deep-bodied semionotid holostean; *C*, *Proscinetes* (*Microdon*), a pycnodontid holostean; *E*, *Caturus*, a fusiform, somewhat central holostean with amioid affinities (see also Figure 130); *F*, *Amiopsis*, an *Amia*-like holostean; *G*, *Hypsocormus*, a somewhat specialized holostean with ossified vertebral centra; *H*, *Ophiopsis*, an early semionotid holostean; *I*, *Aspidorhynchus*, an elongated, armored halecostome; *J*, *Leptolepis*, a Jurassic halocostome, teleost in some classifications; *K*, *Pholidophorus*, a central halecostome. (*A*, *B*, *D E*, *G*, *I*, and *J* after Woodward, 1895 (1891–1895); *C*, *F*, *H*, and *K* after Woodward, 1918 (1916–1919).)

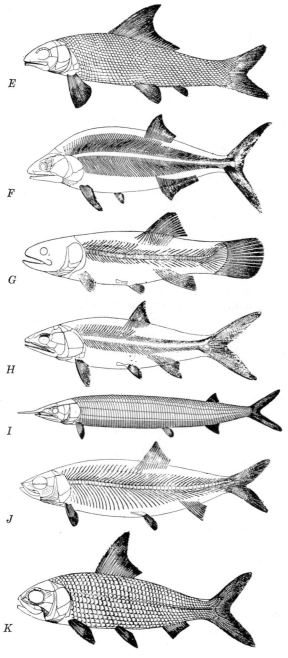

Figure 129 (continued)

569

SURVIVING PRIMITIVE STOCKS. The basic paleoniscoid pattern was preserved nearly intact through the Permian and into the Mesozoic, dying out only in the early Cretaceous. If the conservative classification of Romer (1966a) is followed, a number of the families known in the Carboniferous continued into the late Permian. Even if a greater splitting of genera is employed, the fact that very similar forms persisted for long spans of time is evident.

Late in the Permian the family Palaeoniscidae, more or less the central model of the paleoniscoid structure, came into being. It continued into the late Triassic, with the development of 10 or more genera. In basic structure members of the family are very close to the acrolepids and elonichthyids, differing mostly in some features of the skull roof, nares, and the extent of modification of the cheek region. Typical is *Pteronisculus* (Figure 125C). As this genus shows, the bodies were fusiform, the tail was strongly heterocercal, the suspensorium was oblique, strong fulcral scales were present, the branchiostegals were numerous, the maxilla was tightly joined to other elements of the side of the skull, and the paired fins were broad based and carried a large number of lepidotrichia, exceeding the number of radials. In the early Jurassic the Centrolepidae and Cosmolepidae continued this conservative line, with only moderate changes, and in the lower Cretaceous the family Coccolepidae brought this particular radiation to a close. This family was marked by many primitive features, although it was modified by strong reduction of the scales and a reduced rostrum (Figure 125D).

The majority of these fishes were marine in habitat, but some middle Triassic genera from Australia are from sediments deposited in fresh water. Very little or no change in basic osteological features, in body form, in feeding mechanisms, or in the placement and structure of fins accompanied the shifts of environments that must have occurred in various branches of this stock. There appears to have been considerable radiation, but it was not accompanied by marked changes in the structures that are preserved in the fossil record. Similar types as far as body shape, dentitions, and fin placement are found among holosteans, as shown in Figure 129, but these are accompanied by vast differences that mark the distinct grade levels of these superficially similar fishes.

PERSISTENT, MODERATELY PROGRESSIVE LINES. In contrast to the paleoniscids, which changed but little from the ancestral condition, were several developing lines of the Mesozoic, the Triassic in particular, in which some primitive characters persisted with little change, whereas others showed notable modification, partly related to specific adaptive changes and partly interpretable as expressions of the general tendencies toward the "holostean" grade found widespread among subholosteans.

Certain of the deep-bodied fishes of the time fall into this general category. They have taken on many of the special adaptive features associated with disk-shaped fishes, have advanced somewhat in basic structure over their presumed amphicentrid ancestry, but have retained features that are basically central to paleoniscoids. Best known is *Bobasatrania* (Figure 127*D*), a strikingly adapted, deep-bodied genus, characterized by strong crushing teeth. Probably not distantly related is the deep-bodied genus *Dorypterus* from the upper Permian of Germany. Both, in different ways, show marked modifications associated with the body form, convergent both with more distinctly paleoniscoid fishes, such as *Platysomus*, a deep-bodied, persistent type from a separate source, and with more advanced fishes at both the holostean and teleostean levels.

Bobasatrania and *Dorypterus*, in somewhat different ways, show a mixture of specialized features, retention of paleoniscid features, such as a heterocercal tail, and development of holosteanlike structure, in particular in the median fins. As far as is known, they did not leave any descendants.

The Redfieldiformes form a second line of somewhat conservative sub-holosteans, which includes a wide array of freshwater fishes ranging through the Triassic. They appear to have been adapted primarily to feeding on detritus. Although they remained conservative in many respects, the skull patterns show rather wide modifications (Figure 127*A*). This group probably arose from the central paleoniscid stock and retained not only the general body shape but also various other features characteristic of this stock. Unlike the paleoniscids, however, in the course of their evolution during the Triassic members of the redfieldiforms attained various holostean features. All have a hemiheterocercal caudal fin, but there is considerable difference in the degree of departure from the true heterocercal condition. The suspensorium tends to approach the vertical, the number of lepidotrichia is reduced, and the paired fins have somewhat narrow bases. Fulcral scales, however, are retained in association with the median fins.

A third, moderately conservative line, which has persisted to the present in the gar pike, *Lepisosteus*, and the paddlefish, *Polydon*, includes Saurichthyiformes and Chondrosteiformes of the Mesozoic. Loss of scales, and their replacement by large bony plates, modifications of the skull and jaws, especially in the rostral region, and other adaptive changes, are accompanied by the retention of basically primitive features of the appendages and skeletal organization of the trunk.

Of very uncertain relationship and history are the polypteriforms, known only from two extant genera. In many respects they are paleoniscoid, but there are many departures from the normal chondrostean pattern, for example, in the paired and median fins. The group appears to include much modified survivors of another persistently primitive but adaptively altered line.

PROGRESSIVE LINES. The major radiation and progressive lines of the subholosteans appear to have stemmed from an ancestry close to the Gonatodidae of the Carboniferous (Figure 115). Some concept of the scope of the radiation as far as adaptive types is concerned is evident from the members illustrated in Figure 127. In each of the several developing lines, along with evident adaptive changes, the mosaic pattern of differential acquisition of holosteanlike features is readily apparent. It is largely the distinctive associations of chondrostean and holostean characters in the various genera that demonstrate that many of them must have evolved separately, approaching the holostean grade in general, reaching it in some features, falling short in others, and with little or no constancy in the rates of change of the same characters in the different forms.

Numerically very important among these progressive lines are three families grouped into the Perleidiformes by Gardiner (1967a) (Figure 127). Together the three—Perleididae, Colobodontidae, and Cleithrolepidae—comprise over 20 genera that inhabited both marine and nonmarine waters. Ganoid scales were retained, trending somewhat toward the holostean type in late members, whereas the unpaired fins, the suspensorium, and opercular structures trended toward the holostean condition. The deep-bodied *Cleithrolepis* (Figure 127G) represents a sharp modification of the more common fusiform shape but has many fundamental features that suggest its inclusion within this group.

Peltopleuriforms (Figure 127I), deep-bodied cephaloxeniforms, with heavy-boned skulls and crushing teeth, the very advanced luganoiiforms (Figure 127H), which fall just short of the holostean grade in the somewhat primitive structure of their prearticular, and ptycholepiforms (Figure 127B), with essentially holostean postcranial features but a chondrostean maxilla and preopercular, represent a number of the variations on patterns of modification found in these progressive lines.

Most important of all from an evolutionary point of view are the parasemionotiforms, which include fishes from the marine beds of the lower Triassic. In almost all respects members of this order appear to lie close to the ancestry of one stock of the holosteans proper, the amiiforms, and to the halecostomes (pholidophoriforms and leptolepiforms), which lie at the base of the teleostean grade.

This line of development, the parasemionotiforms, attained a number of holostean features during the early Triassic, a sufficient number for it to be sometimes classified with the holosteans. It fell short of this grade primarily in the condition of the preopercular, which is large and broad medially, and in the absence of a true suborbital series of bones. Clearly any major distinction in taxonomy at this level becomes largely arbitrary. The postcranial structure is holostean and halecostome in grade level. Within this group nearly all of the essentials of structure fundamental to the holostean radiation are present, both in the feeding apparatus and in the agility and mobility made possible by the level of development of the body and fins. Similar levels were reached by members of some of the other lines, noted above, but the combinations of characters and the specific structure of many features of the postcranium and cranium strongly indicated that it was from among the parasemionotiforms that both the amiiform holosteans and the halecostomes took origin.

Initially the differences between the amiiforms and the halecostomes were very slight, relating primarily to differences in the preopercular, in its form and relationships to the suborbital and opercular structure, and in the caudal structure, which approaches a homocercal condition in the halecostomes, attaining it fully in leptolepids and just falling short of full attainment in the pholidophorids. The differences, although initially very slight, were of great importance both in the immediate future and at later stages. The slight reduction of the dorsal process of the premaxilla noted in the pholidophorids (see page 544) contrasts with the strong process present in both semionotiforms and amiiforms, and presumably was attained by modification of a strong process in the ancestral parasemionotiforms. This feature, though seemingly unimportant in the cranial anatomy, later was to prove a critical item in the complex series of changes that led to a great reorganization of structures associated with feeding in the course of radiation of the teleost fishes. It did not, as far as can be told, assume a special adaptive role at the time of its initiation, because the critical steps at the base of initiation of the teleost grade primarily involved the attainment of locomotor advantages.

Quite distinct from the early Triassic radiation of subholosteans was the development of the order Semionotiformes (see the Lepidotidae and the derived families, Figure 116). This group first appeared in the late Permian in the form of the fully holostean *Acentrophorus* (Figure 124*J*), probably derived from a stock that included the earlier aeduellids. Ultimately both this line and that leading to the parasemionotiforms probably were derived from the Gonatodidae, but the split seems to go well back into the Permo-Carboniferous. This then represents a very distinct line of development, but one that at a somewhat earlier date reached a level of organization very much like that of the amiiforms and underwent an extensive radiation on its own. Deep-bodied forms with strong teeth were developed from this base as well as the present-day predacious gar pike, *Lepisosteus*.

The holostean radiations that followed the attainment of the holostean grade were complex and extensive, and their members dominated the waters of Jurassic times. With the onset of the Cretaceous, replacement by teleosts

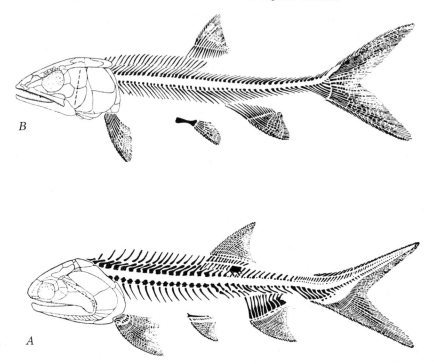

Figure 130. A comparison of the skull and skeleton of *A*, a Triassic paleoniscoid, *Glaucolepis*, and *B*, a Jurassic holostean, *Caturus*, showing the major differences between the two levels. Note especially the differences in the angle of the suspensorium, in the median and paired fins, and, in particular, between the caudal fins. The data analyzed by Schaeffer and shown in Table 50 were based on the skeletal features of these two genera as illustrated. (After Schaeffer, 1956.)

began. Modifications of the locomotor system, with increased mobility of the fins, reduction of the epibatic characteristics of the caudal fin, and reduction of scales undoubtedly were important. Reorganization of the feeding apparatus, as reflected in changes of the bones of the skull and jaws and inferred modifications of the muscle, produced a stronger and more flexible complex, readily subject to considerable adaptive modification, Figure 130. Undoubtedly these changes evident in the hard anatomy were accompanied by modifications of body systems not available to analysis among the fossils. Together they provided the basis for the holostean radiation. Although not approaching that of the later teleosts, this radiation did have considerable success, producing a large array of adaptive types. Only rather specialized and somewhat isolated members have persisted to the present, because as the teleosts arose in the Cretaceous the holostean radiation was brought to a rather sudden halt.

The Teleosts

Teleost fishes provided one of the most striking, complex, and instructive examples of adaptive radiation known among animals. Its very complexity has made for immense difficulties in sorting out lines of development and establishing a satisfactory taxonomy of the group. Two points are of particular interest within the context of the aims of this chapter: the initiation of the grade known as teleost, and the basis of the changes that lay behind the great radiation. As pointed out in the preceding section, the shift from the holostean level, as expressed in primitive halecostomes, to the incipient teleost level, seen among leptolepids, was basically locomotor. Attainment of the full teleost grade involved development of a completely homocercal tail, ossified vertebral centra, reduction of the lepidotrichia to equal the internal elements in number, reduction of scales with loss of ganoine and cosmine, and development of mobile fins. The holostean feeding mechanism was carried over with little change; in fact early halecostomes seem to have had somewhat less progressive adductor musculature than the early holosteans. There were minor differences, such as those of the premaxilla, already noted, but these were of little adaptive significance at the time of separation of lines leading to the pholidophoriforms and the amiiforms.

The teleosts long were considered to be a coherent, monophyletic group, but it now appears that two or more groups arose independently from the pholidophorid sources. One line gave rise to the leptolepiforms, classed as teleosts by some and as halecostomes by others, and their probable descendants the distinctly teleost clupeomorphs. This group, though remaining generalized in many features, had great success from the late Jurassic to the present, and today is represented by a vast array of herringlike fishes.

Although the teleosts are recorded from the late Triassic, there was a notable lag in their development, because it was not until the Cretaceous that their radiation got under way. Such a lag after initiation is characteristic of various major groups—for example, the reptiles and the mammals—but explanations are primarily speculative since the records during the period of lag are characteristically poor. A result in the case of the teleosts is much confusion and difference of opinion on the relationships of the basic groups. In the early Cretaceous the elopomorphs, another moderately primitive group of teleosts, appeared. The tarpon and some deep-sea fishes represent this group today, but its primary success was in the Cretaceous. The source probably was in an unknown line, stemming from the pholidophorids and passing unknown through the Jurassic, but this matter is far from settled as yet.

Elopomorphs were generally primitive in skull features—for example, possessing maxillary teeth and an articulation of the lower jaw posterior to the orbit—but they appear to have initiated the ball-and-socket joint between the maxilla and premaxilla, to have had a somewhat free dorsal process on the premaxilla, and a tandem rather than serial arrangement of the maxilla and premaxilla (see, for example, Figure 120). Palatal and dermal elements of the skull proper were somewhat reduced in ossification.

Within these early and rather primitive groups modest changes of feeding structure did take place, and some of their descendants that came to occupy special environments, such as the eels, attained distinctive special adaptations.

Originating in the upper Cretaceous was a group called Protacanthopterygia, from which, it would appear, stemmed the teleost lines that have undergone extensive and progressive radiations characterized in particular by modifications of the feeding mechanisms. Most spectacular and carrying the feeding mechanism to its highest development are the members of the Acanthopterygia, the spiny-finned fishes. The changes of jaw structure leading to the realization of the acanthopterygian level are those outlined on page 544 and illustrated in Figure 120. The superorder Acanthopterygia comprises some 11 orders, of which one, Perciformes, includes 15 suborders (conservatively) and over 100 families. The adaptive radiation that followed attainment of the advanced feeding apparatus has been truly fantastic and has brought to its peak the long course of actinopterygian evolution.

It was not, of course, this one modification that made such a radiation possible, but the combination of features of the body systems as a whole, fitted gradually into the mosaic of evolving parts, that provided for the attainment of an organizational threshold from which departure along a multitude of adaptive corridors was possible.

EXPLANATIONS

The course of evolutionary events suggested by the record of the actinopterygians, reviewed in the foregoing section, shows a succession of radiations, each following the attainment of an organizational level that made the particular radiation possible. This pattern, which involves extensive parallelism leading to similar thresholds, themselves preludes to radiations, appears to represent the most reasonable interpretation of the form of the very complex history of the actinopterygians. It shows many resemblances to patterns deduced from less good evidence for such groups as the labyrinthodont amphibians and the synapsid reptiles as they evolved from primitive pelycosaurs through successive levels of therapsids to the mammals. Wherever this pattern has been discerned, it has given rise to considerable speculation on the causes and to many efforts at explanation. The idea of some sort of orthogenetic force has, of course, often been invoked but has been abandoned either as a mere restatement of what has been observed or as being beyond the scientific ken.

As far as the actinopterygians are concerned, greatest attention has been paid to the evolutionary patterns at the more primitive levels, involving the paleoniscoids and the subholosteans.

Brough (1936), in exploring the concept of the subholostean level, suggested that repeated appearance of similar features in different lines might have resulted from similar mutational changes in stocks with common genetic heritages that dated far back in their separated lines. That a common genotype may to some extent be involved is by no means impossible. Westoll (1944), however, countered this concept, questioning the extent of the common pattern change suggested by Brough and maintaining that the only persistent common change was the reduction of the heterocercal tail, which could be related to locomotion. He then proposed that the major change from the paleoniscoid to the subholostean grade, followed by various lines and recognized mainly by the reduction of the heterocercal condition of the tail, was related to an environmental shift from fresh to saline waters. This poses an interesting problem, because during this shift loss of ossification suggests that the specific gravity decreased, even though the fishes were passing from a medium of lower to one of higher specific gravity. Westoll's point, however, is that in the transition, as buoyancy increased, it was necessary for the tendency to rise produced by the caudal and pectoral fins to be reduced and for more active swimming to be developed. The changes that are seen fall into a rational pattern as he describes it.

The problem that arises in this hypothesis is that it is far from certain that any such transition from fresh to salt water actually took place. Among the fishes of the Devonian and Mississippian, some occur in deposits formed in fresh water and others occur in those formed in salt water. Some genera occur in beds formed in both environments. It would appear that at least part of the early history of the actinopterygians was in marine waters, and it may well be that many early forms were euryhaline. Many of the Pennsylvanian and Permian paleonisciforms are known from freshwater deposits, but a considerable number also seem to have been marine. It cannot be demonstrated that the various lines of subholosteans had either a marine or nonmarine ancestry. Thus it is certainly at least questionable that the acquisition of similar features in various different lines can be related to some general shift in life environment, such as from one type of water to another.

More recent studies, particularly those of Schaeffer, have developed a complex hypothesis that brings into play many facets of evolutionary theory in explaining the development of higher categories under the concept of adaptive experimentation (Schaeffer, 1956c and earlier). The basic idea is that a transition from one organizational level to another involves some biological improvement and that this is attained, at least in part, by a number of lines with common heritage, developing similar adaptations that are expressed in various associations of characters in "experimental" mosaics in which the old and the new are associated in different arrays. The similarities of the changes are imposed by the common ancestry, by the genetics and epigenetics, somewhat reminiscent of Brough's concept, by canalization of development, and by the actions of selection, which, induced by changes of environment, often tend to be directional. Once the actinopterygian pattern was set, by whatever innovation may have been basic to it, the directions of changes that could lead to more effective locomotion and more effective feeding, in a very broad sense, were to a considerable extent limited. Boundary conditions on future modifications within the habitus were set, and only those suites of modifications that could together eliminate the restrictions, partially at the most, could lead to radiations beyond the limitations of the established structure.

Adaptive experimentation within these limits, exploiting the variability of the populations, is presumed to have produced the many paleonisciform patterns that are known. In some patterns of change modifications of feeding mechanisms were prominent, with little change in the locomotor structures; in others the opposite was the case. Each such new experiment had some adaptive superiority over its predecessor within the environmental circumstances in which it came to be. Some of the particularly favorable combinations, opportunistically favorable in terms of their coincidence with particular life conditions, persisted, resulting in the lines that led to the subholostean radiations. These forms retained some of the old features but had many of the new. All of them, while at the chondrostean level, retained one special chondrostean feature, an undivided neurocranium. Once this was lost, the lines so affected were at a new grade level, that of the holosteans, because only with the attainment of otherwise full suites of holostean characters did this modification come into existence.

The key to such modifications as exhibited in this series, as well as in various others at more advanced phylogenetic levels, lies in general adaptations, those that increase efficiency as a whole for life under a wide variety of circumstances. These form a focal point about which a great many other changes may be organized and give considerable constancy, directionalism, and coherence to an evolutionary series that follows their appearance. Such an initial feature in the Osteichthyes, as noted, may possibly have been the introduction of the hydrostatic potential with the development of lungs. In the particular line leading to actinopterygians minor differences—expressed in the scales, the form of paired fins, the nature of the jaws, and dentitions—may have been sufficient to differentiate the course of development from that of other Osteichthyes and to produce the small, heavily scaled, broad-finned, strong-jawed, sharp-toothed predator that lay at the base of the whole radiation.

Once this pattern was established, successive improvements of the feeding and locomotor mechanisms, in various combinations within the limitations imposed by inherited structure, became key general adaptations. These eventually led to a new organizational level, undoubtedly in concert with other less evident features of morphology and physiological changes. At each new level the stage was set for a new radiation. Thus in succession the lines led to the current culmination in the vast array of teleosts, climaxing in the acanthopterygians.

4 Transitions to land

INTRODUCTION

During the time that the actinopterygians were undergoing the initial differentiations that eventually led them into full exploitation of aquatic environments, other branches of the Osteichthyes followed courses that set the stage for the occupancy of land. The two radiations of Osteichthyes that followed the initial separation had many common aspects, but one exploited the medium in which vertebrates had originated, whereas the other eventually succeeded on land. The special features of the transition to land, contrasted with the radiations considered in the last chapter, are the focal point of the descriptions and discussions that follow.

Presumably the fishes that attained capabilities for radiations onto land, the crossopterygians and the dipnoans, had a common origin with the actinopterygians, but at the time of first appearances the separation lay far in the past and nothing is known of any common stock. Similarly the dipnoans and crossopterygians are structurally quite separate at all known stages. The usual practice has been to consider them to be fairly closely related, stemming from a common stock that had already diverged from the line leading to actinopterygians. This is formalized by their association in a single subgroup of Osteichthyes, variously called Choanichthyes, Sarcopterygia, or Crossopterygia.

The evidence of such a close relationship is tenuous, and many authorities (e.g., Berg and more recently Jarvik) have felt that a wide separation actually existed. If this is a proper assessment and the possibility that the amphibians took origin from both Dipnoi and Crossopterygii is granted, much greater convergence than is generally accepted must have taken place.

What induced the divergence of the actinopterygians and dipnoans and crossopterygians is not known. The last two clearly became adapted to life in shallow waters in which oxygen deficiencies periodically developed. The first well-known assemblages of actinopterygians, in the Mississippian, show that they too had come to live in this sort of environment. The critical evidence on habitats during the time of initial differentiation has not been found.

Later, as is known from living fishes, many lines of actinopterygians became adapted to survival in waters low in oxygen, with some species able to persist for extended periods out of water and to make their way over land. Some teleosts, such as the mudskipper *Periophthalmus*, cross the strand line from marine waters to penetrate tidal flats and mangrove swamps in search of food. Except in extreme cases of modification of appendages, such as in *Periophthalmus*, structures necessary for life under such conditions leave little imprint on the skeletons. Whatever its extent among the actinopterygians, this sort of adaptation did not result in a permanent and effective transition to land.

In the search for ancestors of land vertebrates the spotlight first focused on the Dipnoi, because, of the living fishes, they are in many ways closest to the amphibians. *Polypterus*, a paleoniscoidlike fish, long considered a crossopterygian, and increasing knowledge of fossil fishes directed attention to the crossopterygians. Late in the nineteenth century and during the early part of the present one, through the studies of Cope, Baur, Dollo, and later W. K. Gregory, Goodrich, and Watson, the rhipidistian crossopterygians came to be generally recognized as the most probable ancestors of the tetrapods.

The idea that lungfishes may have been ancestors, however, has been revived sporadically, primarily in the context of a diphyletic origin of urodeles and anurans, with the former from dipnoans and the latter from rhipidistians. During the 1930s Säve-Söderbergh proposed a complex vertebrate phylogeny based on this concept, and as late as 1949 the dual origin was given strong support by Holmgren. Even later Lehman (1956) and, implicitly, Jarvik (1967b) have given some support to the ideas of Säve-Söderbergh.

Lungfishes have at least some potential for terrestrial life. They breathe air and are structurally able to move about on land in a somewhat eel-like fashion, with anguilliform body motions supported by paired appendages. Paleozoic lungfishes appear to have met severely deteriorating water conditions by aestivation. Some Paleozoic lungfishes—for example, *Sagenodus* and similar genera—appear to have remained entirely aquatic, and their remains occur in rocks that were deposited in swamps and ponds where waters, though probably low in oxygen, were persistent. *Lepidosiren*-like fishes, such as *Gnathorhiza*, were aestivators and had developed this habit as early as the Pennsylvanian.

Aestivation as an accommodation to low oxygen or periods of drought appears to be a dead end in the course to life on land. Once aquatic animals have adopted it, their chances for further exploitation of the potentials of land life seem to be slight.

Rhipidistians did not, as far as can be told, become aestivators. Possibly their predatory habits were responsible. During their development some of them, like the teleosts, must have made their way onto land. This step followed the acquisition of air breathing and of the ability to move about on land. Air breathing is found in many groups of fishes, but a locomotor system based on a tetrapodlike fin appears to have been unique to the rhipidistians, although Holmgren has argued to the contrary. Whether still fully aquatic animals became tetrapods, while most other features remained rhipidistian, or whether mature tetrapod limbs arose after land occupancy is not certain, although logic and a few hints from the fossil record tend to support the first interpretation.

The course to life on land even after the introduction of the basic requirements was a long and tortuous one, much of which left no record in known fossil deposits. It probably began early in the Devonian, possibly even earlier, and reached full fruition only in the Permian, when some reptiles and apparently amphibians as well broke their final ties to water. Because the early record is poor, late animals have supplied most of the information about what happened. Many morphological details and information on embryology, physiology, and behavior have come from moderns only.

Progress toward a more complete understanding of the actual course of change has been slow, depending largely on new data from the fossil record. Lack of this basic information has contributed to persistent disputes centered around the phylogeny of the tetrapods and their predecessors, particularly with reference to the role of dipnoans in tetrapod ancestry and the origin of urodeles and anurans from two distinct groups of rhipidistians. Although proponents of one or another form of diphyletic origin of the amphibians are in the distinct minority, their effects on studies in this area have been immense. One favorable result has been that supporters of the various hypotheses have unearthed a great deal of factual information. On the other hand, the use of assumed phylogenies as bases for deductions concerning almost all of the problems of tetrapod origins have introduced such a mixture of objective and subjective analyses of data that dispassionate assessments are difficult and too infrequently encountered.

It is currently possible to gain a fair idea of the general course of the transition from water to land—to know what stocks were involved, the patterns of the major trends, something about the environments in which changes took place, and a great deal about the morphology of animals fairly closely related to those directly involved in the transition. The evidence, however, is far from sufficient to resolve many of the controversies about the details of change or even to eliminate effectively even some of the most speculative hypotheses about very broad relationships. It is to be hoped that accumulation of more information will serve to remedy this situation.

PROTOTETRAPODS: THE BASIC ADAPTATIONS

All major groups of living Osteichthyes include fishes that breathe air. The evidence that this was true for rhipidistians is of course indirect. *Polypterus*, the living paleoniscoidlike fish, dipnoans, and amphibians all have lungs developed embryologically as diverticuli from the digestive tract and certainly homologous. This leaves little doubt that such a lung was the common property of all of the Osteichthyes, including rhipidistians, which, as ancestors of the tetrapods, must have been the source of the amphibian lung.

What the functions of the primitive lung may have been and how it came into existence are less clear. If it was a structure primitively adapted to air breathing, it seems likely that the ancestral Osteichthyes lived in fresh water. The lung, in this case, probably developed in response to need for supplemental oxygen. Perhaps, however, the initial function was hydrostatic, providing greater buoyancy to the heavy-boned and heavy-scaled early fishes and supplementing the lift of the epibatic tail and pectoral fins. In this event the lung might have developed in either fresh or salt water. Not out of the question is the possibility that a lung developed in several lines of Osteichthyes, perhaps as each shifted from an initially marine to a freshwater habitat. Still another possibility is that the early Osteichthyes were euryhaline, living part of their life in fresh water and part in marine, with the lung developed to meet the needs of the freshwater phase.

Geological evidence about the early environment of the Osteichthyes is not definitive. The actinopterygians are unknown before the middle Devonian, except for the scales. They are found at that time in both marine and nonmarine deposits. The earliest Dipnoi, from the early Devonian, are found mostly in marine rocks, but some, from Beartooth Butte, Wyoming, are from beds that may be marginally marine or possibly freshwater in origin.

Rhipidistians occur at several sites in the lower Devonian, both in marine and nonmarine deposits. The earliest record is a scrap from the fluviatile beds of Podolia. Remains are fairly abundant in a slightly later freshwater deposit at Wood Bay, Spitzbergen. However, remains occur in marine deposits at Grey Hoek, Spitzbergen, in several places in Germany, in the Taimer Peninsula of Siberia, and in the Water Canyon formation of Utah (Denison, 1956).

The early fossil remains are from primitive rhipidistians, the Porolepi-formes (Holoptychoidea). It seems quite possible that members of this group were euryhaline. If this rather scant record can be considered at all indicative, it seems to suggest that the early rhipidistians lived under a wide range of environments.

By mid-Devonian, however, the majority inhabited fresh water. Many of the occurrences of this age, and of the late Devonian as well, indicate that the sediments were formed in shallow waters characterized by a high organic content, warm temperatures, and periods of drying up. Oxygen deficiencies might well have occurred, and air-breathing fishes probably were the rule. This is the sort of environment often pictured as one of which early transitions to land may have taken place.

Limbs, girdles, and axial skeletons of rhipidistians of the middle and upper Devonian were competent for locomotion on a solid substrate underwater and probably would have served for clumsy progress on land. The paired fins of the more primitive forms, porolepiforms, have long, slender, fleshy, scale-covered lobes much like those of lungfishes. They probably were less capable of terrestrial locomotion than those of either the Osteolepidae or Rhizo-dontidae. In these the lobes were shorter and broader and the internal skeletons somewhat resembled tetrapod limbs. Only the porolepiform group, however, is known from the early Devonian. As illustrated by representative genera in Figure 131, the trunks of all the three main families of rhipidistians were fusiform, with those of the two osteolepiform families being somewhat less deep. Each group was varied, and parallel progressive evolution of the median fins and scales indicates trends toward more active swimming. Except in some of the porolepiforms (e.g., *Onychodus*), these changes seem to have been in adaptation to life in fresh water.

Paired fins of all of the rhipidistians, in particular those of the osteolepi-forms, could have been used for "bottom walking" and might have been particularly useful as the fishes made their way through weedy, stagnant water. An alternative suggestion is that the fins were developed for burrowing, perhaps in connection with aestivation, but there is little in the fossil record to support this.

Rhipidistians appear to have been active, predacious carnivores that fed on fish and active invertebrates. Rapid swimming must have been important to feeding. It is possible that the initiation of the paired-fin pattern was related not to "walking" but to mobility and flexibility during rapid maneuvering in pursuit of prey. The fins of *Polypterus*, *Latimeria*, and Dipnoi are extremely flexible and have highly developed muscular mobility.

Once developed, such fins could have taken on various roles. It has been suggested that the pectorals may have acted to raise the anterior part of the body above the substrate, perhaps for gulping air in shallow water and

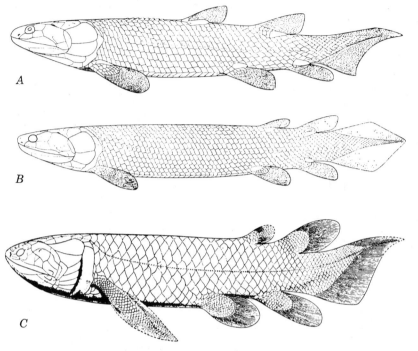

Figure 131. Body form of three distinctive crossopterygian rhipidistians. *A, Osteolepis; B, Eusthenopteron; C, Holoptychius.* (*A* and *B* from Jarvik, 1948; *C* from Lehman, 1966.)

concurrently freeing the throat for easier buccal breathing. Persistence of this habit on land would have an obvious advantage.

Axial support is important on land, but much less so in water. The vertebral centra of teleost fishes are strongly ossified, probably in relationship to the active and intricate swimming maneuvers characteristic of this group. In the few cases in which it is known the vertebral column of the rhipidistian has well-ossified neural arches and disk-shaped or crescent-shaped central ossifications. Probably, as in teleosts, these were developed as adaptations to locomotion in water, perhaps as responses to the needs of increased speed and maneuverability in developing predators. Once present, vertebral ossifications gave a base for the kind of support needed on land.

With respect to breathing and locomotion, the rhipidistians were appropriate prototetrapods. The recognition of this relationship to amphibians rests not so much on these potentials, because other fishes have them, but on the fact that many skeletal features not directly related to terrestrial life resemble those found in the amphibians and their descendants. The fin-limb resemblance is important, but so are features of the dermal roof of the skull, the endocranium, the nares, the auditory region, and the axial skeleton. Thus many of the amphibian characteristics resulted from adaptations of the rhipidistians to their special ways of life in water as altered to meet the amphibian mode of life.

ATTAINMENT OF THE TETRAPOD LEVEL

The Limb: A Basic Criterion

Its Source

The rhipidistians were distinctly fishes, well adapted to aquatic life. None that are known can be considered to have been actual tetrapod ancestors. It seems probable, however, that the initial tetrapods similarly were aquatic creatures, fishlike in their habits, and that much of what we think of as amphibian was developed after the realization of this condition.

Technically the limb defines the tetrapod level, and its existence in an anamniote defines the amphibian. Whether or not the amphibian status was attained more than once cannot now be ascertained.

Schaeffer (1965a) has argued that there was but a single origin of the tetrapod limb and hence of amphibians. Resemblances of all limbs are so close, he felt, that more than a single source seemed very unlikely. The acquisition of a tetrapod limb, under his hypothesis, was an initial event in the establishment of the amphibian habitus. Thomson (1967a) suggested, on the basis of evidences of mosaic evolutionary progress among the rhipidistians, that the limb may well have developed more than once. However, he also envisages a restricted ancestral group from which origins occurred very early in the course to amphibians, so that his position is not very different from Schaeffer's. The dual origin of urodeles and anurans from lungfishes and rhipidistians is at a different level. Although this was originally proposed on the basis of skull morphology, Holmgren used the evidence of limbs to support his case (see Figure 137 and pages 595–597).

Paleontological Evidence from the Amphibians

The most complete data on the limb and foot structure of extinct amphibians have come from the Permian genera *Trematops* and *Eryops*. Many other labyrinthodonts and lepospondyls from the Pennsylvanian and Permian have added to this, but the details of the carpus and tarsus and phalanges are incomplete. Earlier evidence goes back to the uppermost Devonian, but it is quite meager.

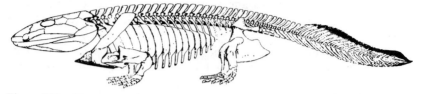

Figure 132.   Reconstruction of *Ichthyostega* (from Jarvik, 1955.)

DEVONIAN AMPHIBIANS.   By late Devonian true amphibians had come into existence. Except for some rather dubious tracks (see, for example, Willard, 1935), no earlier indications have yet been found. Part of the skull roof of an animal called *Elpistostege*, from the lower part of the upper Devonian at Scaumanac Bay, Nova Scotia, has been referred to the Amphibia by Westoll (1938). Its proportions are somewhat intermediate between those of *Eusthenopteron* and primitive amphibians, and it appears to lack dermal expression of the joint between the anterior and posterior parts of the endocranium. The age is definitely Devonian, but assignment to the Amphibia is necessarily somewhat tentative.

From east Greenland have come definite remains of amphibians, including three or four genera that represent at least two families of ichthyostegal labryinthodonts. These are fully developed if somewhat fishlike tetrapods (Figure 132). The deposits in which they occur are generally placed in the upper Devonian, although some have argued that they are earliest Mississippian. These animals give the earliest clear evidence of a tetrapod limb (Figure 133). Not all of the details are clear, but the general pattern is established in both the fore and hind leg. These amphibians are not quite on the line to other labyrinthodonts and are rather far from the base of the lepospondyl

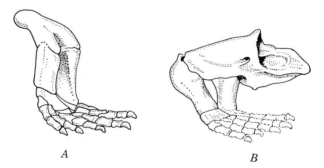

A   B

Figure 133.   The limbs and feet of *Ichthyostega*, showing the completely tetrapod nature of the appendages. *A*, hindlimb; *B*, forelimb. (Redrawn from Jarvik 1955a (see Figure 132) to show limb structure in greater detail.)

				Mauch Chunk (North America)	*Spathicephalus* "trimerorhachid"
Mississippian	Upper	Carboniferous	Limestone Series	Loanhead (No. 2) ironstone (Scotland) Gilmerton ironstone (Scotland)	*?Papposaurus* *Spathicephalus* *Pholidogaster* *Loxomma* *Crassigyrinus*
	Lower	Calciferous	Sandstone Series	Upper oil-shale group (Scotland) ⎰ Dunnet shale	*Adelogyrinus*
				Upper oil-shale group (Scotland) ⎱ Burdiehouse limestone	*Otocratia* *Dolichopareias*
				Lower oil-shale group (Scotland) ⎰ Curley shale	*Palaeomolgophis*
				Lower oil-shale group (Scotland) ⎱ Wardie shale	Aistopod
Devonian	Upper			(Greenland) "Old red sandstone"	*Ichthyostega,* *Ichthyostegopsis* *Acanthostega*
				(Scaumenac Bay, Nova Scotia)	*Elpistostege*

Figure 134.   The Devonian and Mississippian occurrences of amphibians, showing the order of appearance and the great diversity at these early times. Most groups are represented by but one or a few individuals.

stocks. Throughout the world at this time environments suitable for amphibians probably existed, and it is likely that a diverse array of tetrapods had already come into existence.

MISSISSIPPIAN AMPHIBIANS.   The oldest Mississippian amphibians have come from the oil-shale groups of the calciferous sandstone series of Scotland (Figure 134). This is the series that has produced many of the early Carboniferous actinopterygians. The beds, being of Visean age, are considerably later than those from which the east Greenland amphibians were obtained.

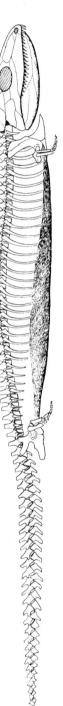

Figure 135. The skeleton of *Pholidogaster*, the oldest well-known anthracosaur (see Figure 134). Total length a little over 1 meter. Note the small limbs, contrasting with limb proportions in most other anthracosaurs. A, remains as preserved; B, restoration. Note also the well developed otic notch and structures indicating full development of hearing and breathing in the atmosphere. (From Romer, 1964b.)

The oldest known specimen, from the Wardie shale, is an aïstopod, sometimes called *Ophiderpeton*. Slightly later, from the Curley shale, has come a better preserved amphibian named *Palaeomolgophis* and placed by Brough and Brough (1968) in that somewhat heterogeneous assemblage of lepospondyls called microsaurs. Some features have suggested that it might in fact be an adelogyrinid. Two other microsaurs, *Dolichopareias* and *Adelogyrinus*, both adelogyrinids, have come from the upper oil shale, the Burdiehouse and Donnet shales, respectively. From the former has come the ichthyostegal *Otocratia*.

In the Lower Carboniferous limestone of Scotland, younger than the oil shales but still Visean, is the Gilmerton ironstone from which a number of labyrinthodonts have come: *Pholidogaster*, an anthracosaur; *Loxomma*, a rhachitome; and, tentatively assigned to the Gilmerton, an anthracosaur, *Crassigyrinus*. Much of the described material consists of skulls and jaws, but the skeleton of *Pholidogaster* is quite well known (Figure 135, after Romer, 1964*b*).

Deposits of upper Mississippian age (Namurian) in Scotland have produced skulls of a loxommid, *Spathicephalus*, and part of a skull of this genus has come from Nova Scotia. Recent work in the upper Mississippian is revealing a varied amphibian array. Two families of anthracosaurs and two of temnospondyls have tentatively been identified. Much is to be expected from these beds. The currently scant remains of the Mississippian are followed by an increasingly good record from the Pennsylvanian.

Data on the limbs and feet of the very early amphibians are as follows: the first known labyrinthodonts, ichthyostegals, had well-developed limbs and girdles that included most of the structures found in later temnospondyls. *Pholidogaster*, an anthracosaur of the middle Mississippian, had very small limbs, but as far as known they were of a standard tetrapod organization (see Figure 135). *Gephyrostegus*, a rather reptilelike seymouriamorph from the Pennsylvanian, had a normal tetrapod limb. Although rhachitomous amphibians occur early, in the upper Mississippian, the limbs from this age have not been worked out. The limbs of several Permian labyrinthodonts— *Trematops*, *Eryops*, *Archegosaurus*, and dissorophids—have been the main basis for detailed comparisons of tetrapod and rhipidistian structure, along with those of the problematic reptile or amphibian *Diadectes*.

Late microsaurs, from the Pennsylvanian and Permian, have limbs with a general tetrapod cast, although most details of the more distal portions are not available. *Microbrachis* of the late Pennsylvanian seems to have but three digits in the manus, but otherwise the limb is orthodox. Gymnarthrids had well-developed, typically tetrapod fore and hind limbs. Other microsaurs—the molgophids, adelogyrinids, and lysorophids—had long elements similar to those of other tetrapods. The appendages were proportionately very small, perhaps absent in some, and little is known of the distal parts or of the girdles.

Nectrideans of the family Keraterpetontidae had small but typically tetrapod limbs. In other nectrideans limbs were reduced or absent. Finally the earliest known lepospondyls, the aïstopods, had no limbs whatsoever.

There is then no evidence of more than one type of tetrapod limb. All amphibians that had appendages at all conformed to the same pattern in both the limbs and the girdles. Within this pattern there probably was considerable diversity, but the record is not sufficient for full evaluation. It does indicate, however, that a great deal of adaptive modification had taken place even before the first traces of amphibians appear in the rocks. Various interpretations of limb phylogeny are possible from the scant data, but the very close morphological resemblances of all that are known leave little doubt that origin was within some taxonomically very compact group.

The Development of Tetrapod Structures

The Evolutionary Mosaic of the Structural and Functional Systems

The transition from rhipidistians to tetrapods has been treated within a basically typological framework, with separate analysis of the various systems. Of course, coordinated changes did take place, and modifications of each of the major systems were reflected in adaptive and structural changes of others. How this occurred can be judged only from animals in which many of the fundamental transitions had already been accomplished.

Different rates of change in the separate systems through experimental adaptation probably were characteristic of separate lines of rhipidistianlike prototetrapods. Initial key modifications in habit and function likely set the stage. Schaeffer (1965a) implied that the attainment of a tetrapod limb was such a change, with resulting evolution, as shown in Figure 136.

Given such a beginning, other related modifications involving feeding, respiration, and sensing, may have taken place, probably in more than one line. Modifications of the skulls and jaws are most readily integrated by correlation with structural changes of skull kinetics and proportional alterations related to incipient modifications of feeding habits. As far as can be ascertained, however, the particular feeding modifications were not per se necessary to the attainment of a tetrapod status or invasions of terrestrial habitats. If they were central—that is, modifications about which other parts of the changing organisms were integrated—many features of the skulls and jaws of tetrapods were first initiated not in direct adaptive response to a major environmental change but to a modification attained through adaptation

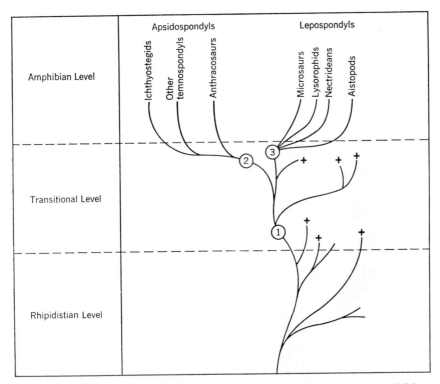

Figure 136.   Diagram of a possible phylogeny of the development of amphibians from the rhipidistian level based on the concept of evolutionary experimentation. In this various developments toward an amphibian-tetrapod limb structure took place, but only one was a successful experiment. This is based on the proposition that the consistency of the structure of tetrapod limbs indicates but a single origin (after Schaeffer, 1965*a*). +, unsuccessful experiment in amphibian direction; 1, successful fin-to-limb experiment; 2, apsidospondyl radiation, double centra, well-developed limbs, moderately short bodies; 3, lepospondyl radiation, single centrum, small limbs, long body.

to somewhat different ways of life in water. Each system, however, also altered under selective pressures related to its own particular functions.

Some such complex history as this is suggested by the record, which shows great diversity among the early amphibians, but a diversity superimposed on basically similar patterns of the tetrapod limbs and generally comparable respiratory and feeding mechanisms. If remnants of the actual changes are ever found, they are likely to show a diversity of adaptive systems comparable to that seen in the early Mississippian paleoniscoids and their Devonian forerunners.

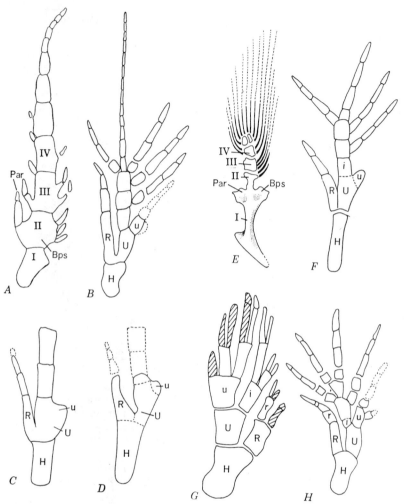

Figure 137.   Illustrations of Holmgren's (1933, 1949) concepts of the derivation of the pectoral appendage of urodeles from a different source (dipnoan or coelacanth) from that of the appendage of anurans and of other amphibians, reptiles, birds, and mammals. Urodeles are considered to have a biserial arrangement. Pairs *A, B;* *C, D; E, F;* and *G, H* are to be compared. *A, Ceratodus,* pectoral based on an 8-month larval stage; *B,* urodele, forelimb showing the biserial archipterygium of the *Ceratodus* type; *C* and *D,* schematic comparison of basal part of pectoral fin of *Ceratodus, C,* and forelimb of the urodele, *D;* compare also with *A* and *B. E,* pectoral of *Laugia,* a coelacanth; *F,* urodele forelimb as archipterygium of *Laugia* type; *G,* Jolmgren's interpretation of the pectoral fin of *Sauripterus* (see Figure 138 for other interpretations) in relation to tetrapod limbs, except urodeles; *H,* schematic representation of anuran forelimb as dichotomous. Bps, basal postaxial segment; H, humerus; i, intermedium; Par, preaxial ray; R, radius; r, radiale; U, ulna; u, ulnare; I–IV, axial elements.

Structural Modifications of Locomotor Structures

THE TETRAPOD LIMB. The fins and limb girdles of the rhipidistians and dipnoans are the only ones among vertebrates from which the structures of tetrapod appendages might have been derived. Except for a few proponents of the dipnoan source (Säve-Söderbergh, Holmgren, and Lehman), by changes such as shown in Figure 137, students have agreed that only the rhipidistian limbs and girdles lie close to the source of the tetrapod limbs. The girdles pose little difficulty in the transition, and most studies have concerned the free

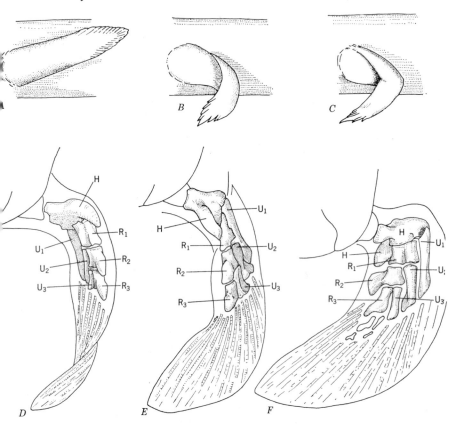

Figure 138. The origin of the forelimb and the homologies of the elements under the theory of Romer and Byrne (1931), in which a turnover and twisting of the limb is involved. A through C, three stages showing a reconstruction of the external appendage as the transition from fish to tetrapod took place; D through F, the osseous elements followed through the critical stages from fish to pretetrapod condition. H, humerus; R_1, equivalents of radius; R_2, equivalents of radiale; R_3, equivalents of metacarpal of prepollex; U_1, equivalents of ulna; U_2, equivalents of intermedium; U_3, equivalents of centrale 4. (A, B, and C after Romer and Byrne, 1931; D, E, and F after W. K. Gregory, 1935.)

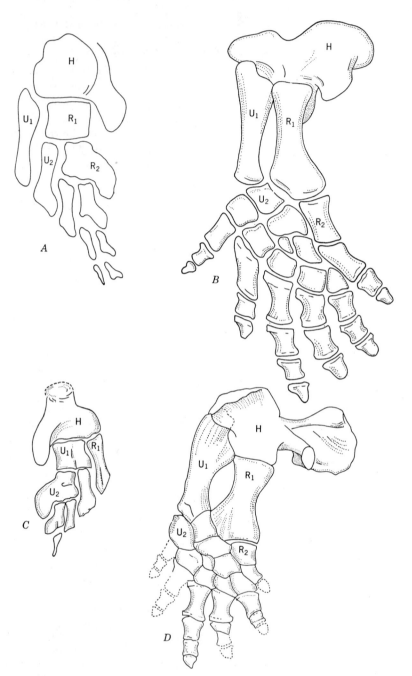

Figure 139. Two possible patterns of homologies of the elements in the forelimbs of rhipidistians, *A* and *C* (here *Sauripterus*), and amphibians, *B* and *D*. *A* and *B* are based on theory proposed by Romer and Byrne, 1931; *C* and *D* are based directly on W. K. Gregory, also comparable to the concepts of others (e.g., Westoll, 1943*c*). Lettering as in Figure 138. (After Gregory, 1935, somewhat modified.)

appendages, especially the pectoral appendage, for which the record is the best.

Internal structures of the fins are well known in *Eusthenopteron* and *Sauripterus* (Figure 64). They are for the most part very similar. Less-well-known fins, such as those of *Ectosteorhachis* and *Megalichthys*, indicate that the general pattern was common among both osteolepids and rhizodontid rhipidistians. The internal structures of the porolepiform fin are poorly known. Probably the fin had a long central axis and preaxial and postaxial radiales, as do some lungfishes. Most analyses have been based on the rather narrow fin of *Eusthenopteron*, but W. K. Gregory and Raven made use of the broader fin of *Sauripterus* in their studies.

Given this basic material and no intermediate stages, it is natural that many patterns of change have been proposed. These fall into two general groups. One, suggested by Romer and Byrne (1931), recognizes a twisting and turning over of the crossopterygian fin to produce the tetrapod condition (Figures 138 and 139). The theory of W. K. Gregory and Raven, augmented by W. K. Gregory and elaborated and somewhat modified by Westoll (1943c), is one of a group in which no twisting is postulated. The variants of this pattern differ in interpretations of the ways in which the fins were brought into contact with the substrate, the position of the main mesomeral axis, and the sources of the individual carpal and tarsal elements, the metapodials, and the digits.

Westoll's theory took into account the probable motions of the rhipidistian fin and explained the origin of the tetrapod limb as a consequence of the potentials of these movements. Rotation of the proximal segment of the fin was compared to the rotation of the proximal long element of the tetrapod limb during striding. The fore and hind limbs were considered to have been very similar initially. Westoll, like W. K. Gregory and Raven, concluded that the digits were new developments, not present in the rhipidistian paddle.

These interpretations are functional and are based on the best structural evidence available. The transition, as interpreted, occurred rather directly from a fin to a limb capable of fairly effective terrestrial locomotion. If, as is rather widely held, the tetrapod limb was an initial amphibian feature, followed by most other terrestrial adaptations, it would appear that the limb was first an aquatic specialization and had a long aquatic history after its origin. This does not dovetail closely with the structural interpretations in which, among other things, the role of contact with a terrestrial substrate is considered important. The adaptive motivations are less easily derived with origin under aquatic conditions.

THE AXIAL SKELETON.   Vertebrae and ribs are known in a few genera of rhipidistians. Those of *Eusthenopteron* (see Figure 72) are the best known.

The vertebrae of *Ichthyostega* and *Eusthenopteron* are very similar, and if *Eusthenopteron* is considered representative of the source, little change took place in the transition to tetrapods. The centra in *Ectosteorhachis* and *Lohsania* (Thomson and Vaughn, 1968), rather unlike those of *Eusthenopteron*, seem to indicate a considerably greater spread than had been envisaged earlier.

Development of vertebrae characteristic of the different phyletic lines of labyrinthodonts evidently took place after the emergence of full-fledged representatives of this large group. These derived structures are not relevant to the initial stages of transition from rhipidistians to tetrapods.

The differences between the vertebrae of lepospondyls and labyrinthodonts are found in the earliest members of the two groups. They have raised doubts about the derivation of the two from common amphibian ancestors. Recently, however, it has become increasingly clear that overextension of the Gadovian system of vertebral ontogeny has produced unnecessary difficulties. The form of the adult centrum now seems to have much less phylogenetic significance than had been accorded it earlier. The spool-shaped centrum of the lepospondyls and lissamphibians gives little information on the developmental role of embryological structures. The ontogeny of the urodeles and anurans (see, for example, Schmaulhausen, 1964, 1968; and Williams, 1959) shows that very similar adult structures can arise through different developmental processes that involve unlike contributions from the ontogenetic components.

Following this general line of reasoning, Thomson (1967a) suggested that the labyrinthodonts and lepospondyls may represent two different adaptive courses, stemming from similar rhipidistian ancestors. The labyrinthodont evolution led toward a generally terrestrial or semiaquatic existence, exploiting the potentials of the tetrapod limb structure and the development of centra, vertebral arches, and zygapophyses suitable for axial support on land. The long-bodied, small-headed, small-limbed lepospondyls emphasized aquatic and semiaquatic locomotion, taking advantage of the subanguilliform basis of rhipidistian locomotion in which the limbs were of little importance. Development of locomotion requiring effective differential contraction of axial musculature may have been accompanied by formation of solid spool-shaped centra. If this is true, a logical case for the monophyletic origin of labyrinthodonts and lepospondyls can be made. Also, however, it leaves the door open to the possibility that the lepospondylous centrum may have developed more than once, being a functional requirement, and that the homogeneity of the lepospondyls can be questioned.

Modifications of the Skulls and Jaws

GENERAL PATTERNS OF CHANGE.   Skulls and jaws participated directly in many of the modifications in the transition from water to land and were intimately related to soft structures that must have altered drastically. The senses of smell, hearing, and vision all changed, and the first two left good records on the skeleton. Respiration altered from predominantly gill breathing to air breathing, with profound changes in associated structures. The functions of feeding and respiration tended to be intimately related in fishes but became separated in the development of tetrapods.

The skull and jaws of *Eusthenopteron* (Figure 102) represent an initial morphostage. *Palaeogyrinus* (Figure 140) represents a stage in which basic amphibian characters have been fully developed. Comparisons of the illustrations of these two will make clear many of the gross modifications.

Superficially the proportionately longer facial region and shorter posterior platform and temporal region of the amphibian are readily apparent. This reflects enlargement of the jaws and increase in dentition related to increasing effectiveness of the predatory way of life initiated among the rhipidistians. The fundamental mechanical system remained the kinetic-inertial type (see Figure 57), but its capabilities were extended and presumably enhanced by decreasing cranial kinesis, which accompanied the enlargement of jaws, and by increasing size and differentiation of the adductor musculature. Increased mobility between the head and body was accompanied by migration of the axial musculature onto the posterior margin of the dorsal platform and its concentration on the posterior oto-occipital surfaces. Concurrently elements of the shoulder girdle, the cleithrum and clavicle, were freed from the skull.

Accompanying the proportional modifications was a series of correlated events that affected all of the structures and functions of the head. The mobile joint between the ethmoidal and oto-occipital parts of the ossified endocranium was lost, along with the complementary joint in the dermal roof and reduced mobility in the cheek region. The large notochordal pit in the basioccipital was reduced and eventually lost. The anterior moiety of the endocranium became proportionately long, and the posterior moiety was reduced.

The massive palatoquadrate of the rhipidistians lost part of its role in the feeding system as the hyostylic jaw suspension was eliminated, freeing the hyomandibular from its supporting role. The suspensorium, at least in the line leading to labyrinthodonts, moved posteriorly, modifying the spiracular and opercular region and producing a large otic notch. A new depressor musculature for the greatly enlarged lower jaws was developed. With the reduction of the opercular, and its eventual loss or conversion into a tympanic membrane, the hyomandibular assumed the role of an ear ossicle. The visceral arches took on new functions as air breathing replaced gill breathing.

The net result of these and other similar modifications was an animal with very different skull proportions, with reduced kinesis in the head, separation of the functions of feeding and breathing, and altered sensory functions. All went to make effective life on land possible.

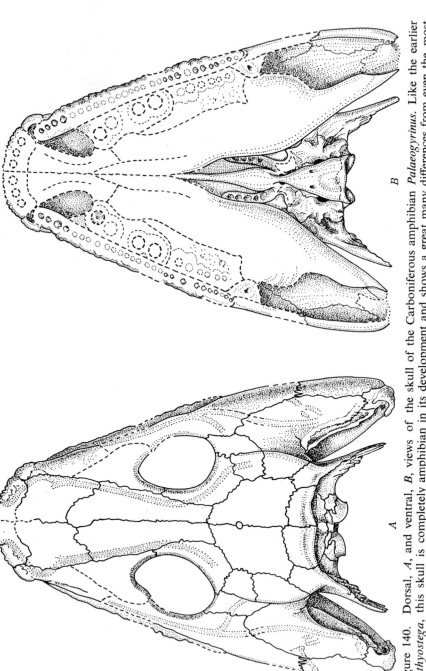

Figure 140. Dorsal, *A*, and ventral, *B*, views of the skull of the Carboniferous amphibian *Palaeogyrinus*. Like the earlier *Ichthyostega*, this skull is completely amphibian in its development and shows a great many differences from even the most amphibian like of rhipidistians. The otic notch is deep and presumably totally involved in hearing. The basicranium is short and without evident intimate association with hyoidean elements. Air breathing by lungs seems to have been the primary mode of respiration. The structures shown here may be compared with those of a rhipidistian in Figure 102: see also

The details of these changes have been treated at great length, most recently by Westoll, Jarvik, Parrington, Panchen, Schmaulhausen, Szarski, Romer, and Thomson. Analyses of the olfactory, respiratory, and lateral-line acoustical systems along with some aspects of visual structures are particularly important. Changes of the dermal elements and problems of their homologies have been matters of continuing debate.

RESPIRATORY AND OLFACTORY MODIFICATIONS. The narial region has received special attention, partly because of its functional importance relative to the olfactory–respiratory system but also because of the results of a challenging study made by Jarvik in 1942. In this he proposed his well-known thesis that the urodeles and anurans were derived from the porolepiform and osteolepiform rhipidistians, respectively. Although later studies on this matter have focused on other parts of the skulls and jaws, the narial region still is basic to his concept.

The distinctions between porolepiforms and osteolepiforms as seen by Jarvik are shown in Figure 141. Interpretations by various students have differed widely, strongly affecting their phylogenetic conclusions. The extensive accounts and analyses by Schmaulhausen (1968), Thomson (1967a), and Jarvik (1967b) can be consulted for recent expositions on both sides of this question.

Even if two widely separated rhipidistian sources of amphibians are discounted, as they are by most students, more than one line of rhipidistians may have been undergoing changes similar to those that led to the amphibians. Considerable diversity in external proportions is known among rhipidistians, and coordinated changes of related systems presumably were being initiated. Various lines appear to have approached an amphibian level in the structure of the skulls and jaws. Variations in skull proportions are shown in Figure 142.

Air breathing was well established in the earliest ichthyostegals. True choanae were present in the rhipidistians, but gill breathing remained the dominant mode of respiration. Thus a vast structural gap appears between the known stages. Some efforts to bridge this theoretically show the possible courses that modifications might have taken.

It is not certain whether the choanae of rhipidistians were used for air breathing or were adapted primarily to olfaction, assuring free passage of water over sensitive membranes of the nasal capsule. In either event the emergence of choanae provided a potential if not realized link between the respiratory and olfactory systems and laid the groundwork for later development.

The evolutionary source of the choanae and related structures has not been certainly determined. The internal nares of Dipnoi, which open in the

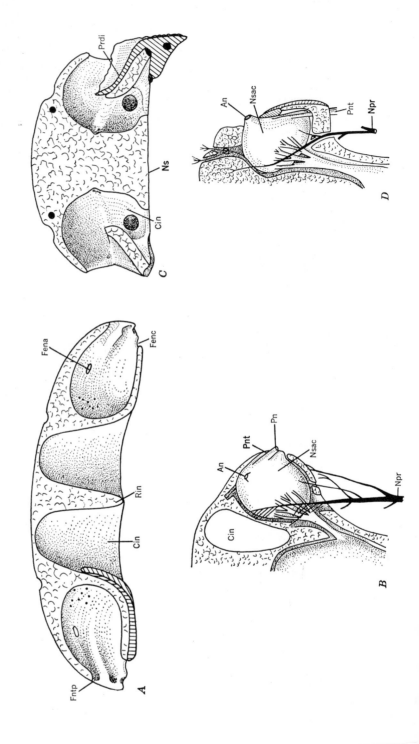

Figure 141. Comparisons of the snout region of the porolepiform, *A* and *B*, and osteolepiform, *C* and *D*, rhipidistians as made by Jarvik (1955*a*). *A*, porolepiform snout seen from behind, showing narrow internarial ridge; *B*, porolepiform, showing the nasal sac and related features; *C*, osteolepiform, showing snout from behind and emphasizing the broad nasal septum; *D*, osteolepiform nasal sac and related structures. Note especially the differences in comparable structures, but also the overall resemblances, especially as shown in *B* and *D*. The comparisons of such differences with those found between urodeles and anurans among living amphibians are considered to strengthen their significance. The labels are for the most part after Jarvik, with some additions and deletions. Differences in the labels, as used, tend to emphasize the differences. An, anterior (incurrent) external nostril; Cin, internal nasal cavity; Fena, opening for anterior (incurrent) external nostril; Fenc, endochondral fenestra; Fntp, opening for posterior nasal tube; Npr, nervus profundus; Ns, nasal septum; Nsac, nasal sac; Pn, posterior (excurrent external nostril; Pnt, posterior external nasal tube; Prdi, processus dermintermedius of lateral rostral; Rin, internasal ridge. (Modified after Jarvik, 1955*a*.)

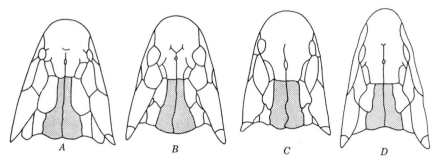

Figure 142. Dorsal view of the skulls of four rhipidistian crossopterygians, showing the differences in proportions relative to orbits, pineal openings, and length of snout as brought out by the postparietal element, shaded for emphasis. *A, Glyptomus; B, Osteolepis; C, Gyroptychius; D, Eusthenopteron.* (After Thomson 1966*a*.)

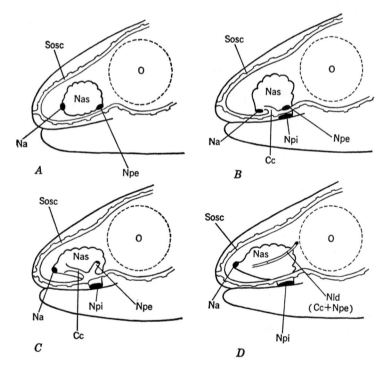

Figure 143. Illustration of the theory of origin of the choana based on Schmaulhausen (1958, 1964, 1968) and Szarski (1962). For an alternative hypothesis see Panchen (1967*a*). *A*, condition in primitive fish; *B*, hypothetical transitional stage; *C*, rhipidistian; *D*, tetrapod. Cc, connecting branch of suborbital canal; Na, anterior nostril; Nas, nasal sac; Nld, nasolachrymal duct; Npe, external section of posterior portion of nostril; Npi, internal (choanal) part of posterior nostril; O, orbit; Sosc, suborbital sensory canal.

oral cavity, are considered homologous to choanae by some, but not by others, one of the most recent dissenters being Panchen (1967a). One theory of the origin of the choanae of rhipidistians is that they developed from the posterior, excurrent narial opening, such as occurs in porolepiforms, coelacanths, and actinopterygians. A possible course of this change is that shown in Figure 143, which illustrates the interpretation of Schmaulhausen (1968). Panchen, on the other hand, has argued that the choanae probably had an independent origin, arising from the accessory nasal sac. Such an origin is known in the teleost *Astroscopus*.

The evidence from the rhipidistians is not conclusive. The precise origin, although intrinsically interesting, is not critical to a theory of acquisition of air breathing. Osteolepiforms have but a single external naris on each side and in this regard resemble the amphibians more closely than the porolepiforms, which have two. Also present in the osteolepiforms is the processus derminternus of the laterorostral bone. This process is assumed to have carried muscles that served to close the narial opening, as it does in the presumed homolog of the laterorostral, the septomaxilla, in amphibians. Panchen (1967a) deduced from these conditions that the osteolepiforms maintained an air-filled, dry buccal cavity. The presence of such a buccal cavity in plethodont salamanders and the functioning of a nasolabial groove to induce closure in the presence of water were cited as structural justifications for this interpretation. Panchen found, furthermore, that there is evidence for such a nasolabial groove in the coal-measures amphibians. If his interpretation is correct, the osteolepiforms were efficient air breathers, possibly in contrast to other rhipidistians. Also, however, they were gill breathers, and this probably was their principal source of uptake of oxygen and elimination of carbon dioxide, except at times when the oxygen content of the water was very low.

If restricted environmental circumstances were to arise frequently, perhaps periodically, in the normal habitats of such fishes, selective pressures toward increased efficiency of this system would exist. This might be accompanied by reduction and eventual loss of the gills. Modifications of feeding habits quite surely accompanied these changes in respiration and olfaction. Among them was the separation of function, so that the gills and associated pumping mechanisms were no longer part of the feeding structures. If, as some have suggested, feeding on land was a motivation for leaving the water, then the relationship of changes in the feeding and respiratory structures would be obvious. There is, however, as yet little evidence for this interpretation.

The loss of gills, closing of the opercular region, and other related modifications also were functionally and topographically related to the auditory structures, which were becoming adapted to sensitivity to the pressure waves in the air. During the development of an air-filled buccal cavity and closure of the opercular region, as proposed in Panchen's theory, breathing by lungs became more and more efficient. Adequate exchange of carbon dioxide between the blood and air became a necessity, requiring a relatively large respiratory surface area relative to body weight. Many living amphibians and some fishes utilize dermal respiration effectively, and it has been suggested, because of evidence of extensive vascularization of the cranial region, that this might have been the case in some of the rhipidistians (Schmaulhausen, 1968). The scale-covered body found among rhipidistians would certainly restrict the extent of dermal respiration. All of the early amphibians, both labyrinthodonts and lepospondyls, in which the condition is known were covered by overlapping dermal scales (see Figure 22), and later ones had various patterns of dermal ossifications. It seems unlikely that dermal respiration played a major role in their respiratory functions.

Such features as pharyngeal diverticuli, branchial labyrinths, or respiratory capacities of the lining of the digestive tract could leave no traces. They may have been developed in rhipidistians as in some modern fishes. If the lung had been effective, however, it seems improbable that such supplementary areas for gaseous exchange would have become prominent. The modifications of the skull and branchial arches produced a structure suitable to buccal breathing, and it seems likely that respiration by means of a buccal pump did occur in the transition to land. A rib-operated pump could not come about until modifications of the osteolepiform axial structure had taken place, probably long after air breathing had become well established.

THE LATERAL-LINE ACOUSTICAL SYSTEM. The characteristics of the lateral-line acoustical system in aquatic animals have been taken up in some detail on pages 488–490. There it was noted that the lateral-line components received near-field (seismic) sounds and the otic portion received far-field (pressure-wave) sounds. It is the latter that underwent strong modifications as it became capable of hearing in air.

No fundamental changes in the sensory components of the lateral-line system occurred in the transition from fishes to amphibians. In both the receptors function only in water and in much the same ways. Many extinct amphibans reveal the presence of this system by grooves in the dermal bones, but in ichthyostegals, as in rhipidistians, the nervous structures were housed in canals in the bone, with pores leading to the surface (see Figure 40, showing this condition). The change to a more superficial position appears to relate to the development of a thicker dermis and epidermis in amphibians and probably had little functional significance in itself. In some amphibians the lateral lines have lost contact with the dermal bones and are not impressed on them. Considerable modification of the system did occur within the evolution of the amphibians, but little during the time of their origin from fishes. The lateral-line system, by virtue of this stability, has figured prominently in discussion on the homologies of dermal elements of the skulls and in this role is considered later in this chapter.

The changes of the acoustical system are quite another matter. Hearing of far-field sounds in water and air requires structures with very different mechanical characteristics. As far as the inner ear is concerned, little gross morphological change took place in the transition, although there may have been modifications of the nervous structures not preserved in the fossils. The primary innovation was the development of the fenestra ovalis over which the foot of the stapes (hyomandibular) lay. At just what stage this developed is uncertain, but it appears to be present in the very early amphibians and absent in the rhipidistians.

Changes of the structures that led to formation of the middle ear of the amphibians were extensive. The hard parts are well known in both the rhipidistians and amphibians, although the stapes is not fully preserved in any of the early tetrapods. The essential features are shown in Figure 140, illustrating otic features, except for the stapes, which is shown in Figure 144.

The nature of sound reception in the rhipidistians is not entirely clear. It is possible that they did not hear far-field sounds in the water. If they did, there must have been a structure, presumably an enclosed air bubble, that was opaque to pressure waves in water and responded to pressure changes. The gas bladder of various modern fishes acts in this way. Although such a structure and some device for transmission to the inner ear may have been present in rhipidistians, there is no evidence of it. Possibly there was some other skeletal structure that interrupted the waves and acted as a transducer, but again none is known.

Thomson (1966a), after considering the various possibilities, outlined two conceivable ways in which an air bubble in close association with the hyomandibular might have been formed. One was by air trapped under the operculum, and the other by an air bubble in the spiracular slit. Favoring the latter, he proposed that the spiracle might possibly have been used for the intake of air, as in *Polypterus*, and thus have the potential of trapping air bubbles. If this was the case, selective pressure might have favored development of a diverticulum in the recess of the spiracular organ, but not incorporating this organ. This diverticulum, lying adjacent to the hyomandibular, could have formed a mechanism that would have been capable of at least some degree of far-field hearing in water.

Present-day amphibians do receive airborne sound waves by a variety of means. Generally their receptor systems are so modified that they contribute but little to an understanding of the origin of hearing. All amphibians, even the earliest, differ sharply from the rhipidistians in the structure of the middle ear, although homologous structures can be identified. The two major groups of extinct amphibians differ significantly in their otic regions and seem to have followed rather different evolutionary courses (Figure 145).

No otic notch is present in most of the lepospondyls; the stapes, where known, projects posterolaterally, without a dorsal-lateral tympanic process.

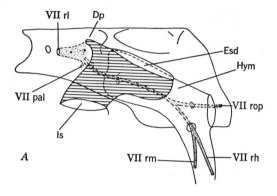

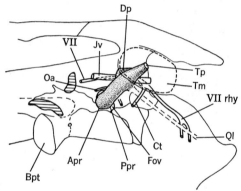

Figure 144.   Lateral views of the otic regions in a rhipidistian, *Ectosteorhachis, A,* and a primitive amphibian, *Palaeogyrinus, B.* The latter shows a restored stapes (which is unknown) and an inferred condition of the vestibule and the cranial nerves. In comparisons note particularly the dispositions of the main stapedial axes, resemblances and differences in the inferred conditions of cranial nerve VII, and the relationships of other soft structures to the stapes as shown in *B.* Apr, anterior process of stapedial foot; Bpt, basipterygoid process; Ct, chorda tympani; Dp, dorsal process; Esd, external opening of spiracular diverticulum; Fov, fenestra ovalis; Hym, hyomandibular, (= stapes in tetrapod); Is, internal opening of spiracular cleft; Jv, jugular vein; Oa, orbital artery, (= stapedial artery); Ppr, posterior process of stapedial foot; Ql, quadrate ligament of stapes; Tm, position of tympanic membrane; Tp, tympanic process; VII, facial nerve, pal, palatine ramus, rh, hyomandibular ramus, rhy, hyoidean ramus, rl, lateral ramus, rm, mandibular ramus (= chorda tympani), rop, opercularramus. (After Thomson, 1966*a.*)

The middle ear is well known in very few lepospondyls. It has been described in two microsaurs, *Cardiocephalus* and *Goniorhynchus*, in which the stapes appears to be preserved. A puzzling situation, currently lacking an explanation, is that the stapes appears very different in this closely related pair.

Most attempts to explain the origin of the middle ear of amphibians from a rhipidistian base have used the labyrinthodont pattern as a major point of reference. Some workers, however, have felt that the typical stegocephalian, with its deep otic notch and long tympanic process, represented not an intermediate stage but rather a special condition within the amphibians (see, for example, Parrington, 1949*b*).

Even the most primitive labyrinthodonts had a large otic notch and an open space in which the stapes lay, probably housed in an open tympanic cavity. The preserved stapes consists of a foot plate, lying over the fenestra ovalis, a dorsal process to the crista parotica, and a long tympanic process that passes to the vicinity of the otic notch. The hyomandibular of rhipidistians has five points of attachment to the adjacent structures. Only three are recorded in the ancient amphibians, but all are present in some living reptiles—for example, some lizards (Figure 145). The three characteristic processes of the labyrinthodonts are readily homologized with comparable attachments in rhipidistians, but a quadrate process (equivalent to the attachment of the hyomandibular to the palatoquadrate) and the hyoidean process are poorly developed or absent, presumably being present in life as cartilage or tendons. Like the skeletal features, the positions of the principal blood vessels and the nerves in this region, indicated by canals and foramina in the fossils, appear to be very similar. Figure 144 shows a reconstruction of the stapes in the ancient amphibian *Palaeogyrinus* and includes a comparison of the stapes and the hyomandibular.

The resemblances are unmistakable and convincing of relationships. Problems do arise, however, when rather primitive reptiles (e.g., *Captorhinus*, Figure 145*C*) and lepospondylous amphibians are considered. The former lacks an otic notch, and the stapes passes laterally toward the quadrate without evidence of a tympanic process. Various other structural differences also occur, but these relate more to evolution within the tetrapods than to origins. If an otic notch is considered to have been present in the transition to tetrapods, the history of such animals as *Captorhinus* must have been quite complicated. The tympanic process (= opercular process of the hyomandibular) appears to have shifted from its primitive, ventral position, first dorsally, and then either to have disappeared, reappearing at least as a functional analog later, or to have migrated once again to a ventral position. Figure 145 (Parrington, 1958*a*) shows the three conditions. Parrington and others have questioned that a stegocephalian stage, with a deep otic notch, was present in the ancestry of the reptilian tetrapods (for an alternative theory see

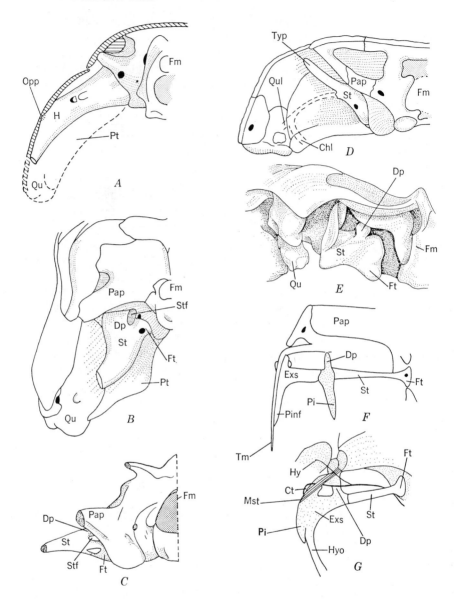

Figure 145.   The hyomandibula in a rhipidistian and adjacent structures in several amphibians and reptiles. All small or moderate in size, but not to scale. These figures illustrate some of the problems in the derivation of the reptilian stapes from rhipidistian and amphibian conditions as known and discussed in the text, pages 603–613. *A, Eusthenopteron,* a rhipidistian; *B, Dimetrodon,* a sphenacodont pely-cosaur; *C, Captorhinus,* a moderately primitive reptile; *D, Lydekkerina,* a typical

Tumarkin, 1955). In this event one of the immediately obvious sources of reptiles is among the microsaurs, particularly the gymnarthid group, and these frequently have been cast in this role. This suggests that the tetrapods that led to the reptiles did not come through the same course as the labyrinthodonts. Many features of the known microsaurs, however, seem to exclude them as ancestors of the reptiles (Romer, 1950). Also there is some evidence that primitive captorhinomorphs do in fact have an otic notch, although the stapes is not known in any of the very primitive genera. It should be noted that the theory of development of the auditory structures by Thomson, outlined below, requires an otic notch in the sequence.

Many unsolved problems of phylogeny remain at this evolutionary level and strongly affect the interpretations of the way in which hearing in air was developed. A big stumbling block, of course, is that both the rhipidistians and amphibians in which the otic structures are well known have departed significantly from presumed common ancestors. This poses a problem comparable in many respects to that of derivation of the mammalian ear structure from that of a contemporary lizard, without knowledge of the primitive reptiles from which both stemmed.

If it is assumed that the structure of a rhipidistian, such as *Eusthenopteron* is a reasonable morphological stage in the origin of tetrapods and that the evolution proceeded through a structural stage represented by ichthyostegals, fairly consistent hypotheses of the nature of origin of air breathing are possible. One such hypothesis, based on the requirements indicated by bioacoustical studies (see, for example, Van Bergeijk, 1966), has been proposed by Thomson (1966a). In developing his hypothesis Thomson rejects another proposed by Van Bergeijk (1966), based on similar mechanical requirements but not conforming to what is known of the fossil record.

Thomson's hypothesis starts with the assumption that far-field hearing in water was established within the rhipidistians (see page 607). Once these fishes left the water, hearing, which had depended on the volume-pressure relationships of an enclosed air space and transmission to the inner ear by the hyomandibular, could occur only if a very different mechanism were present.

---

labyrinthodont amphibian; *E*, *Goniorhynchus*, a gymnarthrid microsaurian lepospondyl (original length of stapes less than 2 mm); *F*, a young lacertilian, showing various parts of the stapes present during development; *G*, adult *Sphenodon*. Chl, ceratohyal ligament; Ct, chorda tympani (VIIth cranial nerve); Dp, dorsal process of stapes; Exs, extrastapedial; Fm, foramen magnum; Ft, foot of stapes; H, hyomandibula; Hy, hyoid branch of cranial nerve VII; Hyo, hyoid; Mst, stapedial muscle; Opp, opercular process of hyomandibula; Pap, paroccipital process of opisthotic bone; Pi, internal process of the stapes; Pinf, inferior part of the stapes; Pt, pterygoid bone; Qu, quadrate bone; Qul, quadrate ligament; St, stapes; Stf, stapedial foramen; Tm, tympanic membrane; Typ, tympanic process of stapes. (*A*, *B*, *C*, and *D* after Parrington, 1958a; *E*, *F*, and *G* after Olson, 1966a.)

The body tissues, transparent to pressure waves in water, would block such waves in air, so that they could not affect the pressure-volume relationships of the air-filled cavity. Part of the tissue thus must have become responsive to pressure waves in the air. Reactivity to relatively slight pressure changes is required. This responsive area must transmit the effects of its mechanical deformation to the inner ear by means of some conductor. The opercular region, connections of the operculum to the hyomandibular, and the abutment of the hyomandibular against the otic capsule could have formed such a mechanism, provided that the opercular became reduced and flexible and the hyomandibular freely suspended in a tympanic cavity.

The physics of hearing in air show that the relationship between the surface area of the tympanum and the mass of the ossicle, along with the nature of its attachment and possible level actions, are critical to the range of frequency received and transmitted to the ossicle for a given intensity of sound. A very large ossicle, such as was present in amphibians, requires a large tympanum in order for sounds of normal intensities to be effective.

From this base, and under the assumption that air hearing was present in the early amphibians, the following hypothesis was developed by Thomson. Changes in skull and jaw structure in the rhipidistians facilitated expansion and modification of a spiracular diverticulum that came to surround the hyomandibular, producing a tetrapod condition (see Figure 144, after Thomson, 1966a). The air cavity was necessarily closed, if there had been hearing in water, and in the late stages contact of the wall of the chamber with the hyomandibular was at the hyomandibular–opercular articulation. Concurrently the gill system was modified, but the rates of development of the different functions and structures are indeterminate. Thomson argued that the tetrapod condition of the middle ear probably was attained as gill breathing was completely lost. This would be almost at the same time as attainment of complete development of the air-filled buccal cavity suggested by Panchen.

As the fish, or tetrapod, came on land, the operculum became a thin, flexible structure that formed the outer part of the tympanic membrane. The tympanic cavity was in existence, and its lining at the opercular contact of the stapes formed the inner part of this membrane. The hyomandibular, freed of its role in the hyostylic jaw suspension, was reduced in size and took on the functions of the stapes. A large otic notch, produced as already described by modifications primarily related to feeding mechanisms, housed the large tympanic membrane.

By such a process, without fundamental change in composition or relationships of any of the structures involved, a system capable of hearing in air could have developed. That some such transition took place to produce the auditory mechanisms of labyrinthodonts is plausible. It must be pointed out, however, that many alternative hypothetical courses have been proposed.

One suite of these involves the proposition that hearing in air developed several times within the amphibians. Among these are ideas that initial hearing was done by bone conduction from contact of the jaws on the substrate. Another suggestion is that hearing in mammals was developed independently from reptilian ancestors that had no functional middle ear. There are many variants of this central theme.

The evidence within the amphibians remains problematical. If the labyrinthodonts and lepospondyls followed an initially similar course, their common ancestry must have been very remote from the time of their first appearance in the record. It might be, however, that the rhipidistians followed a course that paralleled that of the labyrinthodont stock and that the actual common ancestor had a structural pattern more or less common to the crossopterygians and actinopterygians. A simple and plausible hypothesis is that some processes more or less like those envisaged by Thomson produced creatures capable of hearing in air and that these were the central stock of the tetrapods. Later evolution within the amphibians and reptiles probably represents variations on this primitive pattern. The most serious difficulty with such a hypothesis comes from the lepospondyls, since the otic notch is not present even in the earliest of them.

Information on the stapes, however, comes only from late members, in the Permian. It is mostly the otic notch that causes trouble, and even in this area there is some confusion, since some of the nectridean lepospondyls do seem to have a vestige of an otic notch. Perhaps, as Parrington has suggested, the notch was not a feature of the primitive tetrapods. If this was the case, considerable modification of Thomson's hypothesis is necessary. Here the matter must rest until more detailed evidence and information from missing stages become available.

PROPORTIONAL CHANGES AND IDENTITY OF DERMAL ELEMENTS.   The related modifications of feeding mechanisms, changes of the relative proportions of the anterior and posterior portions of the endocranium, and alterations of the respiratory structures were inevitably matched by modifications of the dermal bones covering the internal structures of the head. How they occurred and the implications of these changes on phylogenetic interpretations have been controversial. The most generally accepted interpretation of the course of change from the rhipidistians to the tetrapods views this as a straightforward, graded modification in which the posterior elements of the skull were reduced, some being lost, and the more anterior elements were enlarged proportionately. Initial changes in the anteriormost parts of the skull involved simplification by fusion and loss of some small bones, to produce a stable pattern.

This is the course proposed by Westoll (1938), elaborated by him later and supported strongly by Romer (1945a, 1946, 1956b) Parrington (1956b,

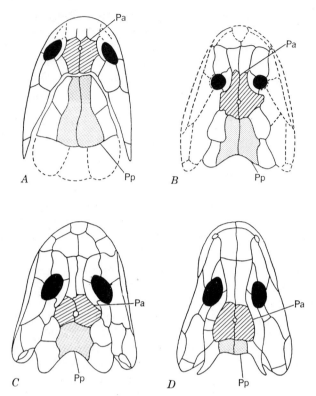

Figure 146.   Westoll's theory of the evolutionary trends of the pattern of the skull roof from rhipidistians to amphibians and reptiles, shown by diagrams indicating the changes in relative sizes of the postparietal and parietal bones. This "orthodox" view has been supported by Romer and Parrington in particular, accepted by many others, and repeatedly contested by Jarvik (see discussions in accompanying text). *A, Osteolepis*, a Devonian rhipidistian, showing a short anterior and long posterior part of the skull roof (matched by underlying structures); *B, Elpistostege*, a presumed amphibian, representing an intermediate stage based on a single skull roof from the upper Devonian; *C, Ichthyostega*, a well-defined uppermost Devonian amphibian; *D, Palaeogyrinus*, a Carboniferous labyrinthodont. The identities of the elements listed as Pa, parietal, and Pp, postparietal, are disputed by Jarvik (see Figure 149). (After Jarvik, 1967*b*.)

1967*a*), and Thomson (1969). Many others, influenced by the logic and simplicity of this system, have fully accepted it. The sequence of changes is illustrated in Figure 146.

The interpretation was based on paleontological evidence, first on the topographic resemblances of elements in a series of graded steps from the record, as shown in Figure 146. These representatives include stages most

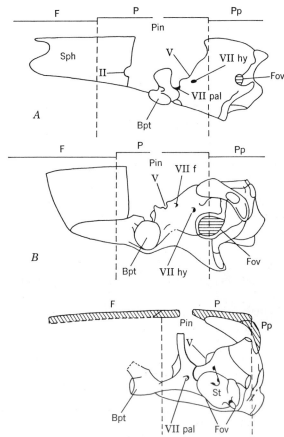

Figure 147. Diagrams of the oto-occipital parts of the skulls of *A, Kotlassia; B, Seymouria; C, Captorhinus,* in lateral aspect with external elements removed. Relationships of the frontal, parietal, and postparietal to internal structures of the skulls are illustrated to show the moderately constant relationships to the major features of the braincase, but differing relationships to such structurally dynamic features as the basipterygoid process. Bpt, basipterygoid process; F, frontal bone; Fov, fenestra ovalis; P, parietal bone; Pin, pineal opening; Pp, postparietal bone; Sph, sphenethmoid bone; St, stapes; II, position of the optic nerve; V, position of the trigeminal nerve; VII, facial nerve; f, facial branch; hy, hyoid branch; pal, palatine branch. (After Parrington, 1967*a*.)

favorable to the interpretation and do not indicate the range of variation that is known. They portray not a true evolutionary series but a morphotypic array, that shows the presumed landmarks in a highly complex transition. This does not detract from their demonstration of a probable course of change, but it does argue for caution in using the series as a basis for dynamic evolutionary interpretations.

In tests of this theory many studies have been made of the consistency of relationships between the superficial dermal bones and the underlying structures of the endocranium. Figure 147, based on Parrington (1967a), illustrates the extent of positional consistency of the frontal, parietal and postparietal, and representative cranial structures, such as the basipterygoid process and the proötic incisure, marking the position of emergence of the fifth cranial nerve. Again, of course, as Parrington emphasized, there is much variation within each of the groups represented.

The patterns of changes under this theory can be explained as the result of modifications in relative proportions of elements and reduction in the numbers of bones by losses. In the anteriormost region, if the osteolepid condition be considered representative of the actual ancestral stage, fusion of small elements to produce the bones found in tetrapods is necessary. The changes of the dermal bones seem to be closely related to the functional adjustments of the various systems of the skull and jaws that occurred in the transition from water to land.

These interpretations stem from a strictly comparative anatomical approach based on the concept that the extent of morphological resemblance in all characteristics of organisms provides the best basis for estimations of relationships. Similarities of a few features in animals that in most respects are very different and thus taxonomically remote are not considered to have major significance in establishment of lineages. Rather, very similar animals whose morphology and geological positions suggest that they represent phylogenetic lines are used for comparisons; where possible, these selected animals are used only if they are reasonably representative of their larger group and somewhat central and primitive within this group. Beyond the usual precepts of evolutionary principles, in particular with regard to consistency of morphological relationships and the adaptive nature of change, little advocacy of particular modes of modification, to the exclusion of all others, is to be found.

There are, however, other approaches, and these have produced very different results. Differences with the foregoing type of analysis of changes of skull elements have centered in particular around the problems of the homologies of the bones. If these were merely restricted to the affixing of particular names to individual bones, they would be trivial in evolutionary considerations. They have, however, entered deeply into problems of the transition to land by becoming a principal basis for phylogenetic interpretation. Parts of the difficulty arose first from the fact that skull bones in fishes and tetrapods were named independently. In some cases the same names were applied to nonhomologous bones and different names were given to bones that have proven to be homologous. For the most part this has caused only minor trouble. Controversy has centered around the equivalency of the "fish frontal" and "tetrapod parietal." In proposing his system of bone homologies, Westoll (1941a) considered these two to be homologous and proposed that the "frontal" in osteolepids be called the "parietal." Westoll's initial study was stimulated by the work of Säve-Söderbergh (1932, 1934, 1935), which presented a very different phylogeny of tetrapods, one involving the multiple origins of the major lines of amphibians. Building on Säve-Söderbergh's work, also incorporating methods used by Stensiö and adding immense amounts of detailed information, Jarvik has carried ideas of independent origins much farther.

The general approach used by these students involves strict adherence to two propositions that relate to the matter of homologies of bones. In addition they make use, with some exceptions, of a general concept of consistency of morphological relationships, much as does the other school. A very important difference also is found in the use of data from selected or chance members of widely separated groups (e.g., two classes) for comparisons that are considered to have phylogenetic significance, with little regard for variation within each of the major groups represented and for all other information bearing on the supposed phylogenetic relationships. These aspects of the approaches used by this school are briefly described below.

1. Patterns of Dermal Bones. Patterns of dermal bones change primarily by fusions of elements, and different patterns result from different courses of fusion. This has been a mainstay of many studies by Stensiö, in particular on primitive fishes. It was applied by Säve-Söderbergh to explain the differences in ichthyostegids and labyrinthodonts when, in his early studies, it was found that the ichthyostegids did not fall, as expected, in a position intermediate between rhipidistian ancestors and later amphibians. From these considerations he deduced two major lines of tetrapods, batrachomorphs, and reptiliomorphs. Jarvik adopted this concept and carried it over into his considerations of tetrapod relationships.

Fusion of elements does occur and seems to have been particularly important in the early evolution of actinopterygian fishes. Problems arise when it is considered to be the only way of change.There is convincing evidence from many sources that bones, in the course of evolution, reduce in size and drop out and that the space which they occupied can be taken over by adjacent elements. There is equally evidence that loss of bones sometimes occurs by fusion of two or more elements. Change occurs by both of these means and by changes in shapes and proportions of elements.

Some bones in adults have formed from two or more centers of ossification. As De Beer (1937) has strongly argued and others too have shown convincingly by experimental work over a wide range of animals, the centers of ossifications in the formation of bones cannot be taken as valid evidence in questions of bone homologies or of phylogenetic histories of bones. Clearly the adult elements in some fishes have arisen by fusion, a process very clear, for example, in some extinct lungfishes (Westoll, 1944, E. I. White, 1965). The concept that this is the sole course of origin and change, however, carries the assumption that size and form are intrinsic properties of bones (Parrington, 1956b), and this quite clearly is not the case.

If the doctrine of development of dermal bones by fusion is carried to its extreme, limitless explanations of the origin of any particular pattern are possible. Unless there is evidence of the actual fusions, which rarely is the case, the patterns can provide no concrete basis for determining that this was the course of its formation.

2. Lateral Lines. Lateral lines have immutable relationships to particular dermal bones. This hypothesis works hand-in-glove with the concept of fusions, since it seemingly offers a way of tracing which bones have become fused, independently of their topographic relationships. It derives from some experimental evidence that neuromasts may act as bone organizers and a derived conclusion that they always induce homologous elements in particular (relative) positions. The work of Allis (1898) and supporting studies by Pehrson (1922) and Severtsov (1926a,b,c) have done much to establish this concept. These studies were based on modern fishes. There can be no doubt that considerable stability exists in the general relationships. That this is an inviolate rule, without exceptions, is not determinable.

In order for this to be established, as Parrington (1956b) has noted, it is necessary to know the identity of the bones independently of the evidence of the lateral-line canals. This is basically a matter of topographic relationships. When, as happens, topographic relationships and the lateral-line canals lead to different interpretations, it is logically inconsistent to follow the evidence of the latter. That such difficulties do arise is shown in Figure 148.

3. Relationships between Sensory Structures and Dermal Elements. Related to the idea of constancy of lateral lines and dermal bones is the concept that there are consistent relationships between various sensory structures of the head and the dermal elements. Once again, this is broadly true, but there are many exceptions. This general principle has been accepted by most workers, but applied in a somewhat irregular way, without dogmatic acceptance. Much attention has been devoted to the relationship of the pineal foramen and the parietal and frontal bones. If Westoll's concept of the parietal in osteolepids and tetrapods is followed, the pineal opening lies in the parietal in both. Westoll did not make important use of this point, but it has figured prominently in support of his position and this has led to commentaries (as in Jarvik, 1967b) on the variable position of the pineal. Also Schmaulhausen (1968), in a rather curious discussion, has argued the case of consistent

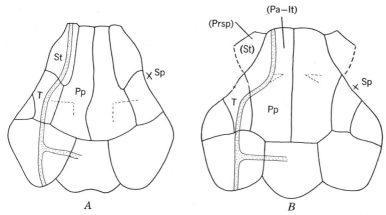

Figure 148. Diagram of the posterior dermal bones of two rhipidistian crossopterygians to show the relationships of these bones to the lateral-line canals. *A*,
Osteolepis; B, Holoptychius. These illustrations have been used by Parrington
(1967*a*) to support the position that the relationships between lateral-line sensory
organs and dermal bones are not always constant and cannot be indiscriminately
used as keys to homologies. The centers of ossification of his postparietals (as
labeled in the figure without parentheses) are at the junctions of the pit lines, well
forward in the broad *Holoptychius.* He considers it likely that these broad elements
have "captured" the lateral-line canals and that the homologies are as labeled.
Jarvik, following the concept of constant relationships of lateral-line sensory organs
and dermal bones, has identified the bone considered tabular by Parrington as
supratemporal, the parietals as postparietals plus intertemporals, and Parrington's
supratemporal as what he has termed a "prespiracular" element. (Pa-It), parietal-
intertemporal element of Jarvik; Pp, postparietal of Parrington; Sp, position of
spiracle; (Prsp), prespiracular plate of Jarvik; St, supratemporal of Parrington;
(St), supratemporal of Jarvik, T, tabular. (Modified after Parrington, 1967*a*; data
from Jarvik in part.)

association of sensory organs and dermal bones, in particular the orbit and the
frontal, to arrive at the concept that fish frontals and tetrapod frontals and
fish parietals and tetrapod parietals are in fact homologous. But he supported this set of homologies by arguing that the pineal opening does not
have constant relationships, in contradiction to his basic doctrine.

4. Anatomical Studies. The use of genera and species that are very widely
separated taxonomically for comparative anatomical studies and conclusions
on phylogenies, without consideration of possible intermediates or the variations within the major groups that they represent seems at best a questionable
practice. Yet it lies at the heart of some conclusions on polyphyletic origins.
The great variability of the vertebrates of each class makes it possible, by
selection of a few features and particular genera from two or more classes, to
support almost any hypothesis of relationship that may be proposed.

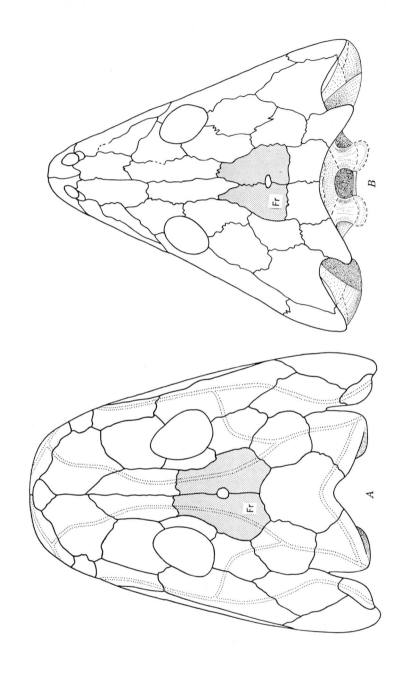

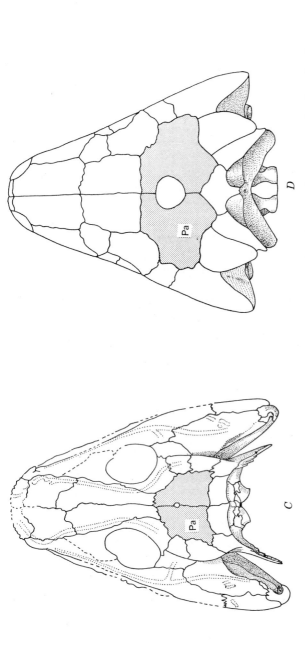

C

D

Figure 149. Four skulls in dorsal view with the pattern emphasizing the element that surrounds the pineal opening. *A, Ichthyostega,* temnospondyl; *B, Lyriocephalus,* temnospondyl; *C, Palaeogyrinus,* anthracosaur; *D, Diadectes,* related to anthracosaurs and close to the reptiles in many features (often classified as a reptile). The labeling of the elements follows Jarvik in keeping with the concept that the temnospondyls and batrachosaurs are latitabular, whereas the anthracosaurs and their reptilian descendants are angustitabular. The former are "batrachomorphs" and the latter "reptilomorphs," with separate origins among the rhipidistians. This is a modification, but generally a continuation of the dual origin proposed by Säve-Söderbergh (1932, 1935). Fr, frontal; Pa, parietal. (After Jarvik, 1967b.)

The use of nasal structures of porolepiforms, osteolepiforms, urodeles, and anurans (see pages 601–603) falls in this category. More ambitious is the comparison of an osteolepid fish and embryonic rabbit (*"Lepus"*) (Jarvik, 1952). In the similarities of the dermal and endochondral structures of the osteolepid and the developing bones in the rabbit Jarvik found evidence for denial of the equivalency of the fish "frontal" and tetrapod "parietal" proposed by Westoll. In refuting this approach Parrington (1967a) pointed out that very different results would have followed had another mammal, *Talpa*, been used. But this to some extent really dignifies the approach beyond what it is worth, because inadvertently it might seem to give some support to the possibility of meaningful results from such practices. A most serious aspect is that no weight is given to any of the abundant evidence available from other sources: other aspects of anatomy, embryology, function, data from related animals in which differences occur, and all of the information on the historical development of vertebrates as seen in the sequential deposits of the fossil record.

Fragmentation of phylogeny is almost certain to result from these practices, because the evidence of continuity is ignored. In his 1967 analysis Jarvik reached some of the following conclusions. Temnospondyls, representing Säve-Söderbergh's Batrachomorpha, stemmed independently from the osteolepiforms. Their parietal, in Westoll's terminology, and that of almost everyone else, is actually the frontal, as illustrated in Figure 149 (data from Jarvik, 1967b). The assemblage including anthracosaurs, seymouriamorphs, and reptiles, the Reptiliomorpha, arose separately from the osteolepiforms, and in them the tetrapod parietal has been correctly named. By this interpretation and by the criteria also noted above, the diphyletic origin of the Eutetrapoda of Säve-Söderbergh has, Jarvik avers, been confirmed.

Other conclusions are that the anurans probably came independently from the osteolepiforms, as did the ichthyostegids. Acanthostegids and perhaps most of the Triassic temnospondyls also represent separate lines. Sauropsid and theropsid reptiles may have evolved independently from an osteolepiform-like ancestry and probably other groups of reptiles as well. Thus the Eutetrapoda are not merely diphyletic but polyphyletic. In this report Jarvik does not mention mammals, whose origin he has also considered to be very remote from other tetrapods.

This very divergent analysis of phylogeny has found few adherents, and its controversial nature tends to cast a shadow over Jarvik's outstanding studies. It seems also to have had the unfortunate effect of making phylogenetic conservatives of many who oppose it, causing the pendulum to swing toward support of positions of simplistic monophyletic origins of all major groups and avoidance of the abundant and intriguing evidence that evolution did not go by simple, directional pathways, with inevitable "progress" and a set of steps well documented by an evident set of stages in the fossil record.

LIGHT RECEPTORS. Perception of images in water and in the air poses rather different problems because of the differences in the indices of refraction of the two media. Comparisons of the structure of the eyes in modern fishes and various tetrapods indicate how accommodations to these differences have been made. Since they involve the crystalline lens, the cornea, and nervous tissues, no record can be expected among fossils. Protection of the cornea from desiccation involved major changes in glands and the muscles of structures adjacent to the eyes proper. These too have left little direct evidence, except in the case of the development of the lacrimal canal and occasional modifications of the shape of the orbital chamber that have been interpreted as being related to glands.

The paired orbits and the pineal foramen, plus occasional impressions of some part of the optic structure on the underside of the dermal surface and in the endocranium, provide much of what can be seen. The size and position of the orbits and the occasional preservation of sclerotic plates tell something about the adaptive nature of the visual system to ways of life, but little else. Most of the recorded changes relate to modifications after the tetrapod stage was well established.

The pineal foramen is of interest primarily because it is persistent and indicates that the organ of the pineal system that it housed retained functional response to external stimulation during the transition from water to land. Its probable functions in water were discussed in Chapter 2 of this section (pages 501–503). Most probably it acted as a light receptor in rhipidistians and in their immediate tetrapod descendants.

Steyn (1961) has suggested that the pineal organ in transitional animals and in the amphibians and reptiles of the Carboniferous and early Permian may have been important in adaptation to land as an indicator of approaching desiccation. This is based on several bits of evidence from a number of sources. Sensitivity to light does exist in various fishes, both those with pineal foramina and those without, and in various tetrapods, both by direct and indirect stimulation (Kelly, 1962). It appears to have a role in thermal regulation in lizards (Stebbins and Eakin, 1958) and possible relationships through its association with the subcommissural organ to thirst and sweating in mammals.

Some such function as Steyn has suggested then is at least plausible. Changes of the pineal during the early evolution of tetrapods are suggested by modifications of the pineal foramen, but these were subsequent to the transition to land and appear to represent adaptive responses among the different groups of amphibians and reptiles.

The Problem of Desiccation

The severity of problems of desiccation faced by the early invaders of the land related both to the conditions in which this transition occurred and to

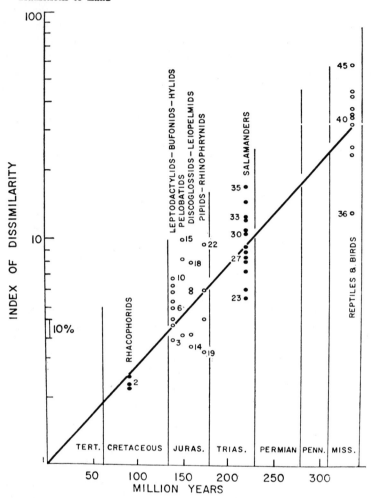

Figure 150.   A graph based on immunological tests dependent on the concept that rates of modifications of the biochemistry tested have been moderately constant over long periods of time. An index of dissimilarity is plotted against a time axis on which are shown the presumed times of origins of the groups tested to assess the least remote possible time of divergences of the stocks. (After Salthe and Kaplan, 1966.)

the structure and physiology of the animals. If there were various emerging lines, they may have faced rather different problems, depending on the conditions under which they existed and their particular level of development of terrestrial adaptations. If the successful invasions occurred where the climate was warm and the humidity high, problems were not severe. If they took place in response to drying environments, the opposite may have been true.

In the course of later evolution tetrapods developed many different ways of maintaining their water balance in a subaerial existence. Even among the existing amphibians, however, the adaptations are so intimately related to their own environments that they do not give a basis for reliable estimates of the remote predecessors.

As noted on page 606, with reference to dermal respiration, the early amphibians appear to have carried a well-developed cover of overlapping, bony, fishlike scales. Later these were modified and sometimes lost; in some instances keratinous outer surfaces of the integumentum were developed. The scales of the early amphibians probably formed an effective shield against desiccation in their usual environments. As in many other features, the principal conditions found in the amphibians were attained not at the time of transition from water to land, but later, in the radiations of the various stocks.

Biochemical Evidence and Amphibian Origins

Relatively little biochemical research has had a very direct bearing on the nature of transition from water to land. It seems unlikely at present that it can be particularly helpful. The only materials available for study are extremely remote from the participants, and their relationships are at best difficult to establish. A study by Salthe and Kaplan (1966), however, does have some bearing on the problem of monophyletic or diphyletic origins of anurans and urodeles from the osteolepiforms.

Their work employed the techniques of immunology and was undertaken to set limits to the date of divergence of the urodeles and anurans. It involves the basic assumption that the extent of differences between homologous protiens is a direct function of the time elapsed since the divergence of the stocks tested. By use of times of "known" divergences this assumption was checked and found to be reasonable. By use of a dissimilarity index applied to a wide variety of vertebrates the diagram reproduced in Figure 150 was obtained.

From their studies it was concluded that divergence of the urodeles and anurans might have been in the early Triassic, and possibly as far back as the early Permian, but no earlier. If the conclusions are accepted as reasonably reliable, the proposed origins of the two from stocks of rhipidistians in the Devonian is untenable.

ECOLOGY OF THE ORIGIN OF TETRAPODS

Scope of the Problem

Frequently the origin of land vertebrates has been treated in a rather simple way, as if it were an event that happened once, at a more or less given time, and consisted of the emergence of aquatic animals onto land, followed by increasingly effective occupation of the new habitat. The ecology of origin becomes the center of dispute. It is now evident that this is far from the full story of the transition. The process began in the middle Paleozoic, probably in the Devonian, and extended into the Permian. It involved development of the capacities to occupy land among aquatic animals, probings of the terrestrial environments, establishment of an amphibious existence, regressions to water, and finally complete freedom from water, followed by effective exploitation of the terrestrial environments.

Some such sequence is now generally recognized by most students. Often, however, specific proposals on one or another aspect of the ecology and the motivations of changes have become oversimplified both with respect to the ecological conditions and to the differences in the ways of development of the various lines of amphibians. No one aspect can stand alone, without consideration of the implications of other phases of the process.

During the 1950s this topic stirred considerable interest and many different points of view were aired. Although they did not settle the matter, these studies brought out many facets of the problems related to the origins of terrestrial adaptations. They have illustrated, among other things, how the use of different sets of criteria may lead to widely different interpretations. Although the pace has lessened somewhat, various aspects of the ecology of the prototetrapods and early tetrapods continue to be studied as new materials are found. Particularly active in investigations of this subject have been the following: Romer (1945a, 1956b, 1958b), Orton (1954), D. W. Ewer (1955), Goin and Goin (1956), Gunther (1956), Inger (1957), Cowles (1958), Eaton (1960), Tihen (1960), and somewhat more recently Carroll (1964, 1967a) and Thomson (1966a).

Suggested Ecologies and Modes of Origin

There have been almost as many variants on the general theme of origin of tetrapods as there have been authors, but these can be grouped into three categories or combinations of these.

1. What may be termed the "classic" theory is that land invasion occurred, probably in the later Devonian, under semiarid conditions when ponds and streams dried out periodically. Their inhabitants, as they were able, crossed over land in search of remaining bodies of water. Those that were successful produced the generations that followed, which became better and better adapted to life on land. Exploitation of land began in this fashion and continued in a more or less straightforward course from amphibians to reptiles. This was the general thesis proposed by Lull (1918) in

coordination with environmental studies of redbeds by Barrell (1918). Romer, in various publications, developed the concept and carried it somewhat farther.

One modification of the thesis that semiarid conditions were involved is that aestivation was a consequence and an intermediate stage. Along these lines Orton (1954) suggested that the tetrapod limb was developed in response to digging activities related to aestivation.

2. In contrast is the theory that occupancy of land took place under warm, moist climates suitable for animals vulnerable to desiccation. Waters of the ponds and streams probably underwent periods of low oxygen tension but they did not dry up. The time has also been suggested as being middle or late Devonian. Variants of this general concept have come largely from students of modern fishes and amphibians (Inger, 1957; Gunther, 1956; Cowles, 1958; D. W. Ewer, 1955; Gordon, 1964). Motivations for leaving the water have been at the heart of differences within the general framework. They involve one or another form of population pressure in the original, aquatic environment, competition for space, for food, for breeding sites, and so forth. Goin and Goin (1956) have proposed a variant in which partial drying is involved, with population pressures growing as animals find less and less space as their habitats dwindle. Intermittent wet periods, during the general drought, promote migrations during the short, favorable times.

Feeding on land has also been considered as a motivation under the theory of origins in a warm, moist climate. Escape from predators has been suggested. Somewhat akin, but under drying conditions, has been the proposal of Romer that the amniote egg was developed to resist desiccation on land, where it would be less vulnerable to destruction by water-feeding animals. A variant of this, by Tihen (1960) discounts the desiccation but incorporates the escape from egg feeders.

3. Still a different theory is that amphibians originated in water and their initial adaptations were to special aquatic circumstances. The limbs, hearing, air-breathing, and feeding modifications all have been related to life in warm, shallow waters, low in oxygen. Terrestrial radiations followed much later and came from various lines of aquatic tetrapods. Eaton (1960) has been the strongest proponent of this concept.

4. Several variants that are combinations of the above explanations have been proposed:

a. The "classic" theory and the idea of subsequent long evolution in water have been combined by Romer (1958*b*). The semiarid Devonian conditions, with markedly seasonal rainfall, produced the conditions necessary for the origin of tetrapod features, related to survival as before. Development of true land animals, or even successful amphibious creatures, took place much later under more favorable, moist, warm conditions. There was a long period of predominantly aquatic life by both amphibians and reptiles. Romer argued against the concept that motivation to land life lay in a search for food, but proposed that in the case of developing amniotes reproduction might be involved.

b. Ideas of polyphyletic origin necessarily tend to lead to the concept that somewhat different ecological conditions attended the different origins. Mostly, however, those who have been concerned with multiple origins have given little attention to this aspect of the problem, so that formal proposals have not been made.

c. The idea expressed by Thomson (1967*b*) that the labyrinthodonts and lepospondyls followed different courses after stemming from a common amphibian source, like polyphyletic theories, involves a combination of items from each of the three basic interpretations. The labyrinthodonts, trending toward a terrestrial life, may have following the combined program indicated by Romer and outlined as (*a*) above. The lepospondyls, however, remaining primarily aquatic conform to the aquatic origin of tetrapods and to a long period of life in the water. Most of them, however, did not give rise to any terrestrial descendants, so this part of the third hypothesis above is not applicable.

The Sources of Evidence

The evidence used for these various interpretations has come from three main sources: the nature of the sedimentary deposits from which fossils have come; the fossils, including vertebrates, invertebrates, and plants; and the structure, physiology, behavior, and environments of present-day fishes, amphibians, and reptiles. Differences in interpretations of the environments of deposition of the sediments and the life habits of some of the extinct vertebrates have contributed to divergences in interpretations of the ecology of the origin of tetrapods. In some instances the tangible evidence is so slight that arguments about its meaning seem somewhat futile. However, this is all there is, and if any conclusions, no matter how tentative, are to be made, they must be based on these scant data. What is most important is recognition that interpretations are tentative and serve in particular as hypotheses whose deductions are to be tested by new information and reassessment of old.

Geological Data

Much of the interest has been directed toward the climates under which nonmarine beds of Devonian age were deposited. Many of these are red beds, red sands, shales, silts, and conglomerates, colored by highly oxidized iron that coats the grains. Red beds have often been cited as evidence for semi-aridity. The fact is, of course, that red sediments can be formed under a variety of conditions and be reworked, transported, and redeposited several times without loss of the red color. What is important to their persistence is that the iron oxide is not subjected to strongly reducing conditions. It is, however, true, as Romer (1958b) has pointed out, that other characteristics of many of the terrestrial red sediments of the Devonian and later Paleozoic do suggest their deposition under conditions in which periodic desiccation took place.

As far as the evidence from sediments of the Devonian is concerned, however, two things are important. First, none of the typical Devonian red beds, except the apparently uppermost Devonian of east Greenland, is known to carry any amphibians, so that it is pure conjecture that they indicate conditions under which terrestrial vertebrates came into being. They do carry rhipidistians. At Escaumanac Bay, Nova Scotia, in nonred deposits, rhipidistians are associated with the possible amphibian *Elpistostege*. Conditions of deposition, except for the lack of red color, do not appear to be greatly different from those of other sites in which rhipidistians occur. In east Greenland amphibians and rhipidistians occur together in red beds, in deposits usually considered to be of Devonian age. If amphibians did in fact arise in red-bed environments, they left no record in these beds, which are a principal source of information on Devonian freshwater vertebrates.

The second point is that the well-known deposits are relatively restricted, mainly to eastern North America, Greenland, Spitzbergen, and northern and western Europe. Some scattered deposits, however, are known in Australia, eastern Asia, and Antarctica. What little has come from these deposits suggests that faunas were similar over very large parts of the earth. Conditions over immense areas, some certainly continental, remain unknown. It is quite possible that a wide variety of environments existed and that terrestrial life, or at least tetrapods, could have come to be under any one or several of these. Unless it is first assumed that red beds represent the environment in which transitions took place, arguments concerning their ecological meanings are futile. To use the evidence that is at hand this assumption must be made, but its tentative nature must be recognized.

Freshwater elements of the vertebrate faunas known from the Mississippian are very restricted, as discussed on pages 591–592. Largely they have come from coal-measure deposits or similar beds formed in association with swampy environments. Some of the facies in eastern United States are in the general red-bed category. The deposits that have yielded amphibians all seem to have been formed in low-lying terrestrial or marginal marine environments. Had well-developed terrestrial organisms existed, their chances of preservation would seem to have been very slight. Geographically sampling is restricted to the British Isles and eastern North America.

The environment of deposition of most Pennsylvanian, freshwater verte-brate-bearing beds was much the same, largely that of coal swamps. In the later Pennsylvanian, however, some red-bed deposits are known. Geographi-cally, Pennsylvanian beds are more widespread, covering large areas in North America and in Europe. In Nova Scotia, at Joggins and Florence, amphibians occur in tree stumps, formed in association with coal measures. Here Carroll (1964, 1967a) has suggested that the environments were some-what more terrestrial than in most coal swamps, but this is based on faunal rather than sedimentological evidence.

Near Danville and Newton, Illinois, and Garnett, Kansas, are upper Pennsylvanian tetrapod-bearing beds not of the typical coal-measures types. All, however, were formed on very low land or in estuaries and do not tap the uplands. The lowest Permian deposits, in east central United States and in the midcontinent continue this, with the sediments formed on delta mar-gins, in streams, ponds, and swamps. In Europe the general coal-measures type of deposition continued. In the later part of the early Permian the sedi-ments give evidence of increasing periodicity in rainfall and some representation of slightly more upland deposition. By the beginning of the upper Permian the geographic scope has increased notably and the environments represented in the still typically red-bed deposits have become much more varied.

The geological evidence on sediments indicates that until very late during the development of terrestrial tetrapods, vertebrate-bearing beds were formed only under aquatic and marginally aquatic circumstances. What must be decided, if interpretations are to be made, is whether or not this sampling is adequate as a broad representation of the record or whether there was a totally unknown radiation in the uplands. Interpretations come largely from what happened subsequently in terrestrial radiations. These seem to show, in general terms, that the types of organisms known from the late Pennsylvanian on do give a moderately adequate basis for later tetrapod evolution. No very extensive, unknown upland radiation need be postulated. They also suggest, however, that within the types of environments tapped by the sediments groups of animals were developing for which little or no known record exists. The broad ecological zone seems to be suitable for the evolution that preceded extensive expansion into the uplands in the very late Paleozoic and early Mesozoic, but only a very small part of its total is known.

Before the late Pennsylvanian, however, the record is less good, for the roots of many later stocks do not appear in the samples. The environments in which the known sediments were formed do not seem to be representative of all of the ecological conditions under which the tetrapods of these times lived. Before the middle of the Pennsylvanian, however, the sampling of even known types of ecologies is so imperfect that it is entirely possible that many of the missing ancestral types did in fact exist in them and have merely not been found among the few specimens that are known.

Fossil Evidence

The fossil record of the Devonian and Mississippian vertebrates has been summarized on pages 590–591. The amphibians of these times generally have aquatic characteristics, and this has been emphasized in many studies. *Ichthyostega* has been interpreted by some as fully aquatic, but its structures cast doubt on this determination. The limbs are quite strong, and the girdles are fully developed. Ossification is moderately complete. The skull is fully separated from the shoulder girdle. Skull and visceral structures were suitable for air breathing, not water breathing by gills. The auditory region has all the characteristics of an animal that could hear only in air.

The subsequent Mississippian record is slight, and the few skeletons that are known are from animals that were mostly strictly aquatic. This was certainly the case for the lepospondyls and for anthracosaurs. The trimerorhachid from the upper Mississippian seems to conform closely to the pattern of later aquatic members of the family. The situation of the ichthyostegid, *Otocratia*, is uncertain. Among the much more highly varied amphibians of the Pennsylvanian, aquatic adaptation appears to be the rule. Within their ranks, however, are creatures that are related to more distinctly semiterrestrial and terrestrial amphibians of the early Permian. Among these are immature individuals, from coal swamps, which seem to be close to possible larval stages of adults known only from more terrestrial beds of later times. The adult aquatic labyrinthodonts are fairly closely related to semiaquatic and terrestrial adults of the Permian. Anthracosaurs show marked aquatic tendencies, but they are not far removed from seymouriamorphs, among which there were terrestrial adults during the Permian. Even the apparently aquatic microsaurian lepospondyls are matched by possibly terrestrial relatives in the Permian.

An interpretation that a trend toward increasing terrestrial life existed during the Pennsylvanian and the Permian is not inconsistent with the evidence, but neither is the interpretation that what is known represents the primarily aquatic segment of a more diverse array that included many semiterrestrial and terrestrial amphibians.

The earliest reptile, *Romeriscus*, occurs in the lower Pennsylvanian, known from one specimen. The state of preservation precludes an interpretation of its aquatic or terrestrial nature. The environment of deposition, fluviatile or estuarine, means little. The fact that this is a reptile on the basis of the skeleton gives no sure indication that it was an amniote, although this is at least a reasonable presumption. The reptile is a captorhinomorph and as such represents a group that seems to have had the earliest success in assuming a truly terrestrial existence.

The coal-measures tetrapods from Joggins and Florence, Nova Scotia, include both amphibians and reptiles. Very few unequivocally aquatic vertebrates are present. Carroll (1964, 1967a) has suggested that these creatures lived in a somewhat terrestrial environment. This is based on the skeletal characteristics of the reptiles, especially their high degree of ossification, and the general nature of the faunas. Both are valid, if somewhat tenuous, bits of evidence.

Some amphibians and reptiles of the early Permian, with increasing numbers from the earliest to latest part of this epoch, were well adapted to life on land. Among the amphibians the dissorophids and trematopsids are outstanding, but some of the gymnarthrids also may have fitted this role. Other amphibians—edopids, eryopids, archegosaurids, and melanosaurids— were amphibious, and many—embolomeres, trimerorhachids, zatrachyids, keraterpetontid nectrideans, and lysorophids—were aquatic. Some seymouriamorphs appear to have been well equipped for terrestrial life, as *Seymouria* itself, but some at least (e.g., *Discosauriscus*) had aquatic larval stages.

Some of the reptiles were well adapted to life on land. Outstanding in this respect are Permian captorhinomorphs. Also large pelycosaurs, such as *Dimetrodon* and *Sphenacodon*, and some of the smaller, more captorhinomorphlike forms, grouped as eothyrids, fall in this category. Among the captorhinomorphs, however, *Limnoscelis*, which is very primitive, has some features indicative of semiaquatic life, and a number of pelycosaurs—such carnivores as *Varanosaurus*, *Ophiacodon*, and *Varanops* and such herbivores as *Edaphosaurus*—clearly were semiaquatic.

The fossil evidence of terrestrial plants of the lowlands during the later Devonian, Mississippian, Pennsylvanian, and early Permian is generally satisfactory for interpretation of possible terrestrial environments of vertebrates as they passed from water to land. Pteridophytes, lycopods, equisetales, and cordaitales all were present in abundance and provided a land cover. There is no evidence that these served as food sources for any tetrapods.

The record of invertebrates on the land is much less good, as is true through much of the Paleozoic. If the few records are taken as representative, it is questionable that foodstuffs for land-living terrestrial vertebrates were present among invertebrates of the Devonian. The variety of later terrestrial invertebrates, of the Carboniferous and Permian, indicates that there was a vast, nearly unrecorded land radiation among a number of groups. This could have been early enough to provide food in areas marginal to water in the late Devonian. That it in fact did or did not is completely open, but the chances that there was adequate animal food on land in the very late Devonian seem to be good.

Evidence from Moderns

Many aspects of the transition to land are understandable only by use of data from living animals. Considerable attention has been paid to their environments, responses to changing conditions, and their habits as indicators of how transitions might have occurred. Dipnoans, actinopterygians, *Latimeria*, urodeles and anurans, and to some extent apodans, have been most commonly cited in explanations. Probably the closest structural analog to the ancestral stocks is *Polypterus*, but its utility has been more morphological than behavioral.

Teleosts, which include many fishes that may be rather close behavioral analogs to prototetrapods, are rather remote structurally from the ancestral rhipidistians. Modern amphibians have departed so far from ancestral tetrapods that it is difficult to equate either their structure or behavior with those of the first amphibians.

Modern air-breathing fishes show two things very clearly. First, air breathing is readily acquired by animals and may take many forms. Carter and Beadle (1930, 1931) recognized six independent forms of air breathing. Second, the majority, although not all, have developed in warm, tropical conditions in fresh waters that have low oxygen tensions at times. Adaptation has been to life in such waters, not to migration out of water. Most air breathers never leave the water, and many fishes, not strictly air breathers, survive in waters low in oxygen by obtaining their supply from the very thin oxygen-rich uppermost layer of water. Some fishes, however, do leave the water, once the potential for this has been well developed.

Today air-breathing fishes are widespread but occur predominantly in southeastern Asia and tropical South America. Lung breathers, including gas-bladder breathers, include fishes of some eight families: the Polypteridae, two holostean families, three among the teleosts, and the two families of lungfishes. They are more diverse environmentally than most other types of air breathers.

Of the modern fishes that come onto the land, the mudskipper, *Periophthalmus*, and the climbing perch, *Anabas*, have essentially terrestrial habits. *Periophthalmus* and *Clarias* feed out of water, and the former leaves the water specifically for feeding. All three have good powers of locomotion on land. *Clarias* appears to leave the water primarily to find new habitats, as is also true of the bettas. Some fishes, such as the snakehead, *Ophiocephalus*, although capable of terrestrial locomotion, aestivate when conditions become harsh, rather than seeking more suitable places for continuing active life. Motivations of behavior are notably difficult to determine, and some fishes do not exhibit any discernible reason for going onto land.

Neither *Polypterus* nor the Dipnoi seem to give particularly useful clues about the behavior of prototetrapods. Eaton (1960) has used *Neoceratodus* as a model for the way in which the limbs might have been used by early tetrapods to prop the body up for air breathing.

Modern amphibians have been cited as models for some behavioral patterns. For example, Panchen (1967*a*) has taken *Plethodon* as a model in the use of an air-filled buccal cavity to indicate the probable nature of the first air breathers. That some amphibians lay eggs out of water, presumably to escape predation, has been noted by Tihen (1960) to show that this could occur without desiccation, contrary to the suggestion of Romer (1957) relative to the amniote egg. Responses of amphibians to drying by aestivation (Orton, 1954) and by clustering around remaining wet areas and then fanning out rapidly during temporary wet times (Goin and Goin, 1956) have been thought to be pertinent. Running through many of the suggestions has been the idea that leaving water, either during favorable times or when conditions became harsh, was a response to population pressure. Feeding similarly has been considered a basic motivation. Each of these ideas has come from observations of the ways that some groups of amphibians have reacted under various environmental conditions.

Interpretations of Origins of Terrestrial Vertebrates

Few of the several themes of the ecological circumstances of the origin of tetrapods have considered more than limited parts of all of the evidence. None has made a thorough and objective evaluation of its authenticity or completeness. For the most part those that have given detailed attention to morphological changes have paid little attention to the ecological evidence or attempted to fit their data to it. Any attempts at such a synthesis reveal the very great difficulties of coordinating information from many sources, especially when much of it is subject to more than a single interpretation.

All of what is known cannot be more than suggestive of a possible course. As far as evidence of geological history is concerned, the probable biases in sampling make it impossible to know whether the data are in fact actually pertinent to the actual course of events. Modern analogies can suggest many possible courses but do not point directly to one as much more likely than another. If, in spite of these difficulties, it is assumed that what is known is pertinent, although with recognition of many gaps, the following events and environmental interpretations appear to follow from selection of the most probable phylogenies, coordination of all the data at hand, and selection of what appear to be the most probable interpretations of detailed bits of evidence in their relationship to the hole.

1. Rhipidistians were the predecessors of tetrapods. They lived in warm, shallow waters, rich in vegetation and subject to periods of low oxygen content. They were efficient air breathers. Complete drying up of the waters in which they lived did not occur.

2. In adaptation to these conditions air breathing by lungs became increasingly efficient. Gill breathing in water was retained during the non-tetrapod stages, and these fish and their early tetrapod descendants breathed by external gills during very immature or larval stages.

3. Early during evolution in this environment a particular pattern of the internal skeleton of the paired appendages arose, probably in response to initiation of locomotion on the substrate. As air breathing became more efficient, the pectoral fin was employed to raise the anterior part of the body off the substrate, bringing the head out of water in a passive position and facilitating buccal breathing with the opercular region closed. This fin became a tetrapod limb with the loss of its rays (a totally unknown transition, perhaps taking place under similar conditions in various lines).

4. From this stage two, possibly more, adaptive lines developed, as follows:

a. One exploited the aquatic medium, with development of an elongated body, small limbs, small head, rather weak, short jaws, and a closed otic region, lacking an otic notch. Vertebrae acquired spool-shaped centra. Arches and zygopophyseal articulations were adapted to sinuous body actions rather than to support on land with the body suspended between limbs. This line produced the lepospondyls. It was successful in the water, and later some members, gymnarthrid microsaurs, of the Pennsylvanian and Permian became partially successful on land.

b. The other branch led to the labyrinthodonts. Increase in the predatory capacities of the jaws, developed along with feeding on increasingly large, active aquatic prey, was a key to their evolution, coupled with development of features suitable to land life, especially the limbs and respiration. The tetrapod limb became larger and stronger, possibly in relationship to its combined use in bottom locomotion and in active maneuvering during rapid swimming. This radiation took place mostly in shallow waters, in environments probably little different from that in which the prototetrapod first lived. The tetrapods so developed were fully capable of life on land, and their immediate descendants, known from ichthyostegids, spent part of their time on land. The association of so many features suitable for terrestrial life in ichthyostegids argues strongly against their having been strictly aquatic, in particular the facts that they were air breathers only and had large, well-ossified, and fully developed tetrapod limbs and girdles.

This concept differs from one that supposes an origin and persistence (in water) in this medium throughout much of the Mississippian and Pennsylvanian. It depends primarily on assessment of the morphological features of the ichthyostegids, which seem surely to have developed adaptations to partially terrestrial ways of life.

The motivations for initial occupancy of land must remain uncertain. The presumed environment, based on known habitats of rhipidistians and ichthyostegids, suggests impetus related to the seeking of new aquatic abodes, either because of intolerable conditions that could not be survived or because of increased population pressures of various possible sorts during times of environmental stress. Feeding out of water is a possible motivation, but emphasis on the aquatic predatory habits suggests the contrary. The initial change, in any event, seems to have been carried out under conditions of ecological stress rather than under equable circumstances.

The terrestrial habitus in adults was attained at this stage, whether through a sequential development of characters or through more or less simultaneous coordinated modifications of the systems that united to produce it. As far as structures of the head are concerned, the latter seems more probable.

5. From an ichthyostegidlike base, which is not precisely known, radiation of labyrinthodonts occurred in two main branches—rhachitomes and anthracosaurs. Members in both of these lines remained amphibious, but most of those known were highly aquatic, with small limbs and rather elongated bodies. This represents a regression from the more terrestrial ancestry required to produce the many terrestrial structures that persist throughout regardless of subsequent habitats.

The temnospondylous rhachitomes appear to have lived mostly in shallow waters, but very little is known about them before the Pennsylvanian. Some, as their descendants indicate, probably were distinctly amphibious, a habit that was established early and probably was not subsequently lost and then regained. Life on land was in low-lying, swampy, humid areas. Invertebrates were abundant both on land and in water. These creatures, along with small fish, probably were the important elements of the diet. The known amphibians, however, were predominantly aquatic and lived in the waters of the coal swamps and probably under various other similar circumstances. These habits persisted into the Permian, exhibited by such well-known genera as *Trimerorhachis* and *Zatrachys*.

From the persistently amphibious rhachitomes arose edopids and eryopids. Among the latter, under increasingly harsh climatic conditions, were developed highly terrestrial amphibians, such as the dissorophids and trematopsids. Although the strictly amphibious labyrinthodonts flourished in a warm, moist climate, they became selectively reduced where increasing seasonality with periodic fluctuations of rainfall and temperatures occurred. Such a sequence is recorded in the early Permian of Texas. It is under these harsher conditions that the acceleration of terrestrial adaptations of amphibians is found.

From the general eryopid base came increasingly aquatic amphibians, represented in the Triassic by trematosaurs, stereospondyls, and plagiosaurs.

These also represent a reversion from the semiaquatic to a more strictly aquatic way of life. That this was a response to increasing terrestrial competition seems plausible, but there is very little direct evidence from detailed faunal analyses that can be brought to bear on the matter.

6. Anthracosaurs, whose semiaquatic heritage is suggested by the combined presence of many terrestrial characteristics—of limbs, hearing, and respiration in particular—trended toward a predominantly aquatic life throughout the Mississippian and Pennsylvanian. During the Pennsylvanian or earlier there arose the seymouriamorph stock, in which the trends were partly in the opposite direction. Some members tended toward aquatic life, but others were terrestrial during adult life, much like the dissorophids and trematopsids with which they characteristically are found in fossil deposits.

7. Sometime during the late Mississippian or very earliest Pennsylvanian reptiles arose from an unknown stock close to anthracosaurs and possibly represented by the somewhat later seymouriamorph *Gephyrostegus*. The earliest reptile gives little clue to the habitat of origin. By middle Pennsylvanian both captorhinomorphs and primitive pelycosaurs were in existence, apparently living on the land, but adjacent to coal swamps. The climate was apparently warm and humid, and it may well be that in *this* instance transference to land took place under conditions that put less environmental stress on the participants. The motivation for the change, of course, is conjectural. At this time there almost certainly was adequate food on land, and these small creatures, which have capabilities for insect feeding, may have exploited this food supply and the favorable climates. This migration to land, under these conditions, following on the initial development of terrestrial features in the manner suggested by ichthyostegids, probably was very different from that either of the ichthyostegid ancestors or of the later course of some amphibians.

The role of the amniote egg, however, is uncertain. By definition, of course, it should be present in reptiles. Most students have considered it unlikely that this type of egg developed more than once. But just when it developed and in what groups is currently uncertain. Romer (1957) suggested that laying of eggs on land, for protection, may have provided a motive for the occupancy of land. This is, of course, possible, but so are many other interpretations, between which there is very little basis for choice. Studies of this problem now being carried out give reason to hope for more definitive conclusions in the near future.

The slight evidence from the middle Pennsylvanian suggests that the initial reptiles may have been moderately terrestrial. That some of the larger, later reptiles, both pelycosaurs and captorhinomorphs, were partially aquatic may

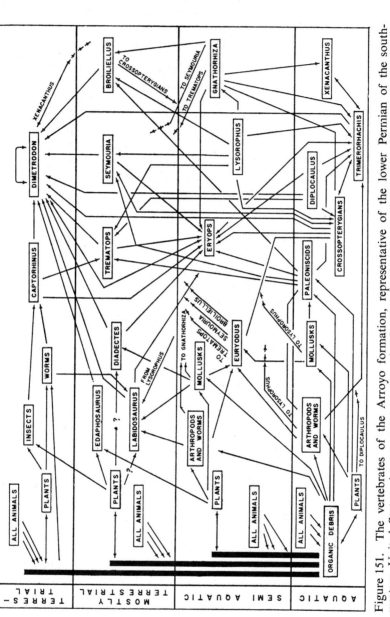

Figure 151. The vertebrates of the Arroyo formation, representative of the lower Permian of the southwestern United States. An attempt to assess the feeding relationships with the flow of energy indicated by the arrows is shown on the diagram. It is of course much oversimplified. Fishes: *Xenacanthus* (a predacious shark), *Gnathorhiza* (a lungfish), crossopterygians, and paleoniscoids; amphibians: *Lysorophus, Diplocaulus, Trimerorhachis, Euryodus, Trematops, Broiliellus, Seymouria, Diadectes*; reptiles: *Edaphosaurus, Labidosaurus, Captorhinus, Dimetrodon*. (From Olson, 1966b.)

be the result of a reversion to this medium as size increased and a terrestrial diet was no longer satisfactory.

Late Pennsylvanian and early Permian red beds contain remains of a faunal and floral complex that for the first time appears to give some idea of a complete ecological system, incorporating into its structure both terrestrial and aquatic elements, related through a complex series of food dependencies. One part of this system, the small captorhinomorphs and small pelycosaurs, appears to have been strictly terrestrial and probably derived directly from the type of complex seen briefly in the mid-Pennsylvanian. Sources of the other elements seem fairly evident, although in some instances their immediate predecessors are not well known. The large predators *Dimetrodon* and *Sphenacodon*, in contrast to the aquatic fish eaters *Varanops* and *Ophiacodon*, seem to have developed primarily as terrestrial animals, but ones that found their prey in the ponds and streams among the fish and the aquatic and amphibious amphibians. Probable food relationships within this complex are shown in Figure 151 (Olson, 1966b). The only elements of this fauna able to penetrate beyond its water-limited confines were those represented by small captorhinomorphs and small pelycosaurs. The former underwent a fairly significant radiation within the complex, but the latter dwindled in numbers and were almost absent in the late part of the early Permian.

A potential for terrestrial life existed by middle Pennsylvanian in the small captorhinomorphs. As far as the record shows, it was not exploited at that time, although the chance that there were totally unknown upland faunas cannot be discounted. Among the known faunas, the first full emancipation from water took place when terrestrial carnivores found food sources among herbivores that were in no way water dependent either for their life cycle or for food. The two earliest groups to supply this potential within the framework of the known faunas were the captorhinomorphs, as they radiated in the early Permian and the *Cotylorhynchus*-like caseids of the late early and early upper Permian.

5 Patterns of evolution in the attainment of the terrestrial way of life

INTRODUCTION

By the late Pennsylvanian, possibly earlier, some reptilian tetrapods seem to have developed the physical capacity to live without dependence on water for feeding or reproduction. From these animals came the successful terrestrial radiations. Indications are, however, that acquisition of full independence from water was a slow affair. All of the vertebrate-bearing deposits of the Pennsylvanian and the Permian appear to have been formed in lowlands, close to open waters. Unless it is assumed that the mainstream of evolution was in totally unknown regions, the sources of terrestrial vertebrates must have existed within the faunal complexes known from these beds.

It is not, of course, improbable that critical evolution did occur elsewhere. If so, all that can be done to understand the origin of terrestrial radiations is to follow them from records that are distinctly tangential to the actual course of events. Two features of the known records suggest that this is not entirely the case. First, within the lowland faunas there are some elements that existed without direct dependence on an aquatic environment. Second, the later faunal complexes, which attained a large measure of independence, do not require a completely different course of evolution from that evident in the lowland faunas.

Sites where Pennsylvanian and Permian vertebrate remains are known to exist have been collected intensively, and the vertebrates that they have yielded have been very thoroughly studied. What is available is well known as far as the morphology of the animals and the general relationships of lines of descent are concerned. Less well known are the details of faunal compositions, distributions of the known ecological types, and, of course, the nature of vertebrate life throughout the vast areas of the earth for which we have no records. The development of truly terrestrial faunas seems to be most profitably studied in the context of faunal evolution, because it was complexes of associated animals, not individual phyletic lines, that underwent major shifts in ecologies. Recent studies, as paleoecology has become increasingly prominent, have been slanted more and more toward such faunal analyses.

Ecological evolutionary studies demand a detailed knowledge of faunas and of the conditions of deposition of the sediments in which the remains occur. This must come from sequences of deposits that encompass periods of time long enough for evolutionary aspects of change to be recognized. The Permo-Carboniferous deposits, more than most terrestrial beds formed prior to the Cenozoic, include the necessary sequences. What they fail to provide are samplings from a wide array of environmental circumstances and from widely dispersed geographic realms.

The present chapter is based on the known materials, and conclusions depend on the assumption that these give information pertinent to the problems of the origin of terrestrial vertebrate lines. It will be apparent in context that such an assumption is not always justified. The data are necessarily very complex since they involve faunal interrelationships of animals from single sites, faunal successions, and environmental information from both physical and biological sources. Only a skeleton of the total is appropriate to our objectives here, but cited references present both the detailed data and methods of study (see especially Hotton, 1967; Olson, 1952, 1961a, 1962, 1966b; and Olson and Beerbower, 1953).

The most successful terrestrial radiations among the vertebrates were those of the amniotes, since amphibian successes early in the Permo-Carboniferous and later, among the anurans, were sporadic. Amniote radiations took place in four major lines: captorhinomorph–procolophon–?chelonian, pelycosaur–therapsid–mammalian, lepidosaurian (including lacertilian and ophidian), and archosaurian–avian. Roots of the first two clearly can be found in the early semiterrestrial complexes of the Pennsylvanian and very early Permian. Whether the course of actual ancestry can be followed is another matter, but the morphotypic sources, at least, are evident. The bases of the other two are less secure, and this applies as well to the bases of the Mesozoic aquatic-reptile radiations. Elements of the semiaquatic complex could have been ancestral to both the lepidosaurian and archosaurian lines, but the inter-mediate steps to the radiations, even in the event that this was true, remain quite uncertain.

PERMO-CARBONIFEROUS FAUNAS

Geological and Ecological Distributions and
Their Consequences

The semiterrestrial and terrestrial faunas of the Permo-Carboniferous are known best from the midcontinent of the United States, from Texas in particular but extending into adjacent Oklahoma, north into Kansas, and westward into Colorado, New Mexico, and Utah. Most of the areas include deposits of Wolfcampian age (Figure 152), but some older vertebrate-bearing beds are known in Colorado. In Texas and Oklahoma sections continue through the Leonardian and into the later Guadalupian. Essentially all of the deposits that include vertebrates were formed marginal to seas and, in large part, in deltas that maintained their surfaces at about sea level over long periods of time. The faunas were made up primarily of freshwater sharks, crossopterygians, paleoniscoids, labyrinthodont and lepospondylous amphibians, and captorhinomorph and pelycosaurian reptiles. In addition there were minor elements that do not fit well into any known categories. The percentages of the major constituent groups varied considerably through time and in different environments of deposition, but the general cast is unmistakable throughout.

Elsewhere in the United States late Pennsylvanian members of this same general faunal complex are known from Illinois and Kansas, and from sites in Pennsylvania, West Virginia, and Ohio. In the eastern area the principal source is the Dunkard formation, which is largely Permian and contains a faunal complex that, though distinctive in some features, is organized in a manner similar to that of the midcontinent. Some even earlier remains from the Pennsylvanian of Nova Scotia may possibly represent the same type of complex.

Records from other parts of the world are meager. A few sites are known in central and eastern Europe, mostly in beds formed under aquatic conditions or with a coal-measures cast. Two localities are reported from Siberia and one, with a few aquatic and semiaquatic genera, has been found in Brazil. The only other early Permian occurrences are in beds of South Africa and Brazil in which the aquatic reptiles *Mesosaurus* and *Stereosternum* are present.

As in the well-known North American localities, the early Permian terrestrial and semiterrestrial vertebrates from elsewhere appear to have lived in lowland areas. Nowhere, except in the *Mesosaurus* and *Stereosternum* sites, is there any evidence of marked departure from the kind of faunal complexes found in the Texas–Oklahoma region of the United States.

The early Permian sequence of Texas is generally taken to be typical as far

G u a d a l u p i a n	P e a s e r i v e r	North Central Texas	Oklahoma	Utah, Colorado, New Mexico, Arizona
		White Horse group	Rush Springs sandstone Marlow sandstone	
		Dog Creek shale	Dog Creek shale	
		Blaine shale, gypsum	Blaine shale, gypsum	
		Flowerpot shale 320-650 ft	Flowerpot shale 650 ft Chickasha formation	
		San Angelo sandstone, shale	Duncan sandstone	
L e o n a r d i a n	C l e a r f o r k	Choza shale 850-1000 ft	Hennessey shale 350-700 ft	
		Vale shale, sandstone, conglomerate 350-500 ft	Hennessey-Garber shale, sandstone (interfingering) 50 ft	
		Arroyo shale 260-700 ft	Garber sandstone, shale 400-600 ft·	
		Lueders limestone 120 ft		
		Clyde shale 180-200 ft	Undifferentiated Wellington—Garber	
W o l f c a m p i a n	W i c h i t a	Belle Plains shale (Bead Mountain limestone) 300-350 ft	North W e l l i n g t o n	South W i c h i t a
		Admiral (Elm Creek limestone in south) 250-350 ft		
		Putnam shale, sandstone (Coleman Junction limestone in south) 175-200 ft		C u t l e r
		Moran shale, sandstone (Sedwick limestone in south) 200-360 ft·		DeChelly sandstone Organ Rock shale Cedar mesa sandstone
		Pueblo shale, sandstone Saddle Creek limestone near base 200-350 ft		Halgaito sandstone, shale

Figure 152. Section of three of the four Permian epochs in the southwestern United States, with the Ochoan, which has no known vertebrates, not included. The highest known terrestrial vertebrates yet found have come from the Chickasha formation of Oklahoma, about equivalent to the middle Blaine formation of the Guadalupian epoch.

as vertebrates are concerned. It is only in this region that the continuous succession of beds with well-preserved remains makes possible an understanding of the faunal composition at many stratigraphic levels and the general nature of the evolutionary changes. Throughout the following discussions this complex will be termed the Permo-Carboniferous chronofauna. It seems to be fairly representative in composition and ecological relationships of major elements to what is known to have existed elsewhere in the world at about the same time. It would be, however, dangerous to think that it was the only source of later terrestrial faunas and that all, or even any, of the later radiations stemmed directly from it. It is a special case and probably one of many, with somewhat differing compositions and ecological structures, that occupied the continental margins in various areas of the earth. As such, it may form an analog to the "morphotypic" genus in a phylogenetic line. In some respects the chronofauna appears to lie close to the main line of development, but as will be seen, some later events cannot be explained in terms of it alone.

Composition, Characteristics, and Evolution of the Permo-Carboniferous Chronofauna

The fauna of the deltaic deposits of the early Permian of Texas is the one on which the concept of the chronofauna was based (Olson, 1952). Its primary characteristics are its continuation with little change in basic structure over a long period of time, certainly several millions of years, and its geographic restriction. It seems to have extended over much of the midcontinental region and somewhat to the west, but not far from the margins of the midcontinental Permian sea. The general stratigraphy is shown in Figure 152. The thickness of the total section runs between 4000 and 4500 feet. Through this section extends a nearly continuous series of vertebrate remains, broken in the lower parts by short periods of marine deposition, but thereafter essentially unbroken into the late Leonardian.

Only the Leonardian faunas have been studied from the chronofaunal point of view, but the fundamental structural features appear to continue to the base of the section. An assemblage from the Arroyo, shown in Figure 151, is typical of this stage, whereas earlier portions were somewhat more aquatic and later ones slightly more terrestrial. Four more or less distinct ecological types, which combine the occupancy of different environmental subzones and feeding habits, can be identified throughout:

1. Purely aquatic elements, including fish and amphibians (typical members: xenacanth sharks, rhipidistians, paleoniscoids, *Sagenodus*-type lungfishes, *Archeria*, *Trimerorhachis*).

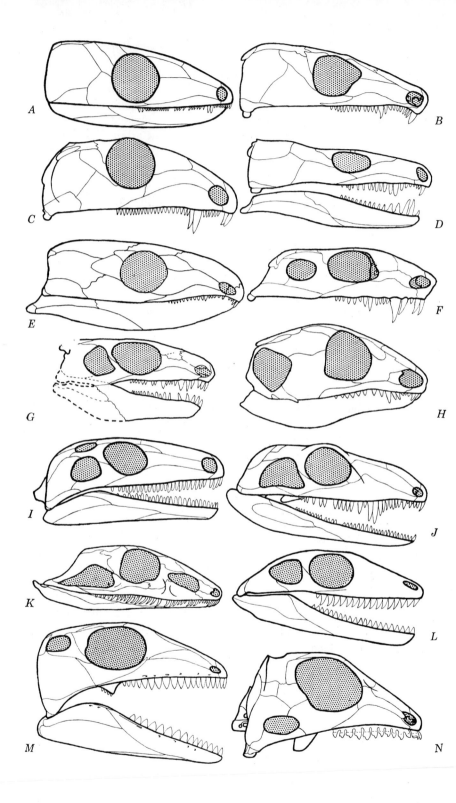

2. Semiaquatic elements, living and feeding mostly in water but able to move about on land or to aestivate (typical members: *Gnathorhiza*-type lungfishes, edopid and eryopoid labyrinthodonts, *Lysorophus*).

3. Terrestrial animals that lived mainly on land but fed primarily in water (typical members: large predacious pelycosaurs, such as *Ophiacodon* and *Dimetrodon*, and possible herbivores, such as *Edaphosaurus* and *Diadectes*).

4. Terrestrial animals that lived and fed on land (typical members: captorhinomorphs, small pelycosaurs, and possibly some amphibians, such as dissorophids and gymnarthrids).

Elements of the last two groups (Figure 153) could have occupied land as their sole habitat except insofar as food sources may have kept them dependent on water. As known in the Permo-Carboniferous chronofauna, of course, they are part of an interacting network. In order to become completely independent of any ties to water they had to break out of this framework. If the sources of terrestrial radiations lay within this chronofauna or within other similar ones, there would appear to have been three main pathways by which this break might have been accomplished:

1. By internal evolution of the structure of the chronofauna, which when well along would result in modifications sufficient for a new complex to result.

2. By penetration of a strictly terrestrial environment by some of the elements, which would have broken their dependence on the chronofaunal structure.

Figure 153.   Skulls of representative Carboniferous and early Permian small to moderate-sized reptiles, including captorhinomorphs, pelycosaurs, and some forms of uncertain affinities. Except for the varanopsids, included to indicate a contemporary semiaquatic element, these creatures are primarily members of the early radiative dispersal of the reptiles. *A, Hylonomus; B, Romeria; C, Protorothyris; D, Paracaptorhinus; A* through *D* are captorhinomorphs, the last three, members of the Romeriidae. *E, Captorhinus,* an advanced captorhinid captorhinomorph; *F, Eothyris; G, Oedaleops; H, Bayloria; F, G,* and *H,* are primitive pelycosaurs. *I, Petrolacosaurus,* shown as reconstructed by Peabody with two temporal fenestrae, but considered an edaphosaurian pelycosaur by Romer, presumably with only a lower fenestra; *J, Varanops; K, Varanodon; L, Elliotsmithia; J, K,* and *L* are varanopsid pelycosaurs showing special adaptations that occurred simultaneously with later parts of the primitive radiation. *M, Araeoscelis,* an araeoscelid, but of uncertain affinities; *N, Bolosaurus,* a small reptile of uncertain affinities. All but *J, K,* and *L* are less than 3 inches long, but are not shown to scale. (*A* after Carroll, 1964; *B* through *F* and *N* after Watson, 1954; *G* after Langston, 1965*b; H* after Olson, 1941; *I* after Peabody, 1952; *J* and *L* after Romer and Price, 1940; *K* after Olson, 1965*d; M* after Vaughn, 1955.)

3. By modifications resulting from interpenetrations and intersections of elements from two or more adjacent chronofaunas to produce a food balance suitable to life on land away from water.

The only abundant and strictly terrestrial types, which a priori seem to be the most likely source of land radiations, are captorhinomorphs, romeriids early and captorhinids later. Their structure indicates that they could have fed on small animals, both invertebrates and vertebrates, and it seems probable that insects formed an essential part of their diet. They were not dependent on a water-based, herbivorous source of food. Less abundant in the collections are small pelycosaurs, such as eothyrids and nitosaurids. These could have had much the same habits. Numerically more important within the faunal complex were large pelycosaurs, structurally capable of life on land, but seemingly finding their primary food among aquatic animals so that their food chains were ultimately based in water (Olson, 1961a, 1966b). If this is correct, they could have left the chronofauna only with the advent of a source of food among herbivores that fed on land. Such do not seem to have existed in significant numbers in the Permo-Carboniferous chronofauna. Finally, among the potential terrestrial animals were small, poorly known genera, typified by *Araeoscelis* and *Bolosaurus*. Their roles in the total complex are most uncertain.

Internal Evolution and Stability of the Chronofauna

Over the period of its existence the chronofauna by definition maintained a strong structural consistency, although important ecological roles may have been played by different species and even genera in succeeding times. The climatic conditions under which the complex lived seem to have been moderately constant up to the beginning of the Leonardian, but thereafter there was an increasing seasonality of rainfall and general reduction in the amount of moisture (Figure 154).

During the transition to this later phase and in the course of its development proportionate representation of the major elements underwent modest changes. The main types of fishes persisted, but aestivating lungfishes became dominant. Strictly aquatic amphibians, *Trimerorhachis* and *Diplocaulus*, remained in numbers, and some of the amphibious types, such as *Eryops*, persisted. The number of kinds was reduced. Among terrestrial water-feeding reptiles *Dimetrodon* became dominant and more strictly aquatic types, such as *Ophiacodon*, disappeared. The terrestrial creatures, primarily captorhinomorphs, underwent a fairly extensive adaptive radiation, with development of larger genera and species, some of which may have moved into a herbivorous subzone.

FM	Sedimentation	Topography	Rainfall
.2200'	Extensive, even clay, sand and evaporite beds. Few channels and ponds. No known life record.	Same	Low rainfall
Choza	Predominantly flood plains. Channels small, few, widely spaced. Evaporite basins markedly increased. Some Playa, fresh water ponds.	Local relief low. ·Land near sea level.	Continued decrease in total rainfall.
	Few channels, not clay-pebble type. Initiation of evaporite beds. Scattered fresh water ponds.	Local relief moderately low. Basins with fresh water and evaporite deposits.	Decrease in total. Not torrential. Seasonality persistent.
1200'	Channel and lag deposits all of clay-pebble conglomerates, sediment locally derived. Ponds scattered, predominantly persistent. Ponds show seasonality in nature of deposition.	Local relief increased.	"Monsoonal" type fully developed. Torrential seasonal rains.
	Same, but beginning of clay-pebble conglomerates, derived from local sediments.	Local relief same. Source area high.	Initiation of "Monsoonal" type.
Vale	Marked channel development. Coarse, dipping, marginal flood plain deposits. Few but extensive ponds. Coarse materials of streams derived from source area.	Moderate local relief. Marked increase in relief in source area of sediments.	Increase in seasonality in source area of streams.
	Initiation of large stream channels with conglomeratic deposits. Ponds widely spaced.		
700'	Even red shale, with broad linear belts of silt and fine sand. Deposited by very slowly flowing water. No definite channels and few ponds.	Local relief very low. Area near sea level. Possibly tidewater estuaries.	Same.
Arroyo	Slight increase in ponding, especially temporary ponds.	Same	Slight increase in seasonality.
	Flood plain deposits predominant, a few scattered ponds. Divides low, streams small.	Local relief low. Source area of streams low.	Moderate, evenly distributed throughout year.

Figure 154. Sedimentation and changing topographic and climatic conditions during Arroyo times as recorded in deltaic deposits in north central Texas. This is the time through which the Permo-Carboniferous chronofauna has been most thoroughly studied.

The end result of this evolution, as seen from the early Choza, was a somewhat impoverished, but nevertheless basically unchanged, faunal structure. Genera were for the most part the same as those at the beginning of the Leonardian. Most striking is the persistence of the large predator *Dimetrodon* without species change. Only the radiation of the captorhinids

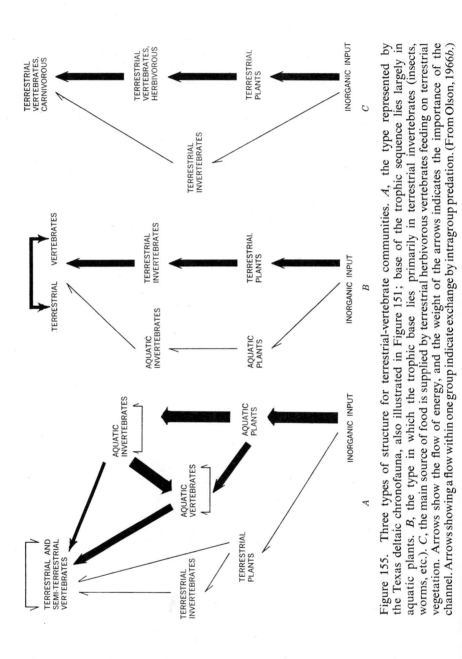

Figure 155. Three types of structure for terrestrial-vertebrate communities. *A*, the type represented by the Texas deltaic chronofauna, also illustrated in Figure 151; base of the trophic sequence lies largely in aquatic plants. *B*, the type in which the trophic base lies primarily in terrestrial invertebrates (insects, worms, etc.). *C*, the main source of food is supplied by terrestrial herbivorous vertebrates feeding on terrestrial vegetation. Arrows show the flow of energy, and the weight of the arrows indicates the importance of the channel. Arrows showing a flow within one group indicate exchange by intragroup predation. (From Olson, 1966b.)

suggests a possible basic structural change, one that could have provided an enlarged terrestrial food source for *Dimetrodon*, resulting in a movement of this genus away from water. How far this actually developed is uncertain. It would appear, however, that internal evolution did not produce a fauna that had shifted markedly from its original aquatic-terrestrial characteristics.

This description, of course, gives a much oversimplified picture. Floras changed, primarily with an increase in the proportion of conifers and losses of more aquatic elements. Furthermore, and perhaps very important, the area over which these changes have been recorded is local, much more local than the full extent of the chronofauna. The early portions are known over a rather broad area and do show considerable constancy. Areas as much as 100 miles to the south of the best known sites show no significant differences and thus provide some idea of the spread. In Oklahoma, to the north, much the same is the case, but the trend toward seasonality appears to have begun earlier, and changes of faunas in part reflect this by initiation of modifications somewhat earlier than to the south. Also, deposits in New Mexico, restricted sites in the Texas area, and some features of the Hennessey formation in Oklahoma, suggest the presence of another chronofauna, as discussed below.

Penetrations Beyond Chronofaunal Limits

The chronofauna itself cannot, of course, give direct evidence of penetration of any of its elements beyond its geographic or structural limits. The morphology and apparent feeding habits of the captorhinomorphs and small pelycosaurs gave them the capacity to exist beyond the limits of the central complex. If such a penetration beyond the limits were made, it probably involved the general food-chain relationships shown in Figure 155*B*.

Evidences of such penetrations are very meager and tenuous, but they do exist. Such a step appears to have been at least possible from the time of middle Pennsylvanian assemblages of Joggins and Florence, Nova Scotia. Only much later is there any tangible evidence. From fissure fills in the Arbuckle limestone north of Fort Sill, Oklahoma, have come remains of an assemblage that could be of this type. The source appears to have been the "highlands" of the Wichita Mountains, flanked, partly at least, by the seas in Arroyo times. *Captorhinus* and *Labidosaurus* are the dominant members of this assemblage, which also includes common gymnarthrid amphibians, the small temnospondyl *Doleserpeton*, and remains of various small, apparently insectivorous, pelycosaurs. In addition, a parietal bone indicates the existence of a possible diapsid (Carroll, 1968). Evidence of any aquatic components is very slight, although a few traces have been recovered. This is the sort of assemblage that could have made a break from the more aquatic mother complex. On the other hand, since it is found in fissures, its composition probably has been influenced by selective preservation, which could have altered the cast by removing large components.

Some deposits of the early part of the late Permian in the Soviet Union were noted as possibly having this type of structure (Olson, 1962). These are early examples of the "cotylosaur" complexes of Efremov and Vjushkov (1955) and include primarily procolophons and very primitive carnivorous or insectivorous therapsids. Within one of the collections is *Mesenosaurus*, which has been called a primitive eosuchian. Once again, this is a possible indication of a breaking away and segregation of the two types of faunal complexes. Evidence, however, is very tenuous, and the number of cases and their distribution in time and space are at best suggestive.

Interactions of Two Developing Systems

Circumstances in which two or more faunal systems overlap or in which one comes to impinge on another by moving into its zone of habitation, either in part or in toto, can be readily envisaged, and many cases are known within the history of modern organisms and from the geological history of the recent past. Interactions between the mammals of North and South America, once migrations became possible in the late Tertiary, provide one of the best examples known among extinct animals. For definitive interpretations the evidence must be extensive with respect to the composition and geographic spread of faunas at given times and must have a considerable span of time. Thus, for the more remote past, interpretations of the recorded events as representing such interactions become more difficult and the results more open to question.

In older deposits there are two sorts of evidence that aid in the identification of adjacent chronofaunas. Direct indications, of course, come from adjacent contemporary deposits that carry complexes of different compositions. This situation is usually extremely difficult to document. The other is the presence of uncharacteristic or "erratic" fossils in collections of a well-known complex.

The collections of the Permo-Carboniferous chronofauna, like most extensive collections, do include species represented by one or a very few individuals. These are the "erratics," which may be explained in a number of ways, depending particularly on the nature of the unusual animal and the conditions of its occurrence.

Among the explanations is the possibility that the erratics may have been members of another faunal complex that, in one way or another, found its way into the depositional area of the principal fauna. This could have occurred by actual overlap of the living faunas or as a result of taphonomic processes acting on faunas that were reasonably close geographically but quite distinct ecologically.

Collections of beds of the Wolfcampian and Leonardian give some hints of a separate chronofauna in the general midcontinental area. The caseid and varanopsid pelycosaurs and possibly *Cacops* and *Cacops*-like dissorophids show repeated joint occurrences as erratics in the Permo-Carboniferous chronofauna and adjacent to it. On the assumption that they do represent a second, if poorly known chronofauna, the name "Caseid chronofauna" will be used for reference to the assemblage.

At three places in the Texas sequence—at the base of the Vale, in the uppermost Vale, and in the middle Choza—caseids are present. In the earliest instance several specimens occur along with *Varanops* and *Cacops*. Except for *Captorhinus*, most other elements of the Permo-Carboniferous chronofauna are missing. In the upper Vale two specimens of *Casea* occur, not in association with other tetrapods. In the middle Choza a single isolated specimen of *Casea* has been found. These erratics are not found in the common assemblages that have come from many sites.

In New Mexico, in beds formed during the later part of the Wolfcampian, occurs a faunal complex that seems to have had little contact with the more usual chronofauna. It extends over a considerable period of time. *Dimetrodon* is absent, but *Sphenacodon* occurs along with a number of genera not found elsewhere and with a few that are common to the Permo-Carboniferous chronofauna. The reptile *Aerosaurus*, probably congeneric with *Varanops*, is present. No caseids are known, although Langston (1965*b*) has suggested that *Oedaleops* (see Figure 153) may be close to this family. This may represent a distinct chronofauna and it is possible, although far from demonstrated, that it is antecedent to that which appears to have been tapped at three levels in the Texas Leonardian. Elsewhere in New Mexico and Utah typical examples of the Permo-Carboniferous chronofauna are found.

The caseid *Cotylorhynchus* occurs in abundance in parts of the Hennessey formation of Oklahoma (see Figure 51). This formation is an age equivalent of the Choza of Texas, which shows the Permo-Carboniferous fauna in a stage of rapid decline over the known area. Associated with *Cotylorhynchus* is *Captorhinikos chozaensis*, which is also known from Texas. In the Hennessey, associated with *Captorhinikos chozaensis* but not directly with *Cotylorhynchus*, is an assemblage of small animals, including another species of *Captorhinikos*, an abundant small gymnarthrid, a labyrinthodont, numerous specimens of *Lysorophus*, and the lungfish *Gnathorhiza*. The only known large carnivores from the Hennessey also come from this small assemblage, represented by a few teeth of sphenacodontids and possibly varanopsids.

From the Guadalupian of Oklahoma and Texas have come less ancient assemblages in which caseids are the dominant animals and elements of the Permo-Carboniferous chronofauna are rare. A varanopsid occurs in the Oklahoma assemblage. These, and the Hennessey vertebrates, may possibly have been developed by interpenetration and development of ecological interrelationships of the two chronofaunas, the Permo-Carboniferous and the caseid. The food-chain relationships appear to be of the type illustrated in Figure 155C.

Later Permian Faunas

United States

The known post-Leonardian Permian vertebrate faunas of the United States cover only a short time span, but two faunal complexes are present. In spite of their limitations, they appear to have considerable bearing on the early exploitation of land and fall somewhere intermediate between the better known earlier faunas and those of later deposits in eastern Europe and South Africa.

One of the complexes comes from the Duncan–Chickasha formation of Oklahoma, formed during the early Guadalupian just marginal to evaporite basins (Olson, 1965d). The faunal constituents, which are far from fully known, include abundant large captorhinomorphs, *Rothia*, caseids, a varanopsid, a dissorophid amphibian, and sparse remains of gymnarthrids, paleoniscoids, and xenacanth sharks. No traces of *Dimetrodon* have been found.

The other complex, from the San Angelo formation of Texas, is essentially contemporary but more complicated. It includes very large caseids; large captorhinomorphs, somewhat like the Oklahoma counterpart and the early Hennessey; *Dimetrodon*, presumably from the Permo-Carboniferous chronofauna; and, in addition, a number of elements not known from any earlier beds. These comprise primitive therapsids; carnivores, such as *Eosyodon* and *Steppesaurus; Dimacrodon*, somewhat venjukovid in aspect; and several very large, poorly known animals that most closely resemble primitive dinocephalians. No ancestors for members of this component have been found as yet.

The Duncan–Chickasha fauna appears to be a continuation of the Caseid chronofauna, substantiating its existence in earlier beds. It lacks large amphibians, and the only strictly aquatic elements are the rare sharks and paleoniscoids, both of which may have been able to live in highly saline waters. *Varanodon*, the only known reptilian carnivore, may have been an aquatic feeder, but the fauna as now known does not include what appears to have been an appropriate aquatic food source.

The San Angelo complex seems to have had three sources, one of which is not known. It is not at all certain, however, that all three derivatives formed a single complex, because the extent of sampling necessary to ascertain this has not been possible in the relatively restricted and sparsely fossiliferous beds of this formation. As far as can be told, the complex was independent of a water-based food supply. Like the Duncan–Chickasha array, it lived near evaporite basins and lacked large amphibians and typically aquatic forms, except rare paleoniscoids and sharks.

The Soviet Union

The copper sandstones of the Soviet Union have yielded fossils from many somewhat scattered sites (Efremov, 1954; Olson, 1957a, 1962). The isolated remains and the ways the early specimens were collected have made faunal conclusions very difficult. Two sites somewhat to the west of the copper sandstones proper, however, have produced large arrays of associated vertebrates. One near Ocher (Ezhovo) is very early Kazanian and appears to include two units from slightly different times. The other, near Isheevo, is considerably later. Because other sites do not tend to have good associations, it is difficult to judge the extent to which these two represent life assemblages or taphonomic accumulations. It is assumed that both sample fairly coherent faunal zones, although all the elements may not have interacted directly.

The Ocher site (Figure 156) has yielded a number of primitive therapsid carnivores, phthinosuchids, or eotitanosuchids, titanophonids, a venjukovid,

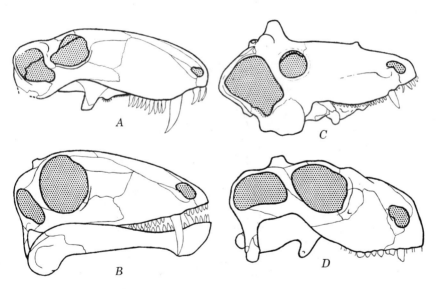

Figure 156.   Skulls of four reptiles from the late Permian site at Ezhovo (Ocher) in the Soviet Union. Not to scale. A, *Eotitanosuchus*, length about 30 cm; B, *Biarmosuchus*, length about 13 cm; C, *Estemmenosuchus*, length about 60 cm; D, *Otsheria*, length about 10 cm. (A, B, and D after Chudinov, 1960; C after Olson, 1962.)

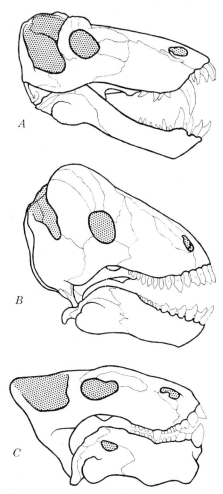

Figure 157.   Skulls of three reptiles from the Isheevo locality, uppermost Permian of the Soviet Union. Not to scale. *A*, *Titanophoneus*, a primitive dinocephalian; *B*, *Moschops* (= *Ulemosaurus*); *C*, *Venjukovia*, a presumed dicynodont ancestor. (*A* after Orlov, 1958; *B* after Efremov, 1940*b*; *C* after Efremov, 1938.)

and very odd "horned" dinocephalians, the best known of which is *Estemmenosuchus*. The last in size and body structure seems to have been a herbivore, but its teeth give little evidence of the nature of the herbivorous diet.

This assemblage is of general interest to the problem of origin of terrestrial faunas. First of all, if it was a coherent assemblage, it was one capable of existing without a water-based food chain and could have had the structure shown in Figure 155*C*. In this respect it is similar to the caseid chronofauna of the United States. Its constitution, however, was very different. The principal elements are close to the poorly known, "ancestorless" elements of the San Angelo faunal complex, which there occur along with the caseids, sphenacodonts, and captorhinomorphs. The phthinosuchids and titanophonids undoubtedly had a sphenacodont ancestry, but neither *Dimetrodon* nor *Sphenacodon* can fill this role. This assemblage suggests that a complex much like the Permo-Carboniferous chronofauna was developing in lowlands. It differed enough in its evolutionary course for somewhat more "advanced" synapsids to become characteristic. This unknown faunal complex appears to have supplied the "ancestorless" portion of the San Angelo fauna. It is not known whether the large herbivores that made possible emancipation from water, if this in fact occurred, developed within this structure or externally, as seems to have been true for the caseids.

In the Soviet Union, in deposits slightly younger than those at Ocher, occur both caseids and captorhinomorphs, but they are known from very few sites and do not have meaningful associations. As far as known, they did not enter into the Ocher-type complexes.

The Isheevo site has yielded an assemblage of remarkably-well-preserved animals, including a large dinocephalian, *Moschops* (*Ulemosaurus*); a number of titanophonid carnivores; *Venjukovia*, presumably ancestral to anomodonts; many amphibians; and fishes (Figure 157). The site was a trap and specimens were rafted in. Some rare elements—presumably fragments of anomodonts, therocephalians, and gorgonopsians—suggest that some of the creatures came from other than the main sources, represented by the common, well-preserved animals. The assemblage is reminiscent of the ecological structure of the Permo-Carboniferous chronofauna, with its food base primarily in the water. One bit of evidence is that large coprolites, apparently from the common, large amphibians, include remains of ganoid fishes, amphibians, and the large titanophonids, suggesting that the last, as their skeletons also indicate, were partially aquatic. If this interpretation is correct, the Isheevian faunal complex represents the last known example of the terrestrial and semiaquatic faunal structure that began sometime during the middle of late Pennsylvanian.

South Africa

The Permo-Triassic Beaufort series of South Africa has been divided into faunal zones based on the contained vertebrates. Hotton (1967) has clarified the stratigraphy with respect to these zones and emphasized the marked ecological differences between different zonal groups. The lowest vertebrate-producing beds contain the fauna that identifies the *Tapinocephalus* zone. The animals lived in lowlands and seem to have been in part at least buried in place. The fauna consists predominantly of large dinocephalians, pareiasaurs, primitive anomodonts, gorgonopsians, and therocephalians. The numerical balance between herbivores and carnivores suggests that the food-chain structure was that shown in Figure 155C.

In many respects the *Tapinocephalus* complex resembles that from Isheevo, but with the large herbivores in considerable numbers rather than rare. Pareiasaurs occur in the Soviet section after Isheevian times. In South Africa they were important herbivores of the *Tapinocephalus* zone and provided part of a terrestrial food base for the carnivores of the time.

Collation of Data: Sources of Captorhinomorph–Procolophon and Pelycosaur–Therapsid Radiations

The fragmentary evidence leaves many questions open or subject to tentative answers. There seems to emerge a picture of two types of lowland faunal complexes, one tied closely to water by a food chain based on aquatic animals and the other, probably stemming from the first, with a potential independence from ties to water made possible by the development of abundant large herbivores. The first seems to represent an earlier stage of transition from water to land. It was very persistent and includes within its limits many creatures that seem to lie fairly close to the ancestors of later terrestrial radiations. Both probably stemmed from an earlier, little known complex, that included small insectivorous terrestrial reptiles.

These two types of faunal complexes, which in their development through time have the characteristics of chronofaunas, seem to have overlapped and merged at various times, with the second type supplying an herbivorous base for the large carnivores of the first. Parallel development of both types occurred in adjacent areas in North America, and there is some indication of the two types, with somewhat different constituents, in Europe, in the Ocher and the Isheevo complexes of the Soviet Union. These geographic distributions may be real, or they may merely reflect the very incomplete record, because some traces of "North American" elements are found in Europe and some "European" elements in North America.

In addition to these two types of lowland faunal complex, there are indications of separate terrestrial complexes, perhaps more upland, in which invertebrates, particularly insects, formed the herbivorous base of the food pyramid. The captorhinomorph and small pelycosaur complex in the early Permian of North America and the cotylosaur complex of Russia are of this sort.

How many other types of faunal complexes may have existed cannot be estimated. Those that are known are, for the most part, adequate to explain the bases of radiation of the captorhinomorph–procolophon and the pelycosaur–therapsid lines. In none of the known samples are all of the actual ancestral genera and species known, but these could have developed under circumstances very similar to those in which known morphotypic ancestors evolved.

The source of the captorhinomorph–procolophon radiation lay in the very early Pennsylvanian, and the radiation seems to have developed its two main branches by early Permian. The captorhinomorph ramus follows a rather well-known course through the lower Permian into the base of the upper Permian, after which it is unknown, perhaps extinct. The procolophon branch has a spotty record from its possible first occurrence in the middle of the early Permian, in *Acleistorhinus* (Daly, 1969), through the Permian and into the late Triassic, with the pareiasaurs probably stemming from it in the middle part of the Permian.

The general course of the pelycosaur–therapsid radiation is fairly clear, starting in mid-Pennsylvanian and continuing with increasing complexity until near the end of the Triassic. From the carnivorous lines of this radiation the mammals eventually arose. Repeatedly, within the synapsid complex, large herbivores arose—edaphosaurids, caseids, dinocephalians, anomodonts, and gomphodont cynodonts. These provided, at various stages, a terrestrial base for the food chain, along with captorhinomorphs and pareiasaurs, but until late in the history all of the faunal complexes existed in lowlands close to water.

Although the general courses of change seem clear, very large gaps occur in the actual lines. *Dimetrodon* and *Sphenacodon* are in general excellent forerunners of the primitive therapsids, but they have many features of the skull and skeleton (e.g., the elongated neural spines) that eliminate them as actual ancestors, which remain unknown. This may merely indicate that similar chronofaunas, developed where no records were preserved, carried the actual ancestors. Or, more drastically, it might indicate that what is actually being sampled are faunal complexes that were tangential to the actual courses of development, which were taking place in more terrestrial environments. Only later faunas from more upland circumstances can indicate which of the alternatives may be true. They show that the types of animals that developed in lowland environments are structurally suitable ancestors and that no special features calling for adaptations to uplands are necessary in their ancestry. Such radiations are not required, but clearly this is no demonstration that they did not exist.

ORIGINS OF OTHER LINES

General Circumstances

By the late Permian eosuchians were well established, and by the beginning of the Triassic archosaurs were widely distributed around the earth. Most

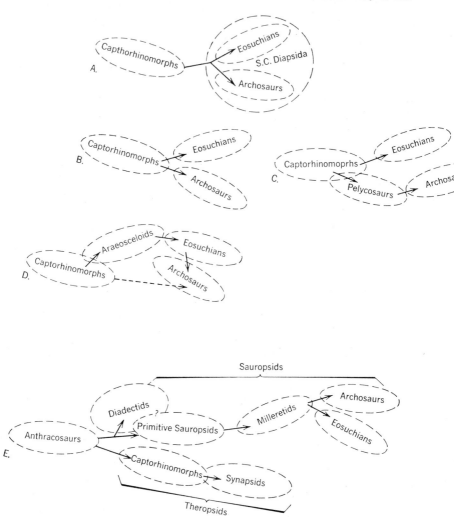

Figure 158. Five possible phyletic schemes of the origin and relationships of eosuchians and archosaurs. Envelopes of dashed lines include the groups and arrows show the lines of descent. Where problems of differentiation or assignment occur, the envelopes intersect.

early remains of the latter have come from the Lower Triassic, but one specimen is known from the late Permian of the Soviet Union. The problems of determining both the sources of these two groups and the evolutionary patterns involved in their origins are much greater than for the captorhinomorph–procolophon and pelycosaur–therapsid lines. This results either from a lack of the actual ancestors in the record or from an inability to determine with any assurance the identity of ancestors among known forms.

Under these circumstances no clear decision can be made as to whether the events leading to the terrestrial radiations were enacted within the known lowland faunas or whether they took place in more upland terrestrial environments. An additional complication arises in the likelihood that the two stocks followed quite different paths.

No positive answers to questions of the ancestry or the ecological settings of the origins are presently possible. The rest of this chapter is devoted to considerations of the possible modes of origin and interpretations of the implications of the more probable courses with regard to evolutionary patterns.

Possible Patterns of Origin

Figure 158 shows five different phylogenetic patterns in which the possible relationships of the eosuchians and archosaurs to each other and to earlier animals are portrayed. In four of these the captorhinomorphs are considered ancestral to both, which conforms to the orthodox thinking of the present time. In the last the two lines are separated, so that the captorhinomorphs are ancestral only to other anapsids and synapsids, whereas the lines in question stemmed from a separate base among the amphibians. Each of these arrangements is cast within the framework of currently accepted groupings of the early reptiles. As will be noted later, this tends to cloud the issue and some of the differences appear less important if considered without this restriction. First, however, some brief comments will set the various patterns in perspective within the taxonomic framework.

1. Figure 158A. This is a more or less orthodox concept in which the eosuchians and archosaurs are considered to belong in a single subclass, the Diapsida, which arose from captorhinomorphs. A common base, which could be defined as pertaining to one of the groups, or one that is neither but is distinct from the captorhinomorphs, is implied by this pattern. No such group has been specifically identified in the Permo-Carboniferous chronofauna, and the captorhinomorphs within this complex, though rather eosuchianlike, had already begun radiations based on their own special characters. Some of the small pelycosaurs could fill an ancestral position, but positive evidence has not been found. The gaps suggest that under this interpretation the evolution of the diapsids must have taken place beyond the limits of the chronofaunas as we now know them.

2. Figure 158B. Eosuchians and archosaurs stemmed from the captorhinomorphs but from different branches within this category, without an eosuchian stage, or any other common noncaptorhinomorph line, ancestral to both. No captorhinomorphs show any trends in an archosaur direction; hence, once again, under this hypothesis, the archosaurs must have arisen beyond the limits of what is known. Eosuchians could, however, have stemmed from some of the early captorhinomorphs. This evolutionary sequence permits two very different courses of evolution for the two groups.

3. Figure 158C. Eosuchians arose from captorhinomorphs, as in Figures 158*A* and *B*, but archosaurs arose from pelycosaurs. This conforms to the suggestion of Reig (1967) as far as archosaurs are concerned. In Reig's proposal, and implied in the distance of the point of derivation of archosaurs from pelycosaurs in the diagram, archosaurs arose from rather advanced pelycosaurs. Specifically, according to Reig, the ancestry is to be found among the varanopsids, which were partially aquatic animals. Thus the origin of archosaurs is presumed to have had an aquatic base within the Permo-Carboniferous types of fauna, particularly within the Caseid chronofauna in which varanopsids seem to have been fish-eating carnivores.

A variant of this origin could be that the archosaurs originated as outlined, but that the eosuchians came not from a captorhinomorph base but from a primitive pelycosaur base, technically placing the origins of both groups in the same subclass.

4. Figure 158D. The archosaurs might have originated in one of the ways shown in Figures 158*A*, *B*, and *C*, but the eosuchians arose from the areoscelids (protorosaurs). These are presumed to represent a stage between the captorhinomorphs and eosuchians. The strongest support for this came from Williston (1925), who saw the origins of lacertilians among the areoscelids. Evidence of a diapsid ancestry, with lizards passing through a younginid stage, showed his concept of origin of the lizard temporal region by ventral emargination of the large lateral squamosal to be invalid. It is not, however, impossible that a lateral fenestra developed to produce a diapsid, although the construction of this region in the known genera seems to make this improbable.

5. Figure 158E. This diagram shows one of several possible ways of origin of eosuchians and archosaurs that follow if it be considered that the two did not arise from a common reptilian stock but had common ancestors only among the amphibians. The figure shows the approximate consequences of a proposal made by Watson (1954), based on the separation of reptiles into two major groups, the sauropsids and theropsids (see Section III, Chapter 8). A major problem in tracing such a course is the absence of early sauropsids in the record. *Diadectes* has been suggested as a side branch that might show the existence of these, and possibly *Bolosaurus* can fit in. If this approximates the actual course of events, then its unfolding took place almost completely outside any known ancient environments. Although by no means impossible, the course leaves the whole matter up in the air and is not subject to any sort of testing.

Each of these courses has variants and totally different ones are not out of the question. If it be granted that the records do show at least something about the ancestry of the eosuchians and the archosaurs, then the origins probably took place along the lines indicated by one of these phylogenies or some combination of parts of two or more. Examination of the evidence will aid in determining which is most likely and direct attention to where information critical to the selection of one or another must be sought.

The Evidence

Archosaurs and Eosuchians

OCCURRENCES.   All of the known early representatives of the archosaurs and eosuchians necessarily occur in lowland faunal complexes, associated with members of the other two major lines, captorhinomorph–procolophon and pelycosaur–therapsid. As such they were constituents of the complexes that were derived from the early Permian chronofaunas.

The earliest eosuchians proper occur in the upper Permian *Cistecephalus* zone of the Karroo basin in South Africa (Figure 159). Apparently somewhat earlier in the Soviet Union is *Mesenosaurus*, originally called a pelycosaur but later tentatively placed with the eosuchians or millerettids. Peabody (1952) suggested that *Petrolacosaurus* from the upper Pennsylvanian of Kansas had eosuchian affinities. Although this has been discounted by most students, it points up one of the major difficulties in this problem—that of the great resemblance of all early reptiles. The younginids (Figure 159C) epitomize the primitive eosuchians and quite clearly lie at or near the base of the lepidosaurian terrestrial radiation. It is their origin that must be explained.

Hotton (1967) concluded that the sediments of the *Cistecephalus* zone, in which younginids are present, were deposited in open water and that the vertebrates were floated in. In this event the known assemblages appear to be taphocoenoses and not necessarily representative of biocoenoses. In earlier zones, the *Endothiodon* and especially the *Tapinocephalus*, burial areas seem to have been close to actual habitats. No eosuchians have been found in these deposits, which seem to have been formed in lowlands by stream deposition. There is thus very little positive information on the actual life associations or ecological circumstances of existence of the early eosuchians. It is, of course, quite possible that the well-known *Cistecephalus*-zone vertebrate assemblage does represent a coherent faunal unit, with eosuchians a part of it, and as such it was not very different in general structure from some of the earlier assemblages of the caseid chronofaunal type.

The only Permian archosaur is from the uppermost Permian of the Soviet Union, in zone V (P_2) in Tatarian beds. This is *Archosaurus* (Tatarinov, 1960*b, c*), which is very close to the better known *Chasmatosaurus*. The latter

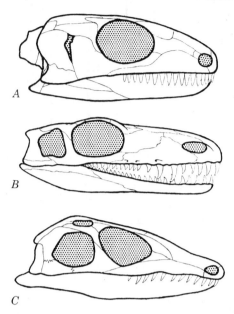

Figure 159. Eosuchians and possible eosuchians. *A, Milleretta,* assigned to Eosuchia by some authors, to Captorhinomorpha by others, and considered intermediate between either Captorhinomorpha and Eosuchia or primitive sauropsids and Eosuchia by still others. *B, Mesenosaurus,* first classified as a pelycosaur and later as an eosuchian but not a diapsid. *C, Youngoides,* a central eosuchian. *A, B,* and *C* all about 3 inches long. (*A* after Watson, 1957; *B* after Efremov, 1940*a*; *C* after Broom, 1914, Olson and Broom, 1937, and Olson, 1936*b*.)

(see Figure 160) is widely distributed in lower Triassic beds. It has been identified in South Africa, China, India, and the Soviet Union, and fragments of possibly related forms have been found in South America. The typical occurrence is with the aquatic anomodont *Lystrosaurus.*

The other very primitive and early archosaurs, which with *Chasmatosaurus* form the Proterosauria, fall in the family Erythrosuchidae. They too are widespread in the early Triassic, but in beds a little younger than those containing *Chasmatosaurus.* They occur in the *Cynognathus* zone of South Africa and in equivalent beds of zone VI in the Soviet Union. Remains are known from China and possibly from western United States and South America from very fragmentary specimens. In addition a possible representative has been found in the upper Triassic Elgin sandstone of the British Isles.

Erythrosuchids occupied more terrestrial environments than did chasmatosaurids, being associated with large cynodonts and anomodonts. In South

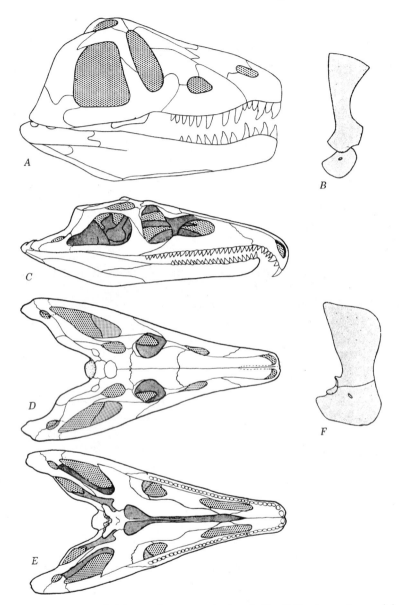

Figure 160.   Two primitive archosaurs. *A*, the skull of *Erythrosuchus*, original
length about 80 cm; *B*, scapulocoracoid of *Erythrosuchus; C*, *D*, and *E*, skull of
Chasmatosaurus, lateral dorsal and ventral views, respectively; original length
about 45 cm. *F*, scapulocoracoid of *Chasmatosaurus*. (*A* after Von Huene, 1911; *B*
and *F* after Hughes, 1963; *C*, *D*, and *E* after Broili and Schroeder, 1934.)

Africa a somewhat more advanced genus, *Euparkia*, sometimes associated with the Pseudosuchia, occurs at the same stratigraphic level as *Erythrosuchus*.

MORPHOLOGY. The skulls and postcrania of the central members of the eosuchians and proterosuchian archosaurs are quite well known. The skulls of *Youngoides*, *Chasmatosaurus*, and *Erythrosuchus*, illustrated in Figures 159 and 160, show the major resemblances and differences. Except for the diapsid condition and the retention of primitive features, there are few positive resemblances. Eosuchians possess a reptilian otic notch similar to that of lizards and *Sphenodon*, and this is not present in the most primitive archosaurs, although it is found in more advanced forms. Eosuchians lack the special archosaur features found in *Chasmatosaurus* and *Erythrosuchus*, in particular the strongly ossified laterosphenoid and the antorbital fenestra. Because they are basically primitive, younginids could be morphologically ancestral to archosaurs, but there is nothing that compels the establishment of such a relationship.

The story of the postcranium is much the same. Eosuchians are lizardlike and not far from primitive reptiles. They lack the broad vertebral arches and holocephalous ribs of captorhinids, but otherwise, although lighter, are similar in general structure. *Chasmatosaurus* and *Erythrosuchus* are basically archosaurian, with a somewhat triradiate pelvis, but most of the archosaurian characteristics are incipient. The femora have remained quite primitive, with a terminal head and a strong internal trochanter, but the vertebrae show the beginning of archosaurian features in which the transverse processes have become closely spaced in the trunk region and the capitulum has swung posteriorly to lie beneath the tuberculum in the cervical region. The hind limbs are somewhat longer than the front ones, but not sufficiently to suggest bipedalism. The shoulder girdle has a reduced coracoid, a feature more strongly developed in *Erythrosuchus* than in *Chasmatosaurus* (Figure 160). The essentially primitive postcrania of younginid eosuchians do not bar them as ancestors to archosaurs, but similarly there are no structural trends to suggest that they in fact were.

Possible Ancestors

Possible ancestors, as far as they can be known, must occur among the constituents of the Permo-Carboniferous and caseid chronofaunas of the late Pennsylvanian and early Permian, and among later complexes known from the Soviet Union and the upper part of the Permian of South Africa. Some of the abundant constituents of these complexes—captorhinomorphs and pelycosaurs—could have been ancestral, as well as some of the less common and generally less-well-known creatures.

Figure 153 shows some of the possible ancestors, several of which are known from very few but well-preserved specimens. In addition there are similar genera known only from rather incomplete specimens. The role of the anapsid captorhinomorphs, which most students concede to have been on the ancestral line, will be noted specifically later. Of the others, none are diapsids but each, by one or another modification of the temporal region, could have attained the diapsid condition.

Petrolacosaurus, as noted above, was assigned to the eosuchians by Peabody (1952), but Romer has called it an edaphosaurian pelycosaur. The upper part of the temporal region and dorsal platform of the skull are not well preserved, so that it is not known whether there was an upper temporal fenestra. This is a lightly structured, primitive reptile of late Pennsylvanian age. It could very well have been ancestral to any of the lines of terrestrial radiation except the captorhinomorph–procolophon line.

Araeoscelis was Williston's candidate for the ancestry of lizards, but Vaughn (1955) considered it to be a lizardlike experiment of the early reptilian radiation, without descendants. Romer (1966a) considered *Araeoscelis* to be a euryapsid, related to the very puzzling group of Protorosauria, or Araeoscelidia (see Section III, Chapter 8). It had very long limbs and a lightly built skeleton, but the vertebrae, except for absence of the very broad arches, were captorhinomorphlike. The skull is somewhat similar to that of some pelycosaurs, except for the temporal region. Other similar animals have been found in the Old World, and a possible representative has been identified in the middle Permian of Africa.

Bolosaurus is an enigmatic genus whose position has never been satisfactorily determined. It may be a somewhat aberrant offshoot of the captorhinomorphs, or it may somehow be related to the diadectids. How it fits into the ancestry of either eosuchians or archosaurs is, as of now, a wide-open question.

Small pelycosaurs, such as *Eothyris*, *Bayloria*, and *Oedaleops*, are essentially captorhinomorphs with lower temporal fenestrae. Similarly they are very close structurally in general features to eosuchians but have only a ventral temporal fenestra. They are not abundant in the lower Permian assemblages but occur with some degree of regularity throughout the range. Their roots seem to go well back into the Pennsylvanian, possibly to such creatures as *Petrolacosaurus*.

Millerosaurs are from considerably later beds, and the time of occurrence has given a somewhat different direction to their study. In many respects they are captorhinomorph, and Broom (1938) assigned the first representative to the cotylosaurs. They have either an incipient or well-developed lower temporal fenestra, quite pelycosaurlike, but such features as their small reptilian otic notch resulted in their assignment elsewhere. Such a notch, however, is not unknown among pelycosaurs, being well developed, for example, in the caseids (Olson, 1968).

All of these primitive animals, including the captorhinomorphs, have common features that can be used to cast them in the role of ancestors of the eosuchians. The structure of the middle ear, and the stapes in particular, has raised serious questions about this relationship. The stapes passes laterally or ventrolaterally to the quadrate, is very heavy, and shows no clear evidence of a tympanic process. This is true whether or not an otic notch is present (see Figure 76). D. M. S. Watson (1954) made a very strong point of this structure in his separation of the therapsids and sauropsids, although he did not at that time have the needed information on millerettids to see the dilemma that they posed, placed as they were in his Sauropsida.

Barry (1963) pointed out the great variability in this region in both living and extinct tetrapods, and Olson (1966a) indicated that the condition in these various possible ancestors represented a generally primitive condition among reptiles, which probably had little to do with differences in the lines of descent in which it underwent various modifications.

Ancestors to the archosaurs have been sought among the varanopsid pelycosaurs by Reig (1967) based on the resemblances shown in Figure 153. Most of Reig's arguments apply to *Chasmatosaurus*, but the antorbital fenestra is found in *Erythrosuchus* as well as other archosaurs. Resemblances clearly exist, but how significant they are phylogenetically is a moot point.

Possible Patterns of Change

When all of the possible ancestors of later lines are viewed together, an evolutionary pattern not unfamiliar in the origins of major groups seems to emerge. By mid-Pennsylvanian there seems to have been a fairly effective break with aquatic environments by primitive, limnoscelid captorhinomorphs. From this base a rather rapid differentiation of a number of somewhat modified small terrestrial stocks took place. Probably there were many branches, although only a few are preserved in the record. Initially, it appears, there was a small otic notch, but this was lost or modified in later lines. The most immediately successful line was that which continued the general captorhinomorph habitus, including members of the families Romeriidae and Captorhinidae. These were common constituents of the Permo-Carboniferous chronofaunas. They were characterized by maintenance of an anapsid temporal condition, loss of the otic notch, and broad-arched vertebrae.

In other groups the temporal regions developed some sort of fenestration. Some retained otic notches, others did not. The vertebral arches, as far as is known, did not assume the structure found in captorhinomorphs. Because of similar temporal fenestration, many of these animals have been grouped as synapsids and, being primitive, as pelycosaurs. They lasted well into the lower Permian and are best known from end members of the lines. Very little different, but with upper temporal fenestration, were such forms as *Araeoscelis* and *Dictybolos*. The latter, like members of some other early lines, was partially aquatic.

All of these creatures had similar middle ears, with the stapes passing laterally or lateroventrally and making a strong contact with the quadrate. How this condition was derived from the presumably ancestral amphibians is not certain, because neither the temnospondyls nor the microsaurs include appropriate ancestral conditions.

Representative skulls are those illustrated in Figure 153. All of these animals represent variations on a common pattern, probably developed during experimental probing of the new environments opened by the threshold passed by the first land occupants. Insects probably formed the initial food base on land, although other invertebrates also may have been important. From these initial probings a variety of lines were sorted out. Among these were the well-known captorhinids and the three major branches of the pelycosaurs—ophiacodont, primarily ophiacodontids; edaphosaurian, including edaphosaurids and caseids; and sphenacodonts, including sphenacodontids and varanopsids. It is not necessary to the concept of origins that these lines had a common synapsid base. Most of the larger pelycosaurs tended to find their food sources in water, perhaps as large size made this possible and reduced the effectiveness of invertebrates as a primary terrestrial source. Caseids were an exception, being terrestrial herbivores.

The smaller pelycosaurs represent stocks that did not change their feeding habits appreciably. Whereas the larger animals were very successful members of the water-related Permo-Carboniferous complex and appear abundantly within it, the smaller, more terrestrial, insectivorous pelycosaurs are less common, their potential zone being mostly occupied by captorhinids. The same appears to apply to areoscelids.

Two related interpretations emerge from such a view of these early events. One is that temporal fenestration was not related to the adaptive shifts inasmuch as experimental probings were carried on equally in similar environments by creatures with different patterns of fenestration. Thus anapsid forms, limnoscelids, various synapsids, and euryapsids, with *Dictybolos* as an early example, all became somewhat adapted to aquatic life, and all three have comparable terrestrial phases.

Through these terrestrial probings, various lines approached the lizardlike habitus. Among these were the millerettids and younginids, which could have attained their structures within the lowland chronofaunas or might have developed external to them by virtue of their potential independence of the chronofaunal structure. The pattern of temporal fenestration has become the basis for the placement of early genera into subclasses, inasmuch as later distinctive radiations, stemming from one or another of the successful early lines, had in their heritage one particular type of temporal region. The imposition of this separation on the early experimental lines clouds the picture of the multiplicity of the early experimental patterns by creating apparent limiting envelopes, such as those in Figure 158. As far as can be told, the early temporal fenestration had no direct relationship to the adaptive deployment of the early lines.

Stemming from the basically insectivorous members of the early differentiation, along with terrestrial and aquatic predators, were lines that began to exploit the plants of the terrestrial environment directly as a food source. *Edaphosaurus* represents a very early example; its foodstuffs probably were the soft plants of the water and water margins, judging from its structure and the general occurrence in association with aquatic animals. The larger captorhinids—*Captorhinikos, Labidosaurikos, Hecatogomphius, Kahneria,* and particularly *Rothia*—exemplify a later case. Caseids, from a synapsid base, represent a third instance, and later procolophons, pareiasaurs, some dinocephalians, anomodonts, and some cynodonts represent still others.

The origin of archosaurs remains the biggest question, and the concepts of the early radiations suggested here do not aid particularly in the solution. The two proterosuchians do not point either to a particular common ancestor or to a common ecological condition of origin. If it be considered that the *Chasmatosaurus* type is primitive and that erythrosuchids came from it, an aquatic-based origin along the lines suggested by Reig seems quite reasonable. However, the resemblances of the varanopsids and chasmatosaurs, beyond the common possession of an antorbital fenestra, whose meaning is uncertain, can well be explained as aquatic specializations superimposed on basically primitive characters.

If erythrosuchids are thought to represent the primitive habitus, with chasmatosaurs an aquatic offshoot, then no known animals offer a particularly close source, although many lines of speculation are possible. Dual origin from different pelycosaur bases is not out of the question. If a nonpelycosaurian base is sought, the eosuchians seem to be the most likely candidates. The absence of the reptilian otic notch in the primitive archosaurs is the most compelling contrary evidence, aside from the fact that no intermediates are known.

Even within the limits of very broad speculations, as made in this chapter, it is not possible to arrive at a consistent picture of how the later terrestrial radiations came into existence from their aquatic and partially aquatic ancestors. What has been taken up in this chapter is a survey of the problems from the standpoint of faunal analysis. They can be viewed strictly from a phylogenetic point of view, without analysis of the possible dynamic aspects of origin, but, although this is simpler, the missing actual and morphotypic links are such that even within this limited framework no more plausible interpretations emerge.

6 Evolutionary patterns in the origin of mammals

INTRODUCTION

In the world of today mammals are readily distinguished from all other vertebrates, but the situation is quite different in the transitional stages from reptiles to mammals recorded in the rocks of Permian and Triassic age. A clear separation of the very advanced therapsids and the primitive mammals can be made only if some arbitrary criteria are adopted. The problems raised by this situation are important in systematics (see Section III, pages 369–377), but affect understanding of the origin and evolution of mammals only indirectly, through the medium of classification. The well-preserved record of this closely graded transition and the detailed information about the reptiles that led to it are important in their own right, but even more so in documenting the general nature of change from one major adaptive level to another.

The evolutionary events that produced the great variety of therian mammals in existence today may be usefully visualized as a series of adaptive radiations, each following on the passage of some critical organizational threshold. The last major radiative phase, that of the therian mammals, fits nicely into the system without need for special treatment. It has been the most extensive and successful of the several postcaptorhinomorph radiations, but only because the level of organization that marked its inception opened ecological vistas greater than those that had been available previously.

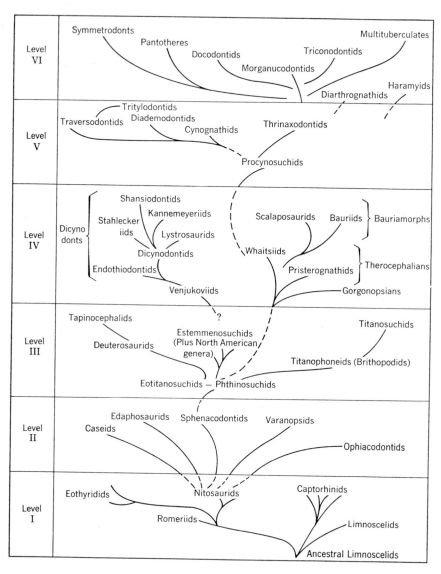

Figure 161.   Diagram of "phylogeny" from primitive reptiles to primitive mammals, showing suggested six levels of radiation used as a framework for discussion in the text. The horizontal lines separating successive radiation levels are not time lines, although in most instances the roughly approximate temporal divisions with time passing from older to more recent from the base to the top.

Starting with romeriid captorhinomorphs, the course passed through successive radiations involving primitive pelycosaurs, more advanced members of this group, primitive carnivorous therapsids, advanced therapsids, the initial mammals, and then the massive radiations of therian mammals. The process commenced in the early Pennsylvanian and began to attain full fruition in the late Cretaceous.

In the course of these changes extensive reorganization of all body systems took place. These did not occur in simple ordered sequences or at constant rates. During each of the radiations new features were introduced and, variously, became part of the structural mosaic of the members of the different radiating lines. The modifications, in context of the radiations, were adaptations to particular circumstances, but some were common to all stocks regardless of their ways of life and thus must be viewed in a more general adaptive context. Coordination and integration of such modifications brought some members of a particular radiation to the threshold of a new radiation.

It is the features that are carried over from one radiation to the next, as "advancements," that have received the most attention in studies of the origin of mammals. Short-term adaptations that left no mark on the general course of change have generally been ignored. Many of the modifications can be followed quite well through successive levels. Although problems of identification of ancestral lines have not been fully solved, morphotypic stages are well known. From studies of direct descendants where possible and morphotypic series where necessary a reasonably adequate concept of the evolutionary events leading from primitive reptiles to mammals has emerged. Detailed knowledge of the changes in some systems is available, whereas others are known mainly by inference. Some understanding of the relationships between the modifications and the environments in which they took place, and at least working hypotheses concerning the causal factors that have contributed to the course of evolution have been developed. These are the matters with which this chapter is concerned.

THE COURSE OF THE RADIATIONS

General Matters

Figure 161 shows a diagrammatic "phylogeny" involving the captorhinomorphs, synapsids, and primitive mammals. The sequence is divided into six levels, labeled I through VI, based on successive series of radiations. Included are some of the families of reptiles that were considered in earlier chapters in a faunal rather than a phyolgenetic context, which here casts them in a different role.

The initiations of the radiations are successive in time, and to some extent this is true for the radiative expansions as well. Thus in Figure 161 the levels separated by horizontal lines are more or less sequential in time, but the boundaries are not time boundaries, and the events depicted at one level may temporally overlap some of those that are depicted in one or more that follows. The controlling factors are the levels of development, which mark the initiation of a particular radiation, and the exploitation of this organization in radiations that it made possible.

Within each level several adaptive types were developed, and in general similar types occur in each. They differed somewhat with the biological potentials of the animals and the physical environments and faunal associations at the different levels. At some stages, for example, the herbivorous component of the vertebrate faunal complex came largely from nonsynapsid groups of organisms, whereas at others it came from the members of the radiation. At each level diet was a basic aspect of the radiation, acting, it would seem, as the primary factor in the divergence of the stocks.

The use of dashed lines and incomplete connections in Figure 161 indicates some of the problems of phylogenetic associations, but the full extent of these cannot be simply portrayed. The interpretations as made are conservative in that where alternatives exist the simpler is used. Thus single origins of major radiations are shown where multiple origins might have occurred. The transition from level II to III, for example, is tentatively indicated as passing through the sphenacodont line only, whereas at least a fair case can be made for dual origin, involving the caseids as well as the sphenacodonts (Olson, 1962).

Patterns of Morphological Change

Perspectives of Analysis

The patterns portrayed in Figure 161, like those of all radiations, may be viewed in two distinct ways, and interpretations based on the two perspectives give emphasis to distinct aspects of evolution, both of which are important in the assessment of transitions in continuing series of organisms.

Each of the radiations may be looked at as an elaboration of the potentials of the basic stock from which it arose. This is a joint function of the biological capacities of the stocks at the particular organizational level and of the ecological opportunities provided for their exploitation during the radiation. Interest focuses on the morphological changes or modifications of the primitive pattern and, to a greater or lesser degree, on the adaptive significance of the changes. The radiating complex itself is the main point of interest, with little or no concern with what preceded the radiation and none at all in what followed, which, while potentially knowable for extinct radiations, was not pertinent to the radiation itself.

"Review of the Pelycosauria" by Romer and Price (1940) exemplifies many aspects of this approach. Studies of modern faunal complexes, as far as aspects of the radiations enter in, must fall within this framework. It is only to the extent that any such studies give bases for prediction that they may have concern with the putative future. Predictive properties are of very low order at best.

The important aspect of this perspective in evolution is that it centers attention on the dynamics and motivations of change. These relate the events directly to the processes of organic change which, as they are known from studies of living organisms, have resulted in the formulation of evolutionary theory based on natural selection.

Lacking any significant predictive properties, however, this point of view, for all of its power, does not advance our major aim of understanding the course and process of change in which successive radiations are involved. It provides no means for more than a vague understanding of the significance of the various aspects of the radiation in the development of new levels of organization. Even if the data on the structure of the animals were adequate for this purpose, which they are not, the fact that realization of a particular potential depends on the future characteristics of the physical and biological environment makes any prediction extremely precarious.

It is necessary then for interpretations such as those made in this chapter to view the total course of events in retrospect with respect to some end actually attained through the evolutionary changes. Each of the several radiations then appears as a span in the bridge between a primitive condition and an end stage. In the present case the primitive captorhinomorphs and pelycosaurs, central in level I, and the mammalian condition attained within level VI are the defined ends under consideration. Once the course of change, through actual lines of descent or through suitable morphotypes is known, attention within the several radiations becomes focused primarily on the stock or stocks that formed the bridge to the next radiation, or in short, was "on the line" from primitive reptiles to mammals. Not only are particular stocks selected from the many, with the "unsuccessful" being largely ignored, but also attention is directed primarily to those structures that are considered especially important in the transition. A very strong selectivity is thus introduced. This diverts analyses of the nature and causes of change away from the intimate relationship to the basic factors of evolution, which can be more readily comprehended from changes within radiations.

Many things, of course, determine the selection of the morphological features to which special attention is devoted. Most important is that the structures show distinct and easily definable states at the two ends of the series. Such features may be considered in the context of this chapter as discretely "reptilian" at one end and discretely "mammalian" at the other.

If such features can, in addition, be traced through their modifications of state, their use is enhanced, and intermediate states may be expressed either by some identifying designation (e.g., example sphenacodont or bauriamorph) or by some numerical designation, 30 percent mammalian, indicating progress about one-third of the way along the course of total change.

A second basis for selection lies in the interpretability of the meaning of the features. Some, such as the jaws and dentitions, have evident adaptive aspects that are both readily recognized and understood. Others—for example, the structure of the middle ear—may be traced through the sequences, but the adaptive significance may be less clear.

A third basis, which leads into very speculative areas, lies in the relationship of structures to modifications of systems or processes that are of particular significance in the sequence. The many features of mammals that are related to the development of homoiothermy or to the parental care of the young impel this sort of interest. Development of homoiothermy, increased duration of activity, and the many accompanying phenomena cannot be directly known and must be inferred from known conditions in existing end members, both reptilian and mammalian, and evidence from preserved structures in the series under study. Because of their very great significance, such items do figure importantly in analyses in spite of the obvious hazards of interpretation.

Various studies of the origin of mammals have been made from the vantage view of this second perspective (e.g., Olson, 1944; Brink, 1956). Like the study of individual radiations, it contributes an essential aspect to understanding of the evolution of major categories, but, similarly, it clarifies only a part of the picture. Both approaches must be used if such evolutionary events are to be understood in the context of modern evolutionary theory. Use of the retrospective point of view alone gives relatively little insight into the evolutionary mechanisms of change and may, in seeming to so do, engender concepts that appear unrealistic in the other context. Some hypotheses of linear evolution—for example, teleological concepts or orthogenetic ideas involving the inevitable realization of innate potentials—stem from this one-sided point of view. The interpretations following from either point of view can be considered "correct" or "incorrect," but those that are little or not at all influenced by information available only in retrospect maintain the closest ties to the mechanisms of evolution by natural selection and are least likely to induce faulty applications of its precepts.

Levels of Radiation

Level I

Primitive captorhinomorphs, limnoscelids, romeriids, and primitive pelycosaurs of the Pennsylvanian and early Permian are the members of the earliest reptilian radiation, level I. Within it appear various branches that presumably resulted from adaptations to somewhat different environmental circumstances in the lowland terrestrial environment. Evidence on the precise nature of such adaptations is, however, relatively slight.

During the Pennsylvanian from some of these lines came the basic stocks from which later reptiles arose, as discussed in the preceding chapter. The captorhinomorphs themselves underwent a modest radiation, but it was from one or several of the branches in which a lower temporal fenestra had arisen that the second level of radiation in the course to mammals developed. In retrospect it is principally the temporal fenestra, lying below the postorbital and squamosal, that establishes this relationship, for most other features are present in all primitive reptiles. Even the temporal fenestra may be suspect, since it may have arisen in similar position in lines that had no descendants and in lines that were at the bases of other radiations. It is clear, however, that it was a common feature among the ancestors of the radiation of level II.

None of the earliest and most primitive members of level I are fully known; hence the primitive reptilian condition must be reconstructed as a composite from a number of genera. Much of the detail, especially of the braincase and otic region, comes from *Captorhinus* and *Labidosaurus*, members of the Captorhinidae that departed far from the primitive condition.

Many of the general features of the skulls at this level are shown in Figure 153 in the preceding chapter. The outer surface of the skull was fully covered by closely applied skin, as indicated by its pitted surface, and the lateral surface of the lower jaws, as exposed during occlusion, was similarly covered. There were no origins or insertions of the jaw musculature over these regions. The adductor musculature of the lower jaw consisted largely of equivalents of the external and posterior adductors of the lizards and of a large reptilian pterygoid muscle. The former were housed within the temporal adductor chamber and inserted primarily in and around the adductor fossa of the lower jaw. The pterygoid (or mandibulopterygoid) originated on the strong transverse process and the quadrate ramus of the pterygoid bone and inserted along the ventral margin of the angular of the lower jaw (see Figure 59). The depressor musculature appears to have originated in the fascia of the neck and possibly on the posterior margin of the skull. It inserted on the retroarticular process of the lower jaw. The temporal region was closed in the most primitive genera, but in some others it carried a small temporal opening below the postorbital and squamosal.

The orbits were moderately large, the snout fairly short, and the flat dorsal platform carried a medium to large parietal foramen. Laterally the platform was joined to the temporal region, but some movement between these two portions may have been possible in primitive members of the group. Anteriorly small paired nares were present. The maxilla, which bore a single row of little-differentiated, conical teeth, occupied only a small part of the prefacial region of the skull, being little expanded dorsally.

The dentary covered about half the outer surface of the lower jaw. No coronoid process was developed. Lower teeth, like the uppers, were little differentiated. The articular and quadrate bones formed the sole articulation of the jaw. Both were large and carried well-developed articular surfaces that were rounded and little subdivided into separate portions. Anteriorly was the large adductor fossa and posteroventrally, the retroarticular process.

The palate included a large pterygoid with a strong transverse process, a long quadrate ramus, and a slender palatine process. A moderate interpterygoidal vacuity was present. The parasphenoid floored the basicranial axis with its somewhat expanded posterior, or alar, portion, and a slender cultriform process passed forward into the interpterygoidal vacuity. Extending laterally from the basisphenoid was a strong basipterygoid process that formed a movable joint with the epipterygoid–pterygoid complex. Internal nares lay far forward, lateral to the vomers. The dermal elements of the palate, except for the parasphenoid, carried small, simple teeth.

The cranial cavity was floored by the basicranial elements and walled by the small oto-occipital complex plus the laterosphenoid. Anteriorly lay the sphenethmoid, which carried the nasal tracts and was Y-shaped in cross-section. Lateral to the well-developed laterosphenoid, forming the outer margin of the cavum epipterycum, was a slender, rodlike epipterygoid (Figure 162).

The otic capsule was ossified by an opisthotic and proötic, which along with the basioccipital enclosed a large fenestra ovalis. The opisthotic projected laterally in a strong paroccipital process. The medial walls of the otic part of the braincase carried a widely open internal auditory meatus and anterodorsally to it was a very shallow subarcuate fossa. The anterior and posterior semicircular canals were about equal in length and curvature, and the osseous conduit of the utricular sinus was essentially normal to the horizontal plane of the basicranial axis.

The stapes was the sole ossicle of the middle ear. Its large foot overlay the fenestra ovalis. The shaft was heavy and carried a strong dorsal process that made osseous contact with the paroccipital process. The quadrate process was strong, but the distal portion was not always fully ossified. When complete, it abutted strongly against the quadrate. No evidence of a tympanic has been certainly identified.

Many of the features of the postcranium are similar to those of *Sphenodon* and generalized lizards. Vertebrae were but little differentiated in the presacral region and all carried ribs. The femur carried both an internal and fourth trochanter, and the head of both the femur and the humerus occupied the proximal end of the elements. The head of the humerus was somewhat

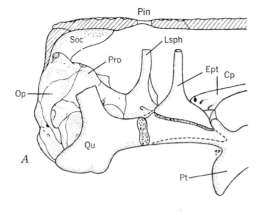

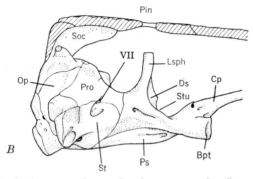

Figure 162. The braincase and associated structures in *Captorhinus* in lateral aspects. *A*, with the lateral dermal bones removed showing the pterygoid-epipterygoid-quadrate complex; *B*, with the pterygoid, epipterygoid, and quadrate removed showing the deep structures. Bpt, basipterygoid process; Cp, cultriform process of parasphenoid; Ds, dorsum sellae; Ept, epipterygoid; Lsph, laterosphenoid; Op, opisthotic; Pin, pineal opening; Pro, proötic; Ps, parasphenoid; Pt, pterygoid; Qu, quadrate; Soc, supraoccipital; St, stapes; Stu, sella turcica; VII, foramen for facial nerve. Redrawn from *Vertebrate Paleozoology*, 3rd edit. by A. S. Romer. Copyright © 1966 by the University of Chicago Press.

S-shaped and made articulation with the screw-shaped glenoid fossa of the scapulocoracoid. Two coracoids were present in those forms in which the shoulder girdle is well known. The ilium carried a strong caudal process, related to heavy caudal musculature. A single sacral rib was strongly bound to the ilium by ligaments, and the anterior caudal vertebrae carried large recurved ribs that had ligamentous attachments to the caudal musculature.

Limbs varied from rather short and heavy to long and slender within the group. The animals that lay close to the ancestry of the pelycosaurs of level

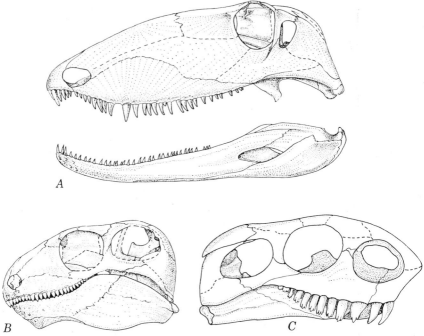

Figure 163.   Representatives of three important pelycosaurian types of level II. *A*, left-lateral view of *Ophiacodon*, an advanced, semiaquatic ophiacodont; original length about 30 cm; *B*, left-lateral aspect of *Edaphosaurus*, the best known edaphosaurid, original length about 17 cm; *C*, right-lateral view of *Casea*, a primitive member of the caseids, original length about 7 cm. (*A* and *B* after Romer and Price, 1940; *C* after Olson, 1968.)

II appear to have had moderately short, stocky limbs. The phalangeal formula was 23453 and 23454 in the fore and hind limbs, respectively.

It is from animals with these features that all of the reptiles likely arose. Special affinities with pelycosaurs of the second level are indicated primarily by the temporal fenestra and the double coracoid. The other characters that are shared are those common to primitive reptiles as a whole.

Level II

RADIATION AND MORPHOLOGICAL FEATURES.   Level II comprises the pelycosaurs of the late Pennsylvanian and early Permian deltas. Its members probably were widespread over the semitropical lowlands of the continental margins of the earth, but the best known representatives come from North America. What is known undoubtedly constitutes a very small part of the total.

Five families of reptiles (Figures 163 and 164) include the principal lines of

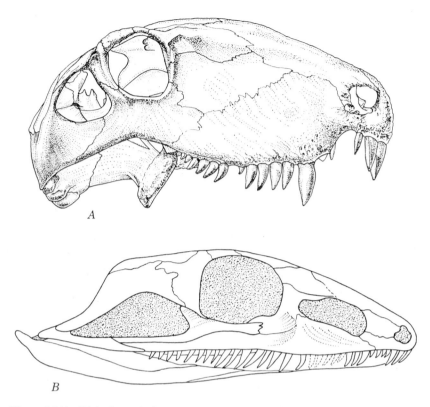

A

B

Figure 164.   Right-lateral view of the skull of *Dimetrodon*, *A*, central member of the family Sphenacodontidae; original length about 32 cm. *B*, *Varanodon*, one of the varanopsids, original length about 22 cm (*A* after Romer and Price, 1940; *B* after Olson, 1965c.)

development, these forming three suborders: Ophiacodontia, including very primitive pelycosaurs and Ophiacodontidae; Sphenacodontia, including Sphenacodontidae and Varanopsidae; and Edaphosauria, including Edaphosauridae and Caseidae. Caseids were herbivorous, as were some of the edaphosaurids; ophiacodontids and varanopsids were primarily piscivorous; and sphenacodontids were primarily carnivorous-piscivorous.

Radiations were characterized by increase in size and various departures from the presumably insectivorous diets of level I. None of the lines, once it had left the primitive insectivorous mode of life, became adapted to it secondarily. The greatest changes, of course, were in the herbivorous lines, whereas the carnivores, preserving many aspects of the primitive diet, showed less pervasive structural modifications. The herbivorous caseids were the last to radiate and ecologically were associated primarily with faunas composed of members of the third level of radiation.

With increase in size, even among the carnivores, there arose modifications related both to locomotion and to feeding. A striking, and probably very significant, aspect of these modifications was that even with the increase in size the animals remained highly active. This is in direct contrast to the herbivores, in particular the caseids.

As far as the record shows, there were few large terrestrial plant-eating reptiles or amphibians associated with the pelycosaurian carnivores. The probable fish-feeding habits of the varanopsids and ophiacodonts presumably required them to be quick, agile creatures. Sphenacodonts show less clear indications of aquatic adaptations, but their food appears to have included small, active, terrestrial animals, primarily captorhinomorphs and aquatic animals including fishes and amphibians. The latter, from the composition of the faunas and the evidence of coprolites, probably formed a major part of the diet (Olson, 1966b). Their activity as well was probably related to dietary necessities. Thus the basis of many mammalian features, which seem to be intimately tied to activity, may be found in the feeding habits and ecological adaptations developed during early radiation, not far removed from the source of reptiles.

The fundamental morphological structures, if modifications related to size and special functions are set aside, are in general similar to those in the ancestral stocks of level I. The adductor temporal muscles lay completely within the temporal chamber and inserted largely in and around the adductor fossa of the lower jaw and not on the outer surface (Barghusen, 1968). The pterygoids were primitive and lizardlike. There was no significant development of the coronoid process, and the elements of the lower jaw retained fairly primitive relationships in size and disposition.

The braincase had changed but little, and there is no indication that any significant enlargement of the brain had occurred. The internal auditory meatus remained widely open, and the subarcuate fossa (= floccular fossa) was very shallow. The anterior and posterior vertical semicircular canals remained of about the same proportions as in primitive reptiles, indicating that the head was still carried with the basal axis more or less parallel to the ground, rather than at an angle, with the anterior end deflected, as in members of later radiations.

The quadrate and articular bones formed the sole articulation between the skull and lower jaws. Within the various lines of the radiation the articular surface became modified. It lies well below the level of the tooth line in sphenacodonts and edaphosaurs, although not in ophiacodonts. In the first two as well, the glenoid fossa is divided into two distinct concave areas, with a strong ridge between. The face of the articular surface of the articular bone tends to be turned somewhat medially, meeting the oppositely deflected quadrate articular surface. In edaphosaurids the articular surfaces are somewhat elongated, suggesting the existence of some anteroposterior motion. These modifications appear to have been adaptations to the special feeding habits of the different groups, each being a somewhat special departure from the basic pattern found among primitive ophiacodonts.

Other skull features underwent rather modest modifications, with specializations related mostly to feeding and superimposed on the primitive pattern. In all the basipterygoid-palatal articulation remained movable, the epipterygoid slender, the basic pattern of dermal elements little modified, and the internal nares well forward.

Limbs retained their primitive lateral projection and foot motions were induced largely by a rotation of the distal ends of the upper limb bones. The phalangeal formula remained primitive, except in caseids, in which considerable modification of the locomotor system occurred. Within each line, however, limb bones have distinctive features. The sphenacodonts have some features suggesting that they were forerunners of primitive therapsids, but they did not depart far from the basically primitive form and posture.

SPECIAL FEATURES. Each of the lines of pelycosaurs has many special features related to its own adaptive courses. Most of these, although of interest in studies of the nature of the radiation, are not directly pertinent to later radiations. Some, however, give possible insights into ways of life or physiological processes that may be pertinent to the origin of mammals.

Neural Spines. Very long neural spines developed in several lines of pelycosaurs. They are best known in *Dimetrodon* and *Edaphosaurus*. The spines of *Sphenacodon* are long, but not excessively so. Broad flat elongated spines occur in *Ctenospondylus*, and *Lupeosaurus*, though an edaphosaur, has spines very like those of some species of *Dimetrodon*. These appear to have developed independently in different lines. It is thought that the spines carried a web of skin between them, both because they appear to have been highly vascular and because it is not uncommon for spines, broken in one way or another, to have retained their position and healed during life.

The long spines have been explained in many ways, ranging from reasonable to ridiculous. Among these is the suggestion that they served to regulate body temperature by acting as a heat receptor and radiator. If this is accepted, as it more or less has been by various students, it might be viewed as an early tendency toward temperature control. This, in turn, might then be considered an expression of a feature that is important in mammalian development, although perhaps carried out in a very different way.

Elongated spines are found elsewhere among the reptiles—for example, in the dinosaur *Spinosaurus* and in such lizards as *Basiliscus* and *Lophura*, in which a web of skin is supported by the neural arches. Also in the amphibian *Platyhystrix* very long, knobby neural spines are developed at least on some of the vertebrae. The long neural spines appear to have had some compelling adaptive advantage, and this seems to have been most frequently expressed in the pelycosaurs. It is, of course, not established that long neural spines all served the same ends. Evidence that they have relationship to temperature control is emerging from recent studies of *Basiliscus*. Another suggestion has been that both the modern and ancient developments may relate to sexual activities, in particular to courtship or territorial practices. In view of the occurrence of the long spines over a wide taxonomic range, it is at least questionable to seek in any of them an indication of the initiation of temperature control of the type found in mammals.

Parietal Openings. A parietal foramen was inherited by all members of this radiation and persists throughout. Only in very advanced members of the therapsid stocks did it finally disappear. Its persistence suggests a functional significance, and, as discussed on pages 501 and 623, one primary function of the organ associated with the foramen is photoreception. Various consequent activities may have developed, including control of the amount of radiation received, possible control of reproductive cycles, and avoidance of desiccation. One or all of these could have been significant in the initial stages of development of primitive reptiles and the synapsid stocks. The evolutionary patterns of the distribution of the parietal foramen are interesting from this point of view.

Some of the small pelycosaurs, most conveniently associated with level I (e.g., *Oedaleops*) have quite large foramina and presumably very-well-developed pineal systems. The foramen is moderately large in many ophiacodontids, but relatively small in the large *Ophiacodon* and in *Sphenacodon*. The small skull of *Edaphosaurus* carries a fairly large parietal opening, but it is only in caseids that it is truly immense. For the most part the size of the parietal opening appears to have been much the same, regardless of the size of the skull, creating an impression that it was smaller in large skulls than in small ones. All of the pelycosaurs, regardless of their environments, must have had a highly functional pineal system. If, as seems most likely, it acted as a photoreceptor, it presumably related to some general activity, common to all. What this may have been is pretty much an open question.

The very large opening in caseids seems to represent an actual increase in the size of the associated organ, for the absolute size of the foramen is considerably greater than that in other lines. A similarly large foramen is present in *Diadectes*, a large, heavy creature, and in pareiasaurs, whose structure somewhat resembled that of the caseids. In addition, a large parietal foramen occurs in procolophons. Some members of the third level of radiation, taken up in the next section, also had very large foramina—particularly dinocephalians, which also were large, ponderous creatures. This extra-large structure seems to be present primarily in herbivores and for the most part in large, slow-moving creatures. Some of the procolophons are the principal exceptions as far as size is concerned. It seems to make a certain amount of sense that pineal systems in such creatures may have been related to temperature control, in view of the usual large size, but again, this is extremely speculative and cannot be a basis for any interpretation of their role in, or tangential to, the lines of evolution leading toward more advanced therapsids and the mammals.

Dentitions. Modifications of the dentitions from the more or less isodont ancestral conditions took several directions, each of which appears to have been strictly adaptive. In various captorhinomorphs and small pelycosaurs enlargement of one or more anterior maxillary teeth produced a somewhat caninelike pattern. This became emphasized in the sphenacodont line and was accompanied by the development of a step in the upper jaw at about the canine level. Other carnivores developed modestly differentiated dentitions, consisting of "incisors," one or several "canines," and a series of postcanines. Strong palatal teeth usually were present. Dentitions appear to have been adapted to seizing, holding, positioning, and swallowing prey, but not to any extent to mastication.

Herbivorous adaptations of two very different sorts are present (see Figure 47). One is found in *Edaphosaurus*, which had crushing palatal plates, and the other in the caseids, in which a cutting, tuberculated crown developed on the cheek teeth, as the anterior teeth assumed a somewhat rakelike aspect, slightly reminiscent of that in diadectids. *Edaphosaurus*-like adaptations occur in some of the larger captorhinids but are not present elsewhere among the members of the synapsid radiations. The general caseid type of postcanine tooth occurs widely throughout the reptiles, in other synapsids—for example, *Estemmenosuchus*–procolophons and pareiasaurs, some dinosaurs, and some lizards.

It was from the sphenacodont type of dental pattern that the lines that eventually led to mammals stemmed.

Extremities. Of the various pelycosaurian lines, only the caseids underwent notable modifications of the extremities. In this line the phalangeal formula was reduced to 22332 and possibly even more, in both the fore and hind feet. Very strong clawlike unguals were present, apparently in adaptation to digging. The bones of the short, massive feet have very heavy rugosities and prominences to which tendons were attached, and the modifications appear to be strictly adaptive. In the varanopsid line the phalangeal formula remains primitive, but there is considerable elongation of some of the toes, probably related to the aquatic habits of these creatures.

Various later therapsid stocks attained phalangeal formulas of 23333, and it is tempting to look at the caseid pattern as an early expression of some "tendency" to go in this "mammalian" direction. This sort of reduction, however, occurs in many lines, often in association with heavy limbs (e.g., in pareiasaurs), and its development in the caseids and several later therapsid lines cannot legitimately be considered as anything more than similar ends attained in the course of adaptive responses to both the internal and external circumstances of the evolving animals.

Other structural changes of the limbs among the pelycosaurs were moderate. In none are the limbs modified in such a way that they could have been carried completely under the body. The differences in limbs and girdles resulted from modifications of the primitive ancestral ground plan related to increases in size and the locomotor demands of the various environments occupied as the adaptive radiations took place.

Level III

RADIATION AND THE GENERAL MORPHOLOGY.   In many respects level III is the least known and most difficult of the levels of radiation to assess. Its most primitive members lie on the border line between the pelycosaurs and therapsids, but all others are generally considered full-fledged therapsids. Clearly there was a primitive therapsid radiation that included animals often grouped as dinocephalians. Near the base are the carnivorous eotitanosuchians and phthinosuchids, which are very pelycosaurlike but seem to have led to the dinocephalians on one hand and the gorgonopsians on the other.

The phylogenetic problems arise in part from the accidents of geographic distribution and discovery, and from the rather sudden appearance of the new types in the Old World, as discussed in Section III (pages 320–325). The simplest phylogeny on the basis of what is known assumes that the sphenacodonts led to the eotitanosuchians and phthinosuchids, and that all other therapsids stemmed from this base. None of the known sphenacodonts is an entirely suitable ancestor, and members of the primitive stocks, presumed to have an eotitanosuchian base, first appear along with their supposed ancestors. The best known forms are early Permian in age, but distinctly sphenacodont reptiles are known from the later Pennsylvanian as well.

The key feature that supports the concept of a sphenacodont ancestry for the therapsids as a whole is the inflected angular process of the lower jaw (Figures 59 and 74). All well-known therapsids have it, and it occurs in the pelycosaurs only among the sphenacodonts. Its function has been the subject of considerable debate, and this has added to the difficulties of interpretation.

If a single origin from the sphenacodonts is accepted, the radiation here termed level III represents a distinct and recognizable phase in the pelycosaur–therapsid evolutionary sequence. From near its base the radiation of level IV also originated, producing the gorgonopsians, therocephalians, bauriamorphs, and dicynodonts. This assessment of the radiations poses two principal problems. The first is that both levels III and IV stemmed, nearly simultaneously it would appear, from the same base, eotitanosuchians, but level IV had a very successful diversification, whereas that of level III was much more restricted. Second, the position of the venjukoviids as members of one or the other radiation is a problem. This is partly because their origin is obscure, but more importantly it depends on whether a basic adaptive shift, which led to the dicynodont radiation to which they were antecedent, had taken place. Whether or not the pair of curved articulating surfaces of the jaws, which permitted free anterior-posterior movement in the dicynodonts (Crompton and Hotton, 1967), had developed cannot be definitely ascertained from known venjukoviids. *Otsheria* has no lower jaw and the quadrate region is not well preserved. In *Venjukovia* the articular surface of the lower jaw is suggestive, but the posterior part of the skull is undescribed, although present in a recently discovered skull. The skull shapes of these two genera suggest that appropriate musculature could have been present. For this reason the venjukoviids and dicynodonts are grouped together as a herbivorous group in the fourth radiation level, along with the carnivorous gorgonopsians, therocephalians, and bauriamorphs.

The stocks of level III were highly diverse, as shown in Figures 156, 165 and in the phylogenetic chart (Figure 158). *Eotitanosuchus* was adaptively close to the sphenacodonts, as were titanophoneans, but both the large carnivorous and herbivorous dinocephalians had departed far from this base. Adaptively this level represents a major expansion over anything known earlier in the synapsid radiation, but most changes involved variations on the basic morphological patterns developed within the level II. No particular adaptive shift can be pinpointed, and from this standpoint level III merely represents a successful exploitation of the potentials of level II. This marks the greatest contrast with the carnivores and herbivores of level IV, because in these there were basic modifications both in feeding and locomotion.

SPECIAL FEATURES. *Feeding Mechanisms.* Skull and tooth morphology is quite varied within this level of radiation, and presumably habits varied accordingly. Although it seems probable that omnivores, carnivores, and herbivores all developed, it is difficult to assess the diet of many of the members of the radiation because there are few reasonable living analogs. It does not appear that there were basic changes in either the jaw musculature or in the general ways in which food was treated. In none of the different types is there any evidence of oral preparation of food prior to swallowing. Palatal teeth remained an important part of the dental structures, related, it appears, to positioning and swallowing of large pieces of food.

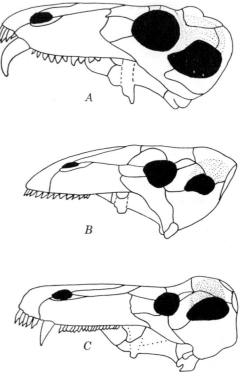

Figure 165.   Three dinocephalians representing three distinct families. *A, Syodon,* an anteosaurid (titanophoneid or brithopodid by some authors); *B, Moschosaurus,* a titanocephalid; *C, Jonkeria,* a titanosuchid. Not to scale, all moderately large. (After Boonstra, 1963*b*.)

The temporal and pterygoid musculature seems to have retained a fundamentally reptilian composition. The temporal adductors, however, became greatly enlarged over those of their ancestors. Even in *Eotitanosuchus* the temporal fossa was somewhat extended in a dorsoposterior direction (Figure 156A), and among the titanophoneans (Figure 157A) the origin of adductor musculature had crept over the postorbital bar, passing out of the temporal fossa onto the outer surface of the skull. Primitively the temporal fossa faced laterally, but in various dinocephalians it opened dorsally as well. Much stronger, more effective adduction was attained, in conjunction with development of strong jaws and jaw articulations.

Both the transverse process and quadrate ramus of the pterygoid remained large and strong, indicating continuation of typically reptilian pterygoid musculature. This muscle appears to have been powerful and to have played an important role in jaw movements. The anterior parts of the quadrate processes of the pterygoid tend to be closely appressed along the midline, forming a strong median keel from which part of the pterygoid musculature took origin.

The jaws had a simple vertical motion, with very small lateral or fore-and-aft components. Teeth were somewhat differentiated into incisors, canines, and postcanines. The crowns of the incisors and canines were mostly simple cones. Those of the postcanines varied from such cones to peglike, cuspidate teeth, reminiscent of those of some odobenids, and leaf-shaped, serrated teeth, not unlike those of pareiasaurs. The teeth were adapted to a variety of foods, probably ranging from both vertebrate and invertebrate animals to various kinds of plants.

Braincase and Parietal Opening.   In those animals in which it is known the braincase was much the same as in the sphenacodonts. Only in a few types is anything known of the internal structure. In these there is no evidence of outpouching of the floccular portions of the brain. Anterior and posterior vertical semicircular canals are similar in length and curvature as in sphenacodonts. The pituitary fossa is large, like that of the sphenacodonts and gorgonopsians. The wall of the braincase is formed posteriorly by the otic elements, the laterosphenoid appears to have been little ossified, and the sphenoid carrying the nasal tracts is at the sphenacodont level of organization. The epipterygoid is a slender rod and did not provide a site for origin of major musculature or form a secondary osseous wall lateral to the cavum epipterycum.

The parietal opening is always present and tends to be large. In primitive members of the radiation, as in sphenacodonts, it lies far back on the skull. In titanophoneans and some of their descendants the posterior position is exaggerated. Where the skull bones are thick, the canal leading from the braincase to the parietal opening is long, and the foramen may be set in a craterlike protuberance on the outer surface of the skull.

There can be no doubt that the organ housed by this canal and foramen was highly functional and that its exposure at the surface of the skull was critical to this function. The way is open, of course, to speculation about how its function, presumably closely allied to reception of radiation, might be related to temperature control in large terrestrial animals that have at least some relationships to stocks from which mammals arose. There are, however, few solid facts to which such speculations can be related.

Locomotor Features.   In attaining large size many of the members of this radiation departed significantly from the sphenacodont posture. The limbs were brought somewhat more under the body, and the knee and elbow became directed more anteriorly and posteriorly, respectively. Both limb bones and girdles showed modifications related to changes in posture, but these were not drastic (Figure 166). The rotary actions of the distal ends of the humerus and

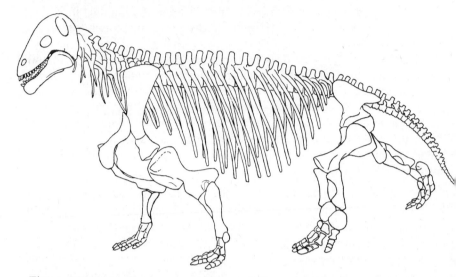

Figure 166. *Moschops*, a large dinocephalian, showing the proportions and structure typical of this level of development, level III. Based on a specimen from South Africa mounted in the American Museum of Natural History, New York. (After W. K. Gregory, 1926.)

femur were partially replaced by a fore-and-aft swing. The feet were applied to the ground more directly in the fore-and-aft action and were turned somewhat more forward. With this came the development of a reduced phalangeal formula, reaching the typical mammalian formula of 23333 in various large dinocephalians.

Many of the locomotor changes in various therapsids seem to have been related to increasing speed and agility. In this phase of radiation, however, the changes, much like those in the pareiasaurs, appear to be primarily in response to the problems of increase in size.

Level IV

GENERAL EVOLUTIONARY PATTERNS. The most extensive radiations recorded by the therapsids lay at this developmental level, which includes the carnivorous gorgonopsians, therocephalians, and bauriamorphs, and the herbivorous dicynodonts. From near the base of this radiation originated the cynodonts, which exploited an adaptive modification of the feeding mechanisms to deploy in a separate and in some respects more advanced radiation, here called level V. Levels III, IV, and V all may have stemmed from the common morphological base exemplified by the sphenacodonts. Level III produced diverse adaptive types but departed only moderately from the essential sphenacodont heritage. The other two, though but little removed in their initial stages, showed progressive departures from this heritage. The cynodonts seem to have attained features that eventually resulted in a thorough reorganization of most of the morphological systems.

The directions followed are clear from intermediate and end members of each of the radiating lines, but it is far from obvious what triggered the several divergences from the ancestral stocks. The ostensive initial differences between the radiations are in the feeding systems. Discussion of the special aspects of the system in the cynodonts will be deferred for the present. The primitive carnivores of level IV, the gorgonopsians and therocephalians, followed a course in which emphasis was placed on the incisor and canine dentitions, with the postcanine teeth becoming of reduced functional significance. This is in contrast to the primitive carnivores of level III, the titanophoneans, whose postcanine dentitions remained moderately prominent, with the teeth being stout and somewhat bluntly crowned (Figure 167). It is possible that these differences were related to minor ecological shifts, involving somewhat different foods and that this may have been sufficient to initiate the shift.

Such an explanation does not take into account the dicynodonts, unless it be assumed that their basic adaptations were subsequent to some such initial separation among carnivorous ancestors. The key to their origin seems to relate to the attainment of a capacity for fore-and-aft motion of the lower jaw relative to the upper. Thus, although all of the lines of level IV attained similar developmental levels, it is far from certain that they had a common heritage in an ancestor that originated by a key shift from the sphenacodont condition.

Within the several radiating lines trends that in retrospect appear to be "toward mammals" appear repeatedly. Among these, in the skull, are the development of a secondary palate, reduction or loss of the kinetic relationship of the basicranium and palate, development of a double occipital condyle, thinning of the bones of the basicranium, reduction in the relative size of the sella turcica, development of a deep floccular fossa, modification of the vertical semicircular canals, reduction and loss of pineal opening, loss of the skull platform with great enlargement and dorsal expansion of the temporal fossae, and development of a complex pattern of foramina and canals for blood vessels and nerves in the labial region. Lower jaws were modified by an increase in the proportionate size of the dentary bone, development of a strong coronoid process, and reduction of the articular elements. The postcranium shows comparable, often commensurate, changes, but these are for the most part less readily definable. Many of them were related to the change of posture in which the limbs came to lie more under the body and to act in a more nearly fore-and-aft plane.

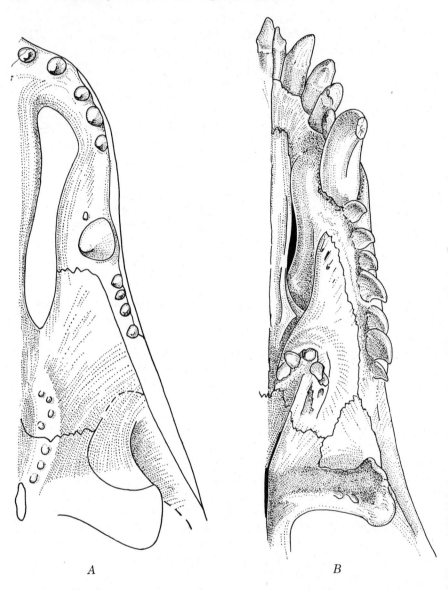

A   *B*

Figure 167. Comparisons of the marginal and palatal dentitions of *A, Arctops,* a gorgonopsian, and *B, Titanophoneus,* a dinocephalian, showing the differences that may have been to some extent responsible for the divergent evolutionary paths in the two groups and the differences observed between levels III and IV. (*A* after Sigogneau, unpublished; *B,* after Orlov, 1958*a*.)

Not all of these trends occur in each of the lines, and the degree of development and the associations are varied within particular lines. Some are found in at least incipient states in the first known members of each of the major groups and may be supposed to be a common heritage from some unknown ancestor. Others, however, were initiated within the radiating lines after their establishment. Both in the initiation and development of such similar features the different lines show striking cases of parallel development, which have occasioned considerable speculation with respect to their causes. Discussion of these will be deferred until the relevant evidence has been presented.

EXTENT OF THE RADIATIONS. Primitive gorgonopsians are very close to *Eotitanosuchus* and presumably stemmed from the group of which this genus is representative. *Phthinosuchus* has been classified as an eotitanosuchid by some students and placed in the Gorgonopsia by others. Although over 70 genera and 17 or 18 families have been named, most of the supposed genera represent relatively small variations on the common pattern. Sigogneau (1966) recognized only two families with about 20 genera. The examples shown in Figure 168 illustrate common patterns and the range of variation. In all the anterior elements of dentition are dominant, with postcanine teeth being small, and all seem to have been adapted to seizing and tearing prey and swallowing large pieces without additional oral preparation. Gorgonopsians range in Africa from the Tapinocephalus zone through the Cistecephalus zone and in the Soviet Union occur in zone IV and possibly in zone II as well.

Therocephalians and bauriamorphs form a more or less coherent array of carnivores, including quite primitive therapsids with a rather sphenacodont cast, the pristerognathids; advanced families of therocephalians (e.g., whaitsiids); and a number of more advanced families generally placed among the bauriamorphs (e.g., the scaloposaurids and silpholestids). This array includes not only a broad spectrum of modifications of basic structures but also a variety of dietary adaptations. The most primitive forms, many of which are quite large, appear to have fed, much like the gorgonopsians, on large prey, which were reduced to manageable pieces and swallowed with little oral preparation. Among the more advanced lines, scaloposaurids in particular, insectivorous diets are suggested by the size and patterns of the teeth. The bauriids (Figure 169), closely related forms, appear to have initiated oral food preparation, involving the marginal postcanine teeth, tongue, and palate.

The therocephalian–bauriamorph assemblage ranges from the Tapinocephalus into the Cynognathus zone in Africa and occurs in zone IV, and probably also in zone II in the Soviet Union.

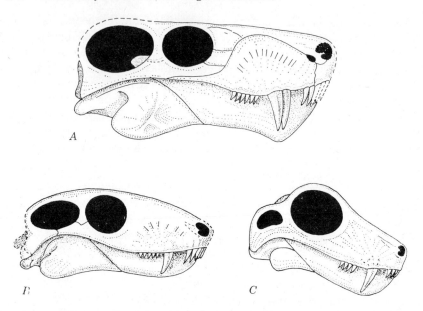

Figure 168.   Three gorgonopsians indicating the approximate extent of morphological variation in the form of the skulls and jaws. Dentitions show somewhat greater variation than that in the illustration, both with more and less cheek teeth than shown. There is also great variation in size. *A*, a more or less central type; *B*, a moderately primitive gorgonopsian; *C*, showing about the maximum of modification of skull shape. (After Broom, 1933.)

Dicynodonts, here very broadly conceived to include venjukoviids, occur throughout the later Permian and well into the middle Triassic. They were widespread and are known in all continents in the later parts of their record. Although they are highly varied in size and in special features, all conform closely to a central pattern. End members and aquatic forms, the lystrosaurs, departed farther than most from the usual dicynodont form, but even in these the basic pattern in evident. The feeding mechanisms,

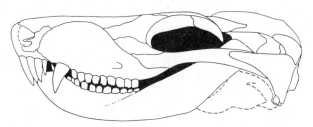

Figure 169.   The skull of *Bauria* (After Brink, 1963*a*.)

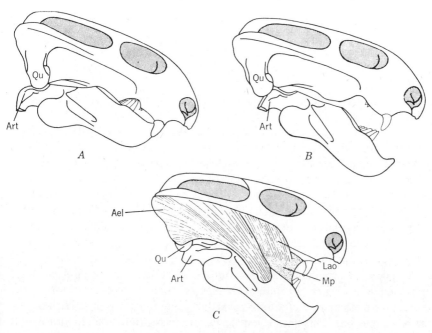

Figure 170.   Three illustrations of the dicynodont *Emydops*. *A*, with the lower jaw in completely retracted position, representing the beginning or end of the masticatory cycle; *B*, the position of full protraction and the beginning of elevation. Note the changes in the articulation between *A* and *B*. *C*, the superficial musculature as reconstructed, indicating its fiber directions and presumed role in elevation during retraction. Ael, musculus adductor externus lateralis; Art, articular bone; Lao, musculus levator angularis oris; Mp, Mundplatt; Qu, quadrate. (After Crompton and Hotton, 1967.)

with fore-and-aft motion of the jaws highly developed (see Figure 170), are highly distinctive, but along with these occur various "progressive" trends not unlike those found among the gorgonopsians and therocephalians.

SPECIAL FEATURES.   *Feeding Mechanisms.*   None of the members of any earlier radiations, with the probable exception of some edaphosaurids, undertook any extensive oral preparation of food. This holds for most members of the fourth level of radiation as well, with advanced members of the therocephalian–bauriamorph complex providing the exceptions. The basic reptilian jaw musculature—with strong external and posterior adductors, no masseteric component, and a well-developed reptilian pterygoid—seems to have persisted throughout. In the gorgonopsians the temporal fossa enlarged, extending posteriorly, and a prominent coronoid developed on the dentary bone. The dorsal, parietal platform, however, persisted. The effect of the changes was to strengthen the adductor action through the agency of a longer lever arm and more powerful musculature. In particular this provided for resistance to forces at the anterior end of the lower jaw, where tendencies for excessive depression must have developed in the course of feeding on large prey. The system is "improved" over that of the sphenacodont but is not altered in a fundamental way.

Both the therocephalians and bauriamorphs retained this basic pattern but carried it further, with a loss of the parietal table and the attainment of at least a superficially temporalislike muscle. The origin was probably largely from a temporal aponeurosis. In none, as far as can be told (see Barghusen, 1968) was a proper masseteric muscle developed, and no masseteric fossa occurs on the coronoid in either gorgonopsians or therocephalians–bauriamorphs. In the former, however, there is some evidence of a specialized condition in which a slip of the adductor musculature passed down onto the angular and reached the upper part of the surface of the inflected lamina.

The absence of masseteric musculature limited the possibilities of lateral jaw motions, and in most genera the deep transverse processes of the pterygoids posed a mechanical restriction on lateral motion when jaws were approaching occlusion. In spite of these restrictions, food preparation by the postcanine teeth and other mouth parts does seem to have been carried out by the bauriamorphs. This has extended to its limits the functional modifications of the "reptilian" type musculature of the jaws among therapsids. Only with the development of a masseter muscle, known only among cynodonts, was alteration to the "mammalian" pattern initiated.

Dicynodonts, as shown in Figure 170, followed a different direction of adaptation, keyed by the jaw motions made possible by the elongated, rounded articular surfaces of the jaws. In dicynodonts, however, the jaw served to break up food and position it for swallowing, but they did not accomplish additional preparation by chopping or grinding. The mechanism may have been quite turtlelike in action. The musculature, though different in form and action, is readily derived without major reorganization from the general reptilian condition exemplified by sphenacodonts and more primitive reptiles.

A strong tendency toward development of a secondary palate exists throughout this level. Osseous palates are found in bauriamorphs and dicynodonts, and are incipient but not developed in therocephalians or gorgonopsians. Among the ictidosuchids this feature is exaggerated, and in such forms as *Ericiolacerta* and *Bauria* (Figure 171C and D) an osseous palate including the maxilla and palatine is fully formed. In the former the vomer plays a role in the palatal surface.

Although neither gorgonopsians nor therocephalians have osseous secondary palates, the formation of the bones of the palates and the positions of the internal nares indicate development of a soft, secondary palate in many genera.

The secondary palate formed in dicynodonts, but in a very different way, apparently related to the great enlargement of the premaxilla and modifications of the maxilla relative to its cutting functions in the jaw. In a sense, palatal development in the theriodont and dicynodont lines may be thought of as parallel, because generally similar structures emerge. In both cases there are stong adaptive reasons for the development, but these were quite different in the herbivores and carnivores.

The emphasis on the incisors and canines and reduction of postcanines in gorgonopsians and therocephalians have been indicated as related to feeding on large prey. The edentulous or nearly edentulous condition of the dicynodonts clearly relates to the substitution of horny plates for teeth in perfection of procurement and reduction of the plant materials that formed the foodstuffs of this group. Canines are variously retained and in some form strong tusks.

Radical departures from these established patterns occur in the advanced therocephalians and bauriamorphs. In some of the very small forms the postcanine teeth are relatively large, number up to 12 or more, and have sharp, slightly recurved crowns. *Silphedocynodon* and *Silpholestes* have postcanine teeth that have a three-cusped pattern and appear to have had well-developed shear between uppers and lowers. These small reptiles probably fed on insects, the first clear instance of this among synapsids since the small pelycosaurs of level I.

The bauriids and some advanced thorocephalians had strongly developed postcanine teeth, in both the upper and lower jaws. These were rather blunt, and their crown surfaces were opposed during occlusion. Food preparation seems to have been primarily by chopping or crushing, and not by grinding, which would require jaw movements that seem impossible in view of the structures of the skulls, jaws, and inferred musculature.

Data on tooth replacement within this level are not abundant, but where known, as in *Ericiolacerta* (Crompton, 1962) give no indications of departure from the general reptilian pattern.

Braincase and Associated Structures. In each of the lines the posterior part of the braincase is formed by the periotic, including ossifications of the proötic and opisthotic, generally fused together. The anterior margin of these elements carries a fairly distinct proötic incisure that accommodated the exit of the mandibular and perhaps maxillary rami of the fifth cranial nerve. In most respects the braincase is similar to that of the sphenacodonts, but with two major differences. Most important is the occurrence of a deep floccular fossa around which the anterior vertical semicircular canal swings. The second involves the relative lengths and curvature of the vertical semicircular canals. Changes in these canals appear to relate to modifications of the angle at which the basicranial axis was oriented relative to the substrate.

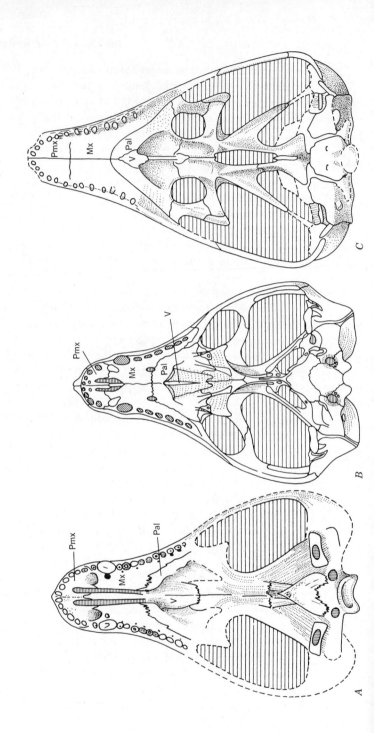

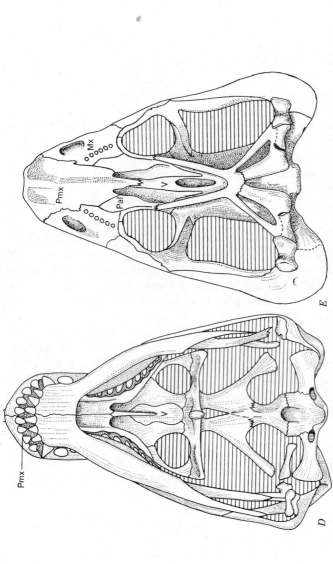

Figure 171. Palatal view of various therapsids to show the ways of development of the secondary palate. *A*, primitive cynodont, *Procynosuchus*, with secondary palate only incipient; *B*, intermediate cynodont, *Thrinaxodon*, with palate well formed, involving both maxillary and palatine elements; *C*, the bauriamorph, *Ericiolacerta*, with the vomer entering into the osseous secondary palate; *D*, *Bauria*, another bauriamorph, with secondary palate highly developed. *E*, the dicynodont, *Eurychororhinus*, showing the extensive involvement of the premaxilla and the unique method of formation of the secondary palate. Mx, maxilla; Pmx, premaxilla; Pal, palatine; V, vomer. (After Piveteau, 1961, with data for *A* from Broom, for *B* from Brink and Kitching, for *C* from Brink, for *D* from Parrington, and for *E* from Broili and Schroeder.)

The floccular fossa (Figure 49) is deep in all gorgonopsians and therocephalians in which it has been studied. Because it lies enclosed within the cranial cavity, this structure is not often revealed in prepared materials. Its condition in bauriamorphs and most advanced therocephalians is not known, but it can fairly safely be assumed to have been well developed. The fossa ranges from almost absent in primitive dicynodonts to moderately deep in many genera. The existence of the deep fossa suggests not only that the floccular region of the brain was well developed but also that the brain itself was enlarged sufficiently to fill the ventral portion of the braincase. Impressions of cranial structures on the base of the cavity (Olson, 1944, see Figure 41), give corroborative evidence. The deep fossa appears to have developed independently in the dicynodonts, assuming that its absence in a very primitive genus means that it was not present at the outset. It is present in the most primitive known forms of the other phyla at this level, but it is absent in sphenacodonts and dinocephalians in which this region is known. The condition of *Eotitanosuchus* has not been determined.

In primitive dicynodonts and gorgonopsians the anterior and posterior vertical semicircular canals are both strongly curved and more or less equal in length. The utricular sinus passes ventrally more or less normal to the horizontal plane of the basicranial axis. In therocephalians and more advanced dicynodonts the posterior canal is short and straight, the anterior canal long and curved, and the utricular sinus passes anteroventrally to make an acute angle with the horizontal plane of the basicranial axis. These structures suggest that in these two lines there was a tendency for the basicranial axis to slope anteroventrally reflecting change in the way that the head was carried.

The sella turcica in sphenacodonts was very large, lying in the deep basicranial complex. This is continued in the gorgonopsians with little change. The chamber is relatively large in primitive therocephalians, but the encompassing basicranial elements are less thick. The sella turcica is small in the dicynodonts. In both the gorgonopsians and therocephalians as in sphenacodonts, the bridge above the sella, equivalent to the dorsum sellae, is formed by the proötic rather than by the basisphenoid. No dorsum sellae is present in the dicynodonts.

A distinct basipterygoid process of the basisphenoid is present in gorgonopsians and primitive therocephalians, and an at least partially open movable joint with the pterygoid and epipterygoid is preserved. In more advanced therocephalians and in the bauriamorphs the motion is reduced or lost and the joint is replaced by sutural connection or by fusion in the tightly knit skulls of dicynodonts.

The epipterygoid is rodlike, as in the sphenacodonts, in the gorgonopsians, pristerognathid therocephalians, and most dicynodonts. Some of the more advanced therocephalians and the bauriamorphs have a somewhat expanded epipterygoid, alisphenoidlike in its general characteristics. The trigeminal exit, or foramen pseudoovale, is isolated between its posterior margin and the anterior margin of the periotic. The expansion of the epipterygoid appears to relate to an increase in the areas of origin of parts of the temporal adductor musculature.

Information on the anterior parts of the braincase is scant. All of the members of the radiation appear to have retained a generally reptilian sphenethmoid, solid in anomodonts but composed of two or more separate portions in some gorgonopsians and therocephalians. Ethmoid nasal scrolls may have been present in advanced members of the therocephalian–bauria-morph line.

Other Skull Features. Parietal openings are present in the gorgonopsians and range from moderately large to small. In some primitive forms the opening rests in a crater, much as in some of the dinocephalians. The parietal foramen also is persistent throughout the dicynodonts. It is large in *Otsheria* and modest to large in other genera. The opening is small among primitive therocephalians and absent in some of the advanced genera. The condition is varied among the bauriamorphs, but in most the parietal opening is absent.

The upper jaws of a wide variety of reptiles carry small foramina in the labial region, on the maxilla above the tooth margin. There is fair development, for example, in various lizards. Among the pelycosaurs such foramina are present in the caseids, but otherwise they appear to be absent. Where present, they transmitted the blood vessels and nerves that supplied the skin of the labial region. In mammals, in which the nerves and blood vessels are very well developed, with a well-defined infraorbital canal, they supply soft tissue of the labial and snout areas and the vibrissae.

The foramina and canals tend to be well developed in members of this level of radiation (Figure 172) and in some of the therocephalians and bauria-morphs are sufficiently mammal-like for the presence of vibrissae to be suggested by several writers (D. M. S. Watson, 1931; Brink, 1956; Tatarinov, 1967). The fairly deep impressions of the canals for the nerves and vessels in the bones in these animals suggest that soft, labial tissues were not developed. The existence of vibrissae is necessarily quite problematical, but it is almost surely true that the lips and snout were highly sensitive areas.

Postcranial Structures. The postcranium within this level is pelycosaur-like in the primitive representatives and changed markedly only in some of the end members. Dicynodonts, as herbivores commonly do, show marked departures from the ancestral condition.

The changes in the gorgonopsians were related largely to a rotation of the

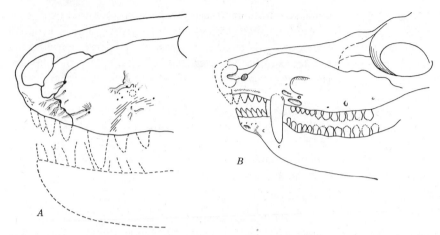

Figure 172.   Foramina, channels, and pits apparently related to blood vessels and nerves in the upper-lip and snout region of *A, Moschowhaitsia*, a therocephalian, and *B, Permocynodon*, a cynodont. Note the lesser expression of the channels on the cynodont skull. (After Tatarinov, 1967*a*.)

limbs that brought them under the body, to function in a more nearly fore-and-aft plane. The most marked change was in the shift of heads of the humerus and femur, which came to lie at an angle rather than normal to the main axes of the bones. Accompanying this change were modifications of the shape of the bones, including a general lengthening and lightening. A tendency toward reduction of digital elements existed but as reduction of the size of particular phalanges rather than their outright loss.

Primitive therocephalians had already reduced the phalangeal formula to 23333 but were otherwise at much the same level as the gorgonopsians. Only among very advanced genera do marked changes take place, and these are known from only a very few specimens. The bauriamorph *Ericiolacerta* had a very long and slender scapula, with the glenoid fossa turned sharply back and down. The feet appear to have been very mammalian, but these are better known in *Bauria* itself. The hindfoot had a well-developed astragalus and calcaneum that are mammalian except that the astragalus has over-ridden the calcaneum very little laterally.

In moderately advanced forms as well there is some evidence of differentiation between the thoracic and lumbar parts of the rib cage, as seen in the therocephalian *Aneugomphius*. The situation in the most advanced therocephalians and the bauriamorphs is not known.

Dicynodonts present a conflicting array of features, some of which are primitive, (e.g., the presacral column and ribs) and others that are advanced, as the 23333 phalangeal formula and modifications of the shoulder girdle.

The five or six sacral vertebrae and ribs are a specialization. As in the other lines, there was a strong tendency for the limbs to assume a fore-and-aft motion, with the knee pointing forward and the elbow backward. For the most part dicynodonts were rather slow-moving creatures, and this is emphasized in large genera. There was little lightening of the skeleton.

Level V

THE RADIATION. Level V includes the cynodonts and some presumably descended genera that approached quite closely to the mammalian grade, the tritylodonts and possibly some obscure genera known mainly from teeth. From this radiation arose many, possibly all, of the mammals. Origin of some mammals from the therocephalian–bauriamorph line has long been supported by many students (e.g., Olson, 1944, 1959; Crompton, 1958; and Brink in various publications). This hinges in part on the source of *Diarthrognathus*, which is essentially mammalian in many features, sometimes placed with the mammals, and has been thought to have a bauriamorph source. Broadening of the concept of cynodont to include all such animals that have an expanded zygomatic arch and presumably a mammal-like masseter muscle, on the assumption that this arose but once, brings the source of at least most of the mammals within the cynodont stock. On this basis, although the bauriamorphs showed many similar trends, the radiation of the cynodonts and their immediate descendants is distinct and at an organizational level not reached by other nonmammalian groups.

At its beginning this radiation was not far removed from the morphological stages found in primitive and intermediate therocephalians. The first representatives, procynosuchids of the Cistecephalus zone of South Africa and zone IV of the Soviet Union, resemble some of the therocephalians in many features.

Barghusen (1968) has indicated that procynosuchids, like all other cynodonts, have a bowed zygomatic arch. Accompanying this is the development of a space between the arch and jaw, and of a masseteric fossa on the coronoid process of the lower jaw (Figure 173) and presumably a true masseteric muscle. As a preliminary to this interpretation, Barghusen rejected, for what appear to be valid reasons, the concept that a superficial masseter passed from the zygomatic arch to the reflected process of the angular among the sphenacodonts and derived synapsids. This idea had been proposed and supported by Parrington (1959) and was an important aspect of the interpretations made by Crompton (1963*b*).

The change as interpreted by Barghusen appears to be a key innovation basic to cynodont radiation and probably central to the development of many skull features generally considered to be mammalian. With it more

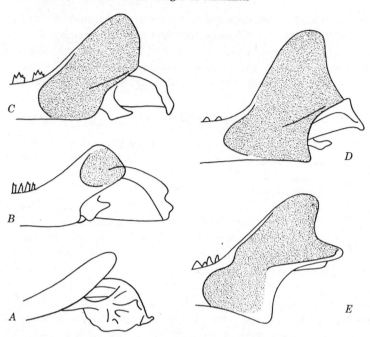

Figure 173.   Lateral aspect of the posterior portion of the left lower jaw of several therapsids showing the extent of the masseteric fossa (stippled). *A*, a scaloposaurid therocephalian in which no fossa is present; *B*, a procynosuchid primitive cynodont, in which a small fossa is present; *C*, *Thrinaxodon*, an intermediate cynodont with a well-formed, large fossa; *D*, *Trirachodon*, an advanced cynodont with a very large fossa and strongly reduced nondentary elements; *E*, *Diarthrognathus*, a very mammal-like therapsid with a large fossa and very small nondentary elements. (After Barghusen, 1968.)

complex motions of the jaw and new ways of food preparation in the mouth prior to swallowing could be developed.

Once the shift had occurred, a radiation that exploited a wide variety of feeding habits took place. Not only were many features of the skulls and jaws altered, affecting areas only remotely connected with mastication, but also the dentitions underwent many changes. Throughout there is a clear separation of incisors, canines, and postcanines. In some instances the last underwent additional differentiation. The primitive stock, procynosuchids, includes a number of genera, most of which appear to have been either small carnivores or insectivores. Moderately advanced was the line represented by *Thrinaxodon*, which retained much the same habits. The cynognathids were large animals, apparently for the most part carnivorous, although very possibly somewhat

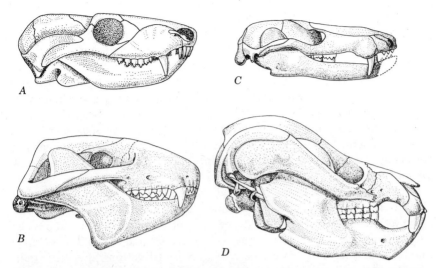

Figure 174.   Skulls of a cynodont and several animals very close to the reptilian-mammalian transition. *A, Thrinaxodon,* a cynodont which may lie fairly close to the line leading to mammals; original length about 5 cm; *B, Diarthrognathus,* an ictidosaurian "reptile," original length about 3.8 cm; *C, Sinoconodon,* a very primitive "mammal," original length about 4 cm; *D, Bienotherium,* a tritylodont "reptile," original length about 12 cm. (*A* after Broom, 1933; *B* after Crompton, 1958; *C* after B. Patterson and Olson, 1961; *D* after Hopson, 1964*a*.)

omnivorous. Other lines, diademodonts and traversodonts, developed complex postcanine dentitions, adapted to crushing, and some of these quite certainly had a herbivorous diet. This mode was highly developed in the tritylodonts. Outside of the cynodonts and their descendants, the only approach to dental modifications related to oral food preparation was in the bauriids.

The cynodonts portray a coherent radiation and the tritylodonts, although late in time, fit into it well (Figure 174). There is, however, a problem of the most constructive way of viewing the animals presumably derived from this radiation, but not directly traceable to it. These are the "mammals" and "near-mammals": *Diarthrognathus, Sinoconodon,* morganucodonts (eozostrodonts), docodonts, triconodonts, symmetrodonts, and pantotheres. *Diarthrognathus* is often considered a therapsid, but Crompton (1963*b*), who has studied it in detail, has agreed with some others that it might just as well be placed with the mammals. The tritylodonts, here placed in the earlier radiation, might also be considered with this "mammalian" group, and where convenient will be used in comparisons.

What the record shows is a series of mostly very small animals, known principally from teeth and jaws, and skulls or skull parts in a few genera.

The dentitions have a mammalian cast, and the jaws had a squamosal-dentary articulation, usually accompanied by the quadrate-articular joint. These are the controversial forms involved in the problem of the origin and constitution of the class Mammalia (see Section III, Chapter 10).

A sharp break in continuity occurs in the record, with intermediates between the well-known cynodonts (or bauriamorphs) and these very mammalian creatures poorly known. The late cynodonts, from the middle Triassic, are predominantly herbivores and their descendants in the later Triassic were the tritylodonts. The most likely antecedents of at least many of the small "mammals" of the later Triassic and Jurassic are therapsids of the *Thrinaxodon* type, of the late Permian and Early Triassic. One possible representative is known from the middle Triassic.

With this record it is at least convenient to consider the fifth level of radiation as encompassing only the cynodonts and their direct descendants the tritylodonts. The latter, like the caseids before them and the multituberculate mammals at a later time, were herbivores that thrived long after their radiation had otherwise waned to overlap, as an ecological herbivorous associate, the deployment of the following radiation.

SPECIAL FEATURES.   The cynodonts underwent many changes similar to those that have been noted among the bauriamorphs. There was an extensive adaptive evolution within the radiation, but the similar basic patterns were superimposed on the different adaptive modifications relating mostly to feeding and locomotion.

Feeding Mechanisms.   The dentary increased greatly in its size relative to the other elements of the lower jaw. In late members the articulation of the upper and lower jaws, characteristically performed by the quadrate and squamosal in reptiles, may have been partially taken over by the dentary and squamosal. Differentiation of a masseteric muscle accompanied enlargement of the dentary bone and the acquisition of a masseteric fossa on the coronoid process. More complex lateral movement, different from that of the bauriamorphs, thus became possible.

Many changes of the skull and jaw musculature accompanied this modification of the lower jaws. Temporal musculature was greatly enlarged, and an essentially mammalian condition appears to have been attained. No parietal platform was present, and the adductor muscles seem to have originated largely from a temporal aponeurosis. The increase in mass of musculature, modifications in the directions of force, modifications in skull proportions, and an increasingly higher coronoid process resulted in a more mobile, more efficient, and more powerful adductor system. Changes also took place in the pterygoid and depressor systems, but controversy persists concerning the history and development of these structures. Crompton (1963B) suggested the following homologies for the pterygoid complex:

Reptile	Therapsids and mammals
Capiti mandibularis profundus	External pterygoid
	Internal pterygoid
Anterior pterygoid	Tensor tympani

This area is one in which much remains to be done with functional interpretation. Studies of the sort made for the adductor musculature are needed. Irrespective of the details of change, it does appear that the late members of the fifth radiation level had attained pterygoid musculature very similar to that of mammals.

The development of a hard palate was only incipient in the primitive cynodonts (Figure 171). In these an extensive soft palate probably was developed, but only in the more advanced cynodonts did the premaxillae, maxillae, and palatines close to form a well-defined structure. This was very similar to the secondary palate of some of the bauriamorphs, although it differed in details, but it was very different from that of the dicynodonts. The palate, perhaps among other functions, provided a surface against which the tongue could act in positioning and holding food during its preparation. Palatal teeth, of course, were absent, having been last present in the gorgonopsians and therocephalians of the preceding radiation. The use of jaws, tongue, palate, and hyoid apparatus in the preparation and swallowing of food had become vastly modified from that of the predecessors of the cynodont radiation.

Strong incisors and well-developed canines were characteristic and did not change materially in the radiation. Postcanines, however, were modified not only in form but also to some extent in replacement. The basic postcanine tooth form, as seen in procynosuchids and retained in thrinaxodontids, was a narrow crown with three longitudinally placed cuspules (Figure 175A). Cynognathids developed postcanines with somewhat serrated crown margins, and diademodonts, traversodonts, diarthrognathids, and tritylodonts developed broad crowns, with various cusp patterns. Some of the patterns resemble those of some primitive mammals, but as yet close cross-ties between types have not been discovered.

The postcanines of most therian mammals consist of a premolar series, in which teeth are replaced once and a molar series, in which there is no replacement. Many studies have been made of the replacement patterns of cynodonts in efforts to trace the origin of the mammalian condition. The primitive forms and thrinaxodontids have normal reptilian replacement by successive waves of *Zahnreihen*. The postcanine dentition of *Diademodon* (Figure 176) consists of a few anterior conical teeth, a molariform series, and several posterior, more sectorial teeth. Changes in the numbers of these types

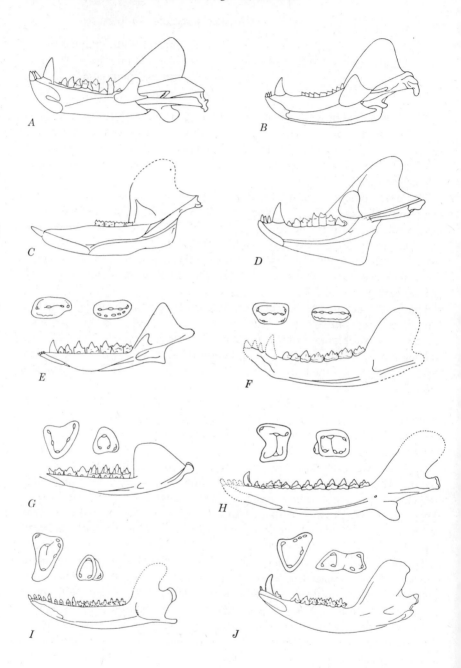

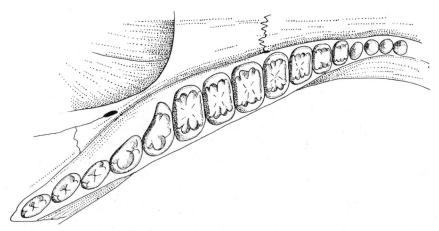

Figure 176.   The left-upper dentition of a gomphodontlike cynodont, *Diademodon*, in crown view. Note the differences between the three types of cheek teeth—anterior, middle, and posterior. (After Brink, 1956; see also Ziegler, 1969.)

take place during growth. The nature of replacement, however, has remained something of a puzzle. There appears to have been at least a single replacement of most, or perhaps all, postcanine teeth, and more than one may have taken place. Eruption of teeth was slow, and the molariform teeth have a somewhat sectorial cast when but partially erupted. To some extent at least initial sectorial teeth appear to have been replaced with age by molariform postcanines, and there is some evidence that some of the posterior teeth were not replaced. If the reptilian pattern of successive waves of replacement did hold, as has been suggested, the waves must have come in slowly and

Figure 175.   The right lower jaws of various advanced therapsids and primitive mammals seen from the inner side to show the degree of development of the non-dentary elements and the recesses on the inner surface of the dentary for their reception. For *E* through *J* a representative upper and lower molar tooth, to the left and right, respectively, are shown diagrammatically above the tooth row of the lower jaw. These teeth are in all cases from the right upper and lower jaws with the labial margin to the top of the page. *A*, *Thrinaxodon*, moderately advanced cynodont; *B*, *Diademodon*, an advanced cynodont with rather complex molar teeth; *C*, *Oligokyphus*, a tritylodont. For *B* and *C* note the diastema between the anterior and cheek teeth, related to the replacement series in which anterior teeth tend to be lost as posterior teeth are added; see also *C* and *D* in Figure 174. *D*, *Diarthrognathus*, an ictidosaur; *E*, *Morganucodon*, a very primitive mammal associated with doco-donts by some authors; *F*, *Priacodon*, a triconodont mammal; *G*, *Tinodon*, a sym-metrodont mammal; *H*, *Docodon*, a docodont mammal; *I*, *Laolestes*, a pantothere; *J*, *Didelphis*, a primitive marsupial. (*A* through *E* after Crompton, 1963*b*; *F* through *I* after Simpson, 1929.)

with several teeth intervening between replacing members of the new series.

In some gomphodonts and in the tritylodonts no successional replacement occurred. Teeth were lost at the anterior end of the series, and new teeth were added to the posterior end. This led to development of a diastema between the canines and postcanines (Figures 174C, D and 175). This pattern is the same as that found in the primitive "mammal" *Sinoconodon*.

No known member of this radiation had developed the type of tooth replacement found in mammals. Either they had a typical reptilian replacement pattern, sometimes in quite modified form, or else the pattern of eruption at the posterior end of the series and loss at the anterior end had become the mode. These adjustments occurred in the forms with dentitions suitable for crushing and grinding food, and seem to have been a response to the need for a continuously acting grinding series.

Auditory Region.   The stapes in most therapsids is a robust element that passed from the fenestra ovalis to the quadrate and showed little or no evidence of a tympanic process. This condition has led to speculation concerning how primitive therapsids heard, and even whether they heard at all. The matter has not been satisfactorily resolved. There is little question that the stapes gave rise to the mammalian stirrup and that the malleus and incus came from the articular and quadrate, respectively. These changes did not, however, occur within level V, because even in its most advanced members, as well as some members of the next radiation, the articular and quadrate continued to function in the jaw articulations. The stapes remained the only auditory ossicle, becoming proportionately smaller in the course of evolution within this level. With the modifications of the skull, the middle ear came to lie beneath the surface, and sound reception required a moderately long tube from the exterior to the tympanic membrane. The bony impression of such a conduit is present in various cynodonts and leaves no question of the existence of an effective sound-conducting mechanism at this level. The nature of the tympanic membrane remains obscure and the sources of the detailed structures of the membrane and associated soft parts must come from embryology rather than from fossils.

Hearing that utilizes three free auditory ossicles is a mammalian trait, not one that was developed before the origin of the mammals. It was attained only after considerable evolution at the "mammalian" grade had taken place. None of the members of either the fourth or fifth radiations approached this condition at all closely.

The Snout and Nasal Capsule.   The facial regions of some cynodonts show evidence of grooves leading forward from the foramina in the orbital region and presumably acting as conduits for nerves and blood vessels serving the snout. These, however, are much less strongly developed than they are in some of the bauriamorphs and therocephalians. In the latter they were interpreted as indicating a highly sensitive area, perhaps vibrissae, on the snouts. The lack of prominence in the cynodonts can be taken as evidence that these features were less developed, but Tatarinov (1967) has suggested the contrary—namely, that they were highly developed and that the lack of channels supports the interpretation that labial tissues were fleshy, rather than thin and inflexible, as suggested for therocephalians and bauriamorphs.

The nasal capsules were highly modified, and evidence of turbinals has been found in the anterior part of the nasal passage. On the facial bone of some cynodonts, in addition to conduits for the nerves and blood vessels, is a series of deeply impressed channels that pass forward from shallow pockets to the region of the external nares. These have been interpreted as ducts leading from moisture-producing glands, perhaps some sort of sweat gland, which maintained moisture on the anterior region of the snout.

The existence of these various soft structures, inferred from the rather slender evidence of the surface of the bone, is necessarily speculative. The interpretations as made suggest distinctly mammalian structures in this general region.

The Braincase.   The brain, although undoubtedly relatively enlarged over that of the primitive reptiles, still did not fill the braincase completely (see Figure 41). The base was somewhat impressed on the ventral osseous surface, and the lateral portions were evident posteriorly in the well-formed floccular fossae. The ear opened widely into the cranial cavity in a large, unossified internal auditory meatus. The anterior vertical semicircular canal was long and curved and the posterior was short and straight, suggesting that the head was carried with the snout deflected ventrally.

The anterior margin of the proötic made a sutural connection with the epipterygoid, which was expanded to form an alisphenoidlike element. A trigeminal foramen, or foramen pseudo-ovale, carrying the mandibular and probably the maxillary rami in the fifth nerve, lay between the two bones. In advanced forms (tritylodonts) the proötic sent processes lateral to the cavum epiptery cum (Figure 177).

There have been two rather different interpretations of this region, and to some extent these affect interpretations of the members of the next level of radiation as well. In cynodonts the cavum epiptery cum lay between the alisphenoid and the persistent pila antotica. In tritylodonts the anterior lamina of the periotic, perhaps an intramembranous ossification, passed forward and the trigeminal foramen (foramen pseudo-ovale) is partially enclosed above and below. K. A. Kermack (1962, 1963) and Hopson (1964a) have given alternative interpretations (Figure 178A,B), placing the cavum epiptery cum differently relative to the periotic. In Hopson's interpretation the anterior lamina is lateral and at the level of the epipterygoid (alisphenoid),

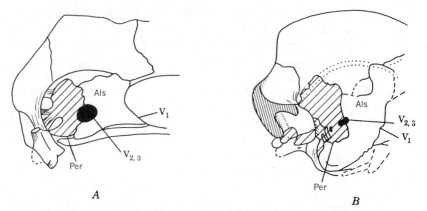

A

B

Figure 177.   Lateral view of the orbitotemporal region of *A*, *Thrinaxodon* and *B*, *Bienotherium*, a cynodont and tritylodont, respectively, showing an interpretation of the positions of exits of the trigeminal nerve. Als, alisphenoid (= epipterygoid); Per, periotic; V_1, exit of profundus branch of the trigeminal nerve; $V_{2,3}$, exit of maxillary and mandibular branches of trigeminal nerve. (After Hopson, 1964.)

with the cavum epipterycum between the pila antotica and the periotic–alisphenoid complex. In K. A. Kermack's interpretation the cavum lies between the periotic and the alisphenoid. The sections of *Bienotherium* described by Hopson appear to confirm his position, at least as far as this genus is concerned. Tatarinov (1968*a*) has made a thorough study of this region in *Dvinia* (Figure 178*C*), revealing a condition that resembles that described by K. A. Kermack.

In cynodonts and tritylodonts the parietal sent flanges ventrally to form part of the dorsolateral wall of the posterior braincase. The floor of the braincase in cynodonts was thin and the sella turcica was shallow, with the dorsum sellae little developed. The tritylodonts *Bienotherium* and *Oligokyphus* have deep basicranial bones and a large sella turcica. A parietal foramen was present in many members of this radiation, usually a small opening located rather well forward. It was very small in advanced members and absent in some.

Postcranium.   Postcranial changes were in general less definitive than those of the skull and jaws, and followed for the most part a course from a relatively primitive to advanced state, much as in the therocephalians and bauriamorphs. The shoulder and pelvic girdles became increasingly adapted to serving as areas for muscles to limbs that were carried below the body and acted mostly in a fore-and-aft direction. Even in the most advanced genera, however, the principal reptilian bones of the shoulder girdle were retained, with coracoids and an interclavicle well developed and, in many at least, a cleithrum still

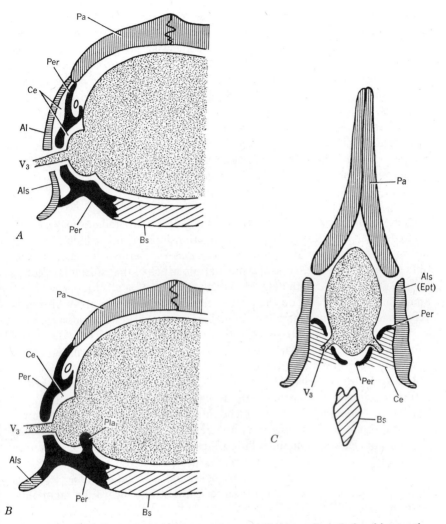

Figure 178. Interpretations of the cavum epipterycum in relationship to the alisphenoid (= epipterygoid), periotic, and its anterior lamina. *A* and *B*, Hopson's comparison of his interpretation based on *Bienotherium* with that of K. A. Kermack (1962, 1963). *C*, Tatarinov's (1968*a*) diagrammatic interpretation of the condition in *Dvinia*, a cynodont, modified to bring it into conformity of presentation with *A* and *B*. Als, alisphenoid (= epipterygoid); Bs, basisphenoid; Ce, cavum epipterycum; Ept, epipterygoid (= alisphenoid); Pa, parietal; Per, periotic; Pla, pila antotica; V_3, mandibular branch of the trigeminal nerve. (*A* and *B* after Hopson, 1964; *C* after Tatarinov, 1968*a*.)

present. As viewed from the right side, the pelvis showed considerable clockwise rotation from the primitive condition, the ilium lying somewhat anteriorly and the pubis and ischium somewhat posteriorly, in a rather mammal-like fashion. Limb elements modified toward a mammalian state, but no more so than in bauriamorphs. Epiphysial growth was not developed. The phalangeal formula remained primitive, with reduction in size of some elements, except in some very advanced members of the radiation.

Vertebrae displayed increasing differentiation along the column, and possibly of considerable significance was the differentiation of the dorsal and lumbar vertebrae and ribs, suggesting possible involvement of the anterior elements in breathing, accompanied by a muscular diaphragm (Brink, 1956).

Level VI

THE VALIDITY OF THE RADIATION. Unlike the two radiations just considered, level VI is not clear-cut. The relationships of the included lines are not at present fully understood, and the materials are such that it is very difficult to estimate the scope of the radiation. Much of the information comes from jaws and teeth, with data on skulls and parts of skulls for a few genera and some scattered bits on the postcranium. Included is an array of mostly very small animals of generally mammalian cast from the late Triassic and the later nontherian mammals—that is, those mammals that lack fully developed tribosphenic dentitions. Ancestries of some of the lines are known, but others have very uncertain backgrounds.

The array, so conceived, includes some of the creatures often grouped as therapsid ictidosaurs (diarthrognathids and haramyids, or microcleptids), the "Mesozoic mammals" morganucodonts (eozostrodonts), triconodonts, symmetrodonts, pantotheres, multituberculates, and the monotremes. The problems of classification and relationships of these groups were discussed in Section III, Chapter 10, and differences in opinions on their associations necessarily result in different terminologies and associations. The terms as used here are applied in a general sense, without fine regard for taxonomic relationships.

The matter of origins and relationships is critical in questions of whether this array comprises a single radiation or whether its various lines, in part or in whole, were merely end branches of the cynodont and perhaps bauriamorph radiations. If there was a single source, such as a limited part of the cynodont or bauriamorph complex, then clearly these animals do represent the type of radiation that arises after a threshold has been passed. If there were no such threshold and no limited source, the various lines could be considered as merely continuing the trends that were established in the cynodonts and possibly the bauriamorphs. Neither concept, of course, need be completely right or wrong, since some major part of the array may have had a single source, whereas some groups (e.g., diarthrognathids or multituberculates) may have come from another therapsid base.

The matter of monophyletic or polyphyletic origin of the mammals is, of course, wrapped up in this question, as well as in the problems of class boundaries. Currently opinion is swinging to a rather broadly conceived monophyly. Once this is accepted, a rationale for making the necessary grouping may become an objective of research and the framework strengthened by the very natural process of discounting earlier evidence that had seemed to point in other directions. However, as recently as 1965 K. A. Kermack, who has made detailed studies of the involved animals, concluded that the evidence of polyphyly in the origin and early evolution of mammals was hardly questionable and that it implied that the initiation of major taxonomic categories more generally must, among other things, be polyphyletic, with several origins occurring at about the same time through parallel development of evolving lines. The differences of opinion at the present really relate not to whether or not the grade seen in level VI arose more than once, because quite clearly it did, but how divergent were the lines that crossed between levels V and VI, and how much parallelism is tenable in the development of many of the definitive mammalian characters.

No final answers can be given to these problems at the present because the critical evidence is not available. The temporal and morphological gaps between possible ancestors among therapsids and the earliest groups of level VI are large. In some respects the most likely candidates seem to lie among the thrinaxodontids, which are best known from the late Permian Cistecephalus zone of Africa, with a few genera from the early Triassic, and one possible incompletely known representative from the middle Triassic (Hopson, 1967). The middle Triassic cynodonts, which have become well known in recent years from studies in South and East Africa and in South America, are predominantly large, herbivorous creatures of the gomphodont type. Although these were probably ancestral to tritylodonts and conceivably to multituberculates (although this is stretching the evidence very thin), they cannot be considered to be closely related to the majority of lines of level VI.

It seems probable that connecting links were small creatures as yet essentially unknown. The situation thus is somewhat like that between members of the second and third phases, where the actual ancestry is not known and interconnecting lines include hypothetical animals that hopefully will sometime be discovered.

GENERAL EVENTS. Because the information about level VI comes largely from teeth and jaws, its interpretation must be cast in a mold quite different from that which has applied earlier. Recently, however, some information on skull morphology has come from new finds and from detailed studies of older materials, and, concurrently, more attention has been paid to teeth, jaws, and the masticatory patterns and functions among the advanced therapsids. This has closed the gap somewhat, but at best, even now, the extent of commensurate information for members of the two levels is derived from a relatively few individuals and gives a very spotty taxonomic sampling.

Ictidosaurs. Diarthrognathids, haramyids, and possibly some very poorly known genera—such as *Dromatherium, Microconodon, Tricuspes,* and *Trithelodon*—of the upper Triassic represent an uncertain array that may be considered either as end members of the reptilian radiation exemplified by cynodonts (and possibly bauriamorphs) or "mammals," comparable to *Morganucodon* and similar genera in which a squamosal-dentary jaw articulation had been established. The term "ictidosaur" has been used in many senses and perhaps might best be dropped. As used here it excludes the tritylodonts, although they have been included, even as the major constituent.

Of the ictidosaurs mentioned above, only *Diarthrognathus* is known from more than teeth and, in a few cases, jaws. This genus has been described in detail by Crompton (1958 and 1963b). The squamosal and dentary made contact in the jaw articulation. Cheek teeth are sharply cusped but have broad, transversely widened crowns. The angle of the jaw is deep and far forward from the articular condyle (Figure 174B). The skull, although different in many particulars, is near the level of development of the tritylodonts (Figure 174D).

Little can be said about the other ictidosaurs except that their teeth suggest that they belong at about this level. None, including *Diarthrognathus*, can be clearly related to any of the Mesozoic mammals.

Morganucodontids (Eozostrodontids). A problem of terminology arises with reference to these small very later Triassic and Jurassic forms. A tooth named *Eozostrodon parvus* Parrington was described in 1941, earlier than *Morganucodon.* It has been suggested that the two are synonyms, and hence the name *Eozostrodon* is proper. Parrington (1967b) follows this practice, but D. Kermack, K. Kermack, and Mussett (1968) reject it, on the basis that the affinities of *Eozostrodon* are indeterminate. Because *Morganucodon* is a much better known term and because of the uncertainty of synonymy, it is being used here as the basis for the family encompassing *Morganucodon* and similar forms.

These very small creatures are known from Wales, South Africa (Crompton, 1964a), and Yunnan, China (Rigney, 1963). The Chinese and Welsh specimens are referred to *Morganucodon,* but the African one is named *Erythrotherium* Crompton. In all cases teeth, jaws, and partial to nearly complete skulls are known. The cheek teeth are basically three cusped, with cusps in longitudinal lines and accessory cingular cusps. The lower jaw has accessory elements, and both the quadrate-articular and dentary-squamosal contacts existed in the articular region (Figure 175*E*).

The skull has been studied in some detail, especially the periotic, and along with the teeth has been a basis for several suggestions about the relationships of *Morganucodon*. The most common association, based on teeth and jaws, is with the Jurassic docodonts, but the group has also been assigned to triconodonts, called ancestral to triconodonts proper, considered ancestral to multituberculates, and closely allied to monotremes. At present its position is a matter of considerable controversy.

Docodonts.   Several upper Jurassic genera pertain to this group, often given ordinal status with the morganucodonts included. Teeth and jaws are known. The upper cheek teeth are somewhat square and have a prominent lingual and labial cusp, plus accessories. Lowers are broad crowned and have a complex cusp pattern. A prominent sulcus is present on the inner surface of the lower jaw, indicating the presence of accessory elements, and the probable existence of a Meckelian cartilage suggests the presence of an articular in the jaw articulation. A process similar to the angular process of therian mammals is present, but it has been argued (B. Patterson, 1956) that it is a pseudo-angular process not homologous to the true angle of therian mammals (Figure 175*H*).

Triconodonts.   Two rather distinct groups—Amphilestidae of the middle and late Jurassic and Triconodontidae of the late Jurassic and early Cretaceous—make up the orthodox order Triconodonta. In addition, a late Triassic form, *Sinoconodon*, has been included in the family Triconodontidae and, as noted, morganucodonts have been called triconodonts. The molar teeth form a sharp, continuous shearing battery, with adjacent teeth interlocked. They are basically three cusped, with very small accessory cusps. The lower jaws lack an angular process, and there is a splenial groove on the inner surface, suggesting the presence of accessory elements. The groove is absent or vestigial in the Cretaceous genus.

In many respects members of this group are remote from all others. *Sinoconodon* (see page 379 and Figure 174*C*) has been the subject of controversy. Its dentition is of the triconodont type, but of course relationships are hard to determine. It is known from skull parts and lower jaws. The cheek teeth were not replaced but erupted successively at the back of the tooth row as they were lost anteriorly and formed a diastema between the strong canine and the postcanine tooth row.

The braincase of *Sinoconodon* is known, although the specimen has been said to pertain not to *Sinoconodon* but to *Morganucodon*. Also parts of the periotic and associated structures are known for two upper Jurassic triconodonts, *Trioracodon* and *Triconodon*. Although points of controversy on exact structures, particularly relating to the cavum epipterycum and alisphenoid, have arisen, as they have for tritylodonts (see pages 711–712), resemblances among these structures in the Jurassic forms *Morganucodon*, *Sinoconodon*, and tritylodonts strongly suggest that the braincases were very similar among the earliest mammals.

Symmetrodonts. Here also a modest problem of relationships and assignment exists. Symmetrodonts have been considered as independent, forming a separate order, but more often they have been related in one way or another to pantotheres. These two are considered together here, on the assumption that a close relationship does exist (Figure 175*G*).

Symmetrodonts and Pantotheres. The earliest known genus is *Kuehneotherium* from the Rhaetic of Wales. It is a possible ancestor to both the Jurassic–Cretaceous symmetrodonts and the pantotheres proper (Figure 175*I*). If *Kuehneotherium* occupies this ancestral position, the symmetrodonts must be considered somewhat regressive in their tooth patterns (D. Kermack, K. Kermack, and Mussett, 1968). Crompton and Jenkins (1967) have argued, on the basis of wear patterns, that *Kuehneotherium* should be called a symmetrodont, which was its initial disposition, whereas D. Kermack et al. have contended that it is a pantothere.

Symmetrodonts proper, from the Jurassic and early Cretaceous, had lower molars with three cusps and no talonid, and uppers with a five-cusp pattern (Figure 175*G*). The lower jaws lacked a well-formed angular process. Pantotheres range from the Jurassic through the early Cretaceous and extend back to the later Triassic if *Kuehneotherium* is included. The lower jaws of the later genera had a well-formed angle, and their molar dentitions attained an incipient tribosphenic pattern, with the lower molars having a talonid. Although it is agreed that the tribosphenic dentition of therians arose from that of the pantotheres, the precise homologies of the cusps of the upper teeth have remained controversial (see pages 373–374).

Multituberculates. These animals, the Allotheria, form a very isolated group, ranging from upper Jurassic into the Eocene. Superficially they resemble tritylodonts, but these is no solid evidence that they were derived from this source. They are fully mammalian in that the articulation is formed only by the dentary and squamosal. Skulls are known for later members, and under one interpretation (D. Kermack, K. Kermack, and Mussett, 1968), they have a very primitive cast, although Simpson (1937) believed them to have a therian cast. It has been suggested on very tenuous dental grounds that multituberculates were ancestral to monotremes, and their ancestry has been sought among various nontherian mammals. At present the ancestry is unknown, and some students (e.g., K. A. Kermack, 1965) have recently maintained that they probably arose independently of other mammalian groups from reptilian predecessors.

PATTERNS OF CHANGE. The rather meager information available for the sixth level of radiation allows only limited and rather uncertain interpretations of the patterns of change. One problem comes from the lack of a substantial phylogeny. Teeth, as the principal structures known for many of the groups, have been used both for the study of evolutionary change and for the establishment of relationships, bringing any conclusions on these two matters into intimate, dependent relationships.

Teeth and lower jaws give a fair concept of the extent of adaptive radiation. The most extensive departures, if the molar teeth of *Thrinaxodon* are taken as a morphological base, are found in the herbivores, principally the multituberculates. *Diarthrognathus* also has a distinctive pattern, but little can be said about its adaptive significance. The basically three-cusp pattern of a number of types of Mesozoic mammals is somewhat similar to that found in *Thrinaxodon*. Presumably all creatures with this sort of dentition were carnivores or insectivores whose cheek teeth had primarily a shearing function, with some possible crushing action where cingular cusps were well formed.

The triconodonts, *Sinoconodon* and triconodonts proper, emphasized shearing action in their interlocking array of narrow teeth. In such forms jaw action probably was fairly simple. No angular process is developed on the jaws.

Morganucodonts had somewhat similar cheek dentitions, but additional, cingular cusps were prominent. They too lacked a therianlike angular process on the jaw. Symmetrodonts and pantotheres appear to have developed more complex jaw actions, with the predominantly vertical shearing action being supplemented by crushing and grinding. The relationships of the upper and lower jaws were complex and permitted a variety of possibilities for occlusion. Angular processes were well developed in the pantotheres but not in the symmetrodonts. Docodonts had developed crushing molar teeth, but information is insufficient to show clearly how these were used. The presence of an angle or pseudoangle suggests a fairly high level of development, and relationships of the upper and lower dentitions probably were complex.

Thus there were moderate adaptive divergences, but with only a limited exploitation of the available environmental spectrum. All of the animals were small, with the "giants" being the multituberculates, which, like the caseids and gomphodonts before them, persisted to become associated with members of later evolutionary radiations.

All of the fairly-well-known stocks of the late Triassic—diarthrognathids, morganucodontids, pantotheres (or symmetrodonts), and tritylodonts (as known from the earlier radiation)—had attained many features common to mammals.

The dentary-squamosal articulation was present in the diarthrognathids, morganucodontids, and probably in the pantotheres and *Sinoconodon*. It was accompanied by the older reptilian articular-quadrate articulation. Occipital condyles were double, as in the cynodonts and tritylodonts, and presumably some of the modifications of the atlas–axis complex found in mammals had occurred.

A secondary palate was present in all and in some at least (*Sinoconodon*, *Diarthrognathus*, tritylodonts, and morganucodonts) was essentially mammalian in cast.

Cheek teeth had developed double roots, and oral preparation of food seems to have been general. The masseteric and the adductor musculature appears to have attained a mammalian condition, but the situation with respect to the pterygoid musculature and the depressors of the lower jaw is somewhat uncertain. Tooth replacement is not well understood. In *Sinoconodon*, as in some of the gomphodonts and tritylodonts, there was no replacement of cheek teeth and new teeth were added at the posterior end of the row, as old were lost anteriorly. Replacement did occur in morganucodonts and may have been of the mammalian pattern, although this is not certain.

None of the Triassic members of this radiation had lost the accessory elements of the lower jaws, and thus none had developed the three-ossicle middle-ear structure of the mammals. All had retained a cavum epipterycum, but as discussed on page 711, there has been argument about the nature of this extracranial space.

If these various groups, excluding the tritylodonts, are presumed to have a common origin, which a concept of monophyly would require, the common features may be thought to have been inherited directly. Most of those that are known represent extensions of the changes that were occurring in the cynodonts and bauriamorphs. It is not difficult to envisage an unknown representative of one of these lines in which the common mammal-like features of the triconodonts, morganucodonts, symmetrodonts, pantotheres, and multituberculates were all expressed. On the other hand, diarthrognathids and tritylodonts do not seem to have come through such a path, although the former conceivably might have. Their mammal-like features, which are at about the same level, most probably were attained somewhat independently.

The descendants of the upper Triassic forms, in the Jurassic and Cretaceous, carried many of the trends farther. Irrespective of what detailed early relationships are drawn, it does seem clear that separation of the various lines occurred in the Triassic, prior to the time that full expression of mammalian features had been attained. Triconodonts, pantotheres, and multituberculates must have been long separated, and the later part of the evolution of the symmetrodonts must have been independent of other lines. If this is the case, loss of the accessory elements of the lower jaws, at least, and probably attainment of the mammalian pattern of auditory ossicles of the ear were independent in these lines. Whether the ossicles came to function in a mammalian way is not, of course, known from the fossils, for they have not been preserved, but they are present in monotremes, which seem to have had a course long separated at least from the lines leading to therians. The existence of very extensive parallelism in what seem to be fundamental, highly complex structures is hard to deny.

The braincase, especially the periotic and alisphenoid and related structures, has been intensively studied. Except for the possible changes in the cavum epipterycum, as proposed by K. A. Kermack (see page 712), the braincase was similar to that in the cynodonts. Although other features were changing rapidly during the evolution of levels IV and V, the braincase seems to have remained stable.

This region is known for tritylodonts, *Diarthrognathus*, *Sinoconodon*, morganucodonts, triconodonts, multituberculates, and monotremes. It is unfortunate but true that information for the presumed ancestors of the therian mammals is almost nonexistent.

At the time of writing no general agreements on either the structures or significance of the posterior part of the braincase had been reached. The differences of Hopson, K. A. Kermack, and Tatarinov on the nature of the cavum epipterycum illustrate one area of disagreement. Hopson and K. A. Kermack, however, appear to favor the concept that both the mandibular and maxillary rami of the trigeminal nerve penetrated the periotic through the foramen pseudo-ovale, so named to distinguish it from the foramen ovale in the alisphenoid of mammals, which carried V_3. Whether the nerves V_2 and V_3 both passed through the foramen pseudo-ovale, or only V_3, cannot be determined from the direct evidence. In primitive reptiles the proötic incisure, essentially a predecessor of the foramen pseudo-ovale, served as an exit of part of nerve V from the braincase, and on the basis of the condition in *Sphenodon* and lizards in which the osteology is similar, it has been supposed that both V_2 and V_3 passed from the ganglion in the vicinity of the incisure and *posterior* to the epipterygoid. In monotremes only V_3 emerges from the foramen pseudo-ovale and thus poses something of a puzzle. The structure and presumably the relationships of nerves and hard parts appear to have been very similar in all known nontherian mammals. If the monotremes are included in this array, there seems to be some justification for assuming that only V_3 was carried by the foramen in these stocks.

The conclusions of K. A. Kermack and of Hopson that the foramen pseudo-ovale penetrated an anterior extension of the periotic suggest a condition very different from that found in the therian mammals. It raises the question of whether the lines in which this sort of structure is known might have been well off the line to the therians.

MacIntyre (1967), appealing to studies by Van Bemmelen (1901), Watson (1916), Vandebroek (1964), and his own independent research, denied categorically that the foramen pseudo-ovale in the monotreme *is* formed within the periotic. In *Ornithorhynchus*, following Watson and Vandebroek, the foramen for the mandibular nerve (V_3) lies in a suture between the alisphenoid and periotic (as in cynodonts). In *Tachyglossus* it appears to pass between the true pterygoid and a membrane bone of somewhat uncertain affinities. The latter has been identified as an anterior intramembranous ossification of the periotic, the lateral lamina already cited in discussing the cavum epipterycum, mainly because it does not form in conjunction with the alisphenoid. It is a highly variable bone of uncertain derivation. Irrespective of the source of this element, periotic or otherwise, MacIntyre's interpretation casts the whole matter of the trends of development of the braincase and associated structures into a different perspective.

MacIntyre suggested tentatively that all of the nontherian mammals, which he calls quasi-mammals, had a general resemblance to *Ornithorhynchus* and had changed very little in fundamental structure from the cynodonts as far as the foramen pseudo-ovale is concerned. The tentative interpretation of B. Patterson and Olson (1961) of a braincase believed to pertain to *Sinoconodon* portrayed a condition much like that suggested by MacIntyre. In this specimen, as in most, however, two serious difficulties plague interpretations. One is poor preservation of minute structures; the other is the tendency for all elements in this region to be fused in the adults. This occurs in the monotremes and has, it would seem, been responsible for the wide differences in interpretation. The possibility of fusion in the various fossil forms, such as *Morganucodon* and tritylodonts, necessarily adds to uncertainties in interpretations.

What occurred in the various developing lines leading to therians, except modifications of the teeth and jaws, is not certain. By the late Cretaceous a true foramen ovale penetrating the alisphenoid was present (Sloan, 1964; Sloan and Van Valen, 1965). Just when this came into being is uncertain. It is also clear that by this time, and apparently even during the early Cretaceous, standard mammalian adductor and depressor musculatures of the lower jaw had been developed.

Monotremes, however, have a different depressor mechanism, with the detrahens present rather than the digastric of therian mammals. Parrington (1967*b*) has taken this as definite evidence of a very early split and feels that the difference must have arisen during modifications leading from the reptilian to the mammalian jaw structure. The musculature of the lower jaws of those Mesozoic mammals that lack a true angle of the jaw has been interpreted in different ways, but at present evidence favoring any particular hypothesis is slight.

The bits and pieces of data, based on very few, mostly fragmentary specimens, seem to add up to a picture in which complex patterns of parallel development of similar features have prevailed. These include independent loss of accessory elements of the lower jaw and probably independent acquisition of the three ossicles of the mammalian middle ear. All of the nontherians showed considerable stability of the braincase, and what changes there were seem to have been similar in all. An angle of the lower jaw arose in the therian lines and a somewhat similar structure among nontherians. It is not certain that these were actually related to the same musculature, and they may represent different solutions of similar problems.

What took place in the pantothere lines and among the symmetrodonts, as they led toward therian mammals, remains largely uncertain. Whether or not the early course of the braincase was like that of the nontherians is undetermined. If there was a similar beginning, then divergence took place somewhere between the early pantothere stage and the true therians of the late Cretaceous.

Once the fully matured mammalian pattern of the therians had come into existence, the stage was set for the radiations of the next phase, level VII. It was to this threshold that the complex evolutionary modifications outlined in this chapter led.

INTERPRETATIONS

The Problems and the Evidence

Many questions are raised by the patterns of evolution outlined in the preceding parts of this chapter. Most fundamental is whether or not it is possible to explain the changes that took place in the successive levels of adaptive radiation and the origin of new levels strictly on the basis of the events of adaptive radiation within each of the levels. This kind of knowledge of change makes possible an integration of the processes of evolution known from studies of modern organisms and the temporally more extensive events portrayed by the fossil record. Departures into other sorts of interpretations, if necessary, tend to carry studies away from the juxtaposition of these aspects of evolutionary change.

The origin of common features in discrete lines and the resemblances of many single features to those associated in definitive suites in later, derived organisms seem to require either a "goal directedness" or the existence of common integrating factors that impart selective advantages to particular sorts of changes. Genetics, epigenetics, ontogeny, adult morphology and behavior, and the biological and physical factors of environment all impose boundary conditions on the course of evolution. These limiting conditions, though not actively directing change, do exert a strong canalizing force. This undoubtedly has been an important factor in the extensive parallelism seen in the mammal-like reptiles.

Many common or parallel modifications occurred independently of the particular adaptations of diverging lines, occurring alike in herbivores and carnivores, for example. Many of these changes are those later interrelated in the morphological patterns we associate with mammals. Each new radiation, while repeating some of the adaptive patterns of the older ones, moved in new directions as well, indicating some relaxation of the canalization effective at earlier times. Boundary conditions were to some extent altered by the innovations at the new level of organization.

The culmination of the successive radiations was the condition we call mammalian. Whether it was attained fully by one or several of the evolving lines has not been finally determined. That the divergent lines independently approached more and more closely the mammalian grade, while becoming adapted to a wide spectrum of environmental conditions, is undeniable and is an evolutionary phenomenon requiring explanation.

Early Phases

Within the radiations of the first three levels relatively few changes can be considered "progressive" if the mammalian condition is viewed as the end result. The sphenacodont morphological heritage is never completely obliterated in the successive descendants, but it cannot of itself be seen to presage the later developments. Extrapolation of the changes that took place in the evolution from the primitive pelycosaurs to the sphenacodonts does not predict the majority of patterns of changes seen in the development of levels. IV, V, and VI.

A few features, such as the reduction of the phalangeal formula, are found in some lines in levels II and III, and are "in retrospect" mammalian. This, along with modifications of the limbs and girdles, seems to be related to the development of large, heavy animals in which the limbs were brought somewhat under the body and the feet directed so that the digits were directed forward. Similar phalangeal changes occurred in nonsynapsid reptiles under similar circumstances and also in some of the small, active therapsids as their limbs came to lie under the body in association with increasing rapidity of locomotion. Such changes are strictly adaptive.

Attention has been given to the development of thermal control within the early levels of radiation. The knowledge that this was present in mammals and probably in some of their therapsid ancestors gives special significance to any trends toward thermal control. The long neural spines of some of the pelycosaurs and the very large parietal foramina are most commonly cited as possible evidence. Even if both are conceded to be related in some way to thermal control, which is probable, two things raise doubts that they represent tendencies toward homoiothermic adjustments of the mammalian type. First, both occur in a number of different groups of reptiles not on the line to mammals. Presumably they represent some common adaptation. Second, neither can clearly be related to physiological thermal control. Neural spines could have acted only as radiatorlike structures, using external energy for temperature control. The pineal system could possibly have acted by triggering physiological adjustments in response to radiation, but its mere existence is certainly not evidence that the necessary system for physiological response was present. Modern reptiles do not give much support to such an interpretation. The positive aspects of such devices, if they had any functions related to thermal control, is that they might be taken to indicate that a thermal control system, should it come into existence, would be selectively advantageous.

The nature of the skin is not well known for the members of the first two levels of development. What evidence there is suggests the existence of reptilian scales. *Estemmenosuchus* suggests that a different type of skin may have been present in level III. Only the caseids and some of the dinocephalians show any notable development of foramina and canals in the suborbital and facial regions. These structures, however, are rather similar to those of lizards and could well have accompanied a reptilian type of skin.

The feeding mechanisms in all members of these levels, although adapted to a wide variety of diets, were based on typically reptilian patterns. The structure of the sphenacodont is a good morphotypic ancestor for those of the eotitanosuchids, phthinosuchids, gorgonopsians, and therocephalians. In these, as well, the structure is strictly reptilian, adapted to dealing more or less effectively with large prey, which could be reduced by biting and tearing to moderately large pieces and then positioned in the mouth and swallowed. This is a common adaptation in large predacious reptiles. Its particular expression in the sphenacodonts set the stage for the later modifications in the therapsids, to some extent limiting the courses they might follow. Only in this sense can it be considered to be "oriented toward a mammalian end."

Throughout these three levels, then, there seems to be nothing that cannot be explained as simple, straightforward adaptive evolution. No indications of innate directedness or of any control by central morphological or physiological processes emerge from study. Resemblances between various members are genetic, expressing a common origin, and giving features by which these animals may be related in a coherent phylogeny. The occasional "mammal-like" features are all explicable as adaptations in particular lines and occur in similar guise in other reptiles. No evidence of related suites of "mammalian" features is to be found.

Later Phases

The picture is quite different for radiation levels IV, V, and VI. In all of these, and in most of the included subgroups, very extensive parallelism does occur, expressed in various characters and combinations of characters that are in retrospect "mammalian" or "incipiently mammalian." Although the extent of parallelism may be interpreted differently under different assessments of phylogenetic relationships, the many instances that arise under any plausible concept leave little doubt that this phenomenon is ubiquitous and basic to the evolutionary patterns at these levels.

These kinds of changes occurred early within the developing lines of level IV and were continued in later phases. In a broad sense the parallel modifications that are observed must be considered adaptive, but the repeated occurrences of similar modifications of fundamental features in lines in which feeding and locomotion were becoming adapted to very different circumstances suggest that they should be viewed in a special sense of adaptation.

Explanations

Attempted explanations of this phenomenon of parallelism, excluding those that involve some orthogenetic or teleological principle, have centered around two general ideas: (a) adaptation to increased activity and (b) initiation of homoiothermy. The two are not, of course, discrete, but the thinking about them has emphasized either activity as producing an environment in which homoiothermy could thrive or the initiation of temperature regulation as a general "improvement" that acted as a key innovation to give selective advantage to particular suites of characters as they appeared.

At what stage in the evolution arose a capacity for sustained activity, requiring a continuing high level of metabolism, is not known. Under the first interpretation this could have been quite late, whereas under the second it must have been much earlier. Mere increase in activity, which is evident in the skeletons of many therapsids, does not imply the development of a physiology in which high temperatures are maintained by body processes, as is shown by snakes and lizards among living reptiles. Very likely in the course of therapsid evolution complex interactions of function and physiology were important and undoubtedly strongly selective feedback between physically adaptive functions and the physiology occurred, tending to increase the effectiveness of each with respect to the other and to both internal and external environmental circumstances. In a symposium in Volume 14 of *Systematic Zoology* these relationships were extensively explored in general considerations of evolution of higher categories as summarized by Olson (1965a).

Although many of the changes that developed independently in the different lines of therapsids were advantageous for an active life, some of them also arose in rather obviously slow, clumsy animals, such as the large dicynodonts, therocephalians, and gorgonopsians. It is of course possible that all such forms were derived from small, active ancestors who had already developed the pertinent features associated with active life. Although this may hold for the cynodonts, it is less tenable for the others because their most primitive known forms are large, slow-moving animals.

A basic change took place at about the time that the gorgonopsian, therocephalian, and dicynodont stocks were initiated. The titanophoneans and dinocephalian lines, which appear nearly simultaneously, show no evidence of these changes. If it is assumed that all of these lines came from a sphenacodont ancestry, it is possible that some key modification occurred within the sphenacodonts. If so, it must be assumed that it had little effect on the members of radiation level III, which continued to carry out the general development of the pelycosaurs, but did affect the others, including the dicynodonts, whose sources are far from clear.

The principal gross difference between basal members of the titanophonean–dinocephalian radiation and the primitive gorgonopsians and therocephalians, both of which may have had ancestry in the eotitanosuchians, is in the dentitions. In both there was modest modification of the temporal region that resulted in a somewhat more effective adductor musculature. Titanophoneans, however, retained moderately strong postcanine teeth, which tended to become somewhat blunt, whereas gorgonopsians and early therocephalians had poorly developed, small postcanine teeth. This presumably indicates a difference in diets or in the ways in which food was treated. The earliest and most primitive known gorgonopsians and therocephalians had a deep floccular fossa and an incipient, soft secondary palate. These are not known to have been developed in the titanophonean–dinocephalian lines, but members of these lines did develop modifications of the adductor musculature that departed much farther from the sphenacodont condition than did the gorgonopsians.

What was responsible for the divergence from the eotitanosuchian stage is uncertain. All that can be determined from the morphology and ecological associations is that there may have developed some difference in habits, which set in motion the wide range of other changes that eventually occurred. Such a modification would be in keeping with the suggestions of Von Wahlert in which change of behavior is seen as the initial step in adaptation to new circumstances (Von Wahlert, 1965). Such a change, it can be imagined, might be sufficient for divergence of the titanophoneans and the gorgonopsians, and the development of the latter might have been sufficient to produce a base for the later radiations of level IV.

The geological evidence does not give much support to such a simple, straightforward interpretation. At the time of the earliest appearance of the abundant therapsids in South Africa, therocephalians were common and gorgonopsians rare. In the Soviet Union there is no evidence to indicate which might have come first. The dicynodonts were well developed in the early African deposits, along with therocephalians, and a presumed ancestor, *Otsheria*, occurs in one of the oldest deposits of the Russian upper Permian, of early Kazanian age. Most early representatives of these stocks of radiation level IV were large, somewhat comparable to the dinocephalians in this regard. In North America, in the San Angelo as well, there are very large, primitive therapsids of rather uncertain affinities. All this seems to indicate that there may have been a moderately long history and deployment considerably earlier than seen in the actual records.

Although it is quite possible to assume some adaptive shift as the basis for the dichotomy, with one branch leading to the more advanced therapsids, it is not possible from the available evidence to support this idea with any concrete information on what such a shift might have entailed. The basis of this divergence cannot now be determined.

Once the separation had been accomplished, whether it was by a series of simple or very complex events, and the developing lines had been well established, an increasing number of "progressive" features with a mammalian cast appeared in each. If these lines were as sharply separated as the evidence suggests, their strong resemblances seem to call for some central, integrating feature. This must have been related to increasing, sustained activity, but not to activity alone, because the modifications continued to develop even in the face of other adaptive trends that indicate a decrease in the intensity of activity, although not necessarily a decrease in the advantages of maintaining continuing high-level metabolic activity.

Such considerations have led to the concept that homoiothermy must have been incipient at or near the base of the radiation of gorgonopsians, therocephalians, and dicynodonts. Most features of the central nervous system and the feeding and respiratory mechanisms that developed would have a strong selective advantage once a capacity for sustained activity was initiated (Olson, 1959). Other features—such as a glandular, soft skin or parental care, including nursing, which have been suggested on tenuous evidence in the bauriamorphs and cynodonts (Brink, 1965, 1963d; Tatarinov, 1967)—also can readily be related to homoiothermy.

The homoiothermy fully expressed in mammals probably originated as a very imperfect system, with gradual integration of mechanisms of various systems to produce the final level of perfection. Just when in the evolution of mammals it arose and to what degree it was actually involved in the parallelism must remain conjectural. The evidence discussed above has convinced some that homoiothermy was deeply involved. For example, Brink (1963d) speaks of this evolution as breaking the "thermal barrier." He has tentatively maintained that many mammalian features had developed in both the bauriamorphs and cynodonts, including diaphragm breathing and care of young. Others have taken a more conservative outlook, but the accumulation of evidence seems to point strongly to a physiological basis of many of the features of the therapsid evolution.

The point of view is not important as far as the mere existence of parallelism in the steps leading to mammals is concerned, because this is well demonstrated below the mammalian threshold. It is, however, critical to the understanding of the nature of mammals to know whether they arose from a very restricted source, such as the thrinaxodontid cynodonts; from several cynodont lines; or from cynodonts, bauriamorphs, and possibly some of the therocephalians. There are no certain answers at present. Currently it seems most likely that most of the "Mesozoic mammals" had their source among cynodonts and that many of these may have come from a restricted group within the cynodonts.

The key to this, as conceived by Barghusen (1968), lies in the development within the cynodonts of a mammalian adductor system in the skull and jaws, involving in particular the development of a masseteric muscle. This likely set the stage for the evolution of a mammalian masticatory structure, and without it, Barghusen has argued, no such development could have occurred. A primary result was the development of capacities for the oral preparation of food, related to increasing frequency of feeding and a continuous supply of energy.

The known bauriamorphs did not develop this type of adductor musculature and are thus considered to be removed from the ancestry. On this basis, such animals as *Diarthrognathus* could not have come from bauriamorphs, thought until recently to have been their source. In view of the common mosaic patterns of evolution, there is no particular reason to suppose that bauriamorphs, which had attained dental and jaw features that were in many respects like those of the cynodonts, could not subsequently have developed a mammalian adductor system, but it is certainly true that there is no tendency in this direction in any of the known genera. We may, then, for the time being accept the cynodont origin of mammals, or at least of most mammals, as being the most probable.

After development of the cynodont type of adductor system, probably superimposed on an earlier shift involving incipient homoiothermy, many lines of radiation developed. In these, irrespective of their specific adaptive features, the patterns of parallelism continued. They appear to have exploited the two basic innovations in a way not found, and perhaps not possible, among the bauriamorphs and to have developed a variety of means of obtaining and preparing food, dependent on the new potentials of the masticatory system. Several parallel lines, with very different feeding habits, attained a dentary-squamosal contact at the jaw articulation and again, within radiation level VI, independently attained an articulation in which only the dentary and squamosal participated. Some groups, such as the tritylodonts and diarthrognathids, approached the mammalian threshold, crossing it or not as it is defined arbitarily. They seem to have left no descendants. Others, such as the docodonts, went somewhat further but also failed to attain a fully mammalian jaw articulation. They too left no descendants. Symmetrodonts and triconodonts, in their Jurassic and Cretaceous members, seem to have attained an exclusively squamosal-dentary articulation but were abortive in that no radiations occurred after their passage over this threshold. The ancestors of the multituberculates, which are unknown, presumably made the shift, since all multituberculates in which the condition is known have a strictly mammalian jaw articulation. The same is presumably the case among the ancestors of the monotremes.

The shift in jaw articulation, being one thing that is fairly well known, has received a great deal of attention and has, of course, loomed as very significant in the crossing of the mammalian threshold. There can be little doubt that feeding was important, but it may well be that because of its availability for study its importance has been overestimated.

Successful exploitation of rather limited environments by the nontherian mammals, irrespective of the precise condition of their jaw structures, persisted into the early Cretaceous. One herbivorous group, the multituberculates, competed successfully with therian radiations into the Eocene. Except for this group, however, the radiations were limited, and relatively few basic changes took place in the known features. Whether the limitations were related to feeding or to other parts of the organisms cannot, of course, be determined.

The symmetrodont–pantothere line gives some insight into the matter. It is assumed to be ancestral to the true therian mammals and to have led to a very extensive radiation in which many environments were exploited, partly by extensive dental modifications. The two main things known about the pantotheres are that they possessed a lower jaw with a true angle and that their dentitions trended toward the development of the tribosphenic pattern that lay at the base of the therian radiation. Without this dental evolution, of course, the radiations of the therians as we know them could not have occurred. This does not imply that the feeding mechanisms were the key to the success of this group in contrast to the others. They may have been, but, on the other hand, it could have been that this particular group for the first time attained a complete complement of mammalian structures and functions and that association with the particular type of dentition was fortuitous. This is not an attractive hypothesis, because it cannot be tested except very indirectly by assessment of the "mammalness" of the monotremes.

As matters now stand, the only evidence on the evolution of the stocks leading to the therians and the changes before the late Cretaceous comes from jaws and dentitions. Their feeding mechanisms are different from those in other lines, and they appear to give evidence of a third important step in the attainment of full "mammalness." This followed early steps involving acquisition of incipient homoiothermy and the expansion of the zygomatic arch, leading to the development of a mammalian type of jaw musculature. Once this full array of changes had been accomplished, the great radiation of the therian mammals, which can be termed level VII, was initiated.

7 Biogeography and patterns of vertebrate evolution

GENERAL CONSIDERATIONS

The analyses of the origin of mammals in the preceding chapter carried events to the threshold of the adaptive radiation of the marsupials and placentals. The base of this radiation lay among the first mammals with tribosphenic dentitions, which, along with their descendants, are called therian mammals in this chapter. These radiations are more suitable for biogeographic analyses than are those of any other terrestrial vertebrates and provide a basis for research into patterns of vertebrate evolution within this framework. This field has been of major interest to many paleontologists, and among those who have dealt primarily with mammals W. D. Matthew and G. G. Simpson stand out. The present chapter is devoted to discussion of some biogeographic principles, especially with reference to mammals and the way that these principles apply to therian origins and radiations. This is followed by a brief consideration of the problems that are involved in biogeographic studies of earlier terrestrial vertebrates.

Studies of the biogeography of fossil vertebrates have taken two distinct directions. On the one hand they have been concerned with distributions of animals in the past and how these distributions came to be, and on the other with major topographic features of the earth as they can be interpreted from distributional patterns of animals. Often the two aims are pursued simultaneously, with information from one played against that from the

733

other so that objectives and principles become intimately interwoven. In the first part of this chapter the emphasis is on distributions of organisms with the shifting physical parameters as a background for the changing patterns. In the later parts more attention is directed to patterns of distribution as evidence of physical conditions and the role that various concepts of evolution play in such interpretations.

Biogeographic interpretations deal with biological events cast in a framework provided by the continents and ocean basins, which form the stage on which they were enacted. Evolutionary theory is a basic element in all interpretations, but both physical and biological explanations are essential. The physical setting and biological characteristics of any given time are the product of the interdependent development of the continents and ocean basins and the evolution of life, but the roles of these two are not commensurate.

During the early parts of the evolution of the earth physical processes alone modified its surface. Later, as organisms appeared, increased in volume and diversity, and occupied a greater variety of environments, they came to play a significant role as a source of sedimentary materials and to exercise control over such physical parameters as rainfall, erosion, and patterns of deposition of sediments. In these ways they had important effects on the development of surface configurations in their own right, but their most pervasive effects were through their indirect influences on the course of physical events. The latter are detectable and interpretable by geological and geophysical studies without direct involvement of biological principles. Much of the information on distributions of landforms in ancient times has come from this sort of analysis. Biological events, important as they have been in some cases, take a secondary position throughout much of geological time.

Interpretations of relationships of the earth's major features based on analyses of distributions of organisms require a set of general biogeographic principles. Early studies dealt largely with data on spatial distributions of taxa and the history of how these came to be. Studies by Darwin and Wallace vital to their development of the theory of natural selection were deeply rooted in interpretations of the distributions of living plants and animals. Wallace in particular became a pioneer in the field of biogeography, and many of its general concepts stemmed from his work. Gradually during the last century principles related to distributions and the histories of development have matured. A landmark was the comprehensive work *Climate and Evolution* by Matthew (1915) and more recently the studies of Simpson have broadened immensely the conceptual basis as it applies to both fossils and recent animals.

Attention has come to focus on the development of a comprehensive theory of biogeography with principles that have universal application. One of the major difficulties, much as in the formulation of a general theory of organic evolution, has been the complexity of the subject and its intimate relationship and dependency on principles of related disciplines, notably the concept of natural selection, systematics, ecology, population genetics and population structure, paleontology, and geology and geophysics.

PRINCIPLES OF BIOGEOGRAPHY

Many books and papers expressing many viewpoints and interpretations have been written about the principles of biogeography and their application to various problems of the distribution of plants and animals. Here we are concerned largely with those that have a direct bearing on the problems faced by the paleozoologist. The interested reader can find basic and supplemental information in the following references and their bibliographies and reference lists: Allee et al. (1949); Baker and Stebbins (1965); De Beaufort (1951); Briggs (1966); Cole (1954); Darlington (1957, 1959, 1935); Ekman (1953); Ford (1964); Gadow (1913); Hesse, Allee, and Schmidt (1951); Lack (1964); Levins (1966); MacArthur (1962); MacArthur and Levins (1964); MacArthur and Wilson (1967); Margalef (1959); Matthew (1915); Schmidt (1943); and Simpson (1940A, B, 1953A).

In *The Theory of Island Biogeography* MacArthur and Wilson (1967) identified the fundamental processes of biogeography as dispersal, invasion, competition, adaptation, and extinction. Each is treated in a detailed and partially quantitative manner in their study, and each, at a cruder level, has been the subject of study in analyses of biogeography of extinct vertebrates. The bridge between the detailed studies of the modern situation and the necessarily broader studies of the fossils can never be completely crossed because the data are very different and methods of approach must be suited to the particular situation. Nevertheless, as MacArthur and Wilson point out, the property of insularity with which they are specifically concerned is basic to all biogeography and principles developed in controlled studies of islands have much broader applicability.

At a less refined level, more in the tradition of the historical approach, Simpson (1940B) has stated a general tenet of biogeography that, although generally taken as understood, is so basic that it bears repeating. This is the principle that life tends to expand by evolutionary radiation and in so doing tends to fill all possible roles of the biota. A more explicit statement, not quantitative in the sense of MacArthur and Wilson but seeking to establish both operational and general principles, is to be found in Darlington's exposition of his "evolution-dispersal" hypothesis, succinctly summarized in Chapter II of *Biogeography of the Southern End of the World* (Darlington,

1965) and more fully presented in somewhat less mature form in an earlier article (Darlington, 1959). Briefly this states that the number of species and patterns of dispersal are observed to be correlated with area and climate. These observations suggest the existence of a worldwide pattern of evolution and dispersal of organisms from large to small areas and from favorable to unfavorable climates.

In the development of this hypothesis Darlington stressed the role of what he called evolution of general adaptation as contrasted to adaptation to local conditions or a special function. General adaptation, with somewhat varied meaning, has been applied in studies of phyletic evolution and in the development of major vertebrate categories (see, for example, W. L. Brown, 1958). Here it is conceived as the source of the dominant organisms that comprise the principal dispersing elements. Briggs (1966) refined Darlington's concept, making a distinction between successful and unsuccessful evolution, with the former that of general adaptation in major centers of dispersal and the latter evolution in peripheral areas, with its elements drawn largely from the centers of major dispersal. General adaptive evolution, or phyletically successful evolution, thus is conceived to take place in large areas under tropical or subtropical conditions and provide the basis for major dispersals.

This hypothesis, which incorporates the principle noted in reference to Simpson's studies, contrasts sharply in some respects with the earlier hypothesis of Matthew (1915). Both conceived of dispersal from large areas, but Matthew suggested that the centers were climatologically unfavorable areas from which the more primitive rather than the "most adapted" dispersed. The general theory of dispersal itself has been questioned, in particular by Rosa (1931), who postulated that there was no such thing as a center of origin but that new life forms appeared throughout the full range of antecedent forms and that the initial distributions were essentially those that persisted. Between these extremes, touching on many details of dispersion, are various intermediate concepts. Most, in one way or another, hold to the principle of centers of origins and dispersal from them.

Each of the five somewhat interdependent processes of biogeography of MacArthur and Wilson has been studied with fair intensity, most particularly in ecological contexts. Dispersal is especially subject to analyses in paleozoological contexts and has received the most attention. Many of its features have been described fully by Simpson (1940a,b; 1947a,b; and 1953a) and Darlington (1957, 1965) and are generally well known. Dispersal is, of course, far from simple in itself and its course is affected in each individual case by all of the other processes. It seems generally to have been somewhat cyclical in nature, with successive expansions and contractions playing very significant roles in the establishment of patterns of distribution. It may involve the spread of but a single species but equally may be the property of a whole community. If only area were involved, under the usual concept of a center of origin, dispersal would have a radial pattern, but the effects of climate on the centers of origin and dispersal routes tend to produce a zonation as well.

Tendencies toward radial movement and the effects of zonation are modified by the relative positions of potential reception areas and the centers, and by the physical nature of their connections and their positions within the complete system. Establishment of secondary centers and tendencies for reversals of initial directions of dispersal complicate matters, as do changes in the nature of both the dispersal and reception areas during the course of interchange.

Invasion, a consequence of dispersal, generally involves a confrontation of the expanding and endemic biotas. Competition is inevitable except in the rare instances in which invaders in no sense have an effect on the niches of the endemics. The records of fossil vertebrates are replete with cases of invasions and replacements following on competition. The dynamics of the processes are not evident but must be inferred from the results seen in the subsequent records.

Competition takes place over the area of confrontation and is also an important factor in the establishment of capacities to expand from the central area (see, for example, G. Cox, 1968). Darlington in one way and MacArthur and Wilson in another have related competition in the center of dispersal to the climate. Darlington's hypothesis also considered it to be related to the dimensions of the area. Direct paleontological information on competition is generally poor, but its effects are evident in the changes of the biotas.

The term "adaptation," as used by biogeographers, often refers to the development of fitness of organisms for some way of life or some particular function. The dominant life forms in Darlington's scheme are the best adapted members of their community, expressing a general capability for existence rather than something specific. Adaptation in the sense of MacArthur and Wilson concerns the capacity of organisms to meet the conditions encountered and to adjust to new circumstances. Most such concepts express a state of being, rather than a process of evolutionary change, but the latter is implicit. In all such considerations criteria of the state of adaptation include survival and numbers of individuals, but beyond such similarities there are wide differences. The general concept of adaptation has so many meanings that unqualified statements concerning it are usually uninterpretable. Nevertheless statements about adaptation, sometimes well defined and sometimes not, figure prominently in all studies of biogeography based on the fossil record. As a rule success in terms of numbers and in production of new life forms, along with the ability of units to meet new circumstances without change, is the criterion used to estimate the extent of adaptation.

Extinction has always been in important aspect of paleozoological studies.

It has been related to biogeography mostly with respect to major fluctuations of the continents and ocean basins, extensive modifications of habitats related to climatic changes, and major competition in broad confrontations between different communities (see, for example, Newell, 1967). A number of more specific and carefully analyzed cases are found in the recent literature; for example, in Axelrod's analyses of extinctions of Quaternary mammals and Cretaceous dinosaurs (Axelrod, 1967; Axelrod and Bailey, 1968) or in the study by Van Valen and Sloan (1966) of the extinction of multituberculates. These studies depend on biogeographic analyses, although such explanations are not their goals.

Extinctions such as those studied by MacArthur and Wilson are on a scale so different from those made by paleontologists that relationships between the two are difficult to establish. It is likely that the sorts of basic interpretations that can be made at the very detailed level will eventually become applicable to broader problems, and it is at the former level that experimental testing of principles must be carried out.

Throughout the known history of organisms the *processes* of biogeography seem to have altered but little. As different life forms have successively exploited new major areas and habitats of the earth, the ways in which the processes have come into play have changed. Dispersal from an aquatic habitat onto land and dispersal from original terrestrial bridgeheads to continental interiors, when the invaded areas were not occupied by terrestrial communities, were very different from dispersals from a well-established center into a fully occupied area. Very early occupations of land, presumably by plants, did not involve competition between invaders and invaded. The initial terrestrial animals met only plants. As vertebrates first penetrated the land their invasion and competition was very different from that of the various fishes that today seem to be making a generally parallel transition. Thus, although processes may remain much the same, different hypotheses or models are required to cover radically different circumstances.

The concepts of centers of origins and dispersal are deeply ingrained in biogeographic thought and supported by so much evidence in the best known cases that other concepts have received little attention. Yet in many specific cases the nagging questions of what the center was or whether there was a center do arise. What is commonly seen in the fossil record seems to suggest that evolution always occurred somewhere else. For two reasons this may well have been the case. If most samples come from peripheral regions, it would be expected that little general evolution would be recorded. Many, at least, do come from just such areas. Also, random sampling would only very occasionally tap the centers, which occupy but a relatively small part of the whole surface of the earth.

Taxonomic and historical approaches to biogeography have treated such a variety of situations that only the very broadest and indefinite sorts of generalizations have emerged. The development of present-day plant and animal distributions is reasonably well explained by Darlington's hypothesis of the relationship of the nature of the centers of origin and dispersal to area and climate, in an appropriate land–ocean setting. That the hypothesis may be inadequate for all vertebrates of the past must remain open. Whether patterns of origin and dispersal of the early therians can be explained by it is not immediately evident.

If there were centers of origin of new major groups in the past, meeting the requirements of the Darlington model, these groups may have arisen within them in one of several ways. Each mode would result in a somewhat different pattern of dispersal. Little concrete consideration has been given to the precise way of origin of such groups, within the context of a large, thriving community. There are various possibilities, each with biogeographic implications. It may be that origin took place from a single species in toto, or that some isolated part of a species lay at the base. A small, rapidly changing succession of populations could have evolved, fitting the pattern of Simpson's "quantum evolution" (Simpson, 1944) or various modifications of it (e.g., Simpson, 1953b).

Origin, however, may have been from a widespread species composed of partially isolated subgroups, demes, or perhaps subspecies, each of which evolved somewhat independently, as in the general models of Wright (1940, 1960). From several of these, over a fairly broad area, the new level of adaptation could have come into being. All descendants could have been so similar that they would be placed within a single genus. Still another origin might have involved several closely related species, each of which moved in the direction of the new level. In such an event the derived animals might perhaps be grouped at a familial level.

Each of these modes of origin presupposes a major, broad center of origin, and each involves evolution of general adaptations. The distribution of the newly emerged group, within the community of its center of origin, would be quite different under the different circumstances, and of course dispersal would follow somewhat different patterns. Under the last model, in addition, it is possible that the source species did not live in a single broad region but occurred separately in two or more major land areas. Very similar animals might then have come into being in these separate regions. As far as any particular group under consideration is concerned, the development of biogeographic patterns would differ immensely under the alternative models. In effect, however, the hypothesis of multiple regions of origins merely pushes the center of origin and dispersal back in time, since presumably the source species had to attain the conditions of dispersion requisite to the supposed pattern of origin. In fact, no matter what pattern of polyphyletic origin is proposed for members of a particular distributional pattern, it is necessary to infer an earlier dispersal from some fairly localized source or to be faced with a regression of independent origins of major stocks that stretches back to the very beginning of life, or beyond if the initial life stock is not considered to have had some sort of limited dispersal. Various polyphyletic concepts have trended in this direction, but not on biogeographic grounds. The extent to which such hypotheses of origins may be tenable depends largely on the credence given the roles of convergence and parallelism in evolution.

In the following sections the origins and early dispersals of the therian mammals are considered in light of the sorts of problems posed above. The data are such that no final answers are to be expected, but they are also sufficient for the problems to be meaningfully studied and for relevant alternative hypotheses to be examined. This is a field in which new information is rapidly accumulating, and this will undoubtedly alter details between the time of writing and publication of the analyses that follow. A major study of the recently discovered late Cretaceous Bug Creek assemblage of therians is now under way by Van Valen and Sloan. This already has added information that alters some earlier concepts and more is to be expected. For our purposes the detail is generally less important than the nature of analyses. For this reason most of the analyses are based on the data presented by Romer (1966a) and organized taxonomically under his classification. Where items of special importance that postdate Romer's book are critical, they have been used. His classification, as noted in Chapter 10 of Section III, has been viewed differently by different workers, and its use masks taxonomic distinctions that some consider important. Both the data and classification, however, provide a stable, basically conservative, and fully accessible basis for analyses, and conclusions drawn from them will not be severely altered by changes necessary to fit the particular predilections of one student or another or by additions of information during the last few years.

THE ORIGIN AND EARLY DISPERSAL OF THE THERIANS

Relationships to Earlier Radiations

In several chapters of this section the origins and radiations of selected groups of vertebrates were analyzed. Each had its own biogeographic patterns, but for the most part these are obscure. Once large continuous continental areas had come to host a coherent and extensive series of communities of organisms, the establishment of their distributions could have proceeded in ways not different in kind from those that have produced present-day distributions.

As far as terrestrial vertebrates are concerned, this pattern may have been established as early as the beginning of the Triassic period, perhaps even earlier, and certainly the necessary conditions had been attained by the late Triassic. By this time very mammal-like therapsids and their immediate mammalian relatives were in existence and widely spread over the earth, with some representatives being known from North America; the British Isles; the continent of Europe; South, East, and West Africa; and Southeast Asia. How similar the several late Triassic assemblages and their early Jurassic equivalents were is uncertain because of their incompleteness, but in all cases in which members of the same major groups have been found in two or more places they have been proven to be very similar. It would appear that there was a very widespread community type in which small mammals and near mammals existed as a significant unit.

In African and Asiatic deposits archosaurian reptiles are closely associated with these primitive mammals, and in the British Isles lizards appear to have been part of the general faunal complex. Very broadly the early mammals seem to have assumed a role as part of the general lepidosaurian–archosaurian complex. It was presumably from these mammals, as discussed in the preceding chapter, that the therians arose. Only in the fissure fills of the British Isles does the assemblage of Triassic mammals and mammal-like creatures include traces of probable therian ancestors. The common features of the nearly worldwide assemblage, however, suggest that similar animals may have been present wherever the complex existed. If this were the case, there were several potential sources for therian mammals.

How the dispersal that led to this widespread upper Triassic faunal complex took place raises many interesting questions, but ones that are not strictly germane to the central focus of this discussion. We will start with this distribution without concern for how it came to be.

The Jurassic mammal-bearing sites add relatively little to biogeographic interpretations. A middle Jurassic site in the British Isles has produced symmetrodonts and pantotheres, and upper Jurassic sites in the British Isles and the United States have produced triconodonts, symmetrodonts, pantotheres, and multituberculates. Lower Cretaceous mammals have been found in the Wealden beds of Britain (multituberculates and a possible therian, *Aegialodon*), in Spain (therian), in eastern Asia (therian, originally placed in the upper Jurassic), and in Texas, from the Trinity formation of Albian age (triconodonts, symmetrodonts, pantotheres, therians, and multituberculates). These scattered finds point to a very wide distribution of mammals, including therians, in the lower Cretaceous and by Albian times possibly both placentals

TABLE 51 North American and Asian Distributions of Upper
Cretaceous Mammals

	North America	Asia
Order Marsupialia	#	
Polyprotodonta	×	
Didelphidae	+	
Alphodon	0	
Boreodon	0	
Campodus	0	
Delphodon	0	
Diaphorodon	0	
Didelphodon	0	
Ectonodon	0	
Eodelphis	0	
Pediomys	0	
Thalaeodon	0	
Order Insectivora	#	#
Proteutheria	×	×
Leptictidae	+	
Procerberus	0	
Zalambdalestidae		+
Zalambdalestes		0
Liptotyphla	×	
Adapisoricidae	+	
Gypsonictops	0	
Incertae sedis	×	
Glasbius	0	
Order Creodonta (Deltatheridia)	#	#
Deltatheridia	×	×
Deltatheridiidae	+	+
?Cimolestes	0	
Deltatheridium		0
Deltatheridoides		0
Hyotheridium		0
Sarcodon		0
Order Primates	#	
Plesiadapoidea	×	
Paramomyidae (Phenacolemuridae)	+	
Purgatorius	0	

TABLE 51 (continued)

Order Condylarthra	#	
Arctocyonidae	+	
Protungulatum	0	
Order Multituberculata	#	#
Ptilodontoidea	×	
Ectypodidae	+	
Cimexomys	0	
Mesodma	0	
Cimolodon	0	
?Essonodon	0	
?Liotomus	0	
Taeniolaboidea	×	×
Cimolomyidae	+	
Cimolomys	0	
Meniscoessus	0	
Eucosmodontidae	+	+
Eucosmodon	0	
Proliotomus	0	
Stygimys	0	
Djadochtatherium		0
Incertae sedis	×	
Paronychodon	0	

Key: #, order; ×, suborder; +, family; 0, genus.

and marsupials were present (Slaughter, 1968). Unfortunately the known sites give no information about the southern continents.

Upper Cretaceous Therian Mammals

Table 51 shows the distributions of upper Cretaceous therian mammals by order, family, and genus, along with distributions of multituberculates and Table 52 shows the distribution of dinosaurs. It is only from eastern Asia, North America, and Europe that mammal remains have come. The only specimen yet described from Europe (France) is probably a eutherian (Lecoux et al., 1966).

A striking feature of the distributions of upper Cretaceous mammals is that marsupials form much the largest and most varied element in North America and that this group is completely absent from Asia. This may, of course, be due to sampling, but the fact that marsupials do not appear later

TABLE 52   Genera of Dinosaurs, Numbers in Principal Groups of Saurischians and Ornithischians in Continental Areas

	N.A.	E.As.	S.As.	Eur.	S.A.	N.Af.	Af.	Genera Common to N.A. and E.As.
Coelurosauria	8	4	4	1	0	1	0	2 + ?1
Carnosauria	4	8 + 1?	2	1?	1	4	0	2
Sauropoda	1 + 1?	1	3	2	6	1	4	0
Ornithopoda	23	7	0	4	0	0	1	1
Ankylosauria	10	8	1	4	2	0	0	0
Ceratopsia	14	3	0	0	1	0	0	?1

Abbreviations: N.A., North America; E.As., East Asia; Eur., Europe; S.A., South America; N. Af., North Africa; Af., Africa (excluding North Africa).

in the record in Asia lends some support to the idea that they were not present. Clemens (1968) and others earlier have not considered the evidence of absence as particularly significant.

Insectivores and creodonts (following Romer's use of the order) are present in North America and Asia, but condylarths are absent from the latter. Until very recently, however, condylarths were known in North America from a very few specimens, and the lack in Asia, again, may be due to sampling. A single specimen of a primate has been described from the upper Cretaceous of North America. A considerable change in outlook on the faunas of the upper Cretaceous has followed the discovery of the Bug Creek assemblage in Montana. In contrast to other North American assemblages, condylarths form the dominant therian element and marsupials are relatively few in number.

The few samples that are known show that the radiation of both placentals and marsupials had begun in earnest by this time, with the dispersal of marsupials probably somewhat greater than that of the placentals. Less certain is the possible interpretation that the faunas of North America and Asia were similar only in very general ways.

The meager evidence on distributions of therians can be somewhat augmented by information on other groups that appear to have been part of the general faunal complex. Multituberculates are the most prominent element of the mammalian fauna of North America at this time. They were present in eastern Asia, but known only from a single genus. Additional collecting as new sites are found may well turn them up in greater numbers and variety. They do show that the faunas of the two areas had this common element, but from what is known the resemblance can only be deemed very general.

Evidence from dinosaurs, Table 52, must be based on the proposition that associations of mammals with dinosaurs indicated by their occurrences in the same bed and in beds of equivalent ages in North America and eastern Asia show that they were parts of the similar dinosaur complexes elsewhere. If this was the case, therians were probably much more widespread than the direct evidence indicates.

In a limited sense the dinosaurs from North America and Asia can give additional information on the intimacy of relationships of the faunal complexes of the two areas. The few common genera from the two regions suggest that similarities were only general and that there was little likelihood of close relationships. Dinosaurs, however, are taxonomically difficult because of the practical problems posed by their large size, which, among other things, leads to very small samples. It could thus be that the resemblances were somewhat closer than the data in Table 52 indicate.

More broadly the dinosaurs show certain important facts about the upper Cretaceous:

1. Similar reptiles were widespread over the continents (North America, Asia, Europe, South America, and Africa), with some genera ranging very widely.

2. Environments in the various areas of deposition differed markedly, and these are strongly reflected in the different characteristics of the preserved faunas.

The dinosaurs of the Northern and Southern Hemispheres have notable differences, with the former being dominated by ornithischians and the latter by sauropods, akin to types found in the Jurassic of North America and East Africa.

During the Jurassic pretherian mammals, the pantotheres, existed in proximity to sauropod faunas, although how closely they were associated in the ecological systems is not certain. They were, it appears, able to exist under the general environmental circumstances in which sauropods flourished. Thus it may be assumed that climates over major portions of the earth during the late Cretaceous were suitable for therians. The dispersal of some common genera of dinosaurs indicates that there were no insurmountable barriers to their movements during the middle and late Jurassic and probably during much of the Cretaceous. Thus it seems probable that there were no barriers to the types of mammals found along with the sauropods in the Jurassic sedimentary sequences at Como Bluffs, Wyoming (triconodonts, symmetrodonts, pantotheres, and multituberculates). A single, edentulous jaw has come from the similar Tendaguru beds of East Africa.

These very tenuous bits of evidence indicate that therian mammals may have been widespread in the upper Cretaceous. It is not known that they in fact were, but there appear to have been no major barriers to their dispersal, and climates suitable to their needs were widespread.

Early and Middle Paleocene

Mammal deposits formed during the early and middle Paleocene are currently known only in southwestern and northwestern United States. Lower Paleocene deposits are more restricted than those of the middle part of the sequence. Table 53 shows the orders of the upper Cretaceous and Paleocene therians and the number of genera of each at the various time levels. The fauna of the Paleocene can be seen to be essentially a continuation of that of the Cretaceous, and this is very generally true of the sediments as well. Dinosaurs, however, are absent. A strong reduction of marsupials has occurred, and the condylarths have increased. The middle Paleocene shows a sharp increase in major groups, and in part gives a true picture. At this level of evolution of the placentals, however, members of different groups were very similar and there continues to be controversy about the placement of various genera and families in the major categories. Continuing reassessment of the placement of several families among the Insectivora and Primates (see Chapter 10, Section III) is the most graphic example, but also the creodont–condylarth–deltatheridian controversy is far from settled. Thus, even though more orders are present in the middle Paleocene, the changes from a morphological point of view are not great and the taxonomic arrangement can be changed so as to alter the picture considerably. Nevertheless, there can be no doubt that adaptive radiation was very active during this time.

This radiation, displayed in North America during the late Cretaceous and the early and middle Paleocene, is one that could have been at the base of the whole marsupial and placental radiation of later times. If this were the case, Darlington's hypothesis would fit the conditions extremely well, for evidently there was a large area, or center, and this occurred in essentially tropical conditions—that is, in an area with a favorable climate.

Late Paleocene and Eocene

Upper Paleocene mammal-bearing deposits occur in North America, South America, Europe, and Asia. Table 53 shows the marked increase of genera from middle to upper Paleocene, and Table 54 gives a summary of orders, suborders, and families through the Eocene. Africa remains a blank during the early Eocene, but beds of the Fayum region have yielded a few terrestrial placentals from the middle Eocene. Upper Eocene and Oligocene beds of this region have produced a good fauna. Australia's record of Cenozoic mammals is, as well known, notably poor, although recent work headed by

TABLE 53   Ranges of Orders of Therian Mammals from the Upper Cretaceous through the Upper Paleocene[a]

Order	Cretaceous			Paleocene		
	Lower	Middle	Upper	Lower	Middle	Upper
Marsupialia	?1		10	1	0	14[b]
Insectivora	?1	?1	3	1	16	26
Tillodontia						1
Taeniodontia				2	3	1
Primates			1[c]	1[c]	10[c]	10
Creodonta			5	1	4	5
Carnivora					2	3
Condylarthra			1 (?2)	19	32	32
Amblypoda					2	10
Notoungulata						9
Astrapotheria						2
Litopterna						3
Edentata						1
Rodentia						1
Lagomorpha						1

[a] Range shown by bar graph; the number of genera of each order for subepochs given below the bar. Data from Romer (1966a) unless otherwise indicated. A genus that extends over more than one time unit is shown as a member of each unit in which it occurs. New, as yet undescribed, materials from Mongolia will eventually modify the figures for the lower Cretaceous.

[b] Thirteen South American, one North American.

[c] After Van Valen and Sloan (1965).

TABLE 54 Orders (Suborders, Infraorders) and Families of Mammals from the Upper Paleocene through the Upper Eocene (Data from Romer 1966a)

	Upper Paleocene						Lower Eocene						Middle Eocene						Upper Eocene					
	N.A.	Eur.	E.As.	S.As.	Af.	S.A.	N.A.	Eur.	E.As.	S.As.	Af.	S.A.	N.A.	Eur.	E.As.	S.As.	Af.	S.A.	N.A.	Eur.	E.As.	S.As.	Af.	S.A.
Marsupialia																								
Didelphidae	x	?x					x	x					x	?x					x	x				
	O	?O					O						O	?O					O	O				
Carloameghinidae						O						O O O						x						x
Borhyaenidae						O																		O
Caenolestidae												O												
Polydolopidae						O						O												
Insectivora																								
Proteutheria	x +	O					x +	O					x +	O		x +			x +	x +				
Leptictidae		O						O						O		O			O	O				
Anagalidae			O						O															
Paroxyclaenidae								O						O										
Tupaiidae	O							O																
Pantolestidae	O						O						O							O				
Pentacodontidae	O						O						O							+				
Apatemyidae	+	O					+	O					+	O						O				
Dermoptera																								
Mixodectidae	+	O																	O	+				
Plagiomenidae																				O				
Picrodontidae	+	O					+	O					+	O					+	O				
Litoptyphla																								
Adapisoricidae																			+	O				
Erinaceidae																				O				
Talpidae																				O				
Plesiosoricidae														O O						O				
?Apternodontidae																				O				

TABLE 54 (continued)

Tillodontia								
Esthonychidae	×○		×○	×○	×○	×○		×○
Taeniodontia								
Stylinodontidae	×○		×○	×○	×○	×○		×○
Chiroptera								
Microchiroptera								
Emballonuridae	×+○○○		×+○○○+○+○○ ○	×+ ○ ○○				×+○○○○○○ ○
Megadermatidae								
Rhinolophidae								
Hipposideridae								
Phyllostomatidae								
Vespertilionidae								
Palaeochiropterygidae								
Archaeonycteridae								
Incertae sedis								
Incertae sedis				○ ○○	○			○
Primates								
?Plesiadapoidea	×+ ○○		×+ ○ +○○	×+ ○ +○○	×+ ○ +○○	×+○○○+○+○○ ○		× +○ ○
Phenacolemuridae								
Carpolestidae								
Plesiadapidae								
Lemuroidea								
Adapidae	+○		+○	+○	+○	+ ○		+○ ○
Tarsioidea								
Anaptomorphidae								
Omomyidae								
Tarsiidae								
?Microsyopsidae								

TABLE 54 (continued)

	Upper Paleocene						Lower Eocene						Middle Eocene						Upper Eocene					
	N.A.	Eur.	E.As.	S.As.	Af.	S.A.	N.A.	Eur.	E.As.	S.As.	Af.	S.A.	N.A.	Eur.	E.As.	S.As.	Af.	S.A.	N.A.	Eur.	E.As.	S.As.	Af.	S.A.
Creodonta				(UK)																				
Deltatheridia																								
Deltatheridiidae	×	+	○				×						×						×	×				
?Didymoconidae																			○					
Micropternodontidae	+			○			+	○	○				+	○					+	○	○			
Hyaenodontia																								
Hyaenodontidae	×	+	○				×	+	○				+	○					×	+	○	○	○	○
Oxyaenidae	×	+	○				×	+	○				×	+	○				×	+	○			
Carnivora																								
Fissipedia																								
Miacidae	×	+	○				×	+	○				×	+	○				×	+	○	○	○	○
Viverridae																			○	○				
Felidae																			?○	○	?○			
Mustelidae																				○	○			
Canidae																			○	○				
Procyonidae																				?○				
Condylarthra																								
Arctocyonidae	×	○	○	○	○		×		○			○	×		○	○			×	○	○			
Mesonychidae	×	○	○	○	○	○	×	○	○	○	○		×	○	○				×	○	○			
Hyopsodontidae							×	○				○	×	○	○									
Meniscotheriidae							×	○					×	○										
Pteryptychidae																								
Phenacodontidae							×	○				○	×	○										
Didolodontidae						○						○						?○						

TABLE 54 (continued)

Taxon	Distribution
Amblypoda	
Pantodonta	
Pantolambdidae	× +
Barylambdidae	+
Titanoideidae	○ ○
Coryphodontidae	○ +
?Pantolambdodontidae	○
Dinocerata	
Prodinocertidae	+ ○
Uintatheriidae	○
Gobiatheriidae	○
Xenungulata	
Carodniidae	+ ○
Pyrotheria	
Pyrotheriidae	+ ○
Proboscidea	
Moeritherioidea	
Moeritheriidae	× + ○ + ○
Barytherioidea	
Barytheriidae	○
Sirenia	
Dugongidae	× ○
Notoungulata	
Notioprongia	
Arctostylopidae	× +
Henricosborniidae	○ ○
Notostylopidae	+ ○
Toxodontia	
Oldfieldthomasiidae	+ ○

TABLE 54 (continued)

	Upper Paleocene						Lower Eocene						Middle Eocene						Upper Eocene					
	N.A.	Eur.	E.As.	S.As.	Af.	S.A.	N.A.	Eur.	E.As.	S.As.	Af.	S.A.	N.A.	Eur.	E.As.	S.As.	Af.	S.A.	N.A.	Eur.	E.As.	S.As.	Af.	S.A.
Archaeopithecidae						O						O						O						O
Isotemniidae												O						O						
Homalodotheriidae																		?O						
Notohippidae						?+O						+O						+						
Typotheria																								
Interatheriidae						O						O						O						
Hegetotheriidae						?O												O						
Archaeohyracidae						?O												O						?O
Astrapotheria																								
Trigonostylopidae						×O																		×
Astrapotheriidae												×OO						×O						O
Litopterna																								
Protorotheriidae						×OO						×OO						×O						×O
Macraucheniidae																								
Perissodactyla							×+O O+O+OO	×+O +O	×				×+O O +OO	×+OO +O+	×+ ?O				×+ O+O+O OO	×+OOO +	×+O O+O+OO			
Hippomorpha																								
Equidae																								
Palaeotheriidae																								
Brontotheriidae																								
Ancylopoda																								
Eomoropidae																								
Ceratomorpha																								
Isectolophidae																								
Helatelidae																								
Lophialetidae																								
Deperetellidae																								

TABLE 54 (continued)

Lophiodontidae				○		○○		○○○	
Hyracodontidae							○	○	
Amynodontidae						○○		○○○	
Rhinocerotidae									?×
Artiodactyla									
Palaeodonta									
Diacodectidae		×+○○		×+		×+	×+	×	×
Homacodontidae							+		
Dichobunidae				○		○		○	○
Achaenodontidae									
Suina									
Choeropotamidae						+	+○	○	
Cebochoeridae							○	+○○	○+○
Entelodontidae						○	○		○
Suidae									?○
Anthracotheriidae				○		○+	○○+○○○○		
Ruminantia									
Cainotheriidae						○		+	
Anoplotheriidae									
Xiphodontidae									
Amphimerycidae									
Agriochoeridae									
Oromerycidae						○○○	○○○○		
Camelidae							○		
Hypertragulidae									
Gelocidae									
Edentata									
Palaenodonta		×+○		×	+○	?○	×	+○	×	+○○
Metacheiromyidae									
Epoicotheriidae									
Xenarthra				×+○○		×+○○			
Dasypodidae									
Palaeopeltidae									
Megatheriidae									
Glyptodontidae				○		○		○	

TABLE 54 (continued)

	Upper Paleocene						Lower Eocene						Middle Eocene						Upper Eocene					
	N.A.	Eur.	E.As.	S.As.	Af.	S.A.	N.A.	Eur.	E.As.	S.As.	Af.	S.A.	N.A.	Eur.	E.As.	S.As.	Ad.	S.A.	N.A.	Eur.	E.As.	S.As.	Af.	S.A.
Tubulidentata							?×												?×					
Orycteropodidae							?○												?○					
Cetacea								×					?×			×			×	×	×	×		
Protocetidae								○					?○			○			○		○	○		
Dorudontidae																			○			○		
Basilosauridae																○			○			○		
Agorophiidae																								
Rodentia	×						×	×					×	×					×	×				
Sciuromorpha	+						+	+					+	+					+	+				
Paramyidae	○						○	○					○	○					○	○				
Sciuravidae							○						○						○					
Cylindrodontidae																			○					
Protoptychidae													○						○					
Aplodontidae																			○					
Sciuridae																				○				
Myomorpha																			+	+	+			
Eomyidae																			○	○	○			
Geomyidae																			○		○			
Gliridae																				○				
Pseudosciuridae																				○				
Theridomyidae																				○				
Lagomorpha			×																×	×				
Eurymylidae			○																					
Leporidae																			○	○				

Key: ×, order; +, suborder; ○, family; N.A., North America; Eur., Europe; E.As., East Asia; S.As., South Asia; Af., Africa; S.A., South America.

the late R. A. Stirton (see especially 1955, 1957, 1961 with Tedford and Miller) has improved the situation somewhat. The recent fauna and what is known of the fossils show a history of mammals very different from that on any other continent, with marsupials as the sole therian element in at least its early parts and probably well into the Cenozoic (see Keast, 1968, for an excellent review).

It is clear that by the late Eocene, and much earlier wherever evidence is available, several distinct centers of mammalian evolution had been established. Some of these—South America and Australia—were essentially isolated centers, whereas others—North America, Asia, and Europe—evidently had faunal interchanges from time to time. Africa had its own faunal complex, but some elements had come in by faunal exchange by the late Eocene. It is this general distribution that biogeographic analyses of the origin and early dispersals of therian mammals must explain. The Eocene is an evolutionary crossroads in placental evolution, a time in which ancient lineages, with their roots in the Cretaceous, still persisted to live side by side with the beginnings of the newer lines from which present-day radiations stemmed. The following subsections present the basic data for the Paleocene and Eocene and are followed by an examination of possible interpretations of the Cretaceous and early Tertiary.

North America, Asia, and Europe

These three major areas had many common features during the early Tertiary and are conveniently treated together. North America and Europe have been intensively collected and studied. Work in Asia has little more than scratched the surface. This results in strong discrepancies in the comparisons, especially for the earlier parts of the record. Adding to the problems are the different taxonomic practices that have been applied to the collections from different areas and to all collections with the passage of time. The use of Romer's compilation tends to reduce these problems but cannot eliminate them.

The basic data are given in Table 54. Eastern and southern Asia are entered separately, but they have been lumped in Table 55, in which similarity indices are used. This is done to enlarge samples at the expense of combining somewhat different provinces.

The index $(C/N_1) \times 100$ (see Simpson, 1960b) has been used for the determination of similarities and differences. It is probably the best index available for this purpose but, like all such indices, leaves much to be desired. The very small samples of the Paleocene produce the least meaningful results. Few patterns that appear to have validity show up in the figures of Table 55. The families of North America and Europe show a decrease in the similarity index from an initial high in the Paleocene through the Eocene. This is not

TABLE 55   Numbers of Orders, Families, and Genera of Mammals from the Upper Paleocene and Eocene of North America, Europe, and Asia, and Indices of Similarity

Region	Number of Orders				Number of Families				Number of Genera			
	U. Pal.	L. Eoc.	M. Eoc.	U. Eoc.	U. Pal.	L. Eoc.	M. Eoc.	U. Eoc.	U. Pal.	L. Eoc.	M. Eoc.	U. Eoc.
N.A.	10	14	14	13	29	39	34	43	73	81	81	92
Eur.	5	9	9	11	13	17	28	48	20	29	49	54
As.	3	4	7	9	7	4	8	27	7	8	13	60

	Common Orders				Common Families				Common Genera			
N.A. Eur.	5	8	9	9	12	13	9	14	1	11	2	7
N.A. As.	1	4	6	8	2	5	4	15	1	3	2	10
Eur. As.	0	3	6	6	0	2	6	8	0	3	0	7

	$(C/N_1) \times 100$, Orders				$(C/N_1) \times 100$, Families				$(C/N_1) \times 100$, Genera			
N.A. Eur.	100	89	100	82	92	96	32	32	5	38	4	13
N.A.-As.	33	100	85	90	50	70	50	55	14	27	15	17
Eur.-As.	0	75	85	66	0	29	75	29	0	37	0	12

[a] The data are primarily from Romer (1966a). Orders, families, and genera common to three regions and similarity indices determined on the basis of the index $(C/N_1) \times 100$, where C = number of common taxons and N_1 the sample size of the smaller of the two samples from the pertinent areas. Abbreviations: U, L, M—upper, lower, and middle, respectively; Pal., Paleocene; Eoc., Eocene; N.A., North America; Eur., Europe; As., Asia.

matched by the other comparisons, which show irregular fluctuations. The genera from North America and Europe reveal no clear trends. The similarities between North America and Asia are somewhat greater than those between Europe and Asia, both in families and orders.

Many pertinent facts, of course, are not revealed in the figures, especially where small samples are involved. Even a single specimen—for example, the one individual of a notoungulate in the lower Eocene of North America—can add a full order, family, and genus and carry great weight in the index. Five of the 10 orders from the Paleocene of North America are found in Europe, but only 3 of the 14 orders of the lower Eocene of North America are found in Asia. The three, however, represent 75 percent of the full number known from Asia. Very important differences are hidden by the resulting high indices. In the Paleocene, for example, where the index for North America and

Europe is 100, tillodonts, taeniodonts, amblypods (which many recognize as including two orders), and edentates are present in North America and absent in Europe. These typically New World orders have only very meager Old World representation or none at all in later times. Those that do appear have come from migration out of North America.

The essentially simultaneous appearance of perissodactyls in Europe and North America is of much greater biogeographic significance than mere addition of a common order and common families. It bears significantly on the location of the center or centers of origin of these advanced ungulates and how this was related to the areas in question.

The summary of data in Table 54 supplies part of the information needed for qualification of the information in the indices, but even beyond this the full effects of taxonomic practices, the numbers of individuals of species and genera, conditions of occurrences, and understanding of the nature of both human and natural sampling are important. Such data may be obtained in part from the literature, but intimate contact with the materials in the field and laboratory is indispensable for full insight into the meaning of the data. Lack of this experiential information inevitably must hamper the efforts of all who attempt to draw very broad biogeographic interpretations, and this looms as one of the major barriers to effective analysis.

Another important aspect of the data is their significance relative to the places of origin of major groups. Here faunal similarity is not the central problem. The earliest record, even of a single genus or species, may confer special significance on the area in which it was found. The Paleocene no-toungulate genus of Asia, known from many specimens, and the single lower Eocene notoungulate in North America are of immense importance in understanding this group, which is known otherwise only from South America, where it underwent its major radiation. Such finds often, as in this case, disturb simpler biogeographic analyses that follow from apparent patterns of dispersal of the common and well-known groups of animals.

North America, South America, and Australia

These three continents, strange bedfellows in some respects, are alike in that all include marsupials both today and in the past. Marsupials also occur in the Tertiary of Europe, but quite clearly these were derived from North America long after the radiation was under way. Only didelphoids ever lived in North America. They were relatively abundant during the upper Cretaceous, when there was a wide variety, present sporadically into the Miocene, and, after a gap in the record, present in the late Tertiary, Quaternary, and Recent. Marsupials first occur in the upper Paleocene of South America, the earliest producing beds of that continent. By that time they had undergone considerable differentiation. They are characterized by development of the borhyaenids, the principal carnivores of the Tertiary faunas before the Pliocene, and caenolestoids, both of which are exclusively South American. Australian Tertiary terrestrial mammals occur in deposits that seem to go back as far as the Oligocene. All that have been found to date, before the late Tertiary, are marsupials, and these all show affinities with modern or recently extinct families.

The placental faunas of the Tertiary of North and South America for the most part are very different, as shown in Tables 53 and 54. Condylarths of the family Hyopsodontidae occur in both, and rare notoungulates and xenarthrans are known in North America. Litopterns, astrapotheres, and pyrotheres are unique to South America.

Some of the clearest insights into distributions and the effects of intermigrations come from these two continents and their faunas have played important roles in the development of the concepts of biogeography. As far as evidence pertinent to early therians is concerned, however, the most important is that which indicates an early isolation of the two continents, the presence of marsupial carnivores and absence of placental carnivores in South America, and the very limited occurrences of typically South American placentals in North America and Asia. The possibilities of connections between South America and Australia to explain, among other things, the distribution of marsupials is a matter that is still subject to debate and is considered in more detail later on.

Africa

The middle and late Eocene and Oligocene deposits of the Fayum region of Egypt are the only sources of the information on therians in Africa before the Miocene. Late as they are, the faunas indicate that a separation of this part of Africa from other major centers of placental evolution existed during the early Tertiary. This, it would seem, went well back into the Paleocene. The most distinctive elements of the fauna are the proboscideans and their "subungulate" relatives. These along with the sirenians and hyracoids appear to have developed somewhere in Africa. Information from the Eocene is meager, and it is primarily the Oligocene that has produced adequate faunas. By this time some routes of migration appear to have been established between Africa and Europe, and also rodents and primates, not known previously in the Fayum, arrived from unknown areas, apparently African rather than European (A. E. Wood, 1968).

Interpretations

For a complete explanation of the distribution of the therian mammals of the late Eocene, definitive information on phylogenetic origins, the patterns of adaptive radiations, the site or sites of origin, and the modes and paths of dispersal is needed. Evidence is far from sufficient, and in the end we must arrive at one or more series of conditional, or "if," statements with the most likely event selected at each level at which definition is required. Biological conditions and physical circumstances at each stage dictate the choice, but at any given level the available choices have already been limited by earlier decisions. One or many contingency series may be erected with at least the possibility of an exhaustive treatment in which all possible series have been determined. What follows is something of a compromise between possible limits.

Shortly after or perhaps at the time of origin of the therians the marsupials and placentals were differentiated. Thereafter they underwent somewhat independent radiations and attained distinct distributions. One of the major aspects of the total distribution to be explained is this difference. Each of the two major groups also has its own peculiar patterns that relate to its origins and the accessibility of dispersal paths to creatures endowed with the general potentials of its own group. Important for any interpretation is an understanding of the phylogenetic origins of the therians and of the place or places in which origins occurred. All that follows is of course dependent on these initial events. Data are inadequate for a confident interpretation, but what is known is sufficient to impose some general limits on the possibilities and to suggest a limited number of likely events.

The majority of biogeographic interpretations of terrestrial vertebrates, and especially those of the late Mesozoic and Cenozoic, have been made under the influence of the concept of the northern continents as centers of dispersal, following especially from the extensive analyses of Matthew (1915). The data for these interpretations came largely from Eocene and later deposits and from distributions of living mammals. Simpson, more than others, considered the implications of the earlier distributions. The most satisfactory evidence comes from the middle Tertiary on, and from this a strong case can be made for dispersals from the northern continents. It has been a common tendency to apply these same patterns to earlier times, where the evidence is less good and where theoretical guidance is necessary. Although this practice has some justification, it may, nevertheless, lead to conclusions for which the evidence of the time in question is far from sufficient.

The interpretations in the rest of this section are directed to an examination of the consequences of the hypothesis of northern continental dispersal centers and exploration of alternatives. In the allotted space, of course, extensive analysis is impossible. Rather, attention is focused on groups of animals and patterns that seem to be the most definitive. The objectives, as throughout the book, are not full explanations, which are impossible, but a study of methods of research and their consequences.

Origins and Very Early Distributions

Both marsupials and placentals were present in Albian times, in the Trinity beds of Texas, if the interpretations of Slaughter (1968) are credited. It must of course be recognized that this, and nearly all assessments of Cretaceous mammals, are based on teeth, usually single teeth or reconstructed series, and occasionally jaws and jaw fragments with teeth still in place. The probable source of all therians lies among the paurodontid pantotheres known from the Jurassic of North America and Europe. Of all the early Cretaceous therians, from England, Spain, eastern Asia, and North America only *Clemensia* and *Pappotherium* have been interpreted as marsupial and placental, respectively.

Marine deposits of the late Jurassic and early Cretaceous are fairly widespread, but continental deposits are very restricted. Some advocates of continental drift have argued that the breakup of a major, single continental mass began in the late Jurassic or early Cretaceous. Other timetables place this earlier. Both, of course, are countered by the idea of long-separate Gondwanan and Laurasian landmasses and of course that of relatively stable continental positions. These uncertainties raise difficulties in efforts to solve the problems of the site or sites of origin of therians and whether there was a large center fragmented by later continental movements or several separate centers on already discrete continents.

What is known of the climates (see Schwarzbach, 1963; Axelrod, 1967) suggests that temperatures were generally high in the late Jurassic and early Cretaceous and tended to become somewhat reduced and perhaps more zoned during the Cretaceous. The data for the early Cretaceous, once again, are not extensive. The distributions of plants and animals in the late Jurassic and early Cretaceous seem to show that climates in the known areas were suitable to habitation by terrestrial organisms that required warmth and moisture, and that there were ready avenues of communication between them for land animals. Routes of dispersal are uncertain, but considerable attention has been focused on the possibilities of a southern (Antarctic) center of dispersal, as taken up in more detail in the consideration of the marsupials.

Therians may have originated in a single stock from pretherian lines, and both marsupials and placentals may have arisen from this stock after it had attained a tribosphenic dentition. It seems quite possible as well that the marsupials and placentals may have come independently from the pretherians, and of course it is not impossible in either instance that very similar animals arose in different places more or less simultaneously.

The simplest hypothesis in keeping with the data is as follows. There was a single stock from which marsupials and placentals arose. This came into being in the late Jurassic or the very early Cretaceous and dispersed rapidly, at least over the northern continents. From this stock the marsupials and placentals diverged at a time no later than Albian. This took place in North

America, which was probably subtropical and which formed the center of dispersal as well as the main center of progressive evolution.

Nothing known is contrary to such an interpretation. It is, however, at least suspect. There can be no question that sampling is insufficient to ensure that it represents distributions at these times. It is highly unlikely that the meager information available does in fact give an insight into what the distribution actually was. The hypothesis accepts the coincidence that scanty sampling has produced a valid basis for estimation of distributions. With so few data points many hypotheses are possible and several have equal chances of being correct. For example, the therians may have originated in Asia, with primitive members dispersing to North America and Europe in the late Jurassic or early Cretaceous. Marsupials then arose from North American stocks, but placentals arose in Asia, sending successive waves of more advanced forms to Europe, Africa, and North America, from which they passed to South America.

We may go still farther afield and set up the hypothesis that the placentals arose in the Northern Hemisphere, in one or more centers, whereas marsupials arose in the Southern Hemisphere, dispersing to South America, North America, and then to Europe, and to the Australian region. Finally an extremely polyphyletic interpretation might suggest that the placentals and/or marsupials originated separately in each of the major landmasses from elements dispersed in the late Triassic and more or less separate since that time. In some areas, such as Australia, only marsupials developed. In South America and North America both marsupials and placentals arose. In Europe and Asia only placentals came into being. The late Cretaceous mammals thus might be the result of parallel development, but thereafter limited interchanges began, and these increased somewhat during the early Eocene, thereafter decreasing again.

If the information from dentitions and fragments of skulls and jaws is extrapolated to imply the nature of the total organisms, creating the image of marsupials and placentals with many of the distinctive features possessed by the living representatives, the amount of parallelism required by the last hypothesis seems somewhat staggering, although not impossible. The inference that the animals involved were fully developed marsupials and placentals may be valid, but it is not necessarily so, and it may be that only in the observed characters or in some limited suites of characters were the characteristic conditions realized.

The first hypothesis is undoubtedly most comfortable, but it actually has little beyond comfort to recommend it over some of the others. The last hypothesis is probably the least comfortable and depends on what may be considered unusual evolutionary events. They are not needed to explain what is known and perhaps, should be discounted. On the other hand, where similar concepts of parallelism have been needed in explanation of observations of some later mammalian dispersals—for example, those of the hystricomorph and caviamorph rodents or the Old and New World monkeys —they have been called on, reluctantly but with confidence, once other solutions seemed to be inadequate.

The point that is most pertinent to understanding of later distributions of therians is that any one of the sequences outlined, as well as others, *could* have occurred and that there is no very sound basis for choosing which one *did* occur. The clues must lie in what happened afterward, for this is strictly dependent on origins and initial dispersals. If, as is often done, these later distributions are predicated in part on what it is believed to have gone before, the chances of proper solutions are seriously reduced. The evidence of later events within therian radiations is not sufficiently complete to be definitive but, added to that from other organisms and physical events, it is sufficient to at least limit the probable sequences of events appreciably.

Problems of Marsupial Distributions

The facts may be briefly reviewed as follows. No Cretaceous marsupials are certainly known outside the United States. By the late Cretaceous an appreciable radiation had occurred. Most sites of this age contain a preponderance of marsupials, relative to placentals. At Bug Creek, Montana, however, the placentals outnumber marsupials. During the early and middle Paleocene in North America marsupials are uncommon and few kinds are present. When first encountered in the Paleocene of South America (upper Paleocene), three families of marsupials were in existence. *Peratherium* of the Eocene of Europe is a genus also known from North America and probably derived from there. From the upper Paleocene of Cernay, France, has come a single upper molar that is possibly a didelphid (D. E. Russell, 1964). Australian faunas of the Tertiary were predominantly marsupial, with placentals appearing sporadically, probably introduced by overwater transport, and mostly late. The fossil record is poor, so that the time relationships are most uncertain. By the time the first marsupials are encountered, in the Oligocene, they were diverse and related to living or recently extinct families of Australian marsupials. The families in Australia and South America are different, but there are close resemblances between some of the genera.

Clemens (1968) presented a careful survey and conclusions about the origin and dispersal of marsupials. He assumed North America to have been the center of radiation, with very early entry, perhaps by island hopping along with a few placentals, to South America. This presumably was during the late Cretaceous and involved didelphid marsupials and insectivorous and condylarthran placentals. Thereafter there was only very occasional crossing between North and South America, and as far as is known, marsupials were not involved until the late Tertiary, when they passed from South America north-ward. Many aspects of the history of this continent have been presented by B. Patterson and Pascual (1963, 1968), Harrington (1962), and Hershkovitz (1969).

The major problem, of course, is the population of Australia by marsupials and the absence of placentals. Present distributions of mammals and presumed conditions during colonization have been summarized recently by Keast (1968). As far as the records of Australia and adjacent lands of New Guinea and New Zealand are concerned, the arrival of marsupials could have been at any time between early Cretaceous and the Oligocene. It is not impossible, although unlikely, that a radiation of the marsupials had occurred before their arrival in Australia. Also it is not demonstrable that placentals did not reach Australia and then become extinct in the early Tertiary. Although this is not usually considered to have been the case, the same hypothesis, but involving marsupials, is necessary to explain an Asian route of entry into Australia. The main difference is the relative "survivability" of members of the two groups.

Most paleontologically oriented biogeographers—for example, Clemens, Darlington, Keast, and Simpson—have argued that marsupials reached Australia by some sort of sweepstakes pattern. Ride (1964) has held to the position that a southern dispersal is possible, considering some of the evidence on continental drift. Asiatic origin has been carefully reasoned, as far as possible on the basis of phyletic relationships and probable land connections during the Cretaceous and early Tertiary. It may well be correct.

If this is accepted, the absence of marsupials in the Asiatic record must be considered unimportant. It could well be merely the result of the very incomplete sampling. Marsupials may turn up at any time as the increasingly intensive studies of the early Tertiary and late Cretaceous deposits continue. Some North American elements do occur in Asia, suggesting an interchange between the areas. Chow Li (1965) has reported two "North American" genera, *Homogalax* and *Heptodon*, in early Eocene deposits of coastal Asia, suggesting that early Tertiary mammals from North America may have invaded this region. If so, as Clemens has argued, the coastal region may have formed a corridor for marsupials, who may never have reached the continental interior. It is not certain just how long this corridor, if it existed, was in operation and just when. Sometime after the marsupials had made at least a start in their crossing to Australia they must have died out on the continent. They are not known in any Tertiary deposits and inasmuch as late Eocene mammal-bearing beds are fairly well known, it is probable that they died out in the early Tertiary.

An alternative source is Europe and this too is possible, since the European forms were very similar to those of North America and could have been a

Figure 179.   A polar projection of the northern hemisphere illustrating a pattern of land distribution and connections consistent with the distribution of early Tertiary mammals. (Based mostly on Kurtén, 1966.)

source of the morphological types found in Australia. The strong resemblances of the faunas of North America and Europe during the late Paleocene and early Eocene indicate some type of direct connection. Kurtén (1966) has argued strongly for this and also has cited evidence for the existence of a northern arm of the Tethyan sea separating Europe from eastern Asia (see Figure 179). This second circumstance, if the sea was pervasive during the early Tertiary, casts some doubt on a European source but does not rule it out. If there was a European source, the invasion of Australia presumably occurred during the Eocene and radiation must have been very rapid, if it occurred in Australia as these hypotheses suggest.

The alternative to an Asian route, of course, is a South American–Antarctic passageway, either with North America as a source or with a radiation center in Antarctica. Neither the idea of origin in North America and passage through South America and Australia nor that of an Antarctic center has had much support among vertebrate biogeographers (e.g., see Darlington, 1965; and Simpson, 1940a, 1961b). The evidence from botanical and invertebrate studies, as interpreted by some students in these fields, is such, however, that southern dispersals cannot be rejected out of hand. The physical evidence is likewise of greatest importance and rapid strides are being made in studies of Antarctica and of continental and polar movements. These may become highly significant, although to date they are far from definitive and have served more to renew interest in possible connections and interchanges than to give a firm basis for analyses of biological events.

Darlington (1965) concluded from a review of the evidence of the southern beech, *Nothofagus*, which has figured in many discussions, and of the carabid beetles, that explanations based on northern dispersal routes are satisfactory. He argued further that, even were this not the case, the vertebrate patterns, based mostly on fish, amphibians, and reptiles, are distinctive in their own right and do not in any way correspond to those to be expected if there had been a southern dispersal route. Presumably marsupial distributions also fall into this category, but the fossil evidence is such that other interpretations are at least possible.

Strong arguments favoring a southern dispersal of organisms have been based on studies of distributions of the plants, *Nothofagus* in particular, and on various terrestrial invertebrates. Summaries of various of these are to be found in papers in Gressit (1963). Also involved are physical determinants that are related to continental positions and continental connections. From the summaries of King (1962) and Adie (1963) and some paleomagnetic data, such as that analyzed by Irving, Robertson, and Stott (1963), the possibilities of Antarctic land connections do seem to exist. But currently there is so much controversy and continuing modifications of interpretations that it is possible to find support for many different patterns of continental distributions and from these to muster support for a variety of biogeographic hypotheses.

Evidence about climates also is not satisfactory. It does seem from what is known of the flora that Antarctica had a climate suitable for appropriate plants and animals along its margins and that the interior could have been climatically suitable to serve as a center of radiation. Both geographic and climatological evidence is permissive of a southern dispersal under some interpretations, but the former at least is doubtful under others.

Perhaps the strongest argument for southern dispersals by way of land connections in recent years has been made by Brundin (1965) based on chironomid midges. Using Hennig's (1957, 1960) phylogenetic systematic methods (see also Hennig, 1966), Brundin arrived at what he considered indisputable evidence that the relationships within this group were developed during a period when the southern lands were directly connected. Antarctica was considered to be part of a temperate center of evolution. South Africa and shortly thereafter New Zealand became separated from Antarctica in early Cretaceous. Australia separated from it later, and South America and Antarctica were connected until "probably not very long ago."

If we make the assumption that the idea of an Antarctic dispersal is valid, we arrive at the following possibilities:

1. Marsupials used it as a bridge from South America to Australia.
2. Marsupials originated in the Antarctic center and dispersed from there to South America to North America and to Australia.
3. Marsupials originated in Australia and radiated from there.

Africa, it would appear, became detached before these events. As far as the timing is concerned, if Brundin's scale is used, no serious problems arise, although it is necessary to have migrations to North America by Albian times. Most data suggest that North America and South America had some interchange as recently as late Cretaceous. If Australia and Antarctica were separated somewhat earlier than suggested, an overwater route poses no more problems than have been suggested for other hypotheses.

Reversals of the usually accepted directions of interchange of marsupials between North and South America pose no problems. If Antarctica was a center of marsupial dispersal, however, it would appear that this occurred early enough for the land connections with Australia supported by Brundin to have been still in existence. This would not alter the marsupial pattern but would suggest that placentals probably were not present in the Antarctic center.

The hypothesis of an Antarctic center has been stressed here because it has tended to be downgraded by most vertebrate biogeographers, in particular by those who have been most influential. Strong arguments against it can and have been made, most recently by Darlington (1965). They are not, however, to be considered as conclusive any more than are the contrary arguments. The important matter is that the problem of marsupial distributions is far from settled and that it will remain so until much more evidence is available.

Problems of Placental Distributions

The data summarized in Tables 51 through 55 show a contrast between marsupial and placental distributions in the late Cretaceous, marked by the absence of marsupials in Asia. If this is not the result of faulty sampling, which is far from certain, then at this early date the patterns of origin and dispersal of the two may be thought to have been quite different. No other basis for comparisons is available after the late Cretaceous until the late Paleocene. The presence of only marsupials, condylarths, and ancestral South American ungulates at this time in South America suggests a sweepstakes pattern of migration to South America (but see Hershkovitz, 1966). The lack of any therians except marsupials in the initial Tertiary deposits of Australia and the modern composition of the fauna indicate that placentals were not among the early immigrants to that continent. If it is assumed that the various stocks did not develop de novo in these two southern continents, then the data, as far as they go, seem to rule out a southern origin and dispersal of the early placentals, a pattern at least possible for the marsupials, as already discussed.

Just as for marsupials, the record of origin of placentals is very uncertain. Romer (1966a) and Slaughter (1968) tentatively place *Pappotherium* with the placentals, although B. Patterson (1956) did not make a definite assignment of the therians from the Trinity beds. *Endotherium* from the middle Cretaceous has also been tentatively assigned to the placentals by Romer. It is, however, only in the late Cretaceous that unquestionable placentals are known. By that time a considerable adaptive radiation had occurred, producing a number of distinct, although fairly closely related, lines. Just how these are to be classified is somewhat open to question, but as many as four orders—Insectivora, Primates, Condylarthra, and Deltatheridia (Creodonta) —have been designated. There is no question that there had been considerable radiation of placentals by this time.

As in the case of the marsupials, the simplest and in some ways most satisfying explanation of distributions compatible with the evidence envisages origin in North America in the early Cretaceous, with dispersal to Asia and South America during the late Cretaceous. In addition, however, dispersal also took place to Asia and Europe, and probably to Africa, where a separate radiation of placentals south of the Tethyan seaway took place during the very early Tertiary. Coordinated with the equally simple explanation of marsupial origin and radiation, this hypothesis envisages the origin of both groups in North America, a common migration to South America and Asia, and thence, as far as marsupials are concerned, to Australia across the Indonesian archipelago. Placentals reached Europe, but there is no certain evidence in the deposits or in biogeographic requirements that marsupials also arrived in the Paleocene. Additional finds may either support or reject this simple hypothesis, for it is based on extremely slender evidence that is equally subject to other interpretation.

Without considerably more information on distributions in the late Cretaceous and the early and middle Paleocene, various other models are possible, with only a southern origin and dispersal being extremely doubtful. If the hypothesis of a single area of origin of marsupials and placentals is accepted, from either a common therian or pretherian stock, Europe, Asia, or even Africa could have been the center. Or, assuming more than one center, the undifferentiated therians or pretherians in North America, Asia, and Europe might have given rise to placentals and marsupials in each of the areas, or any two of them, with marsupials failing to leave an early record except in North America. Other patterns are equally possible—for example, one in which the marsupials arose in the south, in Antarctica, and the placentals arose in the north, in North America, Asia, or Europe. Only by invoking some overriding conceptual basis, such as the proposition that mammalian radiations all originated in a large northern landmass, for example, Asia, is a definitive interpretation possible. This then, of course, is not based on the primary data, which impose very few limits.

By late Paleocene, for which fairly substantial records are available from several parts of the world, so much had transpired that the threads of the history cannot clearly be sorted out to lead back to any one pattern of origin and distribution. Here, as before, it is possible to adopt the "comfortable" hypothesis that the evolution known from the early and middle Paleocene in North America is in fact the evolution in the major dispersal area. This could have been the case for both the marsupials and placentals; but if it was, it is necessary to assume that for some reason the marsupials left but a sporadic record. Granted this, the radiation patterns of the marsupials and placentals from the same general center were different, perhaps in part reflecting presumed differences in their ways of life.

Patterns of the late Paleocene are consistent with the idea of a North American center, but by no means preclude other interpretations. If the North American center is accepted, it follows that the insectivores, primates, condylarths, and deltatheridians probably originated there and spread to Europe. If, as A. E. Wood (1962) has suggested, rodents arose from among the Primates, perhaps from a *Plesiadapis*-like form, during the middle Paleocene, then this order could have come into existence in this center as well. Dinocerata and Pantodonta (Amblypoda of Romer) may also have originated in North America, spreading to Asia but not to Europe. Whether the xenungulates and pyrotheres of South America are related, as their inclusion in the Amblypoda implies, is less certain. Deltatheridians were present in both Asia and North America, and either of these could have been the source of the European deltatheridians. Fissipedes (miacids) are known at this time only in North America and may have spread to Asia and Europe in the late Paleocene or earliest Eocene. Tillodonts and taeniodonts seem to have been strictly North American in origin and early development.

This primitive placental radiation, the earliest phase of adaptive radiation of the stocks of the late Cretaceous, forms a pattern consistent with Darlington's hypothesis of a major climatologically favorable center of origin and dispersal. The data indicate a very strong linkage between North America and

TABLE 56   Early Eocene Distributions of Perissodactyls, Artiodactyls, and Rodents in North America, Europe, and Asia

Taxon	North America	Europe	Asia
Perissodactyla			
Hippomorpha	×	×	
Equoidea	x	x	
Equidae	+	+	
Hyracotherium	○	○	
Propachynolophus		○	
?Propalaeotherium		○	
Brontotheroidea	x		
Brontotheriidae	+		
Eotitanops	○		
Lambdotherium	○		
Palaeosyops	○		
Ancylopoda	×	×	
Chalicotheroidea	x	x	
Eomoropidae	+	+	
Lophiaspis		○	
Paleomoropus	○		
Ceratomorpha	×	×	×
Tapiroidea	x	x	x
Isectolophidae, etc.	+	+	+
Homogealax	○		○ᵃ
Heptodon	○		○ᵃ
Hyrachyus	○	○	
Chasmotherium		○	
Lophiodochoerus		○	
Lophiodon		○	
Artiodactyla	#	#	
Palaeodonta	×	×	
Diacodectidae	+	+	
?Bunophorus	○		
Diacodexis	○		
?Eohyus		○?	
Homacodontidae	+		
Hexacodus	○		
Dichobunidae		+	
Protodichobune		○	
Rodentia	#	#	
Protorogomorpha (Sciuromorpha)	×	×	
Ischyromyoidea	x	x	
Paramyidae, etc.	+	+	
Decticadapis		○	
Franimys	○		
Leptotomus	○		
Lophiparamys	○		
Microparamys	○		○
Pseudotomus	○		
Reithroparomys	○		
Thisbemys	○		
Dawsonomys	○		
Knigthomys	○		
Mysops	○		
Sciuravus	○		

Key: #, order; ×, suborder; x, superfamily; +, family; ○, genus.
ᵃ Listed by Romer (1966), but not by Radinsky (1965). There is but one specimen of early Eocene tapiroid known from Asia (Radinsky).

Europe, more extensive than merely a narrow land-bridge corridor (see Kurtén, 1966), and a lesser linkage to Asia, perhaps a filter-bridge type (see Figure 179). Various problems do arise, among them the stratigraphic and geographic distributions of the notoungulates, edentates, and lagomorphs, early mammals that like rodents may have had their roots in this primitive placental radiation. We shall return to these later, since they are more or less perturbations to be explained away, rather than carriers of massive information that can cast doubt on the whole hypothesis. Of more immediate interest are some of the events of the lower Eocene, involving members of a more advanced phase of adaptive placental radiation—perissodactyls and artiodactyls—and the rapid development of the rodents, which played an important role in later Tertiary faunas.

Perissodactyls, Artiodactyls, and Rodents

When information from the lower Eocene of North America, Asia, and Europe is added to what is known of the late Paleocene, it introduces complications that seriously disturb any hypothesis involving a single major center of placental dispersal. These data do not necessarily alter the hypothesis of a North American center for the more primitive adaptive phase, but they do indicate that during at least its late phases there were other centers in which more advanced animals came into being. Tables 54 and 55 show quite clearly that Europe and North America were closely connected at this time (late Paleocene and early Eocene), perhaps in the way indicated in Figure 179. This is shown both by the members of the primitive radiation phase and even more graphically by those groups that appeared on the scene near the beginning of the early Eocene. Among these, the perissodactyls and artiodactyls are of special interest. Important as well during the Eocene is the fact that North America and Europe, after the early Eocene, show increasing diversity of faunas, whereas Asia and North America on the whole show increasing resemblances. Beyond these general changes, of course, the particular animals that occur variously in the different areas have much to tell about the nature of migration routes. These are not considered here because our interest is in the initial establishments of centers and early dispersals rather than developments that followed these events.

Data on the perissodactyls of the early Eocene are summarized in Table 56. Important recent studies of this group have been made by Kitts (1956) and Radinsky (1963–1967, inclusive). *Hyracotherium*, the basal equid, appeared nearly simultaneously in Europe and North America. Thereafter equid evolution in the two areas proceeded independently, slowly in North America and rapidly in Europe. An eomoropid appeared in both, but the more advanced European genus seems to have been somewhat later in appearance. Of the tapiroids, *Hyrachyus* is present both in Europe and North

America. Three of the four families of perissodactyls known at this time are thus common to these two regions, with only the brontotheriids, which are quite diverse in North America, absent in Europe. Clearly there was a common source area, but almost certainly this was not the center identified as a possible site for the development of the early placental adaptive radiation.

Radinsky (1966c) has identified the phenacodontid condylarth *Tetra-claenodon* as a likely ancestor of the perissodactyls. This middle Paleocene genus lived about 5 million years before the appearance of the perissodactyls. The presence of four major types of perissodactyls early in the Eocene suggests a rapid rate of evolution. Kitts (1956) saw this radiation as beginning no later than the early Paleocene. Origin in a single area, adaptive deployment in this area, and dispersal from it seem to fit well the patterns of appearances of perissodactyls in Europe and North America. Other interpretations are possible, but here the data are such that this seems much the most likely hypothesis. Radinsky (1966c) summarized the origin as follows:

> The early perissodactyls may have originated isolated from, and probably under different selective pressures than, other descendant lines of the middle Paleocene *Tetraclaenodon* stock.

The "other lines" could have developed in the areas of the known Paleocene series in North America, but the perissodactyls apparently did not. It is of course by no means demonstrated that the North American area was a center of origin and dispersal of even the early radiation phase of placentals. Even if this were shown to be so, evidence is not only insufficient to demonstrate that there were not other centers but rather indicates that such centers probably existed in the Northern Hemisphere and certainly did in South America.

The early Eocene of Asia reveals a pattern quite different from that of North America or Europe, with perissodactyls represented by but a single specimen of a tapir (Radinsky, 1965). In the late Eocene, in contrast, many genera of tapiroids, brontotheres, chalicotheres, and rhinoceroses are present and widespread. Kitts (1956, p. 55) notes the source of evolution of perissodactyls in the Paleocene as "undoubtedly" Eurasian. This determination seems to follow primarily from the general concept that origins of most advanced placental groups were Eurasian, rather than from any specific data pertinent to the equids as such. The early Eocene of Asia has not been thoroughly studied. The perissodactyl record of one tapiroid may be a matter of sampling, but even so it is difficult now to consider Asia, except possibly Far Western Asia, as the source of perissodactyls. The distribution of landmasses, connections, and seaways in Figure 179 fits the picture of early Eocene mammal distributions well, with a discontinuity between much of Asia and Europe. If this is the case, either the large landmass of North America and related territories or that of Europe, as extended in Figure 179, would seem to have been the most appropriate centers under the hypothesis that a large landmass was a requisite. Of these two, the preponderance of varied perissodactyls in North America might, perhaps, be taken as favoring the center in North America.

The nature of the climate in these areas is unknown, of course, because the suggested centers are not recorded by any deposits. How much zonation there may have been also remains uncertain, although it has been suggested on the basis of tapirs by Dehm and Oettingen-Spielberg (1958) that there was zonation in the late Eocene of Asia.

Data on early Eocene artiodactyls are summarized in Table 56. They are meager, because the origin and deployment of this group of ungulates seem to have come a little later than that of the perissodactyls. What little is known is compatible with the idea that artiodactyls and perissodactyls may be the product of a single large area, but has little that can argue against a concept that they originated in separate areas, either within a large single land area or in two different major areas.

Unlike perissodactyls and artiodactyls, rodents may have come into existence from members of the primitive phase of placental evolution within the general region in which this evolution is known. As for any such case, including the supposition that the perissodactyls and artiodactyls did not arise in this area, the conclusions must be very tentative. A. E. Wood (1962) suggested that origins may have taken place elsewhere, with migration into the fossiliferous area, or else, if they were present before the first occurrence (late Paleocene), that they were in niches that tended not to contribute to the fossil record. He felt that they probably arose in the early or middle Paleocene, whereas Simpson (1953a) has suggested a somewhat later Paleocene origin. If the origin was from plesiadapids, as A. E. Wood and others have speculated, then the source at least was present in the known areas.

The early Eocene rodent genera of North America greatly outnumber those of Europe, but those present in Europe are very close to North American genera, and there seems no question that there was intimate contact of the continents that permitted rodent interchange.

Notoungulates, Edentates, and Lagomorphs

These three groups, which do not fit well either into the initial insectivore–condylarth–creodont–primate radiation or that of more advanced placentals, pose some interesting biogeographic problems, each of which has received considerable attention. Notoungulates are predominantly South American, where they form an important element of the ungulate radiation of the Tertiary. They appear there in the upper Paleocene and had initiated a significant adaptive radiation by that time. A small notoungulate, *Palaeostylops*, is abundant in the Paleocene of Asia. Associations are with pantodonts and dinoceratans, which are very similar to North American genera.

In North America a single specimen of a notoungulate has been found in the lower Eocene. On the basis of the general idea of Holarctic origins of major groups of mammals, with Asia as the center, it has been supposed that the notoungulates originated in Asia and passed through North America to South America, leaving a scant record along the way. This is not impossible in view of the evidence and not even particularly unlikely. If this was the case, however, in the light of the evidence of close connections between Europe and North America and the presumption of centers of origin of such groups as perissodactyls outside eastern Asia, then the notoungulates must be considered to give evidence of another separate center. This too is not impossible, but inasmuch as the evidence can be interpreted in a number of ways, even under the restricting hypothesis of large areas and favorable climates, there is little to justify such a conclusion.

Of the several other possibilities, either a North American origin and spread to both Asia and South America or a South American origin with spread through North America to Asia is not unreasonable. Any of the three hypotheses requires a passage between North and South America at a time when few other animals, if any, made this transition. The history of South America, however, seems to have been one of population by occasional immigrants—marsupials and condylarths very early, and primates and rodents later in the Tertiary. So perhaps this too fits the pattern. An alternative to any of the above, of course, is that the morphological resemblances are the result of parallelism. The assignment of the North American and Asiatic specimens is based in large part on dental features, especially the existence of an isolated entoconid in the lower molars. Certainly the possibility of parallelism cannot be ruled out, although it has not received serious consideration in this biogeographic context. Accepted, it completely alters the likelihood of the several possible interpretations noted above.

Edentates pose somewhat similar problems, being predominantly South American but represented in North America by a few specimens ranging in age from the late Paleocene to the Oligocene. In this case resemblances between North American and South American forms are based on much more information, including features of skulls and especially very distinctive brain characteristics. The chances of parallel development seem much more remote. A very questionable edentate, *Chungchienia*, has been found in the Asian Eocene. The edentates of North America are primitive and cannot be placed in any of the known South American families. The first of the latter, Dasypodidae, occurs in the early Eocene, represented by five or more genera. Origin of edentates in North America with a chance crossing to South America sometime during late Paleocene is not unlikely, but in view of the absence of any record before the upper Paleocene, the opposite could apply, and of course an Asiatic origin and dispersal through North America to

South America, fitting a scheme of Old World, Holarctic origin for major placental groups, cannot be demonstrated to be false.

Lagomorphs pose some interesting biogeographic problems, partly because of the scant record before the Oligocene, when their remains first become abundant. Three phylogenetic sources have been proposed in recent years: periptychid condylarths (A. E. Wood, 1957a), zalambdodont insectivores (L. S. Russell, 1959), and *Pseudictops*, an anagalid leptictoid (Van Valen, 1964a). The earliest known fossil is *Eurymylus* from the late Paleocene of Mongolia. *Pseudictops* occurs in the same beds. *Mimolagus* has come from China and probably is somewhat later. Thereafter there is no record until the late Eocene of North America.

Asiatic occurrences in the late Paleocene and suggested origin from an Asiatic insectivore give some evidence, however shaky, of an Asiatic origin, possibly along with the notoungulates. The long break, however, means that there was a period of development in an unknown area lasting until the late Eocene. This is a time for which many mammal-producing series are known throughout the world, and thus it would seem that the lagomorphs must have developed away from the "mainstream" of evolution of the well-known lines, or possibly were present in the depositional areas but not in niches that were contributing to the suite of preserved fossils.

In many respects this history resembles that of some artiodactyls, the merycoidodonts, cameloids, and hypertragulids in particular. Whether or not this was of biogeographic significance is uncertain, but each of these groups does suggest a complex picture of origins and dispersals in the early Tertiary, much more complex than can be encompassed in any simple model.

Implications of the Interpretations

This rather brief exploration of possible explanations and interpretations of the events leading to the Eocene dispersal of therians after their origin in the late Mesozoic shows, above all, that even at this relatively recent stage of vertebrate history the raw data are far from sufficient for interpretations that are not readily countered by equally likely alternatives. This emphasizes the very great need for universal principles and strongly established theories of biogeography. All of the interpretations, of course, have been made within a framework consistent with known biological and physical principles, with evolutionary "laws" as the main source of guidance. If Darlington's general hypothesis is adopted, as it was in outlining the simplest and most "comfortable" interpretations, then limits are imposed that reduce the options in decision making. If the limit of possible parallelism is kept minimal, possibilities once again are limited. The principles relating the nature of corridors of communication, as treated extensively by Simpson, coupled with geological data on the positions of continents, their connections and climates, also form restricting guides. If particular interpretations drawn from each of these areas are combined into a single hypothesis—for example, if the concept of Holarctic origins of major groups is coupled with the notion of favorable climates over a large area as requisite to origin and dispersal, in a setting of stable continents and ocean basins, with continental connections by isthmian links, and the concept of constant polar position is added—fairly confident interpretations become possible.

Whether or not this practice is effective at the present state of development of theory is uncertain. Were the data of the late Mesozoic and early Tertiary sufficient, as they begin to be in the Eocene, they could be used for testing the hypothesis and principles that have proven valid in explanation of many later Tertiary and Recent distributions. For the time before the late Eocene this in general is not the case. The capacity to explain distributions during these times thus depends heavily on the degree of acceptance of principles and hypotheses. As was pointed out repeatedly in preceding sections, applications of these concepts as now known from studies of late Cenozoic distributions, including moderns, do produce a reasonable interpretation of the data. The sample points, however, are so few that such interpretations must be considered suspect in view of the serious questions concerning the generality of current concepts of the development of major patterns of distributions.

BIOGEOGRAPHY AND PALEOGEOGRAPHIC INTERPRETATIONS

General Matters

The problems of arriving at definitive conclusions about the origin and dispersal of mammalian groups, even in the later parts of the Tertiary, give little basis for optimism about the successes of studies of this sort for times before the Cenozoic. The problems that plague such studies increase, although irregularly, through increasingly remote times, as the sampling problems treated in the first section of this book become more intense. Data necessary for biogeographic interpretations suffer particularly from increasing restrictions of sampling that affect the absolute area sampled, the diversity of environments, and the taxonomic properties of the materials with respect to their representation of actual diversity and to the problem of accurate determinations. Time relationships between fossil-bearing beds also become less precise for older deposits, adding to the difficulties of interpretation of the relationships of geographically separated events.

The result has been that relatively few studies of the origin and dispersal of major groups of vertebrates have been made on the nonmammalian terrestrial vertebrates of the Mesozoic and Paleozoic. Analyses have been directed more toward solutions of problems of ancient climates and of the positions, relationships, connections, and movements of continents and ocean basins. It is mostly to these aspects of paleogeography that the concluding portions of this chapter are devoted.

The same biogeographic processes and the same general concepts of origin and dispersal are applicable, although possibly in considerably modified form in the early parts of the record, before the full occupancy of the continents by terrestrial vertebrates. These are assumed as known in studies whose aim is primarily determination of physical conditions, whereas in investigations of biological aspects of biogeography the physical setting is usually assumed to be known and the biological events the object of investigation. Studies of both types, of course, have been carried out for the Cenozoic, and very commonly the two go hand in hand. For earlier times this is possible to a much reduced extent, because the possible instability of the continental masses and poles introduces imponderable variables in any interpretations of the specifics of development of patterns of distribution.

As in other parts of this book, analysis of the problems of study will focus on a few instances in which careful study has been made and tentative conclusions have been based on well-analyzed data. No effort at coverage is intended. Two studies, the Permo-Carboniferous by Westoll (1944) and by Romer (1945b), will be used. Following these, brief considerations of the distributions of other times are presented, noting in particular the current status of interpretations.

The restriction to cases involving terrestrial vertebrates—that is, creatures living on land or in continental waters—does not imply that these have some special significance in studies of ancient biogeography. They are merely most appropriate in the context of the earlier parts of this section. Marine vertebrates, to some extent, and marine invertebrates, in such studies as those of Stehli (1957; with Helsley, 1963), have figured prominently in efforts to determine the nature of continent–ocean basin relationships. Studies of plants, especially those of the *Glossopteris* flora of the Southern Hemisphere have played a prominent role in the establishment of such concepts as that of Gondwanaland.

Special Studies of the Upper Carboniferous and Early Permian

The analyses considered here, one by Westoll (1944) and the other by Romer (1945b), are especially significant in that each gives serious attention to the problems involved in biogeographic interpretation, rather than making the more usual somewhat casual remarks on the meanings of distributions on the basis of one or a few genera. Both reach only very tentative conclusions after consideration of the evidence. These are but two of many fairly detailed studies of this general time (e.g., Nopsca, 1934, or Olson, 1955, 1962), but as representative of two of the most important ways of analysis they are of particular interest here.

Westoll (1944) included his study in an extensive treatment of the haplolepid paleoniscoid fishes. He based his conclusions in particular on the haplolepids and two associated amphibian groups, nectrideans and aïstopodans, forming a haplolepid–nectridean–aïstopod complex. This association was found in Ohio, Ireland, England, and Bohemia (Czechoslovakia). The complex occurs in these places in coal swamps, in what Westoll interpreted as a special environment characterized in particular by low oxygen tension. At the various sites the other fishes, amphibians, and reptiles that occurred along with this suite were somewhat different.

Westoll considered the haplolepid–nectridean–aïstopod assemblage most important because it appeared to exist only in this restricted environment; thus, he argued, such an environment must have been, at least over a period of time, continuous between the various localities. The other creatures could probably have made their way along fairly restricted paths, perhaps across narrow land bridges, along streams, coastlines, and even in some cases through shallow marine waters. This, Westoll felt, would not have been possible for the special complex. After his analysis, he concluded that the pattern of distribution is more readily explained by an assumption of broad continental connection across which the coal basins were spread, implying later continental drift, than by land bridges. He stated very explicitly that this was not definitive, but merely the most likely explanation from the data in hand.

Some new finds of both nectrideans and aïstopods under a wider variety of circumstances than were known to Westoll have the effect of somewhat weakening his argument, but the particular association with which he works has not been found under other circumstances.

Romer (1945b) made a very thorough analysis of Stephanian deposits at Kounova, Czechoslovakia, and the early Permian Wichita beds in Texas. He made special note of a matter that is of great importance in the formulation of biogeographic conclusions from extinct organisms, this being that the published faunal lists for two widely separated areas may be extremely misleading. Students are generally reluctant to assign specimens from two regions as far apart as Texas and Czechoslovakia to the same genus or species even though there are no demonstrable morphological differences. If the faunal lists available for these two areas in 1945 had been used, very little faunal similarity would have been evident. Romer, with his expert knowledge of the Texas vertebrates, reached the conclusion that they were extremely similar to those from Kounova. To illustrate this he went into considerable detail to formulate a comparison of similarity relative to that of modern European and American amphibian and reptile genera and families. Comparisons were based on the percentage of common taxa at the level in question. From this he concluded that the resemblances of the two Paleozoic faunas were much greater than those of the modern faunas of comparable animals.

This resemblance, Romer considered, was most readily explained by assuming the existence of a very direct avenue of interchange. He explored as alternatives an Asiatic center of origin, with dispersal to the two areas, and worldwide distributions of very similar animals. Although these were not ruled out, Romer stated that

In the late Carboniferous time the possibilities of intermigration between Europe and North America were much greater than was the case during the Cenozoic *via* the Alaskan-Siberian bridge and that, therefore, some more intimate type of connection was present.

Romer then, as Westoll had done in his study, added confirming evidence from earlier and later in the Paleozoic, based on less full or less trustworthy analyses. Romer felt that the evidence did not establish whether this avenue was provided by continental connections, with later separation by drift, or a massive land bridge that later foundered.

These two studies bring out a number of items of considerable interest in the use of data from the relatively remote past. First of all, they illustrate two very different approaches, each of which has merit and may be used separately or together. Westoll's approach is essentially paleoecological and rests heavily on the occurrence of a suite of animals in a particular environmental situation. The critical feature is that the physiological limitations of the members of this group did not permit them to migrate beyond the boundaries of a restricted environment. This then required the existence of such conditions between the populated areas, either at one time or passing across from one to another area through a relatively limited time. The coal-swamp conditions with low oxygen content of the water, he then concluded, indicated a broad, continental type of connection.

Romer, on the other hand, argued from a faunal analysis, the sort of analysis that has been most commonly used in biogeographic studies. He established the necessity for easy access routes on the basis of a large number of similar or identical genera.

Neither Westoll nor Romer used taxonomic differences as a prime source of evidence, as is often done. In ignoring this aspect they eliminated the very significant dangers of relying on absences, or so-called negative evidence. Westoll mentioned this aspect but concluded that either sampling or ecological differences of the areas supplying the other elements of the collections probably were involved.

Both of the studies, but Romer's in particular, emphasize the role of taxonomy. Even in the case of modern animals for which complete studies are possible taxonomic problems can be very serious; for the remote past they often are insurmountable. Romer's study relies on an assumption that the taxa of the two faunas are well sampled, so that his percentages have some validity, and also on his capacity to make judgments on the relationships, often on the basis of fairly fragmentary, and at best somewhat distorted, skeletal remains. Anyone who has had to make such judgments is fully aware of the necessary reservations. Furthermore, as he noted, the effects of both geographic and temporal spread does have an effect on assignments. Without question, the most reliable assessment is that made by an expert, such as Romer in this case, but two equally able experts may arrive at somewhat different conclusions, and these are certain to be influenced by their whole general outlook on the very problems for which they are seeking data.

Recognition of these difficulties, which Romer freely notes, is to be found in his estimates of generic similarity between Texas and Kounova as being at least 25 percent and possibly as high as 75 percent. Familial problems are less difficult, but the resemblances that they reveal are certainly less significant.

Another taxonomic problem that arises is in the establishment, as a scale for comparison, of an analog that has well-known biogeographic properties. The best, from a taxonomic point of view, is the modern situation, which Romer used. A very real problem, however, is involved in the necessary assumption that the criteria for genera, families, or species, as the case may be, of the extinct and living animals are commensurate.

Probably the most difficult aspect of all such studies is the very limited areas from which the samples are drawn. Both Westoll and Romer recognize this, and it is one of the reasons their conclusions are stated with such caution. If studies in, say, central Asia or Australia were to turn up Westphalian beds in which suites similar to those used by Westoll occurred, the picture would be completely altered. As far as is known this is unlikely, not, however, because beds of this age with other characteristics are known in such critical places, but largely because nothing at all is known. Even when wide distributions are known, as for the upper Triassic discussed below, far less than 1 percent of the total earth's surface produces the fossils. The rest can be known only by extrapolation.

These fairly obvious problems are not such that interpretations lack value. They do, however, argue for the kind of caution that both Romer and Westoll exercised in drawing their conclusions. Outside of this necessarily tentative aspect of answers, perhaps the most pertinent point emphasized in these studies is that it is only through the judgment of an expert that the basic data can be brought into usable form. In effect this eliminates repeatability and makes attempts at general syntheses difficult since they must include separate arrays of data that cannot be tested to determine whether or not they are commensurate.

Other Pre-Cretaceous Circumstances

Terrestrial deposits of Jurassic age as currently known offer little that is appropriate to biogeographic interpretation, and in this they are much the same as the beds of early Cretaceous age. The Morrison in western United States and the Tendaguru of east Africa have produced very similar remains of sauropod dinosaurs, and the Khota beds in India, which are still being explored, seem to have a similar fauna. This may indicate a wide dispersal of the dinosaurs at this time but gives no additional clues on continental relationships. Before the Jurassic, as far back as the Devonian and even the uppermost Silurian, the rocks have yielded vertebrates that have been interpreted as indicating patterns of continental relationships. Most interpretations have been of a rather casual nature and not based on analyses that could produce definitive results.

During the last decade, with increased interest in continental movements and polar shifting and as new physical data have become available, additional attention has been focused on the use of organisms to support various hypotheses. In particular these have related to the concept of a large northern continental mass, Laurasia, and a southern mass, Gondwanaland, separated by the Tethyan seaway. The nature of the evidence from some of the critical times and the interpretations that they have suggested are treated briefly below to show the current status of investigations and the problems that attend such interpretations.

The Triassic

Among those who have accepted in one form or another the idea that there once existed a single large continental mass that fragmented to form the present continents there have been great differences of opinion as to when the disruption began. Initiation has been placed in the late Paleozoic, the early Mesozoic (probably the most popular), and carried all the way into the early Cenozoic. If there was separation not long before the Triassic or during the Triassic and if there resulted from this, among other things, at least partially isolated northern and southern continental masses, the distribution of Triassic terrestrial vertebrates might be expected to have been affected by these events and give evidence of them. As studies of the Tertiary have shown, a vast amount of detailed taxonomic, geographic, and stratigraphic information is necessary for such studies. The Triassic, more than any other pre-Cenozoic time seems to offer the possibility of providing the kinds of data that are required.

Extensive studies of Triassic terrestrial vertebrates are currently under way. They involve deposits in North America, South America, England, the Soviet Union, South and East Africa, Southeast Asia and Antarctica.

Deposits in central Europe, though yielding predominantly marine verte-brates, add to the picture. So much is being done at present that an attempt at synthesis is impractical, but it may be anticipated that within the next several years one will be possible. In the meantime we may look at some of the things that are going on and consider their potentials.

Romer (1966*b*) has suggested the existence of three successive major faunal complexes in the Triassic, without specifying them as strictly lower, middle, and upper. These are as follows:

1. An early complex characterized especially by carnivorous cynodonts and including dicynodonts, a few gomphodonts, archaic archosaurs, and a few primitive rhynchosaurs.

2. A complex characterized by abundant gomphodonts, with many rhynchosaurs, some thecodonts, and a few dinosaurs.

3. A complex with few therapsids, advanced thecodonts, and dinosaurs, with few rhynchosaurs.

As Romer noted, the succession is actually more complex, with intermediates, and with distinctive faunal complexes.

During the time of existence of the latest fauna of which the Rhaetic component is most characteristic, deposition was taking place over a wide range of the earth, and preservation of fossils deposits has been extensive. Representatives are known from North America (Kayenta formation, the Newmark basin), British Isles (fissure fills), central Europe (Rhaetic), East Africa (upper red beds), and China (Lu Feng series). Contemporaneity of the deposits is not completely established, and current collections and studies are insufficient for faunal analyses of the type carried out by Romer for the Kounova–Texas complexes. These should, however, become possible as study goes on and taxonomy of the components from the different sites is worked out.

Currently the tritylodonts and the morganucodontids, of which one or both are present in the British Isles, central Europe, East Africa, southern China (Yunnan), and North America are the most instructive. Very similar representatives of both groups have come from widely separated regions. Different generic names have been given to the creatures from the scattered areas, but whether this is justified in all cases is open to question. Adding to the evidence of cosmopolitanism are the dinosaurs, whose taxonomy, as usual, causes trouble. In addition, from the British Isles and the Newark basin of North America have come specimens of gliding lizards, *Kuehneosaurus* and *Icarosaurus*, respectively. These too may not in fact be generically distinct, and Robinson (1967) has argued that their distribution could be well explained by the concept of a single continental mass and a later separation.

These data suggest that there was widespread faunal interchange during the Triassic, and that this persisted to the very latest Triassic. There is no distinction between northern and southern continents as far as known elements are concerned. At the current stage of study not much more than this can be said, because several hypotheses pertaining to origins and dispersals are equally plausible. Close comparisons of the faunal complexes from the various sites must precede any broad interpretations on the patterns of continental relationships at this time.

Romer's latest Triassic fauna includes, in the Northern Hemisphere, a phytosaur–metoposaur complex. This, as he notes, results in complications, since the complex (often referred to as the Maleri complex) may span a considerable period of time and is distinctly older than the Rhaetic faunas. Colbert (1958) analyzed the Maleri fauna as then known, following on Von Huene's earlier study (1940). Additional study (see, for example, Jain, Robinson, and Roy Chowdhury, 1964) has augmented the data, strengthened Colbert's summaries, and added to the concepts of the widespread nature of this type of fauna. This phytosaur–metoposaur complex is clearly a facies assemblage and probably covers a considerable period of time. It varies in its components over the different areas of occurrence, with rhynchosaurids present in some places and not in others. Elements of the fauna are known in India, continental Europe, the British Isles, East Africa, western and eastern United States, and South America. Continuing studies indicate increasingly that this was a coherent lowland-fauna complex, with extensive inter-communication between its widespread segments. A great deal of uncertainty remains, however, and only continuing study will clarify many of the problems relating to the consistency of the complex, the temporal relationships of its segments, and the taxonomy of the constituents. In 1958 Colbert reached the conclusion that the Maleri fauna was essentially a northern assemblage that pushed southward into India but did not penetrate Africa. At that time he argued that the evidence strongly favored the existence of continents in positions not greatly different from those of today.

Additional information, both biological and physical, has paved the way for some changes in this position. Recently, for example, a fragment of a labyrinthodont amphibian was found in Antarctica (Barrett, Baillie, and Colbert, 1968), implying some sort of connection of this area with a continental source of amphibians. Colbert, like most vertebrate paleontologists, has come to accept the likelihood of continental drift and to study the new information in the light of this probability. Confidence in the theory has been increased by a very recent, but only briefly described, discovery of a "bone bed" of Triassic age in Antarctica. Among the vertebrates present is a fragment of the aquatic therapsid *Lystrosaurus*.

Romer's middle Triassic fauna, including gomphodonts and rhyncho-saurids, has recently been found in profusion in South America. Elements of this fauna are extremely close, perhaps generically identical, with some in its African counterpart. Extensive study is now being carried out under Romer and Crompton on the faunal complexes of these two regions, and one of the finest opportunities for paleobiogeographic comparisons will emerge from this work. At first glance the common aspects of South America and Africa lead to a conclusion that they must have formed part of a single continent, with free interchange of creatures that were part of a single, probably continuous faunal complex. The added information from the Antarctic assemblage noted above seems to bring this landmass into possible relationships as well. This ties in well with a considerable body of physical and botanical evidence of a great southern continent. It also shows how it is possible to overstate probabilities on the basis of quick judgments or of inadequate samples.

Baird and Take (1959) reported on a reptilian fauna in the basal upper Triassic of Nova Scotia. In it have been found rhynchosaur remains and, of great significance, a gomphodont cynodont. In Arizona, near St. John's (Camp and Wells, 1956), a single quarry yielded abundant remains of a large dicynodont, *Placerias*. In addition some fragmentary remains of large dicynodonts are known from Wyoming. Without these scant and scattered finds it might be possible to conclude with confidence that there had been a southern dispersal. With them it is not possible to conclude that this was not the case, but other possibilities open up. More than anything else, these few northern remains emphasize very strikingly the need at all times to consider the possible effects of lack of samples from areas that might have figured in the development of dispersals.

From their information, Camp and Welles, after a thorough analysis of dicynodonts from other regions, concluded that a northern dispersal offered the most reasonable explanation of the distributions of large dicynodonts, but they did not rule out southern routes, merely stating that what was known did not conclusively support southern intercontinental connections. Large dicynodonts of the family Kannemeyeriidae, of which *Placerias* is a member, are known from East Africa, South America, eastern Europe (Soviet Union), and North America. All except *Placerias*, which is upper Triassic, are from the earlier part of the Triassic, lower or middle. The same applies to members of the family Stahleckeriidae known from eastern Asia, East Africa, and South America. Here, as before, both the time and the faunal associations introduce complex factors that must be clarified before any firm faunal resemblances can be determined and used for interpretation of continental relationships.

Romer's earliest Triassic fauna is less widespread, unless some of the dicynodonts noted above are considered to be parts of it. It is best known in South Africa and Shansi, China, with some elements from the Soviet Union as well. Lower Triassic beds that have produced terrestrial and semiaquatic vertebrates are restricted in comparison with those of later parts of the period. In North America, for example, only the Moenkopi of Arizona and Utah has produced a fauna of presumed lower Triassic age. The vertebrates are largely amphibians, and there is some question as to whether they may be of middle rather than lower Triassic age.

Freshwater fish faunas from northern Europe, Greenland, and northern North America have very similar elements. These add something to the data, but compared with later Triassic information, that of the early Triassic must be considered to have very limited biogeographic significance. On the other hand, it must be noted that, deficient as it is, it is better on the whole than the information available for Paleozoic times.

This short survey of some of the emerging information about the Triassic shows the directions that studies are taking and indicates that in time they may be sufficient for conclusions that have some high degree of likelihood. What has emerged to date suggests that there was a worldwide distribution of several faunal complexes in the middle and late Triassic, and by inference this may well have been the case in the early Triassic as well. There are few hints concerning how these distributions came to be. Opinion has been shifting toward an interpretation consistent with concepts of continental drift. Easy solutions to distributional problems are available if this is assumed. But the evidence at hand is little more consistent with such an interpretation than it is with other continental patterns and dispersal routes. This does not imply that the concept of continental drift is invalid but merely that evidence of vertebrates as now known is not definitive with respect to the disposition of continents and their connections at the times in question.

The Permian

Records of the Permian, like those of the lower Triassic, are not satisfactory for definitive analyses of the nature and positions of the continental masses. Basically the problem is simply that there is an insufficient geographic and ecologic spread of the fossil-producing beds. Almost all, if not all, terrestrial fossil-bearing beds were deposited near the margins of seas, and many were formed on large deltas. Some concepts of possible connections, such as those treated by Romer, can be obtained from the data, but no picture of general, world-wide significance, such as required to support or contradict the hypothesis of continental drift, is possible.

Most of our knowledge of the upper Permian comes from deposits in South Africa and the Soviet Union. Faunas with very similar elements, some generically identical, occur in the two regions. Intermigration must have taken place during part of the time that the continent of Gondwanaland existed, if it did, and the northern and southern landmasses cannot have been fully isolated. Vertebrates are known from limited areas during early Kazanian and early Guadalupian times in the Soviet Union and North America, respectively. Comparable faunas are not known elsewhere. There are some resemblances between the two regions. The fauna from the Soviet Union is predominantly composed of members of the dinocephalian complexes, related closely to those in slightly later beds in South Africa. Very fragmentary remains of this complex have been found in North America, with the extent of similarity difficult to judge because of the incompleteness of the materials. The main part of the North American fauna consists of elements derived from the caseid chronofauna, caseids and large captorhinids. These elements, represented by closely related genera are known from a few specimens from two sites in the Soviet Union. Intimate connections seem to have existed, with easy migration from one region to another. Two faunas from distinct facies, however, seem to be represented. Whether these followed similar or different paths of exchange cannot be determined.

The vertebrates of the early Permian are known very largely from the United States, from the midcontinental regions and from the east central part of the country. Only sporadic finds have been made in other parts of the world, central Europe, the western part of the Soviet Union, Siberia, and Brazil. These are largely from the very early parts of the Permian, comparable to the vertebrates from Kounova and Texas compared by Romer. Beyond the conclusions that he reached, they show very little of biogeographic significance. One interesting occurrence is a specimen of an archegosaur amphibian in South America, a type known from Europe, but not from North America. Equally intriguing is the occurrence of *Mesosaurus*, an aquatic reptile, in Africa and South America, but not in the Northern Hemisphere. Although interesting, such isolated instances of distribution are insufficient for definitive conclusion.

Earlier Paleozoic

The vertebrates, both terrestrial and freshwater, have been the basis for various statements about continental relationships. The most definitive study is that of Westoll reported earlier for the Westphalian. Essentially all finds on which any suggestions have been based have come from North America and Europe. They suggest, on the basis of common genera from the two areas, that close connections and easy migration routes were available for freshwater vertebrates. This holds particularly for the upper Devonian, when such fishes as *Eusthenopteron* and *Cheirolepis* occur in the deposits on both continents. In addition Australia has yielded similar vertebrate remains, suggesting that the distributions were very widespread but not, of course, giving any information on how this may have been accomplished. What little is known for the upper Silurian gives much the same picture for Europe and North America. At this time, and for earlier times, the marine invertebrates assume a much more important role as indicators of biogeography than the very few, scattered samples of vertebrates.

SUMMARY

From the early Cretaceous to the beginning of their record, vertebrate remains give tantalizing glimpses into faunal distributions, and these in turn hold the possibility of supplementing other evidence related to the paleogeography of these times. Paired comparisons of well-preserved faunas from sites on two continents, if carried out either on a faunal or a faunal-ecological basis, can give insight into the intimacy of relationships of the lands involved. They have not proven suitable for definitive interpretation of the nature of continental connections. For most of the Paleozoic only such paired comparisons can be made.

The Triassic record, which is developing rapidly, holds out the promise of eventually providing information that may be moderately definitive. As yet, this has not been realized. Even when the currently known regions have been well sampled and their faunas worked out and compared, results will be based on coverage that leaves most of the area of the continents unknown. Until the last decade the majority of interpretations of this record were cast within a framework influenced by concepts of continental stability and a denial of continental drift. This was less true among European than American students. As late as 1956 and 1958 Camp and Welles, and Colbert, respectively, interpreted the evidence of Triassic faunas as showing a northern source of dispersals as most likely. Undoubtedly the influence of Matthew's work was very strongly felt in all such decisions.

During the last decade, with the development of studies of paleomagnetism and new ideas about the physical mechanisms for continental drift, this hypothesis has gained much support in the United States and increased its popularity in Europe. As this has happened, tendencies to interpret evidence in accord with this hypothesis have increased. The importance of the physical framework cannot be ignored in evaluating interpretations. It is notable, however, that both Westoll and Romer, in the middle 1940s, after very careful examination of their limited evidence felt that the concept of drift offered an acceptable, and probably the most most likely, explanation of what they found.

The major problems in studies directed toward understanding paleogeography and the biological aspects of biogeography from fossils lie in three areas: the inadequacy of samples, the lack of satisfactory taxonomies, and the difficulties with time relationships of the beds from different areas. Each of these is subject to great improvement, but for each period there exist finite limits of the areal extent of samples and the types of environment that they represent. This must necessarily place a limit on the success of analyses, a limit that becomes increasingly restrictive in more and more ancient times.

References

Most of the references in the following list are cited in the text figures, or tables. The few that are not have been included because they give information on the subject of a cited reference by the same author.

Abel, O. 1912. *Grundzüge der Palaeobiologie der Wirbeltiere.* Stuttgart. P. 708.
———— 1919. *Die Stämme der Wirbeltiere.* Berlin and Leipzig. P. 914.
Adams, L. A. 1919. A memoir on the phylogeny of the jaw muscles in recent and fossil vertebrates. *Ann. N.Y. Acad. Sci.,* **28**: 51–166.
Adie, R. 1963. Geological evidence of possible Antarctic land connections, in *Pacific basin biogeography,* J. L. Gressit, Ed., Honolulu. Pp. 455–463.
Ager, D. V. 1963. *Principles of paleoecology.* New York. P. 371.
Aldinger, H. 1937. Permische Ganoidfische aus Ostgrønland. *Medd. Grønland,* **102**: 1–392.
Alexander, R. McN. 1965. The lift produced by the heterocercal tails of Selachii. *J. Exptl. Biol.,* **43**: 131–138.
———— 1967. *Functional design in fishes.* London. P. 160.
Allee, W. C., A. Emerson, O. Park, T. Park, and K. P. Schmidt. 1949. *Principles of animal ecology.* Philadelphia and London. P. 837.
Allis, E. P. 1889. The anatomy and development of the lateral line system in *Amia calva. J. Morphol.,* **2**: 463–540.
———— 1898. On the morphology of certain bones of the cheek and snout of *Amia calva. J. Morphol.,* **14**: 425–466.
Ameghino, F. 1906. Les formations sédimentaires du Crétacé supérieur et du Tertiare de Patagonie avec un parallèle entre leurs faunes mammologiques et celles du l'ancien continent. *Anal. Mus. Nac. Hist. Nat. Buenos Aires,* **15**: 1–568.
Andrews, C. W. 1907. Notes on the osteology of *Ophthalmosaurus icenicus* Seeley, an ichthyosaurian reptile from the Oxford clay of Peterborough. *Geol. Mag.,* **4**: 203.
Appleby, R. M. 1961. On the cranial morphology of ichthyosaurs. *Proc. Zool. Soc. London,* **B137**: 333–370.

Atz, J. W. 1952a. Internal nares in the teleost, *Astroscopus. Anat. Rec.*, **113**: 105–115.

———— 1952b. Narial breathing in fishes and the evolution of internal nares. *Quart. Rev. Biol.*, **27**: 366–377.

Axelrod, D. 1967. Quaternary extinctions of large mammals. *Univ. Calif. Publ. Geol. Sci.*, **74**: 1–42.

Axelrod, D., and H. Bailey. 1968. Cretaceous dinosaur extinction. *Evolution*, **22**: 595–611.

Backhouse, K. 1960. Locomotion and direction finding in whales and dolphins. *New Scientist*, **7**: 26–28.

Bader, R. 1965a. Fluctuating asymmetry in the dentition of the house mouse. *Growth* **29**: 291–300.

———— 1965b. A partition of variance in dental traits of the house mouse. *J. Mammology*, **46**: 384–388.

———— 1965c. Reliability of dental characters in the house mouse. *Evolution*, **19**: 378–483.

———— 1965d. Phenotypic and genotypic variation in odontometric traits in the house mouse. *Am. Midland Naturalist*, **74**: 28–38.

Baier, J. J. 1708. Oryctographia Norica. Nuremberg. p. 95.

Bainbridge, R. 1958. The speed of swimming of fishes as related to size and to the frequency and amplitude of the tail beat. *J. Exptl. Biol.*, **35**: 109–133.

———— 1961. Speed and power in swimming. *Symp. Zool. Soc. London*, **5**: 13–32.

———— 1963. Caudal fin and body movement in the propulsion of some fish. *J. Exptl. Biol.*, **40**: 23–56.

Baird, D. 1957. Rhachitomous vertebrae in the loxommid amphibian *Megalocephalus* (abstr.) *Bull. Geol. Soc. Amer.*, **68**: 1698.

———— 1964. The aïstopod amphibians surveyed. *Breviora*, **206**: 1–17.

———— 1965. Paleozoic lepospondyl amphibians. *Am. Zoologist*, **5**: 287–294.

Baird, D., and W. Take. 1959. Triassic reptiles from Nova Scotia. *Bull. Geol. Soc. Amer.*, **70**: 1565–1566.

Baker, H. G., and C. L. Stebbins, Eds. 1965. *The genetics of colonizing species.* New York, P. 588.

Barghusen, H. 1968. The lower jaw of cynodonts (Reptilia, Therapsida) and the evolutionary origin of mammal-like adductor jaw musculature. *Postilla*, **116**: 1–49.

Barrell, J., et al. 1918. *Evolution of the earth.* New Haven. P. 200.

Barrett, P., R. Baillie, and E. Colbert. 1968. Triassic amphibian from Antarctica. *Science*, **161**: 460–462.

Barry, T. H. 1963. On the variable position of the tympanum in recent and fossil tetrapods. *S. Afr. J. Sci.*, **59**: 160–175.

———— 1968. Sound conduction in the fossil anomodont *Lystrosaurus. Ann. S. Afr. Mus.*, **50**: 275–281.

Beckner, M. 1959. *The biological way of thought.* New York. P. 200.

Beerbower, J. R. 1963. Morphology, paleoecology and phylogeny of the Permo-Carboniferous amphibian *Diploceraspis. Bull. Mus. Comp. Zool.*, **130**: 31–108.

———— 1968. *Search for the past.* 2nd ed. New Jersey. P. 512.

Bellairs, D'A. A., and G. Underwood. 1951. The origin of snakes. *Biol. Rev.*, **26**: 193–237.

Belon, du Mans, P. 1555. L'histoire de la nature des oyseaux. Chapt. 12: L'anatomie des ossemens oyseaux. Paris.

Bendix-Almgreen, S. E. 1967. The bradydont elasmobranchs and their affinities: a discussion, in *Current problems of lower vertebrate phylogeny*, T. Ørvig, Ed. Stockholm. Pp. 153–170.

Berg, L. S. 1940. Classification of fishes both recent and fossil. *Trans. Inst. Zool. Acad. Sci. USSR*, **5**: 7–517 (Russian and English).

—— 1955. *Classification of fishes, living and fossil.* 2nd ed., revised by E. N. Pavlovsky. *Trans. Inst. Zool. Acad. Sci. USSR,* **20:** 1–286 (Russian).

Berg, L. P., D. Obruchev, et al. 1964. Actinopterygii, Systematic part, in *Fundamentals of Paleontology,* Vol. II, Y. A. Orlov, Ed. Moscow, pp. 336–472 (*in Russian*).

Bergounioux, F. M. 1938. *Archaeochelys Pougeti* nov. gen., nov. sp. Tortue fossile du Permien de l'Aveyron. *Bull. Soc. Geol. Fr.,* **8:** 67–75.

—— 1955. Chelonia, in *Traité de paléontologie,* vol. 5, J. Piveteau, Ed. Paris. Pp. 487–544.

Bertin, L., and C. Arambourg. 1958. Super-ordre de teleostéens (Teleostei), in *Traité de zoologie,* vol. 13, fasc. 3, P.-P. Grassé, ed. Paris. Pp. 2204–2500.

Bertmar, G. 1966. The development of the skeleton, blood vessels, and nerves in the dipnoan snout, with a discussion on the homology of the dipnoan posterior nostril. *Acta Zool.,* **47:** 81–147.

—— 1968. Lungfish phylogeny, in *Current problems of lower vertebrate phylogeny,* T. Ørvig, Ed. Stockholm. Pp. 255–283.

Bigelow, H., and W. C. Schroeder, 1948–1953. Fishes of the western North Atlantic. *Sears Foundation for Marine Research, Mem.* 1, Pt. 1. New Haven. P. 576.

Bloom, W., and D. Fawcett. 1962. *A textbook of histology.* 8th ed. Philadelphia. P. 720.

Blot, J. 1966. Étude des Palaeonisciformes du Bassin Houiller de Commentry, *Cahiers de Paléontologie,* ed. du C.N.R.S. P. 99.

Blumenbach, J. F. 1779. *Handbuch der Naturgeschichte.* Göttingen. P. 559.

Bock, W. J. 1959. Preadaptation and multiple evolutionary pathways. *Evolution,* **13:** 194–211.

Bohlin, B. 1945. The Jurassic mammals and the origin of the mammalian molar teeth. *Bull. Geol. Inst. Uppsala,* **31:** 363–388.

Bolt, J. 1968. The osteology and relationships of *Doleserpeton annectens,* a new rhachitomous amphibian from the lower Permian of Oklahoma. Ph.D. thesis, University of Chicago. P. 176.

Boonstra, L. D. 1963a. Early dichotomies in the therapsids. *S. Afr. J. Sci.,* **59:** 176–195.

—— 1963b. Diversity within the South African Dinocephalia. *S. Afr. J. Sci.,* **59:** 196–206.

Boulenger, G. A. 1884. Synopsis of the families of existing Lacertilia. *Ann. Mag. Nat. Hist.,* ser. 5, **14:** 117–122.

—— 1893–1896. *Catalogue of snakes in the British Museum* (Natural History). 3 vols.

Brandt, J. 1855. Beiträge zur nähern Kenntniss der Säugethiere Russlands. *Mem. Acad. Imp. Sci. St. Petersbourg,* ser. 6, **7:** 1–365.

Breder, C. M., Jr. 1926. The locomotion of fishes. *Zoologica,* **4:** 159–312.

Briggs, J. 1966. Zoogeography and evolution. *Evolution,* **20:** 282–289.

Brink, A. S. 1956. Speculations on some advanced mammalian characteristics in the higher mammal-like reptiles. *Palaeont. Africana,* **4:** 77–96.

—— 1958. On the skeleton of *Aneugomphius ictidoceps* Broom and Robinson. *Palaeont. Africana,* **5:** 29–36.

—— 1960. A new type of primitive cynodont. *Palaeont. Africana,* **7:** 119–154.

—— 1963a. On *Bauria cynops* Broom. *Palaeont. Africana,* **8:** 39–56.

—— 1963b. A new skull of the procynosuchid cynodont, *Leavachia duvenhagei* Broom. *Palaeont. Africana,* **8:** 57–75.

—— 1963c. Two cynodonts from the Ntawere formation in the Luahgwa Valley of Northern Rhodesia. *Palaeont. Africana,* **8:** 77–96.

—— 1963d. The taxonomic position of the Synapsida. *S. Afr. J. Sci.,* **59:** 153–159.

—— 1963e. Notes on some new *Diademodon* specimens in the collection of the Bernard Price Institute. *Palaeont. Africana,* **8:** 97–112.

Brinkmann, R. 1959. *Abriss der historische Geologie.* Vol. 2. Stuttgart. P. 360.

Brodkorb, P. 1963. Catalogue of fossil birds. Pt. 1, Archaeopterygiformes through Ardeiformes. *Bull. Fla. State Mus. Biol. Sci.*, **7**: 179–293.

—— 1964. Catalogue of fossil birds. Pt. 2, Anseriformes through Galliformes. *Bull. Fla. State Mus. Biol. Sci.*, **8**: 195–335.

—— 1967. Catalogue of fossil birds. Pt. 3, Ralliformes, Ichthyornithiformes, Charadriiformes. *Bull. Fla. State Mus. Biol. Sci.*, **11**: 99–220.

Broili, F., and J. Schroeder. 1934. Beobachtungen an Wirbeltieren der Karrooformation, V. Über *Chasmatosaurus van koepeni* Haughton. *Sitz. Ber. Akad. Wiss. München*, 225–264.

Broom, R. 1914. A new thecodont reptile. *Proc. Zool. Soc. London*, 1702–1777.

—— 1933. *The mammal-like reptiles of South Africa and the origin of mammals.* London. P. 376.

—— 1938. On a new type of primitive fossil reptile from the upper Permian of South Africa. *Proc. Zool. Soc. London*, **B108**: 535–542.

Brough, J. 1935. On the structure and relationship of the hybodont sharks. *Mem. Proc. Manchester Lit. Philos. Soc.*, **69**: 35–49.

—— 1936. On the evolution of bony fishes during the Triassic period. *Biol. Rev.*, **11**: 385–405.

—— 1939. *The Triassic fishes of Besano, Lombardy.* London. P. 117.

Brough, M. C., and J. Brough. 1967. Studies on lower tetrapods. I. The early Carboniferous microsaurs. II. *Microbrachis*, the type microsaur. III. The genus *Gephyrostegus. Philos. Trans. Roy. Soc. London*, **B252**: 107–165.

Brown, M. E., Ed. 1957. *The physiology of fishes.* New York. P. 526.

Brown, W. L. 1958. General adaptation and evolution. *Syst. Zool.*, **7**: 157–168.

Brundin, L. 1965. On the real nature of transantarctic relationships. *Evolution*, **19**: 496–505.

—— 1967. Application of phylogenetic principles in systematics and evolutionary theory, in *Current problems of lower vertebrate phylogeny*, T. Ørvig, Ed. Stockholm. Pp. 473–495.

Bruno, E. 1929. Die Goboiidenflosse und ihre Anpassung an das Landleben. *Z. Wiss. Zool.*, **133**: 411–446.

Busnel, R. G., Ed. 1965. *Acoustic behavior of animals.* New York. P. 933.

Butler, P. M. 1939. The teeth of the Jurassic mammals. *Proc. Zool. Soc. London*, **B109**: 329–356.

Bystrov, A. P. 1938. *Dvinosaurus*, als neotenische Form der Stegocephalen. *Acta Zool.*, **19**: 209–295.

—— 1939. Zahnstruktur der Crossopterygier. *Acta Zool.*, **20**: 283–338.

—— 1955. The microstructure of the shield of Agnatha, the jawless vertebrates from the Silurian and Devonian. *Acad. Sci. USSR*, 472–523 (Russian).

Cabrera, A. 1927. Datos para el conocimiento de los Dasiuroideos fossiles Argentinos. *Rev. Mus. La Plata Buenos Aires*, **30**: 271–315.

Cain, A. J. 1959. Deductive and inductive methods in post-Linnean taxonomy. *Proc. Linn. Soc. London*, **170**: 185–217.

Camp, C. L.1923. Classification of the lizards. *Bull. Am. Mus. Nat. Hist.*, **48**: 289–481.

—— 1942. California mosasaurs. *Mem. Univ. Calif.*, **13**: 1–68.

—— 1945. *Prolacerta* and the protorosaurian reptiles. *Am. J. Sci.*, **243**: 17–32, 84–101.

Camp, C. L., H. J. Allison, and R. H. Nichols. 1964. Bibliography of fossil vertebrates, 1954–1958. *Geol. Soc. Amer. Mem.*, **92**: 1–647.

Camp, C. L.,and S. P. Welles. 1956. Triassic dicynodont reptiles. Pt. I,The North American genus *Placerias.* Pt. II (by C. Camp) Triassic dicynodonts compared. *Mem. Univ. Calif.*, **13**: 255–348.

Carlson, K. J. 1968. The skull morphology and estivation burrows of the Permian lungfish, *Gnathorhiza serrata*. *J. Geol.*, **76**: 641–663.

Carroll, R. L. 1964. The earliest reptiles. *J. Linn. Soc. London* (Zool.), **45**: 61–83.

———— 1966. Microsaurs from the Westphalian B of Joggins, Nova Scotia. *Proc. Linn. Soc. London*, **177**: 63–97.

———— 1967*a*. A limnoscelid reptile from the middle Pennsylvanian. *J. Paleont.*, **41**: 1256–1261.

———— 1967*b*. An adelogyrinid lepospondyl amphibian from the upper Carboniferous. *Can. J. Zool.*, **45**: 1–16.

———— 1968. A ?diapsid (Reptilia) parietal from the lower Permian of Oklahoma. *Postilla*, **117**: 1–7.

———— 1969. A middle Pennsylvanian captorhinomorph and the interrelationships of primitive reptiles. *J. Paleont.*, **43**: 151–170.

———— 1969. Problems of the origin of reptiles. *Biol. Rev.*, **44**: 393–432.

———— 1970. The ancestry of reptiles. *Philos. Trans. Roy. Soc. London*, **B257**: 267–308.

Carroll, R. L., and D. Baird. 1968. The Carboniferous amphibian *Tuditanus* (*Eosauravus*) and the distinction between microsaurs and reptiles. *Am. Mus. Novitates*, No. 2337: 1–50.

Carter, G. S. 1931. Aquatic and aerial respiration. *Biol. Rev.*, **6**: 1–35.

———— 1967. *Structure and habit in vertebrate evolution.* Seattle. P. 520.

Carter, G. S., and L. C. Beadle. 1930. The fauna of the swamps of the Paraguayan Chaco in relation to its environment. I. Physico-chemical nature of the environment. *J. Linn. Soc. London* (Zool.), **37**: 205–258.

———— 1931. The fauna of the swamps of the Paraguayan Chaco in relation to its environment. II. Respiratory adaptations in the fishes. *J. Linn. Soc. London* (Zool.), **37**: 327–368.

Chase, J. 1965. *Neldasaurus wrightae*, a new rhachitomous labyrinthodont from the Texas lower Permian. *Bull. Mus. Comp. Zool.*, **133**: 153–225.

Chow, M. and C. Li. 1965. *Homogalax* and *Heptodon* of Shantung. Vert. Palasiatica, **9**: 15–22.

Chudinov, P. 1960. Upper Permian therapsids of the Ezhovo locality. *Paleont. J. Acad. Sci. USSR*, **4**: 81–94 (Russian).

———— 1968. On the structure of the skin in theromorph reptiles. *Acad. Sci. USSR*, **179**: 207–210 (Russian).

Clark, J., J. R. Beerbower, and K. K. Kietzke. 1967. Oligocene sedimentation, stratigraphy, paleoecology, and paleoclimatology in the Big Badlands of South Dakota. *Fieldiana: Geology Memoirs*, **5**: 1–158.

Clemens, W. 1968. Origin and early evolution of marsupials. *Evolution*, **22**: 1–18.

Colbert, E. 1941. A study of *Orycteropus gaudryi* from the island of Samos. *Bull. Am. Mus. Nat. Hist.*, **78**: 305–351.

———— 1942. An edentate from the Oligocene of Wyoming. *Notulae Naturae*, No. 109: 1–16.

———— 1955. Scales in the Permian amphibian *Trimerorhachis*. *Am. Mus. Novitates*, No. 1740: 1–17.

———— 1958. Relationships of the Triassic Maleri fauna. *J. Paleont. Soc. India*, **3**: 68–81.

———— 1963. Relationships of the Triassic reptilian faunas of Brazil and South Africa. *S. Afr. J. Sci.*, **59**: 248–253.

———— 1966. A gliding reptile from the Triassic of New Jersey. *Am. Mus. Novitates*, No. 2230: 1–23.

Cole, L. C. 1954. The population consequences of life history phenomena. Quart. Rev. Biol. **29**: 103–137.

Cope, E. D. 1868. Synopsis of the extinct Batrachia of North America. *Proc. Acad. Nat. Sci. Phila.*, 208–221.

—— 1871. On the homologies of some cranial bones of the Reptilia, and on the systematic arrangement of the class. *Proc. Am. Assoc. Adv. Sci., 19th meeting.* Pp. 194–297.

—— 1891–1898. *Syllabus of Lectures on Geology and Paleontology.* Philadelphia. P. 90.

—— 1898. *Syllabus of lectures on the Vertebrata.* Philadelphia. P. 135.

Couper, R. 1960. Southern Hemisphere Mesozoic and Tertiary Podicarpaceae and Fagaceae and their paleogeographic significance. *Proc. Roy. Soc. London*, **B152**: 491–500.

Cowles, R. B. 1958. Additional notes on the origin of the tetrapods. *Evolution*, **12**: 419–421.

Cox, C. B. 1967. Cutaneous respiration and the origin of the modern Amphibia. *Proc. Linn. Soc. London*, **178**: 37–48.

Cox, G. 1968. The role of competition in the evolution of migration. *Evolution*, **22**: 180–192.

Cranwell, L. M. 1964a. *Nothofagus*, living and fossil, in *Pacific basin biogeography*, J. L. Gressit, Ed. Honolulu. Pp. 387–400.

——, Ed. 1964b. *Ancient Pacific floras: the pollen story.* Honolulu. P. 114.

Crompton, A. W. 1955a. On some Triassic cynodonts from Tanganyika. *Proc. Zool. Soc. London*, **B125**: 617–669.

—— 1955b. A revision of the Scalaposauridae with special reference to kinetism in this family. *Res. Nat. Mus., Bloomfontein, S. Afr.*, **1**: 149–183.

—— 1958. The cranial morphology of a new genus and species of ictidorsaurian. *Proc. Zool. Soc. London*, **B130**: 183–216.

—— 1962. On the dentition and tooth replacement in two bauriamroph reptiles. *Ann. S. Afr. Mus.*, **46**: 231–255.

—— 1963a. Tooth replacement in the cynodont *Thrinaxodon liorhinus* Seeley. *Ann. S. Afr. Mus.*, **46**: 479–521.

—— 1963b. On the lower jaw of *Diarthrognathus* and the origin of the mammalian lower jaw. *Proc. Zool. Soc. London*, **B140**: 97–753.

—— 1964a. Preliminary description of a new mammal from the upper Triassic of South Africa. *Proc. Zool. Soc. London*, **B142**: 441–452.

—— 1964b. On the skull of *Oligokyphus*. *Bull. Brit. Mus. (Nat. Hist.) Geology*, **9**: 69–82.

—— 1968. The enigma of the evolution of mammals. *Optima*, 137–151 (September).

Crompton, A., and A. Charig. 1962. A new ornithischian from the upper Triassic of South Africa. *Nature*, **196**: 1074–1077.

Crompton, A., and F. Ellenberger, 1957. On a new cynodont from the Molteno beds and the origin of the tritylodontids. *Ann. S. Afr. Mus.*, **44**: 1–14.

Crompton, A., and N. Hotton, III. 1967. Functional morphology of the masticatory apparatus of two dicynodonts (Reptilia, Therapsida). *Postilla*, No. 109: 1–51.

Crompton, A., and F. Jenkins. 1967. American Jurassic symmetrodonts and Rhaetic "pantotheres." *Science*, **155**: 1006–1009.

—— 1968. Molar occlusion in late Triassic mammals. *Biol. Rev.*, **43**: 427–458.

Crusafont-Pairo, M., and R. Adrover. 1966. El primer Mammifero del Mesozoico Espanol. *Fossilia*, **5–6**: 28–33.

Cuvier, G. 1800. *Leçons d'anatomie comparée.* Vol. 1. Paris. P. 521.

Czopek, J., and H. Szarski. 1954. The cutaneous respiration of amphibia and its bearing on the evolution of the class. *Kosmos* (Lwow), **A3**: 256–267 (Polish).

Daly, E. 1969. A new procolophonoid reptile from the Lower Permian of Oklahoma. *J. Paleont.*, **43**: 676–687.

Darlington, P. J., Jr. 1957. *Zoogeography: the geographical distribution of animals.* New York. P. 675.

—— 1959. Area, climate, and evolution. *Evolution*, **13**: 488–510.

———— 1965. *Biogeography of the southern end of the world.* New York. Paperback ed., 1968. P. 238.

Davis, D. D. 1964. The giant panda. *Fieldiana: Zoology Memoirs,* **3**: 1–339.

Dawson, J. W. 1860. On a terrestrial mollusk, a chilognathus myriapod, and some new species of reptiles from the coal-measures of Nova Scotia. Quart. Jour. Geol. Sci. **16**: 268–277.

Dean, B. 1895. *Fishes, living and fossils; an outline of their forms and probable relationships.* New York and London. P. 300.

De Beaufort, L. F. 1951. *Zoogeography of the land and inland waters.* London and New York. P. 208.

De Beer, G. R. 1937. *The Development of the vertebrate skull.* Oxford. P. 522.

———— 1954. *Archaeopteryx lithographica,* a study based upon the British Museum specimen. *London, Brit. Mus. (Nat. Hist.)* P. 1–68.

De Blainville, H. M. D. 1816. Prodrome d'une nouvelle distribution systématique de règne animal. *Bull. Sci. Soc. Philom. Paris,* ser. 3, **3**: 105–124.

———— 1834. Classification adopted in 1834. Published in *Dictionaire pittoresque d'histoire naturelle,* vol. 4, 1936. Gervais. P. 619.

———— 1839. Nouvelle classification des mammifères. *Ann. Franc. Étrang. Anat. Physiol.,* **2**: 1–268.

Dechaseaux, C. 1955. Lepospondyli, Urodèles, Ichthyopterygia, in *Traité de paléontologie,* vol. 5, J. Piveteau, Ed. Paris. Pp. 275–305, 306–318, and 376–408.

Dehm, R., and T. zu Oettingen-Spielberg. 1958. Palaontologische und geologische Untersuchungen im Tertiär von Pakistan. 2, Die mittleocänen Säugetiere von Ganda Kas bei Basal Nordwest Pakistan. *Abh. Bayerische Akad. Wiss. Math. Nat. n.s.* **91**: 1–54.

De Lapparent, A. F., and R. Lavocat. 1955. Dinosauriens, in *Traité de paléontologie,* vol. 5, J. Piveteau, Ed. Paris. Pp. 785–962.

Denison, R. 1941. The soft anatomy of *Bothriolepis. J. Paleont.,* **15**: 553–561.

———— 1947. The exoskeleton of *Tremataspis. Am. J. Sci.,* **245**: 337–365.

———— 1951. Evolution and classification of the Osteostraci; the exoskeleton of early Osteostraci. *Fieldiana: Geology,* **11**: 157–196, 197–218.

———— 1956. A review of the habitat of the earliest vertebrates. *Fieldiana: Geology,* **11**: 359–457.

———— 1961. Feeding mechanisms of Agnatha and early gnathostomes. *Am. Zoologist,* **1**: 177–181.

———— 1966. The origin of the lateral line sensory system. *Am. Zoologist,* **6**: 369–370.

———— 1967a. The evolutionary significance of the earliest known lungfish, *Uranolophus,* in *Current problems of lower vertebrate phylogeny,* T. Ørvig, Ed. Stockholm. Pp. 247–257.

———— 1967b. A new *Protaspis* from the Devonian of Utah, with notes on the classification of Pteraspididae. *J. Linn. Soc. London* (Zool.), **47**: 31–37.

———— 1967c. Ordovician vertebrates from western United States. *Fieldiana: Geology,* **16**: 131–192.

Devillers, C. 1943. Nerfs craniens et circulation céphalique de *Plioplatecarpus Marsh Ann. Paleont.,* **30**: 47–59.

Devillers, C., and J. Corsin. 1967. Les os dermiques craniens des poissons et des amphibians: points de vue embryologiques sur les "territoires osseux" et les "fusions," in *Current problems in lower vertebrate phylogeny,* T. Ørvig, Ed. Stockholm. Pp. 414–428.

Dijkgraaf, S. 1963. The functioning and significance of the lateral-line organs. *Biol. Rev.,* **38**: 51–105.

Dollo, L. 1904. Les mosasauriens de la Belgique. *Bull. Soc. Belge Géol. Paléont. Hydrol.* **18**: 207–216.

——— 1905. Un nouvel opercule tympanique de *Plioplatecarpus*, mosasaurien plongeur. *Bull. Soc. Belge Géol. Paléont. Hydrol.*, **19**: 125–131.

Eaton, T. H., Jr. 1951. Origin of tetrapod limbs. *Am. Midland Naturalist*, **46**: 245–251.

——— 1959. The ancestry of modern Amphibia: a review of the evidence. *Univ. Kansas Publ. Mus. Nat. Hist.*, **12**: 155–180.

——— 1960. The aquatic origin of tetrapods. *Trans. Kansas Acad. Sci.*, **63**: 115–120.

——— 1962. Teeth of edestid sharks. *Univ. Kansas Publ. Mus. Nat. Hist.*, **12**: 347–362.

Eaton, T. H., Jr., and P. I. Stewart. 1960. A new order of fish-like Amphibia from the Pennsylvanian of Kansas. *Univ. Kansas Publ. Mus. Nat. Hist.*, **12**: 219–240.

Edgerton, H. E., and Breder, C. M. 1941. High speed photographs of flying fishes in flight. *Zoologica*, **26**: 311–314.

Edinger, T. 1942. The pituitary body in giant animals, fossil and living: a survey and a suggestion. *Quart. Rev. Biol.*, **17**: 31–45.

——— 1948. Evolution of the horse brain. Geol. Soc. Amer. Mem. **25**: 1–174.

——— 1955*a*. Hearing and smell in cetacean history. *Monatsh. Psychiatr. Neurol.*, **129**: 37–58.

——— 1955*b*. The size of parietal foramen and organ in reptiles. A rectification. *Bull. Mus. Comp. Zool.*, **114**: 3–34.

——— 1956. Paired pineal organs, in *Progress in neurobiology, Proc. First Intern. Meeting of Neurobiologists.* Pp. 121–129.

——— 1964. Recent advances in paleoneurology, in *Topics in basic neurology*, W. Bargmann and J. P. Schadé, Eds. *Progress in brain research*, vol. 6. Amsterdam. Pp. 147–160.

Efremov, I. A. 1938. On some new Permian reptiles of the USSR. *Compt. Rend. (Doklady) Acad. Sci. USSR*, **19**: 771–775.

——— 1940*a*. Die Mesen-Fauna der permischen Reptilien. *Neues Jahrb. Mineral., Geol., Palaeont.*, **B84**: 379–466.

——— 1940*b*. *Ulemosaurus svijagensis* Riab., ein Dinocephale aus den Ablangungen des Perm der USSR. *Nova Acta Leopoldina*, **9**: 12–205 (German, English summary).

——— 1946. On the subclass Batrachosauria—a group of forms intermediate between amphibians and reptiles. *Izv. Biol. Div. Sci., Acad. Sci. USSR*, 616–638 (Russian, English summary).

——— 1950. Taphonomy and the geological record. *Bk. I, Paleont. Inst. Acad. Sci. USSR*, **24**: 3–107 (Russian).

——— 1954. The fauna of terrestrial vertebrates in the Permian copper sandstones of the western Cis-Urals. *Tr. Paleont. Inst. Acad. Sci. USSR*, **56**: 1–416 (Russian).

Efremov, I. A., and B. P. Vjushkov. 1955. Catalog of localities of Permian and Triassic terrestrial vertebrates in the territories of the USSR. *Tr. Paleont. Inst. Acad. Sci. USSR*, **46**: 1–185 (Russian).

Ekman, S. 1953. *Zoogeography of the sea.* London, p. 417.

Ellerman, J. R. 1940–1941. The families and genera of living rodents. 2 vols. London. Pp. 689 and 690.

Estes, R. 1965. Fossil salamanders and salamander origins. *Am. Zoologist*, **5**: 319–334.

Ewer, D. W. 1955. Tetrapod limb. *Science*, **122**: 467–468.

Ewer, R. F. 1965. The anatomy of the thecodont reptile *Euparkeria*. *Phil. Trans. Roy. Soc. London*, **B248**: 379–435.

Felts, W. J. L. 1966. Some functional and structural characteristics of cetacean flippers and flukes, in *Whales, dolphins and porpoises*, K. S. Norris, Ed. Berkeley and Los Angeles. Pp. 255–276.

Flower, W. H. 1869. On the value of characters of the base of the cranium in the classification of the order Carnivora and the systematic position of *Bassaris* and other disputed forms. *Proc. Zool. Soc. London*, 4–37.

———— 1883. On arrangements of the orders and families of existing Mammalia. *Proc. Zool. Soc. London*, 178–186.

Ford, E. B. 1964. *Ecological genetics* London. p. 335.

Fourie, S. 1963. Tooth replacement in the gomphodont cynodont, *Diademodon. S. Afr. J. Sci.*, **59**: 211–212.

Fox, H. 1965. Early development of the head and pharynx of *Neoceratodus* with a consideration of its phylogeny. *J. Zoology.*, **146**: 470–554.

Fox, R. C. 1964. The adductor muscles of the jaw in some primitive reptiles. *Univ. Kansas Mus. Nat. Hist. Publ.*, **12**: 657–680.

———— 1968. Therian and quasi-mammals. *Evolution*, **22**: 839–840.

Fraser, F. C., and P. E. Purves. 1960a. Hearing in cetaceans. Evolution of the accessory air sacs and the structure and function of the outer and middle ear in Recent cetaceans. *Bull. Brit. Mus. (Nat. Hist.) Zool.*, **7**: 1–140.

———— 1960b. Anatomy and function of the cetacean ear. *Proc. Roy. Soc. London*, **B152**: 62–77.

Fürbringer, M. 1888. Untersuchungen zur Morphologie und Systematik der Vögel, zugleich ein Beitrag zur Anatomie der Stutz- und Bewegungsorgane. *Bidj. Dierk., K. Zool. Genoot. Nat. Art. Mag. Amsterdam*, Afl. 15, pt. 2, 1–1751.

Gadow, H. 1896. On the evolution of the vertebral column of Amphibia and Amniota. *Philos. Trans. Roy. Soc. London*, **B187**: 1–57.

———— 1913. *The wanderings of animals*. New York. p. 150.

Gardiner, B. 1960. A revision of certain actinopterygian and coelacanth fishes chiefly from the lower Lias. *Bull. Brit. Mus. (Nat. Hist.), B, Geol.*, **4**: 239–382.

———— 1963. Certain palaeoniscoid fishes and the evolution of the snout in actinopterygians. *Bull. Brit. Mus. (Nat. Hist.) Geol.*, **8**: 258–325.

———— 1967a. Further notes on palaeoniscoid fishes with a classification of the Chondrostei. *Bull. Brit. Mus. (Nat. Hist.) Geol.*, **14**: 145–206.

———— 1967b. The significance of the preoperculum in actinopterygian evolution. *J. Linn. Soc. London* (Zool.), **47**: 197–209.

Garman, S. 1913. The Plagiostomia. *Mem. Mus. Comp. Zool.*, **36**: 1–515.

Garstang, W. 1931. The phyletic classification of Teleostei. *Proc. Leeds Phil. Lit. Soc., Sci. Sect.*, **2**: 240–260.

Geoffroy, S.-H., and G. Cuvier. 1795. Mémoire sur une nouvelle division des mammifères et sur les principes qui doivent servir de base dans cette sorte de travail. *Magazin Encyclopédique*, **2**: 164–190.

Gero, D. R. 1952. The hydrodynamic aspects of fish propulsion. *Am. Mus. Novitates*, No. 1601: 1–32.

Gill. T. N. 1871. On the relations of the orders of mammals. *Proc. Am. Assoc. Adv. Sci., 19th meeting.* Pp. 267–270.

———— 1872. Arrangement of the families of mammals and synoptical tables of characters of the subdivisions of mammals. *Smithsonian Misc. Coll.*, No. 230: 1–98.

———— 1893. Families and subfamilies of fishes, *Mem. Natl. Acad. Sci. U.S.*, **6**: 125–138.

Gilmore, C. W. 1905. Osteology of *Baptanodon* Marsh. *Mem. Carnegie Mus.*, **2**: 77–129.

Gilmour, J. S. L. 1940. Taxonomy and philosophy, in *The new systematics*, J. S. Huxley, Ed. Oxford. Pp. 461–474.

———— 1951. The development of taxonomic theory since 1851. *Nature*, **168**: 400–402.

———— 1961. Taxonomy, in *Contemporary biological thought*, A. M. McLeod and L. S. Colby, Eds. Edinburgh and Chicago. Pp. 27–45.

Gislén, T. 1930. Affinities between the Echinodermata, Enteropneusta, and Chordonia. *Zool. Bidr. Uppsala*, **12**: 199–304.

Glickman, L. C. 1964. Elasmobranchii, in *Fundamentals of paleontology*, vol. 11, Y. A. Orlov, Ed. Moscow. Pp. 196–237. (Russian).

Goin, C. J., and O. B. Goin. 1956. Further comments on the origin of tetrapods. *Evolution*, **10**: 440–441.

Goodrich, E. S. 1909. Vertebrata Craniata. First fascicle: cyclostomes and fishes, in *A treatise on zoology*, pt. 9, E. R. Lankester, Ed. London. Pp. 1–518.

———— 1916. On the classification of the Reptilia. *Proc. Roy. Soc. London*, **B89**: 261–276.

———— 1928. *Polypterus*, a palaeoniscoid? *Palaeobiologica*, **1**: 87–92.

———— 1930. *Studies on the structure and development of vertebrates*. Dover Press. ed., vol. 1, 1958. New York. Pp. xv–xxx.

Gordon, M. S. 1964. Animals in aquatic environments: fishes and amphibians, in *Handbook of physiology*, sect. 4, vol. 1, *Adaptation to the Environment*, D. B. Dill, E. F. Adolph, and C. G. Wilber, Eds. New York Pp. 697–713.

Gosline, W. A. 1965. Teleostean phylogeny. *Copeia*, 186–194.

Gray, J. E. 1844. *Catalogue of the tortoises, crocodiles, and amphisbaenians in the collection of the British Museum*. London. P. 80.

Gray, J. 1933. Studies in animal locomotion. I. The movements of fish with special reference to the eel. II. The relationship between waves of muscular contraction and the propulsive mechanism of the eel. III. The propulsive mechanism of the whiting (*Gadus merlangus*). *J. Exptl. Biol.*, **10**: 88–104, 386–390, and 391–400.

———— 1936. Studies in animal locomotion. IV. The neuromuscular mechanism of swimming in the eel. *J. Exptl. Biol.*, **13**: 170–180.

———— 1946. The mechanism of locomotion in snakes. *J. Exptl. Biol.*, **23**: 101–120.

———— 1949. Aquatic locomotion. *Nature*, **164**: 1073.

———— 1953a. *How animals move*. London. P. 114.

———— 1953b. The locomotion of fishes, in *Essays in marine biology*. Edinburgh. Pp. 1–16.

———— 1957. How fishes swim. *Scientific American*, **197**: 48–54.

Greenwood, P. H., D. E. Rosen, S. H. Weitzman, and G. S. Myer. 1966. Phyletic studies of teleostean fishes with a provisional classification of living forms. *Bull. Am. Mus. Nat. Hist.*, **131**: 339–455.

Gregory, J. T. 1948. The structure of *Cephalerpeton* and affinities of the Microsauria. *Am. J. Sci.*, **246**: 550–568.

———— 1950. Tetrapods from the Pennsylvanian nodules from Mazon Creek, Illinois. *Am. J. Sci.*, **248**: 833–873.

———— 1951. Convergent evolution: the jaws of *Hesperornis* and the mosasaurs. *Evolution*, **5**: 345–354.

———— 1952. The jaws of the Cretaceous toothed birds, *Ichthyornis* and *Hesperornis*. *Condor*, **54**: 73–88.

———— 1965. Microsaurs and the origin of captorhinomorph reptiles. *Am. Zoologist*, **5**: 277–286.

Gregory, J. T., F. E. Peabody, and L. I. Price. 1956. Revision of the Gymnarthridae, American Permian microsaurs. *Bull. Peabody Mus. Nat. Hist.*, **10**: 1–77.

Gregory, W. K. 1910. The orders of mammals. *Bull. Am. Mus. Nat. Hist.*, **27**: 1–524.

———— 1911. The limbs of *Eryops* and the origin of paired limbs from fins. *Science*, **33**: 508–509.

———— 1913. Crossopterygian ancestry of the Amphibia. *Science*, **37**: 806–808.

———— 1915a. Present status of the problem of the origin of Tetrapoda, with special reference to the skull and paired limbs. *Ann. N.Y. Acad. Sci.*, **26**: 317–383.

—— 1915b. On the classification and phylogeny of the Lemuroidea. *Bull. Geol. Soc. Amer.*, **26**: 426–446.

—— 1916. Studies on the evolution of Primates, pts. 1 and 2. *Bull. Am. Mus. Nat. Hist.* **35**: 239–355.

—— 1926. The skeleton of *Moschops capensis* Broom, a dinocephalian reptile from the Permian of South Africa. *Bull. Am. Mus. Nat. Hist.*, **56**: 179–251.

—— 1931. Studies on the body-forms of fishes. *Zoologica*, **8**: 325–421.

—— 1933. Fish skulls. A study of the evolution of natural mechanisms. *Trans. Am. Philos. Soc.*, **23**: pt. 22, 75–479.

—— 1934. A half-century of trituberculy: the Cope–Osborn theory of dental evolution, with a revised summary of molar evolution from fish to man. *Proc. Am. Philos. Soc.*, **73**: 169–317.

—— 1935. Further observations on the pectoral girdle and fin of *Sauripterus taylori* Hall, a crossopterygian fish from the upper Devonian of Pennsylvania, with special reference to the origin of the pentadactylate extremities of Tetrapoda. *Proc. Am. Philos. Soc.*, **75**: 673–690.

—— 1947. The monotremes and the palimpsest theory. *Bull. Am. Mus. Nat. Hist.*, **88**: 1–52.

—— 1951. *Evolution emerging: a survey of changing patterns from primeval life to man* 2 vols. New York. Pp. 736 and 1013.

Gregory, W. K., and C. L. Camp. 1918. Studies in comparative myology and osteology, No. III. *Bull. Am. Mus. Nat. Hist.*, **38**: 447–563,

Gregory, W. K., and G. M. Conrad. 1942–1943. The osteology of *Luvarus imperialis*, a scombroid fish: a study in adaptive evolution. *Bull. Am. Mus. Nat. Hist.*, **81**: 225–283.

Gregory, W. K. and H. C. Raven. 1941. Studies on the origin of paired fins and limbs, pts. I–IV. *Ann. New York Acad. Sci.* **42**: 273–360.

Gressit, J. L., Ed. 1964 (dated 1963). *Pacific basin biogeography.* Honolulu. P. 563.

Gromova, V. I., et al. 1962. Mammalia, in *Fundamentals of paleontology*, vol. 13, Y. A. Orlov, Ed. Moscow. Pp. 61–340 (Russian).

Gross, W. 1935. Histologische Studien am Aussenskelett fossiler Agnathen und Fische. *Palaeontographica*, **83A**: 1–60.

—— 1937. Die Wirbeltiere des rhenischen Devons, Teil II. *Abh. Preuss. Geol. Landesanstalt*, **176**: 1–83.

—— 1951. Die paläontologische und stratigraphische Bedeutung der Wirbeltierfaunen des Old Reds und der marinen altpaläeozoischen Schichten. *Abh. Dtsch. Akad. Wiss. Berl. Math. Nat. Kl.* 1949, **I**: 1–130.

—— 1958. Über die älteste Arthrodiren-Gattung. *Notizbl. Hess. Landesamt. Bodenforsch.*, **86**: 7–30.

—— 1966. Kleine Schuppenkunde. *Neues Jahrb. Geol. Paleont. Abh.*, **125**: 29–48.

Grove, A. J., and G. E. Newell. 1936. A mechanical investigation into the effectual action of the caudal fin of some aquatic chordates. *Ann. Mag. Nat. Hist.*, ser. 10, **17**: 280–290.

—— 1939. The relation of the tail form in cyclostomes and fishes to specific gravity. *Ann. Mag. Nat. Hist.*, ser. 11, **4**: 401–430.

Gunther, G. 1956. Origin of the tetrapod limb. *Science*, **123**: 495–496.

Haeckel, E. 1866. *Generelle Morphologie der Organismen*, vol. 2. Berlin. P. 462.

—— 1895. *Systematische Phylogenie der Wirbeltiere (Vertebrata)*. Dritter Thiel des Entwurfs systematischen Phylogenie. Berlin. P. 660.

Hardy, A. C. 1954. *Evolution as a process.* London. Pp. 122–142.

Harrington, H. J. 1962. Paleogeographic development of South America. *Bull. Am. Assoc. Petrol. Geol.*, **46**: 1773–1814.

Harris, G. G. 1964. Considerations on the physics of sound production by fishes, in *Symposium on marine bio-acoustics*, W. N. Tavolga, Ed. New York. Pp. 233–247.

Harris, G. G., and W. A. van Bergeijk. 1962. Evidence that the lateral-line organ responds to near field displacements of sound sources in water. *J. Acoust. Soc. Amer.*, **34**: 1831–1841.

Harris, J. E. 1936. The role of fins in the equilibrium of the swimming fish. 1. Wind-tunnel tests on a model of *Mustelus canis* (Mitchell). *J. Exptl. Biol.*, **13**: 476–493.

—— 1937. The mechanical significance of the position and movements of the paired fins in the Teleostei. *Publ. Carnegie Inst. Washington*, No. 475: 171–189.

—— 1938. The role of the fins in the equilibrium of swimming fish. 2. The role of the pelvic fins. *J. Exptl. Biol.*, **15**: 32–47.

—— 1953. Fin patterns and mode of life in fishes, in *Essays in marine biology*. Edinburgh. Pp. 17–28.

Hasse, C. 1879–1885. *Das natürliche System der Elasmobranchier auf Grundlage des Baues und der Entwickelung ihrer Wirbelsäule.* 3 vols. Jena. Pp. 76, 284, and 27.

Haughton, S. H., and A. S. Brink. 1954. A bibliographic list of Reptilia from the Karroo beds of Africa. *Palaeont. Africana*, **2**: 1–187.

Hay, O. P. 1902. Bibliography and catalogue of the fossil Vertebrata of North America. *Bull. U.S. Geol. Survey*, No. 179: 1–868.

—— 1930. Second bibliography and catalogue of the fossil Vertebrata of North America. *Publ. Carnegie Inst. Washington*, No. 390, **II**: 1–1074.

Hecht, M. 1962. A reevaluation of the early history of the frogs, pt. I. *Syst. Zool.*, **11**: 39–44.

Heilmann, G. 1926. *The origin of the birds.* London. P. 208.

Heintz, A. 1932. The structure of *Dinichthys*, a contribution to our knowledge of the Arthrodira. Dean memorial vol. Pp. 115–224.

—— 1958. The head of the anaspid *Birkenia elegans* Traq, in *Studies on fossil vertebrates*, T. S. Westoll, Ed. London. Pp. 71–85.

Hempel, C. G. 1965. Fundamentals of taxonomy, in *Aspects of scientific explanation*. New York. Pp. 137–154.

Henderson, J. L. 1913. *The fitness of the environment.* New York. P. 317.

Hennig, W. 1957. Systematik und Phylogenese. *Ber. Hundertjahr, Deutsch. Ent. Ges. Berlin*, 50–71.

—— 1960. Die Dipteren-Fauna von Neuseeland als systematisches und tierographisches Problem. *Beitr. Ent.*, **10**: 221–329.

—— 1966. *Phylogenetic systematics.* Transl. by D. Davis and R. Zangerl. Urbana, Ill. P. 263.

Hershkovitz, P. 1966. Mice, land bridges, and Latin American faunal interchange, in *Ectoparasites of Panama*, R. L. Wenzel and V. J. Tipton, Eds. Chicago. Pp. 725–751.

—— 1969. The evolution of mammals on southern continents. VI. The recent mammals of the neotropical region: a zoogeographic and ecological review. *Quart. Rev. Biol.*, **44**: 1–70.

Hertel, H. 1966. *Structure, form, movement.* Transl. from German, issued 1963. New York. P. 251.

Hesse, R., W. C. Allee, and K. P. Schmidt. 1951. *Ecological animal geography.* 2nd ed. New York. P. 715.

Hildebrand, M. 1966. Analysis of the symmetrical gaits of tetrapods. *Folia Biotheoretica*, No. 6: 9–22.

Hoerner, S. F. 1958. *Fluid dynamic drag.* Author's publ. Pp. 1–28.

Hoffstetter, R. 1955. Squamates de type moderne—Thecodontia, in *Traité de paléontologie*, vol. 5, J. Piveteau, Ed. Paris. Pp. 606–664, 665–694.

—— 1962. Revue des recentes acquisitions concernant l'historie et le systematique des Squamates, in *Problems actuels de paléontologie*. Colloq. Intern. Cent. Nat. Rech. Sci. Paris. Pp. 243–279.

Holmgren, N. 1933. On the origin of the tetrapod limb. *Acta Zool.*, **14**: 185–295.

—— 1939. Contribution to the question of the origin of the tetrapod limb. *Acta Zool.*, **20**: 89–124.

—— 1949. On the tetrapod limb problem—again. *Acta Zool.*, **30**: 485–508.

—— 1952. An embryological analysis of the mammalian carpus and its bearing upon the question of the origin of the tetrapod limb. *Acta Zool.*, **33**: 1–115.

Hopkins, D. M., Ed. 1967. *The Bering land bridge*. Stanford. P. 495.

Hopson, J. A. 1964a. The braincase of the advanced mammal-like reptile *Bienotherium*. *Postilla*, No. 87: 1–30.

—— 1964b. Tooth replacement in cynodont, dicynodont, and therocephalian reptiles. *Proc. Zool. Soc. London*, **B142**: 625–654.

—— 1966. The origin of the mammalian middle ear. *Am. Zoologist*, **6**: 437–450.

—— 1967. Mammal-like reptiles and the origin of mammals. *Discovery*, **2**: 25–33.

Hotton, N., III. 1960. The chorda tympani and middle ear as guides to origin and divergence of reptiles. *Evolution*, **14**: 194–211.

—— 1967. Stratigraphy and sedimentation in the Beaufort series (Permian Triassic), South Africa, in *Essays in paleontology and stratigraphy*, C. Teichert and E. Yochelson, Eds. Lawrence, Kansas. Pp. 390–428.

Hotton, N., III, and J. W. Kitching. 1963. Speculations on upper Beaufort deposits. *S. Afr. J. Sci.*, **59**: 254–258.

Hough, M. J. 1948. The auditory region in some members of the Procyonidac, Canidae and Ursidae: its significance in the phylogeny of the Carnivora. *Bull. Am. Mus. Nat. Hist.*, **92**: 67–118.

—— 1953. Auditory region in North American fossil Felidae: its significance in phylogeny. U.S. Geol. Survey Professional Paper 243-G. Pp. 95–115.

Howell, A. B. 1930. *Aquatic mammals*. Springfield. P. 338.

Huber, O. 1901. Die Kopulationsglieder der Selachier. *Z. Wiss. Zool.*, **70**: 592–674.

Huena, E. von 1944. *Cymatosaurus* und seine Beziehungen zu anderen Sauropterygiern. *Neues Jahrb. Min. Geol. Palaeont. Monatsh.*, B, 192–222.

Huene, F. von 1911. Über *Erythrosuchus*, Vertreter der neuen Reptilordnung Pelycosimia, *Geol. Paleont. Abh.*, **10**: 1–60.

—— 1922. *Die Ichthyosaurier des Lias und ihre Zusammenhänge*. Berlin. P. 114.

—— 1936. The constitution of the Thecodontia. *Am. J. Sci.*, **32**: 207–217.

—— F. 1937. Die Frage nach der Herkunft der Ichthyosaurier. Bull. Geol. Inst. Univ. Uppsala **27**: 1–9.

—— 1940. The tetrapod fauna of the upper Triassic Maleri beds. *Mem. Geol. Survey India*, **32**: 1–42.

—— 1944. Die Zweiteilung des Reptilstemmes. *Neues Jahrb. Min. Geol. Palaeont. Abh.*, **B88**: 427–440.

—— 1952a. Skelett und Verwandschaft von *Simosaurus*. *Palaeontographica*, **A102**: 163–182.

—— 1952b. *Die Saurierwelt und ihre geschichtlichen Zusammenhänge*. Jena. P. 64.

—— 1960. Ein grosser Pseudosuchier aus der Orenburger Trias. *Palaeontographica*, **114**: 105–111.

Hughes, B. 1963. The earliest archosaurian reptiles. *S. Afr. J. Sci.*, **59**: 221–241.

Huxley, T. H. 1867. On the classification of birds. *Proc. Zool. Soc. London*, 415–472.

—— 1880. On the application of the laws of evolution to the arrangement of the Vertebrata, and more particularly of the Mammalia. *Proc. Zool. Soc. London*, 649–662.

Ihle, J., P. van Kampen, H. Nierstrasz, and J. Versluys. 1927. *Vergleichende Anatomie der Wirbeltiere*. Berlin. P. 906.

Illiger, C. 1811. *Prodromus systematis mammalium et avium additis terminis zoographicis utriudque classis*. Berlin. P. 301.

Inger, R. F. 1957. Ecological aspects of the origins of the tetrapods. *Evolution*, **11**: 373–376.

Irving, L. 1966. Effective regulation of the circulation in diving animals, in *Whales, dolphins and porpoises*, K. Norris, Ed. Berkeley and Los Angeles. Pp. 381–393.

Irving, E., W. Robertson, and P. Stott. 1963. The significance of the paleomagnetic results from Mesozoic rocks of eastern Australia. *J. Geophys. Res.*, **68**: 2313–2317.

Jain, S., P. Robinson, and T. K. Roy Chowdhury. 1964. A new vertebrate fauna from the Triassic of the Deccan, India. *Quart. J. Geol. Soc. London*, **120**: 115–124.

Jarvik, E. 1942. On the structure of the snout of crossopterygians and the lower gnathostomes in general. *Zool. Bidr. Uppsala*, **21**: 235–675.

——— 1944. On the dermal bones, sensory canals and pit lines of the skull in *Eusthenopteron foordi* Whiteaves, with some remarks on *E. sävesöderberghi* Jarvik. *Kungl. Sven. Vet.-Akad. Handl.*, **21**: 3–48.

——— 1948. On the morphology and taxonomy of the middle Devonian osteolepid fishes of Scotland. *Kungl. Sven. Vet.-Akad. Handl.*, **25**: 300.

——— 1952. On the fish-like tail in the ichthyostegid stegocephalians. *Medd. Grønland*, **114**: 1–90.

——— 1954. On the visceral skeleton in *Eusthenopteron* with a discussion of the parasphenoid and palatoquadrate in fishes. *Kungl. Sven. Vet.-Akad. Handl.*, **5**: 1–104.

——— 1955a. The oldest tetrapods and their forerunners. *Sci. Monthly*, **80**: 141–154.

——— 1955b. Ichthyostegalia, in *Traité de paléontologie*, vol. 5, J. Piveteau, Ed. Paris. Pp. 53–66.

——— 1960. *Theories de l'evolution des vertèbrés*. Paris. P. 104.

——— 1963. The composition of the intermandibular division of the head in fish and tetrapods and the diphyletic origin of the tetrapod tongue. *Kungle. Sven. Vet-Akad. Handl.*, **9**: 1–74.

——— 1964. Specializations in early vertebrates. *Ann. Soc. Roy. Belgique*, **94**: 11–95.

——— 1965a. On the origin of girdles and paired fins. *Israel J. Zool.*, **14**: 141–172.

——— 1965b. Die Raspelzunge der Cyclostomen und die pentadactyle Extremität der Tetrapoden als Beweise für monophyletische Herkunft. *Zool. Anzeiger*, **175**: 101–143.

——— 1966. Remarks on the structure of the snout in *Megalichthys* and certain other rhipidistid crossopterygians. *Ark. Zool.*, **19**: 41–98.

——— 1967a. On the structure of the lower jaw in dipnoans: with a description of an early Devonian dipnoan from Canada, *Melanognathus canadensis* gen. et. sp. nov. *J. Linn. Soc. London* (Zool.), **47**: 155–183.

——— 1967b. The homologies of the frontal and parietal bones in fishes and tetrapods. *Colloq. Intern. Cent. Nat. Rech. Sci. Paris*, No. 163: 181–213.

——— 1968a. The systematic position of the Dipnoi in *Current problems of lower vertebrate phylogeny*, T. Ørvig, Ed. Stockholm. Pp. 224–245.

——— 1968b. Aspects of vertebrate phylogeny in *Current problems of lower vertebrate phylogeny*, T. Ørvig, Ed. Stockholm. Pp. 497–527.

Jeffries, R. P. S. 1968. The subphylum Calcichordata (Jeffries 1967). Primitive fossil chordates with echinoderm affinities. *Bull. Brit. Mus. (Nat. Hist.) Geol.*, **16**: 245–339.

Johnson, R. G. 1960. Models and methods for analysis of the modes of formation of fossil assemblages. *Bull. Geol. Soc. Amer.*, **71**: 1075–1086.

Johnson, R. G., and E. S. Richardson, Jr. 1966. A remarkable Pennsylvanian fauna from the Mazon Creek area, Illinois. *J. Geol.*, **74**: 626–630.

Jones, F. W. 1923–1925. The mammals of South Australia. Adelaide. P. 458.

——— 1930. A reexamination of the skeletal characters of *Wynyardia bassiana*, an extinct Tasmanian marsupial. *Papers and Proc. Roy. Soc. Tasmania*, 96–115.

Jordan, D. 1963. *The genera of fishes, and a classification of fishes.* Stanford. P. 800.

Kalin, J. 1955. Crocodilia, in *Traité de paléontologie*, vol. 5, J. Piveteau, Ed. Paris. Pp. 695–784.

Keast, A. 1968. Evolution of mammals on southern continents. IV. Australian mammals: zoogeography and evolution. *Quart. Rev. Biol.*, **43**: 373–408.

Kelly, D. E. 1962. Pineal organs: photoreception, secretion, and development. *Am. Scientist*, **50**: 597–625.

Kermack, D., K. Kermack, and F. Mussett. 1968. The Welsh pantothere *Kuehneotherium praecursoris*. *J. Linn. Soc. London* (Zool.), **47**: 407–423.

Kermack, K. A. 1948. The propulsive powers of blue and fin whales. *J. Exptl. Biol.*, **25**: 237–240.

——— 1962. Structure cranienne et évolution de mammifères Mesozoïques, in *Problems actuels de paléontologie.* Colloq. Intern. Cent. Nat. Rech. Sci. Paris. Pp. 311–317.

——— 1963. The cranial structure of the triconodonts. *Philos. Trans. Roy. Soc. London*, **B246**: 83–103.

——— 1965. The origin of mammals. *Science J.*, 66–72.

Kermack, K. A., P. Lees, and F. Mussett. 1964. *Aegialodon dawsoni*, a new tuberculosectorial tooth from the lower Wealden. *Proc. Roy. Soc. London*, **B162**: 535–554.

Kermack, K. A., and F. Mussett. 1958a. The jaw articulation in Mesozoic mammals. 15th Intern. Congr. Zool., Sect. V, paper 8, 1–2.

——— 1958b. The jaw articulation of the docodonts and the classification of Mesozoic mammals. *Proc. Roy. Soc. London*, **B149**: 204–215.

——— 1959. The first mammals. *Discovery*, **20**: 144–153.

King, L. C. 1962. The morphology of the earth. A study and synthesis of world scenery. Edinburgh. p. 699.

Kitts, D. 1956. American *Hyracotherium* (Perissodactyla, Equidae). *Bull. Am. Mus. Nat. Hist.*, **110**: 7–60.

Konzhukova, E. D. 1964. Apsidospondyli, Batrachosauria in *Fundamentals of paleontology*, vol. 12, Y. A. Orlov, Ed. Moscow. Pp. 60–82 and 133–144 (Russian).

Kramer, M. O. 1930a. Boundary layer stabilization by distributed damping. *J. Am. Soc. Nav. Engrs.*, **72**: 25–33.

——— 1960b. The dolphin's secret. *New Scientist*, **7**: 1118–1120.

Krogh, A. 1939. *Osmotic regulation in aquatic animals.* London. P. 242.

Kuhn, O. 1964. Ungelöste Probleme der Stammesgeschichte der Amphibien und Reptilien. Jahrb. Ver. Vaterl. Naturkde. Württemberg, **118/119**: 293–325.

——— 1965. *Die Amphibien.* München. P. 102.

Kuhn-Schnyder, E. 1952. *Askeptosaurus italicus* Nopsca. *Schweiz. Palaeont. Abh.*, **69**: 1–73.

——— 1961. Der Schädel von *Simosaurus. Palaeont. Z.*, **35**: 95–113.

——— 1963. Wege der Reptilien Systematik. *Palaeont. Z.*, **37**: 61–87.

Kühne, W. G. 1949. On a triconodont tooth of a new pattern from a fissure filling in south Glamorgan. *Proc. Zool. Soc. London*, **B119**: 345–350.

——— 1956. *The Liassic therapsid* Oligokyphus. London. P. 149.

——— 1958. Rhaetische Triconodonten aus Glamorgan, ihre Stellung zwischen den Klassen Reptilia und Mammalia und ihre Bedeutung für die Reichart'sche Theorie. *Palaeont. Z.*, **32**: 197–235.

—— 1968. Origin and history of Mammalia, in *Evolution and environment*, E. J. Drake, Ed. New Haven. Pp. 109–123.

Kulczycki, J. 1960. *Porolepis* (Crossopterygii) from the lower Devonian of the Holy Cross Mountains. *Acta Palaeont. Polonica*, **5**: 65–104.

Kurtén, B. 1966. Holarctic land connexions in the early Tertiary. *Commentationes Biologicae, Soc. Sci. Fennica*, **29**: 1–5.

—— 1967. Continental drift and the paleogeography of reptiles and mammals. *Commentationes Biologicae, Soc. Sci. Fennica*, **31**: 1–8.

Lacépédé, B. G. E. 1799. Tableau des divisions, sous-divisions, ordres et generes des mammifères, in *Histoire naturelle*, vol. 14, G. L. L. de Buffon, Ed. Paris. Pp. 144–195.

Lack, D. L. 1954. The natural regulation of animal numbers. Oxford. P. 343.

Lagler, K., J. Bardach, and R. Miller. 1962. *Ichthyology*. New York. P. 545.

Lambrecht, K. 1933. *Handbuch der Palaeornithologie*. Berlin. P. 1024.

Landry, S. O., Jr. 1957. The inter-relationships of the New and Old World hystricomorph rodents. *Univ. Calif. Publ. Zool.*, **56**: 1–117.

Langston, W. 1965a. Fossil crocodilians from Columbia and the Cenozoic history of the Crocodilia in South America. *Univ. Calif. Publ. Geol. Sci.*, **52**: 1–157.

—— 1965b. *Oedaleops campi* (Reptilia: Pelycosauria), a new genus and species from the lower Permian of New Mexico, and the family Eothyrdidae. *Bull. Texas Mem. Mus.*, **9**: 5–47.

Latreille, P. A. 1825. *Familles naturelles du règne animal, exposées succinctment et dans un ordre analytique, avec l'indication de leurs generes*. Paris. P. 570.

Lavocat, R. 1951. *Revision de la faune des mammifères Oligocènes d'Auvergne et du Velay*. Paris. P. 153.

—— 1955. Quelques progrès récentes dans la connaissance des rongeurs fossiles et leurs conséquences sur divers problems de systematique de peuplement et d'évolution. *Colloq. Intern. Cent. Nat. Rech. Sci. Paris*, **60**: 77–85.

—— 1956. Reflexions sur la classification des rongeurs. *Mammalia*, **20**: 49–56.

Ledoux, J., J. Hartenberger, J. Michaux, J. Sudre, and L. Thaler. 1966. Dècouverte d'un Mammifère dans le Crétacé superieur à Dinosaures de Champ-Garimond près de Fons (Gard). Comp. Rend., Acad. Sci. Paris **262**: 1925–1928.

Lehman, J. P. 1947. Description de quelques exemplaires de *Cheirolepis canadensis* (Whiteaves). *Kungl. Sven. Vet.-Akad. Handl.*, **24**: 1–40.

—— 1952. Étude complementaire des poissons de l'Eotrias de Madagascar. *Kungl. Sven. Vet.-Akad. Handl.*, **2**: 1–201.

—— 1955. Amphibiens, Amphibia Linné, généralités. Rhachitomi. Anthracosauria, Phyllospondyli. In *Traité de paléontologie*, vol. 5, J. Piveteau, Ed. Paris. Pp. 3–62, 67–125, 173–224, and 227–249.

—— 1956. L'évolution des dipneustes et l'origine des urodéles. *Colloq. Intern. Cent. Nat. Rech. Sci. Paris*, **60**: 69–76.

—— 1959. *L'évolution des vertébrés inférieurs*. Paris. P. 188.

—— 1966. Actinopterygiens, crossopterygiens, dipneustes, in *Traité de paléontologie*, vol. 4, pt. 3, J. Piveteau, Ed. Paris. Pp. 1–422.

Lehman, J. P., et al. 1958. Super-ordre des chondrostéens (Chondrostei), formes fossiles, in *Traité de zoologie*, vol. 13, fasc. 3, P.-P. Grasse, Ed. Paris. Pp. 2130–2164.

Leigh-Sharp, W. H. 1920–1926. The comparative morphology of the secondary sexual characters of elasmobranch fishes. The claspers, clasper siphons, and clasper glands. *J. Morphol.*, **34**: 245–265; **35**: 359–380; **36**: 191–198 and 199–220; **39**: 553–566 and 567–577; **42**: 307–320.

Levins, R. 1966. The strategy of model building in ecology. *Am. Scientist*, **54**: 421–431.

Liem, K. F. 1967. Functional morphology of the integumentary, respiratory, and digestive systems of the synbranchoid fish *Monopterus albus. Copeia*, 375–388.

Lighthill, M. 1960. Notes on the swimming of slender fishes. *Fluid Mechanics*, **9**: 305–317.

Lillegraven, J. A. 1969. *Latest Cretaceous mammals from the upper part of the Edmonton formation of Alberta, Canada, and review of marsupial-placental dichotomy in mammalian evolution.* Univ. Kansas Publ., Paleont. Contr. Art. 50 (Vertebrate 12). P. 122.

Linnaeus, C. 1758. *Systema naturae per regne tria naturae, secondum classes, ordines, genera, species cum characteribus, differentiis, synonymis, locis.* Editio decima, reformata, vol. 1. P. 824.

Lowe, P. 1933. On the primitive characters of penguins and their bearing on the phylogeny of birds. *Proc. Zool. Soc. London*, 483–538.

——— 1939. Some additional notes on Miocene penguins in relationship to their origin and systematics. *Ibis*, **3**: 281–294.

——— 1944a. Some additional remarks on the phylogeny of Struthiones. *Ibis*, **86**: 37–43.

——— 1944b. An analysis of the characters of *Archaeopteryx* and *Archaeornis*. Were they reptiles or birds? *Ibis*, **86**: 517–543.

Lull, R. S. 1918. The pulse of life, in *The evolution of the earth and its inhabitants*, by J. Barrell et al., New Haven. Pp. 109–146.

Lull, R. S., and N. Wright. 1942. *Hadrosaurian dinosaurs of North America.* Geol. Soc. Amer. Special Paper No. 40. P. 242.

Lund, R. 1967. An analysis of the propulsive mechanisms of fishes, with reference to some fossil actinopterygians. *Ann. Carneg. Mus.*, **39**: 195–218.

MacArthur, R. 1962. Some generalized theorems of natural selection. *Proc. Natl. Acad. Sci. U.S.*, **48**: 1893–1897.

MacArthur, R., and R. Levins. 1964. Competition, habitat selection, and character displacement in a patchy environment. *Proc. Natl. Acad. Sci. U.S.*, **51**: 1207–1210.

MacArthur, R., and E. Wilson. 1967. *The theory of island biogeography.* Princeton. P. 203.

MacIntyre, G. T. 1967. Foramen pseudoövale and quasi-mammals. *Evolution*, **21**: 834–841.

Margalef, R. 1959. Mode of evolution in species in relationship to their places of ecological succession, in *Proc. 15th Intern. Congr. Zool.*, H. R. Hewer and N. D. Riley, Eds. London. Pp. 787–789.

Marples, B. 1952. Early Tertiary penguins of New Zealand. *New Zealand Geol. Survey Paleont. Bull.*, **20**: 1–66.

Marsh, O. C. 1872a. Notice of a new and remarkable fossil bird. *Am. J. Sci.*, **3**: 1–344.

——— 1872b. Preliminary description of *Hesperornis regalis*, with notices of four other new species of Cretaceous birds. *Am. J. Sci.*, **3**: 360–365.

——— 1875. Odontornithes, or birds with teeth. *Am. J. Sci.*, **10**: 624–631.

——— 1880a. *Odontornithes: a monograph on the extinct toothed birds of North America.* Report of the geological exploration of the 40th parallel. Washington. P. 201.

——— 1880b. Notice of Jurassic mammals representing two new orders. *Am. J. Sci.*, **20**: 235–239.

Matthew, W. D. 1915. Climate and evolution. *Ann. N.Y. Acad. Sci.*, **24**: 171–318.

Mayr, E. 1963. *Animal species and evolution.* Cambridge. P. 797.

——— 1965. Numerical phenetics and taxonomic theory. *Sys. Zool.*, **14**: 73–100.

McDowell, S. B., and C. M. Bogert. 1954. The systematic position of *Lanthanotus* and the affinities of the anguinimorphan lizards. *Bull. Am. Mus. Nat. Hist.*, **105**: 1–142.

McKenna, M. 1960. A continental Paleocene vertebrate fauna from California. *Am. Mus. Novitates*, No. 2024: 1–20.

——— 1963. Primitive Paleocene and Eocene Apatemyidae (Mammalia: Insectivora) and the primate-insectivore boundary. *Am. Mus. Novitates*, No. 2160: 1–39.

—— 1966. Paleontology and the origin of the primates. *Folia Primat.*, **4**: 1–25.

Menzbier, M. 1887. Vergleichende Osteologie der Pinguine. *Bull. Soc. Imp. Nat. Moscow* **1**: 483–587.

Merrem, B. 1813. Tentamen systematis naturalis avium. *Abh. Akad. Wiss. Berlin*, 1812–1813 (publ. 1816): 237–259.

Meyer, H., von. 1861. Letters directed to Prof. Bronn. *Neues Jahrb. Min. Geol. Palaeont.* 561, 678–679.

Michener, C., and R. Sokal. 1957. A quantitative approach to a problem in classification. *Evolution*, **11**: 130–162.

Miles, R. S. 1964. A reinterpretation of the visceral skeleton of *Acanthodes*. *Nature*, **204**: 457–459.

—— 1965. Some features in the cranial morphology of acanthodians and the relationships of the Acanthodii. *Acta Zool.*, **46**: 233–255.

—— 1968. Jaw articulation and suspension in *Acanthodes* and their significance, in *Current problems of lower vertebrate phylogeny*, T. Ørvig, Ed. Stockholm. Pp. 109–127.

Miller, G. S., Jr., and J. W. Gidley. 1918. Synopsis of the supergeneric groups of rodents. *J. Wash. Acad. Sci.*, **8**: 431–448.

Mills, J. R. E. 1964. The dentitions of *Peramus* and *Amphitherium*. *Proc. Linn. Soc. London*, **175**: 117–133.

Miner, R. W. 1925. The pectoral limb of *Eryops* and other primitive tetrapods. *Bull. Am. Mus. Nat. Hist.*, **51**: 145–312.

Monath, T. 1965. The opercular apparatus of salamanders. *J. Morphol.*, **116**: 149–170.

Montagna, W. 1959. *Comparative anatomy*. New York. P. 397.

Mook, C. 1934. The evolution and classification of the Crocodilia. *J. Geol.*, **42**: 259–304.

Moss, M. L. 1961. The initial phylogenetic appearance of bone: an experimental hypothesis. *Trans. N.Y. Acad. Sci.*, **23**: 495–500.

—— 1967. The origin of vertebrate calcified tissue, in *Current problems of lower vertebrate phylogeny*, T. Ørvig, Ed. Stockholm. Pp. 360–371.

Moy-Thomas, J. A. 1934. The structure and affinities of *Tarassius problematicus*, Traquair. *Proc. Zool. Soc. London*, 367–376.

—— 1939a. *Paleozoic fishes*. London and New York. P. 149.

—— 1939b. Early evolution and relationships of the elasmobranchs. *Biol. Rev.*, **14**: 1–26.

—— 1942. Carboniferous palaeoniscoids from East Greenland. *Ann. Mag. Nat. Hist.*, ser. 11, **9**: 737–759.

Moy-Thomas, J. A., and M. B. Dyne. 1938. The actinopterygian fishes from the Lower Carboniferous of Glencartholm, Eksdale, Dumfriesshire. *Trans. Roy. Soc. Edinburgh*, **59**: 437–480.

Müller, J. 1845. Über den Bau und die Grenzen der Ganeiden und über das natürliche System der Fische. *Abh. Akad. Wiss. Berlin*, 117–216.

Müller, J., and J. Henle. 1841. *Systematische Beschreibung der Plagiostomen*. Berlin. P. 204.

Nelson, G. J. 1968. Gill-arch structure in *Acanthodes*, in *Current problems of lower vertebrate phylogeny*, T. Ørvig, Ed. Stockholm. Pp. 129–143.

Newell, N. D. 1967. Revolutions in the history of life, in *Uniformity and simplicity*, C. Albritton, Ed. Special Paper, Geol. Soc. Amer., **89**: 63–91.

Nielsen, E. 1942. Studies on Triassic fishes from East Greenland. I. *Glaucolepis* and *Boreosomus*. *Palaeozool. Greenlandica*, **1**: 1–403.

—— 1949. Studies on Triassic fishes from East Greenland. II. *Australosomus* and *Birgeria*. *Palaeozool. Greenlandica*, **3**: 1–309.

—— 1954. *Tupilakosaurus heilmani* n.g., n.s.p., an interesting batrachomorph from the Triassic of East Greenland. *Medd. Grønland*, **72**: 1–33.

Noble, G. K. 1931. *The biology of the Amphibia*. New York. P. 577.

Nopsca, F. 1934. The influence of geological and climatological factors on the distribution of non-marine fossil reptiles and Stegocephalia. *Quart. J. Geol. Soc. London*, **90**: 76–104.

Norris, K. S. 1964. Some problems of echolocation in cetaceans, in *Marine bio-acoustics*, W. N. Tavolga, Ed. New York. Pp. 317–336.

———, Ed. 1966. *Whales, dolphins, and porpoises*. Berkeley and Los Angeles. P. 789.

Novitskaya, L. J., and D. V. Obruchev. 1964. Acanthodei, in *Fundamentals of paleontology*, vol. 11, Y. A. Orlov, Ed. Moscow. Pp. 175–195 (Russian).

Nursall, J. R. 1956. The lateral musculature and the swimming of fish. *Proc. Zool. Soc. London*, **126**: 1–127.

——— 1958. The caudal fin as a hydrofoil. *Evolution*, **12**: 116–120.

——— 1962. Swimming and the origin of paired appendages. *Am. Zoologist*, **2**: 127–141.

Obruchev, D. 1943. A new restoration of *Drepanaspis*. *Compt. Rend. Acad. Sci. USSR*, **41**: 268–271 (Russian).

——— 1964. Agnatha. Placodermi. Holocephali, in *Fundamentals of paleontology*, vol. 11, Y. A. Orlov, Ed. Moscow. Pp. 1–175 and 238–267 (Russian).

Oehmichen, E. 1938. Essai sur la dynamiques des Ichthyosauriens *longipinnati* et particulierement d'*Ichthyosaurus burgundiae* (Gaud.). *Ann. Paleont.*, **27**: 91–114.

Ohlmer, W. 1964. Untersuchungen über die Beziehungen zwischen Körperform und Bewegungsmedium bei Fischen aus stehenden Binnengewässern. *Zool. Jahrb. Anat.*, **81**: 151–240.

Olson, E. C. 1936a. The dorsal axial musculature of certain primitive Permian tetrapods. *J. Morphol.*, **59**: 265–311.

——— 1936b. Notes on the skull of *Youngina capensis* Broom. *J. Geol.*, **44**: 523–533.

——— 1939. Fauna of the *Lysorophus* pockets in the Clear Fork Permian, Baylor County, Texas. *J. Geol.*, **47**: 389–397.

——— 1941. New specimens of Permian vertebrates in Walker Museum. *J. Geol.*, **49**: 753–763.

——— 1944. *Origin of mammals based upon the cranial morphology of therapsid suborders*. Special Paper Geol. Soc. Amer., **55**: 1–136.

——— 1947. The family Diadectidae and its bearing on the classification of reptiles. *Fieldiana: Geology*, **11**: 2–53.

——— 1952. The evolution of a Permian vertebrate chronofauna. *Evolution*, **6**: 181–196.

——— 1957a. Catalogue of localities of Permian and Triassic terrestrial vertebrates of the territories of the USSR. *J. Geol.*, **65**: 196–226.

——— 1957b. Size-frequency distributions in samples of extinct organisms. *J. Geol.*, **65**: 309–333.

——— 1958. Fauna of the Vale and Choza: 14. Summary, review, and integration of the geology and the faunas. *Fieldiana: Geology*, **10**: 397–448.

——— 1959. The evolution of mammalian characters. *Evolution*, **13**: 344–353.

——— 1961a. Food chains and the origin of mammals. Intern Colloq. on the evolution of lower and unspecialized mammals. *Kon. Vlaamse Acad. Wetensch. Lett. Schone Kunsten Belg.*, pt. 1: 97–116.

——— 1961b. Jaw mechanisms: rhipidistians, amphibians, reptiles. *Am. Zoologist*, **1**: 205–215.

——— 1962. Late Permian terrestrial vertebrates. USA and USSR. *Trans. Am. Philos. Soc.*, **52**, pt. 2: 3–224.

——— 1965a. Summary and comment. *Syst. Zool.*, **14**: 337–342.

——— 1965b. Relationships of *Seymouria*, *Diadectes*, and Chelonia. *Am. Zoologist*, **5**: 265–307.

―――― 1965c. Introductory remarks to "Evolution and relationships of the Amphibia." *Am. Zoologist*, **5**: 263–265.

―――― 1965d. *Vertebrates from the Chickasha formation. Permian of Oklahoma.* Okla. Geol. Survey, Circular 70. P. 70.

―――― 1966a. The middle ear–morphological types in amphibians and reptiles. *Am. Zoologist*, **6**: 399–419.

―――― 1966b. Community evolution and the origin of mammals. *Ecology*, **47**: 291–308.

―――― 1966c. Relationships of *Diadectes. Fieldiana: Geology*, **14**: 199–227.

―――― 1967. *Early Permian vertebrates from Oklahoma.* Okla. Geol. Survey, Circular 74. P. 111.

―――― 1968. The family Caseidae. *Fieldiana: Geology*, **17**: 225–349.

―――― 1970. New and little known vertebrates from the early Permian of Oklahoma, *Fieldiana: Geology*, **18**: 395–444

Olson, E. C., and J. R. Beerbower. 1953. The San Angelo formation, Permian of Texas, and its vertebrates. *J. Geol.*, **61**: 389–423.

Olson, E. C., and R. Broom. 1937. New genera and species of tetrapods from the Karroo beds of South Africa. *J. Paleont.* **11**: 613–619.

Olson, E. C., and R. L. Miller. 1951. A mathematical model applied to a study of the evolution of species. *Evolution*, **5**: 325–338.

―――― 1958. *Morphological integration.* Chicago. P. 317.

Orlov, Y. A. 1958. The carnivorous dinocephalians of the Isheevo fauna (Titanosuchians). *Tr. Paleont. Inst. Acad. Sci. USSR*, **72**: 3–113 (Russian).

Orton, G. L. 1954. Original adaptive significance of the tetrapod limb. *Science*, **120**: 1042–1043.

Ørvig, T. 1960. New finds of acanthodians, arthrodires, crossopterygians, ganoids, and dipnoans in the upper calcareous flags (oberer Plattenkalk) of the Bergisch-Gladbach-Paffrath Trough. *Palaeont. Z.*, **34**: 295–335.

―――― 1962. Y a-t-il une relation directe entre les athrodires ptyctodontides et les holocephales? *Problèmes Actuels de Paléontologie*, No. 104: 49–58.

Osborn, H. F. 1903. The reptilian subclasses Diapsida and Synapsida and the early history of the Diaptosauria. *Mem. Am. Mus. Nat. Hist.*, **1**: 451–507.

―――― 1910. *The age of mammals in Europe, Asia, and North America.* New York. P. 635.

―――― 1929. *The titanotheres of ancient Wyoming, Dakota, and Nebraska.* U.S. Geol. Survey Monograph 55, vol. 2.

―――― 1936. *Proboscidea*, vol. I. New York. P. 802.

Osgood, W. H. 1921. *A monographic study of the American marsupial* Caenolestes. Publ. Field Mus. Nat. Hist., Zool. Ser., **14**: 1–156.

Owen, R. 1839. On the bone of an unknown struthious bird from New Zealand. *Proc. Zool. Soc. London*, 169–170.

―――― 1841. Outlines of a classification of the Marsupialia. *Trans. Zool. Soc. London*, **2**: 315–333.

―――― 1863. On the fossil remains of a long-tailed bird *Archaeopteryx macrura* from the lithographic slate of Solnhofen. *Proc. Roy. Soc. London*, **12**: 272–273.

―――― 1866. On the anatomy of vertebrates. II. Birds and mammals. London. p. 586.

―――― 1868. On the anatomy of the vertebrates. III. Mammals. London. p. 915.

Panchen, A. L. 1964. The cranial anatomy of two coal measure anthracosaurs. *Philos. Trans. Roy. Soc. London*, **B247**: 593–637.

―――― 1967a. The nostrils of choanate fishes and early tetrapods. *Biol. Rev.*, **42**: 374–420.

―――― 1967b. The homologies of the labyrinthodont centrum. *Evolution*, **21**: 24–33.

Panchen, A. L., and A. D. Walker. 1960. British coal measure labyrinthodont localities. *Ann. Mag. Nat. Hist.*, ser. 13, **3**: 321–332.

Parker, T. J., and W. A. Haswell. 1963. *A textbook of zoology*, 7th ed., vol. 2. Revised by A. J. Marshall. London and New York. P. 952.

Parrington, F. R. 1937. A note on the supratemporal and tabular bones in reptiles. *Ann. Mag. Nat. Hist.*, ser. 10, **20**: 69–76.

———— 1941. On two mammalian teeth from the lower Rhaetic of Somerset. *Ann. Mag. Nat. Hist.*, ser. 11, **8**: 140–144.

———— 1949*a*. A theory of the relations of lateral lines to dermal bones. *Proc. Zool. Soc. London*, **119**: 65–78.

———— 1949*b*. Remarks on a theory of evolution of the tetrapod middle ear. *J. Laryngol. Otol.*, **63**: 580–595.

———— 1956*a*. A problematic reptile from the upper Permian. *Ann. Mag. Nat. Hist.*, ser. 12, **9**: 333–336.

———— 1956*b*. The patterns of dermal bones in primitive vertebrates. *Proc. Zool. Soc. London*, **B127**: 389–411.

———— 1958*a*. The problem of classification of the reptiles. *J. Linn. Soc. London* (Zool.), **44**; Botany, **56**: 99–115.

———— 1958*b*. On the nature of Anaspida, in *Studies on fossil vertebrates*, T. S. Westoll, Ed. London. Pp. 108–128.

———— 1959. The angular process of the dentary. *Ann. Mag. Nat. Hist.*, ser. 12, **13**: 505–512.

———— 1967*a*. The identification of the dermal bones of the head. *J. Linn. Soc. London* (Zool.), **47**: 231–239.

———— 1967*b*. The origins of mammals. *Adv. Sci.*, **24**: 165–173.

Parsons, T. S. 1967. Evolution of the nasal structure in lower tetrapods. *Am. Zoologist*, **7**: 397–413.

Parsons, T. S., and E. E. Williams. 1962. The teeth of Amphibia and their relation to amphibian phylogeny. *J. Morphol.*, **110**: 375–390.

———— 1963. The relationships of modern Amphibia. *Quart. Rev. Biol.*, **38**: 26–53.

Patterson, B. 1956. Early Cretaceous mammals and the evolution of mammalian molar teeth. *Fieldiana: Geology*, **13**: 1–105.

Patterson, B., and E. C. Olson. 1961. A triconodontid mammal from the Triassic of Yunnan. Intern. colloq. on the evolution of lower and unspecialized mammals. *Kon. Vlaamse Akad. Wetensch., Lett. Schone Kunsten Belg.*, pt. 1: 129–191.

Patterson, B., and R. Pascual. 1963. The extinct land mammals of South America, in 16th Intern. Zool. Congr. program vol. Pp. 138–148.

———— 1968. Evolution of mammals on southern continents. V. The fossil mammal fauna of South America. *Quart. Rev. Biol.*, **43**: 409–451.

Patterson, C. 1965*a*. A review of Mesozoic acanthopterygian fishes, with special reference to those of the English chalk. *Philos. Trans. Roy. Soc. London*, **B247**: 313–482.

———— 1965*b*. The phylogeny of the chimaeroids. *Philos. Trans. Roy. Soc. London*, **B249**: 101–219.

———— 1967. Are teleosts a polyphyletic group? *Colloq. Intern. Cent. Nat. Rech. Sci. Paris*, No. 163: 93–109.

———— 1968*a*. The caudal skeleton in lower Liassic pholidophorid fishes. *Bull. Brit. Mus. (Nat. Hist.) Geology*, **16**: 203–239.

———— 1968*b*. *Menaspis* and the bradydonts, in *Current problems of lower vertebrate phylogeny*, T. Ørvig, Ed. Stockholm. Pp. 171–205.

Peabody, F. E. 1952. *Petrolacosaurus kansensis* Lane, a Pennsylvanian reptile from Kansas. *Univ. Kansas Publ. Paleont. Contr.*, Art. 10 (*Vertebrata* 1). P. 41.

Peach, B. N., and J. Horne. 1903. The Canonbei coalfield. Its geological structure and relations to the Carboniferous rocks of the north of England and central Scotland. *Trans. Roy. Soc. Edinburgh*, **40**: 835–877.

Pehrson, T. 1922. Some points in the cranial development of teleostomian fishes. *Acta Zool.*, **3**: 1–63.

Persson, P. 1963. A revision of the classification of the Plesiosauria with a synopsis of the stratigraphical and geographical distribution of the group. *Lunds Univ. Arssk.*, N. F. Adv. 2, **59**: 1–60.

Peyer, B. 1955. Die Trias Fauna der Tessiner Kalkalpen. XVIII. *Helveticosaurus zollingeri* n.g. n.sp. Schweiz. *Paleont. Abh.*, **72**: 1–50.

———— 1968. *Comparative odontology*. Chicago. P. 347.

Peyer, B., and E. Kuhn-Schnyder. 1955a. Placodontia, in *Traité de paléontologie*, vol. 5, J. Piveteau, Ed. Paris. Pp. 459–486.

———— 1955b. Squamates du Trias, in *Traité de paléontologie*, vol. 5, J. Piveteau, Ed. Paris. Pp. 578–605.

Pincher, C. 1948. *A study of fishes*. London. P. 302.

Piveteau, J. 1955. Anoures, Mesosauria, in *Traité de paléontologie*, vol. 5, J. Piveteau, Ed. Paris. Pp. 250–274 and 409–411.

———— 1961. Marsupiaux, in *Traité de paléontologie*, vol. 6, pt. 1, J. Piveteau, Ed. Paris. Pp. 585–640.

Pompeckj, J. 1922. Das Ohrskelett von *Zeuglodon*. *Senckenbergiana*, **4**: 43–100.

Purves, P. E. 1966. Anatomy and physiology of the outer and middle ear in cetaceans, in *Whales, dolphins, and porpoises*, K. Norris, Ed. Berkeley and Los Angeles. Pp. 320–376.

Quinn, J. H. 1955. *Miocene Equidae of the Texas Gulf Coastal Plain*. Univ. Texas Bur. Econ. Geol. Publ. No. 5516. Pp. 1–102.

Radinsky, L. 1961. Tooth histology as a taxonomic criterion for cartilaginous fishes. *J. Morphol.*, **109**: 73–81.

———— 1963. Origin and early evolution of North American Tapiroidea. *Bull. Peabody Mus. Nat. Hist.*, No. 17: 1–106.

———— 1964. *Paleomoropus*, a new early Eocene chalicothere (Mammalia, Perissodactyla), and a revision of the Eocene chalicotheres. *Am. Mus. Novitates*, No. 2179: 1–28.

———— 1965a. Evolution of the tapiroid skeleton from *Heptodon* to *Tapirus*. *Bull. Mus. Comp. Zool.*, **134**: 69–103.

———— 1965b. Early Tertiary Tapiroidea of Asia. *Bull. Am. Mus. Nat. Hist.*, **129**: 181–264.

———— 1966a. The families of the Rhinocerotoidea (Mammalia, Perissodactyla). *J. Mammal.*, **47**: 631–639.

———— 1966b. A new genus of early Eocene tapiroid (Mammalia, Perissodactyla). *J. Paleont.*, **40**: 740–742.

———— 1966c. The adaptive radiation of the phenacodontid condylarths and origin of the Perissodactyla. *Evolution*, **20**: 408–419.

———— 1967. *Hyrachyus*, *Chasmotherium*, and the early evolution of helaletid tapiroids. *Am. Mus. Novitates*, No. 2313: 1–23.

———— 1968. A new approach to mammalian cranial analysis, illustrated by examples of prosimian Primates. *J. Morphol.*, **124**: 167–179.

Ray, J. 1678. *The ornithology of Francis Willughby*. Translated from 1676 ms., Roy. Soc. London, in three books.

———— 1693. *Synopsis methodica animalium quadrupedum et serpentini generis*. London. P. 336.

Rayner, D. H. 1941. The structure and evolution of the holostean fishes. *Biol. Rev.*, **16**: 218–237.

———— 1951. On the cranial structure of an early palaeoniscid, *Kentuckia* gen. nov. *Trans. Roy. Soc. Edinburgh*, **62**: 53–83.

———— 1958. The geological environment of fossil fishes, in *Studies on fossil vertebrates*,

T. S. Westoll, Ed. London. Pp. 129–156.

Reed, C. 1960. Polyphyletic or monophyletic ancestry of mammals, or: what is a class? *Evolution*, **14**: 314–322.

Regan, C. T. 1906. A classification of the selachian fishes. *Proc. Zool. Soc. London*, 722–758.

—— 1909. The classification of the teleostean fishes. *Ann. Mag. Nat. Hist.*, ser. 8, **3**: 75–86.

—— 1929. Fishes, in *Encyclopedia Britannica*, 14th ed., vol. 9. London and New York. Pp. 305–328.

Reig, O. 1958. Proposiciones para una nueva macrosistematica de los anuros. *Physis*, **21**: 109–118.

—— 1964. El problema del origen monofilético o polifilético de los Amfibios, con consideraciones sobre las relaciones entre anuros, urodelos y apodos. *Ameghiniana* **3**: 191–211.

—— 1967. Archosaurian reptiles: a new hypothesis on their origins. *Science*, **157**: 565–568.

Reinhart, R. H. 1959. A review of the Sirenia and Desmostylia. *Univ. Calif. Publ. Geol. Sci.*, **36**: 1–146.

Reysenbach de Haan, F. W. 1957. Hearing in whales. *Acta Otolaryngol. Suppl.*, **134**: 1–114.

—— 1966. Listening underwater: thoughts on sound and cetacean hearing, in *Whales, dolphins, and porpoises*, K. Norris, Ed. Berkeley and Los Angeles. Pp. 583–595.

Ride, W. D. L. 1964. A review of Australian fossil marsupials. *J. Roy. Soc. W. Australia*, **47**: 97–131.

Rigney, H. W. 1963. A specimen of *Morganucodon* from Yunnan. *Nature*, **197**: 122–123.

Ritchie, A. 1968. New evidence on *Jamoytius kerwoodi* White, an important ostracoderm from the Silurian of Lanarkshire, Scotland. *Paleontology*, **11**: 21–39.

Robertson, J. D. 1954. The chemical composition of the blood of some aquatic chordates, etc. *J. Exptl. Biol.*, **31**: 424–442.

—— 1957. The habitat of early vertebrates. *Biol. Rev.*, **32**: 156–187.

Robinson, P. L. 1962. Gliding lizards from the Upper Keuper of Great Britain. *Proc. Geol. Soc. London*, No. 1601: 137–146.

—— 1967. Triassic vertebrates from lowland and upland. *Science and Culture*, **22**: 169–173.

Romer, A. S. 1922. The locomotor apparatus of certain primitive and mammal-like reptiles. *Bull. Am. Mus. Nat. Hist.*, **46**: 517–606.

—— 1923. Crocodilian pelvic muscles and their avian and reptilian homologues. *Bull. Am. Mus. Nat. Hist.*, **48**: 533–552.

—— 1924. Pectoral limb musculature and shoulder girdle structure in fish and tetrapods. *Anat. Rec.*, **27**: 119–143.

—— 1927. The pelvic musculature of ornithischian dinosaurs. *Acta Zool.*, **8**: 225–275.

—— 1933. *Vertebrate paleontology*. Chicago. P. 491.

—— 1937. The braincase of the Carboniferous crossopterygian *Megalichthys nitidus*. *Bull. Mus. Comp. Zool.*, **82**: 1–73.

—— 1939. Notes on branchiosaurs. *Am. J. Sci.*, **237**: 748–761.

—— 1941a. The skin of the rhachitomous amphibian *Eryops*. *Am. J. Sci.*, **239**: 822–824.

—— 1941b. Notes on the crossopterygian hyomandibular and braincase. *J. Morphol.*, **69**: 141–160.

—— 1945a. *Vertebrate paleontology*, 2nd ed. Chicago. P. 687.

—— 1945b. The late Carboniferous vertebrate fauna of Kounova (Bohemia) compared with that of the Texas red beds. *Am. J. Sci.*, **243**: 417–442.

—— 1946. The early evolution of fishes. *Quart. Rev. Biol.*, **21**: 33–59.

—— 1947a. A review of the Labyrinthodontia. *Bull. Mus. Comp. Zool.*, **99**: 3–352.

———— 1947*b*. The relationships of the Permian reptile *Protorosaurus*. *Am. J. Sci.*, **245**: 19–30.

———— 1948. Ichthyosaur ancestors. Am. J. Sci. **246**: 104–121.

———— 1950. The nature and relationships of the Paleozoic microsaurs. *Am. J. Sci.*, **248**: 628–654.

———— 1955*a*. Fish origins—fresh or salt water, in *Papers in marine biology and oceanography*. London and New York. Pp. 261–280.

———— 1955*b*. Herpetichthyes, Amphibioidei, Choanichthyes, or Sarcopterygii? *Nature*, **176**: 126–127.

———— 1956*a*. *The osteology of the reptiles*. Chicago. P. 772.

———— 1956*b*. The early evolution of land vertebrates. *Proc. Am. Philos. Soc.*, **100**: 157–167.

———— 1956*c*. *The vertebrate body*. Philadelphia. P. 486.

———— 1957. Origin of the amniote egg. *Sci. Monthly*, **85**: 57–63.

———— 1958*a*. The Texas Permian red beds and their vertebrate fauna, in *Studies in fossil vertebrates*, T. S. Westoll, Ed. London. Pp. 157–179.

———— 1958*b*. Tetrapod limbs and early tetrapod life. *Evolution*, **12**: 365–369.

———— 1960. Vertebrate-bearing continental Triassic strata in Mendoza region, Argentina. *Bull. Geol. Soc. Amer.*, **71**: 1279–1294.

———— 1961. Synapsid evolution and dentition. Intern. colloq. on the evolution of lower and nonspecialized mammals. *Kon. Vlaamse Acad. Wetensch. Lett. Schone Kunsten Belg.*, pt. I: 9–56.

———— 1962. The fossiliferous Triassic deposits of Ischigualasto, Argentina. *Breviora*, No. 156: 1–7.

———— 1964*a*. Problems in early amphibian history. *J. Anim. Morphol. Physiol.*, **2**: 1–20.

———— 1964*b*. The skeleton of the Lower Carboniferous labyrinthodont *Pholidogaster pisciformes*. *Bull. Mus. Comp. Zool.*, **131**: 131–159.

———— 1964*c*. *Diadectes* an amphibian? *Copeia*, **4**: 718–719.

———— 1965. Possible polyphylety of the vertebrate classes. *Zool. Jahrb.*, **92**: 143–156.

———— 1966*a*. *Vertebrate paleontology*, 3rd ed. Chicago. P. 468.

———— 1966*b*. The Chañares (Argentina) Triassic reptile fauna. I. Introduction. *Breviora*, No. 274: 1–14.

———— 1967. Major steps in vertebrate evolution. *Science*, **158**: 1629–1637.

———— 1968. *Notes and comments on vertebrate paleontology*. Chicago. P. 304.

———— 1969. An icthyosaur skull from the Cretaceous of Wyoming. *Contr. Geol. Univ. Wyo.*, **7**: 27–41.

Romer, A. S., and F. Byrne. 1931. The pes of *Diadectes*: notes on the primitive tetrapod limb. *Palaeobiology*, **4**: 25–48.

Romer, A. S., and T. Edinger. 1942. Endocranial casts and brains of living and fossil Amphibia. *J. Comp. Neurol.*, **77**: 355–389.

Romer, A. S., and E. C. Olson. 1954. Aestivation in a Permian lungfish. *Breviora*, No. 30: 1–8.

Romer, A. S., and L. Price. 1939. The oldest vertebrate egg. *Am. J. Sci.*, **237**: 826–829.

———— 1940. *Review of the Pelycosauria*. Geol. Soc. Amer., Special Paper, **28**: 1–621.

Rosa, D. 1931. *L'ologenes. Nouvelle theorie de l'évolution et de la distribution geographique*. Revised and translated from the Italian edition, 1918. Paris. P. 368.

Rosen, W. M. 1959. *Water flowing around a swimming fish*. U.S. Naval Ordinance Test Station, Tech. Prod. 2298, unclass. AD-N-238395 (Astia), China Lake, Calif.

Roth, S. 1903. Los ungulados Sudamericanos. *Anales Mus. La Plata, Sec. Paleont.*, 1–36.

Rozhdestvenski, A. K. 1964. Archosauria, in *Fundamentals of paleontology*, vol. 12, Y. A. Orlov, Ed. Moscow. Pp. 493–603 (Russian).

—— et al. (P. Chudinov, B. Vjuskhov, L. Tatarinov, V. Suhkanov, N. Novoshilov, E. Maleev, K. Yurev, L. Khozaitskii, and E. Konzhukova). 1964. Reptilia, in *Fundamentals of paleontology*, vol. 12, Y. A. Orlov, Ed. Moscow. Pp. 191–603 (Russian).

Russell, D. E. 1964. Les mammifères Paléocènes d'Europe. *Mem. Mus. Nat. Hist.*, ser. c, Sci. Terre, 13 (n.s.): 1–324.

Russell, D. E., P. Louis, and D. Savage. 1967. Primates of the French early Eocene. *Univ. Calif. Publ. Geol. Sci.*, **73**: 1–44.

Russell, L. S. 1959. The dentition of rabbits and the origin of the lagomorphs. *Bull. Nat. Mus. Canada*, **166**: 41–45.

Saint-Seine, P., de. 1955. Sauropterygia. Pterosauria, in *Traité de paléontologie*, vol. 5, J. Piveteau, Ed. Paris. Pp. 420–458, 963–990.

Salthe, S. N., and N. C. Kaplan. 1966. Immunology and rates of evolution in the Amphibia in relation to the origins of certain taxa. *Evolution*, **20**: 603–616.

Sand, A. 1938. The function of the ampullae of Lorenzini with some observations on the effect of temperature sensory rhythms. *Proc. Roy. Soc. London*, **B129**: 524–553.

Savage, D., D. E. Russell, and P. Louis. 1966. Ceratomorpha and Ancylopoda (Perissodactyla) from the Lower Eocene Paris basin, France. *Univ. Calif. Publ. Geol. Sci.*, **66**: 1–38.

Säve-Söderbergh, G. 1932. Preliminary note on Devonian stegocephalians from East Greenland. *Medd. Grønland*, **94**: 1–107.

—— 1933. The dermal bones of the head and the lateral line system in *Osteolepis macropelidotus* Ag. *Nova Acta Soc. Sci. Uppsala*, **9**: 1–129.

—— 1934. Some points of view concerning the evolution of the vertebrates and the classification of this group. *Ark. Zool.*, **26A**: 1–20.

—— 1935. On the dermal bones of the head in labyrinthodont stegocephalians and primitive Reptilia, with special reference to Eostriassic stegocephalians from East Greenland. *Medd. Grønland*, **98**: 1–211.

Sawin, H. G. 1941. The cranial anatomy of *Eryops megacephalus*. *Bull. Mus. Comp. Zool.*, **88**: 407–464.

Schaeffer, B. 1941*a*. The morphological and functional evolution of the tarsus in amphibians and reptiles. *Bull. Am. Mus. Nat. Hist.*, **78**: 395–472.

—— 1941*b*. The structure and function of the primitive tetrapod. *Trans. N. Y. Acad. Sci.*, **3**: 158–161.

—— 1953. *Latimeria* and the history of coelacanth fishes. *Trans. N. Y. Acad. Sci.*, **15**: 170–178.

—— 1956. Evolution in the subholostean fishes. *Evolution*, **10**: 201–212.

—— 1961. Differential ossification in the fishes. *Trans. N. Y. Acad. Sci.*, **23**: 501–505.

—— 1965*a*. The rhipidistian-amphibian transition. *Am. Zoologist*, **5**: 267–276.

—— 1965*b*. Evolution of concepts related to the origin of the Amphibia. *Syst. Zool.*, **14**: 115–118.

—— 1965*c*. The role of experimentation in the origin of higher levels of organization. *Syst. Zool.*, **14**: 318–336.

—— 1967*a*. Osteichthyan vertebrae, in *Fossil vertebrates*, C. Patterson and P. H. Greenwood, Eds. *J. Linn. Soc. London* (Zool.), **47**: 185–195.

—— 1967*b*. Late Triassic fishes from the western United States. *Bull. Am. Mus. Nat. Hist.*, **135**: 285–382.

—— 1967*c*. Comments on elasmobranch evolution, in *Sharks, skates and rays*. P. Gilbert, R. Mathewson, and David Rall, Eds. Baltimore. Pp. 3–35.

—— 1968. The origin and basic radiation of the Osteichthyes, in *Current problems of lower vertebrate phylogeny*, T. Ørvig, Ed. Stockholm. Pp. 207–222.

Schaeffer, B., and D. E. Rosen. 1961. Major adaptive levels in the evolution of the actino-pterygian feeding mechanism. *Am. Zoologist*, **1**: 187–204.

Schaub, S. 1953a. La trigonodontie des rongeurs simplicidentés. *Ann. Paléont.*, **39**: 29–57.

—— 1953b. Remarks on the distribution and classification of the "Hystricomorpha." *Verhandl. Naturf. Ges. Basel*, **64**: 389–400.

—— 1958. Simplicidentata, in *Traité de paléontologie*, vol. 6, pt. 2, J. Piveteau, Ed. Paris. Pp, 659–818.

Schmaulhausen, I. I. 1916. On the functions of fins of fishes. *Rev. Zool. Russe*, 185–214.

—— 1958. The nostrils of fishes and their fate in terrestrial vertebrates. *Zool. J.*, **37**: 1710–1718.

—— 1964. *The origin of terrestrial vertebrates*. Moscow. P. 271 (Russian).

—— 1968. *The origin of terrestrial vertebrates*. New York. P. 314 (English).

Schmidt, K. P. 1943. Corollary and commentary for *Climate and evolution*. *Am. Midland Naturalist*, **30**: 241–253.

—— 1950. Modes of evolution discernible in the taxonomy of snakes. *Evolution*, **4**: 79–86.

Schmidt-Nielsen, K. 1960. The salt-secreting gland of marine birds. *Circulation*, **21**: 955–967.

Schmidt-Nielsen, K., and R. Fänge. 1958. Salt glands in marine reptiles. *Nature*, **182**: 783–785.

Scholander, P. F. 1964. Animals in aquatic environments: diving mammals and birds, in *Handbook of physiology*, sect. 4, vol.,1, *Adaptation to the environment*, D. B. Dill, E. F. Adolph, and C. G. Wilber, Eds. Pp. 729–739.

Schwarzbach, M. 1963. Climates of the past. Ed. R. W. Fairbridge. New York. P. 328.

Scott, W. 1937. *A history of land mammals in the Western Hemisphere*, rev. ed. New York. P. 786.

Severtsov, A. N. 1926a. Studies on the bony skull of fishes. I: Structure and development of the bony skull of *Acipenser ruthenus*. *Quart. J. Micro. Sci.*, **70**: 451–540.

—— 1926b. Beiträge zur einer Theorie des Knocheran Schädels der Wirbeltiere. *Palaeont. Z.*, **8**: 42–43.

—— 1926c. Der Ursprung der Quadrupeda. *Palaeont. Z.*, **8**: 75–95.

Shann, E. W. 1924. Further observations on the myology of the pectoral region in fishes. *Proc. Zool. Soc. London*, 145–215.

Shikama, T. 1947. *Teilhardosaurus* and *Endotherium*, new Jurassic Reptilia and Mammalia from the Husin coal-field, South Manchuria. *Proc. Japan Acad.*, **23**: 76–84.

Shishkin, M. A. 1964. Stereospondyli, in *Fundamentals of paleontology*, vol. 12, Y. A. Orlov, Ed. Moscow. Pp. 83–122 (Russian).

Shotwell, J. 1958. Intercommunity relationships in Hemphillian (mid-Pliocene) mammals. *Ecology*, **39**: 271–282.

—— 1961. Late Tertiary biogeography of horses in the northern Great Basin. *J. Paleont.*, **35**: 203–217.

Sigogneau, D. 1963. Remarks on Gorgonopsia. *S. Afr. J. Sci.*, **59**: 207–209.

Simmons, D. 1965. The non-therapsid reptiles of the Lufeng Basin, Yunnan, China. *Fieldiana: Geology*, **15**: 1–93.

Simons, E. 1960. *Apidium* and *Oreopithecus*. *Nature*, **186**: 821–826.

—— 1962a. A new Eocene primate genus *Cantius*, and a revision of some allied European lemuroids. *Bull. Brit. Mus. (Nat. Hist.) Geology*, **7**: 1–36.

—— 1962b. Two new primate species from the African Oligocene. *Postilla*, No. 64: 1–12.

—— 1963. A critical reappraisal of Tertiary primates, in *Evolutionary and genetic biology of primates*, vol. I, J. Buettner-Janusch, Ed. New York. Pp. 66–124.

—— 1964. The early relatives of man. *Scientific American*, **211**: 50–62.

——— 1967a. A fossil *Colobus* skull from the Sudan (Primates: Cercopithecidae). *Postilla* No. 111: 1–12.

——— 1967b. The earliest apes. *Scientific American*, **217**: 28–35.

——— 1968a. Early Cenozoic mammalian faunas, Fayum Province, Egypt. Pt. 1. African Oligocene mammals, introduction, history of study, and faunal succession. *Bull. Peabody Mus.*, **28**: 1–21.

——— 1968b. A source for dental comparison of *Ramapithecus* with *Australopithecus* and *Homo*. *S. Afr. J. Sci.*, **64**: 92–112.

Simpson, G. G. 1928. *A catalogue of the Mesozoic Mammalia in the Geological Department of the British Museum (Natural History)*. London. P. 215.

——— 1929. American Mesozoic Mammalia. *Mem. Peabody Mus. Nat. Hist.*, **3**, pt. 1, 1–171.

——— 1930. Post-Mesozoic Marsupialia, in *Fossilium catalogus, Animalia*, Pars 47, 1–87.

——— 1937. Skull structure of the Multituberculata. *Bull. Am. Mus. Nat. Hist.*, **73**: 727–763.

——— 1940a. Antarctica as a faunal migration route, in *Proc. Sixth Pacific Sci. Congr.*, **2**: 755–768.

——— 1940b. Mammals and land bridges. *J. Wash. Acad. Sci.*, **30**: 137–163.

——— 1941. The affinities of the Borhyaenidae. *Am. Mus. Novitates*, No. 1118: 1–6.

——— 1943. Mammals and the nature of continents. *Am. J. Sci.*, **241**: 1–31.

——— 1944. *Tempo and mode in evolution*. New York. P. 237.

——— 1945. The principles of classification and the classification of mammals. *Bull. Am. Mus. Nat. Hist.*, **85**: 1–350.

——— 1946. Fossil penguins. *Bull. Am. Mus. Nat. Hist.*, **87**: 1–99.

——— 1947a. Holarctic mammalian faunas and continental relationships during the Cenozoic. *Bull. Geol. Soc. Amer.*, **58**: 619–688.

——— 1947b. Evolution, interchange, resemblance of the North American and Eurasian Cenozoic mammalian faunas. *Evolution*, **1**: 218–220.

——— 1949. *The meaning of evolution*. New Haven. P. 364.

——— 1953a. *Evolution and geography*. Condon lectures. Oregon State System of Higher Education. P. 64.

——— 1953b. *The major features of evolution*. New York. P. 134.

——— 1957. Australian fossil penguins, with remarks on penguin evolution and distribution. *Res. S. Aust. Mus.*, **13**: 51–70.

——— 1959a. The nature and origin of supraspecific taxa. *Cold Spring Harbor Symp. Quant. Biol.*, **24**: 255–271.

——— 1959b. Mesozoic mammals and the polyphyletic origin of mammals. *Evolution*, **13**: 405–414.

——— 1960a. Diagnosis of the classes Reptilia and Mammalia. *Evolution*, **14**: 388–392.

——— 1960b. Notes on the measurement of faunal resemblance. *Am. J. Sci.*, **258-A**: 300–311.

——— 1961a. *Principles of animal taxonomy*. New York. P. 247.

——— 1961b. Evolution of Mesozoic mammals. Intern. colloq. on the evolution of lower and unspecialized mammals. *Kon. Vlaamse Acad. Wetensch. Lett. Schone Kunsten Belg.*, pt. 1, 57–95.

——— 1964. Species density of North American recent mammals. *Syst. Zool.*, **13**: 57–73.

——— 1966. Mammalian evolution on the southern continents. *Neues Jahrb. Min. Geol. Palaeont. Abh.*, **125**: 1–18.

Sinclair, W. J. 1906. *Mammalia of the Santa Cruz beds: Marsupialia*. Rept. Princeton Univ. Exp. Patogonia, vol. 4. Pp. 333–482.

814   References

Slaughter, B. 1968. Earliest known marsupials. *Science,* **162:** 254–255.

Slijper, E. J. 1961. Locomotion and locomotory organs in whales and dolphins (Cetacea). *Symp. Zool. Soc. London,* **5:** 77–96.

────── 1962. *Whales.* London. P. 475.

Sloan, R. E. 1964. Paleoecology of the Cretaceous-Tertiary transition in Montana. *Science,* **146:** 1–430.

Sloan, R. E., and L. Van Valen. 1965. Cretaceous mammals from Montana. *Science,* **148:** 220–227.

Smith, H. W. 1953. *From fish to philosopher.* Boston. P. 264.

Sneath, P. 1957. The application of computers to taxonomy. *J. Gen. Microbiol.,* **17:** 201–226.

Sokal, R., and P. Sneath. 1963. *Principles of numerical taxonomy.* San Francisco. P. 359.

Spinar, Z. V. 1952. Revuse nekterych moravskych Diskosauriscidu (Revision of some Moravian Discosauriscidae). *Roz. Ustred. Ustav. Geol.,* **15:** 1–159. (Czech, English summary).

Springer, S. 1961. Dynamics of the feeding mechanism of large Galeoid sharks. *Am. Zoologist,* **1:** 183–185.

Stahl, B. S. 1967. Morphology and relationships of the Holocephali with special reference to the venous system. *Bull. Mus. Comp. Zool.,* **135:** 141–213.

Stebbins, R. C., and R. M. Eakin. 1958. The role of the "third eye" in reptilian behavior. *Am. Mus. Novitates,* No. 1870: 1–40.

Stehli, F. 1957. Possible Permian climatic zonation and its implications. *Am. J. Sci.,* **255:** 607–618.

Stehli, F., and C. Helsley. 1963. Paleontological technique for defining ancient pole positions. *Science,* **142:** 1057–1059.

Stehlin, H., and S. Schaub. 1951. Die Trigonodontie der simplicidentaten Nager. *Schweiz. Palaeont. Abh.,* **67:** 1–385.

Stejneger, L. 1885. Natural history of birds, in *The standard natural history,* vol. 4, J. S. Kingsley, Ed. Pp. 10–95, 368–441, 458–547.

Stensiö, E. 1925. On the head of macropetalichthyids. *Field Mus. Nat. Hist. Publ.* 232. Geol. Ser., **4:** 89–198.

────── 1926. Triassic fishes from Spitzbergen, pt. II. *Kungl. Sven. Vet.-Akad. Handl.,* **2:** 1–261.

────── 1932a. *The cephalaspids of Great Britain.* London. P. 220.

────── 1932b. Triassic fishes from East Greenland, collected by the Danish expeditions in 1929–1931. *Medd. Grønland,* **83:** 1–305.

────── 1958. Les cyclostomes fossiles on ostracodermes, in *Traité de zoologie,* vol. 13, fasc. 1, P. P. Grassé, Ed. Paris. Pp. 173–425.

────── 1959. On the pectoral fin and shoulder girdle of arthrodires. *Kungl. Sven. Vet.-Akad. Handl.,* **8:** 1–229.

────── 1961. Permian vertebrates, in *Geology of the Arctic,* vol. 1, G. O. Raasch, Ed. Toronto. Pp. 231–247.

────── 1962. Origine et nature des écailles placoïdes et des dents, in *Colloq. problèmes actuels de paléontologie.* Colloq. Cent. Nat. Rech. Sci. Paris. Pp. 75–85.

────── 1963a. Anatomical studies on the arthrodiran head. 1. Preface, geological and geographical distribution. The organization of the arthrodires, the anatomy of the head in the Dolichothoraci, Coccosteomorphi, and Pacheostomorphi. Taxonomic appendix. *Kungl. Sven. Vet.-Akad. Handl.,* **9:** 1–419.

────── 1963b. The brain and the cranial nerves in fossil, lower craniate vertebrates. *Norske Vidensk.-Akad., Oslo, Mat.-Nat. Kl., Skr.,* n.s., No. 13: 1–120.

────── 1964. Les cyclostomes fossiles ou ostracodermes, in *Traité de paléontologie,* vol. 4, pt. 1, J. Piveteau, Ed. Paris. Pp. 96–382.

——— 1968. The cyclostomes with special reference to the diphyletic origin of Petromyzontida and Myxinodea, in *Current problems of lower vertebrate phylogeny*, T. Ørvig Ed. Stockholm. Pp. 13–71.

Steyn, W. 1961. Some epithalmic organs, the subcommissural organ and their possible relation to vertebrate emergence on dry land. *S. Afr. J. Sci.*, **57**: 283–287.

Stirton, R. 1955. Late Tertiary marsupials from South Australia. *Rec. S. Aust. Mus.*, **11**: 247–268.

——— 1957. Tertiary marsupials from Victoria, Australia. *Mem. Nat. Mus. Victoria*, No. 21: 121–134.

Stirton, R., R. Tedford, and A. Miller. 1961. Cenozoic stratigraphy and vertebrate paleontology of the Tirari desert, South Australia. *Rec. S. Aust. Mus.*, **14**: 19–61.

Storr, G. C. 1780. Prodromus methodi mammalium. (Excerpts and tables of classification), in *Appendix V*, Th. Gill, *Bull. Philos. Soc. Wash.*, **2** (1875–1880).

Sukhanov, V. B. 1964. Testudinata, in *Fundamentals of paleontology*, vol. 12, T. A. Orlov, Ed. Moscow. Pp. 345–446 (Russian).

Szalay, S. 1968. The beginnings of the Primates. *Evolution*, **22**: 19–36.

Szarski, H. 1962. The origin of the Amphibia. *Quart. Rev. Biol.*, **37**: 189–241.

——— 1964. The functions of the myomere-folding in aquatic vertebrates. *Bull. Acad. Polonica Sci., Ser. Sci. Biol.*, **12**: 305–310.

Tarlo, B. 1960. A review of the upper Jurassic pliosaurs. *Bull. Brit. Mus. (Nat. Hist.) Geol.*, **4**: 147–189.

——— 1961. *Rhinopteraspis cornubica* (McCoy) with notes on the classification and evolution of pteraspids. *Acta Paleont. Polonica*, **6**: 367–402.

——— 1962. The classification and evolution of the Heterostraci. *Acta Palaeont. Polonica*, **7**: 249–290.

——— 1963. Aspidin: the precursor of bone. *Nature*, **199**: 46–48.

Tatarinov, L. P. 1960*a*. Evolution of the structure separating courses of blood in the vertebrate heart. *Zool. J. Acad. Sci. USSR*, **39**: 1218–1231 (Russian).

——— 1960*b*. Discovery of pseudosuchians in the upper Permian of the USSR. *Paleont. J. Acad. Sci. USSR*, No. 4: 74–80 (Russian).

——— 1960*c*. The pseudosuchian material of the USSR. Paleont. J. Acad. Sci. USSR, **1**: 117–132 (Russian).

——— 1964*a*. Ichthyopterygia, in *Fundamentals of paleontology*, vol. 12, Y. A. Orlov, Ed. Moscow. Pp. 338–353 (Russian).

——— 1964*b*. On the anatomy of the head of the therocephalians (blood vessels, nerves, and glands of *Moschowhaitsia*). *Paleont. J. Acad. Sci. USSR*, 72–84 (Russian).

——— 1964*c*. A new locality of Permian seymouriamorphs in the USSR. *Paleont. J. Acad. Sci. USSR*, **1**: 139–141 (Russian).

——— 1967. Development of a system of labial (vibrissial) blood vessels and nerves in the theriodonts. *Paleont. J. Acad. Sci. USSR*, **1**: 3–17 (Russian).

——— 1968*a*. Structure of the braincase of *Dvinia*, and some problems of the evolution of the mammalian endocranium, in *Upper Paleozoic and Mesozoic amphibians and reptiles of the USSR*. Division of general biology, Acad. Sci. USSR. Moscow. Pp. 47–64 (Russian).

——— 1968*b*. Discovery of a primitive tailed amphibian in the upper Permian of the Volge region, in *Upper Permian and Mesozoic amphibians and reptiles of the USSR*. Division of general biology, Acad. Sci. USSR. Moscow. Pp. 7–10 (Russian).

Tatarinov, L. P., and E. D. Konzhukova. 1964. Amphibia, in *Fundamentals of paleontology*, vol. 12, Y. A. Orlov, Ed. Moscow. Pp. 25–170 (Russian).

Tatarinov, L. P., and N. Novoshilov. 1964. Sauropterygia, in *Fundamentals of paleontology*, vol. 12. Y. A. Orlov, Ed. Moscow. Pp. 309–337 (Russian).

Tatarinov, L. P., B. P. Vjushkov, and P. K. Chudinov. 1964. Therapsida, in *Fundamentals of paleontology*, vol. 12, Y. A. Orlov, Ed. Moscow. Pp. 246–298.

—— 1964. Lepidosauria, in *Fundamentals of paleontology*, vol. 12, Y. A. Orlov, Ed. Moscow. Pp. 439–492.

Tavolga, W. N. 1964. *Marine bio-acoustics*. New York. P. 413.

—— 1965. *Review of marine bio-acoustics, state of the art: 1964*. Tech. Rpt. Navtradevcen 1212-1. Port Washington. P. 100.

Taylor, G. 1952. Analysis of the swimming of long and narrow animals. *Proc. Roy. Soc. London*, **A214**: 158–183.

Thomson, K. S. 1962. Rhipidistian classification in relation to the origin of the tetrapods. *Breviora*, No. 177: 1–12.

—— 1964a. The comparative anatomy of the snout in rhipidistian fishes. *Bull. Mus. Comp. Zool.*, **131**: 313–357.

—— 1964b. The ancestry of the tetrapods. *Sci. Progress*, **52**: 451–460.

—— 1965. The nasal apparatus in Dipnoi, with special reference to *Protopterus*. *Proc. Zool. Soc. London*, **B145**: 207–238.

—— 1966a. The evolution of the tetrapod middle ear in the rhipidistian-amphibian transition. *Am. Zoologist*, **6**: 309–397.

—— 1966b. Intracranial mobility in the coelacanth. *Science*, **153**: 999–1000.

—— 1967a. Notes on the relationships of the rhipidistian fishes and the ancestry of the tetrapods. *J. Paleont.*, **41**: 660–674.

—— 1967b. Mechanisms of intracranial kinetics in fossil rhipidistian fishes (Crossopterygii) and their relatives. *J. Linn. Soc. London* (Zool.), **46**: 223–253.

—— 1967c. A critical review of the diphyletic theory of rhipidistian-amphibian relationships, in *Current problems of lower vertebrate phylogeny*, T. Ørvig, Ed. Stockholm. Pp. 286–315.

—— 1969. The biology of the lobe-finned fishes. *Biol. Rev.*, **44**: 91–154.

Thomson, K., and P. P. Vaughn. 1968. Vertebral structure in Rhipidistia (Osteichthyes, Crossopterygia) with description of a new Permian genus. *Postilla*, No. 127: 1–19.

Tihen, J. A. 1960. Comments on the origin of the amniote egg. *Evolution*, **14**: 528–531.

—— 1965. Evolutionary trends in frogs. *Am. Zoologist*, **5**: 309–318.

Traquair, R. H. 1877–1914. The ganoid fishes of the British Carboniferous formations. Pt. I. Palaeoniscidae. *Palaeontogr. Soc.* (*Monogr.*) *London*, **31**: 1–180.

Tullberg, T. 1899. Ueber das System der Nagethiere: eine phylogenetische Studie. *Nova Acta Reg. Soc. Sci. Uppsala*, **18**: 1–514.

Tumarkin, A. 1955. On the evolution of the auditory conducting apparatus: a new theory based on functional considerations. *Evolution*, **9**: 221–243.

Turnbull, W. D. 1970. Mammalian masticatory apparatus. *Fieldiana: Geology*, **18**: 149–356.

Turnbull, W. D., and P. F. Turnbull. 1955. A recently discovered *Phlegethontia* from Illinois. *Fieldiana: Zoology*, **37**: 523–539.

Underwood, G. 1957a. On lizards of the family Pygopodidae. A contribution to the morphology and phylogeny of the Squamata. *J. Morphol.*, **100**: 207–268.

—— 1957b. *Lanthanotus* and the anguinomorph lizards: a critical review. *Copeia*, 20–30.

—— 1967. *A contribution to the classification of snakes*. Brit. Mus. (Nat. Hist.) Publ. 653. P. 179.

Van Bergeijk, W. A. 1964. Directional and nondirectional hearing in fish, in *Symposium on marine bio-acoustics*, W. N. Tavolga, Ed. New York. Pp. 281–289.

—— 1966. Evolution of the sense of hearing in vertebrates. *Am. Zoologist*, **6**: 371–377. Pp. 1–179.

Van Bemmelen, J. F. 1901. Der Schädelbau der Monotremen. *Denkschr. Med.-Naturwiss. Ges. Gena*, **6**: 729–798.

Vandebroek, G. 1964. Recherches sur l'origine des mammifères. *Ann. Soc. Roy. Zool. Belg.*, **94**: 117–160.

Van Valen, L. 1960. Therapsids as mammals. *Evolution*, **14**: 304–313.

—— 1964*a*. A possible origin for rabbits. *Evolution*, **18**: 484–491.

—— 1964*b*. Relative abundance of species in some fossil mammal faunas. *Am. Naturalist*, **98**: 109–116.

—— 1965. Some European Proviverrini (Mammalia, Deltatheridia). *Palaeontology*, **8**: 638–665.

—— 1966. Deltatheridia, a new order of mammals. *Bull. Am. Mus. Nat. Hist.*, **132**: 1–126.

Van Valen, L., and R. Sloan. 1965. The earliest primates. *Science*, **150**: 1–2.

—— 1966. The extinction of multituberculates. *Syst. Zool.*, **15**: 261–278.

Vaughn, P. P. 1955. The Permian reptile *Araeoscelis* restudied. *Bull. Mus. Comp. Zool.*, **113**: 302–467.

—— 1962. The Paleozoic microsaurs as close relatives of the reptiles again. *Am. Midland Naturalist*, **67**: 79–84.

—— 1963. New information on the structure of Permian lepospondylous vertebrae from an unusual source. *Bull. S. Calif. Acad. Sci.*, **62**: 150–157.

—— 1964*a*. Vertebrates from the Organ Rock shale of the Cutler group, Permian of Monument Valley and vicinity, Utah and Arizona. *J. Paleont.*, **38**: 567–583.

—— 1964*b*. Evidence of aestivating lungfish from the Sangre de Cristo formation, Lower Permian of northern New Mexico. *Los Angeles County Mus. Contr. Sci.*, No. 80: 1–8.

Viret, J. 1955. Rodentia fossiles. La denture des rongeurs actuels et fossiles, in: *Traité de zoologie*, vol. 17, fasc. 2, P. P. Grassé, Ed. Paris. Pp. 1526–1573.

Vorobyeva, E. I. 1960*a*. New data on the genus of lobe-finned fish *Panderichthys* from the Devonian of USSR. *Paleont. J. Acad. Sci. USSR*, No. 1: 87–96 (Russian).

—— 1960*b*. Concerning the systematic position of *Eusthenopteron Wenjucovi* (Rohon). *Paleont. J. Acad. Sci. USSR*, No. 2: 121–129 (Russian).

—— 1962. Rhizodont lobe-finned fishes of the principal Devonian fields in the USSR. *Paleont. Inst. Acad. Sci. USSR*, **94**: 1–138 (Russian).

Vorobyeva, E. I., and D. V. Obruchev. 1964. Sarcopterygii, in *Fundamentals of paleontology*, vol. II, Y. A. Orlov, Ed. Moscow. Pp. 268–322 (Russian).

Wahlert, G., von. 1965. The role of ecological factors in the origin of higher levels of organization. *Sys. Zool.*, **14**: 288–300.

Walker, A. D. 1964. Triassic reptiles from the Elgin area. *Ornithosuchus* and the origin of carnosaurs. *Philos. Trans. Roy. Soc. London*, **B248**: 53–134.

Walker, M. V. 1967. Revival of interest in the toothed birds of Kansas. *Trans. Kansas Acad. Sci.*, **70**: 60–66.

Walker, W. F., Jr. 1965. *Vertebrate dissection*, 3rd ed. Philadelphia. P. 374.

Walls, L. W. 1942. The vertebrate eye and its adaptive radiation. *Bull. Cranbrook Inst. Sci.*, No. 19: 1–785.

Walters, V., and R. Liu. 1967. Hydronamics of navigation by fishes in terms of the mucus-water "interface," in *Lateral line detectors*, P. H. Cahn, Ed. Bloomington. Pp. 437–446.

Watson, D. M. S. 1916. The monotreme skull. *Phil. Trans. Roy. Soc. London*, **B207**: 311–347.

—— 1917. A sketch classification of the pre-Jurassic tetrapod vertebrates. *Proc. Zool. Soc. London*, 167–186.

—— 1919. The structure, evolution and origin of the Amphibia. The "orders" Rhachitomi and Sterospondyli. *Phil. Trans. Roy. Soc. London*, **B209**: 1–72.

—— 1921. The basis of classification of the Theriodontia. *Proc. Zool. Soc. London*, 35–98.

—— 1924. The elasmosaurid shoulder-girdle and forelimb. *Proc. Zool. Soc. London*, 885–917.

—— 1925. The structure of certain palaeoniscids and the relationship of that group with other bony fishes. *Proc. Zool. Soc. London*, 815–870.

—— 1926. The evolution and origin of the Amphibia. *Philos. Trans. Roy. Soc. London*, **B214**: 189–257.

—— 1929. The Carboniferous Amphibia of Scotland. *Paleont. Hungarica*, **1**: 219–252.

—— 1931. On the skeleton of a bauriamorph reptile. *Proc. Zool. Soc. London*, 35–98.

—— 1937. The acanthodian fishes. *Philos. Trans. Roy. Soc. London*, **B228**: 49–146.

—— 1940. The origin of frogs. *Trans. Roy. Soc. Edinburgh*, **60**: 195–331.

—— 1942. On Permian and Triassic tetrapods. *Geol. Mag.*, **79**: 81–116.

—— 1951. *Paleontology and modern biology*. New Haven. P. 216.

—— 1954. On *Bolosaurus* and the origin and classification of reptiles. *Bull. Mus. Comp. Zool.*, **111**: 297–449.

—— 1956. The brachyopid labyrinthodonts. *Bull. Brit. Mus.* (*Nat. Histl*), **2**: 318–391.

—— 1957. On *Millerosaurus* and the early history of sauropsid reptiles. *Philos. Trans. Roy. Soc. London*, **B240**: 325–400.

—— 1962. The evolution of the labyrinthodonts. *Philos. Trans. Roy. Soc. London*, **B245**: 219–265.

—— 1963. On growth stages in branchiosaurs. *Paleontology*, **6**: 540–553.

Watson, D. M. S., and A. S. Romer. 1957. A classification of therapsid reptiles. *Bull. Mus. Comp. Zool.*, **114**: 35–89.

Watson, M. 1883. Report on the anatomy of the Sphenisciformidae collected during the voyage of *H.M.S. Challenger*, in Report on the scientific results of the voyage of *H.M.S. Challenger* during the years 1873–1876, *Zoology*, vol. 7, J. Murray, Ed. London. Pp. 1–244.

Weber, M. 1904. *Die Saugetiere. Einführing in die Anatomie und Systematik der recenten und fossilen Mammalia*. Jena. P. 866.

Webster, D. B. 1966. Ear structure and function in modern mammals. *Am. Zoologist*, **6**: 451–466.

Weichert, C. K. 1965. *Anatomy of the chordates*. 3rd ed. New York. P. 758.

Welles, S. 1962. A new species of elasmosaur from the Aptian of Columbia and a review of the Cretaceous plesiosaurs. *Univ. Calif. Publ. Geol. Sci.*, **44**: 1–96.

Wenz, S. 1960. Étude de *Ptycholepis bollensis*, poisson du Lias supérieur de l'Yonne et du Württemberg. *Soc. Géol. France*, **B1**: 916–927.

Westoll, T. S. 1937*a*. On the cheek bones in teleostome fishes. *J. Anat.*, **71**: 362–382.

—— 1937*b*. On a remarkable fish from the lower Permian of Autun, France. *Ann. Mag. Nat. Hist.*, ser. 10, **19**: 553–579.

—— 1938. Ancestry of the tetrapods. *Nature*, **141**: 127–128.

—— 1941*a*. Latero-sensory canals and dermal bones. *Nature*, **148**: 168.

—— 1941*b*. The Permian fishes *Dorypterus* and *Lekovichthys*. *Proc. Zool. Soc. London*, **B111**: 39–58.

—— 1942*a*. Ancestry of captorhinomorph reptiles. *Nature*, **149**: 667–668.

—— 1942*b*. Relationships of some primitive tetrapods. *Nature*, **150**: 121.

—— 1943*a*. The origin of tetrapods. *Biol. Rev.*, **18**: 78–98.

—— 1943*b*. The hyomandibular of *Eusthenopteron* and the tetrapod middle ear. *Proc. Roy. Soc. London*, **B131**: 393–414.

—— 1943*c*. The origin of the primitive tetrapod limb. *Proc. Roy. Soc. London*, **B131**: 373–393.

—— 1944. The Haplolepidae, a new family of late Carboniferous bony fishes. *Bull. Am. Mus. Nat. Hist.*, **83**: 1–121.

—— 1945. The paired fins of placoderms. *Trans. Roy. Soc. London*, **61**: 381–398.

—— 1949. On the evolution of the Dipnoi, in *Genetics, paleontology and evolution*, G. L. Jepsen, E. Mayr, and G. G. Simpson Eds. Princeton. Pp. 121–184.

—— 1958. The lateral fin-fold theory and the pectoral fins of ostracoderms and early fishes, in *Studies in fossil vertebrates*, T. S. Westoll, Ed. London. Pp. 180–211.

—— 1961. A crucial stage in vertebrate evolution: fish to land animals. *Proc. Roy. Inst. Great Britain*, **38**: 600–618.

—— 1962. Ptyctodontid fishes and the ancestry of Holocephali. *Nature*, **194**: 949–952.

—— 1963. The paired fins and axial skeleton of the crossopterygian fish, *Eusthenopteron*. *Proc. 16th Intern. Congr. Zool.*, **1**: 177.

—— 1967. *Radotina* and other tesserate fishes. *J. Linn. Soc. London*, **47**: 83–98.

Wetmore, A. 1930. A systematic classification for the birds of the world. *Proc. U.S. Nat. Mus.*, **76**: 1–8.

—— 1951. A revised classification of the birds of the world. *Smithsonian Misc. Coll.*, **117**, No. 4, 1–22 (Publ. 4057).

—— 1960. A classification for the birds of the world. *Smithsonian Misc. Coll.*, **139**, No. 11, 1–37 (Publ. 4417).

White, E. G. 1937. Interrelationships of the elasmobranchs with a key to the order Galea. *Bull. Am. Mus. Nat. Hist.*, **74**: 25–138.

White, E. I. 1927. The fish fauna of the cementstones of Foulden, Berwickshire. *Trans. Roy. Soc. Edinburgh*, **55**: 255–287.

—— 1935. The ostracoderm *Pteraspis* Kiaer and the relationships of the agnathous vertebrates. *Philos. Trans. Roy. Soc. London*, **B225**: 381–457.

—— 1939. A new type of palaeoniscoid fish, with remarks on the evolution of the actinopterygian pectoral fins. *Proc. Zool. Soc. London*, **B109**: 41–61.

—— 1946. *Jaymoytius kerwood*, a new chordate from the Silurian of Lanarkshire. *Geol. Mag.*, **83**: 89–97.

—— 1950. *Pteraspis leathensis*, a Dittonian zone fossil. *Bull. Brit. Mus. (Nat. Hist.), Geol.*, **1**: 69–89.

—— 1952. Australian arthrodires. *Bull. Brit. Mus. (Nat. Hist.), Geol.*, **1**: 249–304.

—— 1958. Original environments of the craniates, in *Studies on fossil vertebrates*, T. S. Westoll, Ed. London. Pp. 212–234.

—— 1965. The head of *Dipterus valencienesi* Sedgwick and Murchison. *Bull. Brit. Mus. Nat. Hist.* (Geol.) **11**: 1–45.

White, T. E. 1939. Osteology of *Seymouria baylorensis* Broili. *Bull. Mus. Comp. Zool.*, **85**: 323–410.

—— 1949. The endocrine glands and evolution. No. 2: The appearance of large amounts of cement on the teeth of horses. *J. Wash. Acad. Sci.*, **39**: 329–335.

—— 1959. The endocrine glands and evolution. No. 3: Os cementum, hypsodonty and diet. *Contr. Mus. Paleon. Univ. Michigan*, **13**: 211–265.

Wilber, G. 1964. Animals in aquatic environments: introduction, in *Handbook of physiology, sect. 4, vol. 1, Adaptation to the environment*, D. B. Dill, E. F. Adolph, and C. G. Wilber, Eds. Pp. 661–682.

Willard, B. 1935. Chemung tracks and trails from Pennsylvania. *J. Paleont.*, **9**: 43–56.

Williams, E. 1959. Gadow's arcualia and the development of tetrapod vertebrae. *Quart. Rev. Biol.*, **34**: 1–32.

Williston, S. W. 1898. Mosasaurs. *Univ. Geol. Survey Kansas, IV. Paleontology*, 83–221.

———— 1907. The skull of *Brachauchenia*, with observations on the relationships of the plesiosaurs. *Proc. U.S. Natl. Mus.*, **32**: 477–489.

———— 1914. *Water reptiles of the past and the present.* Chicago. P. 251.

———— 1918. The evolution of vertebrae. *Contr. Walker Museum*, **2**: 75–85.

———— 1925. *Osteology of the reptiles.* Cambridge. P. 300.

Wilson, J. A. 1941. An interpretation of the skull of *Buettneria* with special reference to the cartilages and soft parts. *Contr. Mus. Paleont. Univ. Mich.*, **6**: 71–111.

———— 1953. Permian vertebrates from Taylor County, Texas. *J. Paleont.*, **27**: 456–470.

Winge, H. 1895. Jordfundne og nulevende Rovdyr (Carnivora) fra Lagoa Santa, Minas Geraes, Brasilien. *Med. Udsigt over Roudyrenes Indbyrdes Slaegtskab*, **2**, pt. 4, 1–103.

———— 1924. *Pattadyr-Slaegter*, vol. 2, *Rodentia, Carnivora, Primates.* Copenhagen. P. 321.

Winn, H. E. 1964. The biological significance of fish sounds, in *Marine bio-acoustics*, W. N. Tavolga, Ed. New York. Pp. 213–229.

Wischnitzer, S. 1967. *Atlas and dissection guide for comparative anatomy.* San Francisco. P. 178.

Wood, A. E. 1942. Notes on the Paleocene lagomorph, *Eurymylus. Am. Mus. Novitates*, No. 1162: 1–7.

———— 1954. Comments on the classification of rodents. *Breviora*, No. 41: 1–9.

———— 1955. A revised classification of the rodents. *J. Mammal.*, **36**: 165–187.

———— 1957*a*. Speciation and evolutionary rates in Eocene rodents. *Proc. Zool. Soc. Bengal*, Mookerjee memorial volume, 223–227.

———— 1957*b*. What, if anything, is a rabbit? *Evolution*, **11**: 417–425.

———— 1958. Are there rodent suborders? *Syst. Zool.*, **7**: 169–173.

———— 1959. Eocene radiation and phylogeny of the rodents. *Evolution*, **13**: 354–361.

———— 1962. The early Tertiary rodents of the family Paramyidae. *Trans. Am. Philos. Soc.*, **52**, pt. 1, 1–261.

———— 1965. Grades and clades among rodents. *Evolution*, **19**: 115–130.

———— 1968. Early Cenozoic mammalian faunas, Fayum Province, Egypt. Pt. 2, The African Oligocene Rodentia. *Bull. Peabody Mus.*, **28**: 23–105.

Wood, A. E., and Patterson, B. 1959. The rodents of the Deseaden Oligocene of Patagonia and the beginnings of South American rodent evolution. *Bull. Mus. Comp. Zool.*, **120**: 281–428.

Wood, H. E., II. 1924. The position of the "sparassodonts" with notes on the relationships and history of the Marsupialia. *Bull. Am. Mus. Nat. Hist.*, **51**: 77–101.

———— 1934. Revision of the Hyrachyidae. *Bull. Am. Mus. Nat. Hist.*, **67**: 181–295.

———— 1937. Perissodactyl suborders. *J. Mammal.*, **18**: 1–106.

Woodward, A. S. 1889–1895. *Catalogue of the fossil fishes in the British Museum.* Part I (1889). Part II (1891). Part III (1895). London. Pp. 474, 467, and 544.

———— 1898. *Outlines of vertebrate paleontology for students of zoology.* Cambridge. P. 470.

———— 1916–1919. The fossil fishes of the English Wealden and Purbeck formations. *Monograph Palaeontogr. Soc. London.* P. 148.

———— 1924. On a hybodent shark (*Tristychius*) from the Calciferous sandstone series of Eskdale (Dumfries-shire). *Quart. J. Geol. Soc.*, **81**: 338–342.

———— 1932. A Cretaceous pristiophorid shark. *Ann. Mag. Nat. Hist.*, ser. 10, **10**: 476–479.

———— 1942. The beginning of the teleostean fishes. *Ann. Mag. Nat. Hist.*, ser. 11, **9**: 902–912.

Wright, S. 1940. Breeding structure of populations in relationship to speciation. *Am. Nat.* **7**: 232–248.

———— 1960. Physiological genetics, ecology of populations and natural selection. In *Evolution after Darwin.* I. *The evolution of life.* Ed. Sol Tax. Pp. 429–475.

Young, C. C. 1947. Mammal-like reptiles from Lu Feng, Yunnan, China, *Proc. Zool. Soc. London*, **B117**: 537–597.

Young, J. Z. 1962. *The life of vertebrates*, 2nd ed. Oxford. P. 820.

Zangerl, R. 1944. Contributions to the osteology of the skull of Amphisbaenidae. *Am. Midland Naturalist*, **31**: 417–454.

——— 1945. Contributions to the osteology of the postcranial skeleton of the Amphisbaenidae. *Am. Midland Naturalist*, **33**: 764–780.

——— 1953. The vertebrate fauna of the Selma formation of Alabama. Pt. 3. The turtles of the family Protostegidae. Pt. 4. The turtles of the family Toxochelyidae *Fieldiana: Geology*, **3**: 61–277.

——— 1966. A new shark of the family Edestidae, *Ornithoprion hertwigi. Fieldiana: Geology*, **16**: 1–43.

——— 1967. The morphology and the development history of the scales of the Paleozoic sharks *Holmesella?* sp. and *Urodus*, in *Current problems of lower vertebrate phylogeny*, T. Ørvig, Ed. Stockholm. Pp. 400–412.

Zangerl, R., and E. S. Richardson, Jr. 1963. The paleoecological history of two Pennsylvanian black shales. *Fieldiana: Geology*, **4**: 1–239.

Ziegler, A. C. 1969. A theoretical determination of tooth succession in the therapsid *Diademodon. J. Paleont.*, **43** 771–778.

Zittel, K., von. 1890. *Handbuch der Palaeontologie*. I. Abt. *Palaeozologie*, III Bd., Vertebrata (Pisces, Amphibia, Reptilia, Aves). München and Leipzig. P. 900.

——— 1895. *Grundzüge der Palaeontologie*. München and Leipzig. P. 971.

Index

Italics refer to figure references. Author citations are entered only for references that extend beyond mere bibliographic citation and only for the first page in such instances. Taxonomic names are cited only for those pages upon which they form the basis for discussions. Anatomical terms that are grouped in a few pages under considerations of the body systems to which they pertain are not entered individually.